D0081013

Encyclopedia of
WORLD SCIENTISTS

Encyclopedia of

WORLD SCIENTISTS

ELIZABETH H. OAKES

Facts On File, Inc.

Facts On File, Inc.
11 Penn Plaza
New York, NY 10001

Library of Congress Cataloging-in-Publication Data
Oakes, Elizabeth H., 1951–
Encyclopedia of world scientists / Elizabeth Oakes.
p. cm.
Includes bibliographical references and indexes.
ISBN 0-8160-4130-X
1. Scientists—Biography—Encyclopedias. I. Title.
Q141.O25 2000
509.2′2—dc21
[B] 00-029419

Facts On File books are available at special discounts when purchased in bulk quantities for businesses, associations, institutions or sales promotions. Please call our Special Sales Department in New York at 212/967-8800 or 800/322-8755.

You can find Facts On File on the World Wide Web at http://www.factsonfile.com

The author is grateful to Lisa Yount for her permission to adapt and reuse material from Yount's *A to Z of Women in Science and Math* (Facts On File, 1999) for the publication of this book.

Text design by Erica K. Arroyo
Cover design by Cathy Rincon

Printed in the United States of America

VB Hermitage 10 9 8 7 6 5 4 3 2 1

This book is printed on acid-free paper.

CONTENTS

Acknowledgments

I am particularly grateful to the writers and researchers who assisted me on this project—Bill Baue, Barbra Brady, Mariko Fujinaka, and Rebecca Stanfel. Their commitment to excellence in every aspect of their work is always appreciated. For assistance with photographs, I would like to thank those scientists who graciously responded to my requests and the many libraries and archives who helped. I would also like to express thanks to the University of Montana Mansfield Library, where much of the research for this book was done, and to the authors of the many science reference books I consulted, especially Lisa Yount. Finally, I would like to thank my editor, Frank K. Darmstadt, for all his work and support and my former editor, Dr. Eleanora von Dehsen, who first proposed the project.

INTRODUCTION

Starting with the Greek physician Galen's discovery in the second century A.D. that the arteries carry blood and not air (the common belief in his time) and moving up through the ages, the *Encyclopedia of World Scientists* is a comprehensive reference tool for learning about famous scientists and their work. Here readers will find the discovery of radioactivity and the planet Uranus; the beginning of the idea of the Ice Age; the development of the atomic bomb, Boyle's law, and the periodic table of elements; and the invention of the light bulb, the Bunsen burner, and the modern computer. All these scientific achievements and many more are presented in basic, everyday language that makes even the most complex concepts accessible.

THE SCIENTISTS

The *Encyclopedia of World Scientists* includes the well-known scientific greats of history, as well as contemporary scientists whose work is just verging on greatness. Among these are well over 200 women and minority scientists who have often been excluded from books such as this. A majority of the nearly 500 scientists in the book represent the traditional scientific disciplines of physics, chemistry, biology, astronomy, and the earth sciences. A smaller number represent mathematics, computer science, philosophy of science, medicine, engineering, anthropology, and psychology. In addition to the biographical entries, the book contains 100 illustrations.

To compile the entrant list, I relied largely on the judgment of other scientists, consulting established reference works, such as the *Dictionary of Scientific Biography,* and science periodicals, award lists, and publications of science organizations and associations. Despite this process, I cannot claim to present the "most important" historical and contemporary figures. Time constraints and space limitations prevented the inclusion of many deserving scientists.

THE ENTRIES

Entries are arranged alphabetically by surname, with each entry given under the name by which the scientist is most commonly known. The typical entry provides the following information:

Entry Head: Name, birth/death dates, nationality, and field(s) of specialization.

Essay: Essays range in length from 500 to 1,000 words, with most totaling around 750 words. Each contains basic biographical information—date and place of birth, family information, educational background, positions held, prizes awarded, and so on—but the greatest attention is given to the scientist's work. Names set in small capitals within the essays provide easy reference to other scientists represented in the book.

In addition to the alphabetical list of scientists, readers searching for names of individuals from specific countries or scientific disciplines can consult one of the following indexes found at the end of the book:

Field of Specialization Index: Groups entrants according to the scientific field(s) in which they worked.

Nationality Index: Organizes entrants by country of birth and/or citizenship.

Subject Index: Lists page references for scientists and scientific terms used in the book.

LIST OF ENTRIES

Encyclopedia of
WORLD SCIENTISTS

Agassiz, Jean Louis Rodolphe

(1807–1873)
Swiss/American
Ichthyologist, Geologist, Paleontologist

Through meticulous observation of the natural world and exhaustive research, writing, and lecturing, Louis Agassiz established himself as the major opponent to CHARLES ROBERT DARWIN in the debate over the origins of natural history in the mid-1800s. Agassiz's belief in the Platonic notion that behind visible reality resides an unseen reality that controls the world challenged Darwin's evolutionary model for the origin of the universe. Agassiz first proved himself in Europe as one of the foremost ichthyologists of his time, before focusing his attention on glaciers and introducing the idea of the Ice Age, a period when ice sheets covered most of the Northern Hemisphere. In 1846 he accepted an invitation to lecture in the United States, and he remained in the country for the rest of his life, contributing to science education by introducing new pedagogical practices and instituting new learning facilities.

Agassiz was born on May 28, 1807, in Moutier-en-Vuly, Switzerland, a village on Lake Morat. His resistance to evolutionary theories probably traced back to the influence of his mother, Rose Mayor Agassiz, and his father, Rodolphe, a Protestant pastor. Agassiz married twice: first, in 1832 to Cécile Braun, who had three children before her death in 1847; second, to Elizabeth Cabot Cary, the first president of Radcliffe College, Harvard University's extension devoted to women's education.

Agassiz studied at the Universities of Zurich, Heidelberg, and Munich; he earned his Ph.D. from the Universities of Munich and Erlangen in 1829 and his M.D. from the University of Munich in 1830. He then migrated to Paris, where he studied under Baron GEORGES CUVIER and Baron ALEXANDER VON HUMBOLDT between 1830 and 1832. Cuvier secured his pupil a position under C. F. P. von Martius and J. B. von Spix, cataloging the fish they had taken back to Paris from Brazil, a project that Agassiz took over in 1826. He published the results in the 1829 text *Selecta Genera et Species Piscium*.

Louis Agassiz, who in 1840 introduced the idea of the Ice Age. *Smith Collection, Rare Book & Manuscript Library, University of Pennsylvania.*

Von Humboldt recommended Agassiz for a position at the newly established College of Neuchâtel in 1832. During his tenure there as a professor of natural history he published two of his most important works. *Recherches sur les poissons fossiles,* a five-volume series, which appeared between 1833 and 1843, cataloged over 1,700 fossil fish, resurrecting interest in extinct species in general. In 1840 Agassiz published *Études sur les glaciers,* based on his observations begun in 1836 of the movement of the Aar Glacier in Switzerland. He postulated that all of Europe had been covered by a single sheet of ice in recent geological time, but that such ice ages tend to ebb and flow.

In 1846 the Lowell Institute in Boston, Massachusetts, invited him to lecture there; the visit led to permanent residence in the United States. He accepted a professorship of zoology in 1847 at the Lawrence Scientific School of Harvard University, a post he held until his death, and in 1859 he published his *Essay on Classification,* a tract that expressed his belief in special creation as opposed to the evolutionary model proposed by Darwin in *On the Origin of Species.* By the time of his death Darwinism had not yet supplanted the predominance of Agassiz's view of natural history.

Agassiz's more lasting achievements were in education, where he posited the innovative approach of learning natural history by interacting with nature instead of gleaning information from books. Agassiz realized his commitment to education by raising over $600,000 in public and private funds to establish the Harvard College Museum of Comparative Zoology in 1859. In 1873 he also established the Anderson School of Natural History on Penikese Island, off Cape Cod, Massachusetts, a marine biology center and summer school. Unfortunately his belief that there was more than one species of human being fueled the theory used to defend slavery.

Agassiz received the Wollaston Medal of the Geological Society of London and the Copley Medal of the Royal Society of London. He remained active until the end of his life, taking an expedition through the Strait of Magellan just before he died of a cerebral hemorrhage on December 14, 1873, in Cambridge, Massachusetts.

Agnesi, Maria Gaetana

(1718–1799)
Italian
Mathematician

Maria Gaetana Agnesi summarized the mathematics of her day in a textbook that was still popular 50 years after she wrote it. Born on May 16, 1718, in Milan, Italy, she was a child prodigy. Her father, Pietro, a mathematics professor at the University of Bologna, loved to show off her abilities. By the age of 11 she spoke French, Latin, Greek, German, Spanish, and Hebrew.

Agnesi's chief love was mathematics. She could solve difficult geometry problems at age 14. At age 20 she began a 10-year project of writing a two-volume textbook that summarized European mathematical discoveries. The first volume covered algebra and geometry; the second was devoted to calculus, a field of mathematics invented in the previous century by Sir ISAAC NEWTON and Gottfried von Leibniz. According to one biographer, Agnesi worked so hard on the book that she sometimes walked to her desk in her sleep and solved problems that had been troubling her.

Agnesi's book, *Le Instituzioni Analitiche* (Analytical institutions), was published in 1748 to great praise. Empress Maria Theresa of Austria, to whom Agnesi dedicated the book, sent her a diamond ring and a crystal box in recognition of the significance of her work and as a way of expressing her thanks for the dedication. M. Motigny of the French Academy of Sciences wrote to her, "I admire particularly the art with which you bring under uniform methods the divers [various] conclusions scattered among the works of geometers and reached by methods entirely different." Pope Benedict XIV offered Agnesi her father's old post at the University of Bologna in 1750, but she declined because she did not want to leave Milan, where she had lived her entire life. The science historian Margaret Alic calls Agnesi's book "the first systematic work of its kind" and states that "50 years later it was still the most complete mathematical text in existence."

Agnesi had never enjoyed public life, and after her father's death in 1752 she retired from the academic world. She headed a charitable institution and turned part of her home into a hospital. When she died on January 9, 1799, she was as renowned for her good works as she had been for her learning.

In later centuries Agnesi's name was remembered mainly in the form of a strange and ironic mistranslation. Because of a confusion of the Italian words for "curve" and "witch," a curve described in her book, called a versed sine curve, acquired the English name *witch of Agnesi.* As a result of this error in translation this saintly woman was sometimes referred to by the same name.

Agricola, Georgius

(1494–1555)
German
Mineralogist, Geologist, Metallurgist

Though his exhaustive knowledge of diverse subjects earned him the title "the Saxon Pliny," Georgius Agricola was best known as the author of *De re metallica libri XII* (On the subject of metals), a seminal text in the under-

Quid Medici poſſent manibus? quas iungere plagas
Vlceribus ſordes, ſigna mouere loco?
Extitit hic ſolus qui pondera, viſcera Terræ
Rimatus, nobis bella metalla fodit.

H 2

Georgius Agricola, whose great knowledge of metallurgy and mining was commemorated in 1912 by future U.S. president Herbert Hoover, who prepared an English edition of Agricola's masterwork, *On the Subject of Metals. Smith Collection, Rare Book & Manuscript Library, University of Pennsylvania.*

standing of metallurgy and the mining and smelting processes of the time. Living in the mining capitals of St. Joachimsthal in Bohemia (Czechoslovakia) and Chemnitz, Germany, Agricola had extensive exposure to every aspect of mining, including the management of the mines and the machinery used, such as pumps, windmills, and waterwheels, which he incorporated into his books.

Born on March 24, 1494, in Glauchau, Saxony, to Gregor Bauer, a dyer and wool draper, Georg Bauer latinized his name to Georgius Agricola, as was the custom at the time. His youngest and favorite brother followed his footsteps to Chemnitz in 1540 to become a metallurgist, and his oldest brother entered the priesthood in Zwickau. In 1526 Agricola married the widow of Thomas Meiner, the director of the Schneeberg mining district; she died in 1541. In 1542 Agricola remarried, this time to Anna Schütz, the daughter of the guild master and smelter owner Ulrich Schütz, who entrusted his wife and children to Agricola's care upon his death in 1534.

In 1514 at the late age of 20 Agricola entered Leipzig University, where he earned his B.A. in 1515. The university retained him as a lecturer in elementary Greek for the next year, after which Agricola went to Zwickau, where he organized a new Schola Graeca in 1519. In 1520, he wrote his first book, *De prima ac simplici institutione grammatica*, which described new humanistic methods of teaching. Agricola then fled from the radicalism of the Reformation back to Leipzig, where he studied medicine between 1523 and 1526. During this time he served on the editorial board in Bologna and Venice for the Aldina editions of texts by GALEN and HIPPOCRATES OF COS, interests he maintained later in life. After earning his M.D., he left Italy via the mining districts of Carinthia, Styria, and the Tyrol bound for Germany, where he stayed briefly until the mining city of St. Joachimsthal elected him town physician and apothecary in 1527.

In 1534 Agricola departed for Chemnitz, yet another city renowned for its mining, which elected him mayor in 1545. The combination of chronically sick miners and heavy metals allowed Agricola to investigate the pharmaceutical uses of minerals. Agricola published on a wide range of topics, including politics and economics. In 1554 he published *De peste libri III*, based on his experiences administering care to sufferers of the black plague that swept through Saxony between 1552 and 1553.

De re metallica libri XII (On the subject of metals), his crowning achievement, did not appear until 1556, four months after his death. The text surveyed all aspects of mining at the time, from working conditions to metallurgy to smelting processes. Agricola had finished writing it during his return visit to St. Joachimsthal, where he had started drafting it 20 years earlier. While there he met the designer Blasius Weffring, who spent the next three years creating 292 woodcuts to illustrate the text. A year later Phillipus Bech translated the work into Old German but retained the woodcuts, creating an edition so fine that it survived 101 years in seven editions.

Agricola died on November 21, 1555, in Chemnitz. The mining engineer and future United States president Herbert Hoover revived Agricola's legacy in 1912 by preparing an English edition of his masterwork, *On the Subject of Metals*, a testament to the primacy of Agricola's work.

Ajakaiye, Deborah Enilo

(c. 1940–)
Nigerian
Geologist

Deborah Ajakaiye has studied the geophysics of Nigeria, where she was born around 1940. Unlike many traditional Africans, Ajakaiye's parents believed in education for girls as well as for boys and encouraged her schoolwork as well as her career pursuits. She has said that it was a primary school teacher who first awakened her interest in science. She received her bachelor's degree from University College in Ibadan, Nigeria, in 1962. She then went on to graduate training at the University of Birmingham in Britain, from which she received her master's degree, and Adhadu Bello University in Nigeria, from which she received her Ph.D. in 1970. "I chose . . . geophysics because I felt that this field could make possible significant contributions to the development of my country," she wrote in a 1993 paper for the American Association for the Advancement of Science.

Ajakaiye points out that geophysics can help a country identify valuable natural resources. For instance, she says, Africa is rich in several minerals needed by high-technology industries, and some parts of the continent, including Nigeria, possess large deposits of uranium, oil, natural gas, and coal. Selling these resources can give a country the money it needs to feed, house, and educate its people. Geophysics can also identify sources of precious groundwater and help to predict natural disasters.

Ajakaiye has looked for all these resources in Nigeria. In some studies she used a new technique called geovisualization, in which computers produce three-dimensional images of materials below the Earth's surface. Ajakaiye and her students, who included several women, also carried out a survey for a geophysical map of northern Nigeria. "By the end of the survey quite a few Nigerian men had changed their attitudes toward their female counterparts," she noted.

In addition to conducting research, Ajakaiye has taught at Adhadu Bello University and the University of Jos, both in Nigeria. In recent years she has been professor of physics at the University of Jos and the dean of the university's natural science faculty. She was the first woman professor of physics in West Africa, the first woman dean of science in Nigeria, and the first female fellow of the Nigerian Academy of Science.

Alexander, Hattie Elizabeth

(1901–1968)
American
Microbiologist

Hattie Alexander discovered a way to prevent most deaths caused by one form of meningitis, a devastating brain disease. Born on April 5, 1901, she grew up in Baltimore, Maryland. She was the second child of William Bain Alexander, a merchant, and his wife, Elsie.

Alexander preferred athletic activities to studying while at Goucher College in Towson, Maryland, and her grades were only Cs. After her graduation in 1923, however, she began working as a bacteriologist for state and national public health services, and she saved her money so that she could go to medical school. As a medical student at Johns Hopkins University she earned very high grades. She completed her M.D. in 1930. Alexander spent her career at Babies Hospital, part of the Columbia-Presbyterian Medical Center in New York City. Eventually she headed its microbiology laboratory. She also taught in the Columbia University medical school; she became a full professor in 1958. Her specialty was meningitis, a disease of the membranes around the brain that was almost always fatal, especially in children. Several types of microorganisms could cause meningitis, but Alexander concentrated on just one, a rod-shaped bacterium called *Hemophilus influenzae.*

When Alexander began her research, there was no effective treatment for *Hemophilus influenzae* meningitis. In the late 1930s, however, she heard of a technique in which rabbits were injected with bacteria. Reacting to the invaders, the rabbits' immune systems produced substances, collectively called antiserum, that could be used as a treatment for the disease caused by that kind of bacterium.

With Michael Heidelberger, Alexander injected rabbits with *Hemophilus influenzae* from children with meningitis. In 1939 she reported that the resulting antiserum had cured several infants. It lowered the disease's death rate by 80 percent by the end of its second year of use.

In the early 1940s Alexander began treating meningitis with antibiotics, which had just come into use, as well as with her antiserum. She noticed that bacteria sometimes developed resistance to the drugs and was one of the first to conclude that this resistance was due to mutations in the microorganisms' genes. In 1944 the American researcher Oswald Avery claimed that the inherited information in genes was carried in a complex chemical called deoxyribonucleic acid (DNA) and reported that changing a bacterium's DNA changed the characteristics of future bacterial generations. Many researchers doubted these conclusions, but Alexander supported them by producing results like Avery's with *Hemophilus influenzae.*

Alexander won several prizes for her work, including the E. Mead Johnson Award for Research in Pediatrics (1942). She was the first woman president of the American Pediatric Society (1964) and one of the first to head any national medical society. She retired in 1966 but continued to work almost until her death from cancer on June 24, 1968.

Alfvén, Hannes Olof Gösta

(1908–1995)
Swedish
Physicist

Numerous scientific discoveries have been attributed to Hannes Olof Gösta Alfvén in fields such as space physics, astrophysics, and plasma physics, a field that Alfvén pio-

Hannes Alfvén, who was the first space scientist to be awarded the Nobel Prize. *AIP Emilio Segrè Visual Archives.*

neered. Alfvén is recognized as the father of magnetohydrodynamics (MHD), the study of plasmas, or ionized gases, in magnetic fields, and received the 1970 Nobel Prize in physics for this work. He was the first space scientist to be awarded the Nobel Prize.

Born on May 30, 1908, in Norrkoeping, Sweden, Alfvén was the son of Johannes Alfvén and Anna-Clara Romanus Alfvén. Both of his parents were physicians. Alfvén married Kerstin Maria Erikson in 1935, and the couple had five children.

After attending elementary and secondary school in his hometown, Alfvén entered the University of Uppsala; he earned a doctorate in physics in 1934. He then took a position as lecturer in physics at the university, and three years later became research physicist at the Nobel Institute of Physics. In 1940 Alfvén joined the staff of the Royal Institute of Technology in Stockholm, where he stayed until 1967. During the years at the institute Alfvén held a variety of positions, including professor of the theory of electricity, professor of electronics, and professor of plasma physics.

Alfvén's work in the late 1930s and early 1940s dealt with astronomical phenomena, particularly plasma. Plasma existed at extremely high temperatures, which could be found in stars, and was a gaslike blend of positively charged ions and electrons. Atoms and molecules were completely charged at such high temperatures. Alfvén's observations of plasma were a result of his investigations of sunspots, magnetic storms, and other stellar happenings. One of Alfvén's first important findings was his theorem of frozen-in flux, which stated that a plasma is "frozen" to magnetic field lines. Under certain conditions, Alfvén proposed, the magnetic field associated with a plasma moves together. This discovery contributed greatly to the field of space physics. In 1939 Alfvén announced his findings on magnetic storms and light displays in the atmosphere. His theories significantly influenced modern theories of the Earth's magnetic field.

The study of sunspots led to Alfvén's revolutionary discovery of hydromagnetic waves, which were named Alfvén waves in his honor. At the time of his finding, the traditional belief was that electromagnetic waves were not capable of significantly penetrating a conductor. Alfvén suggested, however, that electromagnetic waves were actually very good conductors with the ability to go through highly charged solar gas. He announced his theory in 1942, but it was generally ignored and rejected. During a lecture at the University of Chicago six years later Alfvén presented his theory once again, and it was well received, thanks in part to ENRICO FERMI's acceptance of the concept.

One of Alfvén's goals in his work with plasma and MHD was to gain an understanding of the origin of the universe. Alfvén believed the universe was formed from plasma and was a supporter of "plasma cosmology," which opposed the big-bang theory of the creation of the universe. Plasma cosmology suggested that the universe had no beginning or probable end, and that the universe was formed by the electric and magnetic forces of plasma, which affected the organization of star systems and other observed structures.

Though Alfvén's studies focused on space, many of his findings had practical applications. Attempts to control nuclear fusion, for instance, were based on MHD. During Alfvén's later years he became active in the antinuclear movement, believing that nuclear power was dangerous in any application. Disagreements with the Swedish government over nuclear power policies influenced him to leave Sweden and accept a position at the University of California at San Diego in 1967. Alfvén received a number of honors and awards in addition to the Nobel Prize, which he shared with LOUIS EUGÈNE FÉLIX NÉEL. These included the Gold Medal of the Royal Astronomical Society (1967), the Franklin Medal of the Franklin Institute (1971), and the Bowie Gold Medal (1987). Alfvén died on April 2, 1995, in Stockholm, Sweden.

Altman, Sidney
(1939–)
Canadian/American
Molecular Biologist

For pioneering discoveries on ribonucleic acid (RNA) molecules Sidney Altman received the 1989 Nobel Prize in chemistry. Altman found that RNA molecules, which were thought to act only as carriers of genetic codes among various parts of the living cell, were actually able to act as enzymes, thus dispelling the commonly accepted notion that all enzymes consisted of protein and that only protein molecules were capable of catalyzing chemical reactions in enzymes. Altman shared the Nobel Prize with the University of Colorado's Thomas R. Cech, who independently reached similar results. The pair's discovery laid the foundation for new areas of scientific research and biotechnology and shed light on old theories concerning the functioning of cells.

Altman was born on May 8, 1939, in Montreal, Quebec, Canada. His father, Victor Altman, was a grocer, and his mother, Ray Arlin, had worked in a textile factory prior to her marriage. Altman married Ann Korner in 1972 and had two children.

After graduation from West Hill High School in Montreal, Altman entered the Massachusetts Institute of Technology (MIT), where he studied physics. After earning a bachelor's degree in 1960 he worked as a teaching assistant at Columbia University for two years. It was in the early 1960s that Altman changed his emphasis from physics to the up-and-coming interdisciplinary field of molecular biology. Late in 1962 Altman moved to Boulder, Colorado, where he attended the University of Colorado. He studied under Leonard S. Lerman and researched T4 bacteriophage, a virus that infects bacterial cells. In 1967 he received a doctorate in biophysics.

Altman began his investigations of RNA in the late 1960s while a fellow at the Medical Research Council Laboratory of Molecular Biology in Cambridge, England. Altman focused on the transcription of transfer RNA (tRNA), a component of RNA, from deoxyribonucleic acid (DNA), which was known to transfer genetic information into the nucleus of a cell. During his research Altman worked with the enzyme ribonuclease P (RNase P), which was responsible for catalyzing the processing of tRNA. Altman had returned to the United States; at Yale University in 1972 an experiment demonstrated that RNase P was composed of both RNA and a protein, thereby indicating that the RNA had been involved in the enzymatic activity. Because it was commonly accepted that enzymes were composed of protein and not nucleic acids, Altman's discovery was unique. Continuing his studies of enzymatic activity, Altman in 1984 worked with a Yale colleague, Cecilia Guerrier-Takada, to produce only the RNA portion of RNase P. This artificial RNA still acted as a catalyst though there was no protein present, proving conclusively that RNA could act as an enzyme.

Altman's discoveries did much to advance the understanding of proteins and genetic codes. Scientists had long been baffled by whether proteins or nucleic acids had appeared first in the development of life; proteins were responsible for catalyzing biological reactions, but nucleic acids carried the genetic codes that created the proteins. With Altman and Cech's findings this puzzle was solved—nucleic acids could act as both codes and enzymes. In addition to the 1989 Nobel Prize, Altman received the Rosentiel Award for Basic Biomedical Research. Altman is a member of a number of scientific societies, including the National Academy of Sciences and the Genetics Society of America, and has received numerous honorary degrees.

Alvarez, Luis Walter
(1911–1988)
American
Physicist

In his first years as a research physicist at the University of California at Berkeley in the 1930s Luis Alvarez earned the title "prize wild idea man," which acknowledged both his wide-ranging investigations and his ability to identify important questions in need of solutions. A member of the Manhattan Project, which developed the atomic bomb during World War II, Alvarez followed the *Enola Gay,* the plane that dropped the atomic bomb on Hiroshima, Japan, in another B-29 bomber, from which he witnessed the destruction of Hiroshima. Alvarez also modified the bubble chamber technique invented by the University of Michigan physicist Donald Glaser for use in conjunction with particle accelerators to identify previously unknown elementary particles. This work earned him the Nobel Prize in physics in 1968; he never rested on these laurels, remaining active and innovative in physics for the next 20 years.

Alvarez was born on June 13, 1911, in San Francisco, California, to Dr. Walter Clement Alvarez, a research physiologist at the University of California at San Francisco, and Harriet Skidmore Smythe, an Irish woman whose family instituted a missionary school in Foochow, China. Alvarez had two children, Walter and Jean, with his first wife, Geraldine Smithwick, a fellow University of Chicago student, but their relationship ended as a result of the strain of wartime separation. Alvarez had two more children, Donald and Helen, with his second wife, Janet Landis, whom he married in 1958.

Alvarez entered the University of Chicago in 1928, earned a B.S. in 1932, and graduated in 1936 with a Ph.D. in physics. He appreciated his graduate adviser, the Nobel

A member of the Manhattan Project, Luis Alvarez witnessed the destruction of Hiroshima from a B-29 bomber that followed the *Enola Gay*, the plane that dropped the atomic bomb. *AIP Emilio Segrè Visual Archives, Physics Today Collection.*

laureate Arthur Compton, not for guiding him but rather for staying out of his way while he immersed himself in his research. At UC Berkeley, he conducted research alongside two other Nobel laureates, ERNEST ORLANDO LAWRENCE and FELIX BLOCH.

Alvarez quickly earned his reputation as the wild idea man at UC Berkeley with three significant discoveries in the 1930s. Within his first year there he discovered how atomic nuclei decayed and orbital electrons absorbed them, a process known as K-electron capture. He then invented a mercury vapor lamp with a student, and with Bloch he developed a process for determining the magnetic moment of neutrons by slowing their motion in a beam. In the early 1940s he conducted research for the military at the Massachusetts Institute of Technology on radio detecting and ranging systems, developing three new types of radar in three years.

Alvarez never suffered from guilt over his role in the development of the atomic bomb and continued to support nuclear technology, as he believed the benefits far outweighed the drawbacks. After the war Alvarez devoted his research to particle physics, building larger and faster particle accelerators. Alvarez transformed Glaser's bubble chamber design to use liquid hydrogen instead of diethyl ether as the liquid through which the particles passed, leaving their bubble tracks. By this method he discovered dozens of new elementary particles.

In the final years of his career Alvarez projected high-energy muon rays (the subatomic particles of cosmic radiation) at King Chephren's pyramid in Gaza, Egypt, to determine whether it had secret chambers. The Warren Commission called upon his physics expertise to verify whether a lone assassin had murdered President John F. Kennedy. Finally he teamed up with his son, the UC Berkeley geologist Walter Alvarez, to hypothesize that the dust created by the Earth's collision with an asteroid caused the "winter" that exterminated dinosaurs, a widely accepted theory based on iridium deposits in Italian sedimentary rock.

Alvarez remained on the faculty of UC-Berkeley until his retirement in 1978. The 1968 Nobel Prize and the 22 patents he held on his inventions attest to the significance of his work. He died of cancer on September 1, 1988, in Berkeley, California.

Ampère, André-Marie

(1775–1836)
French
Physicist, Mathematician

A mathematical child prodigy, self-educated according to the principles of Rousseau, André Ampère established his scientific significance after a fit of inspiration in September and October 1820, when he developed the science of electrodynamics. Both Ampère's law, which established the mathematical relationship between electricity and magnetism, and the ampere, or amp, a unit for measuring electrical current, were named after him. Though Ampère maintained wide-ranging interests in mathematics, chemistry, metaphysics, philosophy, and religion, his fame was mainly the result of his work with electricity and magnetism.

Ampère was born in Lyon, France, on January 22, 1775, to Jean-Jacques Ampère, an independent merchant who provided his son with a complete library, and Jeanne Desutières-Sarcey Ampère, a devout Catholic who instilled her faith in her son. Ampère spent the rest of his life reconciling his reason with his faith. His personal life amounted to a series of disasters, starting with the execution by guillotine of his father on November 23, 1793, in the midst of the French Revolution. Tragedy took a respite during his relationship with Julie Carron, whom he courted against all odds and married on August 7, 1799. Tragedy revisited him when she died on July 13, 1803, of an illness contracted during the birth of their son Jean-Jacques on August 12, 1800. Ampère then entered into an

André Ampère, for whom the ampere, the unit for measuring electrical current, was named. *AIP Emilio Segrè Visual Archives.*

ill-advised marriage to Jeanne Potot on August 1, 1806. The only positive outcome of the marriage was the birth of his daughter, Albine.

Though Ampère did not hold a degree, he taught mathematics at the Lyceum in Lyon before his appointment as a professor of physics and chemistry at the École Centrale in Bourg-en-Bresse in February 1802. In 1808 Napoleon appointed him inspector-general of the new university system, a post Ampère held until death. In 1820 the University of Paris hired him as an assistant professor of astronomy, and in August 1824 the Collège de France appointed him as the chair of the experimental physics department.

In early September 1820 François Arago reported to the Académie des Sciences on the discovery by the Danish physicist HANS CHRISTIAN ØRSTED that a magnetic needle was deflected when current in nearby wires varied, thus establishing a connection between electricity and magnetism. In less than a month Ampère presented three papers to the Académie that established the science of electrodynamics by positing that magnetism was simply electricity in motion. Specifically Ampère worked with two parallel wires with electrical current flowing through them: He discovered that the currents attracted each other when heading in the same direction and repelled one another when traveling in opposite directions. The implications of his experiments suggested a whole new theory of matter. Ampère published a comprehensive overview of his findings in 1827 in his *Memoir on the Mathematical Theory of Electrodynamic Phenomena, Uniquely Deduced from Experience.* The scientific community did not embrace his theories until Wilhelm Weber incorporated them into his theory of electromagnetism later in the century.

Ampère died on June 10, 1836, alone on an inspection tour in Marseilles, France. Despite a tragic personal life, he had been successful in his profession, contributing an entire field to the study of science.

Ancker-Johnson, Betsy

(1927–)
American
Physicist, Engineer

Betsy Ancker-Johnson made important contributions to understanding the behavior of plasmas in solids. She also held high-level posts in government and the automobile industry. Born on April 29, 1927, to Clinton J. and Fern Ancker in St. Louis, Missouri, she spent what she has called her "idyllic" years studying physics at Wellesley College and graduating in 1949 with high honors.

The happy times ended when Ancker decided to follow her "love of adventure" and interest in other cultures and do graduate work at Tübingen University in Germany. The German professors "told me that women can't think analytically and I must, therefore, be husband-hunting" rather than seriously pursuing a career, Ancker-Johnson recalled in a 1971 talk. Nonetheless, she obtained her Ph.D. with high honors in 1953.

On returning to the United States, Ancker encountered equal skepticism when she tried to find a job. She discovered that "a woman in physics must be at least twice as determined as a man with the same competence in order to achieve as much as he does." Physicists were much in demand, but she was offered only second-rate jobs. "Not one interviewer ever leveled with me" about the reason, but she believes it was that "I was in a . . . subset that employers had decided was not dependable; i.e., a woman will marry and quit, and what is invested in her goes down the drain."

Ancker finally took a minor academic post at the University of California in Berkeley. There, through the Inter-Varsity Christian Fellowship, for which she did volunteer work, she met a mathematician named Harold Johnson. "My husband is man enough not to be threatened by his wife's awareness of electrons," she said later. They married in 1958, after which she used the name Ancker-Johnson.

Ancker-Johnson did research in solid-state physics for Sylvania Corporation (now GTE) from 1956 to 1958 and for RCA from 1958 to 1961. After her marriage she encountered her employers' fear that she would soon quit to raise a family. She informed them that she did plan to have children but would hire live-in help to care for them while she continued to work. During her first pregnancy, she has said, male executives seemed to view her condition as something like "an advanced case of leprosy." For three months before her first daughter's birth she was not even allowed to enter the laboratory building without the director's permission. By the time she had her second child, Ancker-Johnson was working for Boeing, a "more enlightened" company. This time company officials merely stopped her salary eight weeks before the baby's birth and started it again six weeks afterward, even though she continued working during all but two weeks of that period.

Ancker-Johnson made her chief contributions to plasma and solid-state physics while working for Boeing in Seattle, Washington, from 1961 to 1973. (She was also an affiliate professor of electrical engineering at the University of Washington during this time.) In the early 1970s she was supervisor of the company's solid-state and plasma electronics laboratory and manager of their advanced energy systems. She identified several types of instabilities that can occur in plasmas in solids, including oscillation, pinching, and microwave emission. She produced microwaves by applying an external electric field but showed that the field did not have to be present when they appeared, a new discovery. Building on her work, other scientists have suggested that solid-state plasmas may be useful sources of microwave radiation. Other applications of her work potentially affect computer technology and extraction of aluminum and other elements from low-grade ore.

Betsy Ancker-Johnson holds several patents in solid-state physics and semiconductor electronics. She is a member of the National Academy of Engineering and fellow of several professional societies, including the Institute of Electrical and Electronic Engineers and the American Physical Society. She has won excellence awards from Boeing and the Carborundum Company and the Chairman's Award from the American Association of Engineering Societies. She has been a member of the Board of Directors of the Society of Automotive Engineers, the Motor Vehicle Manufacturers Association, Varian Associates, and General Mills.

From 1973 to 1977 Ancker-Johnson served as the assistant secretary of commerce in charge of science and technology. In this job she controlled six organizations with a $230 million total annual budget. In contrast to the prejudice she had faced earlier—but equally irritating to her—she feels she was hired primarily because she was a woman; she was the first appointed by a president to the Department of Commerce.

After her service with the government ended, Ancker-Johnson worked for 14 months as director for physical research at Argonne National Laboratory, near Chicago. Then in 1979 General Motors made her a vice president in charge of environmental activities. She was the first woman in the auto industry to achieve such a high rank. "Environmental activities" included such considerations as pollution controls, automobile safety, and fuel economy. Ancker-Johnson retired from this job in 1992, but has remained active on many committees and as director of the World Environment Center.

Anderson, Carl David

(1905–1991)
American
Physicist

Carl David Anderson's greatest contributions to physics were his discoveries of the positron, the positive electron, and the muon, a negatively charged particle. Anderson's detection of the positron confirmed the existence of subatomic particles and furthered physicists' understanding of the structure and nature of atoms. Anderson won a Nobel Prize in physics in 1936 for his discovery of the positron.

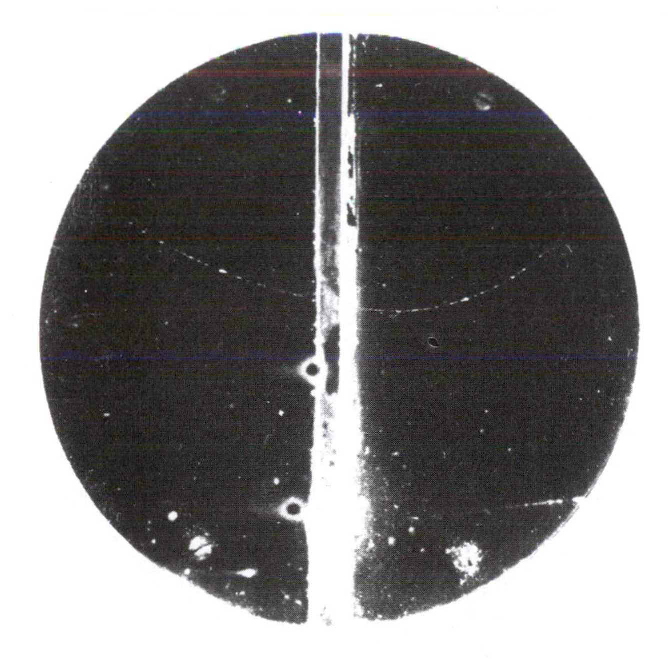

The first image of Carl D. Anderson's positron track, which confirmed the existence of subatomic particles and earned him the Nobel Prize in 1936. *Photo by C. D. Anderson, courtesy AIP Emilio Segrè Visual Archives.*

Born in New York, New York, on September 3, 1905, Anderson grew up in a poor family. His father, who was also named Carl David Anderson, and his mother, Emma Adolfina Ajaxson Anderson, were emigrants from Sweden; his father had arrived in the United States in 1896. Anderson's parents both were from farming families. When Anderson was a child, the family moved to Los Angeles, California, where his father worked as a restaurant manager.

Unable to afford room and board at a faraway college, Anderson enrolled in 1924 at the California Institute of Technology (Caltech), which was located in Pasadena, close enough for a commute. Though initially interested in electrical engineering, Anderson decided to major in physics during his sophomore year. An exceptional student, Anderson received numerous financial grants. After earning a bachelor's degree in physics and engineering in 1927, Anderson remained at Caltech to pursue graduate studies in mathematics and physics. Anderson researched electrons and X rays and received his doctorate in 1930.

After finishing school Anderson stayed at Caltech and worked with the physicist Robert A. Millikan, a Nobel laureate, who had succeeded in measuring the charge of an electron. Millikan was working on measuring the energies of cosmic rays and asked Anderson to assist in the development of a cloud chamber capable of performing such measurements. Anderson and Millikan created a magnetic cloud chamber that could effectively deflect high-energy cosmic particles, and Anderson photographed the tracks of cosmic rays in the chamber. As Anderson analyzed about 1,000 photographs, he noticed that there were as many positively charged particles as there were electrons. Anderson assumed that the positively charged particles were protons, as electrons and protons were the only elementary particles known to exist at that time. As he attempted to prove this, however, he discovered that the mass of the particles was significantly less than the mass of protons. Believing he had found either positively charged electrons or a new particle, Anderson divided the chamber with a lead plate to slow the movement of the particles and allow a better analysis. The resultant photographs confirmed Anderson's suspicion, and in 1932 he announced the discovery of a new particle—the positron, or positive electron.

Continuing his studies of cosmic rays, Anderson discovered yet another new elementary particle in 1936. Using the same cloud chamber in which he had found the positron, Anderson spent four years studying thousands of photographs of the charged particle, which had a mass more than 200 times that of the electron. Anderson named the particle the *mesotron,* a term that was soon shortened to *meson.* The theoretical physicist HIDEKI YUKAWA had predicted the existence of the meson in 1935, and when a particle more similar to the one Yukawa foresaw was discovered in 1947, Anderson's meson was renamed *muon.* He married Lorraine Elvira Bergman in 1946, and the couple had two children.

Anderson remained at Caltech for his entire career, teaching and researching cosmic radiation and elementary particles before retiring in 1976. He succeeded in producing positrons by gamma irradiation in 1933, and during World War II he worked for the U.S. government on rocket research. Besides the 1936 Nobel Prize in physics, which he shared with the Austrian physicist VICTOR FRANCIS FRANZ HESS, Anderson received a number of honors, including medals from the Franklin Institute and the American Institute of the City of New York.

Anderson, Elda Emma

(1899–1961)
American
Physicist, Medical Researcher

Elda Emma Anderson worked on the atomic bomb, then helped to develop a new scientific field, whose goal is to minimize harm from radiation. She was born on October 5, 1899, in Green Lake, Wisconsin, the second of Edwin A. and Lena Anderson's three children. Her older sister, who became a chemistry teacher, was responsible for first igniting her younger sister's interest in science. She received her bachelor's degree from Ripon College in 1922, then attended the University of Wisconsin, where she earned a master's degree in physics in 1924. In 1941 she returned to the university to obtain her Ph.D.

Anderson was dean of physics and mathematics from 1924 to 1927 at Estherville Junior College in Iowa, where she also taught chemistry. In 1929 she became a professor in the new physics department of Milwaukee-Downer College. She became head of the department in 1934.

In late 1941 Anderson took a vacation from teaching to work in the Office of Scientific Research and Development at Princeton University. There she became involved with the Manhattan Project, the code name for the secret project to develop an atomic bomb. She moved to the project's headquarters at Los Alamos, New Mexico, in 1943. She worked—sometimes 18 hours a day—measuring subatomic particles produced in cyclotrons, or atom smashers. This work proved vital to both the development of the bomb and the design of nuclear power reactors.

Anderson returned to teaching in 1947, but her old life seemed dull after the Manhattan Project days. Her research in atomic physics had also stirred her concern about the harm that radiation could do to living things. A new field of science, called health physics, had been established toward the end of the war to discover means to prevent such harmful effects. Anderson left Milwaukee-Downer College in 1949 and devoted the rest of her life to developing health physics and making other scientists recognize its importance.

Anderson became the first chief of education and training for the Health Physics Division at Tennessee's Oak Ridge National Laboratory. She also set up the American Board of Health Physics, a professional certifying agency. Perhaps as a result of her work with radiation Anderson developed leukemia, a blood cell cancer, in 1956. She died on April 17, 1961.

Anderson, Elizabeth Garrett

(1836–1917)
British
Physician

Inspired by ELIZABETH BLACKWELL, Elizabeth Garrett Anderson opened the medical profession to women in Britain, just as Blackwell had in the United States. Garrett was born in Whitechapel, a poor section of London, in 1836. Her father, Newson Garrett, soon became successful in business and moved his family to a large house in the village of Aldeburgh. Elizabeth and her eight brothers and sisters thus grew up in comfort.

In 1859, when Elizabeth Garrett was 23 years old, a friend told her that Elizabeth Blackwell was going to speak in London. She obtained a personal introduction to Blackwell, who assumed that Garrett must be planning to be a physician. That idea had never crossed Garrett's mind until that meeting, but suddenly it began to seem a real possibility. Still, she wrote later, "I remember feeling as if I had been thrust into work that was too big for me."

When Garrett told her parents about her new plans, Louisa Garrett predicted that the "disgrace" of her daughter's action would kill her, and Newson Garrett pronounced the idea of a woman doctor "disgusting." Nonetheless, he agreed to go with her to talk to London physicians.

What Miss Garrett wanted, the doctors said, was impossible. The Medical Act of 1858 said that no physician could be placed on the Medical Register, Britain's list of approved physicians, without a license from a qualified examining board—and no board would allow a woman to take its examinations. Newson Garrett did not take kindly to being told no, whether the refusal was directed at his daughter or at him. The physicians' opposition turned him into Elizabeth's strongest supporter.

Obtaining a physician's license by getting an M.D. degree seemed out of the question, because no British medical school admitted women. However, the charter of the Society of Apothecaries—medical practitioners who made and distributed drugs—said that the society would grant a license to "all persons" who completed five years of training with a qualified doctor or doctors, took certain required classes, and passed its examination. An apothecary's license was not as prestigious as an M.D. degree, but it would get Garrett onto the Medical Register.

Facing the Garrett father-daughter team in August 1861, the society directors had to admit that Elizabeth Garrett was a person and therefore could potentially qualify for a license. They told her to return when she had completed their requirements—hoping, no doubt, never to see her again.

Bit by bit Garrett accumulated the training she needed, and in 1865 she returned to the apothecaries with proof in hand. The society tried to back out of its earlier promise, but after Newson Garrett threatened a lawsuit, it allowed Elizabeth to take the examinations. She found them "too easy to feel elated about" and earned a higher score than anyone else had. By 1866 she had her apothecary's license and her spot on the Medical Register.

Once Garrett set up her medical practice, friends and acquaintances flocked to her. One, the women's rights advocate Josephine Butler, commented, "I gained more from her than [from] any other doctor; for she . . . entered much more into my mental state and way of life than they could." In addition, Garrett opened a small clinic, St. Mary's Dispensary for Women and Children, in a poor section of London. In 1872 the clinic would become the New Hospital for Women. It was renamed the Elizabeth Garrett Anderson Hospital in 1917, after its founder's death.

Meanwhile Garrett wanted to spend more time in hospital work to enhance the practical side of her medical training. In 1869 she applied for a post at London's Shadwell Hospital for Children. One director on the hospital board who was sure he did not want her to work there was James George Skelton Anderson, the Scottish head of a large shipping company. Once he met the young woman doctor, however, he changed his mind. Garrett and Anderson were married in 1871 and later had three children. Their daughter, Louisa, also became a physician and during World War I headed the first group of British women doctors to serve in active duty in wartime.

Elizabeth Anderson was still determined to obtain an M.D. degree. In 1868 the University of Paris had opened its doors to women, and Anderson received permission to take the school's examinations for physicians even though she had not actually studied there. She passed the test and obtained her M.D. at last on June 15, 1870. The British medical journal *Lancet* reported, "All the [French] judges are complimenting Miss Garrett [and have] . . . expressed liberal opinions on the subject of lady doctors."

Working as a physician was only part of Elizabeth Garrett Anderson's busy life. At the London School of Medicine for Women, the teaching arm of her New Hospital for Women, she taught classes and served in the administration as dean; she was president of the school between 1886 and 1902. She was elected to the London school board and at age 71 became mayor of Aldeburgh, where she and her husband had retired in 1902. She was the first woman elected mayor in Britain.

Elizabeth Garrett Anderson died on December 17, 1917, at the age of 81. A fellow physician said of her, "She did more for the cause of women in medicine in England than any other person."

Anfinsen, Christian Boehmer

(1916–1995)
American
Biochemist

For ground-breaking work on enzymes the biochemist Christian Boehmer Anfinsen won the 1972 Nobel Prize in chemistry. Anfinsen investigated the shape and activity of enzymes, specifically the enzyme ribonuclease (RNase), and found that an enzyme's structure was closely tied to its function. He later demonstrated similar behavior in other proteins. Anfinsen's work significantly advanced the understanding of the nature of enzymes.

Born on March 26, 1916, in Monessen, Pennsylvania, near Pittsburgh, Anfinsen was the son of Christian Anfinsen, an engineer who had emigrated from Norway, and Sophie Rasmussen Anfinsen, who was also of Norwegian descent. In 1941 Anfinsen married Florence Bernice Kenenger, and the couple had three children. A year after divorcing in 1978, Anfinsen married Libby Esther Schulman Ely.

For his undergraduate education Anfinsen attended Swarthmore College, where he received a bachelor's degree in 1937. He then went to the University of Pennsylvania to study organic chemistry and completed a master's degree in 1939. After study abroad at the Carlsberg Laboratory in Copenhagen, Denmark, through a fellowship from the American Scandinavian Foundation, Anfinsen returned to the United States and began studying biochemistry at Harvard University in 1940. Anfinsen received his doctorate in 1943. His doctoral thesis concentrated on enzymes, particularly those enzymes in the retina of the eye.

Anfinsen began his investigation of the function and structure of enzymes in the mid-1940s. Enzymes are a type of protein, and proteins are composed of blocks of amino acids. As the amino acid chain folds, the enzymes form three-dimensional shapes, known as tertiary structures. Though there are 100 possible ways for one set of amino acids to bond together for one enzyme, only one arrangement results in an active enzyme. Anfinsen wanted to determine how a set of amino acids knew how to arrange in such a way as to produce an active enzyme. Anfinsen chose to focus on the enzyme RNase, which was responsible for breaking down the ribonucleic acid, also known as RNA, in food. He believed he could gain an understanding of how an enzyme protein is built and is configured into an active form by studying the addition of each amino acid.

A breakthrough in his research occurred during a leave of absence from the National Institutes of Health, his employer. Anfinsen spent 1954 to 1955 studying under the chemist Kai Linderstrøm-Lang, who encouraged him to start with the entire RNase molecule, then break it down block by block, thereby reversing his existing research methods. Anfinsen was able to observe that the severing of particular key bonds brought about the formation of other bonds between the amino acids, which resulted in an inactive form of RNase. By 1962, after Anfinsen had returned to the United States, he had observed that the inactive form of RNase automatically reverted to an active configuration when placed in particular environments. His findings thus confirmed that the configuration for an active enzyme was present within the protein's sequence of amino acids.

In 1972 Anfinsen received the Nobel Prize in chemistry, which he shared with STANFORD MOORE and WILLIAM H. STEIN, who in 1960 had found the precise amino acid sequence of RNase, the first enzyme for which the full sequence was discovered. After being awarded the Nobel Prize, Anfinsen turned his attention to studies of the protein interferon. Throughout his career Anfinsen worked at a number of institutions, including Harvard Medical School, the Medical Nobel Institute in Sweden, the National Institutes of Health, and Johns Hopkins University. Anfinsen received numerous honorary degrees and was a member of several societies, including the National Academy of Sciences and the American Society of Biological Chemists. He died at home on May 14, 1995, of a heart attack.

Anning, Mary

(1799–1847)
British
Paleontologist

Mary Anning discovered fossils that helped give British scientists their first understanding of prehistoric life. Born in 1799, she grew up in Lyme Regis, on a part of Britain's southwest coast that had been a sea bottom 200 million years before. People in Lyme Regis often found bones, shells, or other remains of the creatures that had lived in that long-ago ocean. Now turned to rock, these bones and shells could be found sticking out of the seaside cliffs or washed down onto the beach.

Lyme Regis became a popular resort in the late 18th century, and some townsfolk with an eye for business began collecting fossils to display or sell to visitors. One such businessman was Richard Anning, who earned most of his living from cabinetmaking. He taught his wife,

Molly, and his children, Joseph and Mary, how to look for fossils during walks beside the cliffs. When he died in 1810, leaving his family with little money, they tried to support themselves through their fossil business.

In 1811 Joseph made the family's first important find, a huge skull with a long snout and rows of sharp teeth embedded in a rock on the beach. He thought the skull belonged to a crocodile, but in fact it was that of an ancient dolphinlike sea reptile called an ichthyosaur, or "fish-lizard." A year later 12-year-old Mary found the rest of the animal's 30-foot-long skeleton projecting from a cliff. The two fossils found by the brother and sister added up to one of the first ichthyosaurs ever discovered.

Mary Anning, whom one visitor described as "a strong, energetic spinster . . . tanned and masculine in expression," continued to find and sell prize fossils all her life. In addition to several more ichthyosaurs, she found the first complete skeletons of plesiosaurs, nine-foot-long sea reptiles with small heads, long necks, and paddlelike fins. She found her first plesiosaur in 1821; in 1828 she found the first British pterosaur, a flying reptile.

Anning corresponded with scientists and collectors all over England. Her fossils provided important study material for researchers in the new field of paleontology, and if she had belonged to a time and class in which women were educated, she herself might have become a paleontologist. As it was, she was merely a fairly successful businessperson and a local curiosity. When she died in 1847, a guidebook commented that her "death was in a pecuniary [financial] sense a great loss to the place, as her presence attracted a large number of distinguished visitors." More flattering to Anning, and in greater acknowledgment of her scientific accomplishments, was the dedication of a stained-glass window in her honor, paid for by the scientists whom she had served with her discoveries.

Apgar, Virginia

(1909–1974)
American
Physician

A fellow physician once remarked, "Every baby born in a modern hospital anywhere in the world is looked at first through the eyes of Virginia Apgar." He meant that the quick tests of a newborn infant's health, introduced by Apgar in 1952, are used in hospitals everywhere.

Virginia Apgar was born on June 7, 1909, to Charles and Helen Apgar in Westfield, New Jersey. She shared a love of music with her family, playing the violin during family concerts and even making her own instruments as an adult. Her family also taught her to appreciate science; her businessman father's hobbies included astronomy and wireless telegraphy.

During her undergraduate years at Mount Holyoke College, Apgar helped to pay for her schooling by waiting on tables and working in the school library and laboratories. She also reported for the college newspaper, won prizes in tennis and other sports, acted in plays, and played in the orchestra.

After graduation from Mount Holyoke in 1929 Apgar attended medical school at Columbia University, where she earned her M.D. in 1933. She wanted to be a surgeon, but her professors convinced her that she could never earn a living in that male-dominated field. "Even women won't go to a woman surgeon," she remarked later; when asked why, she sighed, "Only the Lord can answer that one."

Apgar turned her attention to the new specialty of anesthesiology instead. She was only the 50th physician certified as a specialist in administering painkilling and sleep-inducing drugs during surgery. Being a woman was no problem in the field, since anesthesia had previously been given by nurses, most of whom were women. Apgar began teaching at Columbia's medical school in 1936, and she became the school's first professor of anesthesiology and first woman professor in 1949. Beginning in 1938 she was also clinical director of Presbyterian Hospital's anesthesiology department. She was the hospital's first woman department head and helped to establish anesthesiology as a medical specialty.

Within anesthesiology Apgar focused on anesthesia given during birth. After assisting at some 17,000 births during her career, she wrote in the 1972 book *Is My Baby All Right?* "Birth is the most hazardous time of life. . . . It's urgently important to evaluate quickly the status of a just-born baby and to identify immediately those who need emergency care." Yet, she noticed, a newborn baby was often simply wrapped up and hustled off to the hospital nursery. Serious problems sometimes were undetected for hours or days, and by the time they were found, it was too late to treat them.

"I kept wondering who was really responsible for the newborn," Apgar later told a reporter, and apparently she decided that she was. "I began putting down all the signs about the newborn babies that could be observed without special equipment and that helped spot the ones that needed emergency help." The result was the Apgar score system, five tests that a doctor or nurse could perform in a few seconds during the first minute or so after a birth. The tests, which Apgar introduced in 1952, rate a baby's color, muscle tone, breathing, heart rate, and reflexes on a scale of 0 to 2; the combined results are the Apgar score. A combined score of 10 indicates a very healthy baby, whereas a low score warns of problems needing immediate treatment.

After 33 years at Columbia, Virginia Apgar surprised her colleagues by going back to school. In 1959 at the age of 49, she earned a master's degree in public health from Johns Hopkins University. At this same time the charity organization called the National Foundation-March of Dimes, founded to help children with polio, was changing its focus to birth defects, which affect 500,000 children born in the United States each year. The charity asked Apgar to direct its department of birth defects. "They said they were looking for someone with enthusiasm, who likes to travel and talk," Apgar recalled. "I love to see new places, and I certainly can chatter." She knew little about birth defects, but she learned.

Apgar's work for the foundation included writing, distributing research grants, fund-raising, and public speaking around the world. She traveled some 100,000 miles a year for the group. It was said to be largely a result of her efforts that the foundation's annual income rose from $19 million when she joined them to $46 million at the time of her death. She became director of the foundation's basic research department in 1967. In 1965 Apgar also became the first person to lecture on birth defects as a medical subspecialty.

Apgar received many honors for her work, including the ELIZABETH BLACKWELL Citation from the New York Infirmary in 1960 and the American Society of Anesthesiologists' Distinguished Service Award in 1961. The *Ladies' Home Journal* named her their Woman of the Year in science in 1973. The Alumni Association of the Columbia College of Physicians and Surgeons awarded her its Gold Medal for Distinguished Achievement in Medicine in 1973; she was the first woman to win this prize. Apgar died of liver disease on August 7, 1974, at the age of 65.

Archimedes

(287 B.C.–212 B.C.)
Sicilian/Greek
Physicist, Mathematician

KARL FRIEDRICH GAUSS considered Archimedes one of the two greatest mathematicians in history; only ISAAC NEWTON was his equal. Archimedes' estimation for the numerical value of pi survived as the best approximation available into the Middle Ages. However, he was most renowned for his practical applications of mathematical and physical theories. Two of his innovations, Archimedes' principle and the Archimedes screw, involved the displacement of water, and he was considered the founder of hydrostatics.

Though little is known about Archimedes' personal life, his own writings reveal the identity of his father, the astronomer Phidias. Archimedes was born in Syracuse on the island of Sicily in about 287 B.C. Evidence suggests that he traveled to Alexandria, where he studied mathematics under the successors of EUCLID. He returned to Sicily, which was under the rule of King Hieron II, supposedly a relative or at least a friend (Archimedes dedicated *The Sand-Reckoner* to Hieron's son, Gelon).

In this late text Archimedes contrived a notation system for very large numbers, and in another text, *Measurement of the Circle,* he estimated the value of pi as 22/7, a relatively accurate figure. However, his applications of mathematical concepts proved even more profound than his abstract realizations. In *De architectura,* Vitruvius told the dubious story of Archimedes' solution to a problem posed by Hieron—how to test whether a gift crown was indeed pure gold, as the giver claimed, or alloyed with less precious metals, as Hieron suspected. As Archimedes pondered the problem in the bathtub, he noticed that the farther he immersed his body in the tub, the more water spilled over the edge. He hypothesized that the density of the displaced water equaled the density of his submerged body. In his excitement he rushed through the streets naked shouting, "Eureka!" While testing the authenticity of the crown, he noticed that a block of pure gold equal in weight to the crown displaced less water than the crown, thus casting doubt on its composition.

This test, which hinged on relative density and buoyancy, became known as Archimedes' principle. Archimedes described this principle, along with his understanding of buoyancy (or the upward force exerted on solids by liquids), in *On Floating Bodies,* a text that established him as the founder of hydrostatics. Of even greater practical value was his invention of what became known as the Archimedes screw, a device that could draw water along an ascending helix and was used for raising water in irrigation systems.

Archimedes was apparently most proud of his formulation of the volume of a sphere as two-thirds that of the cylinder in which it is inscribed, as discussed in one of his most famous works, *On the Sphere and the Cylinder.* When Cicero was the quaestor, or chief financial officer, of Sicily in 75 B.C., he tracked down Archimedes' grave and verified that it was indeed inscribed with a sphere and cylinder as well as the formula for their intersection.

Hieron called upon Archimedes to invent weapons to stay the Roman invasion of Sicily in 215 B.C., led by Marcellus. Experts doubt that Archimedes invented the weapon of mirrors that ignited distant ships with focused sunlight, though he did invent various catapults. Marcellus ultimately captured Sicily in 212 B.C., and though the general himself admired Archimedes' work, his soldiers put the mathematician to death, supposedly while he was making calculations in the sand.

Aristarchus of Samos

(320 B.C.–250 B.C.)
Greek
Astronomer

Recognized as one of the first to propose and support the theory that the Sun is the center of the universe and that

Earth revolves around the Sun, Aristarchus of Samos was an astronomer and mathematician whose ideas were not universally embraced during his lifetime. Another achievement of Aristarchus was his attempt to calculate astronomical distances by using mathematics, specifically geometry. Aristarchus estimated the relative distances of the Earth from the Sun and the Moon, and though his results were highly inaccurate, his method was correct. It was also the first effort to measure astronomical distances with a method more advanced than mere speculation.

Little is known about the life of Aristarchus. He was born around 320 B.C. on Samos, an island in eastern Greece in the Aegean Sea. The area was the heart of Ionian civilization, where science and philosophy flourished. Aristarchus studied under Strato of Lampsacos, who was the third head of the Lyceum, the school outside Athens, Greece, founded by the philosopher Aristotle. It is believed Aristarchus was taught by Strato in Alexandria rather than at Athens. Aristarchus' colleagues referred to him as "the mathematician," and references by the Roman architect and writer Vitruvius indicate that Aristarchus was among seven who were highly skilled in all areas of mathematics and capable of using their knowledge for practical applications. Vitruvius also attributed to Aristarchus the invention of a commonly used sundial.

Best known for his heliocentric view of the universe, which placed the Sun at the center, Aristarchus faced considerable opposition during his day, despite indications that ideas concerning heliocentrism began as early as the fifth century B.C. with the Pythagoreans in southern Italy. Various theories regarding the rise of heliocentrism exist. Some believe that around 440 B.C. Philolaus proposed that the Earth, Moon, planets, and Sun revolved around a central fire. Philolaus' contemporary Hicetas was credited by others as being the first to ascribe a circular orbit to the Earth. Authorities unanimously, however, attribute the heliocentric theory to Aristarchus. Though only one of Aristarchus' writings, *On the Sizes and Distances of the Sun and Moon* survived, those who followed Aristarchus frequently referred to his theories. The mathematician and engineer ARCHIMEDES, who lived shortly after Aristarchus, discussed in a treatise Aristarchus' opinion that the Sun and stars were stationary.

Aristarchus' achievements in mathematics tend to be somewhat overshadowed by his astronomical advances, but his mathematical skills were considerable, and scholars often rank him with other famous Greek mathematicians such as EUCLID and ARCHIMEDES. In *On the Sizes and Distances of the Sun and Moon* Aristarchus used a geocentric view, which placed the Earth at the center of the universe. To calculate relative distances from Earth of the Sun and Moon, Aristarchus considered the fact that when the Moon is in the second quarter, half-light and half-dark, it forms a right angle with the Earth and the Sun. Aristarchus then calculated the relative lengths of the sides of the triangle with geometrical methods. Though the method was accurate, Aristarchus estimated that the Earth was 18 times farther away from the Sun than from the Moon, a grossly inaccurate result.

Aristarchus' heliocentric theory was not widely recognized or adopted until the 16th century, when NICOLAUS COPERNICUS proposed that the Earth and other planets revolved around the Sun. It is said that Aristarchus' views were considered so contrary to popular beliefs that the philosopher Cleanthes of the Stoic school proposed that Aristarchus be indicted for impiety. Though Aristarchus' advanced beliefs did not enjoy a receptive audience during his lifetime, they contributed greatly to modern astronomical thought. A crater on the Moon is named in his honor.

Aristotle

(384 B.C.–322 B.C.)
Greek
Philosopher of Science

Though Aristotle's theories touched on areas of physics, astronomy, psychology, and biology, his most important contribution was the development of a scientific paradigm for understanding the world that held sway for centuries. It was not until the 16th and 17th centuries that new paradigms transplanted the primacy of Aristotelian science.

Aristotle was born in 384 B.C. in Stagira in Chalcidice. His father, Nicomachus, was a physician at the court of Mayntas II of Macedon, grandfather of Alexander the Great. Both of his parents died when he was a boy. In 367 B.C. at the age of 17 he traveled to Athens, where he remained for the next 20 years, studying under Plato at the Athenian Academy until Plato's death in 347 B.C. Aristotle then traveled for 12 years, teaching at academies he established at Assus in Asia Minor (where he married the daughter of the ruler, Hermeias) and at Mytilene in Lesbos. From 342 B.C. to 335 B.C. Aristotle tutored the young Alexander the Great. Plutarch reported that Aristotle disappointed Alexander by publishing the *Metaphysics,* which, the ruler assumed, was written solely for his own personal benefit. Alexander's ascension to the throne created a favorable situation for Aristotle to return to Athens, where he established his own academy, the Lyceum, in his gardens.

Aristotle's writings were apparently rediscovered and organized by Andronicus of Rhodes in about 50 B.C. Most of these extant writings appear to be lecture notes for courses at the Lyceum. Thomas Aquinas drew upon Aristotelian thought as the foundation for much of his Christian theology. In the sciences Aristotle's biological theories retained the most validity, especially as expressed in *De partibus animalium* (On the parts of animals) and *De gen-*

eratione animalium (On the generation of animals). Aristotle mentioned more than 500 animal species in his writings, and 19th-century zoologists confirmed as true some of Aristotle's observations about animals that had been assumed false by earlier readers. Many of his astronomical theories, based on beliefs such as the centrality of the Earth in the universe, were eventually proved false.

Despite this, the methods by which he arrived at his convictions remain instructive. Aristotle reversed Plato's prioritization of dialectics over mathematics. For Aristotle mathematics and science acted as the organizing principles upon which reality was based. Reality was thus best understood by applying mathematical and scientific logic to the questions that confounded human reason.

Upon the death of Alexander the Great in 323 B.C. Aristotle left Athens for fear that the city, under the sway of anti-Macedonian sentiments, would condemn him to the same fate his mentor, Plato, had suffered: execution for crimes against the state. Aristotle retired to his maternal estate in Chalcis, where he died in 322 B.C.

Arrhenius, Svante August

(1859–1927)
Swedish
Physical Chemist

Svante August Arrhenius posited his theory of electrolytic dissociation despite strong opposition. Almost 20 years after he first defended his conviction to the incredulous doctoral committee at the University of Uppsala, he finally received recognition of the significance of his discovery with the 1903 Nobel Prize in chemistry. His dissertation hypothesized that when salt dissolves in water, its components (sodium and chloride) dissociate into charged ions that can conduct electricity, even though neither dry salt nor pure water acts as an electrical conductor. Doubters resisted his theory in part because it straddled the fence between physics and chemistry; in response Arrhenius became one of the founders of physical chemistry, a scientific discipline that explained chemical phenomena through physical laws.

Arrhenius was born on February 19, 1859, in Vik, in the district of Kalmer, Sweden, to the former Carolina Thunberg and Svante Gustaf Arrhenius, a land surveyor. Arrhenius's first marriage, to Sofia Rudbeck, lasted only from 1894 to 1896, though it produced one child, Olev Wilhelm. In 1905 Arrhenius married Maria Johansson, who bore him three more children, Ester, Anna-Lisa, and Sven.

By the age of three Arrhenius had taught himself to read, and his father's bookkeeping piqued his interest in mathematics. After graduation from the Cathedral School in Uppsala in 1876 Arrhenius attended the University of Uppsala; he received his bachelor's degree in mathematics,

Svante August Arrhenius, one of the founders of physical chemistry. *Smith Collection, Rare Book & Manuscript Library, University of Pennsylvania.*

chemistry, and physics in 1878. He remained at the university to embark upon graduate study in physics; disillusioned with his advisers, he left in 1881, bound for Stockholm, where he continued his doctoral research under the physicist Eric Edlund at the Physical Institute of the Swedish Academy of Sciences. In May 1884 his dissertation on the electrical conductivity of solutions earned a mark of fourth class, the lowest possible passing grade, preventing Arrhenius from teaching.

Arrhenius sent copies of his dissertation to luminary scientists; several supported him by forming a core group of proponents of a new scientific field, physical chemistry. Wilhelm Ostwald demonstrated his faith in Arrhenius's theory by offering him a position at the Polytechnikum in Riga, Latvia. Arrhenius declined, choosing instead the last-minute offer of a lectureship from Uppsala. This decision to prioritize posts in his homeland over foreign offers became a pattern in Arrhenius's life until in 1891 he finally landed a position at the Royal Institute of Technology in Stockholm; he became the rector of the school in 1896. In 1905 the Nobel Institute of the Swedish Academy of Sciences in Stockholm created a chair for him to head the physical chemistry division; Arrhenius maintained his association with the academy until his death.

Arrhenius pursued diverse scientific interests. In the late 1890s he predicted the greenhouse effect by relating increases in atmospheric carbon dioxide to atmospheric warming. In 1908 he published *Worlds in the Making*, in which he conjectured that life traveled through space on spores.

Sweden finally acknowledged the accomplishments of its patriotic son in 1903, when Arrhenius became the first Swede to win the Nobel Prize. The Nobel committee had been undecided whether to award it in physics or chemistry. The previous year, Britain's Royal Society awarded him the Davy Medal, named in honor of the British chemist SIR HUMPHRY DAVY, who discovered nitrous oxide, better known as "laughing gas," in 1800. In 1911 Arrhenius won the first Willard Gibbs Medal from the Chicago section of the American Chemical Society, and in 1914 the British Chemical Society awarded him the Faraday Medal, given in honor of MICHAEL FARADAY, the 19th-century British physicist who first demonstrated that electrical forces could produce motion. Arrhenius died in Stockholm on October 2, 1927.

Aston, Francis William

(1877–1945)
British
Chemist, Physicist

A dedicated and tireless worker, Francis William Aston received the 1922 Nobel Prize in chemistry for his development of the mass spectrograph, a device used to separate atoms of different masses and measure the masses, and for discoveries made with the spectrograph. Aston found that though the atomic weight of each element was a whole number, most elements were mixtures of isotopes, which were atoms of the same element that differed in mass. The spectrograph has wide application in geology, nuclear physics, chemistry, biology, and other fields.

The third of seven children, Aston was born on September 1, 1877, in Harborne, in Birmingham, England. His father, William Aston, was a farmer and metal merchant, and his mother, Fanny Charlotte Hollis, was the daughter of a gun manufacturer. Aston's scientific skills were evident from an early age, and he conducted experiments in a crude laboratory housed on the family farm.

After graduating from secondary school in 1893 with highest honors in mathematics and science Aston attended Mason College, which later became the University of Birmingham. Aston studied organic chemistry and optics under P. F. Frankland but was forced to postpone his studies by a lack of funds. Aston worked as a chemist for a brewing company in the early 1900s and performed experiments in electricity with instruments he designed and built at home. These experiments won him a scholarship in 1903 to the new University of Birmingham. There he researched the flow of electrical currents passing through gases at low pressures. In 1910 Aston began assisting Joseph John Thomson at both the Cavendish Laboratory at Cambridge University and the Royal Institution in London. With Thomson, Aston investigated the gas neon and found the first evidence for isotopes in nonradioactive elements.

In 1919 Aston became a fellow of Cambridge University's Trinity College and perfected the mass spectrograph, which he built himself. From observations taken with his spectrograph Aston successfully calculated the proportion of heavier to lighter atoms in neon and theorized that the atomic weights of elements were always whole numbers and that most elements are combinations of isotopes. Aston was also able to demonstrate that the existence of isotopes was not restricted to radioactive elements.

In the 1920s Aston developed larger spectrographs that were increasingly accurate. Through a series of observations in 1927 Aston discovered some fractional discrepancies with the rule of whole numbers. When particles in the nucleus of an atom were packed more tightly, a greater fraction of the atom's mass became converted to energy devoted to keeping the nucleus intact. Aston used the particles, known as "packing fractions," in calculations that provided important information concerning the stability and abundance of elements.

Aside from the Nobel Prize in chemistry, Aston received a number of awards and honors, including the 1938 Royal Medal of the Royal Society and the 1941 Duddell Medal and Prize of the Institute of Physics. Aston, who remained at Trinity College for the duration of his career, preferred to work alone rather than collaborate. He enjoyed photography and outdoor sports and was an amateur musician. Aston remained single throughout his life. He died on November 20, 1945, in Cambridge, England.

Audubon, John James

(1785–1851)
French/American
Ornithologist, Naturalist

John James Audubon's paintings and drawings of North American birds captured the popular imagination of his time, fostering interest in nature and its inhabitants for years. Though Audubon would not qualify as a hard scientist, he did more for the understanding of ornithology

John James Audubon, the self-educated painter and ornithologist whose paintings and drawings of North American birds have won popular acclaim for almost two centuries. *Smith Collection, Rare Book & Manuscript Library, University of Pennsylvania.*

than many of the field's purists by making it possible for the common person to appreciate birds and their natural surroundings.

Audubon was born Fougère Rabin, or Jean Rabin, in Les Cayes, Saint-Domingue, West Indies (now Haiti), on April 26, 1785, to the French sea captain Jean Audubon and his mistress, Jeanne Rabin. He was baptized Jean-Jacques Fougère Audubon upon his return to France. Anne Moynet, Jean Audubon's wife, graciously adopted her husband's illegitimate offspring in 1794, and in 1803 the younger Audubon was sent to one of his father's American farms, in Pennsylvania, where he first developed his interest in observing and drawing birds. In 1808 he married Lucy Bakewell, though he committed himself more devoutly to his vocation of bird painting, leaving his wife to fend for herself as a governess while raising their two sons, Victor and John.

Having little formal education, Audubon taught himself all he knew about science and painting. The dubious claim that he briefly studied painting under Jacques Louis David in Paris would amount to his only formal training, if it is true. Audubon compensated for his lack of credentials with unflagging energy and confidence in his own powers of observation. In Pennsylvania he tied colored string around phoebes, plain grayish brown and white flycatchers found in the eastern United States, to mark them, antedating the practice of banding birds for tracking purposes by more than half a century. In 1910 he briefly met the preeminent ornithological artist Alexander Wilson; viewing the first two volumes of Wilson's nine-volume *American Ornithology* convinced Audubon of his own superior artistic skills. He reached the height of these skills between 1821 and 1824, painting birds in Louisiana and Mississippi.

After attempting unsuccessfully to find a publisher for his paintings in Philadelphia and New York, Audubon found more favor in London, where the engraver Robert Havell published the four-volume series *The Birds of America,* containing 435 aquatint copperplates of Audubon's ornithological art, between 1827 and 1838. Because of Audubon's limited writing and descriptive skills, William MacGillivray wrote a five-volume accompaniment to *Birds, Ornithological Biography,* published between 1831 and 1839. *A Synopsis of the Birds of North America,* published in 1839, acted as a one-volume index. Audubon continued to conduct fieldwork, traveling through the mid-Atlantic states in 1829; through the Southeast from 1831 to 1832; to Labrador, Canada, in 1833; as far west as Galveston, Texas, in 1837; and as far north as Fort Union in what would become North Dakota in 1843. He collected information on birds as well as on mammals; that formed the basis for the book *Viviparous Quadrupeds of North America,* a three-volume text published between 1845 and 1848. The naturalist John Bachman cowrote the text, and Audubon's sons, Victor and John, assisted in the project.

The Royal Society of London elected Audubon as a member, legitimizing his status as a scientist. His paintings, however, revealed multiple technical mistakes, as well as a tendency to anthropomorphize birds' faces. These errors were nevertheless invisible to all, save ornithologists, and his work was accurate enough to allow nonspecialists to appreciate the grace and beauty of birds. Audubon died on January 27, 1851, in New York City.

Avicenna

(980–1037)
Persian
Physician, Philosopher of Science

Avicenna produced two texts of paramount importance: *The Book of Healing,* a comprehensive treatment of disciplines, ranging from the physical sciences to music and psychology, based on an Aristotelian and Neoplatonic foundation, and *The Canon of Medicine,* the most significant text on medicine produced in the East or the West. Throughout his tumultuous life, living in the midst of

Avicenna, one of the foremost Muslim Aristotelians of his day. *Smith Collection, Rare Book & Manuscript Library, University of Pennsylvania.*

political turmoil, Avicenna retained his faith in a God, which was the basis for much of his writing.

Avicenna was born in 980, in Bukhara, Iran. His father's house served as a kind of intellectual salon, providing learned company for him early on. By the age of 10 he had memorized the Koran, and by the age of 18 his intelligence so surpassed that of all of his teachers that he educated himself through books. At this young age he healed the Samanid prince Nuh bin Mansur, who appointed Avicenna as court physician. This position secured him access to the royal Samanid library, in which he continued his self-education until the Samanid dynasty was overthrown by the Turkish leader Mahmud of Ghazna. The concurrent death of his father cast Avicenna into political and personal exile.

After years of wandering and itinerant work as a physician he landed in Hamadan, where the Buyid prince Shams ad-Dawlah declared him court physician and twice named him vizier. During this period Avicenna commenced writing the two books that established his lasting significance. The five books of *The Canon of Medicine,* containing over a million words, surveyed the entire field of medicine, focusing on anatomy, physiology, etiology, diagnosis, obstetrics, and pharmacology, among other disciplines. Since he had lost his own practice notes in transit, he based his treatise less on personal experience than on the precedents of Greek physicians of the Roman imperial age and Arabic medical documents.

The *Book of Healing* discussed most areas of human knowledge, including logic, psychology, geometry, astronomy, arithmetic, music, and metaphysics. The only two areas that Avicenna did not touch on were ethics and politics. He based the work on a religious foundation, positing the necessary existence of a God. Fulfilling his medical and administrative duties by day, Avicenna devoted his nights to composition and research, which often entailed all-night discussions and general revelry with his students. Both of these seminal texts were translated into Latin in the 12th century, exposing them to a European audience, notably the Franciscan schools, which incorporated Avicenna's thought system with Augustinian theology as a basis for medieval intellectual and religious beliefs.

The death of Shams ad-Dawlah in 1022 prompted Avicenna to migrate to Isfahan, where he enjoyed the favor of the ruler 'Ala 'ad-Dawlah. He continued his prolific output, composing three more major works: the *Book of Salvation,* a summary of *The Book of Healing;* the *Book of Directives and Remarks,* a spiritual map from the initiation of faith to the constant contemplation of God; and *The Arabic Language,* a tome on Arabic philology that remained in draft form at his death. Though Avicenna ranked as one of the foremost Muslim Aristotelians, his true but unfinished goal was to establish an Oriental philosophy. Despite his efforts at self-healing while on a military campaign, Avicenna died of colic and exhaustion in 1037.

Avogadro, Lorenzo Romano Amedeo Carlo

(1776–1856)
Italian
Physicist

Amadeo Avogadro's important hypothesis regarding the atomic weight of gases languished in obscurity for some 50 years, with ANDRÉ-MARIE AMPÈRE its sole supporter, until STANISLAO CANNIZZARO asserted its validity convincingly at the 1860 Chemical Congress at Karlsruhe. What became known as Avogadro's law states that equal volumes of gases at equal temperature and pressure contain equal numbers of molecules. Avogadro's work led to a more informed understanding of atomic molecular theory and, most importantly, distinguished atoms from molecules.

Amedeo Avogadro, whose work distinguished atoms from molecules and led to a more informed understanding of atomic molecular theory. *AIP Emilio Segrè Visual Archives.*

Avogadro was born on August 9, 1776, in Turin, Italy. His parents were Anna Maria Vercellone Avogadro and Count Filippo Avogadro. His father was elected senator of Piedmont in 1768 and appointed advocate general to the senate in 1777. In 1787 Avogadro inherited his father's title, count of Quaregna and Cerreto. He married Felicita Mazzé, with whom he had six children. Like his forefathers, he studied and practiced law until the turn of the century.

In 1800 Avogadro embarked on a private study of mathematics and physics, and by 1806 he taught at the college attached to Turin Academy. On October 7, 1809, the College of Vercelli appointed him as a professor of natural philosophy. In 1820 the Turin Academy established the first chair of mathematical physics in Italy and appointed Avogadro to fill the chair. Political opposition, however, abolished the chair from 1822 until 1832. On November 28, 1834, Avogadro was reappointed to the chair, a position he occupied until his retirement in 1850.

Avogadro conceived of his major hypothesis, discussed in the paper "On the Way of Finding the Relative Masses of Molecules and the Proportions in Which They Combine" and published in an 1811 edition of the *Journal de Physique,* as a mere extension of the work of Gay-Lussac, which was published in 1809. Avogadro's humility prevented his ground-breaking assertion from receiving the audience and recognition it deserved. This oversight of the scientific community not only cost Avogadro personally but also set back the progression of atomic theory by a half-century. As it turned out, Avogadro's law did not apply universally to gases, but specifically to noble gases, among others. However, the basis of his theory encouraged several inferences, namely, that the ratio of the relative molecular weights of two gases is in the same proportion as their densities at the same temperature and pressure. This inference allowed scientists to calculate atomic weights with precision by this ratio. Avogadro also deduced that simple gases such as hydrogen and oxygen are diatomic (H^2, O^2); thus water is H_2O, not HO, as JOHN DALTON had asserted.

On November 21, 1819, Avogadro was elected as a full member of the Academy of Sciences of Turin after 15 years as a corresponding member. A more lasting honor was the naming of the number of particles in one mole as Avogadro's constant: 6.022×10^{23}. Avogadro died on July 9, 1856, in Turin.

Ayrton, Hertha (Phoebe Sarah Marks)
(1854–1923)
British
Physicist, Engineer

One of the first woman electrical engineers, Hertha Ayrton improved the working of the electric arc, used in streetlights and movie projectors of her time, and invented a fan to clear poisonous gases from mines and soldiers' bunkers. She was the first woman to gain an award from Britain's prestigious Royal Society.

Named Phoebe Sarah Marks at her birth, Hertha Ayrton was born in 1854 in Portsea, England, into the large family of Levi Marks, a Jewish refugee from Poland. Marks, a jeweler and clockmaker, died in 1861, leaving his wife, Alice, in poverty with eight children to support. While Alice worked as a seamstress, Sarah, the oldest girl, took care of her brothers and sisters. She did not begin school until she was nine, but eventually she attended the school her aunt ran in London. While there she met Barbara Bodichon, a wealthy women's rights advocate and philanthropist who became her friend and supporter. In 1876 Bodichon helped Sarah enter Girton, a women's college of Cambridge University. Sarah changed her name to Hertha while at college.

After graduating from Girton in 1880, Marks turned a cousin's idea into her first invention, a tool that divided a line into equal parts. She obtained a patent on it in 1884, and it proved useful to engineers, architects, and artists.

Encouraged by this success, Marks began studying at Finbury Technical College, where one of her teachers was the physicist W. E. Ayrton. Ayrton admired Marks's intelligence and energy, and the two married in 1885. They later had a daughter, Barbara. Hertha helped her husband in his work, but he encouraged her to do her own research as well, giving her the use of his laboratory and calling her his "beautiful genius."

Some of Hertha Ayrton's most important work began in 1893 as a continuation of a project her husband was doing on electric arcs, which were used in streetlights, searchlights, and later in movie projectors. The arc, a glowing stream of electrons that flowed between two carbon electrodes separated by a pit or crater, often degenerated into rainbow flickers accompanied by a hissing noise; early movies were nicknamed "flickers" or "flicks" because of this failing. Determined to "solve the whole mystery of the arc from the beginning to the end," Hertha showed that these problems occurred because oxygen in the air entered the crater and combined with the carbon in the electrodes. Drawing on her research, engineers worked out a way to protect the arc from the air and thus increase its power and reliability.

These and other experiments made Hertha Ayrton a national authority on the electric arc. In 1895 the magazine *Electrician* asked her to write a series of articles on the subject, which she expanded into a book in 1902. The Institute of Electrical Engineers was so impressed by her paper explaining the hissing of the electric arc, which she read to them in March 1899, that the group made her its first woman member two months later. A reviewer called the paper "a model of the scientific method of research." Ayrton's paper on the electric arc was also presented to Britain's premier organization of scientists, the Royal Society, in 1901, but this time a man had to read it because the society did not permit women at its meetings.

W. E. Ayrton's health began to fail in 1901, and he and Hertha moved to the coast in an attempt to improve it. Unable to continue her electrical experiments because she now lacked a laboratory, Hertha became curious about the sandy beaches covered with what she later described as "innumerable ridges and furrows, as if combed by a giant comb." To learn how waves shaped the sand, she built glass tanks in her attic, put a layer of sand in the bottom, and filled them with water. She put the tanks on rollers to imitate wave motion and found that when waves moved constantly back and forth over the same spot, they created regular ripples that eventually pushed the sand into two mounds between the crests of the waves. Ayrton believed that this kind of wave action formed both sand dunes on the shore and underwater sandbanks that often wrecked ships. She hoped other engineers could use her research to prevent sandbanks from forming.

Ayrton presented a paper about her sand research to the Royal Society in 1904. This time she was allowed to read it herself, the first woman to read a paper before the group. The society awarded her its Hughes Medal in 1906 for her work on electric arcs and sand.

The *London Times* commented of the award, "It seems that the time has now come when woman should be permitted to take her place in . . . all our learned bodies."

Several of Hertha Ayrton's inventions helped her country during wartime. Her improvements to searchlights made night spotting of aircraft easier. She also used what she had learned in her beach experiments to design a fan that drove poisonous gas out of bunkers and trenches and drew in fresh air. One soldier whom it helped during World War I wrote: "There are thousands and thousands of inarticulate soldier persons who are extremely grateful to you." The Ayrton fan was later modified to drive dangerous gases out of factories and mines.

Hertha Ayrton once said, "Personally I do not agree with sex being brought into science at all. . . . Either a woman is a good scientist, or she is not." She expressed her belief in women's equality in another way by joining the Women's Social and Political Union, one of the most militant organizations seeking votes for women. Ayrton's gender denied her some of the scientific recognition she deserved, but she did live to see British women gain the right to vote in 1918. She died five years later.

B

Babbage, Charles

(1792–1871)
English
Mathematician, Computer Scientist

Charles Babbage's frustration with the persistence of human errors in mathematical tables gave birth to the notion of mechanical computation, an idea that led to modern computer technology. Once Babbage recognized the potential for the mechanization of mathematics, he devoted himself obsessively to its realization. He designed and attempted to build three mechanical computers in his lifetime, though it wasn't until May 1991 that Doron Swade successfully completed the construction of a Babbage computer at the Science Museum in London for a cost equivalent to $500,000. Part of the driving force behind Babbage's project was his belief in the connection between science and culture, especially industry, which stood to progress on the shoulders of scientific discovery.

Babbage was born to affluent parents on December 26, 1792, in Teignmouth, England. He attended Cambridge University from 1810 to 1814, though he found that many of his professors could not match his intellect. In an attempt to bolster British mathematical standards to the level of those in continental Europe, he joined Sir JOHN FREDERICK HERSCHEL and George Peacock in campaigning against British intellectual isolationism. Toward this end he helped found the Analytical Society in 1815. He was instrumental in the founding of many other organizations, among them the Royal Astronomical Society in 1820 and the Statistical Society of London in 1834, a testament to his wide-ranging interests. His theory of "operational research" helped with the establishment of the British postal system in 1840. His diverse scientific interests included cryptanalysis, probability, geophysics, astronomy, altimetry, ophthalmoscopy, statistical linguistics, meteorology, actuarial science, and lighthouse technology.

In 1827 Cambridge University honored Babbage with an appointment as the Lucasian Professor of Mathematics, a position he held through 1839. The position would have served as a perfect platform from which to preach against and perhaps reform the substandard teaching of mathematics in Britain. Ironically, Babbage did not use the position as a bully pulpit because of his preoccupation with mechanized computation and an analytical machine. In about 1830 he commenced work on what became known as his Difference Machine No. One, based on addition rather than multiplication. The construction of this eight-foot-tall machine took a decade and cost the equivalent of $85,000, though he never completed it because of a loss of funding. In 1847 he went on to design his Difference Engine No. Two, a streamlined version of No. One, though he never constructed this design, as the government refused to finance it.

Babbage's concept acted as a harbinger of modern computers, though his engine was an analog decimal machine, as opposed to the modern computer, which is a binary digital machine. Lord Byron's daughter, AUGUSTA ADA BYRON LOVELACE, created programs for the prototypical computer, establishing the field of computer programming in conjunction with Babbage's designs for his computer. These two cohorts devised systems to predict winning horses, but lost much money as the systems were never foolproof.

Although Babbage criticized the Royal Society for its conservatism, he was elected a member in 1816. Babbage died on October 18, 1871, in London.

Bailey, Florence Augusta Merriam

(1863–1948)
American
Zoologist

Many people in the late 19th and early 20th centuries learned about birds through the writings of Florence

Merriam Bailey. Born on August 8, 1863, in Locust Grove, New York, Florence grew up at Homewood, the Merriam family's country estate. Her father, Clinton, a New York businessperson, frequently took his young daughter on camping trips to collect wildlife specimens. She learned about the stars from her mother, Caroline, and about birds and mammals from her older brother, Hart. Of all these she liked birds the best.

Merriam entered Smith College in 1882 but earned only a certificate instead of a degree. (Smith finally awarded her a B.A. in 1921.) She then returned to Homewood, where she observed and wrote about birds by "sit[ting] down, pull[ing] the timothy [a kind of plant] stems over my dress, [and] make[ing] myself look as much as possible like a meadow." Working from the notes she took in the fields, Merriam wrote articles for *Audubon Magazine*, and these in turn grew into *Birds through an Opera Glass*, published as a book for young people in 1889.

Merriam made her first trip to the Southwest in 1893. Although she did so for health reasons, the bird-watching she did there provided enough material for two more books. Meanwhile, her brother, C. Hart Merriam, became the first chief of the U.S. Biological Survey, and when Florence returned to the East in late 1895, she moved to Washington, D.C., to live with him. There she continued to write about birds. Her book *Birds of Village and Field*, published in 1898, became one of the first popular American guides for amateur birders. In it, she said that bird-watchers need only four things: "a scrupulous conscience, unlimited patience, a notebook, and an opera glass [an equivalent of binoculars]."

Merriam's brother introduced her to Vernon Bailey, the Biological Survey's chief field naturalist. Merriam and Bailey married in 1899 and thereafter spent part of each year camping in the West, observing wildlife and gathering specimens. Vernon Bailey specialized in mammals, reptiles, and plants, and Florence continued to focus on birds. Florence Bailey's writing became more scientific after her marriage. In 1902, she published *Handbook of Birds of the Western United States*, which became a standard field guide for the next 50 years. Her most monumental work was *Birds of New Mexico*, published in 1928. Because of this book the American Ornithological Union not only made Bailey its first woman member in 1929 but also awarded her its Brewster Medal in 1931. Bailey's last book, *Among the Birds in the Grand Canyon Country*, appeared in 1939, when she was 76. She died on September 22, 1948.

Banks, Harvey Washington

(1923–1979)
American
Astronomer, Astrophysicist

Harvey Washington Banks broke the color barrier at Georgetown University as the first African American to receive a Ph.D. in astronomy from the institution. Banks focused his career on the study of planetary spectroscopy, or the examination of light emitted by distant sources, and geodetic measurements, using orbiting objects as a means of determining distances in the United States.

Banks was born on February 7, 1923, in Atlantic City, New Jersey, but when he was still young his parents, Nettie Lee Jackson and Harvey Banks, Sr., moved to Washington, D.C., where he attended Dunbar High School. He was married to Ernestine Boykin, and together the couple had four children—Harvey III, Deborah, Dwann, and Darryle.

Banks remained in Washington, D.C., for his undergraduate work at Howard University, where he earned a bachelor of science degree in physics in 1946 and added a master of science degree, also in physics, in 1948. He stayed on at Howard as a research associate in physics until 1952, when he got a job in the private sector as an electronic engineer at National Electronics, Inc. Two years later he left the industry for a job in education, teaching physics and mathematics in the public school system of Washington, D.C. After two years of teaching Banks returned to academia, where he was a research assistant in astronomy at the Georgetown College Observatory while pursuing his doctorate at Georgetown University. In 1961 he became the first African American to receive a Ph.D. in astronomy from Georgetown. His dissertation, "The First Spectrum of Titanium from 6000 to 3000 Angstroms," defined the focus of the rest of his career: He studied the properties of light originating from distant sources, a concentration known as planetary spectroscopy.

Georgetown retained Banks as a fellow for the year after he earned his doctorate, then hired him on as a lecturer and research associate from 1963 through 1967. During this period he also taught at American University, also in Washington, D.C., and at Delaware State College. Then in 1967 Delaware State appointed him as a professor of astronomy and mathematics with a concurrent appointment as the director of the college's observatory. On September 1, 1969, Banks returned to his alma mater, Howard, as an associate professor of astronomy, and two years later the university added an appointment as an associate professor of physics, a position he maintained until his death.

Besides spectroscopy, Banks concerned himself with geodetic measurements, or determinations of distances between two points based on objects orbiting the Earth. He was thus interested in orbits and celestial mechanisms. In the 1970s Banks supervised the construction of an observatory outside Washington, D.C., as a member of the Beltsville Project. Banks also coordinated the Astronomy and Space Seminar for the National Science Teachers' Association. He was also a member of numerous societies, including the American Astronomical Society, the Optical Society, the Washington Academy of Science, the New

York Academy of Science, the Washington Philosophical Society, and the Spectography Society. Banks died in 1979.

Banting, Sir Frederick Grant

(1891–1941)
Canadian
Physiologist

Sir Frederick Grant Banting devised a method for isolating the pancreatic hormone insulin, which regulates blood-sugar levels and thereby controls the disease diabetes mellitus. For this discovery Banting received the 1923 Nobel Prize in physiology or medicine, which he shared with JOHN JAMES RICKARD MACLEOD. The contributions to the research of Charles H. Best and James Bertram Collip were not recognized officially, so Banting shared his prize money with Best and Macleod shared his with Collip. Banting was the first Canadian to receive a Nobel Prize.

Banting was born on November 14, 1891, near Alliston, Ontario. His parents were Margaret Grant, the daughter of a miller, and William Thompson Banting, a farmer of Irish parentage. Banting was engaged to Edith Roach in 1920, but the marriage never materialized. Banting did marry Marion Robertson in 1924, and the couple had one son, but the unhappy marriage ended in divorce in 1932. Banting married Henrietta Ball, a technician in his department, in 1939.

Banting's father sent him to Victoria College of the University of Toronto to enter the Methodist ministry, but in 1912 Banting shifted to studying medicine. With the onset of World War I the university accelerated Banting's course of study, and he received his M.D. in December 1916.

The Canadian Army Medical Corps inducted Banting as a lieutenant immediately upon his graduation. He served as a surgeon at the orthopedic hospital in Ramsgate, England, where he sustained an arm injury that prompted the government to award him a Military Cross in 1919 for gallantry under fire. After the war Banting set up practice in London, Ontario, supplementing his scant income with a position as a demonstrator in surgery and anatomy at the University of Western Ontario.

In May 1921 Banting abandoned his failing practice to conduct an experiment inducing pancreatic ischemia as a means of isolating insulin. Macleod offered him space in his University of Toronto laboratory as well as the assistance of Best while Macleod himself departed for summer vacation. During his absence Banting and Best succeeded in isolating what they called "isletin," named after the islets of Langerhans, the section of the pancreas that produces the hormone, but upon his return Macleod vetoed this name in favor of the traditional name derived from the Greek, *insulin*. By January 1922 the team, joined by Collip to produce a pure extraction, completed successful clinical trials on themselves and on a 14-year-old diabetic.

Recognition of the significance of this discovery was almost immediate: In addition to the Nobel Prize, the University of Toronto established the Banting and Best Chair of Medical Research in 1923, a position that Banting himself filled before it expanded into the Banting and Best Department of Medical Research. That year the Canadian parliament granted Banting an annuity. In 1924 the Banting Research Foundation was established, and in 1930 the University of Toronto named its new medical school buildings the Banting Institute. In 1935 the Royal Society of London inducted Banting as a fellow.

In 1934 King George V, a diabetic, knighted Banting, a symbolic act of appreciation of the man who literally extended the life expectancy of all diabetics through the controlling influence of insulin. Banting's own life was cut short when his plane crashed in Newfoundland on February 21, 1941.

Bárány, Robert

(1876–1936)
Austrian/Swedish
Physician

Robert Bárány received the 1914 Nobel Prize in physiology or medicine for his work on the vestibular apparatus, the mechanism in the inner ear that controls balance. Bárány invented a battery of simple and practical tests that led to the founding of a new field, otoneurology, which investigated the connection between the vestibular and the nervous system.

Bárány was born on April 22, 1876, in Rohonc, Austria, the eldest of six children. His mother, Marie Hock, was the daughter of a prominent Prague scientist; his father, Ignaz Bárány, was a bank official. The intellectual bent on his mother's side influenced Bárány to pursue an academic career, and a childhood case of tuberculosis in the bones pushed him toward medicine. In 1909 Bárány married Ida Felicitas Berger, and all three of their children entered the medical sciences: Ernst became a professor of pharmacology at the University of Uppsala, Franz became a professor of internal medicine at the University of Stockholm, and Ingrid became a psychiatrist in Cambridge, Massachusetts.

In 1894 Bárány entered the University of Vienna medical school, where he received his M.D. at the turn of the century. Between 1900 and 1902 he studied internal medicine, neurology, and psychiatry in Frankfurt, Heidelberg, and Freiberg, before returning to Vienna for surgical training in the hospital there. In 1903 he landed a position at the Ear Clinic in Vienna under the directorship of Adam Politzer. World War I interrupted a period of fertile research, as Bárány served as a medical officer on the Russian front. He was captured as a prisoner of war; at that time he contracted malaria, but his medical skills earned him respect and special treatment. In 1916 Prince Carl of Sweden arranged for Bárány's release. His return to Vienna was marred by infighting among his colleagues, who wanted to claim credit for his Nobel Prize, a development that antagonized Bárány. He accepted a professorship from the University of Uppsala in 1917 and remained there the rest of his career.

The tests that Bárány devised included the Bárány caloric test, whereby one ear is irrigated with hot and the other with cold water, after which the physician observes for nystagmus, or the rapid, involuntary movements of the eyeballs. Bárány also developed the chair test, spinning the patient in a rotating chair and then checking for nystagmus. His noise box isolated hearing on one side by distracting the other side with a constant signal. The pointing test, whereby the patient pointed to a specific position with eyes open and then with eyes closed, tested the connection between the cerebellum and body movements.

In addition to the Nobel Prize, many other distinctions brought recognition to Bárány's work. The Swedish Society of Medicine awarded him the 1925 Jubilee Medal for his development of a surgical technique for curing chronic sinusitis. He also won the Belgian Academy of Sciences Prize, the German Neurological Society's ERB Medal, and the Guyot Prize from the University of Groningen in the Netherlands. After a series of strokes Bárány died on April 8, 1936, in Uppsala. An international meeting celebrating his 60th birthday went ahead as planned a few days after his death. Bárány's achievements were commemorated by the establishment in 1948 of the Bárány medal at the University of Uppsala and by the founding of the Bárány Society in 1960 to advance vestibular research.

Bateson, William

(1861–1926)
English
Geneticist

In coining the term *genetics,* William Bateson established a new branch of scientific inquiry. He disavowed the Darwinian theory of natural selection, which attributed evolution to the slow aggregation of multiple small changes, in favor of the theory of discontinuous variation, which allowed for evolutionary "jumps." Bateson's discovery in 1900 of the 1866 article "Experiments with Plant Hybrids" by the Austrian monk JOHANN GREGOR MENDEL provided support for his own experimentation. Bateson translated Mendel's work into English, using the results from the monk's experiments to prove his own hypotheses.

Bateson was born on August 8, 1861, in Whitby, England. His mother was Anna Aiken Bateson; his father was William Henry Bateson, a classical scholar who became the master of St. John's College in Cambridge in 1857. Though Bateson's academic future seemed unpromising when he was a boy, he nevertheless earned first-class honors in the natural sciences at Cambridge University when he received his B.A. in 1883, having focused his attention on zoology and morphology under the influence of Professors A. Sedgwick and W.F.R. Weldon. He spent the next two years in the United States conducting research on marine organisms at Johns Hopkins University under W. K. Brooks.

In 1885 Bateson returned to Cambridge as a fellow of St. John's College, and in 1887 he received the Balfour Studentship. In 1894 he published his first important paper, "Materials for the Study of Variation," which outlined his theory that evolution was discontinuous and not environmental. The work gained him little recognition, as most scientists rejected his ideas and Cambridge did not appoint him to a teaching position. In 1899 he served as a deputy in zoology to Alfred Newton.

The turn of the century marked a major turning point in Bateson's career as well, as he discovered Mendel's work in breeding pea plants, which corroborated his own breeding experiments. Together with L. Cuénot, he extended his experimentation to animals, focusing on the question of inheritance in the comb shape in poultry. In 1902 Bateson published *Mendel's Principles of Heredity: A Defense,* in which he translated Mendel's work and proposed a Mendelian interpretation of evolution that became the theoretical umbrella for much of the remainder of his career. He coined the term *genetics* in 1906 at the Third Conference on Hybridization and Plant Breeding. In 1907 Cambridge appointed him to a readership in zoology, but in 1908 the university fully legitimized his work by establishing for him a professorship in genetics, the first such chair in Britain. Bateson further validated the new field by founding *Journal of Genetics* with R. C. Punnett in 1910. That same year Bateson departed from his long association with Cambridge to become the first director of the John Innes Horticultural Institution, a position he retained the rest of his life.

Bateson received much recognition for his pioneering work, including the Darwin Medal in 1904, the presidency of the British Association for the Advancement of Science in 1914, and the Royal Medal in 1920. In 1922 the British Museum named him a trustee. He died on February 8, 1926, in London, England.

Becquerel, Antoine-Henri

(1852–1908)
French
Physicist

Henri Becquerel followed in the footsteps of his father, Alexandre-Edmond Becquerel, and grandfather, Antoine-César Becquerel, who were both famous physicists. Henri

Henri Becquerel, whose discovery of radioactivity in 1896 earned him the 1903 Nobel Prize in physics. *AIP Emilio Segrè Visual Archives, William G. Myers Collection.*

established himself in his father's prominent posts at the Museum of Natural History and the National Conservatory of Arts and Crafts and held several other prestigious positions before he made his landmark discovery of radioactivity in 1896. For this work he earned the 1903 Nobel Prize in physics.

Becquerel was born on December 15, 1852, in Paris, France. Eminent scholars in his father and grandfather's circles surrounded him from the very beginning, introducing him at a young age to the world of science, and physics in particular. He began his formal education at the Lycée Louis-le-Grand, from which he graduated to the École Polytechnique in 1872. Upon his departure from the Polytechnique in 1874 he married Lucie-Zoé-Marie Jamin, daughter of the physicist J.-C. Jamin. She died four years later, only weeks after giving birth to their son, Jean, who in turn became a physicist. In 1890, two years after he earned his doctorate from the Faculty of Sciences of Paris, Becquerel married his second wife, the daughter of E. Lorieux, an inspector-general of mines.

Becquerel earned positions from the institutions where he studied soon after his graduation. In 1876 the Polytechnique appointed him as a *répétiteur;* he rose to the position of full professor two decades later in 1895. In 1877 the Administration of Bridges and Highways appointed him as an *ingénieur* after he studied from 1874 to 1877 at the School of Bridges and Highways; he later ascended to the position of *ingénieur de première classe* for the administration. In 1878 the Museum of Natural History appointed him to his father's former position of *aide-naturaliste.* Henri's father, Edmond, died in 1891, and the subsequent year his father's chairs of physics at the museum and at the National Conservatory of Arts and Crafts succeeded to Henri.

In 1895 the German physicist WILHELM CONRAD RÖNTGEN discovered X rays. At that time Becquerel was studying fluorescence, as his father and grandfather had, and so he experimented with the substance he was studying at the time, potassium uranyl sulfate. Hypothesizing a connection between X rays and luminescence, he placed this uranium salt, known to luminesce, on top of photographic plates wrapped in black paper, then exposed this setup to sunlight. On February 24, 1896, he reported his results to the Academy of Sciences: The luminescence of the uranium salt exposed the plates through the black paper, thus suggesting the existence of X rays. Becquerel repeated this procedure twice, but he stored the setups in dark drawers because the Sun was hidden by clouds on February 26 or 27. By happenstance he developed these photographic plates on March 1 and found them fully exposed, suggesting some type of radiation other than X rays, as the lack of exposure to sunlight meant that luminescence could not have been triggered.

The rays that exposed the photographic plates were actually called Becquerel's rays until his doctoral student, MARIE SKLODOWSKA CURIE, named this phenomenon *radiation.* Her research for her dissertation proved so fundamental to the understanding of radiation that she, along with her husband, PIERRE CURIE, shared the 1903 Nobel Prize in physics with Becquerel. Their joint findings shifted the direction of physics in the 20th century. Becquerel died on August 25, 1908, in Le Croisic, in Brittany, France.

Bell Burnell, Susan Jocelyn

(1943–)
British
Astronomer

While she was still a graduate student, Jocelyn Bell Burnell spotted a "bit of scruff" on a paper tape carrying recorded signals from space. Her identification of a new kind of star was a discovery *Current Biography* called "one of the most exciting events in the history of astrophysics."

Susan Jocelyn Bell was born on July 15, 1943, in Belfast, Northern Ireland. She and her sisters and brother grew up in Solitude, their family's large country house. When Jocelyn was about 11 years old, her father, Philip, an architect, helped to rebuild Armagh Observatory. Jocelyn went along, met the observatory's astronomers, and learned to love their science.

At the University of Glasgow Bell was the only woman among her class's 50 physics majors. She earned a B.S. with honors in 1965 and moved on to Cambridge University, where she studied radio astronomy. Radio astronomers map the sky by recording radio waves that stars and other objects in space give off, just as conventional astronomers record and use light. Astronomers can also study the sky through other parts of the electromagnetic spectrum, such as microwaves, X rays, and gamma rays. Each type of radiation provides a different "picture" of the universe.

Bell's first task was to help build a new radio telescope, a job that sometimes meant swinging a 20-pound sledgehammer to drive in the poles that would hold up its antenna wire. Once the telescope was operating, Antony Hewish, the head of the radio astronomy project, gave her the painstaking job of analyzing the 100 feet of tape that its recorders spewed out each day. One day in October 1967, after Bell had been doing this for several months, she saw an unusual signal—what she called a "bit of scruff." As she said later, "I . . . remember[ed] that I had seen this particular bit of scruff before, and from the same part of the sky."

When Bell checked back, she found that the strange signal appeared once every 23 hours and 56 minutes. This indicated that it was keeping sidereal time, or "star time," rather than Sun time. The Earth rotates with respect to the Sun once every 24 hours, but its rotation time with respect to the stars is slightly less. The fact that the signal recurred on a star-time schedule meant that it almost surely originated outside the solar system.

The signals pulsed once every 1.3 seconds. Since no natural object that could make such rapidly pulsing signals was known, Hewish's group began to wonder whether the signals could be communications from another solar system. They joked about "little green men." At the end of 1967, however, Bell disproved this idea by finding another signal, pulsing even faster than the first, in a different part of the sky. "It was highly unlikely that there were two lots of little green men signaling to us from opposite sides of the universe," she concluded. She soon found two more such signals.

Hewish published an article about the discovery on February 9, 1968. The new objects were dubbed pulsars, and astronomers speculated that they might be neutron stars, a strange kind of star predicted by theory but never observed before. They knew that when a large star runs out of nuclear fuel, it blows up in a colossal explosion called a supernova. The core left after a supernova explosion was expected to be only 6.2 to 9.3 miles across, yet heavier than the Sun. Its tremendous gravity would probably smash electrons into protons in the atoms' nuclei, leaving a soup of neutrons. Only something as small and heavy as a neutron star could spin as fast as the pulsars without being torn apart. Astronomers guessed that powerful radio waves streamed from the star's magnetic poles and "flashed" at Earth once each time the object rotated, just as the turning light in a lighthouse seems to flash each time its beam passes an observer.

Jocelyn Bell received her Ph.D. in 1969. She then married Martin Burnell, a government official, and moved with him whenever he was transferred to a new town, as happened often. They had a son in 1973, and Bell Burnell, as she now called herself, decided to work only part-time so she could care for him. "I am very conscious that having worked part-time, having had a rather disrupted career, my research record is a good deal patchier than any man's of a comparable age," she says. On the plus side her peripatetic career has produced a breadth of experience that few other astronomers can equal. For instance, she has done gamma-ray astronomy at Southampton University and X-ray astronomy at the Mullard Space Science Laboratory of University College, London. She has also managed a telescope in Hawaii for the Scottish Royal Observatory.

Antony Hewish was awarded the Nobel Prize in physics in 1974, partly for his "decisive role in the discovery of pulsars." Jocelyn Bell Burnell did not share the award. This angered one of Britain's foremost astronomers, SIR FRED HOYLE, who claimed that Hewish had "pinch[ed] [stolen] the discovery from the girl." He praised Bell Burnell's "willingness to contemplate as a serious possibility a phenomenon that all past experience suggested was impossible." Bell Burnell said that Hoyle's accusation was "overstated," but some other astronomers agreed with it at least in part. In any case, Bell Burnell has received plenty of other awards, including the Franklin Institute of Philadelphia's Michelson Medal (1973), the Herschel Medal of the Royal Astronomical Society of London (1989), and the Jansky Award of the National Radio Astronomy Observatory (1995).

Since 1991 Bell Burnell, now divorced, has been a professor and chairperson of the physics department at the Open University, which offers classes to adults all over Europe through correspondence, television, and computer. She calls herself "a role model, a spokeswoman, a representative, and a promoter of women in science in the U.K." She continues research as well, some of it on pulsars, the mysterious stars she helped to discover.

Benedict, Ruth Fulton

(1887–1948)
American
Anthropologist

Ruth Benedict was one of the first American women to become a professional anthropologist. She proposed the idea that each of the world's cultures has its own "personality." As part of her fight against racism, a term she coined, Benedict wrote books that helped different cultures understand one another and was one of the first to combine anthropology with psychology and sociology to gain a multifaceted understanding of human culture. She also pioneered the use of anthropology to study major modern cultures.

Ruth Fulton was born on June 5, 1887, in New York City. Her childhood was scarred by the death of her father, Frederick, a surgeon, when she was less than two years old. Reaction to this loss and to her mother's continuing grief made her depressed and lonely throughout the first half of her life. Limited hearing, the result of a childhood attack of measles, also tended to isolate her. She and her younger sister, Margery, grew up partly on their grandparents' farm and partly in the cities where their mother, Bertrice, found teaching jobs. Money was always scarce.

Fulton graduated from Vassar College in 1909, but she had little idea of what she wanted to do with her life. She taught school and also wrote poetry, some of which appeared in national magazines under the name Anne Singleton. In 1914 she married Stanley Benedict, a biochemist. They slowly drifted apart, however, and separated permanently in 1930.

To fill time, Ruth Benedict began taking classes in 1919 at the New School for Social Research in New York City. Teachers there introduced her to Columbia University's Franz Boas, the "grand old man" of anthropology. Benedict soon began studying anthropology full-time and earned her doctorate in 1923. Columbia immediately hired her and, according to *Current Biography,* "eventually she became, next to Dr. Boas himself, the key figure in the Department of Anthropology." Her teaching inspired such students as MARGARET MEAD, and her fieldwork among the Native Americans of the Southwest resulted in two books, *Tales of the Cochiti Indians* (1931) and a two-volume work on Zuñi mythology (1935). She also edited *Journal of American Folklore* between 1925 and 1940.

Benedict came to believe that each culture forms a basic pattern into which it tries to integrate all the random details of daily life. It honors only certain human traits, rejecting others that might be respected by other groups. Taken together, she said, the traits a culture honors form a sort of collective personality of that culture. She described these ideas in her best-known book, *Patterns of Culture,* published in 1934. Benedict was one of the first to link culture and individual personality, combining findings from anthropology and psychology. Later scholars have doubted that a single cultural pattern dominates daily life as fully as Benedict suggested, but her book remained a popular introduction to anthropology for more than 25 years.

Although still only an associate professor, a title she had been given in 1931, Benedict became acting head of Columbia's anthropology department after Franz Boas's retirement in 1936. In 1940, when the belief that some races were superior to others was tearing the world apart, she published *Race: Science and Politics* to disprove this poisonous myth, which she called racism. "All the arguments are on the side of the Founding Fathers [of the United States], who urged no discrimination on the basis of race, creed, or color," she wrote. The army's Morale Division arranged for the distribution of 750,000 copies of *The Races of Mankind,* a pamphlet with the same message that she coauthored.

Benedict worked for the Office of War Information in Washington, D.C., from 1943 to 1945, advising the agency about dealing with people in occupied and enemy territories. After the war she extended her idea of cultural patterns into a detailed study of Japanese culture, *The Chrysanthemum and the Sword.* Most Americans thought of Japan only as an enemy they had fought during World War II, but Benedict's book, published in 1946, helped them understand and respect the Japanese.

In 1947 the U.S. Office of Naval Research gave Columbia a grant to carry out research on contemporary cultures and chose Ruth Benedict to head this huge endeavor, the most ambitious anthropology project yet seen in the United States. She also served as president of the American Anthropological Association in 1947–1948. Most people in the field had considered Benedict the leading American anthropologist since Franz Boas's death in 1942, but Columbia waited until 1948 to make her a full professor. Unfortunately, she did not have long to enjoy her new status. Benedict died of a heart attack on September 17, 1948.

Bennett, Isobel Ida

(1909–)
Australian
Marine Biologist

Isobel Bennett gained an international reputation as a marine biologist even though she lacked formal training and never achieved the status of professor at the University of Sydney, where she worked most of her life. She was born in Brisbane on July 9, 1909, the oldest of four children. Her mother died when Isobel was 9 years old, then her family's money ran out when she was just 16, and she had to leave school to earn a living. She worked at secretarial jobs, but in 1932 the Great Depression caused the last of these to vanish.

Refusing to let their spirits be dampened, Bennett and one of her sisters spent the remainder of their savings on a cruise to Norfolk Island. As luck had it, another passenger on the cruise was William J. Dakin, professor of zoology at the University of Sydney. Dakin became friends with the two young women and offered Isobel a job helping with his research on the history of whaling. Bennett took him up on the offer. Dakin's specialty was marine biology, and Bennett too became interested in the subject. Dakin trained her and gave her increasingly challenging assignments. For instance, she became a regular crew member on the university's research ship, *Thistle,* sorting through nets full of plankton (tiny, floating marine life) and later giving informal instruction to the students on the ship's expeditions. She took on other jobs herself, such as cataloging and reorganizing the department's library. "When she saw something that should be done, she simply did it," wrote her biographer, Nessy Allen.

In time Bennett became almost as expert as Dakin. Her specialty was the ecology of the intertidal area. She studied intertidal shore life in Australia and in Antarctica, which she was one of the first four Australian women scientists to visit. She wrote many scientific papers and nine books, some of which became widely used textbooks. *The Great Barrier Reef,* which she wrote after numerous expeditions there between 1948 and 1970, was the first book to give a general picture of the whole Great Barrier Reef. It was published in 1971.

The University of Sydney never paid Bennett what it would have paid someone with a graduate degree, even though she did similar work. In 1962, however, the university did award her an honorary master's degree. Bennett also received other awards, including the Order of Australia and in 1982 the Mueller Medal of the Australia and New Zealand Association for the Advancement of Science. She was only the second woman to receive the latter award. The Royal Zoological Society of New South Wales gave awards to two of her books, in 1982 and 1988.

Although she officially retired in 1971, Isobel Bennett was still working in 1992 at the age of 83. Her peers have called her "one of Australia's foremost marine scientists." One genus and five species of marine animals and a coral reef have been named after her.

Berg, Paul

(1926–)
American
Biochemist

Paul Berg placed his stamp indelibly on the history of science by pioneering the technology to create recombinant deoxyribonucleic acid (DNA). Berg received the 1980 Nobel Prize in chemistry for this innovation, sharing the award with Frederick Sanger and WALTER GILBERT, who worked with DNA and ribonucleic acid (RNA).

Berg was born on June 30, 1926, in Brooklyn, New York, to Sarah Brodsky and Harry Berg, a clothing manufacturer. Berg grew up with two brothers. On September 13, 1947, he married Mildred Levy, and together the couple had one son, John Alexander.

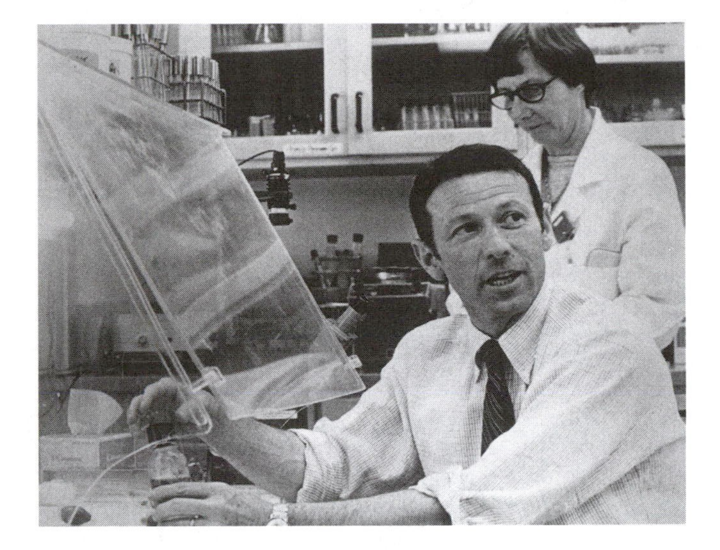

Paul Berg, who received the 1980 Nobel Prize in chemistry for developing the technology to create recombinant DNA. *Department of Special Collections, Stanford University Libraries.*

World War II interrupted Berg's study of biochemistry at Pennsylvania State University, as he served in the United States Navy from 1943 through 1946. He returned to his undergraduate work after the war and received a bachelor's degree in 1948. He went on to pursue a doctorate at Western Reserve University (now Case Western Reserve University), researching under a National Institutes of Health fellowship from 1950 until 1952, when he received a Ph.D. in biochemistry.

Berg conducted postdoctoral work as an American Cancer Society research fellow at the Institute of Cytophysiology in Copenhagen, Denmark, under Herman Kalcker from 1952 to 1953. He continued postdoctoral research the next year under ARTHUR KORNBERG at Washington University in St. Louis, Missouri, where he remained as a scholar in cancer research from 1954 through 1957. Berg served as an assistant professor of microbiology at the Washington University School of Medicine from 1956 to 1959, when the Stanford University School of Medicine recruited him as a professor of biochemistry. At Stanford Berg served as a senior postdoctoral fellow of the National Science Foundation from 1961 through 1968. He acted as the chairman of Stanford's Department of Biochemistry from 1969 to 1974; in 1970 Stanford named him the Sam, Lula and Jack Willson Professor of Biochemistry.

At Stanford, Berg conducted the pioneering research that led to the innovation of the techniques for splicing genes that became known as recombinant DNA technology. However, on the eve of performing an experiment injecting the recombinant DNA of monkey virus SV40 and a complementary bacteriophage into the common intestinal bacterium *Escherichia coli,* Berg realized that the all-pervasive nature of *E. coli* made the combination potentially highly volatile and halted his experiment. He then drafted a letter along with other prominent scientists warning the scientific community of the potential dangers of recombinant DNA technology if wielded unwisely. This missive, known as the "Berg letter," was published in the July 26, 1974, issue of *Science,* and it listed the recommendations arrived at by the group that Berg chaired, the Committee on Recombinant DNA Molecules Assembly of Life Sciences of the National Academy of Sciences. This prompted a meeting of 100 scientists from 16 countries that convened in Pacific Grove, California, on February 27, 1975, to discuss professional standards for the handling of the recombinant DNA question. Federal regulations published by the National Institutes of Health in June 1976 followed the guidelines set at this meeting.

Besides the Nobel Prize, Berg received the 1959 Eli Lilly Prize in biochemistry from the American Cancer Society and the 1980 National Medal of Science. In 1985 he became the director of the New Beckman Center for Molecular and Genetic Medicine, and in 1991 was named head of the National Institutes of Health's Human Genome Project scientific advisory committee.

Berg could be considered the father of the genetic engineering industry, which created new medicines and other chemicals but also created an ethical dilemma in terms of transforming the building blocks of life. He set an important precedent by questioning the implications of his own research, thereby prompting others to proceed into new scientific realms with caution.

Bergius, Friedrich

(1884–1949)
German
Chemist

Friedrich Bergius received the Nobel Prize for discovering a method for converting coal into oil. This so-called Bergius process, which subjected coal to extremely high pressure, was further developed by German industry to provide resources for the German war effort during World War II. Bergius was also a pioneer in cellulose conversion and succeeded in breaking down wood into edible products. Unlike many of his contemporaries, Bergius was not a pure academician; he conducted most of his path-breaking experiments at the Goldschmidt Company, where he served as research director for 31 years.

Bergius was born on October 11, 1884, in Goldschmieden, Germany (now part of Poland), to Heinrich and Marie Hasse Bergius. Later in life Bergius would marry Ottilie Kratzert, with whom he had three children.

In 1903 Bergius enrolled at the University of Breslau, where he studied chemistry and was awarded his doctorate four years later. He then served as an assistant to three renowned chemists—FRITZ HABER, Walther Nernst, and Max Bodenstein. With Bodenstein, Bergius began to explore the uses of high pressure in certain chemical reactions and invented the leak-proof high-pressure apparatus that would figure prominently in his future work.

Bergius accepted a professorship at the Technical University in Hanover in 1909. At his own laboratory there he investigated the effects of high pressure in both the formation of coal from wood and the transformation of heavy oils into lighter ones. By 1913 he was granted his first patent for his insight that adding hydrogen to petroleum to replace hydrogen lost during the refining process, would increase the yield of gasoline. But Bergius found it increasingly difficult to conduct his industrial-scale projects at the Goldschmidt Company in Essen. His position there provided him with an industrial laboratory that could process up to 20 tons of petroleum per day. He remained at the company until 1945.

Bergius's early years at the Goldschmidt Company were devoted to the problem of converting coal into oil. Recognizing that petroleum and coal differ only in their hydrogen content and molecular mass (petroleum has a higher hydrogen content and lower molecular mass), Bergius conceived the conversion process that bears his name. He heated a mixture of coal dust and heavy oil with hydrogen under high pressure in the presence of a catalyst composed of metallic sulfides. The coal took up the hydrogen—that is, became hydrogenated—and could be distilled to yield petroleum. Despite the profound implications of the Bergius process, he was thwarted in his efforts to continue his work after Germany's defeat in World War I and subsequent economic and industrial collapse. Bergius sold the patent rights from his discovery to a large German company, Badische Anilin und Sodafabrik (BASF), in 1926. Although he had not succeeded in making his conversion process economically feasible, BASF expanded his research, and by 1928 the company had constructed a factory that produced gasoline from coal.

Bergius then began to explore the possibility of creating edible products from wood. He found that treating wood with hydrochloric acid and water resulted in the breakdown of cellulose—the fundamental substance of wood and other plant material—into sugar. Throughout the 1930s and 1940s he perfected the hydrolysis (a chemical reaction whereby a substance reacts with water and is converted into other substances) of wood, and established in 1943 a plant in Richau that was essential to the German war effort. Unable to find work in Germany after World War II, Bergius left the country. After a stint in Spain, he moved in 1946 to Argentina, where he served as a scientific adviser to the Argentinean government until his death on March 30, 1949.

Bergius's achievements were recognized in 1931, when he was awarded the Nobel Prize in chemistry (which he shared with CARL BOSCH) for developing chemical high-pressure methods. Bergius was also honored with the Liebig Medal of the German Chemical Society. His work provided the basis for both the coal hydrogenation and cellulose conversion industries. Unfortunately both of these processes helped provide essential resources for Nazi Germany.

Bernard, Claude

(1813–1878)
French
Physiologist

Claude Bernard first intended to distinguish himself as a writer, but a literary critic persuaded him to pursue a stable career first as a safety net, so Bernard chose medicine. Bernard never practiced medicine but distinguished himself through physiological experimentation. He focused his attention on the digestive and vascular systems, making important discoveries on the strength of his objective detachment combined with his ability to see beyond accepted theories. When his health flagged near the end of his life, he devoted himself to the philosophy of science, setting out the key principles of scientific experimentation and hypothesisis.

Bernard honored his modest roots by returning for the yearly grape harvest to his home near the village of St.-Julien, in the Beaujolais region of France, where he was born on July 12, 1813. His parents, Pierre François Bernard and Jeanne Saulnier, were vineyard workers. Bernard shifted the course of his life toward academia through determination and even through calculation. After he failed to earn teaching credentials with the Faculty of Medicine in Paris in 1844, he entered into an unhappy marriage with Fanny Martin for her dowry, which enabled him to continue his beloved experiments. This union yielded three children—a son who died in infancy and two daughters, Jeanne-Henriette and Marie-Claude.

Bernard entered the world of medicine when an apothecary named Millet hired the 19-year-old as an apprentice in his pharmacy in Vaise, a suburb of Lyons. After Bernard's aborted attempt at literature, he passed the baccalaureate in 1834 to enter the Faculty of Medicine in Paris. In 1839 he entered an internship under Pierre Rayer at the Charité, one of Paris's municipal hospitals. He received his M.D. in Paris on December 7, 1843; that same year saw his first publication, "Recherches anatomiques et physiologiques sur la corde du tympan." On March 17, 1853, he received his doctorate in zoology from the Sorbonne on the strength of his thesis, "Recherches sur une nouvelle fonction du foie." In the decade between these two events Bernard made a series of significant contributions to medicine with his physiological experiments.

François Magandie served as Bernard's mentor in the laboratory from 1841 through 1844 and from 1847 through 1852, when Magandie retired from both his academic chair and his laboratory, both of which Bernard inherited. Bernard's discoveries advanced the understanding of the digestive and vascular systems. At the time doctors believed that digestion occurred primarily in the stomach; Bernard discovered that it occurred primarily in the small intestine. He discovered that pancreatic secretions break down fat molecules into fatty acid and glycerin. He also determined that the digestive system is anabolic as well as catabolic; in other words, it both breaks down complex molecules into simpler ones and builds complex molecules out of simpler ones. An example of the latter would be glycogen, a starchlike substance that Bernard discovered in 1856, which the liver produces from sugars and stores as carbohydrates to be broken back down into sugars when necessary. One of Bernard's vascular discoveries involves the climate-controlling mechanisms of the vasomotor system, which dilates and constricts blood vessels on the skin's surface to heat or cool the body.

As Bernard's failing health kept him out of the laboratory, he shifted his focus to the philosophy of science; he published *An Introduction to the Study of Experimental Medicine* in 1865. In 1869 he was elected to the French Academy. Bernard died on February 10, 1878, in Paris, France.

Bernoulli, Daniel

(1700–1782)
Swiss
Mathematician, Physicist

Over four generations the Bernoulli family produced 11 important mathematicians. Daniel Bernoulli was one of the most significant thinkers in his family, affecting the history of science by providing the theoretical foundation for the establishment of the modern science of hydrodynamics. The cornerstone of this new branch of science was Bernoulli's principle, which holds that the pressure in a fluid decreases as its velocity increases. Bernoulli also contributed to the understanding of medicine, mathematics, and the natural sciences, as he held chairs in these disciplines throughout his career.

Daniel Bernoulli was born on February 8, 1700, in Groningen, Netherlands, the second son of Johann I. Bernoulli and Dorothea Falkner Bernoulli, daughter of the patrician Daniel Falkner. The inherent brilliance of the family created competition, expressed in bitterness between father and son. On the eve of Daniel's 1738 publication of

Daniel Bernoulli, who provided the theoretical framework used to establish the modern science of hydrodynamics. *Smith Collection, Rare Book & Manuscript Library, University of Pennsylvania.*

his influential tract *Hydrodynamica* his father, Johann, published a competing text, *Hydraulica,* which he predated to 1737 in order to undercut his son's achievement.

Johann I and Nikolaus II, Daniel's older brother, served as Daniel's early mathematics teachers. In 1713 Daniel commenced studying philosophy and logic, and in 1715 he passed his baccalaureate. After earning his master's degree a year later he embarked on the study of medicine at a series of universities—first in Basel, then in Heidelberg in 1718, and in Strasbourg in 1719, before returning to Basel in 1720. In 1721 he earned his doctorate with the dissertation "De respiratione."

In 1724 Bernoulli published his work on differential equations and the physics of water flow, *Exercitationes quaedam mathematicae,* which prompted a job offer from St. Petersburg, where he remained for eight years. Throughout this period he applied for professorships at the University of Basel without success until 1733, when he received a joint appointment in anatomy and botany. A decade later he exchanged these chairs for one in physiology, which he preferred, and in 1750 he exchanged that for one in physics, which he filled until 1776.

In *Hydrodynamica* Bernoulli discussed the relationships among the properties of pressure, density, and velocity in flowing fluid. He also set the groundwork for a kinetic theory of gases, asserting that an increase in temperature leads to a corresponding increase in molecular pressure and motion, assuming that molecules are in constant but random motion. Bernoulli's contributions to science and mathematics earned him 10 prizes from the Paris Academy of Sciences between 1725 and 1749, with papers on subjects as various as astronomy, gravity, magnetism, navigation, and oceanology. The academy awarded the 1734 prize jointly to father and son for their tandem work on planetary orbits. Believing that he alone deserved the honor, Johann I reportedly threw Daniel out of the Bernoulli house.

Bernoulli assured himself a place in the history of science by utilizing for the first time Leibniz's calculus to solve Newtonian scientific problems. Bernoulli died on March 17, 1782, in Basel, Switzerland.

Bethe, Hans Albrecht

(1906–)
German/American
Physicist

Hans Bethe established his significance in the history of science after his flight from Nazi Germany, where he was persecuted because of his Jewish ancestry. In 1936 he worked with colleagues to create a summary of nuclear physics that became known as "Bethe's Bible," essential reading for physics graduate students. In 1938 he reluctantly agreed to write a paper for an astrophysics conference, then summarily solved the problem of astral energy production. During World War II he helped develop the atomic bomb, as part of the Manhattan Project, an experience that prompted him to promote the necessity of social responsibility in science.

Bethe was born on July 2, 1906, in Strassburg, Germany (now Strasbourg, France), to Albrecht Theodore Julius Bethe, a lecturer in physiology at the University of Strassburg, and Anna Kuhn, the daughter of a professor. Bethe met his future wife, Rose Ewald, while working at the Technical College of Stuttgart under her father, Paul Ewald, who invited his protégé into his home for dinners. The Ewalds also emigrated from Germany, and Rose was attending Smith College when she reconnected with Bethe. The couple married on September 14, 1939, and had two children, Henry and Monica.

In 1924 Bethe entered the University of Frankfurt, and in 1926 he moved to the University of Munich, where

After serving as a member of the Manhattan Project during World War II, Hans Bethe became an advocate for social responsibility in science. *Photograph by Roy Bishop, Acadia University, courtesy AIP Emilio Segrè Visual Archives.*

he earned a doctorate in theoretical physics in 1928. Under Arnold Sommerfeld he wrote his dissertation on the theory of electron diffraction. Over the next three years Bethe held teaching positions at his undergraduate and graduate universities; he also visited Cambridge University on a Rockefeller Foundation fellowship and worked with ENRICO FERMI in Rome. In 1932 the University of Tübingen appointed Bethe as an assistant professor. Over the next year he exchanged knowledge with Hans Geiger, the inventor of the Geiger counter, until the university dismissed Bethe for his Jewish heritage and Geiger dissolved their friendship. After a short stint in exile at the Universities of Manchester and Bristol in Britain, Bethe landed an assistant professorship at Cornell University. In 1937 he earned the title of full professor at the university, where he remained until his 1975 retirement.

Bethe found himself explaining theoretical physics not only to students, but also to Cornell's physics faculty. Bethe, recognizing the lack of publications discussing the basics of nuclear physics, enlisted the help of Robert F. Bacher, M. Stanley Livingston, and his former graduate adviser, Arnold Sommerfeld, to write a series of three articles in 1936 and 1937 that together functioned as a primer on the subject. Two years later GEORGE GAMOW and EDWARD TELLER coerced Bethe to present a paper at their Washington, D.C., astrophysics conference. In a short six weeks he discovered how massive stars generate energy by a six-stage process known as the carbon cycle, whereby a carbon-12 atom reacts with four hydrogen nuclei successively and releases energy in creating a helium nucleus along with the original carbon atom. *Physical Review* published this paper, "Energy Production in Stars," on the first page of a 1939 issue. On the basis of this work Bethe received the 1967 Nobel Prize in physics.

In 1941 Bethe gained U.S. citizenship and subsequently contributed to America's military efforts in World War II by working on the development of radar at the Massachusetts Institute of Technology, where he invented the Bethe coupler to measure increased electromagnetic wave activity. Bethe then traveled to Los Alamos, New Mexico, as a member of the Manhattan Project, in which he headed the Theoretical Physics Division. After contributing to the building of the atomic bombs, Bethe devoted his efforts to educating his profession and the general public about the destructive capabilities of nuclear power. Toward this end he helped launch *The Bulletin of the Atomic Scientist.*

Besides the Nobel Prize, Bethe received a 1946 Presidential Medal of Merit, the 1955 Max Planck Medal of the German Physical Society, and the 1961 Enrico Fermi Award from the U.S. Atomic Energy Commission. As a professor emeritus, Bethe maintained an office at Cornell and continued to educate the public on both social and scientific issues.

Binnig, Gerd

(1947–)
German
Physicist

Gerd Binnig shared the 1986 Nobel Prize in physics with HEINRICH ROHRER as coinventor of the scanning tunneling microscope (STM), a mechanism that allowed the examination of individual atoms on the surface of solids. This pair also shared the prize with Ernst Rohrer, who—some 55 years earlier—had invented the electron microscope, the most powerful microscope until the advent of the STM. Binnig was 39 years old at the time he won.

Binnig was born on July 20, 1947, in Frankfurt am Main, Germany. His parents were Ruth Bracke, a drafter, and Karl Franz Binnig, a machine engineer. Binnig studied at Offenbach and at the Johann Wolfgang Goethe University in Frankfurt, where he received his Ph.D. in 1978. At Goethe University Binnig met Lore Wagler, who was studying to become a psychologist, and the couple married in 1969. Together they had two children—a daughter, born in 1984, and a son, born in the United States in 1986.

Gerd Binnig, who was 39 years old when he won the 1986 Nobel Prize in physics. *AIP Emilio Segrè Visual Archives.*

Immediately upon receiving his doctorate, Binnig went to work for International Business Machines (IBM) in the company's Zurich Research Laboratories, located in Rüschlikon, Switzerland. There he teamed up with Rohrer, who had been working with IBM for 15 years, and together they began to consider the surface of solids at an atomic level, a topic that still perplexed scientists. Binnig and Rohrer hit upon the idea of applying the quantum theory of tunneling that had been experimentally verified by Ivar Giaever in 1960 to the study of solid surfaces. Tunneling is the quantum phenomenon whereby atoms on the surface of a solid escape to form a kind of cloud hovering above the surface; when another surface approaches, the two atomic clouds overlap and tunneling, or an atomic exchange back and forth, occurs.

Binnig and Rohrer applied this phenomenon to their research by attaching a tungsten conducting probe to an instrument that would move the tip of the probe, which was the width of a single atom, into close enough proximity to the surface of the solid to induce tunneling. Since the magnitude of the flow of electrons depended on the distance of the tip from the surface, the mechanism could maintain a constant distance as it scanned the surface, thereby allowing for computerized mapping of the surface of the solid. The STM could achieve a vertical resolution of 0.1 angstrom, or $^1/_{30}$ the size of a single atom, and a horizontal resolution of 6 angstroms. By 1981 Binnig and Rohrer had constructed the first STM, and with it Binnig was the first person to observe a virus escaping a cell.

In 1984 IBM promoted Binnig to the position of group leader and subsequently appointed him as an IBM Fellow. In 1985 Binnig took a leave from IBM to conduct research at Stanford University, where he helped develop the atomic force microscope (AFM) along with Calvin Quate and his IBM colleague Christoph Gerber. The AFM expanded upon the uses of the STM, allowing for the microscopic examination of materials that did not conduct electricity. As an innovator of both the STM and the AFM, Binnig was one of the foremost figures in microscopy in the 20th century.

Blackwell, Elizabeth

(1821–1910)
British/American
Physician

The one medical school that admitted her thought Elizabeth Blackwell's application was a joke, but to her becoming a physician was a "moral crusade." She became the first woman to obtain an M.D. degree.

Elizabeth was born on February 23, 1821, in Bristol, England, the third of 12 children born to Samuel and Hannah Blackwell. Samuel, the well-to-do owner of a sugar refinery, filled his house with visitors who, like him, supported rights for women, the abolition of slavery, and similar social causes. He hired tutors to educate his children and insisted that his daughters be taught the same subjects as his sons.

Samuel Blackwell moved his family to the United States in 1832, when Elizabeth was 11 years old, after a fire destroyed his British refinery. They settled first in New York City, where Samuel established another sugar refinery. He lost most of his wealth in a financial panic in 1837, however. The Blackwells moved to Cincinnati, Ohio, in May 1838, hoping for a new start, but their hopes ended when Samuel died of a fever that August. He left his family with tremendous grief and without money.

Hannah Blackwell and her three oldest daughters, including Elizabeth, tried to earn a living by teaching. Neither teaching nor the idea of marriage appealed to Elizabeth, but she was not sure what else to do with her life until a dying friend, Mary Donaldson, urged her to become a physician. Donaldson said she might have sought treatment sooner if she could have seen a woman doctor.

Most of the physicians Blackwell talked to tried to discourage her, and 29 medical schools turned down her application—but the 30th did not. The dean of Geneva Medical College, a small college in upstate New York, wrote on behalf of his students, "In extending our unanimous invitation to Elizabeth Blackwell, we pledge ourselves that [she shall never] regret her attendance at this institution."

Blackwell did not know it, but the letter had been meant as a joke. When the dean read her application to the students, they thought it was a prank from a rival college, and they replied in the same spirit. They were flabbergasted when a real and serious young woman appeared at their school on November 6, 1847. To their credit, however, they honored the pledge in the dean's letter. The Geneva townspeople were less understanding. They assumed that a woman who said she wanted to be a doctor was either a person of ill repute, an abortionist, or an insane person. They often crossed the street to avoid speaking to her.

Blackwell had her first practical experience in caring for the sick in 1848 at the hospital attached to the huge, grim Blocksley Almshouse in Philadelphia. Her patients were Irish immigrants felled by an epidemic of typhus fever. She wrote her thesis on the illness, emphasizing the importance of sanitation in preventing it.

Elizabeth Blackwell received her M.D. degree on January 23, 1849. "It shall be the effort of my life to shed honor on this diploma," she said. The crowd at the graduation ceremony burst into cheers.

For many years after her graduation she struggled to gain practical experience and establish a clientele, fighting intense social opposition along the way. Finally, in the hope of improving her finances, Blackwell gave lectures in which, among other things, she suggested that girls should take regular exercise in the open air and should be taught how their bodies worked. These novel ideas shocked many, but they appealed to a group of wealthy women who belonged to the Religious Society of Friends, or Quakers. These Quaker women became Blackwell's first patients and provided her with much-needed financial support. Blackwell's lectures were published in 1852 in the book, *The Laws of Life, with Special Reference to the Physical Education of Girls.*

Now that she had a few paying patients, Blackwell also began helping people who could not pay. In March 1853 she set up a one-room clinic for women and children near Tompkins Square, one of the worst slums in New York. The clinic was open just three afternoons a week, but it treated 200 women in its first year.

Two other pioneering women physicians, Blackwell's younger sister, Emily, and a Polish immigrant, Marie Zakrzewska, arrived in 1856 to help her at the clinic. With their labor and with money from Quaker friends, Blackwell transformed a run-down building into a hospital. The New York Infirmary for Women and Children opened on May 12, 1857, the first hospital to be staffed entirely by women.

In 1858 Blackwell went to England, where she gave lectures on medical education for women. One young woman she inspired was ELIZABETH GARRETT ANDERSON, who would later become the first British woman M.D. The Civil War started soon after Blackwell returned to the United States, and Elizabeth and Emily trained battlefield nurses during the war. After the war ended, Blackwell opened the Women's Medical College of the New York Infirmary on November 2, 1868. It had entrance examinations, a rarity in those days. It also offered three years of training instead of the usual two, plus extensive practice in the hospital.

In 1869 Blackwell returned to England, this time for good. In addition to treating a large number of patients, she gave lectures on the need for improved sanitation and wrote two more books, *Counsel to Parents on the Moral Education of Their Children* (1878) and *The Human Element in Sex* (1884). She was ahead of her time in her stress on preventive medicine ("Prevention is better than cure," she wrote) and her willingness to discuss taboo subjects such as sex. She eventually retired to Hastings, on England's southern coast, where she died on May 31, 1910, at the age of 89.

Bloch, Felix

(1905–1983)
Swiss/American
Physicist

Recipient of the 1952 Nobel Prize in physics, Felix Bloch was recognized for the development of nuclear induction, also referred to as nuclear magnetic resonance, which allowed for the measurement of the magnetic fields of atomic nuclei. Though these techniques were initially developed to uncover the magnetic moment of the proton and neutron, they contributed greatly to physics and chemistry and were later used to analyze the structure of large molecules. Bloch also made numerous advances in the field of solid-state physics.

Born on October 23, 1905, in Zurich, Switzerland, to Agnes Mayer and Gustav Bloch, a grain merchant, the younger Bloch showed an early interest in mathematics and astronomy. He married the physicist Lore C. Misch, the German-born daughter of a professor, in 1940, and the couple had four children.

Because of his propensity for the sciences Bloch's family encouraged him to enter the Federal Institute of Technology in Zurich to study engineering. Bloch matriculated in 1924, and soon he became interested in physics. Bloch graduated in 1927 and continued his education at the University of Leipzig. There he studied under WERNER KARL HEISENBERG, one of the founders of quantum

Felix Bloch, who won the 1952 Nobel Prize in physics for the development of nuclear induction, which allows for the measurement of the magnetic field of atomic nuclei. *Department of Special Collections, Stanford University Libraries.*

mechanics. Bloch received his doctorate in physics in 1928. His doctoral dissertation presented a study of the quantum mechanics of solids that significantly extended the understanding of electrical conduction. After earning his doctorate, Bloch studied and conducted research at a variety of institutions, including the University of Utrecht in the Netherlands and the University of Copenhagen, where he worked with NIELS HENDRIK DAVID BOHR. Bloch traveled to the United States after the rise of Adolf Hitler and took a position at California's Stanford University.

Before Bloch became interested in studying the neutron, he had already made many advancements in theoretical physics. With the Bloch-Fouquet theorem he detailed the structure of wave functions for electrons in a crystal. The theorem later became useful for physicists studying the nature of metals. Bloch also worked on the quantum theory of the electromagnetic field. After he began teaching at Stanford, he set out to investigate the magnetic moment of the neutron, the existence of which had been announced in 1933 by OTTO STERN. In 1934 Bloch suggested that proof of the neutron's magnetic moment could be found by splitting a beam of neutrons into two sections that related to polarized neutron beams. Bloch and his collaborator LUIS WALTER ALVAREZ from the University of California at Berkeley used this technique to measure the neutron's magnetic moment in 1939.

After a period during which Bloch worked on atomic energy in New Mexico and countermeasure radar research at Harvard University, he returned to Stanford University and developed the method of nuclear induction with the physicists William W. Hansen and Martin Packard. The technique, which was based on the principle of magnetic resonance, helped determine the relationship between nuclear magnetic fields and the magnetic and crystalline properties of assorted materials. Bloch's method enabled scientists to investigate nuclear particles thoroughly and to measure the nuclear magnetic moment of an individual nucleus accurately.

For his outstanding contribution to physics Bloch received the Nobel Prize in physics in 1952, an award he shared with EDWARD MILLS PURCELL, who had simultaneously discovered a nearly identical technique. The technique later became known as nuclear magnetic resonance (NMR) and enjoyed a wide range of applications not only in physics and chemistry but also in such fields as diagnostic medicine. After 1955 Bloch continued his studies of NMR and also researched superconductivity. He retired from Stanford University in 1971 and became professor emeritus. Bloch was elected to the National Academy of Sciences in 1948 and also served as a fellow of the American Physical Society and the American Academy of Arts and Sciences.

Blodgett, Katharine Burr

(1898–1979)
American
Physicist

Katharine Burr Blodgett was one of the first women scientists to win a major reputation in industry, creating several techniques relating to films on a surface that are still used. She was born in 1898 in Schenectady, New York, home of the giant General Electric Corporation (GE). Her father, who unfortunately died just before Katharine's birth, was GE's head patent attorney. After his death her mother embarked on years of travel, taking Katharine and her older brother to live for a while in France and then in Germany. After graduation from Bryn Mawr in 1917 Blodgett followed her father's footsteps back to General Electric and applied for a research job. She was told that she needed more science background, so she spent a year at the University of Chicago earning a master's degree in physics. Even that might not have been enough if World War I had not taken so many young men out of the country. As it was, however, GE hired her in 1918 as an assistant to IRVING LANGMUIR, who would later (1932) win a Nobel Prize in chemistry. Blodgett was the first woman to work in GE's laboratories and virtually the only woman physicist working in industry at the time.

Langmuir helped Blodgett gain access to the Cavendish Laboratory at Britain's Cambridge University, an almost unheard-of feat for a woman. She earned her Ph.D. in 1926, the first woman to receive that degree from Cambridge. She then returned to Langmuir's laboratory.

Langmuir had invented a way to deposit oil onto the surface of water so that it formed a film just one molecule thick. Blodgett discovered in December 1933 that by dipping a flat surface into such a film, she could deposit an equally thin layer onto the surface. The technique could be repeated to build up as many layers as were desired. "You keep barking up so many wrong trees in research," she later told the writer Edna Yost. "This time I . . . barked up one that held what I was looking for."

Blodgett developed two applications of her technique. First she noticed that when oily layers built up on a surface, they reflected different colors under white light, just as a film of oil on a puddle shows a rainbow reflection. She made a gauge that showed the colors of different thicknesses of film. By matching the color of an unknown film with a color on the gauge, a scientist could measure the thickness of the film with an accuracy of millionths of an inch.

In 1938 Blodgett found that if a film exactly 0.000004 inch thick was put on glass, the light waves reflected from the bottom of the coating and those reflected from the top canceled each other out. As a result, no light was reflected, and the glass in effect became invisible. Her "invisible" coated glass was used to make lenses for cameras, telescopes, and other optical instruments because it stopped the 8 to 10 percent loss of gathered light that was normally caused by reflection. This cut down on the exposure time needed to produce a good image.

Katharine Blodgett received several awards for her work, including the American Chemical Society's Garvan Medal (1951) and the Progress Medal of the Photographic Society of America (1972). She retired from GE in 1963 and died in 1979. The techniques she discovered are still used in physics, chemistry, and metallurgy.

Boden, Margaret

(1936–)
British
Psychologist

Margaret Boden has popularized the idea that the computer programming involved in so-called artificial intelligence (AI) can explain much about how the human mind works. She was born on November 26, 1936, to a lower-middle-class British family. "I'd never expected to go to university," she told the reporter Celia Kitzinger in 1992. "Neither of my parents did." Nonetheless she won a scholarship to prestigious Cambridge University in 1955. She found Cambridge "like being born anew. . . . So many doors opened—intellectual doors, social doors, cultural doors."

Boden studied medicine and philosophy at Cambridge, but she was not happy with her postgraduate experience of working with mental patients and teaching philosophy (from 1959 to 1962 she was a lecturer in philosophy at the University of Birmingham). Soon after entering Harvard University in 1962 to work on a doctorate in social and cognitive psychology she happened to pick up a book, *Plans and the Structure of Behavior.* "Just leafing through it in the bookshop . . . change[d] my life," she recalls. "It was the first book that tried . . . to apply the notion of a . . . computer program to the whole of psychology." This idea struck her "like a flash of lightning."

Boden's doctoral thesis considered how the idea of intention or purpose for actions was applied in various theories of psychology and how it could be understood in terms of actions taken by a computer. This thesis grew into her first book, *Purposive Explanation in Psychology,* published in 1972. She says that book contained all her basic ideas and that her later, more popular books are mere "footnotes" to it.

In 1965 Boden began teaching at Sussex University. She married two years later (she divorced in 1981) and soon had two young children. She wrote her first two books in snatches while they napped. Her second book, *Artificial Intelligence and Natural Man* (1977), used comparisons drawn from subjects including knitting and baking to explain the complex computer programming involved in artificial intelligence and show how it could be used to explore human thinking. She followed this book with one about the Swiss psychologist Jean Piaget, first published in 1979, which discussed his ideas about biology and philosophy as well as his famous theories about children's psychological development. She related his ideas to AI and, in the book's second edition, to artificial life (A-Life). Boden's fourth major book, published in 1990, was *The Creative Mind.* It extended the links between computer programs and human thinking into the realm of creativity. Boden maintained that insights into human creativity can be gained by studying the results of attempts to program computers to be creative.

As of 1998 Boden, a fellow of the British Academy and of the American Association for Artificial Intelligence, was professor of psychology and philosophy at the University of Sussex, where in 1987 she had become the founding dean of the School of Cognitive and Computing Sciences. The department's interdisciplinary courses echo her own ability to combine insights from many disciplines. "I'm interested in the human mind, not computers," she emphasizes, "but I use computers as a way of thinking about the mind."

Bohr, Niels Henrik David

(1885–1962)
Danish
Physicist

The study of physics was transformed from its classical model to a new model founded on quantum theory based on the bold hypothesizing of scientists such as Niels Bohr. Bohr supported his courageous theorizing with careful experimentation, confirming his hunches with physical evidence. He used planetary orbiting as a model for his atomic theory, with electrons circling the nucleus of an atom. Bohr later proposed the liquid drop as a metaphor to visualize atomic motion. The Swedish Academy of Sciences awarded Bohr the 1922 Nobel Prize in physics for his pioneering work in the quantum mechanics of atomic structure.

Bohr was born on October 7, 1885, in Copenhagen, Denmark, to Christian Bohr, a professor of physiology at the University of Copenhagen, and Ellen Adler Bohr, the daughter of the prominent Danish banker D. B. Adler. Bohr married Margrethe Nørlund on August 1, 1912, and the couple had six sons—Hans, Erik, Aage, and Ernst, as well as two others who died young. Aage accompanied his father to Los Alamos, New Mexico, to conduct research on the atomic bomb. Aage won the 1975 Nobel Prize in physics jointly with Ben R. Mottelson and James Rainwater essentially for explicating his father's nuclear model.

Niels Bohr (shown here with his mother), who was awarded the 1922 Nobel Prize in physics for his work in the quantum mechanics of atomic structure. *Niels Bohr Archive, courtesy AIP Emilio Segrè Visual Archives.*

In 1903 Bohr entered the University of Copenhagen. In 1907 he received not only his B.S. in physics but also a gold medal from the Royal Danish Academy of Science for his research on the surface tension of water, as in the vibration of a jet stream. In 1909 Bohr earned an M.S. in physics and in 1911 a doctorate. His doctoral thesis explored the electron theory of metals.

In 1911 Bohr left Denmark to conduct research in the Cavendish Laboratory at Cambridge University under J. J. Thomson but soon realized that ERNEST RUTHERFORD would be a better mentor, so he transferred to Victoria University in Manchester in 1912; he was a lecturer in physics there between 1914 and 1916. Bohr then returned to the University of Copenhagen, which had established a chair of theoretical physics for him. In 1920 he became the director of the Institute for Theoretical Physics (later named after him), a position he held until his death.

As of 1912 atomic physicists realized that electrons orbited the nucleus of an atom; however, the process in which this occurred could not be understood in the terms of classical principles of physics. Bohr attempted a novel approach to solving this problem by applying Johann Balmer's mathematical formula for representing the spectral lines of hydrogen atoms. This formula hedged on two integers whose significance remained a mystery. Bohr simply hypothesized that the integers represented orbital paths where the classical laws of physics were suspended, and some other laws, namely, quantum physics, governed the motion. The hypothesis had no theoretical foundation other than the fact that it worked when tested.

After his experience on the Manhattan Project, Bohr dedicated himself to lobbying for the sane use of nuclear energy by organizing the first Atoms for Peace Conference in Geneva in 1955. In 1957 he won the first Atoms for Peace Award, adding this to his list of other prestigious awards, which included the 1922 Nobel Prize in physics, the 1926 Franklin Medal of the Franklin Institute, the 1930 Max Planck Medal of the German Physical Society, and the 1930 Faraday Medal of the Chemical Society of London. Bohr died on November 18, 1962, in Copenhagen.

Bok, Bart Jan

(1906–1983)
Dutch/American
Astronomer

Bart Jan Bok's name graces the discovery he made, Bok's globules, the dark spots dotting nebulae that he hypothesized to be gaseous clouds gathering energy in the process of becoming stars. This discovery took on added significance as the formation of stars garnered more and more attention in the scientific community as well as in mainstream culture. Bok contributed immensely to the understanding of star formation.

Bok was born on April 28, 1906, in Hoorn, Netherlands. His parents were Gesina-Annetta Van der Lee and Jan Bok. On September 9, 1929, he married Priscilla Fairfield, and together the couple had two children, a son named John Fairfield and a daughter named Joyce Annetta. Bok became a naturalized citizen of the United States in 1938.

Bok attended the University of Leiden from 1924 through 1927 and the University of Groningen from 1927 through 1929; he received his Ph.D. there in 1932. In the meantime he traveled to Harvard University in Cambridge, Massachusetts, as an Agassiz Fellow from 1929 through 1930. He remained at Harvard for almost three decades, first as the R. W. Willson Teaching Fellow from 1930 through 1933. Harvard appointed him as an assistant professor of astronomy from 1933 until 1939, when he was promoted to associate professor. In 1947 Harvard named him the R. W. Willson Professor of Astronomy, a title he held for the next decade.

Also in 1947 Bok observed against the backdrop of luminous gas the small, circular dark patches that would later bear his name. Bok had been studying cosmic evolution, and this phenomenon fit within the theoretical context for the genesis of stars. Bok had already published several books on related astronomical topics, such as *The Distribution of Stars in Space* in 1937 and *The Milky Way,* coauthored by his wife, Priscilla Fairfield Bok, and published in 1941 and again in 1957. Bok also published *Basic Marine Navigation* with F. W. Wright in 1944 and *The Astronomer's Universe* in 1958.

In 1957 Bok accepted an appointment as professor and head of the Department of Astronomy at the Australian National University. He maintained a simultaneous appointment as the director of the Mount Stromlo Observatory near Canberra throughout the next decade. In 1966 he returned to the United States as a professor of astronomy at the University of Arizona and as the director of Steward Observatory in Tucson. Bok maintained this joint appointment until 1974, when he retired to become a professor emeritus at the University of Arizona.

Besides his major discovery, Bok studied the star clouds of Magellan as well as interstellar matter and galactic structure. He conducted radio astronomy to get a clearer map of the Milky Way than simple optical observations could provide. Instead of confirming the optical picture, radio astronomy contradicted it, so Bok revised the existing theories on the Milky Way by fusing the two contradictory hypotheses into one harmonious theory. In 1957 he received the Oranje-Nassau Medal in the Netherlands. He died in 1983.

Boltzman, Ludwig Eduard

(1844–1906)
Austrian
Physicist

Ludwig Boltzman developed statistical mechanics, which describes how atomic properties, such as mass and charge, decide the perceptible properties of matter, such as viscosity and diffusion. Boltzman left the imprint of his name in three places in the annals of science: the Maxwell-Boltzman distribution law, which describes how the energy of a gas is distributed among its molecules; Boltzman's equation, which expresses entropy in terms of probability; and Boltzman's constant, a factor used in his equation.

Boltzman was born on February 20, 1844, in Vienna, Austria, to Ludwig Boltzman, a tax officer, and Katherine Pauernfeind Boltzman. He married Henrietta von Aigentler in 1876; with her he had four children. Boltzman commenced his undergraduate work at Linz and continued his

Boltzman's equation, which Ludwig Boltzman derived in 1896, measures the relationship between entropy and probability. *AIP Emilio Segrè Visual Archives.*

study at the University of Vienna, where he earned his doctorate in 1866 under the tutelage of Josef Stefan. Boltzman held professorships in physics or mathematics at universities in several European cities throughout his lifetime: at Graz from 1869 to 1873 and again from 1876 to 1879; at Vienna from 1873 to 1876, from 1894 to 1900, and from 1902 to 1906; at Munich from 1889 to 1893; and at Leipzig from 1900 to 1902.

The second law of thermodynamics interested Boltzman particularly, and his work focused on applying mechanical and statistical analyses to this law to elucidate its properties further. The work of JAMES CLERK MAXWELL proved instrumental to his own theorizing, in terms of both the kinetic theory of gases and the theory of electromagnetism. Boltzman used Maxwell's kinetic theory as a foundation and a springboard, applying it to conditions in which external forces were present. The fusion of his and Maxwell's theories resulted in the Maxwell-Boltzman distribution law. An offshoot of this finding was the law of equipartition of energy, which stated that the energy created by the motion in different directions in an atom was equally distributed on average. Boltzman called this phenomenon *degrees of freedom*.

Boltzman's equation, which he derived in 1896, measured the relationship between entropy and probability. Through the exploration that led to his assertion of this equation he came to realize that entropy was an expression of the inherent disorder in the atomic system. Boltzman also contributed a theoretical derivation of his former mentor Josef Stefan's law of blackbody radiation.

Prone to depression throughout his life, Boltzman reacted strongly to the consistent criticism of the logical positivist philosophers of Vienna. Although he maintained a close personal relationship with Wilhelm Ostwald, they fought bitterly over their opposing views of science, as Ostwald was an opponent of atomism. Their feuds verged on violence, but Boltzman maintained his confidence privately that the gas theory of science would prevail in the end. This confidence could not overcome his depression in the face of opposition, however, and while on vacation in Duine, on the Adriatic coast near Trieste, Boltzman committed suicide on September 5, 1906. His equation was engraved on his headstone.

Boole, George

(1815–1864)
English
Mathematician

George Boole created a new approach to logic by wresting it from the sole possession of philosophy and applying mathematical precision to the discipline. This new marriage of logic and mathematics allowed for novel applications unimagined by Boole. Boolean algebra, for example, serves as the basis for the design of digital technology that drives computer and telephone circuitry.

Boole was born on November 2, 1815, in Lincoln, in Lincolnshire, England. His father, John Boole, was a cobbler by trade but maintained an interest in mathematics and optical instruments that he passed on to his son. When the Mechanics Institution was founded in Lincoln in 1834, John Boole was named the curator of the reading room, which received publications from the Royal Society. Through this library George Boole continued the self-education that he'd commenced after the decline of his father's business forced him to quit school at the age of 16 to teach in the village schools of West Riding in Yorkshire. On his own Boole worked through complex texts such as Sir ISAAC NEWTON's *Principia* and Lagrange's *Mécanique analytique*. At the age of 20 Boole opened his own school in Lincoln.

In 1839 the newly formed *Cambridge Mathematical Journal* published Boole's paper "Researches on the Theory of Analytical Transformations," the first in a series of original papers Boole submitted to this journal for publication. In 1844 the journal *Philosophical Transactions of the Royal Society* published a paper in which Boole discussed the intersection between algebra and calculus. That same year the Royal Society awarded Boole its Royal Medal for his contributions to analysis, specifically of very large and very small numbers. In 1847 Boole issued a pamphlet, "Mathematical Analysis of Logic," in which he demonstrated the inherent connection between logic and mathematics, thus calling into question the traditional connection between logic and philosophy.

In 1849 the newly founded Queens College in County Cork, Ireland, offered Boole a professorship in mathematics on the strength of his published theoretical works, despite the fact that he held no university degrees. In 1854 he published *An Investigation into the Laws of Thought, on Which Are Founded the Mathematical Theories of Logic and Probabilities*. Along with his 1847 pamphlet this text laid out the underpinnings of Boolean algebra, which uses a two-valued system of classification that acts as the foundation for binary systems of digital technology. In 1855 Boole married Mary Everest, daughter of Sir George Everest (after whom the Himalayan mountain was named) and niece of a Queens College professor of Greek. The couple had five daughters together.

The Royal Society elected Boole as a fellow in 1857. In 1859 and 1860 he published a pair of texts, *Treatise on Differential Equations* and *Treatise on the Calculus of Finite Differences*, that encapsulated his most important ideas. Boole's devotion to his work proved to be his downfall. A hard rain could not deter him from walking to his class, which he taught in wet clothes; the illness that followed caused his death on December 8, 1864, in Ballintemple, in County Cork, Ireland.

Bordet, Jules-Jean-Baptiste-Vincent

(1870–1961)
Belgian
Physician, Immunologist

Recognized as the leading researcher in serology, the study of the properties and reactions of blood serums, Jules Bordet made important contributions in immunology and bacteriology. Bordet uncovered information about the immunity factors of blood serum, a discovery that led to the diagnosis and treatment of numerous diseases. Also credited to Bordet was the development of complement fixation tests, which allowed for the detection of specific antibodies and disease-causing antigens, such as certain bacteria and toxins. For his revolutionary work Bordet was awarded the 1919 Nobel Prize in medicine.

Born on June 13, 1870, in Soignies, Belgium, Bordet was the second son of a schoolteacher, Charles Bordet, and his wife, Célestine Vandenabeele Bordet. In 1874, when Bordet's father gained a position as a teacher at the primary school École Moyenne, the family moved to Brussels. Bordet married Marthe Levoz in 1899, and the couple had three children. Their one son followed his father's career path and became a medical scientist.

An exceptional student, Jules Bordet attended the school where his father taught then went to secondary school at the Athénée Royal of Brussels, where his interest in chemistry began. When he was 16 years of age Bordet enrolled at the Free University of Brussels, where he received his medical degree in 1892. Bordet embarked on his research projects while he was a medical student, and the year he earned his degree he published a paper regarding the reactions of viruses to immunized organisms and the changes to the viruses that resulted. This work earned Bordet a scholarship from the Belgian government to study at the Pasteur Institute in Paris.

Bordet spent 1894 to 1901 at the Pasteur Institute under the zoologist ÉLIE METCHNIKOFF. It was during this period that Bordet made his primary discoveries in immunology. Bordet first investigated bacteriolysis, the destruction of bacteria. The scientist Richard Pfeiffer had found in 1894 that when vaccinated animals were injected with the bacteria against which they were immunized, they would die. Bordet proposed that bacteriolysis was caused by two substances—one was an antibody resistant to heat, which was immune to the specific bacterium, and the other was a heat-sensitive component, found in all animals, which became known as a complement.

Continuing his experiments, Bordet began to introduce red blood cells from one animal into another animal species. The second animal species, Bordet found, destroyed the foreign red cells. Bordet called this process hemolysis and concluded that it worked in a similar manner to bacteriolysis in that it required a complement. This discovery offered insight into how organisms were able to immunize themselves against antigens and led to the development of the complement fixation test, a process that made it possible to detect the presence of specific bacteria in serum.

In 1901 Bordet became the director of the Pasteur Institute of Brussels, a new institution devoted to the study of bacteriology. Collaborating with Octave Gengou, Bordet continued his research in immunology and in 1906 discovered the bacterium that causes whooping cough. Their work also provided for important findings regarding such diseases and infections as typhoid fever and tuberculosis.

Bordet remained at the Pasteur Institute of Brussels until 1940. In 1907 he began teaching bacteriology at the Free University of Brussels. Though administrative duties took time away from his research, Bordet continued his investigations in immunology. He studied blood coagulation from 1901 to 1920 and in 1920 wrote a text on immunology. After 1920 he investigated the bacteriophage, a group of viruses that infected bacteria. When he retired from the Pasteur Institute in 1940, his son assumed the directorship. Jules Bordet died in 1961.

Borlaug, Norman Ernest

(1914–)
American
Agronomist

Norman Borlaug is credited with fathering the "green revolution," a movement in the late 1960s that attempted to create a holistic solution to global hunger and overpopulation problems by increasing crop yields and decreasing the bureaucratic morass that often exacerbated problems instead of relieving them. After working for years with crop management to solve these problems, Borlaug realized that the hunger problem required not only agricultural solutions but political solutions as well.

Borlaug was born on March 25, 1914, in Cresco, Iowa, a farming community populated mostly by Norwegian immigrants who maintained a collective memory of the hunger that drove them from their homeland to the United States. His parents, Henry O. and Clara Vaala Borlaug, farmed 56 acres on the outskirts of Cresco. Borlaug married Margaret G. Gibson on September 24, 1937, and together they had two children, Norma Jean and William Gibson.

At his grandfather's insistence Borlaug attended college instead of going into farming immediately after high school. He graduated in 1937 with a bachelor of science degree in forestry from the University of Minnesota, where he had studied under Elvin Charles Stakman, head of the plant pathology department. Borlaug stayed on at the university to study plant pathology for a master's degree in 1939 and a doctorate in 1942; his dissertation was a study of fungal rot in flax plants. Upon graduation Borlaug worked for E. I. du Pont de Nemours and Company, studying the effects of dichlorodiphenyl trichloroethane (DDT), the chemical pesticide developed by PAUL HERMAN MÜLLER in 1939.

Responding to an appeal from the Mexican Ministry of Agriculture, the Rockefeller Foundation appointed a team of agronomists headed by George Harrar to travel to Mexico to advise the country on methods of improving its crop yields. Harrar chose Borlaug as the director of the Cooperative Wheat Research and Production Program in Mexico. Borlaug initiated the introduction of a new strain of wheat with a taller, thinner stem that could outstretch weeds in competition for sunlight. Whereas Mexico had been importing half of its wheat before Borlaug introduced this new strain, it was self-sufficient in wheat production by 1948. Borlaug navigated a setback in the 1950s, as the successful grain grew too heavy for the thin stalk, which tended to "lodge" or bend over. The solution was a hybrid, blending in a shorter, thicker-stemmed strain; this combination proved twice as productive as the tall-stemmed strain and 10 times as productive as the original strain used by Mexican farmers.

As an outgrowth of this project Borlaug became an associate director of the Rockefeller Foundation, and he also headed the International Center for Maize and Wheat Improvement, applying his expertise to similar situations in Pakistan in 1959 and in India in 1963. Throughout this period Borlaug became increasingly politicized as he realized that agricultural improvements needed to be accompanied by a decrease in the size and scope of bureaucracies managing agricultural affairs.

These tenets of the green revolution earned Borlaug the Nobel Peace Prize in 1970, the first awarded to an agricultural scientist. As the 1970s progressed and the environmental movement gained momentum, Borlaug came under increasing criticism for his overreliance on chemical fertilizers, pesticides, and herbicides, which polluted the environment. Borlaug stood behind his commitment to the belief that the drawbacks of using chemicals in farming did not outweigh the importance of reducing famine.

In 1979 Borlaug retired from the International Center for Maize and Wheat Improvement, and in 1984 Texas A & M University appointed him as a Distinguished Professor of International Agriculture. Borlaug continued to contribute his knowledge to several organizations in his field, including the Renewable Resources Foundation; the United States Citizen's Commission of Science, Law and Food Supply; the Commission on Critical Choices for America; and the Foundation for Population Studies in Mexico.

Born, Max

(1882–1970)
German/English
Physicist

Max Born developed the theory of matrix mechanics, a description of atomic activity that competed with ERWIN SCHRÖDINGER's theory of wave mechanics until Born reconciled the two theories, explaining that they coexist, each applicable in different situations. He won the 1954 Nobel Prize for this work, which clarified scientific understanding of quantum physics.

Born was born on December 11, 1882, in Breslau, Germany, to Gustav Born, an embryologist and professor of anatomy at the University of Breslau, and Margarete Kaufman Born, who died when Born was four years old. In 1890 his father remarried; Born's stepmother was named Bertha Lipstein. On August 2, 1913, Max Born married Hedwig Ehrenberg, the daughter of a law profes-

Max Born, whose theory of matrix mechanics helped to clarify scientific understanding of quantum physics. *AIP Emilio Segrè Visual Archives, Segrè Collection.*

sor at the University of Göttingen. The couple had three children together—Irene, Margaret, and Gustav. Gustav became the head of the department of pharmacology at Cambridge University.

In 1901 Born entered the University of Breslau, studying a cornucopia of fields, including astronomy, chemistry, logic, mathematics, philosophy, physics, and zoology. He studied during the summer semesters of 1902 and 1903 at Heidelberg and Zurich. In 1904 he entered the University of Göttingen, where he studied mathematics under David Hilbert, CHRISTIAN FELIX KLEIN, and HERMANN MINKOWSKI and earned a special assistantship under Hilbert. In 1907 he received his doctorate in physics on the strength of a prizewinning paper on the elasticity of wires and tapes. His mentors had to coerce him into submitting the paper for the prize, as he had already moved on from that topic to studying Minkowski's theories on relativity.

After short stints in the military, serving his compulsory term, and at Cambridge University, studying electrons with J. J. Thomson and Joseph Larmor, Born returned at Minkowski's behest to the University of Göttingen, where he remained from 1909 until 1933, with a hiatus during World War I. In 1915 he worked as an assistant professor at the University of Berlin while fulfilling his wartime military duty there. He took advantage of this opportunity to work with MAX PLANCK and ALBERT EINSTEIN. In 1919 Born taught at the University of Frankfurt on the Main; then he returned to a professorship at Göttingen, and in 1921 he became director of the university's Physical Institute.

He collaborated with the Hungarian aerodynamicist Theodore von Kármán in 1921 to devise a definition of heat capacity of crystals, a mathematical expression of the first law of thermodynamics that became known as the Born-Kármán theory of specific heats. In 1925 he collaborated with his assistants WERNER KARL HEISENBERG and E. P. Jordan on the most significant work of his career: the application of matrix algebra to quantum mechanics, creating matrix mechanics, the first satisfactory explanation of quantum phenomena. A year later Schrödinger suggested a competing explanation of quantum mechanics, known as wave mechanics. Born reconciled the theories, and even elaborated on wave mechanics by suggesting that the square of the wave function indicated the probability of finding a particle on that wave, a formula that became known as Born's approximation. In 1926 PAUL ADRIEN MAURICE DIRAC blended the matrix and wave mechanics into one theory.

Nazi regulations ousted Born from his position at Göttingen on April 25, 1933. Cambridge University offered him the Stokes Lectureship, and during this refuge he composed two influential books, *Atomic Physics* and *The Restless Universe*. In 1936 Born became the Tait Professor of Natural Philosophy at the University of Edinburgh, a post he maintained until his 1953 retirement. The next year he shared the Nobel Prize in physics with WALTHER WILHELM GEORG BOTHE. He also won the 1948 Max Planck Medal of the German Physical Society. Born died on January 5, 1970, in Göttingen; he had the Heisenberg-Born-Jordan equation engraved on his tombstone.

Bosch, Carl

(1874–1940)
German
Industrial Chemist

Carl Bosch transformed FRITZ HABER's elaborate formula for synthesizing ammonia from an experimental process into an industrial process for making fertilizers and explosives. Haber utilized high temperatures and pressure to convert hydrogen and nitrogen into ammonia, a process that Bosch applied to other conversions, such as that of methyl alcohol. Bosch shared the 1931 Nobel Prize in chemistry with FRIEDRICH BERGIUS for their work with large-scale chemical conversions.

Bosch was born on August 27, 1874, in Cologne, Germany. His father and mother were Carl and Paula Bosch, and his uncle was the founder of the worldwide electrotechnical firm that carries the Bosch name. Carl Bosch married Else Schilbach in 1902, and the couple had one son and one daughter. In 1894 Bosch commenced his study of metallurgy and mechanical engineering at the Technical University in Charlottenburg, Germany. In 1896 he moved on to the University of Leipzig, where he earned his doctorate in 1898. He wrote his dissertation on carbon compounds, under the guidance of Johannes Wislicenus.

In 1899 Bosch landed a job with the German dyestuffs company Badische Anilin und Soda Fabrik (BASF) in Ludwigshafen. In 1909 BASF acquired the ammonia conversion patent from Haber, and Bosch began experimenting with ways to increase efficiency and safety while decreasing expenses in the process. The first order of business was to find cheaper catalysts than osmium and uranium. After more than 20,000 experiments Bosch found the combination of iron and blended alkaline material a much cheaper catalyst. Bosch replaced Haber's reaction chamber, which became unstable as the steel lost its carbon to the hydrogen used in the process, with a double-walled chamber. The inner shell, made of soft steel, could leak hydrogen; the outer shell, made of heavy-duty carbon steel, retained its carbon content and hence its strength. This reaction chamber withstood temperatures as hot as 500 degrees Celsius and equally intense pressures. By 1911 BASF had a plant opened near Oppau producing commercial amounts of synthetic ammonia. In 1919 BASF promoted Bosch to the position of managing director as the post–World War I ammonia industry took off in response to increased fertilizer production.

In 1923 Bosch developed a process for converting carbon monoxide and hydrogen into methanol for use in manufacturing formaldehyde. This process relied on high temperatures and pressure, as did ammonia synthesis. In 1925 BASF consolidated with six other German chemical companies to found I. G. Farben, appointing Bosch as president of the new conglomerate. A decade later Bosch ascended to the position of chairman of the board of directors, and in 1937 he filled the highest position for scientists in Germany when he became president of the Kaiser Wilhelm Institute (later known as the Max Planck Society).

Bosch later applied this technique of using high temperatures and pressure to procure urea from ammonium carbamate. He also addressed the problems of carbon hydrogenation and rubber synthesis. Although Bosch never managed to discover a cost-effective means of producing gasoline from the patented coal dust and hydrogen conversion process acquired from Friedrich Bergius in 1925, he and Bergius did jointly win the 1931 Nobel Prize in chemistry. After a protracted illness Bosch died on April 26, 1940, in Heidelberg, Germany.

Bose, Satyendranath

(1894–1974)
Indian
Physicist

Although Satyendranath Bose is much less well known than his collaborator in the creation of Bose-Einstein statistics, Bose managed a feat that even ALBERT EINSTEIN could not achieve: He verified Einstein's quantum theory with mathematical equations, thus reconciling the theory with Planck's law. In honor of Bose's contribution to quantum physics a set of subatomic particles of finite mass were named *bosons* after him.

Bose was born on January 1, 1894, in Calcutta, India. His father, Surendranath Bose, was an accountant and the founder of the East India Chemical and Pharmaceutical Works. His mother was Amodini Raichaudhuri Bose. In 1914 Bose married Ushabala Ghosh; together they had two sons and five daughters. Bose attended Presidency College in Calcutta, where he studied under Jagedischandra Bose, who shared his last name but was not a relative. In 1915 Bose embarked on postgraduate work, earning his master's degree in mathematics and graduating first in his class.

After receiving his degree, Bose became a lecturer in physics at the University of Calcutta's College of Science. In 1921 Bose commenced a relationship of more than 20 years with the newly formed University of Dacca in East Bengal. He became a professor there in 1926; in 1945, the University of Calcutta named Bose the Khaira Professor of Physics, a title he held for 11 years. Visva-Bharati University subsequently appointed him vice-chancellor in 1956, and in 1959 the Indian government designated him as a national professor.

Over the nearly 40 years of his academic career Bose wrote relatively few works, publishing only 26 original papers between 1918 and 1956. His 1924 paper "Planck's Law and the Hypothesis Light Quanta" proved extremely significant, however, to the development of quantum physics. *Philosophical Magazine* rejected the paper in 1923, but Bose maintained his determination by sending a copy of the paper to Einstein. Einstein immediately recognized the importance of Bose's work, which applied a phase-space model to an ideal gas. Bose's paper provided a solution to Planck's law concerning blackbody radiation, while suggesting a solution to Einstein's own theory of electromagnetic radiation. Einstein used his influence to prevail upon the journal *Zeitschrift Für Physik* to publish a German translation of the paper in 1924. Einstein then extended the implications of Bose's work to form Bose-Einstein statistics, which addressed the gaslike qualities of electromagnetic radiation. This system offered an alternative to ENRICO FERMI's approach; hence the name for certain subatomic particles depended on the mathematical method used to derive them—they were either bosons or fermions.

Bose's work also contributed to the understanding of X-ray crystallography, unified field theory, and the interaction of electromagnetic waves with the ionosphere. Bose helped found the Science Association of Bengali in 1948, and he served as a member of the Indian parliament between 1952 and 1958. He received the Padma Vibhushan award from the Indian government in 1954. In 1958 the Royal Society elected him a member, one of the few instances of Bose receiving the international recognition he deserved, though his home country of India held him in highest esteem. Bose died on February 4, 1974, in Calcutta.

Bothe, Walther Wilhelm Georg

(1891–1957)
German
Physicist

The physicist Walther Wilhelm Georg Bothe made significant contributions to science during the period in the early 1900s commonly known as the "Golden Age of Physics." Bothe developed a procedure called the coincidence counting method that was useful in detecting subatomic particles, researched cosmic rays, and advanced nuclear reaction theory. Bothe shared the 1954 Nobel Prize in physics with MAX BORN for his breakthrough work on cosmic radiation.

Born on January 8, 1891, in Oranienburg, Germany, Bothe was the son of Fritz Bothe, a merchant, and Charlotte Hartung. Georg Bothe married Barbara Below, who was from Moscow, in 1920. The couple had two children.

In 1908 Bothe entered the University of Berlin, where he studied physics, mathematics, and chemistry. In 1914 at the age of 23 Bothe was granted his doctorate from the University of Berlin. Bothe worked on his dissertation under the physicist MAX PLANCK, who in 1918 received the Nobel Prize in physics for his discoveries in quantum theory. Bothe's dissertation detailed his study of the molecular theory of reflection, refraction, dispersion, and extinction of light. After completing his doctorate Bothe started work at the Physical-Technical Institute in the laboratory of Hans Wilhelm Geiger, the creator of the Geiger counter, an instrument used to detect and measure radiation.

Bothe's work with Geiger was hindered by World War I, during which Bothe spent five years as a prisoner of war in Siberia. Despite his captivity, Bothe continued his research in physics. Upon his release in 1920 Bothe returned to Germany and resumed his work with Geiger. He also took a position teaching physics at the University of Berlin. Bothe's interest in quantum theory began to develop, and he decided to investigate the subject further. At the time there was little experimental proof of quantum theory, the theory that electromagnetic energy is transmitted in the form of particles and waves. The theory, however, was generally accepted by the scientific community, and advances toward its proof were being made. Arthur Holly Compton in 1923 discovered what became known as the Compton effect, a phenomenon in which electrons scattered X rays as if they were particles, causing them to transfer some momentum and energy to the electrons. This discovery led some scientists to theorize that momentum and energy were conserved in the sum of many interactions between radiation and matter rather than in individual interactions. Bothe and Geiger tested this new hypothesis in 1924 by using a method that would become the coincidence counting method, and their experiments succeeded in demonstrating that energy and momentum were conserved at the level of each interaction.

In the late 1920s Bothe used the coincidence counting method to study cosmic rays. Collaborating with the astronomer Werner Kolhörster, Bothe showed that cosmic rays were not composed of gamma rays only, as had been thought since the discovery of cosmic rays in 1912. While studying radioactivity in 1930, Bothe detected an unusual radiation, which he believed was gamma radiation. This radiation was later discovered by Sir JAMES CHADWICK to be the neutron.

In 1934 Bothe became the director of the Max Planck Institute at Heidelberg and participated in the building of Germany's first cyclotron, an accelerator for charged particles in a constant magnetic field, which was completed in 1943. During World War II Bothe researched atomic energy and worked on Germany's attempts to develop an atomic bomb.

In addition to the 1954 Nobel Prize in physics, Bothe was awarded the Max Planck Prize. He worked at the Max Planck Institute until his death in 1957.

Bovet, Daniel

(1907–1992)
Swiss/Italian
Pharmacologist

For his pioneering work in the science of drugs and the development of synthetic drugs Daniel Bovet received the 1957 Nobel Prize in physiology or medicine. Bovet discovered the first antihistamine, made sulfa drugs a practical commercial reality, and developed medically and financially viable muscle relaxants. Bovet also investigated possible chemical solutions for mental illness.

Born on March 23, 1907, in Neuchatel, Switzerland, Bovet was one of four children of Pierre Bovet and Amy Babut. Bovet's father was a professor of pedagogy and experimental education at the University of Geneva. He also established the Institut J. J. Rousseau. The younger Bovet married a fellow scientist, Filomena Nitti, in 1938, and the two collaborated on research projects, as well as on articles and other publications. They had three children.

After completing his compulsory education in Neuchatel, Bovet attended the University of Geneva, where he studied biology. He received a doctorate in science in 1929 for his studies in zoology and physiology, then accepted a position as an assistant at the therapeutic chemistry laboratory at the Pasteur Institute in Paris. A decade later he was appointed head of the laboratory.

During the 1930s Bovet and his coworkers investigated Prontosil, a dye that had been discovered by the German biochemist Gerhard Domagk to be effective in fighting streptococcal infections such as scarlet fever, pneumonia, and meningitis. Prontosil was prohibitively expensive to produce and was protected by patents. Through a series of experiments on Prontosil, Bovet and his colleagues concluded that sulfanilamide was the ingredient that lent Prontosil its therapeutic powers. Sulfanilamide was easy and inexpensive to produce and soon had mass production. Bovet continued his studies of sulfanilamide and developed many derivatives. These sulfa drugs have saved innumerable human lives.

In the late 1930s Bovet began to look into developing an antihistamine. The human body was not equipped to combat the overabundance of free histamine, which was caused by the presence of an irritant. The effects of this overproduction of free histamine included allergic reactions or swelling. Bovet suggested that a synthetic substance that could block the harmful results of free histamine was needed. He soon developed a rudimentary antihistamine, and from 1937 to 1941 he and his colleagues performed thousands of experiments to find a commercially feasible alternative. Several substitutes were produced, including pyrilamine.

Bovet next turned to muscle relaxants. After moving to Rome in 1947 to assume directorship of the Superior Institute of Health's therapeutic chemistry laboratory, Bovet embarked on a study of curare, a muscle relaxant that had been used by South American Indians for poisonous arrows. At the time curare was used in limited amounts to prepare bodies for surgery. Because pure curare could have unpredictable results, Bovet hoped to develop a synthetic form that would be commercially and medically viable. In eight years of research Bovet produced hundreds of versions of curare, including gallamine and succinylcholine.

Bovet assumed Italian citizenship, and in 1964 he became a professor of pharmacology at the University of Sassari on the island of Sardinia. In 1969 he returned to Rome to head the psychobiology and psychopharmacology laboratory of Italy's National Research Council. He then became professor of psychobiology at the University of Rome in 1971, and an honorary professor after his retirement in 1982. During his later years Bovet studied the effects of various chemicals on the central nervous system in order to learn more about the relationship between mental illness and chemistry. A prolific writer, Bovet published more than 400 articles and numerous books before his death in 1992.

Bowman, Sir William

(1816–1892)
British
Surgeon, Histologist

Bowman is best known for his microscopic investigations of human organs. By this process he discovered that the kidney's blood-filtration system creates urine as its byproduct. He also focused his microscope on the optical organs and discovered the structure and function of the eye as well as the striated muscle. Bowman devoted the first half of his career to histological research; in the second half he developed a very successful private surgical practice in ophthalmology.

Bowman was born on July 20, 1816, in Nantwich, in Cheshire, England. His parents, John Eddowes Bowman and Elizabeth Eddowes Bowman, were first cousins. His father divided his attention between his vocation as a banker and his avocation as a naturalist. He acted as a founding member of the Manchester Geological Society. In 1842 William Bowman married Harriet Paget, the daughter of a Leicester surgeon.

In 1826 Bowman attended the Hazelwood School in Birmingham. In 1832 W. A. Betts, Birmingham Infirmary's resident surgeon, took Bowman on as an apprentice. Bowman also worked under the renowned surgeon Joseph Hodgson. Bowman commenced his path to membership in the Royal College of Surgeons in 1837 by performing his requisite attendance at a London teaching hospital, choosing to work under Richard Partridge, a professor of anatomy in the King's College medical department. In 1838 Bowman served as a prosector under Richard Bentley Todd, a professor of physiology with whom Bowman collaborated extensively during the course of his career. Later that year Bowman embarked on a hospital tour of Europe, mentoring Francis Galton, who later founded the study of eugenics.

Bowman commenced his professional career with an appointment as an assistant surgeon at King's College Hospital in London in 1840. He had already started work, however, on some of the writings that would gain him a lasting reputation. In 1836 Bowman and Todd commenced composition of the *Cyclopaedia of Anatomy and Physiology*. This composition consisted not only of writing but also of microscopic observation and painstaking reproduction of the anatomical samples observed. Bowman's steady hand engraved his drawings directly into wood as his keen eyes stared into the microscope. The pair finished publishing the *Cyclopaedia* in 1852; concurrently Bowman and Todd published the five-volume *Physiological Anatomy and Physiology of Man*, which appeared between 1843 and 1856. These two texts revolutionized the study of anatomy and physiology, mostly as a result of advanced microscopic technology.

Bowman read his most influential paper, "On the Structure and Use of the Malpighian Bodies of the Kidney," to the Royal Society on February 17, 1842. In it he identified what became known as Bowman's capsule, a key component of the kidney that connected to the renal duct, a previously unknown relationship. The Royal Society, which had elected Bowman as a member the previous year, awarded him the Royal Medal in 1842. Thenceforth Bowman focused his career on surgery. He became a member of the Royal College of Surgeons in 1844, and from 1846 to 1876 he worked at the Royal London Ophthalmic Hospital (later known as Moorfields Eye Hospital). Between 1848 and 1855 he also taught at King's College.

Bowman founded the Ophthalmological Society in 1880. In 1884 Queen Victoria knighted Bowman as a baronet. He died on March 29, 1892, near Dorking, in Surrey, having contributed a new method of relating minute anatomical observations to physiological functions.

Boyle, Robert

(1627–1691)
English/Irish
Chemist, Philosopher of Science

Robert Boyle is best known for the law that bears his name. Boyle's law states that the volume of a gas is

Boyle's law, named for Robert Boyle, who developed it around 1660, states that the volume of gas is inversely proportional to its pressure when the temperature is constant. *AIP Emilio Segrè Visual Archives.*

inversely proportional to its pressure when the temperature is constant. Boyle is also known for disavowing the Aristotelian theory of elements in favor of a view that matter consists of primary particles. Boyle was thus a harbinger of the modern theory of chemical elements.

Boyle was born into the Anglo-Irish aristocracy on January 25, 1627, in Lismore, County Waterford, Ireland. He was the youngest son in a family of 14 children, born to Richard Boyle's second wife, Katherine Fenton Boyle. The elder Boyle was the first earl of Cork, residing in Lismore Castle, where Robert commenced his education under a tutor. In 1635 Boyle went away to Eton College. In 1639 he accompanied his brother Francis, who later became Lord Shannon, on a grand tour of Europe with a tutor. While in Florence in 1642 Boyle encountered the text *Dialogue on the Two Chief World Systems* by the recently deceased GALILEO GALILEI, which made a lasting impression on him.

In 1655 Boyle published a collection of essays on morality entitled *Occasional Reflections upon Several Subjects;* supposedly one of these essays inspired Jonathan Swift's *Gulliver's Travels.* From 1656 through 1668 Boyle was in residence at the University of Oxford, where with the assistance of Robert Hooke he constructed an air pump in 1658 that was based on the design developed by Otto von Guericke in 1654. Boyle used this pump to prove for the first time by experimentation Galileo's assertion that all objects fall at the same velocity in a vacuum. He also proved that air is essential for combustion, respiration, and the transmission of sound.

The first edition of Boyle's *New Experiments Physio-Mechanical, Touching the Spring of the Air and Its Effects* appeared in 1660, but it wasn't until the second edition appeared in 1662 that the text became truly groundbreaking. Boyle appended in the second edition his 1661 report to the Royal Society expounding Boyle's law. His discovery occurred during an ingenious experiment that involved tubing shaped into a U and partly filled with mercury to isolate air under atmospheric pressure. Boyle noticed that when he added mercury, the volume halved when the pressure doubled. After experimenting with different pressures and volumes, he noted that the two measurements were inversely related.

Boyle published *The Sceptical Chymist* in 1661; in it he refuted the Aristotelian theory of the four elements (earth, air, fire, and water) and Paracelsus' proposition of three principles (salt, sulfur, and mercury) in favor of a particle theory. In fact, Boyle was staunchly anti-Aristotelian, seeking to replace the primacy of ARISTOTLE's influence with mechanical explanations of physical phenomena. This view remained consistent with his deeply held religious beliefs, which he expressed in the 1690 text *The Christian Virtuoso.* Boyle believed that God set a perfect world in motion, much as a clockmaker sets a clock in motion, then retires to watch its mechanical action.

Boyle was a founding member of the Royal Society of London, which elected him president in 1680, though he graciously declined the offer. Boyle died on December 30, 1691, in London, England.

Brady, St. Elmo

(1884–1966)
American
Chemist

St. Elmo Brady has the distinction of being the first African American to have earned a doctorate degree in chemistry. A lifelong educator and academic with a strong commitment to the progress of others, Brady held professorships at four of the historically black colleges and universities in the southern United States. Interested in agricultural research, Brady studied the native plants of the South for the development of practical chemical products.

Born on December 22, 1884, in Louisville, Kentucky, Brady attended public schools and was an exceptional student. Brady and his wife, Myrtle, had one son, who became a physician. Myrtle was an educated African-American woman who worked in Washington, D.C.

In 1904 Brady entered Fisk University in Nashville, Tennessee. Modern chemistry was a relatively new subject being taught in black colleges, and Professor Thomas W. Talley was among its early teachers. Brady graduated in 1908 and went to Alabama to work at the Tuskegee Institute, now known as Tuskegee University. There he worked with Booker T. Washington, the head of Tuskegee, and the agricultural chemist GEORGE WASHINGTON CARVER. In 1913 Brady left Tuskegee and entered the graduate program in chemistry at the University of Illinois, a significant achievement in that few African Americans held doctoral or other advanced degrees during that era. A year after Brady began the program, he earned his master's degree and received a fellowship that allowed him to pursue his doctorate. Studying under Clarence G. Derick, Brady investigated the divalent oxygen atom for his dissertation research. In 1914 he became the first African American invited into the chemistry honor society Phi Lambda Upsilon. He was also among the first to be admitted into the science honor society Sigma Xi, which he entered in 1915. Brady earned his doctorate in 1916 and returned to Tuskegee to head the science department.

Brady took a job as professor and head of Howard University's chemistry department in 1920 and moved to Washington, D.C. There he spent seven years developing the undergraduate chemistry program and curriculum. Brady then accepted the position of head of the chemistry department at Fisk University and remained there for 25 years until his retirement. In addition to his teaching duties, which included such courses as organic chemistry and general chemistry, Brady carried out studies on castor beans and magnolia seeds. He established the first graduate program in chemistry in a black college and began a series of lectures, named after Thomas W. Talley, to draw great chemists to the Fisk campus. Brady also oversaw the building of a modern chemistry building—the first ever built at a black college. The building was later named in his and Talley's honor. Yet another contribution Brady made was the founding of a summer program in infrared spectroscopy. This program, created in cooperation with faculty members at the University of Illinois, attracted teaching staff from colleges and universities across the United States.

After retiring from Fisk in 1952, Brady moved to Washington, D.C. Retirement did little to slow Brady down, however, and he collaborated in the development of a department of chemistry at a small school in Mississippi, Tougaloo College. Brady also assisted in the recruitment of instructors. He did much to advance opportunities in higher learning for African-American students.

Brahe, Tycho

(1546–1601)
Danish
Astronomer

Tycho Brahe's close observation of the sky yielded accurate calculations of the positions of 777 fixed stars. More importantly, it called into question the prevailing belief in the astronomical ideas of ARISTOTLE, who held that the stars were constant and unchanging. Brahe passed his torch on to his assistant, JOHANNES KEPLER, whose work subsequently laid the foundation for the advancements of Sir ISAAC NEWTON.

Brahe was born on December 14, 1546, in the family seat of Knudstrup, in Scånia, Denmark (now Sweden). His twin was stillborn, but he had five brothers and five sisters. His mother was Beate Billie, and his father, Otto Brahe, was a privy councilor and governor of Hälsingborg Castle. Brahe's childless uncle Jørgen essentially kidnaped him and raised him at his castle in Tostrup, Scånia.

Brahe's uncle's wealth allowed him to study law at the Lutheran University of Copenhagen from April 1559 to February 1562. A celestial event intervened, though, changing the course of Brahe's life from law to astronomy. As predicted, a total eclipse of the sun occurred on August 21, 1660. The 14-year-old Brahe witnessed the event in awe. In an attempt to refocus his mind on law, Jørgen sent Brahe to the University of Leipzig in March 1562. Brahe simply attended law classes in the day and studied astronomy at night while his tutor, Anders Sörensen Vedel, slept. Another celestial event intervened to confirm Brahe's devotion to astronomy. In August 1563 Brahe witnessed a conjunction of Jupiter and Saturn. What piqued Brahe's interest was the fact that all the existing almanacs and ephemerides predicted this event inaccurately; even the Copernican tables were off by a couple of days. Brahe devoted his life to correcting such inaccuracies. His devotion to accuracy extended beyond astronomy. On December 29, 1566, he challenged another Danish nobleman to a duel to decide an argument over a mathematical problem. Brahe lost a slice of his nose in the process, but he affixed a copper replacement to his face.

On November 11, 1572, Brahe observed yet another unusual celestial event—a new star. Later named *Tycho's star,* the nova appeared in the constellation of Cassiopeia, brighter than the rest of the night sky. This observation established his reputation and shook the foundation of the Aristotelian cosmology, which held that the stars were unchanging. Brahe published his findings in *De nova stella* in 1573, the same year he shocked the aristocratic community by marrying a peasant girl named Kirstine. Together the couple had five daughters and three sons.

In order to retain Brahe in Denmark, King Frederick II granted him title to the island of Ven, as well as the financial backing to construct an elaborate observatory that Brahe named Uraniborg after the astronomer's muse, Urania. Here Brahe made the majority of his observations and substantiations of astronomical records. However, when King Frederick's son, Christian IV, succeeded his father in 1588, Brahe fell from favor; he migrated in 1599 to Prague, where he found the financial support of Emperor Rudolph II. Johannes Kepler served as Brahe's assistant, and he later published their joint star catalogue, *Rudolphine Tables,* in 1627. Brahe died on October 24, 1601, in Prague.

Brandegee, Mary Katharine Layne ("Kate")

(1844–1920)
American
Botanist

Katharine Brandegee was curator of botany at the California Academy of Sciences for 22 years and also made major contributions to the University of California at Berkeley's herbarium (dried plant collection). She was born Mary Katharine Layne in Tennessee on October 28, 1844. Her father, Marshall, whom she called "an impractical genius," moved the family from place to place, finally settling near Folsom, California, when Kate, as she was always known, was nine. Having grown up in California's beautiful Gold Rush country, she wrote later, "Biology always attracted me greatly."

In 1866 Kate married Hugh Curran, an Irish police constable, who died of alcoholism in 1874. Seeking a new life, Kate Curran enrolled at the medical school of the University of California at Berkeley in 1875, only the third woman to do so. She earned her M.D. in 1878. She was "not overrun with patients," however, so she decided to pursue an interest in plants that began when she learned about them in medical school as sources for drugs.

Curran studied botany under two experts at the Academy of Sciences in San Francisco and was soon helping them organize the academy's plant collection. When the curator of the collection retired in 1883, the academy, whose charter stated that it "highly approve[d] the aid of females in every department of natural history," took the very unusual step of giving her his job. Her fellow botanist Marcus Jones writes that she "was a model in thoroughness in her botanical work."

Curran fell "insanely in love," as she wrote to her sister, with a plant collector named Townshend Stith Brandegee when he visited the academy in 1886. They married in San Diego on May 29, 1889, and spent their honeymoon walking from there to San Francisco, gathering plant specimens all the way. In 1895 they left the academy's herbarium in the capable hands of ALICE EASTWOOD, whom Brandegee had trained, and moved back to San Diego, where they set up a home, herbarium, and garden that one visitor called a "botanical paradise."

The Brandegees, traveling sometimes together and sometimes separately, made collecting trips all over California, including Baja California, as well as to parts of Arizona and Mexico. In spite of poor overall health, Kate Brandegee enjoyed these arduous journeys. "I am going to walk from Placerville to Truckee," she wrote to her husband in 1908, when she was 64 years old. "I have had considerable hardship in botanizing and perhaps in consequence—I am unusually strong and well."

When the herbarium at the University of California's Berkeley campus was destroyed in the huge earthquake of 1906, the Brandegees not only gave the university their collection (numbering some 100,000 plants) and library but also moved to Berkeley to manage them. They remained there the rest of their lives, working without pay. Brandegee died on April 3, 1920.

Branson, Herman Russell

(1914–1995)
American
Physicist

The physicist Herman R. Branson is known for his contributions in biophysics and his pioneering work in the structure of proteins, which he performed in collaboration with the American chemist LINUS CARL PAULING. His discovery of the spiral structure of proteins did much to further the fields of biochemistry and molecular biology. As important as his scientific discoveries was his dedication to advancing the opportunities of African Americans in the sciences.

Branson was born on August 14, 1914, in Pocahontas, a small town in Virginia. Though his father, a coal miner, had little education, Branson had relatives who were highly educated and worked in the field of medicine. Branson's mother encouraged the children to develop their reading skills from an early age and was known to quiz them about what they had read. The family moved to Washington, D.C., after Branson had started school.

After graduation from high school at the top of his class Branson entered the University of Pittsburgh, his uncle's alma mater. He transferred to Virginia State College in Petersburg, Virginia, after two years and graduated with highest honors in 1936. He immediately entered a graduate program in physics at the University of Cincinnati, where he researched the effect of X rays on worms. When he earned his doctorate in 1939, he became the first African American to receive a doctorate in a physical science at the University of Cincinnati. He married Corolynne Gray in 1939, and the couple had two children, both of whom became physicians.

With his doctorate in hand Branson went to New Orleans, Louisiana, to teach physics, mathematics, and chemistry at Dillard University. In 1941 he returned to Washington, D.C., to teach physics and chemistry at Howard University. He remained there until 1968, making many contributions to the school's science division and, through his research, to the scientific community as a whole. At Howard, Branson successfully established an undergraduate program in physics, an uncommon field of study at historically black colleges and universities. Graduate courses and research fellowships at nearby facilities were added as well. Branson also participated in programs geared toward encouraging young African Americans to take science classes and to pursue careers in medicine and science.

For Branson's own research during his time at Howard University he studied the manner in which the human body uses raw materials, such as phosphorus. After developing a theory about the stages phosphorus undergoes before being used by human cells, Branson tested the theory by tracing radioactive phosphorus in a living animal. Branson's breakthrough work with Linus Pauling on the structure of proteins took place in the late 1940s while Branson was a National Research Council fellow at the California Institute of Technology. Pauling hoped to determine the physical structure of deoxyribonucleic acid (DNA). Branson's investigation of the molecule hemoglobin led to his discovery in 1950 of a helical structural pattern. This advanced Pauling's research immensely and also provided insight into sickle-cell anemia.

In 1968 Branson left Howard to become president of Central State University in Wilberforce, Ohio. Two years later he assumed the presidency of Lincoln University, near Philadelphia, Pennsylvania. After retiring from Lincoln in 1985 Branson went back to Howard to oversee a program designed to encourage promising high school students to pursue careers in science. During his career Branson received numerous honors, including honorary degrees from such institutions as Brandeis University and Drexel University. He died on June 7, 1995.

Braun, Karl Ferdinand

(1850–1918)
German
Physicist

For his pioneering contributions in the development of wireless telegraphy Karl Ferdinand Braun was awarded the 1909 Nobel Prize in physics, a prize he shared with GUGLIELMO MARCONI, the inventor of a long-wave radio transmitting system. Braun's studies of electricity also led to the development of the cathode-ray tube and the creation of the oscilloscope, an electronic device that displays fluctuations of voltage and current on the screen of a cathode-ray tube. Braun also discovered that some crystals could be used as rectifiers to convert current from alternating (AC) to direct (DC) current. This finding led to the development of crystal radios.

Born on June 6, 1850, in Fulda, located in central Germany, Braun was the son of Konrad Braun, a court clerk, and the former Franziska Göhring, whose father was Konrad Braun's supervisor. Braun married Amelie Bühler in 1885, and they had four children.

After completing high school in Fulda, Braun first studied at the University of Marburg and then entered the University of Berlin, where he received his doctorate in 1872. His dissertation focused on the vibrations of elastic rods and strings. Braun worked as an assistant to the physicist Georg Quincke at the University of Würzburg for two years and then held an appointment in Leipzig before accepting a position as professor of theoretical physics at the University of Marburg in 1876.

In the 1870s Braun began studying crystalline materials. He discovered that some semiconducting crystals allowed electrical current to flow in only one direction. In 1874 he published these findings, which later led to the development of crystal radio receivers and advanced the standardization of the measurement of conductivity. Braun left Marburg in 1880 and worked at the University of Strasbourg, the Technical High School at Karlsruhe, and the university in Tübingen before accepting a position as a professor of physics and head of the Physical Institute at the University of Strasbourg in 1895. There he created the first oscilloscope, which he called the *Braun tube* in a paper published in 1897. He developed the oscilloscope by modifying a cathode-ray tube so that the electron beam was deflected by a change in voltage. The fluctuating pattern of the current then appeared on the cathode-ray tube's screen. The oscilloscope was later used in the study of electrical technology and led to the development of the television receiver.

Using his newly developed oscilloscope, Braun began to investigate wireless telegraphy in the late 1890s. He hoped to increase the broadcasting range of Marconi's radio transmitter, patented in 1896, which had a range of about 8 to 12 miles. Braun suggested that the transmitter's range could be expanded by increasing the electrical power of the transmitter. Through his research Braun eventually produced an antenna circuit without sparks. The transmitter's power was conveyed to the antenna circuit through an inductive link—large wire coils converted electricity into magnetic fields. Braun patented his improvements to Marconi's transmitter in 1899, and his sparkless antenna circuit, which greatly increased the transmitter's range, was later used in radio, television, and radar.

Braun's continued research in wireless telegraphy gave rise to improvements in antennas, including the development of unidirectional broadcasting antennas, and led to the employment of radio waves for signaling instruments on boats. Braun traveled to the United States in 1914 to testify in a patent case related to radio broadcasting. When World War I began, Braun was detained in the United States because of his German citizenship. He died in the United States in 1918.

Brewster, Sir David

(1781–1868)
Scottish
Physicist

A thorough experimenter, Sir David Brewster made significant contributions to science through his extensive studies in optics. He made many discoveries about the polarization, reflection, and absorption of light. He also studied spectroscopy and tirelessly promoted and popularized science. Well known for his invention of the kaleidoscope, Brewster made his living not from his scientific endeavors but from his work as a writer and editor of journals and books.

Born on December 11, 1781, in Jedburgh, Scotland, Brewster was the son of James Brewster, the cleric at the local elementary school, and Margaret Key. His interest in science began at an early age, and as a child he studied his father's notes from college science courses. He also assisted his hometown's minister and scholar, taking dictation and copying text, and this helped him hone his writing and editing skills. Encouraged by a local astronomer, James Veitch, Brewster built a number of instruments, including microscopes, telescopes, and sundials.

Intending to join the ministry, Brewster entered the University of Edinburgh in 1794. In 1800 he earned an honorary master's degree, and four years later he received his preacher's license from the Church of Scotland. By then, however, Brewster's interest in science and experimental research was strong, and in the late 1790s he began his study of light. Never ordained as a minister, Brewster worked as a private tutor from 1799 to 1807 and also served as an editor for several magazines and journals while he carried out his experiments.

Brewster's most significant work dealt with the polarization of light. In 1813 he presented Brewster's law, which stated that if a light beam is divided into a reflected ray and a refracted ray when the beam hits a reflective surface, then those rays are polarized. The rays are completely polarized when they are at right angles. The angle at which this occurs was known as the Brewster angle. In other words, the refractive index of the reflective substance equaled the Brewster angle, and the Brewster angle and the refractive angle add up to a right angle.

The kaleidoscope, invented in 1816, was a result of Brewster's investigations in optics and solidified his reputation in popular culture. Brewster was also responsible for founding optical mineralogy and photoelasticity, and in 1819 he categorized hundreds of minerals and crystals according to optical and mineralogical groups. In the early 1820s Brewster began to study spectroscopy and by 1832 had found spectra of gases, colored glass, and the Earth's atmosphere. Brewster improved the stereoscope in the 1840s and improved the lenses used in lighthouses.

Hoping to popularize and elevate the status of science, Brewster worked tirelessly. He was instrumental in the founding of the British Association for the Advancement of Science, the Edinburgh School of Arts, and the Royal Scottish Society of Arts. Brewster published several biographies of scientists and wrote hundreds of articles. Among the honors Brewster received were the Royal Society of London's Copley, Rumford, and Royal Medals and the Royal Society of Edinburgh's Keith Prize. He became a fellow of the Royal Society in 1815 and was knighted in 1831. Brewster served as principal of the University of Edinburgh from 1859. He died in Allerby, Melrose, in Scotland, on February 10, 1868, shortly after contracting pneumonia.

Broca, Pierre Paul

(1824–1880)
French
Surgeon, Anthropologist

Paul Broca helped develop modern physical anthropology with his study of human craniums, comparing features such as the form, structure, and topographical characteristics of skulls belonging to different races and comparing contemporary skulls to prehistoric skulls. Broca also studied living subjects; by that method he developed a theory of cortical localization by finding the seat of articulate speech in the left frontal lobe of the brain, which was named the *convolution of Broca* after him.

Broca was born on June 28, 1824, in Sainte-Foy-la-Grande, France, near the Bordeaux region. His father was Benjamin Broca, a Huguenot doctor, and his mother was the daughter of a Protestant preacher. Broca married the daughter of a Paris physician named Lugol.

Broca received a bachelor's degree from a local college in mathematics and physical sciences before beginning his study of medicine at the University of Paris in 1841. In 1843 he became an extern, or nonresident medical student; the next year he became an intern. By 1848 Broca had become a prosector of anatomy, making dissections for anatomical demonstrations, and in 1849 he earned the M.D. degree, with specialties in pathology, anatomy, and surgery.

In 1847 he was a member of a commission reporting on the findings of the excavations of the cemetery of the Celestins. Broca developed methods for measuring and comparing skulls, and his findings supported CHARLES ROBERT DARWIN's theory of evolution. In 1853 Broca became an assistant professor in the Faculty of Medicine and a surgeon of the Central Bureau; in 1868 the Faculty of Medicine promoted him to professor of clinical surgery.

Broca announced his discovery of cortical localization in 1861 when he identified a brain lesion on the left inferior frontal gyrus, later known as Broca's convolution, of a patient with aphasia, a disease that impairs one's ability to articulate words. Subsequent research revealed that the exact location of the lesion was not the area Broca had surmised; nevertheless, his assertion of hemispheric zones of the brain controlling different functions came to be widely accepted.

Broca also left his stamp on his field by founding organizations that survived him. In 1858 he founded an anthropology laboratory at the École des Hautes Études in Paris. The next year he helped establish the Société d'Anthropologie de Paris. He also acted as a founder of a journal for the field he helped create, *Revue d'anthropologie*. He helped found the École d'Anthropologie in Paris and subsequently was its director.

The five volumes of *Mémoires d'anthropologie,* published between 1871 and 1878, constituted Broca's major written achievement. He also published a two-volume text on tumors, *Traité des tumeurs,* as well as a text in 1853 on strangulated hernias and a text in 1856 on aneurysms. He also contributed a paper, "La splanchnologie," to the *Atlas d'anatomie*. Overall, Broca advanced the understanding of anatomy, pathology, surgery, cerebral functions, and anthropology, publishing a total of 223 papers between 1850 and his death on July 9, 1880, in Paris.

Broglie, Louis-Victor-Pierre-Raymond, prince de

(later seventh duke)
(1892–1987)
French
Theoretical Physicist

Louis-Victor-Pierre-Raymond de Broglie's significance in the development of human understanding of reality cannot be overstated. Physicists acknowledge him as the father of wave mechanics. De Broglie was the first to apply the notion of the dual nature of particle properties and wave properties to matter. At the time experiments showed that both properties applied to light, but it was a vast leap to posit that matter can behave in two different ways. The wave properties of matter apply at the subatomic level, invisible to the eye, so this duality of matter is not apparent. However, the fact that particle properties and wave properties do not always coincide opens the door to the possibility of randomness and unpredictability in matter, a proposition that disturbed de Broglie himself, though he could not find a scientific solution to this conundrum.

De Broglie was born on August 15, 1892, in Dieppe, France. His father, who died when de Broglie was 14, was Duc Victor, and his mother was Pauline d'Armaille. The family carried two titles of distinction—in 1740 Louis XIV granted the title of duc to the family and in the Seven Years War the family earned the title of prinz from the Holy Roman Empire. De Broglie inherited both titles in 1960, when his older brother and scientific collaborator, Maurice, died.

After the traditional familial education de Broglie matriculated at the Sorbonne, where he earned dual baccalaureates in philosophy and mathematics in 1909. He then earned his Licencié ès Sciences from the Faculty of Science at the University of Paris in 1913. He spent the entirety of World War I in the military as a radio operator at the Eiffel Tower. After the war he resumed his scientific research, focusing his doctoral investigation on the question of whether matter might exhibit the same dual properties of waves and particles that light exhibits. De Broglie presented his dissertation, "Investigations into the Quantum Theory," to an advisory committee befuddled by the complexity of the theory. On their advice he sent a copy to ALBERT EINSTEIN, who recognized that de Broglie's theory solved one of the basic mysteries of science. The journal *Annales de Physique* published the paper in its entirety in 1925.

Proof of the theory's validity came from two fronts. Clinton Davisson and Lester Germer located what became known as de Broglie waves by using slow electrons, and George Paget Thomson did the same with fast electrons in 1927. ERWIN SCHRÖDINGER proceeded to formulate the mathematical function known as the Schrödinger equation by which these properties worked, thus helping to establish quantum mechanics. In 1927 de Broglie attended the seventh Solvay conference, where prominent physicists took up the question posed by his theory—Is there determinacy in quantum mechanics? De Broglie defended the position that there must be some organizing principle behind quantum mechanics, proposing his "double solution" of a "pilot wave" as an answer. Further investigation revealed holes in this solution, leaving no option but to accept the random nature of the probabilistic theory.

The Faculty of Science at the University of Paris appointed de Broglie as a professor of theoretical physics in 1928, and in 1933 a special chair in theoretical physics was created for him at the Henri Poincaré Institute. He won the 1929 Nobel Prize in physics. Subsequently many distinguished societies included him in their membership, including the Academy of Sciences in 1933, the Académie Française in 1944, the National Academy in the United States in 1947, and the Royal Society of London in 1953. De Broglie died of natural causes on March 19, 1987.

Brongniart, Alexandre

(1770–1847)
French
Geologist, Paleontologist

Alexandre Brongniart pioneered stratigraphic geology with his study of the Paris Basin when he arranged and chronologically ordered the geologic layers from the Tertiary period, dating as far back as 66 million years ago. Brongniart dated each stratum according to fossils he found in that layer. Interestingly, he found evidence of alternating layers between salt water and fresh water, suggesting an ebb and flow of oceans in that region. He also subdivided reptiles into four classifications.

Brongniart was born on February 5, 1770, in Paris, France. His father was the renowned Parisian architect Alexandre-Théodore Brongniart, and his mother was Anne-Louise Degremont Brongniart. He married the politician and scientist Charles-Étienne Coquebert de Montbret's daughter, Cécile. Their only son, Adolphe-Théodore Brongniart, went on to become a famous botanist, having his father's scientific bent from the very beginning.

Alexandre Brongniart attended the École des Mines and then the École de Médécine. Next he served as an assistant to his uncle, Antoine-Louis Brongniart, a professor of chemistry at the Jardin des Plantes. His next position was as a mining engineer in 1794, and he was later promoted to chief mining engineer in 1818. In 1797 the École Centrale des Quatre-Nations in Paris named him a professor of natural history. Brongniart's final academic appointment lasted the rest of his life. He held the position of professor of mineralogy at the National Museum of Natural History from 1822 until 1847.

Brongniart published his first scientific paper in 1791. In 1800 he published his "Essay on the Classification of Reptiles," in which he broke down the class Reptilia into four groups: Chelonia, Ophidia, Sauria, and Batrachia. The first three groupings survived into modern classification systems, but even at the time Brongniart recognized that the reproductive systems of the members of the fourth group differed significantly from those of the other three groups. In 1804 Pierre Latreille confirmed this distinction by classifying batrachians as a completely separate group, otherwise known as amphibians. Despite this correction, Brongniart's classification system marked an important step in scientific thinking.

In 1804 Brongniart commenced the work that brought him the most recognition, the study of fossil-laden strata in the Paris Basin, which he conducted in collaboration with GEORGES LÉOPOLD CHRÊTIEN FRÉDÉRIC DAGOBERT CUVIER. Together they presented their paper, "Essay on the Mineralogical Geography of the Environs of Paris," on April 11, 1808, and the work was published in June 1808. An appended version of the paper, which added "A Geological Map and Profiles of the Terrain" to the title, was published in 1811. Although Cuvier's name appeared first in the byline, the majority of the geological work reportedly was done by Brongniart.

In 1800 Brongniart was named the director of the Sevres Porcelain Factory, a position he retained the rest of his life, devoting most of his energy in the latter part of his career to the study and perfection of porcelain-making techniques. The Academie des Sciences elected Brongniart as a member in 1815. He died on October 7, 1847, in Paris. It is surmised that Brongniart would have wielded even greater influence in scientific history if he had not been so cautious in his theorizing, but Brongniart refused to venture beyond the boundaries of his own certainty. Even his most cautious work, however, created breakthroughs in the world of science.

Brooks, Harriet

(1876–1933)
Canadian
Physicist

In 1907 the Nobel Prize–winning British physicist ERNEST RUTHERFORD, for whom she had worked, called Harriet Brooks "next to Mme. [Marie] Curie, . . . the most prominent woman physicist in the department of radioactivity." Harriet was born in Exeter, Ontario, on July 2, 1876, the second of George and Elizabeth Brooks' eight children. Her father was a traveling salesman for a flour company, and the family moved frequently during her childhood.

Brooks won a scholarship to McGill University in Montreal, where she studied mathematics, languages, and physics. She graduated with honors in 1898 and began work in the laboratory of Rutherford, who had just immigrated to Canada. She also taught at the Royal Victoria College, a new women's college associated with McGill. In 1901 she earned a master's degree, the first McGill granted to a woman.

Brooks discovered that strange "emanations" given off when radium broke down were not a form of that element but a new element, a gas later called radon. Her paper on the subject, published in 1901, provided the first proof that one element could change into another. She also discovered recoil, a quality of radioactive atoms that aided the identification of the radioactive forms of elements, and she made important studies of the radioactive decay process in radium and actinium.

While spending a year of postgraduate study at the Cavendish Laboratory of Britain's Cambridge University in

1902–1903, Brooks met and fell in love with a physicist from Columbia University named Bergen Davis. She followed him to the United States in 1905 and took a teaching post at Barnard, a women's college at Columbia. She and Davis soon became engaged. When she announced the engagement, she was shocked to receive a letter from Barnard's dean, Laura Gill, saying: "I feel very strongly that . . . your marriage . . . ought to end your official relationship with the college." Brooks snapped back, "I think . . . it is a duty I owe to my profession and to my sex to show that a woman has a right to the practice of her profession and cannot be condemned to abandon it merely because she marries."

As it happened, the engagement broke off, but perhaps not surprisingly, Brooks soon left Barnard. She traveled to Italy, where she met MARIE SKLODOWSKA CURIE; then to France, where she studied the element actinium at the Radium Institute. In 1907 she married Frank Pitcher, a physicist whom she had met earlier at McGill, and they returned to Montreal. Brooks then retired from science to raise two children. She died on April 17, 1933.

Buchner, Eduard

(1860–1917)
German
Chemist

Best known for his breakthrough discovery of cell-less fermentation, Eduard Buchner is recognized as the founder of the chemistry of enzymes. Buchner's findings refuted the generally accepted theory that fermentation of carbohydrates from sugar to alcohol was the result of the presence of live yeast cells, a theory supported by prominent scientists, including LOUIS PASTEUR. Buchner also found that the catalyst for fermentation was an enzyme, thus sparking the field of enzyme chemistry. For his outstanding work on fermentation Buchner was awarded the 1907 Nobel Prize in chemistry.

Born on May 20, 1860, in Munich, Germany, Buchner was a member of an old family of academics. Buchner's father, Ernst Buchner, was a professor of obstetrics and forensic medicine and edited a medical journal. His mother was the former Friederike Martin. Buchner's brother, Hans, became a bacteriologist, and Buchner later worked closely with him in his research.

After attending high school in Munich, Buchner spent some time in the field artillery. He then attended the Technical College in Munich to study chemistry. He took a leave of absence as a result of lack of funds and worked in canneries for four years. When Buchner's brother, Hans, gave him financial assistance, Buchner returned to school; he entered the Bavarian Academy of Sciences in Munich in 1884. There he studied under the organic chemist Adolf von Baeyer. During this time Buchner also worked at the Institute for Plant Physiology under Karl von Nägeli. It was at the institute that Buchner first became interested in alcoholic fermentation. In 1886 he published a paper on the subject in which he diverged from Pasteur's view that an oxygen-free environment was a requirement for fermentation. After Buchner received his doctorate in chemistry in 1888, he became Baeyer's teaching assistant and carried on with his research on fermentation.

Buchner's brother, Hans, who was studying extracts from bacteria, assisted Buchner in his fermentation experiments. After Buchner began teaching at the University of Kiel in 1893, his brother's assistant demonstrated an extraction method in which Buchner would be able to grind down yeast cells with sand to produce a cell-free extract. During the course of his research Buchner found that this extract decomposed quickly, and thus to preserve the extract, he added sugar to the mixture. Buchner, his brother, and his brother's assistant then observed the process of fermentation, which occurred even though there were no yeast cells present. Buchner's experiment successfully disproved the long-standing theory that fermentation required yeast cells. In 1897, after taking a position as professor at the University of Tübingen, Buchner published his findings in the paper "Alcoholic Fermentation without Yeast Cells."

In 1898 Buchner acquired a position at the Agricultural College in Berlin. There he continued his studies of the fermentation process, producing 15 papers on the topic. One paper, "The Zymase: Fermentation," detailed his discoveries and called the catalyst in the cell extract *zymase*. Buchner later discovered that zymase was an enzyme; enzymes often act as biochemical agents. In 1900 he married Lotte Stahl, the daughter of a mathematician, and the couple had three children.

Buchner received the 1907 Nobel Prize in chemistry for his work on fermentation, and in 1909 he took a job at the University of Breslau. He moved once again two years later when he was offered a position at the University of Würzburg. Buchner entered the army during World War I and died in combat in 1917.

Bunsen, Robert Wilhelm

(1811–1899)
German
Chemist

A versatile experimentalist with diverse scientific interests, Robert Wilhelm Bunsen made many contributions to chemistry. He influenced the field of spectrum analysis and discovered, with Gustav Kirchhoff, the elements cesium and rubidium. Bunsen also discovered an antidote to arsenic poisoning and researched the composition of

Robert Bunsen, who developed the well-known Bunsen burner in 1855. *Smith Collection, Rare Book & Manuscript Library, University of Pennsylvania.*

gases. He is most often remembered as an inventor and modifier of numerous devices, including the well-known Bunsen burner, which he developed in 1855.

Born on March 31, 1811, in Göttingen, Germany, Bunsen was the youngest of four children. His father, Christian Bunsen, was a linguistics professor and chief librarian at the University of Göttingen. Bunsen's mother was the daughter of a British-Hanoverian officer. Many of Bunsen's relatives held positions in public office.

After completing high school studies at Holzminden in 1828, Bunsen entered the University of Göttingen, where he studied chemistry, physics, mathematics, and mineralogy. Bunsen's dissertation topic was hygrometers, and in 1830 he received a doctorate in chemistry. In the early 1830s Bunsen traveled around Europe, meeting numerous notable scientists and exploring various scientific topics, including geology. Among the scientists Bunsen worked with during his journeys were the German chemist Justus von Liebig and the French chemist Joseph Louis Gay-Lussac. Bunsen also visited a number of industrial plants and factories. After his return he took a teaching position at the University of Göttingen.

One of Bunsen's first discoveries was a remedy for arsenic poisoning, which he found while investigating metal salts of arsenic acid in 1834. Bunsen observed that hydrated ferric oxide worked as an antidote and noted that it was effective because ferric oxide joined with arsenic to form a mixture that could not be dissolved in body fluids or in water. In 1837 Bunsen embarked on his only work in organic chemistry. Studying the compounds of cacodyl, an organic compound containing arsenic that was very toxic, Bunsen lost the sight in one eye from an explosion and nearly killed himself twice from arsenic poisoning. After obtaining derivatives of cacodyl, such as chloride, iodide, cyanide, and fluoride, Bunsen abandoned organic chemistry for inorganic chemistry. His work, however, influenced his students, including Edward Frankland, who later made important discoveries about organic compounds containing metal.

Beginning in the late 1830s, Bunsen researched the composition of gases emitted from industrial furnaces. He found that charcoal-burning furnaces in Germany lost more than 50 percent of their heat through escaping gases. The coal-burning furnaces in Britain lost more than 80 percent. Bunsen then developed methods for retrieving and using the lost gases and detailed his findings in his only book, *Methods in Gas Measurement,* published in 1857. In the 1840s Bunsen turned toward the modification of batteries. In 1841 he developed a battery using carbon as the negative pole. Carbon cost less than the commonly used platinum or copper. The battery was later known as the Bunsen battery. The Bunsen burner, a laboratory instrument producing a hot flame, was not invented by Bunsen but modified by him to assist in his research identifying metals and their salts. Bunsen and Kirchhoff's studies of light and spectra in the early 1860s proved instrumental in the development of spectroscopy, and spectral analysis led to Bunsen's discovery of cesium and rubidium.

Bunsen also created a grease-spot photometer, the filter pump, and the ice calorimeter. Highly interested in geology, Bunsen went on an Icelandic expedition and collected the gases from volcanic gaps. He also studied geysers. A very popular and charismatic instructor, Bunsen taught at the University of Heidelberg from 1852 until his retirement in 1899. He died shortly thereafter. Among the honors Bunsen received were the Royal Society of London's Copley Medal, the English Society of Arts' Albert Medal, and the first Davy Medal. Bunsen never married.

Burbank, Luther
(1849–1926)
American
Plant Breeder

Luther Burbank developed over 800 new strains of plants, focusing his efforts on those varieties that would prove most successful commercially. For example, of the 113 different varieties of plums and prunes that he bred, 20 strains have retained their commercial viability. His techniques helped establish plant breeding as a branch of science.

Burbank was born on March 7, 1849, in Lancaster, Massachusetts. His parents were Olive Ross and Samuel Walton Burbank. He remained a bachelor until late in life; he married Elizabeth J. Waters on December 21, 1916. He also conducted most of his scientific research with no more than a high-school education from Lancaster Academy; he received his Sc.D. degree from Tufts University late in life, in 1905.

Burbank educated himself, reading CHARLES ROBERT DARWIN's *The Variation of Animals and Plants under Domestication* early in his intellectual development, and this text proved to be particularly influential. In 1870 at the age of 21, when most college students would be completing their undergraduate degrees, Burbank bought a 17.3-acre farm near Lunenberg, Massachusetts, which served as his experiential classroom. Here Burbank commenced his plant-grafting career, which continued for the next 55 years.

Within about a year Burbank had developed a significant new strain of potato (later named the *Burbank potato*) that proved instrumental to countering blight in Ireland. Burbank sold the rights to this potato for $150, money he used to migrate west in 1875 to establish Burbank's Experimental Farm, with a nursery garden and greenhouse, in Santa Rosa, California.

Burbank's methods, combined with the favorable climate of California, yielded much success in developing promising hybrids. In order to maximize his efficiency, Burbank would cross multiple native and foreign strains simultaneously, and then he would graft these seedlings onto mature plants so that he could assess their viability sooner. He developed a sixth sense for spotting characteristics that would prove commercially successful. In fact, this sense led him to refute JOHANN GREGOR MENDEL's 1901 principle of heredity, as Burbank's experience suggested that some characteristics could be acquired rather than inherited.

Besides the eponymous potato, Burbank developed multiple varieties of different fruits: Gold, Wickson, Apple, October, Chalco, America, Santa Rosa, Formosa, Beauty, Eldorado, and Climax plums; Giant Splendor, Sugar, Standard, and Stoneless prunes; and Burbank and Abundance cherries. He even hybridized a new fruit, the Plumcot. He developed not only commercial fruits but also plants, such as Peachblow, Burbank, and Santa Rosa roses. He also served as a special lecturer on evolution at Leland Stanford Junior University.

Burbank published extensively in his career. Between 1893 and 1901 he published descriptive catalogs entitled *New Creations*. From 1914 to 1915 he published the 12-volume series *Luther Burbank, His Methods and Discoveries and Their Practical Applications*. In 1921 he published an eight-volume series, *How Plants Are Trained to Work for Man*. An autobiography coauthored by Wilbur Hall, *Harvest of the Years*, appeared a year after his death. Burbank died on April 11, 1926, in Santa Rosa, California.

Burbidge, Eleanor Margaret Peachey

(1919–)
British/American
Astronomer

Margaret Burbidge helped to explain how chemical elements are created inside stars. She has also headed Britain's most famous astronomical observatory. A 1974 *Smithsonian* article called her "probably . . . the foremost woman astronomer in the world."

Eleanor Margaret Peachey was born on August 12, 1919, in Davenport, England. Her father, Stanley, was a chemistry teacher at the Manchester School of Technology. Marjorie, her mother, had been one of his few woman students. Margaret first became interested in the stars at age four, when she saw them through a porthole during a trip across the English Channel. "They are so beautifully clear at night from a ship," she recalls. When she was 12 she was delighted to learn that astronomy involved not only stars but her other favorite thing, large numbers. "I decided then and there that the occupation I most wanted to engage in 'when I was grown up' was to determine the distances of the stars."

Margaret Peachey majored in astronomy at University College, London; she earned a bachelor's degree in 1939 and a doctorate in 1943. In 1945 she applied for a grant to work at California's Mount Wilson telescope but was turned down because the telescope's administrators did not permit women to use it. She returned to University College two years later to take an advanced course in physics and met another student, Geoffrey Burbidge; they married on April 2, 1948. Geoff, as he was known, started out as a physicist, but he, too, became an astronomer.

After two years of work in the United States the Burbidges returned to Britain in 1953 and began working with the British astronomer Sir FRED HOYLE and the nuclear physicist William Fowler on a theory that explained how elements were made inside stars. The theory came to be called the B^2FH theory from the first letters of its creators' last names. (Other astronomers referred to the Burbidges, who often worked together, as B^2, or "B squared.") It said that as stars age and exhaust their nuclear fuel, they go through a series of reactions that make heavier and heavier elements by fusing the atomic nuclei of the elements made in the previous reaction. Finally, if a star is large, it destroys itself in a violent explosion called a supernova, creating the heaviest elements in the process.

Gas from the supernova, containing all the elements that the star produced, drifts out into space and is eventually captured and reused by other stars and planets.

To gather data to support their theory, the Burbidges braved Mount Wilson again in 1955. They got Margaret in by pretending that she was Geoff's assistant. She actually did most of the telescope work, spending nights high in the unheated observatory dome while pregnant with the couple's only child, Sarah. She photographed the light of stars and analyzed it to show what elements the stars contained. The four astronomers presented the B^2FH theory in 1957, and the Burbidges won the Warner Prize in 1959 for helping to devise and prove it. The writer Dennis Overbye says that the theory "laid out a new view of the galaxy as a dynamic evolving organism, of stars that were . . . an interacting community."

In the late 1950s the Burbidges worked at the University of Chicago's Yerkes Observatory in Wisconsin, studying different kinds of galaxies. The university hired Geoff as an associate professor, but antinepotism rules forbade hiring Margaret as well. (Such rules "are always used against the wife," she says.) The most she could get was a research fellowship. It was little wonder that, when the newly established University of California at San Diego offered to hire both Burbidges in 1962, they accepted. Margaret Burbidge became a full professor of astronomy there in 1964. At San Diego the Burbidges studied quasars, or quasi-stellar radio sources. Astronomers still do not know exactly what these strange starlike objects are.

When the Burbidges visited Britain in 1971, the head of the country's Science Research Council (SRC) asked a startled Margaret to become the director of the Royal Greenwich Observatory, Britain's most famous observatory. No woman had ever held this post. Geoff was offered a job there as well. The Burbidges accepted and moved back to England in 1972.

Unfortunately being head of the Greenwich Observatory entailed as much frustration as honor. For instance, the observatory director was normally given the title of Astronomer Royal, but Margaret, for unknown reasons, was not. The Burbidges also became involved in a dispute over whether the observatory's largest telescope should be moved out of the country. Geoffrey published a blunt letter on the subject in the science journal *Nature,* which angered the SRC. After what Margaret calls a "bitter confrontation," she resigned, after heading Greenwich for just 15 months. The Burbidges then returned to San Diego.

Margaret Burbidge became the first woman president of the American Astronomical Society in 1976. She was also president of the American Association for the Advancement of Science in 1981. She headed a team that designed a faint object spectrograph, one of the instruments attached to the Hubble Space Telescope. She also directed San Diego's Center for Astrophysics and Space Sciences from the early 1980s, when it was founded, until 1988. In 1998 Burbidge had retired as a university professor, but she still worked as a research professor and was investigating quasars, using the spectrograph she helped to design.

Burnet, Sir Frank Macfarlane

(1899–1985)
Australian
Virologist

Sir Frank Macfarlane Burnet made notable contributions to the fields of immunology and virology and received the 1960 Nobel Prize in physiology or medicine for his work. Burnet researched bacteriophages, viruses that attack bacteria, and viruses. Through his studies Burnet developed a method for cultivating viruses in live chick embryos, a practice that later became a common laboratory technique. He also investigated the immune system, formulating a theory about immunological tolerance in living beings.

Born on September 3, 1899, in Traralgon, Australia, Burnet was the son of a Scottish immigrant, Frank Burnet, who worked as a bank manager. His mother was Hadassah Pollock MacKay. Burnet developed an interest in nature when he was a child and spent time studying birds and insects. He married a schoolteacher, Edith Linda Druce, in 1928 and had three children. After the death of his wife in 1973 Burnet married Hazel Jenkin.

After attending Geelong College and studying biology and medicine, Burnet transferred to the University of Melbourne in 1917. There he earned his undergraduate degree in 1922 and his M.D. in 1923. After working in Melbourne for a few years Burnet traveled to London on a fellowship to work at the Lister Institute of Preventive Medicine. There he studied viruses and bacteriophages and received a doctorate from the University of London in 1927. Burnet then returned to Australia to work at the Walter and Eliza Hall Institute of Medical Research at the Royal Melbourne Hospital.

Though Burnet had studied viruses and bacteriophages for some time, it was during a year's leave from the Hall Institute that Burnet made his first significant advance in virology. While working at the National Institute for Medical Research in London as a research fellow from 1932 to 1933, Burnet developed a method for cultivating viruses in chicken embryos. The technique allowed scientists to cultivate viruses readily; the task had previously been complicated and difficult.

While working on the cultivation of viruses, particularly influenza viruses, Burnet became increasingly interested in studying the immune system. He had observed that adult hens infected with the influenza virus could readily develop antibodies against the virus. Chicks born from the eggs in which the virus was cultivated, however, were unable to develop these antibodies. This led Burnet to contemplate how the body was able to recognize foreign substances and build defenses against them. Burnet also examined the development of the immune system to explore when the ability to produce antibodies began. Burnet worked on these questions for more than 20 years, discovering and presenting details about the way the body recognizes the difference between foreign cells and its own cells. Burnet also believed that the body's ability to produce antibodies developed during the fetal stage.

For his pioneering work on immunology Burnet shared the 1960 Nobel Prize in physiology or medicine with Peter Brian Medawar. Burnet became a fellow of the Royal Society in 1947 and was knighted in 1951. He received the Royal Society's Royal and Copley Medals and the Order of Merit. Burnet also studied cancer, autoimmune disorders, and diseases including Murray Valley fever, Q fever, and myxomatosis.

Buys Ballot, Christoph Hendrik Diedrik

(1817–1890)
Dutch
Meteorologist

Christoph Buys Ballot proposed the law that later bore his name, which described the dependence of the motion of wind rotation on its proximity to low- or high-pressure weather patterns and its position in the Northern or Southern Hemisphere. Even more significant than Buys Ballot's law was his influence on the efforts to organize and coordinate meteorological information throughout Europe.

Buys Ballot was born on October 10, 1817, in Kloetinge, Netherlands. His parents were Geertruida Françoise Lix Raaven and Anthony Jacobus Buys Ballot, a Dutch Reformed minister. He attended the Gymnasium at Zaltbommel and the University of Utrecht and received his Ph.D. in 1844. Buys Ballot married twice, and five of his eight children survived him.

After earning his doctorate Buys Ballot remained at the University of Utrecht for the rest of his career, first as a lecturer in mineralogy and geology in 1845, adding duties teaching theoretical chemistry the next year. In 1847 the university appointed him as a professor of mathematics; two decades later the university named him as a professor of physics, a position he held until his retirement in 1888.

From 1851 on Buys Ballot published yearbooks of synoptic weather observations, gathered simultaneously throughout Europe and then compiled onto charts and maps, with shaded areas corresponding to meteorological phenomena such as wind direction and speed and temperature anomalies. This effort was the first of its kind, and Buys Ballot spent the rest of his career organizing and coordinating the field of meteorology. In 1854, for example, he founded the Royal Netherlands Meteorological Institute, where he was later director.

In 1857 Buys Ballot made the observations that led to the law named after him. He noticed that in the Northern Hemisphere wind circulated counterclockwise around low-pressure systems and clockwise around high-pressure systems; in the Southern Hemisphere they rotated in the opposite directions in those situations. These effects resulted from the deflecting force of the Earth's rotation, though Buys Ballot himself did not make this connection, as William Ferrel had previously. He did, however, assert that wind speed is proportional to pressure gradient and that wind blows at right angles to the isobars.

In 1860 Buys Ballot established the first service for weather forecasts and storm warnings. During this period he also invented the aeroclinoscope, an instrument that pinpointed the center of a depression and its pressure gradient. After the 1873 Congress of Vienna, Buys Ballot served as the chairperson of the International Meteorological Committee, charged with the responsibility of making recommendations on which meteorological instruments should be utilized for official observations, the form and format of weather messages and the system whereby these messages would be recorded, and the organization of a system of meteorological observation stations. Although Buys Ballot trained as a theorist and considered himself to be one, he made a much more significant contribution to the field of meteorology with his compilations of information and with his organization and coordination of efforts to record and disseminate meteorological information. Buys Ballot died on February 3, 1890, in Utrecht, Netherlands.

Cannizzaro, Stanislao

(1826–1910)
Italian
Chemist

Amid the turbulence of political strife in his homeland and in his profession Stanislao Cannizzaro discovered an unknown chemical reaction, later known as Cannizzaro's reaction. More importantly Cannizzaro recognized the difference between atomic and molecular weights, a difference that eluded chemists as a result of the limitations imposed by their politicized views of their field. While trying to simplify the contemporary understanding of chemistry for a course he taught, Cannizzaro traced the historical development of atomic chemistry and realized that the distinction between atomic and molecular weights had been made by LORENZO ROMANO AMEDEO CARLO AVOGADRO in 1811 but was unrecognized by all save ANDRÉ-MARIE AMPÈRE, whose sole voice of support was insufficient to save this discovery from obscurity until Cannizzaro asserted its validity convincingly at the 1860 Chemical Congress in Karlsruhe, Germany.

Cannizzaro, the youngest of 10 children, was born on July 13, 1826, in Palermo, Sicily. His father was Mariano Cannizzaro, a magistrate and minister of the Palermo police department, and his mother was Anna di Benedetto, of noble Sicilian lineage. Cannizzaro's family largely supported the Bourbon regime, which ruled Sicily from Naples, but Cannizzaro himself followed in the footsteps of his politically liberal maternal uncles by joining the Sicilian revolution in January 1848. By April 1849 the rebellion had lost its momentum, and Cannizzaro fled from a death warrant to Marseilles and eventually to Paris. In 1856 or 1857 Cannizzaro married Henrietta Withers in Florence. The couple had a daughter and a son who became an architect.

Cannizzaro's education in the sciences began in earnest in 1841, when he entered medical school at the University of Palermo. Michele Foderà introduced Cannizzaro to biological research, as the two collaborated on a study of the differences between centrifugal and centripetal nerves. At

Stanislao Cannizzaro, who successfully asserted the validity of Amadeo Avogadro's discovery of the difference between atomic and molecular weights. *Smith Collection, Rare Book & Manuscript Library, University of Pennsylvania.*

the University of Pisa from 1845 until 1847 Cannizzaro served as a laboratory assistant under Rafaelle Piria, who first prepared salicylic acid.

After Cannizzaro's flight from Sicily he worked in Michel-Eugène Chevreul's laboratory in the Jardin des Plantes in Paris, assisting Stanislaus Cloëz in preparing cyanamide. Also in 1851 the Technical Institute of Alessandria appointed Cannizzaro as a professor of physics, chemistry, and mechanics. Two years later he discovered the reaction named after him by combining benzaldehyde with a concentrated alcoholic hydroxide, creating equivalent amounts of benzyl alcohol and salt of benzoic acid. Cannizzaro moved on in 1855 to a chemistry professorship at the University of Genoa, where he concentrated on teaching because of the school's lack of laboratory facilities. His mind thus occupied, he focused his attention on clarifying the history and present status of chemistry for his course lectures.

In 1858 he wrote a letter to Sebastiano de Luca, attempting to simplify the complexities of the previous half-century of developments in the field of chemistry. Cannizzaro published this letter under the title "Sunto di un corso di filosofia chimica fatto nella Reale Università di Genova" in the journal *Nuovo cimento* that year and as a pamphlet the next year. The subsequent year Cannizzaro expressed the views stated in the letter at the Chemical Congress in Karlsruhe, and copies of the pamphlet, which explicated his argument step by step, were distributed after his departure. His message sank in, and the discipline finally adopted Avogadro's principles after Cannizzaro had championed them so persuasively.

Cannizzaro served as a professor of inorganic and organic chemistry at the University of Palermo from 1861 until 1871, when he moved to the University of Rome as a professor of chemistry. Cannizzaro represented Palermo in the Italian Senate from 1872, and he later became the vice president of the assembly. In 1891 the Royal Society of London awarded him the Copley Medal. He died on May 10, 1910, in Rome, Italy.

Cannon, Annie Jump

(1863–1941)
American
Astronomer

Annie Jump Cannon, whose work formed the core of the *Henry Draper Star Catalogue,* which was published between 1918 and 1949 and is still a primary reference for astronomers. *Harvard College Observatory.*

Annie Jump Cannon studied more stars than any other person—some 350,000 of them. She perfected a system for classifying stars according to patterns in their light and produced a giant star catalog that astronomers still consult. HARLOW SHAPLEY, director of the Harvard Observatory, called her "one of the leading women astronomers of all time."

Annie Cannon, born on December 11, 1863, spent a happy childhood in a large family in Dover, Delaware. Her father, Wilson, was a wealthy shipbuilder, and her mother, Mary, taught her to recognize constellations, using a textbook from her own school days. The two built a makeshift observatory in their attic, and Annie sometimes climbed through the attic's trapdoor to watch the stars from the house's roof. She then returned to bed to read by candlelight, making her father afraid she would start a fire.

Even though college education for women was a new and rather shocking idea, Wilson Cannon recognized his daughter's intelligence and encouraged her to attend Wellesley, a new women's college in Massachusetts. Annie enjoyed her years there, but she was not prepared for the chilly New England winters. During her first year she had one cold after another. The illnesses damaged her eardrums, producing deafness that became worse as she grew older.

After graduation in 1884 Cannon returned to her home in Delaware and led a carefree social life. That life ended abruptly when her mother died in 1893. The two had been very close, and Cannon could not bear to stay in the places they had shared. She decided to go back to Wellesley instead. She took a year of graduate courses there, then enrolled in 1895 as a special student in astronomy at Radcliffe, the women's college connected with Harvard University, so she could use Harvard's observatory.

To Cannon's surprise she found a number of other women working at the observatory, including WILLIAMINA PATON STEVENS FLEMING. Edward C. Pickering, the observatory's director, was making a gigantic survey of all the stars in the sky, and he made a point of hiring women to help him. Publicly he stated that he preferred women because they had more patience than men, a better eye for detail, and smaller hands that could more easily manipulate delicate equipment. However, he also pointed out to the Harvard trustees in 1898 that women "were capable of doing as much good routine work as [male] astronomers who would receive much larger salaries. Three or four times as many assistants can thus be employed . . . for a given expenditure."

Pickering's survey depended on a device called a spectroscope, which converted light from stars or other sources into rainbowlike patterns called spectra. Astronomers had learned to combine a spectroscope, a camera, and a telescope to take pictures of the spectrum made by each star's light. By studying these spectrograms, as the pictures were called, a trained observer with a magnifying glass could find out which elements the star contained, how hot it was, how big it was, and how fast it moved through space.

Annie Cannon joined the Harvard Observatory staff in 1896. Pickering by then had accumulated 10 years' worth of spectrograms, each of which contained the spectra of hundreds of stars, and analyzing them became Cannon's job. No two stars' spectra were exactly alike, so she used each star's spectrum to identify it. She also grouped stars with similar spectra. She refined a classification system that Williamina Fleming had devised, eventually dividing stars into classes that, in order of their surface temperature from hottest to coolest, are designated O, B, A, F, G, K, and M. (Generations of astronomy students have memorized this list by means of the rather sexist sentence "O Be a Fine Girl, Kiss Me.") By 1910 astronomers everywhere were using her system.

Cannon examined the spectra of an unbelievable 350,000 stars during her lifetime. "Each new spectrum is the gateway to a wonderful new world," she once said. She grew so expert that she could classify three spectra a minute. She also calculated each star's position. Her work became the core of the giant *Henry Draper Star Catalogue,* issued in nine volumes between 1918 and 1924, and its two extension volumes, issued in 1925 and 1949. Harlow Shapley, who became director of the Harvard Observatory after Pickering, said that Cannon's contributions to astronomy make "a structure that probably will never be duplicated . . . by a single individual." It is still a standard reference for astronomers all over the world.

Cannon earned her master's degree in 1907 from Wellesley. In 1911 she took over Williamina Fleming's job as curator of photographs at the Harvard Observatory, a post she held for 27 years. She was also named the William Cranch Bond Astronomer at Harvard in 1938, one of the first women to be given a titled appointment by the university. She was elected an honorary member of Britain's Royal Astronomical Society.

Cannon won many awards, including the first honorary doctorate given to a woman by Britain's prestigious Oxford University (1925). She won the Draper Medal of the National Academy of Sciences in 1931 (the first gold medal awarded to a woman by this group) and the Ellen Richards Prize of the Society to Aid Scientific Research by Women in 1932. She used the money that came with the Richards Prize to fund an award of her own, the Annie Jump Cannon Prize, to be awarded every third year by the American Astronomical Society to a woman who had given distinguished service to astronomy. Cannon died of heart disease on April 13, 1941, at the age of 77.

Cantor, Georg Ferdinand Ludwig Philipp

(1845–1918)
Russian/German
Mathematician

Although much of Georg Cantor's work was scorned during his lifetime, his ground-breaking theories are now recognized as the basis of modern mathematical analysis. Cantor was the father of set theory—the study of the relationships existing among sets. Through his research in this area he discovered the theory of infinite sets and transfinite numbers (now termed infinite numbers). Prior to Cantor mathematicians deliberately excluded the concept of infinity from their work. Cantor, though, delved into the study of infinite sets and was the first to distinguish between types of infinity, when he posited that there were countable infinite sets and sets having the power of a continuum. As the founder of the Association of German Mathematicians he also played an essential role in promoting the exchange of ideas among German scientists. The Cantor set and the Cantor function are named in his honor.

Cantor was born in St. Petersburg, Russia, on March 3, 1845, to Georg Waldemar and Maria Böhm Cantor. In 1862 Cantor matriculated at the University of Zurich; he transferred to the University of Berlin after his father's death in 1863. He received his doctoral degree in 1867. Cantor married Vally Guttmann in 1874; with her he had five children.

In 1869 Cantor took a post at the University of Halle, where he remained until his death. His early work dealt with trigonometric series. In 1872 he defined irrational numbers in terms of convergent sequences of rational numbers. The following year he proved that rational numbers are countable—that they can be placed in one-to-one correspondence with natural numbers. In 1874 he published a revolutionary paper in *Crelle's Journal* that was the foundation of set theory. Any collection of objects is termed a set, and Cantor recognized that studying the relationship of the elements within and between sets could define and clarify several aspects of mathematics. This 1874 paper also introduced Cantor's notion of infinite numbers. Infinity was not a concept invented by Cantor (Zeno of Elea had noted the problem of the infinite in B.C. 450), but his genius lay in his careful application of the concept to mathematics. Prior to Cantor's 1874 treatise, infinity was considered a taboo topic among mathematicians. For instance, KARL FRIEDICH GAUSS denied the efficacy of any use of infinity as a mathematical value. Cantor ignored this injunction and proposed two distinct types of infinity. He considered infinite sets not merely as collections of numbers going on "forever" into some nebulous void; instead he viewed them as completed entities, which possessed an actual—though infinite—number of members. He called these actual, infinite numbers *transfinite numbers*. Cantor's first enunciation of these concepts yielded a full professorship at Halle in 1879, though it also won him the undying enmity of Leopold Kronecker, who was so opposed to the ideas that he strove thereafter to stymie Cantor's career.

In a series of papers published between 1874 and 1897 Cantor proved several of his theories. He determined that the set of integers (the positive and negative numbers 1, 2, 3, and so on, or 0) had the same number of members as the sets of even numbers, squares, cubes, and roots to an equation. He proposed that the number of points in a line segment is equal to the number of points in an infinite line, a plane, and all mathematical space. He proved that the number of transcendental numbers (values that can never be the solution to any algebraic equation) was much larger than the number of integers. In addition, Cantor constructed a set (now called the Cantor set) that is self-similar at all scales: That is, a magnified portion of the set is identical to the entire set. He also conceived the Cantor function, a distribution function to define measures of the Cantor set.

From 1884 Cantor suffered bouts of severe depression. He was repeatedly institutionalized, and eventually he died in an asylum in 1918. He received little formal acknowledgment of his work during his lifetime. Nevertheless, his theoretical inventions provide the basis for modern mathematics. He ignited further research into what are now considered fundamental principles. Cantor is considered one of the greatest modern mathematicians.

Cardús, David

(1922–)
Spanish/American
Physician

The physician David Cardús, a professor of statistics at Rice University in Houston, Texas, is known for his work on experimental exercise, sports medicine, and respiratory processes. Cardús specializes in cardiology and biomathematics, the application of mathematical principles to biological functions. He uses computer and mathematical applications to investigate the functions of living organisms.

Born on August 6, 1922, in Barcelona, Spain, Cardús was the son of Jaume and Ferranda Pascual Cardús. In 1951 Cardús married Francesca Ribas. The couple had four children.

Cardús attended the University of Montpellier in France and earned his B.A. and B.S. degrees in 1942. He then entered the University of Barcelona and received his M.D. in 1949. Cardús's medical internship was carried out at the University of Barcelona's Hospital Clínico, and he completed his residency at Barcelona's Sanatorio del Puig d'Olena. Cardús then traveled to Paris as a research fellow funded by the French government. He spent two years in Paris studying cardiology then returned to the University of Barcelona for a diploma in cardiology. Cardús then received another research fellowship—a British fellowship to study at the Royal Infirmary at the University of Manchester.

After working at the Royal Infirmary, Cardús moved to the United States. He began work as a research associate at the Lovelace Foundation in Albuquerque, New Mexico, in 1957. In 1960 Cardús joined the team at Baylor College of Medicine's Institute for Rehabilitation and Research in Houston, Texas. Not only did he teach in the rehabilitation and physiology departments in the medical school, but he also headed the biomathematics department and the exercise and cardiopulmonary laboratories. Cardús made significant advances in the use of computer technology and telecommunications in rehabilitation, staging a demonstration of such applications in 1972 that resulted in a gold medal from the U.S. International Congress of Physical Medicine and Rehabilitation.

Cardús has been an adjunct faculty member in the department of statistics and mathematical sciences at Rice University and a member of the teaching staff at Baylor College of Medicine. He has acted as a planning consultant to the U.S. Public Health Service in the design and construction of health facilities. An active member of a number of professional organizations, he has served as president of the International Society for Gravitational Physiology in 1993 and was the vice-chairman of the Gordon Conference on Biomathematics in 1970. Cardús has received many honors and awards from professional societies in both the United States and his home country of

Spain, including top prizes from the International American Congress of Rehabilitative Medicine and the American Urological Association. He has also been recognized by the American Congress of Physical Medicine and Rehabilitation with a writing award. Though Cardús became a U.S. citizen in 1969, he has remained active in programs with Spain. He has served as the president of Spanish Professionals in America and was the chairman of the board of the Institute for Hispanic Culture in Houston. For his scientific contributions Cardús has received honors from Spain's Generalitat de Catalunya and the Instituto Catalan de Cooperación Iberoamericana Fundación Bertran.

Carnot, Nicolas Léonard Sadi

(1796–1832)
French
Physicist

Sadi Carnot's scientific significance is built upon his only publication, an abstract consideration of an idealized frictionless steam engine that attracted little attention in 1824 but subsequently gained much notoriety in connection with the law of conservation of energy as well as the first and second laws of thermodynamics. Both the Carnot cycle and Carnot's theorem arose from this paper.

Carnot was born on June 1, 1796, in the Palais du Petit-Luxembourg in Paris, France. His father, Lazare Carnot, was known as the "Organizer of Victory" for the French Revolution in 1794. Carnot the elder was named as Napoleon's minister of war in 1799; he resigned from political life in 1807 to devote himself to science and the education of his children, Sadi and his younger brother, Hippolyte, father of France's fourth president under the Third Republic. In October 1815 the elder Carnot was exiled by the Restoration.

After his father's instruction Carnot attended the École Polytechnique from 1812 until October 1814, when he graduated sixth in his class. He then matriculated at the École du Génie at Metz as a student second lieutenant, a rank he attained upon graduation in late 1816. In 1819 Carnot escaped the tedium of his military duties by appointment to the army general staff corps, which allowed him time to attend courses at the Sorbonne, the Collège de France, the École des Mines, and the Conservatoire des Arts et Métiers.

Carnot's reputation rests solely on one work, *Reflections on the Motive Power of Fire,* which was published on June 12, 1824. On June 14 the paper was formally presented to the Académie des Sciences, and on July 26 a favorable review of the text was read to the academy. This review was later published in the August issue of *Revue encyclopédique.* Carnot was inspired as much by nationalist pride as by scientific curiosity, as the British had established themselves as the primary innovators of the steam engine. Carnot recognized the inefficiency and lack of theoretical foundation of the British work on steam engines, so he set out to address both shortcomings in his own treatise.

Sadi Carnot, one of the founders of thermodynamics. *AIP Emilio Segrè Visual Archives.*

Carnot based his hypotheses about heat production in a steam engine on the caloric theory of heat, which was later disproved. This basis did not undermine the validity of his results, however. Most significantly Carnot deduced that the efficiency of the engine depended only on the temperature at the heat source and at the heat sink, and not on the temperature of the working substance. The first deduction amounted to the Carnot cycle; the second was known as Carnot's theorem.

It is not known why Carnot's paper did not generate more initial interest. Émile Clapeyron, a railroad engineer, picked up on Carnot's ideas in 1834, extending his hypotheses. Not until 1849 was the true scientific significance of Carnot's paper fully acknowledged by scientists of the status of William Thomson and Rudolph Clausius. From that time forward Carnot was credited as a founder of thermodynamics. Unfortunately he was not alive to enjoy this recognition, as he had died of cholera on August 24, 1832, in Paris.

Carrel, Alexis

(1873–1944)
French
Surgeon, Biologist

Outspoken and controversial, Alexis Carrel was a pioneering surgeon who developed techniques for suturing together blood vessels. Carrel also studied organ and blood vessel transplantation and researched tissue culture, working toward the artificial maintenance of tissues and organs. His research advanced the study of viruses and facilitated the development of vaccines. For his breakthrough work on vascular ligature and the grafting of organs and blood vessels Carrel was awarded the 1912 Nobel Prize in physiology or medicine.

Born on June 28, 1873, in Sante-Foy-les-Lyon, France, Carrel was the oldest of three children born to Anne-Marie Ricard and Alexis Carrel Billiard, a textile manufacturer. Carrel's family was devoutly Roman Catholic, and after the death of his father when Carrel was only five years old, he and his siblings were raised by their caring mother. Carrel developed an early interest in science and medicine, dissecting birds and conducting chemistry experiments.

After attending Jesuit schools and earning two bachelor's degrees, Carrel entered the University of Lyons in 1891 to study medicine. He spent nine years gaining his education, and in addition to his classwork he gained hands-on experience working as an army surgeon and in local medical facilities. After receiving his M.D. in 1900, Carrel remained at the University of Lyons, hoping to gain a permanent position.

Carrel began investigating methods for suturing severed blood vessels back together shortly after earning his medical degree. After teaching himself how to sew with a needle and thread, he developed ways to minimize the risk of infection and blood clots. Carrel publicized his discoveries in a medical journal in 1902. Because his work was not well received by French colleagues, Carrel traveled in 1904 to Montreal, Canada. Soon he was offered a position in physiology at the University of Chicago, where he worked from 1904 to 1906. During that time Carrel continued his studies of vascular surgery. His suturing methods led him to consider organ transplantation and blood transfusion, and Carrel conducted numerous experiments on dogs, successfully transplanting kidneys. Carrel's recognition grew as a result of these transplants.

In 1906 Carrel accepted a position at the new Rockefeller Institute for Medical Research (now Rockefeller University) in New York City. The facility was dedicated solely to medical research and allowed Carrel to extend his work on blood vessels. At the time, because it was not known how to prevent blood from clotting, storage of blood was impossible; Carrel's methods made blood transfusions feasible. He also began to consider the possibility of keeping human tissues and organs alive outside the body. Of particular interest to Carrel was the process of pumping blood through an organ via artificial means to keep the organ viable, the technique known as perfusion. In 1912 he extracted the heart tissue of a chicken embryo; he kept it alive in the laboratory for 34 years.

During World War I Carrel directed an army hospital for the French government. There he and the biochemist Henry D. Dakin developed a technique for cleaning deep wounds to stop infection. This technique, known as the Carrel-Dakin method, helped stave off gangrene in wounded soldiers. After the war Carrel returned to the Rockefeller Institute to continue his tissue culture research. With the aviator Charles A. Lindbergh, Carrel developed a so-called artificial heart, or perfusion pump, that was able to keep organs alive up to several weeks.

In addition to the Nobel Prize, Carrel was awarded the 1931 Nordhoff-Jung Cancer Prize. After retiring from the Rockefeller Institute in 1939, he returned to France. Carrel was married to a surgical nurse, Anne-Marie Laure de Meyrie, a Roman Catholic widow with one son, from 1913 until his death in 1944. They had no children. Although he led a somewhat controversial life, his contributions to medical science are undeniable. His research resulted in a number of new options for the medical community and greatly advanced vascular surgery

Carruthers, George R.

(1939–)
American
Astrophysicist

The astrophysicist George R. Carruthers is known for his development of a camera/spectrograph that utilized ultraviolet light to photograph the Earth's atmosphere from the Moon's surface. His invention was taken aboard *Apollo 16* in 1972 and traveled to the Moon. The images captured during this mission provided new insight into interstellar gases and the amount of pollution in the Earth's atmosphere. Carruthers has also worked on electronic telescopes and computer-controlled space cameras.

Born on October 1, 1939, in Cincinnati, Ohio, Carruthers showed an early interest in physics and astronomy. His father, whose name was also George, was a civil engineer, and his mother was Sophia Carruthers. The family moved to Milford, Ohio, a suburb of Cincinnati, when George was seven years old. There Carruthers became more interested in space exploration, influenced at first by comic books and later by his father's astronomy texts; it was his father who encouraged his interests and provided additional scientific books for Carruthers to study. Carruthers built his first telescope when he was 10 years old. When Carruthers was 12 his father died, and the family moved to Chicago, Illinois.

In 1957, after excelling in science courses in high school, Carruthers entered the University of Illinois at Champaign-Urbana, where he combined the study of astronomy with aeronautical engineering. He earned his bachelor's degree in physics in 1961 and his master's degree a year later. During his graduate studies at the University of Illinois, Carruthers built a type of rocket engine called a plasma engine and worked on spacecraft design topics. After completing his dissertation on atomic nitrogen recombination, he received his doctorate in aeronautical and astronautical engineering in 1964. That same year he became a National Science Foundation fellow and journeyed to Washington, D.C., to study rocket astronomy at the Naval Research Laboratory.

While working at the Naval Research Laboratory, Carruthers began investigating how to make imaging instruments that would provide detailed information about outer space. He was chiefly interested in spectroscopy and the measurement of ultraviolet light. Carruthers developed a device for detecting electromagnetic radiation, which he patented in 1969, then created the Far Ultraviolet Camera/Spectrograph, an imaging device that could provide information about deep space and about the Earth's atmosphere. Because the camera was designed to be mounted on the Moon, the resulting images would be free of the distortions caused by the Earth's atmosphere. The camera was taken on the *Apollo 16* mission and collected more than 200 images of space. One significant finding was the discovery of hydrogen in deep space. The photographs also revealed information about the amount of pollution in the Earth's atmosphere. The camera was subsequently used on *Skylab 4* and in other missions to measure the thickness of the ozone layer, an important detail in the planning of environmental regulations.

Carruthers continues his research at the Naval Research Laboratory and in past years has worked on the development of electronic telescopes for satellite use, as well as computer-controlled imaging devices. Carruthers teaches occasionally and is active in organizations encouraging young minority students to pursue careers in science and mathematics. He has received numerous honors for his work in astrophysics, including the Arthur S. Fleming Award in 1971, the National Aeronautics and Space Administration's (NASA) Exceptional Scientific Achievement Medal in 1972, the Warner Prize, and the National Civil Service League Exceptional Achievement Award. Carruthers is also a member of a number of professional societies, including the American Astronomical Society. Carruthers married in 1973.

Carson, Benjamin S.

(1951–)
American
Brain Surgeon

Benjamin Carson gained acclaim in 1987 when he performed surgery to separate a pair of Siamese twins connected at the back of the head. These twins shared a common blood supply to support both of their brains, as well as the rest of their vital organs, so the main challenge of the operation was to establish independent cardiac systems in each. With this achievement Carson overcame the prejudice that his African-American heritage might hamper the scope of his career.

Carson was born on September 18, 1951. His parents were divorced when he was eight years old. His father was an automobile factory worker and made a comfortable living, but Carson and his older brother, Curtis, lived with their mother, who could not afford to live in the family house that she had won in the divorce settlement. As a result she had to rent it out to make mortgage payments. She moved the family to Boston for two years to live with her sister before moving her sons back to Detroit to live in a racially integrated neighborhood. After a vision problem was identified, Carson excelled in school.

Like his older brother, Carson joined the Reserve Officer Training Corps (ROTC) in high school, ascending to the highest rank among ROTC students in the Detroit region. Carson turned down offers of financial support and scholarships to West Point and the University of Michigan in order to attend Yale University. Carson balanced the cerebral and sophisticated atmosphere at Yale with summer jobs such as supervising a trash crew for the Wayne County Highway Department, inspecting line work at an auto assembly plant, and operating a crane at a steel mill.

Before graduation in 1973 Carson met Candy Rustin, a fellow Yale student from Michigan, and the couple married in 1976. Carson was then in his third year of medical school at the University of Michigan, when he developed an easier technique that allowed brain surgeons to locate hidden parts of the brain, thus shortening the duration of brain surgery. Carson earned his M.D. in 1977. Although most internships take two years, Carson completed his in one, after which he beat out 130 other applicants for the only available residency in brain surgery, at Johns Hopkins University Hospital in Baltimore, Maryland. In his last year of residency Carson developed a method of inducing brain cancer in rabbits, which facilitated the study of treatments.

Carson turned down an offer to stay on at Johns Hopkins and accepted an offer from Queen Elizabeth II Medical Center in Perth, Australia, where his first child was born. When Carson returned to the United States, he accepted the offer from Johns Hopkins Medical Institution, where he became an associate professor. It was there in 1987 that Carson performed the operation to separate the Siamese twins born earlier that year in Germany. This surgery required consummate skill, as he had to administer 60 blood transfusions and cool the twins' bodies down to 68 degrees Fahrenheit to slow their bodily functions. The operation also required great endurance, as it lasted 22 hours. Carson awakened both babies after 10 days of induced sleep to find them healthy. The success of this operation earned Carson the reputation as a world-class brain surgeon.

Carson, Rachel Louise

(1907–1964)
American
Marine Biologist, Ecologist

Rachel Carson combined a professional knowledge of science (she trained as a marine biologist) with poetic writing skill to create best-selling books about the sea. Her impact on history, however, resulted from a book she did not really want to write: a warning that unless human exploitation of the environment were curbed, much of nature might be destroyed. Carson's book *Silent Spring* introduced the idea of ecology to the American public and almost single-handedly spawned the environmental movement.

Rachel Carson, whose book *Silent Spring* is often cited as the inspiration for the creation of the Environmental Protection Agency. *Beinecke Rare Book & Manuscript Library, Yale University.*

Rachel Carson was born in Springdale, Pennsylvania, on May 27, 1907. Her father, Robert, sold insurance and real estate. Her family never had much money, but the 65 acres of land around their home near the Allegheny Mountains was rich in natural beauty, which her mother, Maria, taught her to love. "I can remember no time when I wasn't interested in the out-of-doors and the whole world of nature," Carson once said.

When Carson entered Pennsylvania College for Women (later Chatham College) on a scholarship, she planned to become a writer. A biology class with an inspired teacher made her change her major to zoology, however. She graduated magna cum laude in 1929 and obtained another scholarship, to do graduate work at Johns Hopkins University. She also began summer work at the Marine Biological Laboratory in Woods Hole, Massachusetts. Carson had always loved reading about the sea, and Woods Hole gave her a long-awaited chance not only to see the ocean but also to work in it. She obtained a master's degree in zoology from Johns Hopkins in 1932.

Carson taught part-time at Johns Hopkins and at the University of Maryland for several years. Then in 1935 her father died suddenly and her mother moved in with her. A year later her older sister also died, orphaning Carson's two young nieces, Virginia and Marjorie. Carson and her mother adopted the girls.

Needing a full-time job to support her family, Carson applied to the U.S. Bureau of Fisheries. In August 1936 she was hired as a junior aquatic biologist—one of the first two women employed there for anything except clerical work. Elmer Higgins, Carson's supervisor, recognized her writing ability and steered most of her work in that direction. He rejected one of her radio scripts, however, saying it was too literary for his purposes. He suggested that she make it into an article for *Atlantic Monthly,* and it appeared as "Undersea" in the magazine's September 1937 issue.

An editor at Simon & Schuster asked Carson to expand her article into a book. The result, *Under the Sea Wind,* appeared in November 1941. Critics liked it, but the United States entered World War II a month later, and book buyers found themselves with little interest to spare for poetic descriptions of nature. The book sold poorly.

Carson continued her writing for the U.S. Fish and Wildlife Service, created in 1940 when the Bureau of Fisheries and the Biological Survey merged. She became editor in chief of the agency's publications division in 1947. In 1948 she began work on another book, drawing on information about oceanography that the government had obtained during the war. That book, *The Sea around Us,* described the physical nature of the oceans. It was more scientific and less poetic than Carson's first book. Published in 1951, it became an immediate best-seller, remaining on the *New York Times* list of top-selling books for a year and a half. It also received many awards, including the National Book Award and the John Burroughs Medal.

Suddenly Rachel Carson found herself famous and, for the first time, relatively free of money worries. In June 1952 she quit her U.S. Fish and Wildlife Service job to write full-time. A year later she built a home on the Maine coast, surrounded by "salt smell and the sound of water, and the softness of fog." She shared it with her mother; her niece, Marjorie; and Marjorie's baby son, Roger. When Marjorie died in 1957, Carson adopted Roger and raised him.

Carson's third book, *The Edge of the Sea,* described seashore life. Published in 1955, it sold almost as well as *The Sea around Us* and garnered its own share of awards, including the Achievement Award of the American Association of University Women.

The book that gave Rachel Carson a place in history, however, was yet to be written. It grew out of an urgent letter that a friend, Olga Owens Huckins, sent to her in 1957 after a plane sprayed clouds of the pesticide dichlorodiphenyltrichloroethane (DDT) over the bird sanctuary that Huckins and her husband owned near Duxbury, Massachusetts. Government officials told Huckins that the spray was a "harmless shower" that would kill only mosquitoes, but the morning after the plane passed over, Huckins found seven dead songbirds. "All of these birds died horribly," she wrote. "Their bills were gaping open, and their splayed claws were drawn up to their breasts in agony." Huckins asked Carson's help in alerting the public to the dangers of pesticides.

Silent Spring appeared in 1962. It took its title from the "fable" at the book's beginning, which pictured a spring that was silent because pesticides had destroyed singing birds and much other wildlife. The health of the human beings in her scenario was imperiled as well. This was the only fiction in the book.

Carson's book did more than condemn pesticides. These toxic chemicals, she said, were just one example of humans' greed, misunderstanding, and exploitation of nature. "The 'control of nature' is a phrase conceived in arrogance born of the . . . [belief] . . . that nature exists for the convenience of man." People failed to understand that all elements of nature, including human beings, are interconnected and that damage to one meant damage to all. Carson used the word *ecology,* from a Greek word meaning "household," to describe this relatedness.

The reporter Adela Rogers St. Johns wrote that *Silent Spring* "caused more uproar . . . than any book by a woman author since *Uncle Tom's Cabin* started a great war." The powerful pesticide industry claimed that if Carson's supposed demand to ban all pesticides—a demand she never actually made—were followed, the country would plunge into a new Dark Age because pest insects would devour its food supplies and insect carriers such as mosquitoes would spread disease everywhere. The media portrayed Carson as an emotional female with no scientific background, ignoring her M.S. degree and years as a U.S. Fish and Wildlife Service biologist. Many scientists took Carson's side, however. President John F. Kennedy appointed a special panel of his Science Advisory Committee to study the issue, and the panel's 1963 report supported most of Carson's conclusions.

Rachel Carson died of breast cancer on April 14, 1964. The trend she started, however, did not die. It resulted in the banning of DDT in the United States and the creation of the Environmental Protection Agency (EPA). Most importantly, it reshaped the way that the American public viewed nature.

Carver, George Washington

(c. 1865–1943)
American
Agricultural Chemist

Born in slavery just before the end of the Civil War, George Washington Carver became an agricultural scientist internationally renowned for his role in the diversification of southern agricultural practices, particularly in the peanut and sweet potato crops. He conducted experiments in crop rotation and the restoration of soil fertility, worked with hybrid cotton, and invented useful products from peanuts, sweet potatoes, and Alabama red clay. His successful testimony before the House Ways and Means Committee in 1921 on the importance of protecting the U.S. peanut industry earned him an identity as the peanut wizard and resulted in a tariff that protected the domestic peanut crop.

Carver's mother, Mary, was a slave on the farm of Moses and Susan Carver when Carver was born. Although his exact birth date is unknown, it is generally thought that he was born in the spring of 1865. His father was said to have been a slave on an adjacent farm, who was killed in an accident soon after Carver's birth. When Carver was still a young child, his mother disappeared, and Carver and his brother were brought up by Moses and Susan. Even as a child Carver developed a reputation as the neighborhood "plant doctor," collecting plants and keeping a small plant nursery.

Leaving home at the age of 12, Carver began a long journey in search of education that would take him through three states, Missouri, Kansas, and Minnesota. In each place Carver found a black family willing to take him in and provide room and board in exchange for chores, while allowing him to attend school during the day. By 1891 Carver was enrolled in the Iowa State College of

Agriculture and Mechanics in Ames, Iowa. Here he found success and recognition for his diverse skills. A painting he did during these years, *Yucca and Cactus,* was exhibited in Cedar Rapids and selected as an Iowa representative for the World's Columbian Exposition in Chicago in 1893. Among his fellow students Carver was known as the "doctor" for his way with plants. Awarded his bachelor's degree in 1894 and his master's degree in 1896 by the college now known as Iowa State University, Carver left Iowa to accept a teaching position at Tuskegee Institute in Alabama, where he spent the next 47 years.

As a researcher at Tuskegee, Carver worked on many projects to improve the prospects for poor southern farmers. He experimented with crop rotation, worked with organic fertilizers, and developed more than 300 products that could be made from peanuts, a staple crop that became even more important when the cotton boll weevil invaded the South, ruining fields of cotton and forcing farmers to grow even more peanuts. He also was instrumental in establishing the first successful agricultural extension programs to teach good farming practices.

Carver received many awards during his life, including an invitation to serve on the advisory board of the National Agricultural Society; the Spingarn Medal, awarded by the National Association for the Advancement of Colored People (NAACP) in 1923; the Franklin Roosevelt Medal in 1937 for "Distinguished Research in Agricultural Chemistry"; and several honorary doctorates. After Carver's death on January 5, 1943, President Franklin D. Roosevelt signed legislation designating Carver's Missouri birthplace a national monument.

Cassini, Giovanni Domenico

(1625–1712)
Italian/French
Astronomer

Cassini's laws, expressed in 1693, proposed several rules of rotation for the Moon, including that its revolution around its own axis coincides with its daily revolution around the Earth and that its tilt remains constant. Cassini also proposed the ovals of Cassini, refuting JOHANN KEPLER's claim of elliptical planetary orbits, and observed a gap in the ring system encircling Saturn, which became known as Cassini's division. Perhaps his most important achievement was his calculation of the astronomical unit, or AU, representing the distance between the Sun and the Earth.

Cassini was born on June 8, 1625, in Perinaldo, in Imperia, Italy. His parents were Jacopo Cassini, a Tuscan, and Julia Crovesi Cassini. Giovanni Cassini's son, Jacques; his grandson, César François; and his great-grandson, Jean-Dominique, all succeeded him as director of the Paris Observatory, a position he held for some 40 years. Cassini was educated at the Jesuit college in Genoa and at the abbey of San Fructuoso. He was persuaded against pursuing astrology by Pico della Mirandola's pamphlet *Disputationes Joannis Pici Mirandolae adversus astrologiam divinatricem.* Cassini was also influenced by Giovanni Battista Riccioli, the Jesuit who published *Almagestum novum* in 1651, and Francesco Maria Grimaldi, whose work, *De lumine,* was published posthumously in 1665. These two thinkers instilled in Cassini the importance of precise observation, as well as an intellectual conservatism.

The construction of a new meridian, an instrument for solar measurements, to replace the one atop the church of San Petronio in Bologna, which had been blocked by new construction, allowed Cassini to make extremely accurate solar observations. He was invited in 1650 by the Bolognese senate to occupy a vacant chair in astronomy at the university. Over the next two decades Cassini made many important astronomical observations. In 1656 he published *Specimen observationum Bononiensium,* in which he recorded information gathered with the new meridian. In 1659 he proposed a planetary system that was consistent with the hypotheses of the Danish astronomer TYCHO BRAHE. In 1661 he developed a way to predict successive phases of solar eclipses by using a method based on the work of JOHANNES KEPLER. And in 1668 he published *Ephemerides Bononieses mediceorem siderum,* charting the movements of the Medici planets, otherwise known as Jupiter's moons, which Galileo discovered. Olaus Römer used these tables in 1675 to calculate the speed of light.

Cassini departed Bologna on February 25, 1669, and arrived in Paris on April 4 to fill his new position as the director of the newly constructed Paris Observatory, a post he maintained for the next 40 years. In 1673 he became a French citizen, cementing his commitment to the observatory. Here he calculated the rotational periods for Jupiter, Mars, and Venus. He took advantage of the powerful new aerial telescopes to locate four new satellites orbiting Saturn—Iapetus in 1671, Rhea in 1672, and Dione and Tethys in 1664. The next year he confirmed the gap in Saturn's rings that William Balle had noticed 10 years earlier. Most importantly, Cassini calculated the astronomical unit, or mean distance between the Sun and the Earth. His figure—87 million miles—gave a much more precise sense of the size of the universe than earlier estimates made by Brahe of 5 million miles and by Kepler of 15 million miles.

Cassini's blindness in 1710 prevented him from making further observations, and he died two years later, on September 14, 1712, in Paris, France. Though he made great contributions to astronomy, many scientists have suggested that had he hypothesized more about the significance of his observations, his contributions might have been even greater.

Cauchy, Augustin-Louis, Baron

(1789–1857)
French
Mathematician

Sixteen mathematical concepts and theorems bear the name of Cauchy, testimony to his enduring effect on mathematics. He contributed to the understanding of calculus, complex functions, error theory, differential equations, and mechanics. He also introduced a commitment to rigor and exactitude that benefited the field far beyond his specific contributions. Though his extreme political and religious conservatism put him at odds with most of his contemporaries, his professional work did not bear any stains from these conflicts.

Cauchy was born on August 21, 1789, in Paris, France. His father, Louis-François Cauchy, was a police lieutenant in Paris and the first secretary to the Senate, and his mother, Marie-Madeleine Desestre Cauchy, bore four sons and two daughters. During the Reign of Terror the Cauchy family escaped to the village of Arcueil, where the founders of the Société d'Arcueil surrounded the young Cauchy. The mathematician the marquis PIERRE SIMON DE LAPLACE, author of *Mécanique celeste,* and the chemist Claude-Louis Berthollet particularly influenced him to pursue study of the sciences and mathematics. After the Terror subsided, Cauchy attended the École Centrale du Panthéon. At the age of 16 in 1805 he matriculated at the École Polytechnique, and in 1807 he moved on to the École des Ponts et Chaussées.

In 1809 Cauchy became an engineer, working on the Ourcy Canal and the Saint-Cloud bridge that year and the harbor of Cherbourg the next year. In 1811 he solved the problem posed to him by Joseph Lagrange as to whether the faces of a convex polyhedron determine its angles. In 1812 he solved Fermat's problem of polygonal numbers. The elegance of his solutions helped establish his reputation as a promising mathematician, so that the École Polytechnique appointed him as a *répétiteur* in 1815; in 1816 he was promoted to full professor of mechanics, and the Faculty of Sciences and the Collège de France followed suit with similar appointments. That year he won the Grand Prix from the Institute of France for a paper on wave propagation.

In 1818 Cauchy married Aloise de Bure, the daughter (or granddaughter) of his publisher. Their two daughters both married counts—the viscount de l'Escalopier and the count of Saint-Pol. Cauchy published a string of important works in the 1820s. In 1821 he published "Courses on Analysis from the École Royale Polytechnique." In 1822 his work set the foundation for the mathematical theory of elasticity, the contribution that carved his name in the history of scientific discovery. In 1823 he published "Résumé of Lessons on Infinitesimal Calculus," and between 1826 and 1828 he published "Lessons on the Applications of Infinitesimal Calculus to Geometry."

In the wake of the July Revolution of 1830 Cauchy exiled himself rather than swear an oath to the monarch who deposed Charles X. The University of Turin offered him refuge with a chair of mathematical physics, which he held only until 1833, when he departed to tutor the duke of Bordeaux, grandson of Charles X. Cauchy returned to France in 1838, and his exemption from the oath, because he was an academician, allowed him to resume his teaching positions. In 1848 he took up a chair at the Sorbonne.

The extremity of Cauchy's political and religious commitments was matched only by the extremity of his production of mathematical papers. The Academy of Sciences limited the length of paper submissions in response to Cauchy's prodigious output. He died on May 22 or 23, 1857, in Sceaux, France, having promised the academy that in his next paper he would explain the meaning of his previous paper at greater length.

Cavendish, Henry

(1731–1810)
English
Physicist, Chemist

Henry Cavendish was a wealthy aristocrat who could afford to be odd. His eccentricity extended to his scientific experimentation in that he did not publish much of his own work, which, it was later discovered, anticipated future scientific developments and discoveries. The Cavendish experiment, which measured the density and mass of Earth, was well publicized and earned him a reputation as an important scientist. He also developed the scientific understanding of the composition of air, the composition of water, the nature and properties of hydrogen, and the properties of electricity, to name a few. His work sometimes confused more than it clarified, though, as he confounded his readers by referring to his own unpublished results in published papers. After his death, scientists such as JAMES CLERK MAXWELL delved into his papers to make known and available his prodigious output.

Cavendish was born on October 10, 1731, in Nice, France. His parents' marriage joined two aristocratic families, as both his grandfathers were dukes. His father, Lord Charles Cavendish, was the fifth son of the second duke of Devonshire, and his mother, Lady Anne Grey, was the fourth daughter of the duke of Kent. She died in 1733 of complications that occurred during the birth of Cavendish's brother Frederick. Cavendish never continued the lineage, however, as he was an infamous misogynist, avoiding contact with women at all costs.

Cavendish matriculated at the Hackney Seminary in 1742. Between 1749 and 1753 he studied at Peterhouse

The Cavendish experiment, which measured the density and mass of the Earth, earned Henry Cavendish a reputation as an important scientist in the late 18th century. *Smith Collection, Rare Book & Manuscript Library, University of Pennsylvania.*

College of the University of Cambridge, but he took no degree, because he would have been required to make a statement of adherence to the Church of England. Cavendish did not necessarily require a degree, however, as he inherited a fortune from his uncle, securing his independence. He devoted much of his wealth to science, acquiring an extensive scientific library and a laboratory filled with scientific apparatus, both of which he made available to other scientists and the general public.

Despite the fact that Cavendish kept many of his important discoveries to himself, he did publish many important works. In 1766 he published "Three Papers Containing Experiments on Factitious Airs," in which he called hydrogen "inflammable air" and carbon dioxide "fixed air," distinguishing these gases as separate from common air. In 1784 he published "Experiments on Air," describing experiments wherein he exploded hydrogen and air, resulting in water with no apparent weight loss. In 1798 he used torsion balance to calculate the mean density of the Earth, a method that came to be known as the Cavendish experiment. The preponderance of Cavendish's work on heat and electricity was not published until 1879, when Maxwell collected an edition of his unpublished experiments and papers.

In 1871 the seventh duke of Devonshire endowed the Cavendish Laboratory, ensuring that Cavendish's legacy would continue well after his death. Cavendish received recognition in his own lifetime as well, having been elected to the Royal Society in 1760 and named one of only eight foreign associates of the Institute of France in 1803. Cavendish died on February 24, 1810, in London, England.

Celsius, Anders

(1701–1744)
Swedish
Astronomer

Anders Celsius made significant contributions to the field of astronomy, but his lasting significance stemmed from his innovation of the thermometric scale that bore his name. This scale was adopted into the metric system of measurement, as it was based on 100 gradations.

Celsius was born on November 27, 1701, in Uppsala, Sweden. His father was a professor of astronomy at the University of Uppsala, where Celsius's uncle and grandfather had held positions. Celsius attended the University of Uppsala himself, studying astronomy, mathematics, and experimental physics. After completing his course of study there, Celsius was appointed secretary of the Uppsala Scientific Society.

After graduation from the University of Uppsala, Celsius remained at the university as a professor of mathematics. In 1730 the university appointed him to the position of professor of astronomy, and Celsius followed in his father's footsteps. That year he published one of his first papers, "A Dissertation on a New Method of Determining the Distance of the Sun from the Earth." In 1732 he embarked on a tour of Europe to visit the centers of scientific inquiry—Berlin, Nuremberg, Italy, Paris, and London.

While in Nuremberg in 1733 Celsius published his collection of 316 observations made between 1716 and 1732 by him and others of the phenomenon of aurora borealis, otherwise known as northern lights. At the end of this tour in 1736 Celsius took part in an expedition to Lapland led by Maupertuis of the Paris Academy of Sciences to measure the arc of the meridian there, in Tornea, Sweden. The guiding principle of this expedition was the goal of verifying Sir ISAAC NEWTON's theory that the Earth flattened at the poles, thereby disproving the opposing Cartesian view of the Earth. Similarly in 1738 Celsius published his "Disquisition on Observations Made in France for Determining the Shape of the Earth," a broadside that argued against the views of Jacques Cassini.

In 1740 Celsius oversaw the construction of the Uppsala observatory, and two years later he supervised the move into the new facility. Celsius subsequently served as the director of the observatory. In 1742 he presented a paper before the Swedish Academy of Sciences wherein he laid out his proposal for his new thermometer, based on a scale of 100 intervals or degrees. This scale became known as the Celsius scale as well as the centigrade scale, because of its 100 gradations.

Celsius died on April 25, 1744, in Uppsala. Five years after his death his colleagues at the Uppsala Observatory inverted the Celsius scale so that 0 degree corresponded to the freezing point of water and 100 degrees corresponded to the boiling point. The innovation of the Celsius scale was a very important development in science; it standardized temperature recordings and provided a base-10 scale, thus representing a simplification over the existing Fahrenheit scale, which ran from the freezing point of water at 32 degrees to the boiling point of water at 212 degrees.

Chadwick, Sir James

(1891–1974)
English
Physicist

Sir James Chadwick, in collaboration with his former mentor and fellow nuclear physicist ERNEST RUTHERFORD, hypothesized the existence of neutrally charged ions orbiting the nucleus of atoms, but only after 12 years of searching did Chadwick make the discovery that confirmed the existence of the neutron. This discovery opened the door for new applications of nuclear power, and with the approach of World War II efforts focused on unleashing that immense power through weaponry. Though Chadwick participated in these efforts to devise atomic bombs, he later conceded that this use of his discovery could not justify its negative consequences. In this sense Chadwick's role as a diplomat and spokesperson proved to be as important as his role as a pure scientist.

Chadwick was born on October 20, 1891, in Bollington, near Manchester, England. His parents were Ann Mary Knowles and John Joseph Chadwick. His father owned a laundry business in Manchester. In 1925 Chadwick married Aileen Stewart-Brown, and together the couple had twin daughters.

At the age of 16 Chadwick entered the University of Manchester on a scholarship. He pursued physics as a result of incorrect paperwork that he was too unassuming to correct—he had intended to study mathematics. He graduated with first honors in 1911 and continued at the university for his master's degree in 1913. He then studied at the Technische Hochschule in Berlin under Hans Geiger, but he was interned for the duration of World War I, during which time he carried out rudimentary experiments to keep his mind occupied. He returned to England to earn his Ph.D. in 1921 from the University of Cambridge.

Rutherford retained Chadwick at Cambridge in the Cavendish Laboratory as a fellow of Gonville and Caius College and later as an assistant director of research. Together they studied the transmutation of elements by bombarding them with alpha particles. Their results always revealed an inconsistency between the atomic number of the element and its atomic weight. The existence of a neutrally charged particle with a mass equal to a proton's would solve the problem, so Chadwick and Rutherford searched for years with no luck. In 1930 WALTHER WILHELM GEORG BOTHE and Hans Becker found unusual radiation emissions, thought to be gamma rays, from beryllium bombarded with alpha particles. Chadwick combined this information with an experiment conducted in 1922 by FRÉDÉRIC JOLIOT-CURIE and IRÈNE JOLIOT-CURIE in which beryllium knocked off hydrogen protons from the absorbing material, paraffin. Chadwick realized that a neutron would have the necessary mass to displace the proton, so he set out to disprove the gamma ray theory and then prove the existence of neutrons. He

James Chadwick, whose work confirmed the existence of the neutron, opening the door to new applications of nuclear power. *AIP Emilio Segrè Visual Archives, William G. Myers Collection.*

published his results and explanation in the journal *Nature* with an article modestly titled "Possible Existence of a Neutron."

Chadwick acted as the spokesperson for the Maud Committee of Britain's Ministry of Aircraft Production, the group attempting to build an atomic bomb. Chadwick then proceeded to the Manhattan Project, and he later negotiated between the United States and Britain over the rights to joint stockpiles of uranium after the war. Chadwick first supported British self-sufficiency with nuclear capabilities, but he later realized the futility of nuclear weapons and worked for disarmament.

Chadwick won the 1935 Nobel Prize in physics for his discovery of the neutron. He also won the Medal of Merit from the U.S. government in 1946 and the Copley Medal from the Royal Society in 1950. He died on July 24, 1974, in Cambridge.

Chain, Sir Ernst Boris

(1906–1979)
German/British
Biochemist

Sir Ernst Boris Chain joined the pathologist HOWARD WALTER FLOREY to isolate, purify, and perform the first clinical trials on penicillin, which had been discovered by Sir ALEXANDER FLEMING in 1928. The Swedish Academy of Sciences awarded the trio of Chain, Florey, and Fleming the 1945 Nobel Prize in physiology or medicine for their work with penicillin. Chain continued to advance the use of penicillin, promoting more efficient means of producing mass quantities and studying related problems, such as bacterial development of immunity to the antibiotic.

Chain was born on June 19, 1906, in Berlin, Germany. His father, Michael Chain, a Russian immigrant who became a chemical engineer, died in 1919. His mother, Margarete Eisner Chain, then supported the family by transforming their home into a guest house. Chain's mother and sister died in Nazi concentration camps. In 1948 Chain married Anne Beloff, a fellow biochemist, and the couple had three children.

Chain graduated in 1930 from Friedrich-Wilhelm University with degrees in chemistry and physiology. He proceeded to work at the Charité Hospital and the Kaiser Wilhelm Institute for Physical Chemistry and Electrochemistry in Berlin from 1930 to 1933, when Hitler came to power. Fearing not for his life, as he believed Hitler would eventually be deposed, but for his professional advancement, Chain immigrated to England. He worked under Sir Frederick G. Hopkins at the University of Cambridge from 1933 until 1935, when Hopkins recommended Chain to Florey, who had just become head of the Sir William Dunn School of Pathology at the University of Oxford.

In collaboration with Norman Heatley, Florey and Chain traced Fleming's work on antibodies back from his discovery of lysozyme to his discovery of penicillin in *Penicillium notatum*, though Fleming had not been able to identify the active agent. With the assistance of a Rockefeller Foundation grant, Chain, Florey, and Heatley began searching in 1938 for this key to penicillin's efficacy, concentrating large amounts of the mold into tiny amounts of penicillin powder—125 gallons of mold broth was required to produce one tablet of penicillin. In 1941 they finally produced enough of the isolated penicillin to conduct an eight-patient trial that yielded promising results. Chain spent the years during and after World War II advocating the unrestricted use of penicillin.

Between 1948 and 1961 Chain worked as the director of the International Research Center for Chemical Microbiology at the Superior Institute of Health in Rome, which gained an excellent reputation during his tenure. Chain then returned to England to become a professor of biochemistry working at the Wolfson Laboratories of the Imperial College of Science and Technology at the University of London from 1961 to 1973, when he retired to emeritus status, though he continued on as a senior research fellow until 1976. In the 1960s Chain discovered penicillinase, an enzyme produced by some bacteria that destroyed penicillin. Throughout his career Chain also studied snake venom, insulin, and tumor metabolism.

Besides the Nobel Prize, Chain won the 1946 Berzelius Medal and the 1954 Paul Ehrlich Centenary Prize. In 1969 Queen Elizabeth II knighted him. He died of heart failure on August 12, 1979, in Ireland.

Chandrasekhar, Subrahmanyan

(1910–1995)
Indian/American
Astrophysicist

Chandrasekhar, or Chandra as he referred to himself, startled astrophysicists in the 1930s with his theory on the development of stars, in which he suggested other evolutionary possibilities besides white dwarfs, such as neutron stars and black holes. Resistance to his revolutionary theory within the field delayed its acceptance by two decades and Chandra's receipt of the Nobel Prize in physics by a half-century; he finally received it in 1983 in conjunction with WILLIAM ALFRED FOWLER. Chandra defined the stars that would not devolve into white dwarfs as those larger than 1.44 times the size of the Sun, a figure also known as the Chandrasekhar limit.

Chandra was born on October 19, 1910, in Lahore, India (now Pakistan). His father was C. Subrahmanyan Ayyar and his mother was Sutakakshmi Balakrishnan. Chandra was the first son in a large family, with two older sisters, four younger sisters, and three younger brothers. Chandra became interested in science early, following the precedent set by his uncle, Sir CHANDRASEKHARA VENKATA RAMAN, winner of the 1930 Nobel Prize for physics. In 1937 Chandra wed Lalitha Doraiswamy, in a rare love marriage of two members of the Brahman caste.

Chandra attended the Presidency College in Madras as an honors physics student, though he also indulged his passion for pure mathematics with the enthusiastic consent of his professors. He earned his M.A. in 1930 and proceeded to Trinity College at the University of Cambridge on a scholarship from the Indian government. Over the next several years, Chandra studied internationally with scientific luminaries: In 1931 he studied at the Institut für Theoretische Physik in Göttingen, Germany, with MAX BORN, and in 1932 he studied in Copenhagen, Denmark, under NIELS HENRIK DAVID BOHR before returning to Trinity, where he earned a fellowship in 1934.

In 1935 he presented a paper to the Royal Astronomical Society outlining his theory on white dwarfs, supernovas, neutron stars, and black holes. The incredulous reception by the prominent astronomer and physicist Sir Arthur Stanley Eddington set the tone for the general rejection of this theory, though other prominent physicists such as Bohr, WOLFGANG PAULI, and PAUL ADRIEN MAURICE DIRAC accepted it. Instead of drawing out that debate, Chandra published his theory in the 1939 text *An Introduction to the Study of Stellar Structure*.

Between 1935 and 1936 Chandra served as a visiting lecturer in cosmic physics at Harvard. In 1937 he commenced a long relationship with the University of Chicago's Yerkes Observatory at Williams Bay in Wisconsin, interrupted only by a brief stint at Aberdeen Proving Grounds, where he performed research for the military during World War II. Chandra was promoted to assistant professor of astrophysics in 1938, associate professor in 1942, professor in 1943, and Distinguished Service Professor in 1946. In 1952 Chandra assumed the title of Morton D. Hull Distinguished Service Professor of Astrophysics in astronomy and physics. In 1953 he became a U.S. citizen.

Chandra served as the editor in chief of *Astrophysical Journal* between 1952 and 1971. He also published many important texts, including *Principles of Stellar Dynamics* (1942), *Radiative Transfer* (1950), *Hydrodynamic and Hydromagnetic Stability* (1961), *Truth and Beauty: Aesthetics and Motivations in Science* (1987), and *The Mathematical Theory of Black Holes* (1983). Chandra's research included work on the system of energy transfer within stars, stellar evolution, stellar structure, and theories of planetary and stellar atmospheres. Besides the Nobel Prize, Chandra received the 1953 Gold Medal of the Royal Astronomical Society and the 1962 Royal Medal of the Royal Society. He died on August 21, 1995, in Chicago, Illinois.

Chang, Min-Chueh

(1908–1991)
Chinese/American
Biologist

Min-Chueh Chang's development of oral contraception with Gregory Goodwin Pincus and John Rock transformed human society by divorcing sex from childbearing more effectively than previous forms of contraception. Chang also worked on in vitro fertilization, which led to the advent of test-tube babies, and on embryo transfer, which allowed for more control over farm-animal reproduction.

Chang was born on October 10, 1908, in Taiyuan, China. His parents were Gen Shu Chang and Shih Laing Chang. Chang married Isabelle C. Chin on May 28, 1948, and the couple had three children.

Chang's postsecondary education began at Tsing Hua University in Beijing, where he graduated in 1933 with a B.S. degree in animal psychology. He then proceeded to Cambridge University in England to work on animal breeding. He earned his Ph.D. in 1941 and stayed on at Cambridge's School of Agriculture for postdoctoral work in John Hammond's research group from 1941 through 1945.

In 1945 Chang immigrated to the United States to work as a research associate at the Worcester Foundation for Experimental Biology in Shrewsbury, Massachusetts, where he remained for the rest of his career. He became a U.S. citizen in 1952. In 1954 the foundation appointed him as senior and principal scientist. Chang had joined the faculty of Boston University in 1951, and in 1961, the university promoted him to the position of professor in reproductive biology.

Chang and Pincus commenced their research on oral contraception for humans in 1951, focusing on the hormone progesterone, which regulates menstruation. They realized that boosting progesterone levels in the blood could prevent ovulation. After experimenting with many different formulations, they arrived at a combination of three steroid compounds, including both progesterone and estrogen derived from the wild Mexican yam. They then called upon the expertise of John Rock of the Rock Reproduction Clinic in Brookline, Massachusetts, to collaborate with them on further research into the issue. Their attempt to eliminate estrogen from the mix failed, but after reintroduction of estrogen, human trials yielded a 99 percent effectiveness rate. The U.S. Food and Drug Administration approved the pill for mainstream use in 1960.

Chang also collaborated with Cyril Adams in the 1950s on the process of embryo transfer, which was ultimately used for fertilizing farm animals. This line of research also led to Chang's innovation of in vitro fertilization, or test-tube fertilization followed by the implantation of the fertile egg into the uterus of an infertile woman. Robert Edwards and Patrick Steptoe performed the first successful implantation of a test-tube baby in 1978.

Chang won multiple awards for his pioneering work. In 1950 he received the Ortho Award, followed in 1961 by the Ortho Medal. In 1954 he won the Lasker Foundation Award. The British Society for the Study of Fertility awarded him its Marshall Medal in 1971, and in 1983 he received the Pioneer Award from the International Embryo Transfer Society. In 1990 he was elected to the U.S. National Academy of the Sciences. Chang died of heart failure on June 5, 1991, in Worcester, Massachusetts.

Charpak, Georges

(1924–)
Polish/French
Physicist

Georges Charpak's invention of the multiwire proportional chamber particle detector allowed many other scientists to research particles much more efficiently than they had been able to with previous devices such as the bubble chamber and the cloud chamber, which relied on photographic tracking instead of computerized tracking. Both SAMUEL CHAO CHUNG TING, who won the Nobel Prize in physics in 1976 for his discovery of the J/psi meson, and Carlo Rubbia, who won the Nobel Prize in 1984 for his discovery of the W and Z particles, used Charpak's multiwire chamber in their experimentation. The Swedish Academy of Sciences finally recognized Charpak with a Nobel Prize in physics in 1992, almost a quarter-century after he first constructed the device.

Charpak was born on August 1, 1924, in Dabrovica, Poland. He migrated to France in 1929 with his father, Maurice Charpak, and mother, Anna Szapiro Charpak. Charpak joined the resistance during World War II, but the Vichy government accused him of terrorism and imprisoned him in 1943, sending him to Dachau, where he remained until the concentration camp was liberated. He became a French citizen in 1946 and married Dominique Vidal in 1953. The couple had two sons and one daughter.

Upon his return to France Charpak completed his degree in civil engineering at the École des Mines. In 1948 he commenced graduate studies in nuclear physics at the College de France in Paris, working in the lab of FRÉDÉRIC JOLIOT-CURIE and earning his Ph.D. in 1955. He worked on nuclear physics at the Centre National de la Recherche Scientifique in France before being recruited in 1959 by the European Organization for Nuclear Research (CERN) in Geneva, Switzerland, where he remained for the rest of his career, with the exception of an appointment as the Joliot-Curie professor at the School of Advanced Studies in Physics and Chemistry in Paris, which he accepted in 1984.

Charpak built his first multiwire proportional tracking chamber in 1968. Previous particle chambers relied on photographic tracking of particles, which handled only a fraction of the output of the new particle accelerators. Charpak designed the multiwire chamber with charged wires separated by 1.2 millimeters and then layered in a gas-filled container. The central wires were charged positively, and the outer wires were charged negatively, so that when particles passed from one charge to another, this activity could be recorded by a computer. The device could handle 1 million nuclear events per second. This chamber spawned several similar designs, such as the drift chamber and the time projection chamber.

Charpak received numerous honors besides the Nobel Prize in his career. In 1985 he was voted a member of the French Academy of Science. In 1989 he received the High Energy and Particle Physics Prize from the European Physical Society. Later in his career Charpak devoted his attention to medical and aerospace research. He also founded the SOS committee at CERN to help those imprisoned, as he was, by repressive governments. His work aided, among others, Andrei Sakharov, whose civil rights had been violated by the government of the former Soviet Union.

Cherenkov, Pavel Alekseyevich

(1904–1990)
Russian
Physicist

In 1934 Pavel Cherenkov observed a faint blue light emanating from water that was absorbing radiation. At first he presumed this to be the well-established phenomenon of luminescence, but on further investigation he discovered by passing gamma rays through different liquids that the light was not a function of the medium, as with luminescence, but rather a function of the radiation. In 1937 ILYA MIKHAILOVICH FRANK and IGOR EVGENIEVICH TAMM ascertained, by developing and testing a mathematical theory as proof, that the light was emitted by radioactive particles moving faster than the speed of light through liquid. The trio shared the 1958 Nobel Prize in physics for their identification and verification of what was called Cherenkov radiation. Understanding of this phenomenon led to the development of the Cherenkov counter, which measures high-energy particles and gained wide usage in physics experiments measuring Cherenkov radiation.

Cherenkov was born on July 28, 1904, in Novaya Chigla in the Voronezh region of Russia. His parents were Russian peasants. He attended Voronezh University and graduated in 1928 with a degree in physics and mathematics. After teaching high school for two years, he commenced graduate studies in physics at the Institute of Physics and Mathematics under Sergei I. Vavilov. The institute relocated to Moscow and was renamed the P. N. Lebedev Institute of Physics. Cherenkov earned his doctorate there in 1940 and remained at the institute for the rest of his life. In 1953 the institute named him a professor of experimental physics, then promoted him to the position of senior scientific officer. He later headed its department of high-energy physics.

Cherenkov gauged the light output from the radiated liquids with unprotected eyes, so he increased his sensitivity to light by blinding his eyes from light for an hour before commencing the experiments, which he performed repeatedly. Using elementary laboratory equipment, Cherenkov exposed weak gamma rays to different liquids and observed the mysterious blue light in every trial, thus verifying the radiation and not the liquids as the source of the light. Furthermore Cherenkov differentiated this radiation from luminescence, which dimmed with additives and was polarized. Frank and Tamm joined Cherenkov in 1937 to explain the source of the blue light, which they hypothesized was the visual equivalent of a sonic boom in sound. In other words, light travels fastest in a vacuum, but in a liquid, light travels slower, and this radiation proved fast enough to overtake light.

The immediate response to their findings was not promising, but with the escalation of World War II and the cold war in its wake interest in nuclear physics grew exponentially. Cherenkov counters were commonly used in satellites and balloons studying cosmic rays and in research on radiation under the polar ice caps. In 1946 the trio of Cherenkov, Frank, and Tamm won the State Prize from the Soviet government before winning the Nobel Prize. In 1970 Cherenkov himself became a member of the U.S.S.R. Academy of Sciences. He died on June 6, 1990.

Cho, Alfred Y.

(1937–)
Chinese/American
Electrical Engineer

During his research in the field of solid-state electronics Alfred Cho utilized what he called his "Oriental patience" to do the painstakingly slow and methodical work behind the technology that drives Western culture. His work affected such everyday inventions as the microwave and the compact disk player. Specifically Cho developed a semiconductor preparation process, labeled *molecular beam epitaxy,* whereby engineers can create synthetic crystalline structures layer by atomic layer.

Cho was born on July 10, 1937, in Beijing, China. His parents were Mildred Chen and Edward I-Lai Cho, a professor of economics. Cho attended Pui Ching high school in Hong Kong until he immigrated to the United States in 1955. That year he commenced studying general science at Oklahoma Baptist University. However, Cho wished to study electrical engineering, so he transferred to the University of Illinois in 1956. There he earned his bachelor's degree in 1960 and his master's degree one year later.

In 1961 Cho moved to Burlington, Massachusetts, to become a research physicist at the Ion Physics Corporation. The next year in nearby Boston he became a naturalized citizen of the United States. That year he moved on to TRW's Space Technology Laboratory in Redondo Beach, California, where he worked until 1965, at which time he returned to the University of Illinois to pursue a doctorate in electrical engineering. Cho received his Ph.D. in 1968, and in June of that year he married Mona Lee Willoughby in Illinois. Together the couple had four children.

Also in 1968 Cho commenced a long-term relationship with AT&T when he joined the technical staff at the AT&T Bell Labs in Murray Hill, New Jersey. He was promoted to the position of department head in 1984, and three years later he was named the director of the Materials Processing Research Lab. In 1990 he made a parallel move within the company to become the director of semiconductor research. Cho also held a position as an adjunct professor at the University of Illinois.

Cho has written over 400 articles about solid-state electronics, most concerning his specialty of molecular beam epitaxy in semiconductors. His periodical publications include "Epitaxy by Periodic Annealing" in *Surface Science* in 1969, "Film Deposition by Molecular Beam Techniques" in *Journal of Vacuum Science and Technology* in 1971, "Growth of Periodic Structures by the Molecular Beam Method" in *Applied Physics Letters* in 1971, and "Growth of III-V Semiconductors by Molecular Beam Epitaxy and Their Properties" in *Thin Solid Films* in 1983.

President Bill Clinton presented Cho with the 1993 National Medal of Science, and the next year he received a Medal of Honor from the Institute of Electrical and Electronics Engineers. In addition Cho has received awards from the Electrochemical Society, the American Physical Society, and the Chinese Institute of Engineers. Cho has served on the board of directors of Instruments SA in Edison, New Jersey, and he holds over 40 patents related to molecular beam epitaxy.

Chu, Paul Ching-Wu

(1941–)
Chinese/American
Physicist

Paul Chu is considered one of the most important superconductivity scientists for his 1987 discovery of a combination of materials that could conduct electricity at temperatures high enough to allow for cheap, efficient energy production. In 1911 Heike Kamerlingh Onnes had discovered superconductivity, or the phenomenon whereby metals lose all electrical resistance as they approach the temperature 0 Kelvin, creating an extremely efficient electrical system. The use of rare and expensive liquid helium as a medium allowed scientists to reach slightly higher temperatures of superconductivity, but the threshold temperature was 77.4 Kelvin, or the temperature at which nitrogen liquefies, since nitrogen is plentiful and thus cheap. Chu's combination of yttrium, barium, and copper proved capable of sustaining superconductivity at 93 Kelvin, well above the threshold temperature for nitrogen, thus ushering in a new age of electrical energy production.

Chu was born on December 2, 1941, in Hunan Province of China. His parents were members of the Nationalist Party and fled to Taiwan in 1949. Chu received his baccalaureate of science in 1962 from the Cheng Kung University. In 1965 he received his master of science from Fordham University, in Bronx, New York. He then studied physics at the University of California at San Diego as a research assistant to Bernard T. Matthias, the acknowledged grandfather of superconductivity. In 1968 Chu received his Ph.D. in physics, and the same year he married May P. Chern; the couple had two children, Claire and Albert.

In 1970 Cleveland State University hired Chu as an assistant professor of physics, promoting him to the position of professor of physics by 1979, when he departed. That year the University of Houston offered him a professorship of physics, and he soon became the director of the Texas Center for Superconductivity. Besides these university appointments, Chu worked for Bell Labs in Murray Hill, New Jersey; the National Aeronautics and Space Administration's (NASA) Marshall Space Flight Center in Huntsville, Alabama; and the Los Alamos Science Lab in Los Alamos, New Mexico.

In 1986 Chu came across a report in *Zeitschrift für Physik* by K. Alex Müller and Johannes Georg Bednorz on a ceramic material composed of barium, lanthanum, and copper oxide that was able to superconduct at temperatures as high as 35 Kelvin. Chu followed this lead in the attempt to break the threshold of 77.4 Kelvin, which would allow nitrogen to be used as the medium for superconductivity instead of helium.

In the March 2, 1987, edition of *Physical Review Letters*, Chu published his discovery of a mixture of rare earth oxide ceramics that could superconduct electricity at temperatures as high as 93 Kelvin. Chu patented what he called "compound 1-2-3," or the mixture of yttrium, barium, and copper, as the most efficient superconductor known to humans at the time. Chu's announcement of his discovery at the March 18, 1987, annual meeting of the American Physical Society in New York City was greeted with enthusiasm.

Chu received the 1988 National Medal of Science and the Comstock Award from the National Academy of Science for his work with superconductivity.

Clark, Eugenie

(1922–)
American
Marine Biologist

Eugenie Clark has done so much research on sharks that she has been called the "shark lady." Sharks, however, are just one form of ocean life she has studied. Clark was born in New York City on May 4, 1922. Her father, Charles, died a year after she was born, and her mother, Yumico, had to work to support herself and her daughter. When Yumico worked on Saturdays, she left Eugenie at the New York Aquarium. "I brought my face as close as possible to the glass [of the largest aquarium tank] and pretended I was walking on the bottom of the sea," Eugenie Clark recalls.

Clark majored in zoology at Hunter College in New York City and graduated in 1942. She then took graduate science courses at New York University at night while working at a plastics factory in the daytime. She finished her master's degree in 1949 and her Ph.D. in 1950.

In June 1949 the U.S. Navy and the Pacific Science Board awarded Clark a scholarship to investigate poisonous fish in the South Pacific. She then went to the Red Sea, between Africa and the Arabian Peninsula, to collect other poisonous fish. She married Ilias Papaconstantinou, a Greek-born physician whom she had met in New York, in June 1951, and they spent their honeymoon diving in the Red Sea. They later had four children.

Clark wrote a book about her experiences in the Pacific and the Red Sea called *Lady with a Spear.* Published in 1953, it attracted a wide audience, including a wealthy couple who asked Clark to start a marine biology laboratory on land they owned in Florida. The facility, which opened in 1955, was called the Cape Haze Marine Laboratory.

One of Clark's chief projects during her years at Cape Haze was a study of the intelligence of sharks. She found that they were much smarter than biologists had thought. She describes them as "magnificent and misunderstood."

Clark and her husband divorced in 1967. She left Florida and joined the University of Maryland the following year. Since then she has done underwater research all over the world. She has studied fish that can change sex in 10 seconds, a flatfish that exudes a poison that makes even a big shark back off in disgust, and sharks that "sleep" in underwater caves while smaller fish pick parasites off their skins. She has also worked to preserve the ocean habitat she loves. She was instrumental in helping to make Ras Muhammad, a favorite diving site in Egypt, the country's first national park. She retired from the University of Maryland in 1992 but remained active there as a senior research scientist and professor emerita. She was also still diving. "I plan to keep on diving and researching and conserving until I'm at least ninety years old," she said.

In the late 1990s Clark was involved in two research projects. One involved studying the reproductive behavior, territoriality, and ecology of tropical sand-dwelling fish in Papua New Guinea, the Caribbean, and the Red Sea. Her other work on sharks dates back several decades and involves studying shark behavior in the deep sea from submersibles at depths of 1,000 to 12,000 feet.

Clark has played a vital role in 24 television specials about marine life, including "The Sharks" (1982), a National Geographic special extremely popular with viewers. She has received numerous awards for her work in marine biology, conservation, and writing, including the Society of Women Geographers' Gold Medal in 1975, the Governor of Red Sea Egypt medal in 1988, election to the Maryland Women's Hall of Fame in 1989, and the National Geographic Society's Franklin Burr Award in 1993.

Cobb, Jewel Plummer

(1924–)
American
Cancer Researcher

Jewel Plummer Cobb has excelled not only as a scientist, conducting research and teaching, but also as a college administrator. Born on January 17, 1924, in Chicago, she grew up in a family who discussed "science things at the dinner table." Her father, Frank, was a physician. Carriebel, her mother, taught physical education. Jewel became interested in biology in high school, when she looked through a microscope for the first time. "It was really awe inspiring," she said.

Plummer earned her B.A. at Talladega College in Alabama in 1944 and her master's degree (1947) and Ph.D. (1950) at New York University. She then joined the Cancer Research Foundation of Harlem Hospital in New York City, where she worked under another African-American woman scientist, Jane Wright. Wright and Plummer tried to develop a way to test anticancer drugs on cells from a patient's tumor in the laboratory to determine the best dose to give to the patient. Plummer did the lab work, studying cells under the microscope and making time-lapse films to show how they changed after drugs were added. Although the project did not succeed, the researchers learned valuable information about how the drugs affected cancer cells.

Plummer left full-time research in 1952 and began teaching at the University of Illinois. In 1954 she married Roy Cobb, an insurance salesman, and they had a son, Jonathan. She moved to Sarah Lawrence College in 1960. There, in addition to teaching, she did research on skin cells, both normal and cancerous, that contain the dark pigment melanin.

Jewel Plummer Cobb, who is renowned as both a cancer researcher and college administrator. *Courtesy Jewel Plummer Cobb.*

The Cobbs were divorced in 1967, leaving Jewel Cobb with a young son to raise alone. In 1969 she became dean of Connecticut College and thus began a third career, that of college administrator. She eventually had to give up research because of the time it demanded.

Cobb became dean of Douglass College, the women's college of Rutgers University, in 1976. Then in 1981 she became president of the California State University campus at Fullerton. No other African-American woman had headed such a large public university on the West Coast. While there she established two new schools—one of communication and one of engineering and computer science—as well as the campus's first residence hall.

Although Cobb retired from Cal State–Fullerton in 1990, she remains a trustee professor of the state university system. From her office in Los Angeles she oversees a center that works to improve science education for minority students. She once told an interviewer that she wanted to be remembered as "a black woman who cared very much about what happens to young folks."

Cohen, Stanley H.

(1922–)
American
Biochemist

Stanley Cohen applied his biochemical expertise to the experiments of the Italian-American neurobiologist RITA LEVI-MONTALCINI to discover and identify both nerve growth factor (NGF) and epidermal growth factor (EGF), substances produced by the body to stimulate the development of nerve and skin tissue. The pair won the 1986 Nobel Prize in physiology or medicine for this pioneering work.

Cohen was born on November 17, 1922, in Brooklyn, New York. His parents were Fannie and Louis Cohen, Russian immigrants both. His father was a tailor. Cohen suffered from polio contracted as a child. He developed a permanent limp at this time but eventually overcame the disease.

Louis Cohen saved enough money to send his four children to high-quality colleges. Stanley attended Brooklyn College; in 1943 he earned a B.A. in chemistry and zoology. He received a scholarship to pursue his master's degree at Oberlin College, where he received a master's degree in zoology in 1945. Cohen then attended the University of Michigan on a teaching fellowship; he received his Ph.D. in 1948 in biochemistry.

Between 1948 and 1952 Cohen worked for the University of Colorado School of Medicine in Denver. In 1952 he was appointed as the American Cancer Society post-doctoral fellow in the radiology department at Washington University in St. Louis, Missouri. The next year he became a research assistant in the zoologist Victor Hamburger's laboratory, where he worked until 1959 alongside Levi-Montalcini. That year he departed for Vanderbilt University in Nashville, Tennessee, where he served as an assistant professor of biochemistry. In 1962 the university promoted him to associate professor status, and in 1967 he attained full professorship.

By the time Cohen arrived in St. Louis, Levi-Montalcini had just proved the existence of NGF by injecting tumor cells of male mice into chicken embryos. By 1956 Cohen had isolated NGF, but he found it a difficult substance to work with. A fellow biochemist, ARTHUR KORNBERG, suggested adding snake venom to the extract; the addition not only made the substance easier to handle but also stimulated nerve growth, accelerating the experimental process. Coincidentally Cohen found that the salivary glands of male mice, which were related to the venom sacs of snakes, harbored copious amounts of NGF. It was not until after Cohen's funding ran out at Hamburger's lab that a full analysis and identification of the amino acid chains of NGF were completed by other researchers at Washington University in 1970. Cohen himself continued his work on growth factors at Vanderbilt, discovering EGF and fully analyzing and identifying its amino acid chains by 1972. EGF proved important for burn healing and skin grafting.

That Cohen and Levi-Montalcini's work was largely unrecognized for decades was both a blessing and a curse. They escaped the limelight that can distract scientific concentration, but they lacked acknowledgment of the importance of their work. This recognition did finally come in 1982, when Cohen won the Alfred P. Sloan Award. Full recognition followed the awarding of the Nobel Prize in 1986, when Cohen also won the National Medal of Science and the Albert Lasker Award. Cohen also received institutional recognition that year, when Vanderbilt named him a distinguished professor.

Conybeare, William Daniel

(1787–1857)
English
Geologist, Paleontologist

William Daniel Conybeare was one of the first scientists to examine cross sections of the Earth to discover the geologic progression of time, as each layer signifies a subsequent historical period. Throughout England and Wales, Conybeare studied the stratigraphic characteristics of the Carboniferous system, so named after the copious amount of carbon deposited in these layers some 280 million to 345 million years ago. Conybeare's overall geologic perspective combined progressionism with catastrophism.

Conybeare's career synthesized his father's legacy in the clergy with his own interest in science. Conybeare was born on June 7, 1787, at St. Botolph, in West Sussex, England. His father, the Reverend William Conybeare, served as the rector of St. Botolph's, exerting a religious influence over his youngest son. Conybeare attended Westminster School, then Christ Church at Oxford University. More important in his intellectual development was his joining of the Geological Society of London in 1811.

In 1814 Conybeare married and also accepted the curacy in Suffolk. This year also marked his first scientific publication, *On the Origin of a Remarkable Class of Organic Impressions Occurring in Nodules of Flint*. This convergence of events set the precedent for Conybeare's subsequent development. His clerical career traced a steady progression that was notable in its own right, but beneath this thrived a scientific life that distinguished him more than his official life in the clergy. In 1821 Conybeare published a tract on the ichthyosaurus, representing the first description of this animal. Three years later he reconstructed a plesiosaur from excavated remains, and he claimed that this skeleton represented the link between the ichthyosaurus and modern crocodiles.

The year 1822 marked another convergence in Conybeare's career, when he accepted the position of rector of Sully in Glamorganshire and published his most influential work, *Outlines of the Geology of England and Wales,* which he composed in collaboration with William Phillips. Conybeare and Phillips dated the geologic layers by identifying fossils embedded in each layer, which allowed them to pinpoint the period when that layer settled. Conybeare planned to follow up on this work with a second study in collaboration with Adam Sedgwick, which never materialized. Conybeare later collaborated with William Buckland on a geologic study of the coal fields surrounding the Bristol area. Conybeare adopted Buckland's view that natural disasters accounted for the disappearance of species, a view that countered Sir CHARLES LYELL's position. In 1829 Conybeare published *Hydrographical Basin of the Thames.*

Conybeare was elected a fellow of the Royal Society of London in 1832. Four years later he became the vicar of Axminster in Devon. In 1839 Oxford named him the Bampton Lecturer, and in 1845 he was appointed the dean of Llandaff in Wales. Conybeare died on August 12, 1857, in Itchen Stoke, in Llandaff, Wales. The scientific world remembers Conybeare for his clear, precise, and accurate descriptions of geological phenomena.

Cooper, Leon Neil

(1930–)
American
Physicist

Together with John Bardeen and JOHN ROBERT SCHRIEFFER, Leon Neil Cooper formulated the path-breaking BCS theory of superconductivity in 1957. Although the trio shared the 1972 Nobel Prize in physics for their work, it was Cooper who conceived a central tenet of the theory. In seeking to explain how electricity flows without resistance through a superconductor, Cooper postulated the existence of what came to be known as Cooper pairs of electrons. Whereas electrons normally repel one another, Cooper posited that at extremely low temperatures two electrons located among positive ions in a metal lattice would develop an attraction to one another and thus bind together. They would then not be affected by electrical resistance as they traveled through the lattice in the same direction. Cooper radically shifted direction after making this influential discovery. He cofounded the Center for Neural Sciences at Brown University and has devoted much of the rest of his career to understanding the human brain.

Born on February 28, 1930, in New York City, Cooper was the son of Irving and Anna Zola Cooper. During his senior year at the Bronx High School of Science, he won the Westinghouse competition for a research project on penicillin-resistant strains of bacteria. Upon graduating from high school, Cooper enrolled at Columbia University, where he earned his A.B. in 1951 and his Ph.D. in 1954. His thesis on the mu-mesonic atom earned him a position as a National Science Foundation Post-Doctoral Fellow at the Institute for Advanced Study in Princeton, New Jersey. He continued his research at Princeton until 1955.

Because of Cooper's growing reputation as a quantum theorist he was invited to join John Bardeen's team at the University of Illinois. From 1955 until 1957 he served as Bardeen's research assistant and also collaborated with John Robert Schrieffer. Together the trio attempted to account for the phenomenon of superconductivity (the fact that certain metals lose all resistance to the flow of electrical current when they are cooled to temperatures approaching absolute zero), which had first been reported in 1911 by the Dutch physicist Heike Kamerlingh Onnes. Cooper posited that because the electrons in superconductive metals are embedded in positive ions, they attract, rather than repel, each other, and form pairs. Cooper's concept of paired electrons—Cooper electrons—was a pillar of the BCS theory of superconductivity proposed by the trio of researchers. They published their results in the December 1, 1957, issue of *Physical Review,* and revolutionized the understanding of superconductivity.

His work at Illinois complete, Cooper joined the faculty of the physics department at Ohio State University in 1957; there he investigated liquid helium-3 and concluded that the substance was a superfluid. Despite his triumphs in the field of physics, Cooper radically changed direction in 1958 when he accepted a post at Brown University. After his appointment as a full professor in 1962 he continued to pursue theoretical physics and published an acclaimed textbook, *An Introduction to the Meaning and Structure of Physics,* in 1968, but he also diversified his areas of research and became increasingly involved in neurology. Cooper married Kaye Anne Allard in May 1969, and the couple had two daughters.

Since 1973 he has cochaired Brown's Center for Neural Sciences; he founded the Institute of Brain and Neural Systems in 1992. At the Center for Neural Sciences he brought together an interdisciplinary team of biologists, linguists, mathematicians, and physicists to study human memory and brain structure and function. As director of the Institute of Brain and Neural Systems he has sought to create intelligent systems for use in electronics and communications. For instance, Cooper helped invent software introduced in 1987 by International Business Machines (IBM) that can convert handwritten letters into typed characters.

Cooper's multifaceted research has been frequently acclaimed. In addition to the 1972 Nobel Prize he shared with Bardeen and Schrieffer, he and Schrieffer were jointly awarded the Comstock Prize from the National Academy of Sciences in 1968. Cooper also received the John Jay

Award from Columbia College in 1985. Cooper's multidisciplinary legacy is considerable. Not only did he increase understanding of the workings of superconductors (helping to increase the practical applications of superconductors), he has also played a role in the development of new cognitive theories.

Cooper has received many other awards, including the 1974 Award of Excellence from the Graduate Faculties Alumni of Columbia University, the 1977 Descartes Medal Academie de Paris from the Universite Rene Descartes, and the 1985 John Jay Award of Columbia College. Cooper's significant publications include *An Introduction to the Meaning and Structure of Physics* (1968), *Introduction to Methods of Optimization* (1970), and *Methods and Applications of Linear Programming* (1974).

Copernicus, Nicolaus

(1473–1543)
Polish
Astronomer

Nicolaus Copernicus helped to found modern astronomy and revolutionize science with his bold assertion that the Earth was not the stationary center of the universe but rather was in motion around a fixed Sun. This heliocentric hypothesis provided a much simpler explanation of astronomical phenomena than the prevailing geocentric theory, which failed to correlate accurately with observations. Copernicus's proposed system created its own inaccuracies because it postulated circular planetary orbits and did not allow for elliptical orbits. However, his theory itself carried enough credence to overturn not only scientific but also religious and social paradigms of the time. The fact that human beings did not occupy the central location in the universe carried far-reaching implications.

Copernicus was born on February 19, 1473, in Toruñ, Poland. His father, a merchant, died in 1483 and Copernicus's uncle, who later became the bishop of Varmia, raised him. In 1491 he entered the University of Kraków. In 1496 he proceeded to the University of Bologna and later to the University of Padua to study canon law, at his uncle's behest. In 1497 Copernicus's uncle plied his influence to secure a canon of the cathedral chapter of Frombork for his nephew. Though this amounted to blatant nepotism, it provided a comfortable income and required only modest administrative duties of Copernicus, freeing him to devote his time to astronomical observations and calculations. On May 31, 1503, Copernicus earned his doctorate in canon law from the University of Ferrara.

As early as March 9, 1497, Copernicus recorded his first astronomical observation, and on November 6, 1500, he recorded a lunar eclipse. He thus had commenced his astronomical career while studying for his career in canon law. It is believed that Copernicus devised his heliocentric theory throughout the first dozen years of the 16th century, but the first evidence of it was dated May 1, 1514, when the anonymous work *De hypothesibus motuum caelestium a se constitutis commentariolus* was cataloged by a Kraków professor. Copernicus distributed this pamphlet discreetly to friends and knowledgeable individuals, but he did not claim it in his own name because of the controversial stance the hypothesis took: In essence it overturned the astronomical views asserted by PTOLEMY in his text *Almagest* from the second century A.D.

Nicolaus Copernicus, whose assertion that the Earth is in motion around a fixed sun revolutionized science in the 16th century. *AIP Emilio Segrè Visual Archives.*

Copernicus delayed the actual publication of his theory until 1543, on the eve of his death. Reportedly he saw a copy of his masterwork, *On the Revolutions of the Celestial Spheres,* on his deathbed. Copernicus died on May 24, 1543, in Frauenberg, East Prussia (now Frombork, Poland). Acceptance of his theory was also delayed. Not until 1609 did GALILEO GALILEI confirm Copernicus's heliocentrism by observation of Jupiter's moons. That same year JOHANNES KEPLER asserted elliptical planetary orbits, refuting Copernicus's theory of circular orbits but allowing the accurate calculations that confirmed heliocentrism. In 1616 the Catholic Church banned Copernicus's book as inconsistent with the Bible and Catholic theology; the ban lasted until 1835. Three years later Friedrich Wilhelm Bessel refuted the last major stumbling block to Copernicus's theory. In Copernicus's time TYCHO BRAHE criticized the heliocentrism theory, arguing that the stars, which appear to be fixed, should actually appear to move if the Earth is in fact in constant motion. Copernicus countered that the stars were too far away for their motion to be observed. Bessel finally measured parallax of the stars in 1838, thus fully confirming Copernicus's theory of heliocentrism.

Cori, Gerty Theresa Radnitz

(1896–1957)
American
Biochemist

The bodies of living things constantly store, use, and recycle the energy that they get from food. Gerty Radnitz Cori and her husband, Carl Cori, worked out the steps in this energy cycle and discovered several chemicals and reactions involved in it. Gerty Cori also showed that certain inherited diseases are caused by the absence of key chemicals in this cycle. This work earned her a share of a Nobel Prize in 1947.

Gerty Theresa Radnitz was born on August 15, 1896, in Prague, then part of the empire of Austria-Hungary and later the capital of Czechoslovakia. The oldest of the three daughters of Otto Radnitz, a well-to-do businessperson and chemist who owned several beet sugar refineries, and his wife, Martha, Gerty enrolled in 1914 in the Carl Ferdinand Medical School in Prague, where she met Carl Cori. Joint work on a research project convinced them that they were ideal partners, and they married on August 5, 1920, two months after they earned their M.D. degrees. They had a son, Tom Carl, in 1936.

After a year of working separately in Vienna the Coris moved to the United States, where they both worked at the New York State Institute for the Study of Malignant Diseases (later the Roswell Park Memorial Institute) in Buffalo. The pair stayed in Buffalo for the next nine years, becoming American citizens in 1928. Gerty later said, "I believe the benefits of two civilizations, a European education followed by the freedom and opportunities of this country, have been essential to whatever contributions I have been able to make in science."

One project the Coris worked on concerned the way cancerous tumors use carbohydrates. Carbohydrates are sugars and starches, the chief foods that plants and animals break down to get energy. The Coris were interested in the way the healthy body uses carbohydrates, and through years of painstaking experiments they worked out the basic cycle of carbohydrate use in the bodies of mammals. They first described this cycle, which came to be called the Cori cycle, in 1929. The two forms of carbohydrate in the Cori cycle are glucose, a simple sugar, and glycogen, the "sugar maker," a complex carbohydrate made of hundreds of glucose molecules bonded together. Glucose is the form that the muscles break down to get energy. Glycogen is the form in which carbohydrate energy is stored.

The Cori cycle goes into action every time a human or animal exercises. When a lion runs, for instance, its muscles break down glycogen stored there to form glucose. They then break down most of the glucose into carbon dioxide and water, releasing energy in the process.

Although the institute in Buffalo gave the Coris a free hand with their work, they decided that they should work for an institution more devoted to basic research. The problem was finding a place that would hire both of them.

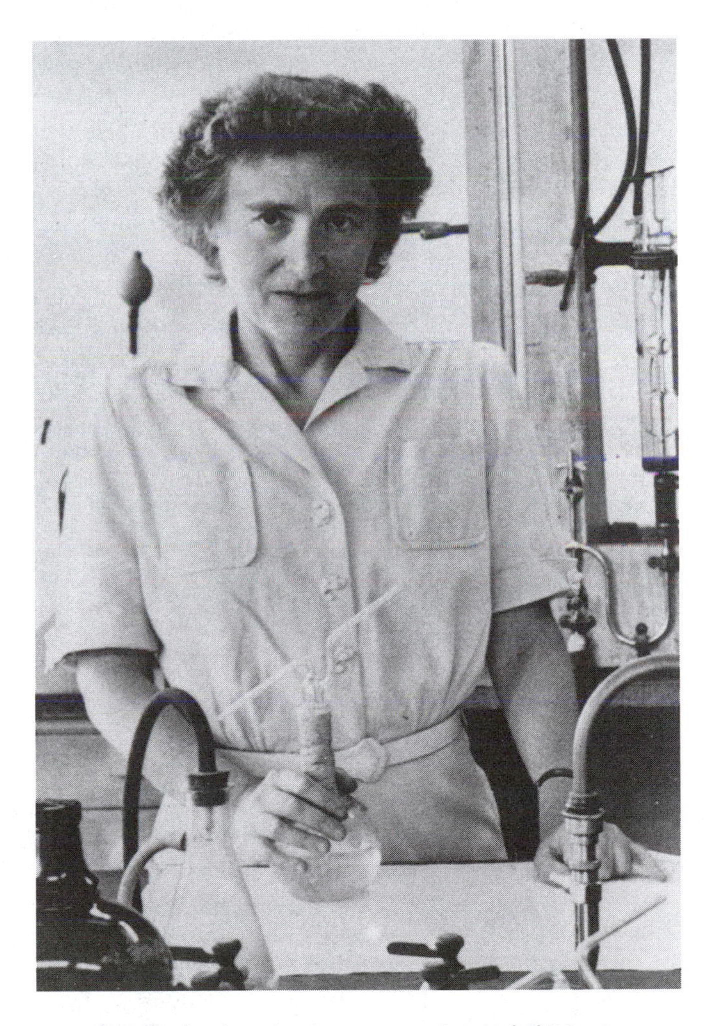

Gerty Cori, the first American woman to win a Nobel Prize in physiology or medicine. *Bernard Becker Medical Library, Washington University School of Medicine.*

Finally in 1931 the Coris found a university that would take both—though not at equal rank. The Washington University School of Medicine in St. Louis, Missouri, accepted Carl as a professor and head of its pharmacology department. It hired Gerty only as a research associate, a sort of glorified lab technician, for a fifth of the pay offered to Carl. She was not made an associate professor until 1944. In 1947 she became a full professor and won the Nobel Prize.

At Washington University the Coris continued their work on the carbohydrate cycle, discovering several key compounds involved in the cycle, including glucose-1-phosphate, a new form of glucose that came to be called the Cori ester. Beginning in 1938 the Coris focused their attention on a poorly understood group of substances called enzymes.

The year 1947 brought both triumph and disaster. On October 24, 1947, they learned that they had been awarded shares of that year's Nobel Prize in physiology or medicine for their discovery of the enzymes involved in the carbohydrate cycle. They shared the prize with Bernardo A. Houssay, a researcher from Argentina. Gerty Cori was the third woman, and the first American woman, to win a Nobel Prize in science.

Before the Coris left for Sweden, they learned that Gerty had an incurable disease of the bone marrow, which makes all the cells in the blood. Over the next 10 years the illness slowly worsened, sapping her strength and requiring blood transfusions to keep her alive.

Gerty Cori died on October 26, 1957. Her memorial service included the playing of a speech she had made on a recording called "This I Believe." She said in part, "For a research worker, the unforgotten moments of his life are those rare ones, which come after years of plodding work, when . . . what was dark and chaotic appears in a clear and beautiful light and pattern." Gerty Cori's work revealed many such patterns.

Coriolis, Gustave-Gaspard

(1792–1843)
French
Engineer, Mathematician

The law known as the Coriolis force explains how a rotating body exhibits the effects of motion differently than a stable body. Though it is essentially a law of physics, its implications spill over into many other fields of science: In meteorology it controls the rotation of storms and the direction of prevailing winds; in oceanography it predicts sea currents; in astrophysics it explains stellar dynamics, such as the direction that sunspots rotate; in ballistics it governs the trajectory of spinning bullets. The practical implications of Coriolis's discovery far outreach its theoretical importance.

Coriolis was born on May 21, 1792, in Paris, France, the son of a loyalist officer of Louis XVI, who later became an industrialist. Coriolis avoided marriage because of his ill health, which plagued him throughout his life. In 1808 he entered the École Polytechnique in Paris, from which he graduated second in his class.

In 1816 Coriolis became a tutor of mechanics at the École Polytechnique, which later appointed him as an assistant professor of analysis and mechanics. He maintained his relationship with this school for the rest of his life. In 1829 he added a chair of mechanics at the École Centrale des Arts et Manufactures. In 1836 he filled the open position in applied mechanics at the École des Ponts et Chaussées, even though this robbed him of time to devote to his own theoretical work. He gave up his teaching duties at the polytechnique in 1838 to take on the position of director of studies there, a role he excelled in by placing utmost importance on the physical condition of the learning environment.

Coriolis's first publication in 1829, *On the Calculation of Mechanical Action,* introduced important terminology into the scientific vocabulary, specifically the terms *kinetic energy* and *work*. This latter term proved particularly important, as it not only traced its roots back to Greek but also created affinities with other areas of theoretical and practical application, namely, economics, in the midst of the rise of industrialism and the response of Marxism. However, Coriolis did not choose the terms simply for their theoretical aptness—he chose them because they were very practical expressions of physical realities. These terms simplified the connection between scientific abstractions and scientific realities.

In 1835 Coriolis published his *Mathematical Theory of the Game of Billiards* as well as his masterwork, *On the Equations of Relative Motion of Systems of Bodies,* in which he proposed the notion of Coriolis force. This theory expounded that inertial force acts upon rotating bodies at a right angle to the direction of motion, resulting in a curved instead of a straight path of motion. The importance of this theory is greatly compounded by the fact that the Earth is a rotating body in motion; thus calculations involving the motion of the Earth must take into consideration Coriolis force.

Coriolis published one other important work, *Treatise on the Mechanics of Solid Bodies,* which appeared posthumously. Coriolis died in Paris on September 19, 1843. His achievements remain notable because they so successfully married theory with practice.

Cousteau, Jacques-Yves

(1910–1997)
French
Oceanographer

Jacques Cousteau popularized the study of the world beneath the water's surface with his books, films, and television series. Though not a scientific expert, Cousteau gained his appreciation of the seas through experience, and he experimented with cinematographic media to try to recreate a hands-on experience for his audience. Later in his career he used his vast influence to raise environmental awareness in hopes of curbing the destruction wrought on the natural environment by its human inhabitants.

Cousteau was born on June 11, 1910, in St. André-de-Kubzac, France. His mother was Elizabeth Duranthon; his father was Daniel Cousteau, a legal adviser. Cousteau suffered from chronic enteritis, a painful intestinal condition, for his first seven years. A 1936 car accident mangled his left arm, but he opted against amputation in favor of rehabilitation, during which time he experienced a spiritual connection with the sea. On July 12, 1937, he married Simone Melchior and had two sons: Jean-Michel was born in March 1938, and Phillipe was born in December 1939. Though both sons worked extensively with their father, Phillipe was slated to inherit his father's role before he died in a plane crash on June 28, 1979, near Lisbon, Portugal.

Cousteau attended the École Navale, graduating second in his class in 1933. After a military stint in Shanghai, China, he returned to the aviation academy and graduated in 1936. He served as a second lieutenant and a gunnery officer in the French navy before World War II. During the war he used his oceanographic experimentation as cover for his participation in the resistance movement, for which he was awarded a Croix de Guerre with a palm after the war.

In 1942 Cousteau produced an 18-minute underwater film entitled *Sixty Feet Down,* commencing his career as an underwater filmmaker. As a pioneer in this field he had to pave his own way. For example, with Emile Gagnan he invented the Aqualung, which they patented in 1943, increasing the mobility and facility of underwater diving. In 1947 Cousteau set the world freediving record at 300 feet. On July 19, 1950, he purchased the *Calypso,* an old U.S. minesweeper that he converted into an oceanic laboratory and film studio. Thus outfitted, Cousteau produced a steady stream of award-winning films from the *Calypso.* In 1955 he filmed a version of his 1953 book *The Silent World,* which sold over 5 million copies worldwide. The 90-minute film won the prestigious Palme D'Or, or Grand Prize, at the 1956 Cannes International Film Festival and a 1957 Oscar from the Academy of Motion Picture Arts and Sciences.

Over the next three decades Cousteau produced a string of successful television programs and series, including ABC's *The Undersea World of Jacques Cousteau,* which ran from 1968 through 1976; PBS's *Cousteau Odyssey,* which commenced in 1977; and TBS's *The Cousteau Amazon,* which ran in 1984. More than 40 Emmy nominations were bestowed on Cousteau's television programs, often for their informational content. The Cousteau Society was founded in the 1970s as a nonprofit peace and environmental awareness initiative.

Several awards honored Cousteau's lifetime achievements, starting with the National Geographic Society's Gold Medal in 1961 and its Centennial Award in 1988. In 1985 the French government awarded him the Grand Croix dans l'Ordre National du Mérite and the U.S. government awarded him the Presidential Medal of Freedom. Though scientific purists sometimes allude to Cousteau's lack of training and qualifications, few scientists have done more to raise worldwide awareness of scientific issues than Cousteau. He died at home, in Paris, on June 25, 1997, after suffering a respiratory infection and heart problems.

Crick, Francis Harry Compton

(1916–)
English
Molecular Biologist

Francis Crick, in conjunction with the American zoologist JAMES DEWEY WATSON, created a model that explained the structure and suggested the replication process of deoxyribonucleic acid, or DNA, the building block of life. Watson and Crick built upon the existing knowledge about DNA, particularly the concurrent work of MAURICE HUGH FREDERICK WILKINS and ROSALIND ELSIE FRANKLIN. Once Watson and Crick had established the model, a flurry of activity in the new field of molecular biology ensued, seeking to ascertain the complete makeup and the replication procedure for DNA. In 1962 Crick shared the Nobel Prize in medicine with Watson and Wilkins.

Crick was born on June 8, 1916, in Northampton, England, to Anne Elizabeth Wilkins and Harry Crick, who ran a shoe and boot factory with his brother. Crick married Ruth Doreen Dodd in early 1940, and the birth of their son, Michael, occurred during an air raid on November 25, 1940. Crick and Dodd divorced in 1947, and Crick married Odile Speed in 1949.

Crick entered University College in London in 1934, and he graduated early, in 1937, with a second-class-honors degree in physics, accompanied by work in mathematics. World War II interrupted his academic progress, but in 1947 he commenced his graduate work, funded by the Medical Research Council at Strangeways Research Laboratory of Cambridge University. In 1949 he transferred to the new Medical Research Council Unit at Cavendish Laboratory of Cambridge University. In 1951 the 23-year-old Watson arrived at the Cavendish as a member of the visiting Phage Group, which was studying bacterial viruses. Watson and Crick clicked immediately and set about finding the foundation of genes.

Whereas Wilkins and Franklin approached the problem of discovering DNA's structure experimentally, Watson and Crick preferred a more theoretical method of model building, based on LINUS CARL PAULING's solution to the structure of the alpha helix in protein. In 1953 they came up with a model of protein strands that intertwined in a spiral, connected by bases of paired nucleotides. Watson and Crick reported their findings and hypotheses in four papers, the first of which was published on April 25, 1953, in the journal *Nature*. Remarkably, Watson and Crick did not brag about the implications of their findings but simply suggested the possibility of genetic replication based on DNA.

Crick could not follow up on this discovery immediately, as he had to finish his doctoral dissertation, which he completed in late 1953. He spent much of the rest of his career tracing the chemical makeup of DNA and deciphering gene replication. In 1957 he announced his "Central Dogma" in the paper "On Protein Synthesis," which stated that genetic transcription occurs from DNA to ribonucleic acid (RNA) to protein. Once a genetic transfer takes place, it cannot be reversed. Crick also suggested that bases group themselves in triplets, known as codons. Crick thus mapped the genetic code whereby life passes unto life.

In 1977 the Salk Institute for Biological Studies in San Diego, California, made Crick a distinguished professor. Crick recorded his impressions of his ground-breaking research in the 1966 book *Of Molecules and Men,* and in 1988 he published an autobiography, *What Mad Pursuit: A Personal View of Scientific Discovery.* Crick's greatest scientific gift was his ability to pierce through irrelevant information straight to the heart of a matter, and to convey this understanding in a clear and uncluttered way.

Crookes, William

(1832–1919)
English
Physicist, Chemist

Sir William Crookes developed excellent experimental techniques that he applied prodigiously throughout his long and wide-ranging career. He also developed a keen sense for synthesizing others' unformed ideas into his own hypotheses and theories. He is best known as the editor of the important journal *Chemical News* and for his discovery of the element thallium. In the process of this discovery he invented the Crookes radiometer, which converted light radiation into rotary motion and illustrated the kinetic theory of gases, though it had little other practical use. The latter half of his career is characterized by controversy, as he tried to apply hard scientific methods to spiritualism and psychic phenomena.

Crookes was born on June 17, 1832, in London, England, the eldest son of 16 children. His father was Joseph Crookes, a successful tailor, and his mother was Mary Scott, the second wife of Joseph. In 1856 Crookes married Ellen Humphry of Darlington, with whom he had 10 children.

In 1848 Crookes commenced studying at the Royal College of Chemistry, where he earned the Ashburton Scholarship. Between 1850 and 1854 he served as A. W. Hofmann's personal assistant. Under the influence of MICHAEL FARADAY, Charles Wheatstone, and George Stokes, Crookes veered away from classical chemistry and toward chemical physics. Wheatstone helped secure Crookes his first position as superintendent of the meteorological department of Radcliffe Astronomical Observatory at Oxford. The next year Crookes secured a position teaching chemistry at the College of Science in Chester. In 1856 he inherited a large sum from his father, freeing him to devote the rest of his life to scientific experimentation and theory.

Crookes established the influential journal *Chemical News* in 1859, serving as its proprietor and editor until 1906. While investigating selenium in 1861, using a spectroscope recently invented by ROBERT WILHELM BUNSEN and Gustav Kirchoff, Crookes discovered a new element, thallium, which he then isolated, noting its properties and calculating its atomic weight in 1873. His experiments on thallium led to the invention of the Crookes radiometer. Many other scientific discoveries and innovations bear his name, including the Crookes tube, an improved vacuum tube; the Crookes glass, used by industrial workers to protect their eyes from radiation; and the Crookes dark space, or the dark area surrounding a cathode when electricity is discharged through rarefied gas.

In general Crookes divided his time and attention between the theoretical and the practical worlds of experimentation. Though he inherited a large sum, his 10 children and avid experimentation required a large income, so he focused a portion of his scientific efforts on experiments that might yield profitable results. The other portion of his time he devoted to pure experimentation, following his instinct. The many subjects of his experiments included the derivation of sugar from beets, the dyeing of textiles, the formation of diamonds, the maintenance of soil fertility, electrical lighting, and sanitation. He also anticipated the discovery of electrons by J. J. Thomson 20 years later. At the time Crookes realized only that he was observing negatively charged particles, an assertion that came under attack by German physicists. Crookes ended his career with a long study of spiritualism and psychic phenomena, a controversial undertaking.

Crookes was knighted in 1897, and in 1910 he received that Order of Merit. He died on April 4, 1919, in London.

Curie, Marie Sklodowska

(1867–1934)
Polish/French
Physicist

Marie Curie was the first woman to win a Nobel Prize, which she shared with her husband, PIERRE CURIE, and

Marie Curie, the first woman to win a Nobel Prize and the first scientist to win the prestigious prize twice. Shown here with her husband, Pierre Curie, with whom she shared the 1903 Nobel Prize in physics. *AIP Emilio Segrè Visual Archives.*

ANTOINE-HENRI BECQUEREL for their joint discovery of radioactivity. Curie followed up on this 1903 Nobel Prize in physics by winning the 1911 Nobel Prize in chemistry for her isolation of pure radium. She was the first scientist to win the prestigious prize twice. She continued her work with radioactivity by seeking medical applications for it.

Curie was born on November 7, 1867, in Warsaw, Poland, the fourth daughter and fifth child of two schoolteachers. The death of her mother, Bronislawa, when Curie was 11 years old, cemented her atheism. She was also an ardent nationalist and educated herself via the radical "Floating University" in Warsaw's underground movement. Curie financed her sister Bronia's study of medicine in Paris in exchange for future assistance with her own education.

In 1891 Curie migrated to Paris and became one of the Sorbonne's few female students. She graduated first in her class in 1893 with a degree in physical sciences, and in 1894 she graduated second in her class with a master's degree in mathematical sciences. Subsequently her monetary concerns were alleviated by the Alexandrovitch Scholarship, which she later repaid to allow others to benefit as she did. That year she met the physicist Pierre Curie, and they married on July 25, 1895. Pregnancy and the births of their daughters Irène in 1897 and Eve in 1904 did not slow her scientific activity.

The discovery of "Becquerel's rays" emanating from uranium in 1896 provided a focus for her doctoral research, which she commenced by renaming the rays *radioactivity.* She used the piezoelectric quartz electrometer, invented by Pierre and Jacques Curie, to identify thorium as the only other radioactive element known at that time. She also noted that excess rays were emitted from pitchblende, so she and Pierre set out to separate its elements. Upon analysis in July 1898 they identified a new radioactive element, named *polonium* after Marie's homeland. They discovered yet another radioactive element—which they named *radium*—six months later.

These discoveries validated Marie's work, and in 1900, the École Normale Supérieure for girls in Sèvres appointed her as its first female teacher. She instituted an innovative style of teaching, focused on experimentation. In June 1903 she earned her doctorate, the first woman to do so in France, summa cum laude. That year held much greater recognition for the couple, as they received the Royal Society's Davy Medal and the Daniel Osiris Prize in addition to the Nobel. With this recognition came financial backing, enabling Pierre to establish a well-appointed laboratory with Marie as his chief assistant. However, on April 19, 1906, a wagon struck him dead.

The University of Paris appointed Curie in her husband's position, granting her full professorship two years later. In 1909 the university announced plans to build laboratories for the new Radium Institute, which would be divided into two branches—one for radioactivity study and one for the study of the medical uses of radioactivity. Curie oversaw this effort, completed in 1914. During this time Curie had managed to isolate pure radium in a metallic state with the help of her assistant, André Debierne, in 1910. In 1911 the Swedish Academy of Sciences awarded her a second Nobel, this one in chemistry.

Curie devoted her later career to finding beneficial uses for radioactivity, as well as for her fame. In 1920 the Curie Foundation was established to fund research, and in 1922 the League of Nations International Commission for Intellectual Cooperation included her on its roster. Her constant exposure to radiation, however, took its toll, and she contracted leukemia. Curie died on July 4, 1934, near Sallanches, France. She was buried in Sceaux beside her husband.

Curie, Pierre

(1859–1906)
French
Physicist

Even before his award-winning work on radioactivity with his wife, MARIE SKLODOWSKA CURIE, and ANTOINE-HENRI BECQUEREL, Pierre Curie had performed important work in physics, studying crystal symmetry. In collaboration with his brother, Jacques, Curie discovered the phenomenon of piezoelectricity and applied it to their invention of the piezoelectric quartz electrometer. Though this work drew some recognition, Curie worked throughout his career in woefully inadequate laboratories, with excessive teaching duties hanging over his head.

Curie was born on May 15, 1859, in Paris, France. His mother was Sophie-Claire Depouilly Curie, the daughter of a prominent manufacturer, and his father was Eugène Curie, a physician. Curie's education began at home, with first his mother, then his father and his brother as teachers. In 1875 at the age of 16 Curie earned his bachelor of science degree. He then entered the Faculty of Sciences at the Sorbonne to earn his master of physical sciences degree in 1877.

The next year he worked as a laboratory assistant to Paul Desains at the Sorbonne. In 1882 he started a 22-year relationship with the new Municipal School of Industrial Physics and Chemistry in Paris as the head of its laboratory. He was later appointed a chair of physics there, but this promotion did not garner him a laboratory of his own or a decent salary. Curie's career was plagued by his disdain for the sordid politics of academia, which he avoided vehemently. This cost him several promotions to higher professional positions. In 1898 the Sorbonne passed him over for a chair in physical chemistry, and in 1903 it rejected his bid for a chair in mineralogy. A year earlier his election bid for membership in the French Academy of Sciences had failed. In 1900 he accepted an assistant professorship at the Polytechnic Institute, though this appointment added little prestige and much more work to his life. That year the Sorbonne did grace him with a chair in physics, largely as a result of the intervention of JULES-HENRI POINCARÉ.

Despite his lack of adequate laboratory facilities, Curie produced an astounding amount of original research. He and his brother published their discovery of piezoelectricity in 1880. In 1884 Curie published his findings on crystal symmetry; in 1885 he published his theory of the formation of crystals. This line of research and publication culminated in 1894 with his publication of a general principle of symmetry.

That year Curie met Marie Sklodowska, a graduate student in physics. His letters to her in Poland when she visited home that summer convinced her of his ardent love, and the couple married on July 25, 1895. Curie earned his doctorate that year with a dissertation on the magnetic properties of substances at varying temperatures. Marie was just commencing her doctoral research under Becquerel, studying his recent discovery, what she later named *radioactivity*. Curie postponed his own return to work on crystals in order to assist with this promising research. Together the Curies discovered two new radioactive elements, polonium and radium, in 1898. They verified the discovery of radium in 1902 by isolating it to determine its atomic weight, thus establishing its status as an individual element. They won the 1903 Nobel Prize in physics in conjunction with Becquerel for their work with radiation.

Also in 1903 they won the Royal Society's Davy Medal, and the Legion of Honor decorated Curie, though he refused the award. Together they also refused to patent their radium-extraction process on moral grounds, though this course would have proved extremely lucrative. Curie, however, did accept a new chair from the Sorbonne in 1904 with the stipulation that the university outfit him in a well-appointed laboratory with Marie as his chief assistant. Curie had scant opportunity to take advantage of this long-awaited luxury, though, as a wagon killed him in a freak traffic accident on April 19, 1906. His wife carried on his legacy by filling his vacant chair at the Sorbonne and continuing their commitment to finding positive applications for radioactivity, lest this potent energy be harnessed toward destructive ends, as he warned in his speech of acceptance for the Nobel Prize.

Cuvier, Georges Léopold Chrêtien Frédéric Dagobert, Baron

(1769–1832)
French
Zoologist, Comparative Anatomist, Paleontologist

Baron Georges Cuvier helped establish the sciences of comparative anatomy and paleontology as distinct fields of study. His scientific methodology prioritized facts over general theories. He eschewed the evolutionary model, supporting instead catastrophism, or the belief that natural disasters controlled the progress of species more than any other factors.

Cuvier was born on August 23, 1769, in Montbéliard, France, the son of a Swiss soldier. Between 1784 and 1788 he studied comparative anatomy at the Académie Caroline in Stuttgart, Germany.

Cuvier served as a tutor for seven years after his graduation, during which time he studied mollusks. He sent some of his original research to Étienne Geoffrey Saint-Hilaire at the Museum of Natural History in Paris, the world's largest scientific research facility at the time, and Geoffrey responded by arranging an assistantship for Cuvier with the professor of comparative anatomy in

Baron Georges Cuvier, who helped establish the sciences of comparative anatomy and paleontology as distinct disciplines. *Smith Collection, Rare Book & Manuscript Library, University of Pennsylvania.*

1795. Initially Cuvier and Geoffrey collaborated, producing a joint tract on the classification of mammals that same year. Eventually, though, their views on anatomical functions diverged.

In 1797 Cuvier published his first book, *Elementary Survey of the Natural History of Animals,* which proved successful. In 1798 Napoleon invited him to fill the position of naturalist on an expedition to Egypt that would last until 1801, but Cuvier chose instead to continue his zoological research at the museum. He accepted a professorship of natural history at the Collège de France in 1799, and in 1802 the Jardin des Plantes granted him a professorship as well.

Cuvier published one of his more important works, *Lessons on Comparative Anatomy,* which he based on his lectures, between 1800 and 1805. In this work he advanced his notion of the "correlation of parts." Cuvier believed that the organs of animals were internally interrelated, each complementing the function of the other. Further, he believed that animals' anatomies suited the functions those animals performed: that form followed function. This theory caused the split between Cuvier and Geoffrey, who believed that function followed form—that animals' anatomies determined what functions they would be able to perform. The rift between the two scientists grew wider as their careers progressed.

Cuvier expanded his work to include educational reform, acting as the imperial inspector of public instruction and helping organize the provincial university system in France. In 1810 he published his assessment of scientific instruction throughout Europe, entitled *Historical Report on the Progress of the Sciences Since 1789.* He also diversified his scientific research, extending his anatomical and zoological classifications to fossils, thereby developing a more comprehensive view of comparative anatomy. Cuvier first published his paleontological findings in 1812 in the text *Researches on the Bones of Fossil Vertebrates.* He later expanded his theory in the 1825 text *Discourse on the Revolutions of the Globe,* in which he expounded his notion of catastrophism, or the belief that natural catastrophes accounted for the progression or extinction of species.

In 1817 Cuvier published *The Animal Kingdom, Distributed According to Its Organization,* in which he extended CARL LINNAEUS's classification system by adding another level, the phylum. Cuvier divided this level into four large groupings—vertebrates, mollusks, articulates, and radiates. This theory exacerbated the antagonism between Cuvier and Geoffrey, which culminated in a public debate in 1830, when Cuvier defended his division of the animal kingdom into four groups, while Geoffrey maintained that all animals are of one class. The issue remained unresolved until the advent of Darwinism.

Cuvier's other important contributions to science included the naming of the pterodactyl and the addition of some 13,000 zoological skeletons to the Museum of Natural History's collection. Cuvier's administrative work with France's educational system earned him the title of chevalier in 1811. Cuvier died on May 13, 1832, in Paris.

D

Dalton, John
(1766–1844)
English
Physicist, Chemist

John Dalton changed the course of the human understanding of physical makeup with his atomic theory of matter, which states that all elements are made up of minute, indestructible particles, called atoms. Dalton arrived at this theory by considering the properties of gases. Dalton's law, or the law of partial pressures, states that the total pressure of mixed gases amounts to the sum of the pressure of each individual gas. Dalton also contributed to the understanding of the aurora borealis, the origin of the trade winds, the barometer, the thermometer, the hygrometer, the dew point, rainfall, and cloud formation. His limited educational background freed him from academic prejudice, and he carefully guarded his freedom of thought against undue influence by accepted theories. Dalton trusted his own observations and experience to guide his scientific research.

Dalton was born on September 6, 1766, in Eaglesfield, in Cumberland, England. His parents were Mary Greenup Dalton and Joseph Dalton, a Quaker weaver. The younger Dalton inherited his family's modest farm in 1834 when his older brother, Jonathan, died. Dalton devoted himself exclusively to his work and his religion, never marrying.

By the age of 12 Dalton had acquired enough education to commence teaching in Cumberland's Quaker School. At age 14 he moved to Kendal, where he taught with his brother at the Quaker School for the next 12 years. In 1793 he moved to Manchester, where he taught mathematics and natural philosophy at the New College, established by the Presbyterians as an alternative to Cambridge and Oxford, which required oaths to the Church of England.

Dalton published his first book, *Meteorological Observations and Essays,* based on his journal of meteorological observations started in 1787, in 1793. The next year he published *Extraordinary Facts Relating to the Vision of Colours,* an issue of personal concern as he was color-

John Dalton, whose atomic theory of matter states that all elements comprise minute, indestructible particles called atoms. *Smith Collection, Rare Book & Manuscript Library, University of Pennsylvania.*

blind. In 1800 the Manchester Literary and Philosophical Society appointed Dalton as secretary, and he read most of his papers there throughout his lifetime. In 1817 the society appointed him president, a position he maintained until his death.

Dalton presented four important papers to the society in 1801. "On the Constitution of Mixed Gases" expounded his law of partial pressures; "On the Force of Steam" discussed the dew point and represented the founding of exact hygrometry; "On Evaporation" proposed that the quantity of water evaporated was proportional to the vapor pressure; and "On the Expansion of Gases by Heat" stated that all heated gases expand equally. This last principle is known as Charles's law, as Jacques Charles discovered the effect in 1787, though Dalton published his statement first.

In 1802 Dalton presented his paper "On the Absorption of Gases by Water," to which he appended the first table of atomic weights. In a December 1803 lecture to the Royal Institution, Dalton explicated for the first time his atomic theory, which held that all atoms of a particular element are alike, having the same atomic weight. Dalton published this theory in his 1808 text *A New System of Chemical Philosophy,* which he revised in 1810 and 1827.

Dalton received a Gold Medal from the Royal Society in 1826, but it was not until after his death that STANISLAO CANNIZZARO in 1858 reasserted LORENZO ROMANO AMEDEO CARLO AVOGADRO's theories from a half-century earlier, which confirmed Dalton's atomic theory undisputedly. Even without this final confirmation Dalton's influence was immense, as evidenced by the 40,000 people who attended his Manchester funeral after his death on July 27, 1844.

Dana, James Dwight

(1813–1895)
American
Geologist, Mineralogist, Zoologist

Much of Dana's most significant work grew out of his role as the team geologist and marine zoologist on the Wilkes Expedition, which circumnavigated the globe from 1838 to 1842. Not known for his seamanship, Dana excelled in observational and descriptive abilities. He also distinguished himself with his coral reef theories, which coincided with those of CHARLES ROBERT DARWIN, and with his geosynclinal theory, which proposed that sedimentary deposits in the Earth's surface could compress into mountain ranges. With this theory Dana introduced the term *geosyncline* into the scientific lexicon.

Dana was born on February 12, 1813, in Utica, New York. His mother was Harriet Dwight Dana, and his father was James Dana, a saddler and hardware merchant. Dana attended Yale College from 1830 to 1833; there Benjamin Silliman took the young Dana under his wing. Silliman directed Dana toward natural history and its more specialized fields of geology and mineralogy. Silliman collaborated with Dana on the latter's first book, *A System of Mineralogy,* published in 1837.

Dana's first job upon graduation foretold his future path, as he secured a position as a teacher to midshipmen aboard the U.S.S. *Delaware.* On this voyage he witnessed volcanic activity that led to his first publication, "On the Condition of Vesuvius in July, 1834." When he returned from this year-long cruise, he was uncertain whether he could sustain a viable career in science. Silliman helped him by using his influence to garner an assistantship for Dana in a chemical laboratory at Yale in 1836.

Between 1838 and 1842 Dana participated in the Wilkes Expedition, which explored the Pacific and Antarctica. Though the voyage lasted only 4 years, processing all of the findings into written form took 10 years. Dana published three monumental books from his experiences with the Wilkes Expedition; these publications established his enduring scientific significance. The first book, *Zoophytes,* published in 1846, along with the second book, *Geology,* published in 1849, advanced the scientific knowledge of coral systems and animals immensely, as very little was known before these books appeared. Cataloging coral zoophytes proved particularly time-consuming for Dana, as he had collected over 400 zoophytes unknown to science. The last book, *Crustacea,* appeared in 1852.

While Dana was compiling his data from the expedition, he also worked on other projects. The most significant of these was his announcement in 1847 of his geosynclinal theory of the origin of mountains from sedimentary deposits that fold up into chains of mountains. In 1856 Yale appointed Dana to a chair in natural history, and in 1864 it appointed him to a chair in geology and mineralogy, which he held until 1890. In 1864 Dana published the comprehensive text *Manual of Geology.* In 1872 he followed up on his work with coral from the Wilkes Expedition with a final statement on the subject in his book, *Coral and Coral Islands.* This book particularly pleased CHARLES ROBERT DARWIN, as it confirmed the subsidence theory he had proposed 30 years earlier. Dana died on April 14, 1895, in New Haven, Connecticut.

Darwin, Charles Robert

(1809–1882)
English
Naturalist

Charles Darwin's name is synonymous with his theory of evolution, known as Darwinism. His assertion and documentation of this theory rocked the world, as it allowed no room for divine intervention in the process of evolution. Theologians rejected the notion, since it explained the historical progression of species in purely scientific terms, suggesting that nature controlled its own progression without intervention by a Creator. However, Darwin's methods were exacting, making it difficult to punch holes in his theory. Darwinism continued to hold sway as an explanation of evolution long after his death.

Darwin was born on February 9 (or 12), 1809, at The Mount, in Shrewsbury, England. His father, Robert Waring Darwin, was a physician and the son of the famous physician ERASMUS DARWIN. His mother, Susannah Wedgwood Darwin, the daughter of the potter Josiah Wedgwood I, died prematurely. Darwin was the couple's fifth child and their second son.

Darwin's elder sisters educated him until he entered the Shrewsbury School in 1818, studying under Dr. Samuel Butler. Darwin moved on to Edinburgh University in 1825 to study medicine, which he found utterly boring. His father made a last stab at salvaging his young son's education and future by enrolling Darwin at Cambridge in 1827 for clerical study. Darwin was not inclined toward academics and barely passed his exams to receive his degree in 1831.

Darwin's education truly commenced when he signed on as the unpaid naturalist for H.M.S. *Beagle*. Both his father and Captain FitzRoy required convincing of the worthiness of this project, but Darwin's uncle, Josiah Wedgwood, supported the 22-year-old's decision. On December 27, 1831, the *Beagle* set sail on its five-year voyage, forever affecting the history of science. The most significant location the ship visited was the Galapagos Islands off the western coast of South America, as these islands were isolated from one another and from the mainland, allowing species to evolve independently. Species that existed on one island did not exist on another, and certain species, for example tortoises, developed distinctly on the different islands.

Though Darwin published the *Journal* of the journey in 1839, he waited another two decades to publish his broad theory of evolution. In the intervening time Darwin married his first cousin, Emma Wedgwood, on January 29, 1839, and together they had 10 children, 7 of whom survived past childhood. A letter to Darwin from ALFRED RUSSEL WALLACE in Malaya in 1858 expressed a similar hypothesis concerning evolution and spurred Darwin into action. The two men presented a joint paper to the Linnean Society in London that year, announcing their coinciding theories. The following year Darwin published his immensely influential text, *On the Origin of Species by Means of Natural Selection*. In this book he set forth the three main principles of Darwinism—variation, heredity, and natural selection.

Darwin's later work, *The Descent of Man, and Selection in Relation to Sex*, published in 1871, followed up on his theory of evolution, but he also made contributions outside the realm of evolution. In 1868 he published *Variation in Animals and Plants under Domestication*. His other work included the study of barnacles, atolls, and earthworms, among other topics. Darwin died on April 19, 1882, at Down House in Kent, England.

Darwin, Erasmus

(1731–1802)
English
Physician

Erasmus Darwin, the grandfather of the naturalist CHARLES ROBERT DARWIN, distinguished himself not so much as a physician, his main occupation, but as a scientific philosopher. In the field of medicine he stood out for his belief in diagnoses that fit the individual, as well as for his recognition of the role of heredity in disease and his advocacy of public health issues. As a writer, however, he truly set himself apart with his versification of scientific philosophy.

Darwin was born on December 12, 1731, at Elston Hall in Nottinghamshire, England. His father was the barrister Robert Darwin. Darwin married Mary Howard in 1757. Together they had three sons—Charles, who died at the age of 20 in 1778 while a medical student; Erasmus, who committed suicide at the age of 40; and Robert Waring, the physician and father of Charles Darwin. After Mary died in 1770, Darwin had two illegitimate daughters, for whom he reportedly wrote *A Plan for the Conduct of Female Education in Boarding Schools* in 1797. Darwin married the widow Elizabeth Chandos Pole in 1781, taking in her two children from her previous marriage. Together they had seven more children.

Darwin attended St. John's College at Cambridge from 1750 to 1754 on the Lord Exeter Scholarship. He then attended the Edinburgh Medical School, purportedly for his M.D., though no evidence exists that he ever earned this degree. In 1755 he returned to Cambridge to complete his education there.

In 1756 Darwin set up a medical practice in Nottingham, but he soon moved it to Lichfield, where he proved more successful after treating an incurable patient. Darwin was elected a member of the Royal Society in 1761. In 1766 he cofounded the Lunar Society with Matthew Boulton and Dr. William Small. Around that time he also helped found the Lichfield Botanical Society. After Elizabeth Darwin prevailed upon him to move from Lichfield to Derby, he founded the Derby Philosophical Society in 1783.

Darwin's lasting fame resulted from four major texts he published late in his life. The first, *The Botanic Garden*, was published in two parts. Darwin published the second part, *The Loves of the Plants*, in 1789; he published the first part, *The Economy of Vegetation*, in 1791. He reversed the order because he felt the second part more accessible, though both were challenging in that they presented scientific ideas in verse form. Though this style was well received in reviews, its radical format laid it open to mockery. In 1798 *The Loves of the Triangles,* a harsh parody, was published, making it difficult to read the original in earnest. Darwin abandoned his style of scientific versification thereafter.

Darwin's three other important books were *Zoonomia or the Laws of Organic Life,* the first volume published in 1794 and the second volume published in 1796; *Phytologia: The Philosophy of Agriculture and Gardening,* published in 1800; and *The Temple of Nature or the Origin of Society,* published posthumously in 1803. Darwin was a radical freethinker whose inclusion of paganism in his writings led to the label of atheist, though his writings reveal a deep belief in a Creator and in biblical ethics. Reconciling these beliefs with rational scientific thought was the challenge Darwin tried to meet in his writings. Darwin died on April 18, 1802, at Breadsall Priory in Derby, England.

Daubechies, Ingrid

(1954–)
Belgian/American
Mathematician

Ingrid Daubechies has worked on mathematical constructs called wavelets, which are used, among other things, to solve the problem of separating useful information—a signal—from surrounding noise, or random data. She was born in Houthalen, Belgium, on August 17, 1954, to Marcel, a civil mining engineer, and Simone, a criminologist. As a child she was always interested in how things worked and enjoyed such hobbies as weaving, pottery, and sewing clothes for her dolls.

Daubechies earned a bachelor's degree in physics from the Free University of Brussels in 1975, followed by a Ph.D. in 1980, and remained at that university, rising to the rank of tenured assistant professor, until 1987. In 1984 she won the Luis Empain Prize for Physics, given once every five years to a Belgian scientist on the basis of work done before age 29. Her early work was on the application of mathematics to physics, especially to quantum mechanics, the laws that govern physics at very small (atomic) scales.

Daubechies's career changed direction in 1985, when she first became interested in wavelets. She was thinking about ways in which to reconcile different requirements for wavelet constructions at the time she attended a conference in Montreal, Canada, in February 1987. She had hoped to tour the city, but it was too cold to go out much. "I was kind of forced to stay in my hotel room—and calculate," she told the *Discover* magazine writer Hans Christian von Baeyer in 1995. "It was a period of incredibly intense concentration." Even her wedding, scheduled to take place in a few weeks, took a back seat to the ideas she was having about ways to process information.

The 19th-century mathematician and physicist Jean-Baptiste Fourier worked out a method of breaking down signals into groups of regular repeating waves in order to analyze them; scientists have used his method ever since. Unfortunately this method, which determines the pitch of a note, cannot at the same time determine when it is struck. Compromise techniques must be used if someone wants to have both types of information, and none of these has been completely satisfactory. During her chilly stay in Canada, Daubechies found a way to construct wavelets that has proved particularly effective in helping computers solve this type of problem. Her method has set wavelets to work in astronomy, physics, computer science, and other fields; for instance, it can be used to analyze the complex patterns in air streaming over a plane's wing.

Daubechies moved to the United States soon after her discovery about wavelets and has since become a naturalized citizen. From 1987 to 1994 she worked primarily at AT&T Bell Laboratories. She then went to Princeton University, where in 1999 she was a full professor and director of the program in applied and computational mathematics. A fellow of the John D. and Catherine T. MacArthur Foundation from 1992 to 1997, Daubechies has won two prizes from the American Mathematical Society and was elected to the American Academy of Arts and Sciences in 1993 and the National Academy of Sciences in 1998. She has continued to research wavelets and their applications, working on such projects as a program to apply wavelets to signals from biomedical devices.

Davy, Sir Humphry

(1778–1829)
English
Chemist

Sir Humphry Davy is best known for his discovery of several chemical elements, including sodium and potassium. He established himself with a study of the effects of nitrous oxide. Later in his career he turned his attention to practical concerns and invented the miner's safety lamp. For his achievements he was knighted on April 8, 1812, and further honored with the title of baronet in 1818.

Davy was born on December 17, 1778, in Penzance, England. His father, Robert Davy, a woodcarver, speculated unsuccessfully in farming and tin mining and died in 1794. After his death Davy's mother, Grace Millett, managed a milliner's shop until 1799, when she inherited a small estate. Davy married Jane Apreece, a widow, on April 11, 1812.

Davy's formal education was quite limited. He attended grammar school in Penzance before transferring to school in Truro in 1793. In 1795 he apprenticed to a surgeon and apothecary, intending to enter the field of medicine. He proved quite adept at chemical science, and in 1798 he was appointed as the chemical superintendent at the Pneumatic Institute in Clifton. His responsibilities included experimenting with gases to understand their effects. Davy worked with nitrous oxide, which came to be known as laughing gas because of the way it released inhibitions in those who inhaled it. Since his name was associated with such a pleasant substance, Davy quickly established a positive reputation. In 1801 the newly organized Royal Institution of Great Britain in London appointed him as a lecturer. His lectures, which became very popular, sometimes included demonstrations of the scientific principles discussed, such as the effects of nitrous oxide. Davy had friendships with the socially elite, including the poets Samuel Coleridge and William Wordsworth. In 1802 he was promoted to the position of professor of chemistry.

Davy subsequently turned his attention to the effects of electricity on chemicals, a discipline known as electrochemistry. In 1806 Davy reported on some of his findings in the paper "On Some Chemical Agencies of Electricity," which won the 1807 Napoleon Prize from the Institut de France despite the fact that England and France were warring at the time. That year Davy used electrolysis to discover sodium and potassium. He followed this with the discoveries of boron, hydrogen telluride, and hydrogen phosphate. In 1812 he published the first part of the text, *Elements of Chemical Philosophy;* however, he never managed to complete another section of the book. In 1813 Davy published a companion piece, *Elements of Agricultural Chemistry.* In 1815 he put his experimental expertise to use by inventing a safe lamp for mining.

Besides his publications and lectures, Davy earned recognition through awards and memberships. The Royal Society elected him a fellow in 1803, appointed him secretary in 1807, and promoted him to president from 1820 to 1827. In 1805 he won the society's Copley Medal. In 1826 Davy suffered a stroke, and he never fully recovered. On May 29, 1829, he died in Geneva, Switzerland.

Debye, Peter Joseph William

(1884–1966)
Dutch/American
Chemical Physicist

Peter Debye was awarded the 1936 Nobel Prize in chemistry for his diverse work on phenomena such as dipole moments, X rays, and light scattering in gases. His name serves as the appellation of the base unit of the electric dipole moment, or the debye (D). It also graces the Debye-Hückel theory of electrolytes, which explains mathematically the discrepancies between the apparent and the actual number of dissolved particles in electrolyte solutions.

Peter Debye, whose name serves as the appellation for the base unit of the electric dipole moment, the debye. *Smith Collection, Rare Book & Manuscript Library, University of Pennsylvania.*

Debye was born on March 24, 1884, in Maastricht in the Netherlands to Johannes Wilhelmus Debije, a foreman at a metalware manufacturer, and Maria Anna Barbara Ruemkens, a theater cashier. He had one younger sister. On April 10, 1913, Debye married Matilde Alberer; the couple had two children. Peter Paul Ruprecht, who became a physicist and collaborated with his father, was born in 1916, and Mathilde Maria Gabriele was born in 1921. Debye became a U.S. citizen in 1946.

In 1905 Debye earned his degree in electrical engineering from the Technische Hochschule in Aachen. In 1910 he received his Ph.D. in physics from the University of Munich, with a dissertation on the effect of radiation on spherical particles with diverse refractive properties. Debye stayed on as a lecturer at Munich for one year.

Debye managed to produce impressive results from his research despite constant professional movement. In

1911 he occupied the chair of theoretical physics previously held by ALBERT EINSTEIN at the University of Zurich. He moved to the University of Utrecht in 1912, and in 1914 he became a professor of theoretical and experimental physics at the University of Göttingen. He returned to the University of Zurich in 1920 as a professor of experimental physics and the director of the physics laboratory at the Federal Institute of Technology. From 1927 to 1934 he taught at the University of Leipzig before moving to the University of Berlin, where he assumed the position of director of the Kaiser Wilhelm Institute for Theoretical Physics. Debye last moved to the United States, where he became a professor of chemistry and the department chairperson at Cornell University from 1940 to 1950.

In 1916 Debye discovered that solid substances could be powdered for use in X rays, a process that circumvented the more difficult process of preparing crystals for that purpose. In 1923 he collaborated with Erich Hückel to transform SVANTE AUGUST ARRHENIUS's theory of the partial ionization of electrolyte solutions to the more precise version of complete ionization of electrolyte solutions by taking into account the fact that charged ions could interact mutually instead of exclusively. In 1923 he also developed a theory that explained the Compton effect mathematically. Later in his career he researched polymers and magnetism.

Besides the Nobel Prize, Debye won a slew of other awards, including the 1930 Rumford Medal of the Royal Society, the 1935 Lorentz Medal of the Royal Netherlands Academy of Sciences, the 1937 Franklin Medal of the Franklin Institute, and the 1949 Faraday Medal. He died of a heart attack on November 2, 1966, in Ithaca, New York.

Delbrück, Max

(1906–1981)
German/American
Molecular Biologist

Max Delbrück is recognized as the founder of molecular biology and molecular genetics. Although he discovered important phenomena, such as the spontaneous mutation of bacteria to become immune to bacteriophages, he exerted even more influence by inspiring other scientists. His collaborative work on bacteriophages, or viruses that infect bacteria, with ALFRED DAY HERSHEY and SALVADOR EDWARD LURIA earned the trio the 1969 Nobel Prize in physiology or medicine.

Delbrück was born on September 4, 1906, in Berlin, Germany, the youngest of seven children. His father, Hans Delbrück, was a professor of history at the University of Berlin and editor of the journal *Prussian Yearbook*, and his mother was Lina Thiersch, granddaughter of Justus von Liebig, who was considered the founder of organic chemistry. Delbrück married Mary Adeline Bruce in 1941, and the couple had four children—two sons, Jonathan and Tobias, and two daughters, Nicola and Ludina. In 1945 Delbrück became a U.S. citizen.

In 1924 Delbrück enrolled in the University of Tübingen, but he transferred first to the University of Bonn before receiving his Ph.D. in physics in 1930 from the University of Göttingen. He started his dissertation on the origin of a type of star, but he then switched topics to the differences in bonding of two lithium atoms, as opposed to the much stronger bonding of two hydrogen atoms. Delbrück performed postdoctoral studies under a research grant at the University of Bristol in England focusing on quantum mechanics. A Rockefeller Foundation postdoctoral fellowship at the University of Copenhagen allowed him to continue his postdoctoral studies under NIELS HENDRIK DAVID BOHR.

From 1932 through 1937 Delbrück served as a research assistant to LISE MEITNER at the Kaiser Wilhelm Institute for Chemistry in Berlin. A second Rockefeller Foundation fellowship allowed him to travel in 1937 to the United States, where he resided for the rest of his life. From 1937 through 1981 he served on the faculty of the California Institute of Technology in Pasadena, with a stint at Vanderbilt University between 1940 and 1947. In 1945 he commenced his tradition of offering annual summer courses on bacteriophages at Cold Spring Harbor Laboratory in New York.

In 1939 Delbrück discovered a one-step process by which bacteriophages multiplied exponentially within one hour. In 1943 he helped organize the Phage Group, which included both Hershey and Luria, and the members began to meet together informally to discuss bacteriophages. That year Delbrück and Luria jointly published "Mutations of Bacteria from Virus Sensitivity to Virus Resistance," an article that represents the genesis of bacterial genetics. The next year the Phage Group drew up the Phage Treaty of 1944, which standardized bacteriophage research. In 1946 he and Hershey independently discovered that different kinds of viruses can combine genetically to create new forms of viruses. In 1947 at the request of George Beadle, head of the biology department at Caltech, Delbrück accepted a position there. By 1950 his interests began to shift away from phage and toward sensory physiology, though he did help launch the next wave of viral genetics—tumor virology. Delbrück died on March 10, 1981, in Pasadena, California.

Descartes, René du Perron

(1596–1650)
French
Mathematician, Philosopher of Science

Although René Descartes is known as the father of modern philosophy, his work in the history of science was equally instrumental. Descartes developed the analytical or

René Descartes, developer of Cartesian geometry, which translates geometrical problems into algebraic problems in order to solve them. *AIP Emilio Segrè Visual Archives.*

coordinate system of geometry known as Cartesian geometry, which transforms geometrical problems into algebraic problems in order to solve them. Descartes also theorized in the opposite direction, applying geometrical solutions to algebraic problems. In a broader sense, he contributed to the sciences in general with his notion of "method," or the systematic doubt of knowledge until it is confirmed as true by a process of consideration that moves from the simplest to the most complex considerations.

Descartes was born on March 31, 1596, in La Haye, Touraine, France. His father was *a conseiller* to the parlement in Brittany. His mother, who died shortly after his birth, left him both her noble name *du Perron* as well as an estate in Poitou.

Between 1604 and 1612 the Jesuits at the Royal College of La Flèche gave Descartes a modern education in physics and mathematics, including the astronomical discoveries of Galileo. In 1616 Descartes graduated from the University of Poitiers with a degree in law, but his independent wealth freed him from the necessity of practicing law for a living, allowing him to travel instead.

Descartes served in the military after his schooling, first under the prince of Orange in Holland in 1618. During this stint he encountered Isaac Beekman, who taught him much about science. In 1619 Descartes served under the duke of Bavaria. It was during this time that Descartes experienced a series of visions that would transform the future of science. He spent the day of November 10, 1619, in a *poêle,* or an overheated room, and came to two realizations: that he must arrive at all knowledge through his own senses, and that he must doubt all in order to prove what is true without learning of prior assumptions. That night he dreamed three dreams that prompted him down the path in pursuit of discovering unshakable truths. Descartes's awakening on this date, which he reported in his 1637 text *Discourse on Method,* is marked as the birth of modernity by many.

Descartes included several appendices that demonstrated his "method"—*Météores, La dioptrique,* and *La géometrie.* This latter appendix proved pivotal to the science of mathematics, as it proposed the Cartesian coordinate system of geometry. He further defined that one can determine the position of a fixed point by the convergence of two or more straight lines. In 1644 he published the *Principles of Philosophy,* which included his theories in physics. Descartes reduced the world to a single mechanistic system in this work. He produced these, his most influential works, while living in Holland between 1628 and 1649.

In 1649 Queen Christina of Sweden enticed Descartes to Stockholm to serve as her private tutor. She forced Descartes, who maintained that he conceived his best ideas while lounging in a warm bed, to rise before five o'clock in the morning for tutorials in the chill of Swedish winter. Descartes died in Stockholm on February 11, 1650, of an illness he contracted on these cold mornings. He is most remembered for his search for certainty, which led him to attempt to apply mathematical method to all knowledge and resulted in his famous *Cogito, ergo sum,* "I think, therefore I am."

De Vries, Hugo

(1848–1935)
Dutch
Plant Physiologist, Plant Geneticist

At the beginning of the 20th century Hugo De Vries recovered from obscurity the laws of heredity that JOHANN GREGOR MENDEL had formulated some 34 years earlier. Independently Karl Correns and Erich Tschermak von Seysenegg rediscovered Mendel's work simultaneously. De Vries also advanced the study of plant physiology by identifying such processes as plasmolysis, or osmosis in plant cells.

De Vries was born on February 16, 1848, in Haarlem, Netherlands. His parents were Maria Everardina Reuvens, who hailed from a family of scholars, and Gerrit de Vries, who was a representative of the Provincial State of North Denmark, a member of the Council of State, and the minister of justice under William III. De Vries studied medicine at the University of Leyden from 1866 until 1870, when he moved to the University of Heidelberg, where he studied under Hofmeister. Reading works by JULIUS VON SACHS and CHARLES ROBERT DARWIN influenced him tremendously, and in 1871, when he moved on to the University of Würzberg, he had the opportunity to study under Sachs with whom he maintained a long-standing professional relationship.

Later that year De Vries accepted a position teaching natural history at the First High School in Amsterdam while continuing to research in Sachs's laboratory in the summers. In 1875 he landed a position with the Prussian Ministry of Agriculture in Würzburg, writing monographs on red clover, potato, and sugar beets, as well as on the processes of osmosis in plant cells. This post lasted two years, until he moved to the University of Halle as a *Privatdozent,* lecturing on the physiology of cultivated plants. De Vries resigned this position in favor of a lectureship in plant physiology at the University of Amsterdam later in the year of 1877. His appointment represented the first academic position in the field of plant physiology in the Netherlands, and De Vries stayed on at the University of Amsterdam for the remainder of his career. In 1878 the university promoted him to the position of assistant professor of botany, and in 1881 he ascended to the status of full professor.

In 1886 De Vries commenced his research in plant genetics when he noticed that some species of evening primrose differed from others and sought to explain this distinction. He reported his early findings in the 1889 book *Intracellular Pangenesis.* He continued to work with plant breeding until 1900, when he formulated the laws of heredity that restated Mendel's work from 1866, though De Vries discovered Mendel's papers only after he had formulated his own version of the same ideas. De Vries developed the theory of mutation, which held that there existed mutation periods, which represented the process of evolution. De Vries described progressive mutants as productive characteristic transformations, whereas retrogressive mutants represented changes that did not benefit the continuation of the species. De Vries published this work in 1901 through 1903 in the book *The Mutation Theory,* which appeared in 1910 and 1911 in an English translation.

In 1896 the University of Amsterdam named De Vries a senior professor of botany, and in 1904 he served as a visiting lecturer at the University of California at Berkeley. In 1912 he visited the Rice Institute in Houston, Texas, to participate in its opening ceremonies. De Vries retired in 1918, though he continued to work in the field that made him famous, plant genetics and plant physiology. De Vries died on May 21, 1935, in Lunteren, Netherlands.

Dicke, Robert Henry

(1916–1997)
American
Physicist

Robert Henry Dicke is perhaps best remembered for his theory that residual radiation from the big bang can still be detected. But just as he was planning an experiment to prove his prediction that microwaves from that cosmic event still echo in the universe, he learned that Robert Wilson and Arno Penzias had inadvertently demonstrated the phenomenon. Dicke was not awarded a portion of their Nobel Prize. Dicke also gained notoriety by challenging ALBERT EINSTEIN's general theory of relativity. In the Brans-Dicke theory (which he formulated with his graduate student Carl Brans) Dicke posited that the gravitational constant so central to Einstein's theory was actually not a constant, but decreased infinitesimally each year. Although the Brans-Dicke theory failed to gain widespread acceptance, Dicke continued to assert its accuracy. Dicke also held at least 50 patents and is credited with verifying Einstein's equivalence principle.

Dicke was born to Oscar and Flora Peterson Dicke on May 6, 1916, in St. Louis, Missouri. In 1942 he married Annie Henderson Currie, with whom he had three children. After graduation with honors from Princeton in 1939 Dicke pursued doctoral work at the University of Rochester. He was awarded his Ph.D. in nuclear physics in 1941.

Dicke's first position was at the Radiation Laboratory at the Massachusetts Institute of Technology, where he investigated radar from 1941 to 1946. He developed a radiometer that could detect weak radio waves. The device not only aided the Allied war effort, but also became an essential tool in the field of radio astronomy. He then accepted a faculty position at Princeton, where he remained for the rest of his career. In 1955 he was named full professor, and he served as the first Albert Einstein University Professor of Science from 1975 until 1984.

In 1961 he proposed the controversial Brans-Dicke theory. Whereas Einstein had maintained that gravity was a constant force, Dicke held that gravity grew weaker with each passing year, at a rate of 1 part in 10^{11}. Dicke advanced this radical reformulation because of what he perceived to be troubling inconsistencies between the calculated age of the universe and the apparent age of objects (such as globular clusters of stars) that seemed much older. Dicke postulated that since gravity had been a stronger force in the past, it had caused these stars to burn faster—and thus age more quickly. This theory has not been widely adopted, however. Dicke tested another aspect of Einstein's theory of general relativity during the 1960s as well. Building on the work of Roland von Eötvös, Dicke conducted experiments to confirm a central tenet of relativity theory—that gravitational mass (measured by weighing) and inertial mass (measured by resistance to acceleration) are equivalent. After painstaking study Dicke concluded that the two types of mass were indeed equivalent.

In 1964 Dicke began to explore the implications of the big bang theory of the origin of the universe. Since that explosion was believed to have produced temperatures above 10 billion degrees, Dicke hypothesized that radiation from this event was still present in space. With his colleague P.J.E. Peebles, Dicke calculated that this vestigial radiation would have a temperature of only about 3 degrees Kelvin (very near absolute zero), would be recognizable only in the form of weak microwaves, and would be constant throughout the universe. Before he could prove his theory, however, he learned that Penzias and Wilson had accidentally done so.

Although Dicke often made grand predictions that defied accepted notions, he remained rooted in the fundamentals of careful experimentation. Not coincidentally, he held over 50 patents for an array of devices, most related to radar. Although he did not share Penzias and Wilson's Nobel Prize, his theory on the echo radiation of the big bang provided a theoretical underpinning that helped establish the big bang as an accepted scientific doctrine. Moreover, even though his work on gravitational theory did not undermine the belief in Einstein's gravitational constant, he spurred further study into the matter. He was honored with the 1970 Medal of Science and the 1973 Comstock Prize from the National Academy of Sciences. An emeritus professor at Princeton since 1984, Dicke died of Parkinson's disease on March 4, 1997.

Dirac, Paul Adrien Maurice

(1902–1984)
English
Theoretical Physicist

Paul Dirac expanded the human view of the universe by positing the existence of antimatter, as well as by suggesting a quantifiable limit to the size of the universe as a ratio with the size of an atom. Dirac's work grew out of the attempt to answer unsolved questions, such as explaining the phenomenon of electron spin, which quantum mechanics at the time did not take into account.

Dirac was born on August 8, 1902, in Bristol, England. His parents were Charles Adrien Ladislas Dirac, a Swiss immigrant who taught French at the Merchant Venturer's Technical College, where Dirac received his early education, and Florence Hannah Holten Dirac. While working at the Institute for Advanced Studies at Princeton University from 1934 through 1935 Dirac met Margit ("Manci") Wigner, the sister of the physicist Eugene Wigner. The couple married in January 1937 and had two daughters, Mary Elizabeth, born in 1940, and Florence Monica, born in 1942.

In 1921 Dirac received his bachelor's degree in electrical engineering from Bristol University and continued there in graduate study on a two-year scholarship. He then transferred to St. John's College at Cambridge University, where he earned his Ph.D. in physics in 1926. His dissertation extended notions of quantum mechanics established by WERNER KARL HEISENBERG. Dirac spent the years 1926 through 1927 traveling, first to Copenhagen, where he entered discussions with NIELS HENRIK DAVID BOHR, and next to Göttingen, where he conversed with MAX BORN and J. ROBERT OPPENHEIMER.

In 1927 St. John's College appointed Dirac a fellow, and in 1929 it promoted him to the status of university lecturer and praelector in mathematical physics, a position that allowed him the flexibility to accept brief teaching stints at the Universities of Michigan and Wisconsin. In 1932 Cambridge named Dirac the Lucasian Professor of Mathematics, a post he held until 1969. In 1971 he round-

Paul Dirac, who first suggested the existence of antimatter, with Richard Philip Feynman at the International Conference on Relativistic Theories of Gravitation in Warsaw, Poland. *AIP Emilio Segrè Visual Archives, Physics Today Collection.*

ed out his career as a professor of physics at Florida State University.

In 1927 Dirac commenced work on a hole in ERWIN SCHRÖDINGER's theory of wave mechanics, which failed to account for electron spin, a phenomenon discovered two years earlier. Dirac introduced relativity into the equation and derived a formula that calculated the correct split energy levels for hydrogen, thus confirming a spinning electron. These results suggested the existence of negative energy, a proposition Dirac pursued in 1930. He concluded that electrons that fall below their lowest possible positive energy level, or the ground state, enter a state of negative energy, otherwise known as antimatter. Dirac published this theory in 1930 in *The Principles of Quantum Mechanics*. Two years later CARL DAVID ANDERSON experimentally confirmed the existence of the positron, as these electrons were called.

In 1937 Dirac published *The Cosmological Constants*, which contained his theory on "large-number coincidences." In his calculations Dirac noticed that the number 10^{40} appeared at vital junctures—as the ratio of the radius of the universe as compared to the radius of an electron, and also as the approximate square root of the number of particles in the universe. Dirac thus posited that this number provided a quantifiable model of the universe.

Dirac received the 1933 Nobel Prize in physics with Schrödinger in recognition of their work on defining quantum mechanics. Dirac's theories in particular carried far-reaching implications, for they suggested physical explanations for some of the mysteries of reality. Dirac died on October 20, 1984, in Tallahassee, Florida.

Doppler, Christian Johann

(1803–1853)
Austrian
Physicist, Astronomer

Christian Doppler, who became famous for formulating the Doppler effect in 1842. *AIP Emilio Segrè Visual Archives.*

Christian Doppler gained his fame by formulating the Doppler effect, which explained the apparent fluctuations in the frequency of light or sound when there is motion between the light or sound source and the eye or ear that acts as the receiver. For example, the pitch of a train whistle seems to get higher as it approaches, then seems to lower as the train whistle moves away. This change in frequency is explained by the Doppler effect. Doppler used this type of acoustic effect to hypothesize a similar optical effect with starlight, which would appear redder when the source and the receiver receded from one another and would appear more violet when they approached one another. The Doppler effect became an important instrument for theorizing and has been useful in many practical applications, such as global positioning systems.

Doppler was born on November 29, 1803, in Salzburg, Austria. His father was a master stonemason. On the advice of the astronomer and geodesist Simon Stampfer, who realized the young man's potential to excel in mathematics, Doppler attended the Polytechnic Institute in Vienna from 1822 to 1825. He then studied privately in Vienna.

Between 1829 and 1833 Doppler worked as a mathematical assistant in Vienna. The State Secondary School in Prague appointed him as a professor of mathematics and accounting in 1835. In 1841 he was named a professor of elementary mathematics and practical geometry at the State Technical Academy. In 1847 he became a professor of mathematics, physics, and mechanics at the Mining Academy at Schemnitz; in 1850 he was named director of the new Physical Institute and full professor of experimental physics at the Royal Imperial University of Vienna.

On May 25, 1842, Doppler read his paper "Concerning the Colored Light of Double Stars," which stated the Doppler effect. CHRISTOPH HENDRIK DIEDRIK BUYS BALLOT experimentally verified the Doppler effect with sound in 1845, though he criticized some of Doppler's astronomical bases for the theory. Doppler refuted these claims vehemently, though it was difficult to prove or disprove his theory when applied to astronomy since the available technology was not sophisticated enough to measure the Doppler effect properly. The Italian astronomer Benedict

Sestini published his observations of star coloration in 1850, confirming within the technological limits imposed at the time that light from a receding star measured at a lower frequency, or a longer wavelength, making it appear reddish, whereas light from an approaching star measured at a higher frequency, or a shorter wavelength, making it appear more violet. Not until 1901 was the astronomical effect fully confirmed by using spectroscopic tools. With the advent of quantum mechanics the Doppler effect started to take relativity into account.

The Doppler effect developed diverse applications, from astronomers' calculating the speed of the Sun's rotation to police officers' measuring the speed of moving vehicles with radar guns. Doppler himself also performed diverse scientific and mathematical experiments in the fields of geometry, optics, and electricity. He died on March 17, 1853, in Venice, Italy, and was survived by his wife and five children.

Dresselhaus, Mildred Spiewak

(1930–)
American
Physicist, Engineer

Mildred Dresselhaus raised herself from poverty to become head of the Massachusetts Institute of Technology's Materials Science Laboratory and one of the university's 12 Institute Professors. She was born Mildred Spiewak on November 11, 1930, in a Brooklyn slum. Her parents were immigrants with no money or education, but they taught their two children that the United States was a land of opportunity. Dresselhaus says, "I found out that opportunities did present themselves [but] one had to take the initiative to find these opportunities and exploit them."

Mildred realized that education was the key to a better life, and she determined to enter Hunter College High School, which prepared talented girls in New York City for college. (The school later became coeducational.) That meant passing a rigorous entrance exam, a task that seemed hopeless. She studied in every spare moment, however, and not only passed but made a perfect score in mathematics. "Passing the entrance examination to Hunter College High School is the greatest achievement of my life," Dresselhaus says. Mildred Spiewak went on to Hunter College, which offered free tuition. At first she planned to become an elementary school teacher, but her physics teacher, later Nobel Prize winner, ROSALYN SUSSMAN YALOW, steered her toward that field. After Spiewak graduated in 1951, Yalow helped her obtain a Fulbright Fellowship to study for a year at Britain's Cambridge University.

Spiewak studied at Radcliffe for her master's degree in physics, which she earned in 1953, and at the University of Chicago for her Ph.D. She decided to specialize in the physics of solids. As her thesis project she studied how magnetic fields affect superconductors, unusual materials that conduct electricity without converting any of it to heat. She married a fellow physicist, Gene Dresselhaus, soon after both received their Ph.D. in 1958.

After two years at Cornell University and the birth of their first child, the Dresselhauses moved to the Lincoln Laboratory at the Massachusetts Institute of Technology in 1960. Mildred switched her research to semiconductors, her husband's specialty. Semiconductors, crystalline materials such as silicon that can be treated so that they conduct electricity in somewhat the way metals do, are used in transistors and computer chips. As she had with superconductors, Mildred studied how magnetic fields affected these materials. She also studied so-called semimetals—elements such as arsenic, graphite (a form of carbon), and bismuth, which act as semiconductors in some ways and superconductors in others.

Dresselhaus says that her seven years at the Lincoln Laboratory were "the most productive years of my research career," even though she had three more children during this time. A live-in nanny cared for the children in the daytime, and the Dresselhauses spent time with them in the evenings. "My children have gained more than they have lost because of my professional career," Dresselhaus believes.

In 1967 the Dresselhauses began working for MIT's National Magnet Laboratory, where Mildred continued her research on the effects of magnetic fields on semiconductors. She became a full professor in the university's electrical engineering department a year later. She was associate head of the electronic science and engineering department between 1972 and 1974, and between 1973 and 1985 the Abby Rockefeller Mauze Professor, an endowed position for a woman professor interested in furthering the careers of women undergraduates. She was director of the university's Center of Materials Science and Engineering from 1977 to 1983. Since 1985 she has been one of 12 Institute Professors, MIT's highest faculty designation—the first woman given this honor. She has also been a visiting professor at universities from Brazil to Japan.

Dresselhaus has won many awards for her work, including the Hunter College Hall of Fame Award (1972), the Society of Women Engineers' Achievement Award (1977), and the National Medal of Science (1990). She was the second woman to become a member of the National Academy of Engineers. In 1984 she became the first woman president of the American Physical Society, and in 1998 she became president of the American Association for the Advancement of Science.

Dresselhaus's work today involves another unusual material—carbon molecules made up of 60 atoms arranged in a structure like that of a soccer ball. These molecules have been nicknamed *buckyballs* because they also resemble the dome-shaped houses popularized by the architect Buckminster Fuller in the 1950s. Buckyballs, part of a larger class of similarly shaped carbon molecules called fullerenes, may prove useful for making industrial diamonds and materials that conduct light. In this work, as in her other research, Dresselhaus has been more interested in discovering the qualities of materials than in developing practical uses for them. "We tend to be . . . 10 years in advance of commercial applications," she told the writer Iris Noble.

Mildred Dresselhaus has made a point of mentoring young women scientists, just as Rosalyn Yalow once did her. She helps them find or make the opportunities they need, as she herself did. "If you go into science and engineering," she tells them, "you go in to succeed."

Dubos, René

(1901–1982)
French/American
Microbiologist

René Dubos transformed the understanding of soil microorganisms by studying them in their natural state, as opposed to the practice common in his time of studying them in laboratory conditions. Following the logic of the Russian bacteriologist Sergei Winogradsky, Dubos believed that the microorganisms themselves were transformed under lab conditions. Dubos isolated several antibacterial substances from soil microorganisms, which led to the discovery of important antibiotics. Later in his career Dubos focused increasingly on ecology, stressing the importance of a holistic view to problem solving, taking into account the complex relationships between humans and the environment.

Dubos was born on February 20, 1901, in Saint-Brice, France. He was the only child of Georges Alexandre Dubos and Adeline Madeleine de Bloedt. When Dubos was 13 his father moved the family to Paris to set up a butcher shop that Dubos and his mother ran while his father fought in World War I. Dubos's father died of injuries sustained in the war. Dubos married Marie Louise Bonnet, and before her death of tuberculosis in 1942, the couple collaborated on *The White Plague: Tuberculosis, Man, and Society,* which was published in 1952. In 1938 Dubos became a U.S. citizen.

In 1921 Dubos graduated with a B.S. in agricultural science from the Institut National Agronomique in Paris. He then moved to Rome to take on the position of assistant editor of the *International Agriculture Intelligence,* an academic journal published by the International Institute of Agriculture, which was part of the League of Nations. The soil microbiologist Selman Waksman convinced Dubos to pursue graduate study in soil microbiology at Rutgers University, helping to secure the French student a research assistantship at the university's State Agricultural Experiment Station. Dubos also served as an instructor in bacteriology while he composed his dissertation on the decomposition of cellulose in paper by soil microorganisms. Rutgers granted him a Ph.D. in 1927.

Dubos jumped at the opportunity to join the Rockefeller Institute for Medical Research in New York City with a fellowship in the department of pathology and bacteriology, where he worked enthusiastically under the bacteriologist Oswald Avery. Except for a two-year stint in the early 1940s at Harvard Medical School as a professor of tropical medicine Dubos spent most of his career at the Rockefeller Institute, which promoted him to professor in 1957 and to professor emeritus in 1971. That year the State University of New York at Purchase named him director of environmental studies.

While at the Rockefeller Institute Dubos made the discoveries that established his scientific significance. In 1930 he isolated an enzyme in a soil microorganism that could decompose the part of the bacillum that causes lobar pneumonia in humans. Nine years later he isolated from *Bacillus brevis* another antibacterial substance that he named *tyrothricin,* which became the first commercial antibiotic; it later proved too toxic for mainstream use as it killed red blood cells. This discovery proved more important, in paving the way for other antibiotics.

In 1968 Dubos published his most important book, *So Human an Animal,* which won the 1969 Pulitzer Prize for general nonfiction. He also won the 1963 Phi Beta Kappa Award for *The Unseen World,* as well as the 1961 Modern Medicine Award and the 1976 Tyler Ecology Award. In 1980 his name graced the René Dubos Center for Human Environments in New York City, which extended his support of grass-roots environmentalism. Dubos died of heart failure on February 20, 1982, in New York City.

E

Earle, Sylvia Alice
(1935–)
American
Botanist, Marine Biologist

Sylvia Earle has spent more than 6,000 hours underwater, including living in an undersea "habitat" for two weeks, and has dived deeper than any other solo diver. Admiring colleagues call her "Her Royal Deepness."

Sylvia was born on August 30, 1935, in Gibbstown, New Jersey, and spent her childhood on a farm near Camden. Her mother, Alice, a former nurse, taught Sylvia and her brother and sister to love nature. "I think I always knew I would work [as a scientist] with plants and animals," Earle once told an interviewer. Her favorite spot was a pond in her backyard.

In 1948, when Sylvia was 12 years old, her father, Lewis, an electrical engineer, moved the family to Dunedin, Florida. Now the "pond" in Sylvia's backyard was the Gulf of Mexico. She made her first ocean dive when she was 17 and "practically had to be pried out of the water."

Earle earned a B.S. from Florida State University in 1955 and an M.S. in botany from Duke University in 1956. She married a zoologist, John Taylor, around 1957 (they divorced in 1966) and had two children, Elizabeth and John (Richie). She began full-time undersea research in 1964. Among other things, she collected algae (seaweeds and related plants) in the Gulf of Mexico for her Ph.D. project at Duke, which she finished in 1966. Unlike most marine biologists of the time, she dived to study undersea life in its own habitat rather than dragging it up to the deck of a ship in nets.

In 1970 Earle lived underwater for two weeks as part of a National Aeronautics and Space Administration–(NASA)-sponsored project called Tektite; the name is taken from a type of glassy meteoric rock often found on the seafloor. She headed a crew consisting of four other women scientists. Their "habitat," 50 feet under the Caribbean Sea, consisted of two tanks connected by a passageway. It included not only beds and a kitchen but even hot showers and television. The women spent up to 10 hours a day in the water, studying ocean life.

The group did nothing that all-male Tektite crews had not also done, but they emerged into a blizzard of publicity, hailed as "aquababes." They also received the Conservation Service Award, the Department of the Interior's highest civilian award. The fuss irritated Earle, who saw it as reverse discrimination, but it also made her realize that as a woman scientist she had a unique opportunity to reach and educate the public. Ironically a greater achievement of Earle's won much less attention. In September 1979 she donned a heavy plastic and metal "Jim suit" (named after a diver who tested an early version of it), a sort of underwater space suit, and dived 1,250 feet into the water near Hawaii. A submarine lowered and then released her. No other diver had gone this deep without being attached to a cable. *Current Biography* called this "possibly the most daring dive ever made." Earle remained submerged for two and a half hours under water pressure of 600 pounds per square inch, observing such creatures as "an 18-inch-long shark with glowing green eyes" and "a lantern fish . . . with lights along its sides, looking like a miniature passenger liner."

While preparing for the Jim suit dive, Earle met the British engineer, Graham Hawkes, who had designed the suit. The two found that they shared a love of diving and a desire to improve diving technology. They formed two companies, Deep Ocean Technology and Deep Ocean

Engineering, in 1981. One of their products was a one-person submersible called *Deep Rover,* which Earle piloted down to about 3,000 feet in 1985, the deepest any solo diver had gone. The couple also married in 1986. (This was Earle's third marriage; from 1966 to 1975 she was married to Giles Mead and had a third child, Gale.) Earle and Hawkes have since divorced. In 1992 Earle founded her own undersea technology company, Deep Ocean Exploration and Research (DOER).

Earle took part in many research projects during the 1970s and 1980s, including diving with humpback whales as part of a study done by Roger Payne and KATHARINE BOYNTON PAYNE. She was also curator of phycology (the study of algae, or seaweeds) at the California Academy of Sciences in San Francisco from 1979 to 1986. Earle has led more than 50 oceanic expeditions and won many awards for her work. They include the John M. Olguin Marine Environment Award (1998), the Kilby Award (1997), the Director's Award of the Natural Resources Council (1992), and the Society of Women Geographers Gold Medal (1990). She is a fellow of the American Association for the Advancement of Science, the Marine Technology Society, the California Academy of Sciences, the World Academy of Arts and Sciences, and the Explorers Club.

Earle served on the President's Advisory Committee on Oceans and Atmosphere from 1980 to 1984. In 1990 President George Bush chose her to be the chief scientist of the National Oceanic and Atmospheric Administration (NOAA). She was the first woman to hold this post. In 1998 Earle's major project was a five-year study of the National Marine Sanctuaries called the Sustainable Sea Expeditions.

Eastwood, Alice

(1859–1953)
American
Botanist

Continuing the work of MARY KATHARINE LAYNE BRANDEGEE, Alice Eastwood was curator of the herbarium, or dried plant collection, of the California Academy of Sciences in San Francisco for 57 years. Alice was born on January 19, 1859, in Toronto, Canada. Her mother died when she was six, and she and her brother and sister were raised by relatives until 1873, when Colin, her father, sent for them to join him at the store he then ran in Denver, Colorado. Her schooling was often interrupted by work, so she graduated from East Denver High School only in 1879, when she was 20.

Eastwood began teaching at the high school, but her happiest times were the summers, when she climbed into the Rocky Mountains to collect plants. By 1890 income from lucky investments let her devote all her time to botany. She visited the California Academy and became friends with Katharine Brandegee, then its curator of botany. In 1892 Brandegee persuaded her to move to San Francisco and become joint curator. When Brandegee moved to San Diego a year later, Eastwood took over her position.

In addition to improving the organization of the plant collection, Eastwood personally added many specimens to it, sometimes hiking 20 miles a day through the Sierras with her heavy wooden plant presses on her back. On April 18, 1906, a huge earthquake shook San Francisco, followed by a citywide fire. Eastwood found the academy building partly in ruins and the fire approaching. She had stored her most valuable plants in her sixth-floor office, but the building's staircase had collapsed. As she wrote later in *Science* magazine, she and a friend "went up chiefly by holding on to the iron railing and putting our feet between the rungs. Porter helped me to tie up the plant types, and we lowered them to the floor . . . by ropes and strings." She got out with her plants just as flames reached the building.

For the next six years the academy had no home, so Eastwood visited herbaria in the East and Europe. She rejoined the academy in 1912, when it built a new headquarters in Golden Gate Park. She spent the rest of her life rebuilding and improving its plant collection, adding over 340,000 specimens to it. Of the work represented by the many plants lost in the 1906 fire, she said only, "It was a joy to me when I did it, and I can still have the same joy in starting it again."

Eastwood finally retired in 1949, when she was 90. The next year the Eighth International Botanical Congress in Sweden recognized her lifetime of work by electing her its honorary president. As a mark of its respect the group seated her in a wooden chair once used by CARL LINNAEUS, the 18th-century Swedish scientist who had designed the system of naming plants and animals that all biologists use. Eastwood died on October 30, 1953.

Eccles, Sir John

(1903–1997)
Australian
Physiologist

Sir John Eccles won the 1963 Nobel Prize in physiology or medicine with ALAN LLOYD HODGKIN and Sir ANDREW FIELDING HUXLEY for their discovery of how impulses are chemically transmitted or inhibited by nerve cells. Eccles hypothesized that this chemical communication expresses itself in electrical charges, with the content of the signal depending on the positivity or negativity of the electrical charge involved, though researchers later demonstrated that the process was more chemical than electrical. Physicians applied Eccles's findings to further research on the kidneys, heart, and brain, as well as to the treatment of nervous disorders.

Eccles was born on January 27, 1903, in Melbourne, Australia. Both of his parents, William James and Mary Carew Eccles, were teachers. Eccles's 1928 marriage to Irene Frances Miller yielded four sons and five daughters but ended in divorce in 1968. Later that year Eccles married Helena Tabarikova, a Czechoslovakian neurophysiologist with whom he collaborated professionally.

In 1919 Eccles commenced medical school at the University of Melbourne, and in 1925 he graduated with highest academic honors and a Rhodes Scholarship. He traveled to England on this scholarship, matriculating in September 1925 at Magdalen College of Oxford University, which named him the Christopher Welch Scholar. Eccles moved to Exeter College at Oxford in 1927 as a junior research fellow under Sir CHARLES SCOTT SHERRINGTON, the founder of cellular neurophysiology. Eccles and Sherrington collaborated on research on neural "motor units," or nerve cells that send messages to muscles, which formed the basis of Eccles's doctoral dissertation that earned him a Ph.D. in 1929. Eccles remained at Exeter as the Staines Medical Fellow from 1932 to 1934.

Eccles returned to his homeland in 1937 as the director of the Kanematsu Memorial Institute of Pathology in Sydney, a position he retained until 1943. He then moved to New Zealand to take a post as professor of physiology at the University of Otago from 1944 to 1951. He filled the same position at the Australian National University at Canberra from 1951 to 1966. Eccles rounded out his career in the United States as the Buswell Research Fellow and a professor of physiology and medicine at the State University of New York at Buffalo from 1968 until 1975.

During his tenure at the Australian National University Eccles performed his most important work on neural excitation and inhibition. He discovered that the release of a chemical, most likely acetylcholine, prompted the passage of sodium or potassium ions to send either a positive or a negative message across the nerve cells. Eccles reported his early findings on neural activity in 1932 in "Reflex Activity of the Spine" and his breakthrough work in 1957 in "The Physiology of Nerve Cells" and in 1969 in "The Inhibitory Pathways of the Central Nervous System." Eccles focused his attention while in the United States on the mind-body problem, publishing his ideas in *The Understanding of the Brain* (1973), *The Self and the Brain* (1977), *The Human Mystery* with Karl Popper (1979), *The Human Psyche* (1980), and in *The Creation of the Self* (1989).

Early in his career Eccles won the 1927 Gotch Memorial Prize and the 1932 Rolleston Memorial Prize. He was knighted in 1958 and later received the 1961 Baly Medal of the Royal College of Physicians, the 1962 Royal Medal of the Royal Society, and the 1963 Nobel Prize in physiology or medicine. He died in Contra, Switzerland, on May 2, 1997, after a protracted illness.

Eddy, Bernice

(1903–)
American
Medical Microbiologist

Bernice Eddy had a knack for discovering facts that her superiors didn't want to know or want others to know. She warned of, but was unable to prevent, a disaster that struck in the early days of making polio vaccines. She also codiscovered the first virus shown to cause cancer in mammals.

Bernice was born in Glendale, West Virginia, in 1903. She grew up in nearby Auburn. Her father, Nathan Eddy, was a physician. Bernice went to college in Marietta, Ohio, where her mother moved after her father's death, and graduated in 1924. She earned a master's degree in 1925 and a Ph.D. in 1927 in bacteriology from the University of Cincinnati.

After remaining at the University of Cincinnati for several years, Eddy joined the Public Health Service in 1930. Five years later she transferred to the Biologics Control Division of the National Institutes of Health (NIH) in Bethesda, Maryland, the department that checks the quality of vaccines that the government distributes. Soon after she joined NIH, Eddy married Jerald G. Wooley, a physician who worked for the Public Health Service. They had two daughters, Bernice and Sarah. Unfortunately Wooley died while the girls were still young. Eddy's mother helped her raise the children. Beginning at the start of World War II and continuing for 16 years, Eddy checked army influenza vaccines. She became chief of the flu virus vaccine testing unit in 1944. In 1952 she also began doing research on possible treatments for polio. She received an NIH Superior Accomplishment Award in 1953 for this work.

In 1954 Eddy was asked to perform safety tests on batches of a killed-virus vaccine for polio that JONAS EDWARD SALK had just invented. A huge national program that would administer the vaccine was about to begin, and Eddy and her staff had to work around the clock. "We had eighteen monkeys," Eddy said later. "We inoculated these monkeys with each vaccine [batch] that came in. And we started getting paralyzed monkeys." This meant that some of the virus in the vaccine was still able to cause disease. Alarmed, Eddy told her supervisors about her results, but they ignored her and went ahead with the vaccination program. Shortly afterward live virus in a few batches of the vaccine produced polio in about 200 children.

Eddy was taken off polio research and assigned to test a vaccine that was supposed to prevent colds. Meanwhile she began working with a fellow NIH scientist, SARAH STEWART, on a virus that Stewart had discovered, which seemed to cause leukemia (a blood cell cancer) in mice. Eddy worked out a way to grow the virus dependably in the laboratory.

Eddy and Stewart began publishing papers about their virus in 1957, reporting that it produced a bizarre collection of tumors in every animal that received it. They named it the SE polyoma virus, *SE* for Stewart and Eddy and *polyoma* meaning "many tumors." Viruses that caused cancer in birds had been known, but polyoma was one of the first viruses shown to cause tumors in mammals and the first to cause the disease in a wide range of animals. "It was a major, major discovery," says NIH's Alan Rabson.

Both the cold vaccine and the polio vaccine contained viruses that were grown in monkey kidney cells in the laboratory. Around 1959 Eddy noticed that the monkey cells sometimes died for no obvious reason. She suspected that they were being killed by an unknown virus. When she injected ground-up monkey cells into newborn hamsters, the animals developed tumors, suggesting that the virus could cause cancer.

Once again Bernice Eddy had disturbing news for her superiors. On July 6, 1969, she told her boss, Joseph Smadel, about the hamster tumors and suggested that steps be taken to keep the virus out of future lots of vaccine. Smadel did not want to hear that vaccines already in widespread use might contain a cancer-causing virus. He dismissed the tumors as mere "lumps."

Undaunted, Eddy described her experiments with the mystery virus at a meeting of the New York Cancer Society in the fall of 1960. Smadel heard about her speech and telephoned her in a fury. "I never saw anybody so mad," Eddy said later. Smadel ordered her not to speak in public again without clearing the content of her speeches with him. Shortly afterward another researcher, Maurice Hilleman, reported the same discovery Eddy had made. He called the virus SV40 (the 40th simian, or monkey, virus to be discovered). It proved to be very similar to the polyoma virus that Eddy and Stewart had described. Fortunately it did not appear to cause cancer in humans.

During her remaining years at NIH Bernice Eddy was pushed into smaller and smaller laboratories and denied permission to attend professional meetings and publish papers. She continued her work as best she could until her retirement in 1973, at age 70. She received several awards at or after the time of her retirement, including a Special Citation from the secretary of the Department of Health, Education, and Welfare in 1973 and the NIH Director's Award in 1977.

Edelman, Gerald

(1929–)
American
Biochemist

Gerald Edelman received the 1972 Nobel Prize in medicine or physiology with Rodney Porter for their work on the chemical structures of antibodies. In the 1960s two competing theories explained the function of antibodies. Dissatisfied with both, Edelman proposed a third theory, which postulated the indeterminacy of chance as a controlling factor in antibody function: Essentially, human immunoglobulin, the serum that carries all antibodies, takes advantage of its own mistaken transcriptions of genetic codes to create a vast variety of antibodies to protect against disease.

Edelman was born on July 1, 1929, in New York City. His parents were Edward Edelman, a physician, and Anna Freedman. Edelman married Maxine Morrison on June 11, 1950, and the couple had three children—Eric, David, and Judith. That same year Edelman received his B.S. in chemistry from Ursinus College in Collegeville, Pennsylvania. Four years later he earned his M.D. from the University of Pennsylvania Medical School, having served one year away as the medical house officer at the Massachusetts General Hospital. Edelman proceeded to seek his doctorate at Rockefeller University in New York City, writing his dissertation on human immunoglobulin. He received his Ph.D. in 1960.

Rockefeller University retained Edelman, and he spent the majority of his career there. In 1966 he attained the status of full professor of biochemistry, and in 1974 the university named him the Vincent Astor Distinguished Professor. The university further promoted him in 1981 to the position of director of its Neurosciences Institute. Edelman finally parted with Rockefeller University in 1992 to direct the Neuroscience Institute at Scripps Research Institute in La Jolla, California.

Rodney Porter had devised a method for splitting immunoglobulins by using enzymes as "chemical knives" as early as 1950. A decade later Edelman used a similar method, known as reductive cleavage, to split immunoglobulin G (IgG) into a light and a heavy chain. The Czech researcher Frantisek Franek added his theory that immunoglobulins trapped antigens within these two chains. This joint determination of the structure of IgG, which Porter announced in 1962, opened up the field to intensive research and sharing of results in global meetings known as "Antibody Workshops," led by Edelman and Porter. Edelman turned his attention to the study of cancer-producing myelomas, and in 1969 he constructed a model of an entire antibody, complete with four chains and 1,300 amino acids. In 1975 Edelman discovered proteins called cell-adhesion molecules, or individual cells that combine to make tissue.

Edelman devoted his later career to understanding the workings of the neural system and the brain. In 1981 he developed a general theory of neural development, and in 1987 he published his most important text, *Neural Darwinism: The Theory of Neuronal Group Selection.* He followed up on this with the 1993 book *Bright Air, Brilliant Mind.* In addition to the Nobel Prize, Edelman has received the Spenser Morris Award, the Eli Lilly Prize of the American Chemical Society, and the Albert Einstein Commemorative Award.

Edinger, Johanna Gabrielle Ottelie ("Tilly")
(1897–1967)
German/American
Paleontologist

Tilly Edinger was the first person to make a systematic study of the brains of long-extinct animals. She was born Johanna Gabrielle Ottelie Edinger on November 13, 1897, in Frankfurt, Germany. Her father, Ludwig, was a wealthy medical researcher who compared the brain structure of different animals. Anna, her mother, was active in social causes. Tilly, as she was called, surely must have become interested in the brain as a result of her father's interest, but he did not believe in careers for women and did not encourage her.

Edinger studied at the Universities of Heidelberg and Munich from 1916 to 1918, then at the University of Frankfurt, where she received a doctorate in 1921. Unlike her father, who examined the brains of animals now living, Edinger studied fossils. She worked at the university for six years as a research assistant in paleontology.

In 1927 Edinger became curator of the vertebrate collection at the Senckenberg Museum in Frankfurt. Her first book, *Die Fossilen Gehirne* (Fossil brains), was published two years later. Edinger essentially invented the field of paleoneurology, the study of fossil brains. Brains themselves are too soft to form fossils, but she discovered that plaster casts of the inside of fossil skulls revealed the shape of the long-vanished organs because mammals' brains fit very tightly against their skulls.

Because Tilly Edinger was Jewish, her quiet days in Frankfurt ended in 1933, when the violently anti-Semitic Nazis took control of Germany. The museum had to make her continued employment a secret. The director removed her name from her office door, and she sneaked in each day by a side entrance. When the secret was revealed in 1938, Edinger decided to leave the country. She fled Germany in May 1939 and entered the United States the next year. She joined the Harvard Museum of Comparative Zoology and became a U.S. citizen in 1945. In her new country Edinger maintained her reputation as one of the top figures in vertebrate paleontology, and the Society of Vertebrate Paleontology made her its president in 1963–1964.

In 1948 Edinger published a second monumental book, *The Evolution of the Horse Brain.* In it she showed that advances in brain structure such as an enlarged forebrain had evolved independently in several groups of mammals. Many scientists pictured evolution as a steady advance along a single "chain of creation," but Edinger showed that evolution was more like a many-branched tree. Edinger died after a car accident on May 27, 1967.

Edison, Thomas
(1847–1931)
American
Inventor

Thomas Edison held 1,069 patents on his inventions, which included such important innovations as the incandescent electric lamp and the phonograph. He combined scientific acumen with an uncanny entrepreneurial sense for identifying the technological advances that would prove profitable. He happened upon his only discovery in the realm of pure science accidentally when he noted what became known as the Edison effect, the emission of electrons from a heated cathode. Edison did not recognize the importance of this discovery, though subsequent scientists used the effect as the basis for the electron tube.

Thomas Edison, who invented the phonograph in 1877 and the incandescent light bulb in 1878. *Smith Collection, Rare Book & Manuscript Library, University of Pennsylvania.*

Edison was born on February 11, 1847, in Milan, Ohio. He was the youngest of seven children born to Samuel Ogden Edison, Jr., and Nancy Elliot. Edison married Mary Stillwell on December 25, 1871, and the couple had three children together—Marion Estell, Thomas Alva, and William Leslie. Edison's first wife died on August 8, 1884; he then married Mina Miller on February 24, 1886. Together they had three children—Charles, Madeleine, and Theodore.

Edison's formal education was limited to home schooling, though he had set up his own chemical laboratory by the age of 10. At age 12 he started developing his entrepreneurial skills by riding trains to sell newspapers. He even set up a makeshift laboratory in one of the trains he rode regularly. He compensated for his lack of education by reading voraciously, and he later characterized his reading of the first two volumes of Michael Faraday's *Experimental Researches in Electricity* in one sitting in 1868 as a seminal event in his scientific life.

Edison became a telegraph operator in 1863, traveling from city to city until in 1868 he reached Boston, where he worked for Western Union Telegraph Company. There he patented his first invention, the electric vote recorder, though the technology proved too efficient for legislators who would rather waste their time in argumentation. In 1869 he moved to New York City, where he landed a $300-a-month job as general manager at Law's Gold Indicator Company. That year he invented a paper tape ticker for standardizing the reporting of stock prices, and he joined Franklin L. Pope and James N. Ashland to establish Pope, Edison and Company, a firm devoted to technological innovation. In 1870 Gold and Stock Telegraphic Company bought out the company, at the same time purchasing the rights to Edison's tape ticker for $40,000. Edison invested this windfall in a scheme that industrialized the invention process, implementing a factory line approach at a plant in Newark, New Jersey. Edison outgrew this facility in 1876 and moved to Menlo Park, New Jersey.

There Edison invented the phonograph in 1877, though he did not immediately recognize its commercial value. He began work on the incandescent light bulb in 1878, and after experimenting with more than 6,000 substances, he discovered a carbon filament from bamboo fiber that lasted more than 1,000 hours in a vacuum. Edison then devoted his efforts to supplying electricity for widespread use, operating the first power station from Pearl Street in New York City as of September 4, 1882. He again outgrew his facilities and moved them to West Orange, New Jersey, in 1887. There he invented the kinetograph, a precursor to modern motion-picture technology.

Edison received numerous honors for his work, both during his lifetime and after his death. He won the 1892 Albert Medal of the British Society of the Arts. In 1927 the National Academy of Sciences elected him a member, and in 1928 the U.S. Congress awarded him a gold medal for his life's work. In 1960 he was inducted into the Hall of Fame of Great Americans. Edison died on October 18, 1931, in West Orange, New Jersey. His prolific output combined with the social significance of his inventions earned him an unquestionable place among the ranks of the most important scientists of the 20th century.

Egas Moniz António Caetano de Abreu Freire
(1874–1955)
Portuguese
Neurologist

António Egas Moniz innovated two important neurosurgical techniques: cerebral angiography in the 1920s and frontal leucotomy in the 1930s. The latter, which he developed with Almeida Lima, was a psychosurgical technique that represented a physical solution to psychological problems by severing the frontal lobes from the rest of the brain to prevent anxiety and neuroses. For this work Moniz shared the 1949 Nobel Prize in physiology or medicine with WALTER RUDOLF HESS.

Egas Moniz was born António Caetano de Abreu Freire on November 29, 1874, in Avança, Portugal. His parents, Maria do Rosario de Almeida e Sousa and Fernando de Pina Rezende Abreu, were members of the aristocracy. A clerical uncle educated Moniz until he entered the University of Coimbra in 1891 to study medicine. He wrote his thesis on diphtheria to receive his M.D. in 1899, after which he studied neurology at the Universities of Paris and Bordeaux. Before departing, he commenced his political career as a liberal republican by serving as a deputy in the Portuguese parliament. While in France he wrote a paper on the physiological pathology of sexual activity, a work that sufficiently impressed the faculty of his alma mater, the University of Coimbra, to offer him a position as a professor upon his return from France in 1902. That same year he married Elvira de Macedo Dias.

Egas Moniz actively opposed the Portuguese monarchy. In fact, his name originated as a pen name on his political tracts as a student, and he retained the designation professionally. His activism got him imprisoned after the revolution of 1908, but by 1910 his cause prevailed, as the monarchy was overthrown. The next year the University of Lisbon hired him as a professor of neurology, a chair he held until his 1945 retirement. His political involvement escalated, peaking in 1917 when he served as Portugal's ambassador to Spain and as the minister of foreign affairs. He thus represented Portugal at the signing of the Versailles Treaty ending World War I. During the war he had conducted research on head injuries and published his findings in 1917 in *A Neurologia na Guerra.* He waged his own battle in 1919, dueling over a political quarrel. This brush with death prompted him to exit politics altogether in 1922.

In 1927 Egas Moniz innovated the technique of cerebral angiography, injecting radioactive solutions into the arteries of corpses, then x-raying the body in order to map blood flow. Egas Moniz practiced this technique on living patients to locate cerebral tumors according to the displacement of arteries. He had the idea for frontal leucotomy after hearing of the technique, which was reported by John F. Fulton and Carlyle G. Jacobsen at the 1935 International Neurological Conference in London. There Fulton and Jacobsen had discussed their removal of the frontal lobes of chimpanzees and observation of their subsequent behavior. Egas Moniz reasoned that the success of the experiment on chimpanzees warranted its use on humans, so he conducted frontal leucotomies on 20 patients, none of whom died and most of whom improved psychologically.

Though the frontal leucotomy earned Egas Moniz his fame, the advent of drug therapy replaced his operation as a solution for severe psychological problems. However, cerebral angiography, which earned him less renown, continues to serve as an important diagnostic technique for doctors. Moniz died on December 13, 1955, in Lisbon.

Ehrlich, Paul

(1854–1915)
German
Medical Scientist

Paul Ehrlich is best known for synthesizing salvarsan, a chemical he proved to be effective against syphilis and other diseases, in 1909, a year after receiving a joint Nobel Prize in physiology or medicine with ÉLIE METCHNIKOFF in recognition of their work on immunology and serum therapy. Ehrlich's belief that chemical reactions influence biological processes led him to experiment with staining techniques to identify aberrant body cell tissues. Ehrlich is considered a pioneer in hematology, chemotherapy, and immunology.

Ehrlich was born on March 14, 1854, in Strehlen, in Silesia, Prussia (now Strzelin, Poland). His father was Ismar Ehrlich, a Jewish distiller, innkeeper, and lottery collector, and his mother was Rosa Weigert. Her cousin, Carl Weigert, was merely nine years older than Ehrlich, and both men became important research physicians.

In 1872 Ehrlich entered Breslau University, though he spent three of his semesters at Strasbourg, where the anatomist Wilhelm von Waldeyer influenced him immensely. He passed his exams in 1874 in Strasbourg and returned to Breslau for medical school. He spent one semester in 1876 at the Physiology Institute of Freiburg im Breisgau and a final term at Leipzig University in 1878; there he submitted his doctoral dissertation on cell staining to differentiate what he termed "mast cells" from plasma cells.

Upon graduation in 1878 Ehrlich was appointed head physician at the medical clinic at the Charité Hospital in Berlin, where he worked and researched for the next nine years. In 1883 Ehrlich married Hedwig Pinkus, and together they had two daughters. In 1884 the Prussian Ministry of Education conferred on him the title of professor. In 1887 he began teaching at the University of Berlin, though the position did not pay him because of the anti-Semitic atmosphere prevalent at the time. He moved on to the Moabit Hospital in Berlin in 1890 to study tuberculin with Robert Koch for the subsequent six years, when he invented a new staining technique for the tuberculosis bacillus discovered by Koch. Then in 1896 the new Institute for Serum Research and Serum Investigation at Steiglitz appointed him as its

Paul Ehrlich, who did pioneering research in hematology, chemotherapy, and immunology. *Smith Collection, Rare Book & Manuscript Library, University of Pennsylvania.*

director. In 1899 he moved to the Institute of Experimental Therapy in Frankfurt, which was later named after him.

Ehrlich published 37 scientific papers in the early period of his career, between 1879 and 1883. Ehrlich considered the last of these, "The Requirement of the Organism for Oxygen," which posited that the rate of oxygen use measures cellular vitality, to be the most significant. Ehrlich also contributed to the understanding of hematology with his extensive work on dye reactions with red and white blood cells. At the Charité he conducted histological and biochemical research on morphological, physiological, and pathological characteristics of blood cells, during which he developed new methods for detecting and differentiating leukemias and anemias. He discovered a use for methylene blue in the treatment of nervous disorders, and his work on antibodies paved the way for modern immunology. Ehrlich suffered a stroke in December 1914, followed by a second stroke the next year. He died on August 20, 1915, in Bad Homburg vor der Höhe, Germany.

Eigen, Manfred

(1927–)
German
Physical Chemist

Manfred Eigen won the 1967 Nobel Prize in chemistry with RONALD GEORGE WREYFORD NORRISH and GEORGE PORTER for their work on fast chemical reactions. Eigen developed a technique for disrupting the equilibrium of the chemical system, which involved bombarding chemical solutions with a burst of high-frequency sound waves, then measuring the time the solutions took to return to the state of equilibrium.

Eigen was born on May 27, 1927, in Bochum, Germany. His father, Ernst Eigen, was a chamber musician, and his mother was Hedwig Feld. Eigen's chemistry and physics studies at the University of Göttingen were disrupted by World War II, when he served in the army with an antiaircraft artillery unit. Afterward he returned to the university to earn his Ph.D. in 1951. The next year he married Elfreide Mueller, and together the couple had two children—Gerald and Angela.

Eigen remained at the University of Göttingen as a research assistant in its Institute of Physical Chemistry for two years after obtaining his doctorate. He moved on in 1953 to the Max Planck Institute for Biophysical Chemistry, where he stayed for the remainder of his career. The institute promoted him to the status of research fellow in 1958. In 1962 he took over the helm of the biochemical kinetics department; in 1964 he became the director of the institute and later was its chairman as well.

In 1954, the year after he arrived at the Max Planck Institute, he employed for the first time what he called "relaxation techniques," referring to the time chemical solutions require to "relax" after they have been disturbed. Eigen commenced the experiment with the solution at a set temperature and pressure. He then disrupted this equilibrium with a pulse of electric current or a burst of sound waves, after which he used absorption spectroscopy to measure in millionths and billionths of seconds the return of the solution to the beginning temperature and pressure. Eigen applied this technique first to simple chemical solutions such as water, later to more complex combinations. Eigen reported his findings in the article "Methods for Investigation of Ionic Reactions in Aqueous Solutions with Half Times as Short as 10^{-9} Sec.: Application to Neutralization and Hydrolysis Reactions," which appeared in the scholarly journal *Discussions of the Faraday Society* in 1954.

Later in his career Eigen applied the same notion of fast chemical reactions to biology in an effort to account for the genesis of life. He hypothesized that chance played an important role in the creation of life and that chance continued to determine much of the reality of the physical as well as the social world. Eigen collaborated with Ruthild Winkler on the 1981 book *Laws of the Game: How Principles of Nature Govern Chance.*

Besides the Nobel Prize, Eigen won numerous esteemed awards. The German Physical Society honored him with its Otto Hahn Prize in 1962. The American Chemical Society granted him its Linus Pauling Medal in 1967. The British Chemical Society awarded him its Faraday Medal in 1977.

Einstein, Albert

(1879–1955)
German/Swiss/American
Physicist

Albert Einstein is perhaps the most important scientist of the 20th century. In one year, 1905, he published four papers that radically revised the scientific understanding of the world. He explained the equivalence of mass and energy and developed both the photon theory of light and the special theory of relativity. He expanded upon this latter theory 11 years later with the general theory of relativity. Once he had established his status as an extremely influential scientist, he used his voice to speak out for Zionism and pacifism, and against Hitler's Nazi regime.

Einstein was born on March 14, 1879, in Ulm, Germany. His mother was Pauline Koch and his father was Hermann Einstein, an electrical engineer. Einstein married twice. When his first marriage ended in divorce, he married his cousin, Elsa, a widow with two daughters.

The collapse of his father's electrochemical business, combined with Einstein's revulsion at all things German, led the family to immigrate to Switzerland, where Einstein gained citizenship in 1901. In 1896 Einstein entered the

Swiss Federal Institute of Technology in Zurich to study physics and mathematics. After failing to secure a teaching position after graduation, Einstein settled into a job with the Swiss Patent Office in Bern as a technical expert, third class, for seven years, starting in 1902. Einstein loved this arrangement, as the light duties allowed him to concentrate his free time on science. This period was the most fertile in his career, as it spawned not only a doctorate, which he earned in 1905 from the University of Zurich with the dissertation "A New Determination of Molecular Dimensions," but also his four papers that same year, which shaped scientific history.

All four papers were published in the journal *Annals of Physics*. The first paper, "On the Motion of Small Particles Suspended in a Stationary Liquid According to the Molecular Kinetic Theory of Heat," concerned Brownian motion, which is the random movement of microscopic particles suspended in liquids or gases as a result of the impact of molecules in the surrounding fluid. The second paper, "On a Heuristic Point of View about the Creation and Conversion of Light," described electromagnetic radiation as a flow of quanta, or discrete particles, now known as photons. The third paper, "On the Electrodynamics of Moving Bodies," contained the special theory of relativity, which stated that the speed of light is constant for bodies moving uniformly, relative to one another. It rejected the notions of absolute space and absolute time, and it asserted the notion of time dilation: that time slows down for a moving body. The final paper, "Does the Inertia of a Body Depend on Its Energy Content?" contained Einstein's famous equation, $E = mc^2$.

A series of university positions followed, commencing in 1908 with a spot at the University of Bern. In 1909 Einstein became an associate professor of physics at the University of Zurich. Two years later Einstein taught at the German University in Prague, and the year after that he returned to the Federal Institute of Technology in Zurich as a professor. Finally in 1914 he gained some stability as a member of the Prussian Academy of Sciences in Berlin and the director of the Kaiser Wilhelm Institute for Physics.

In 1916 Einstein followed up on his special theory of relativity with the publication of his paper "The Foundation of the General Theory of Relativity." In it he extended his special theory of relativity to apply to all situations. When his prediction that a ray of light from a distant star passing near the Sun would appear to be bent slightly, in the direction of the Sun, was observed to be correct during a solar eclipse in 1919, Einstein gained international fame. In 1921 Einstein won the Nobel Prize in physics.

Einstein left Nazi Germany for a position at the new Institute for Advanced Study at Princeton in 1933. He became a U.S. citizen in 1940 and in fact convinced President Franklin Roosevelt of the possibility of the Nazi regime's developing a weapon of mass destruction using nuclear power. Einstein continued to work for peace until an attack due to an aortic aneurysm led to his death on April 18, 1955, in Princeton, New Jersey.

Elion, Gertrude Belle ("Trudy")

(1918–1999)
American
Chemist, Medical Researcher

Although Nobel Prizes in science are usually awarded for basic research, the 1988 prize in physiology or medicine went to three people in applied science—drug developers. "Rarely has scientific experimentation been so intimately linked to the reduction of human suffering," the 1988 *Nobel Prize Annual* said of their work. One of the researchers was Gertrude Elion.

Gertrude, whom everyone called Trudy, was born on January 23, 1918, to immigrant parents in New York City. Her father, Robert, was a dentist. The family moved to the Bronx, then a suburb, in 1924. Trudy spent much of her childhood reading, especially about "people who discovered things."

In 1933, the year Trudy graduated from high school at age 15, her beloved grandfather died in pain of stomach cancer, and she determined to find a cure for this terrible disease. There was no money to send her to college, however, because her family had lost its savings in the 1929 stock market crash. Trudy enrolled at New York City's Hunter College, which offered free tuition to qualified women. She graduated from Hunter with a B.A. in chemistry and highest honors in 1937.

Elion failed to win a scholarship to graduate school, so she set out to find a job—not easy for anyone during the Depression, let alone for a woman chemist. One interviewer turned her down because he feared she would be a "distracting influence" on male workers. She took several short-term jobs and also attended New York University for a year, beginning in 1939, to take courses for her master's degree. She then did her degree research on evenings and weekends while teaching high school and finally completed the degree in 1941.

World War II removed many men from workplaces, making employers more willing to hire women, and in 1944 it finally opened the doors of a research laboratory to Gertrude Elion. Burroughs Wellcome, a New York drug company, hired her as an assistant to a researcher, George Hitchings. Most drugs in those days were developed by trial and error, but Hitchings had a different approach. His lab looked systematically for differences between the ways that normal body cells and undesirable cells such as cancer cells, bacteria, and viruses used key chemicals as they grew and reproduced. The researchers then tried to find or make chemicals that interfered with these processes in undesirable cells but not in normal ones.

Elion at first worked mostly as a chemist, synthesizing compounds that closely resembled the building blocks of nucleic acids. The nucleic acids, deoxyribonucleic acid (DNA) and ribonucleic acid (RNA), carry inherited information and are essential for cell reproduction. Scientists had theorized around 1940 that an antibiotic called sulfanilamide killed bacteria by "tricking" them into taking it up instead of a nutrient that the bacteria needed, thus starving them to death, and Hitchings thought that a similar "antimetabolite" therapy might work against cancer. If a cancer cell took up compounds similar but not identical to parts of nucleic acids, he reasoned, the chemicals would prevent the cell from reproducing and eventually kill it, much as a badly fitting part can jam the works of a machine. He and Elion set out to create such compounds.

In 1950 Elion invented 6-mercaptopurine (6-MP), which became one of the first drugs to fight cancer successfully by interfering with cancer cells' nucleic acid. It worked especially well against childhood leukemia, a blood cell cancer that formerly had killed its victims within a few months. When combined with other anticancer drugs, 6-MP now cures about 80 percent of children with some forms of leukemia.

Another compound Elion developed in her cancer research was called allopurinol. Because it can prevent the formation of uric acid, allopurinol has become the standard treatment for a painful disease called gout, in which crystals of uric acid are deposited in a person's joints.

Work on another breakthrough began in 1969, when Elion sent John Bauer, a researcher at the Burroughs Wellcome Laboratories in England, a new drug she had created that was related to a known virus-killing compound. At Elion's suggestion that he test her drug against a dangerous group of viruses called herpesviruses, he found that it stopped their growth. Elion and her coworkers then launched a search for variants of the drug that would kill herpesviruses even better than the original compound. In 1974 a Burroughs Wellcome researcher, Howard Schaeffer, synthesized a drug called acyclovir, which was 100 times more effective against herpesviruses than Elion's first drug.

As Hitchings and Elion developed drug after drug, they advanced together within Burroughs Wellcome. Finally in 1967 Elion was made head of her own laboratory, the newly created Department of Experimental Therapy. Although she had always enjoyed working with Hitchings, she was glad to have more independence. When Burroughs Wellcome moved to Research Triangle Park, North Carolina, in 1970, Elion moved with it. Her laboratory became a "mini-institute" with many sections.

Gertrude Elion officially retired in 1983, but her scientific legacy lived on. Workers from her team, using approaches she developed, discovered azidothymidine (AZT), the first drug approved for the treatment of acquired immunodeficiency syndrome (AIDS).

Elion's received her greatest honors after her retirement. On October 17, 1988, she learned that she had won the Nobel Prize. In 1991 she was given a place in the Inventors' Hall of Fame, the first woman to be so honored. She also received the National Medal of Science that year. She is included in the National Women's Hall of Fame and the Engineering and Science Hall of Fame as well. In 1997 Elion received the Lemelson/MIT Lifetime Achievement Award. Elion died on February 21, 1999.

Enders, John Franklin

(1897–1985)
American
Microbiologist, Virologist

John Franklin Enders won the 1954 Nobel Prize in physiology or medicine with FREDERICK CHAPMAN ROBBINS and THOMAS HUCKLE WELLER for their cultivation of the polio virus on human tissue in a test tube for the first time. This development led directly to the development of a polio vaccine by ALBERT BRUCE SABIN and JONAS EDWARD SALK. Enders also demonstrated that the polio virus could grow on various tissues, not just on nerve cells, as was previously believed.

Enders was born on February 10, 1897, in West Hartford, Connecticut, to John Enders, a wealthy banker, and Harriet Whitmore. Enders married Sarah Bennett in 1927, and together the couple had two children—John Enders II and Sarah Steffian. His first wife died in 1943, and in 1951 Enders married Carolyn Keane.

World War I interrupted Enders's junior year at Yale University, where he commenced his undergraduate study in 1914. He served as a flight instructor and lieutenant in the U.S. Naval Reserve Flying Corps; he graduated from Yale after the war in 1920. He proceeded to Harvard University to earn an M.A. in English literature in 1922. Enders then shifted direction by enrolling at Harvard Medical School, where he specialized in microbiology under the influence of Hans Zinsser. Enders earned his Ph.D. in 1930 with a dissertation on anaphylaxis, an allergic reaction to foreign protein.

Enders commenced his long career as a member of the Harvard faculty while still a student, with an assistantship in the department of bacteriology and immunology, in 1929. In 1935 Harvard promoted him to the status of assistant professor. Enders attained the level of associate professor in 1942. During World War II he served as a civilian consultant on infectious diseases to the U.S. War Department, a role he retained after the end of the war until 1949. Harvard named Enders a full professor in 1956, and then in 1962 it granted him its highest title, university professor.

Enders happened upon his discovery about the polio virus by accident. His assistant, Thomas Huckle Weller, prepared an excess of cultures of human embryonic tissue for an experiment on chicken pox; instead of letting them go to waste, Enders honored his affiliation with the National Foundation for Infantile Paralysis by introducing the polio virus into the extra cultures, which yielded growth. *Science* magazine reported this development in a 1949 issue, dispelling the myth that the polio virus grew only on nerve cells and revealing that it could grow on human tissue. This revelation facilitated research on the virus immensely, opening the door for Sabin and Salk's development of the polio vaccine. Though Enders did not develop this vaccine, he did develop the measles vaccine in 1957.

Besides the Nobel Prize, Enders won multiple other awards, including the 1955 Kyle Award from the U.S. Public Health Service, the 1963 Presidential Medal of Freedom, and the 1963 American Medical Association Science Achievement Award, one of the few times it was awarded to a nonphysician. Enders died of heart failure on September 8, 1985, at his summer home in Waterford, Connecticut.

Erasistratus of Chios

(c. 304 B.C.–250 B.C.)
Greek
Anatomist, Physiologist

Although none of Erasistratus' prolific writings has survived to this day, his numerous theories were later referenced and often employed by GALEN. Erasistratus made significant contributions to the fields of human anatomy and comparative anatomy. His detailed research, which relied on vivisections of mammals and postmortem examinations of humans, led him to make a number of important discoveries about the structure and function of the cardiovascular, digestive, and nervous systems. Not only did he correctly posit that the heart functioned as a pump, he also demonstrated that both veins and arteries originated at the heart and extended through the entire body. He founded a school of anatomy in Alexandria and was the first well-known proponent of pneumatism, a physiological theory that connected life with a subtle vapor called pneuma.

Erasistratus was born in approximately 304 B.C. in Ilius, Chios (or Ceos), a Greek island. His interest in medicine was surely cultivated at a young age. Not only was his father, Cleombrotus, a doctor, but his mother, Cretoxene, was the sister of the physician Medios. Erasistratus, like his brother, Cleophantus, joined the family profession. After studying medicine with Metrodorus in Athens, he entered the university in Cos in about 280 B.C. The medical school followed the teachings of Praxagoras.

Erasistratus remained in Cos for most of his career. Because the city had close political and cultural ties to Alexandria, Erasistratus was influenced by a number of medical theories and practices that flourished there. Unlike the Greeks, the Ptolemies, who ruled Alexandria (and Egypt) during this period, were not opposed to the notion of human dissection. Erasistratus was thus able to conduct extensive postmortem examinations of human bodies, as well as the dissections of birds and mammals. From these experiments Erasistratus was able to draw a number of conclusions about physiology and anatomy.

In his study of the human digestion system Erasistratus located the epiglottis and accurately described its function as being responsible for closing the larynx during swallowing to prevent food from entering the trachea. With this discovery he disproved the older theories of Plato and Diocles, who had held that it was possible for liquids consumed to enter the lungs. Erasistratus also proposed, contrary to ARISTOTLE's belief that the digestion of food was akin to cooking, that food was broken apart by the peristaltic motion of the stomach muscles. In addition, he correctly described these muscles. Erasistratus also identified significant details about the structure of the brain. He distinguished the cerebellum and the cerebrum and noted the cerebral ventricles within the brain, as well as the membranes that cover the brain. Unlike Aristotle and Praxagoras, Erasistratus held that the brain, not the heart, was the center of intelligence. He also correctly correlated the greater number of convolutions in the human brain to higher brain function and intelligence. Moreover, Erasistratus' theories about the vascular system, including that the heart acted as a pump, were far more advanced than those of his predecessors. His pneumatic theory informed much of what he asserted about circulation. For Erasistratus pneuma entered the body from the outside and infused living creatures with a sort of life force.

Later in his life Erasistratus moved to Alexandria, where he founded a school of anatomy and devoted himself to full-time research. The date and cause of his death are unknown, though legend has it that he committed suicide because of an incurable ulcer on his foot. Erasistratus greatly influenced Galen and is considered to be one of the founders of modern medicine.

Eratosthenes

(c. 276 B.C.–c. 194 B.C.)
Greek
Geographer, Astronomer

Though Eratosthenes was a polymath, contributing to the fields of astronomy, drama, geography, mathematics, and poetry, he is best remembered for his simple but elegant formula for calculating the Earth's circumference. On the basis of the assumption that the Sun is sufficiently far away that its rays strike the Earth in parallel lines, Eratosthenes used the known distance between two cities in conjunction with the angle of the Sun's rays in those locations to calculate the portion of the Earth's circle represented by that distance. Taking into account the imprecision of Eratosthenes' measuring techniques, scientists estimate that his calculation was a mere 50 feet off the modern calculation of the Earth's actual circumference.

Eratosthenes was born in about 276 B.C. in Cyrene, in what is now Libya. Little is known of his history, as most of the knowledge about him derives from secondary sources citing his works, which have since been lost. It is known that he studied in Athens before moving to Alexandria, where he tutored the son of Ptolemy III before becoming the chief librarian at the Alexandrian Museum in about 255 B.C. Scholars surmise that he took full advantage of the academic resources available to him to inform his writings, though it is equally clear that he possessed an independent intellect capable of creating solutions to complex problems.

The Geography, comprising three books, represents Eratosthenes' masterwork. In it he sought to correct the Ionian map. He based his calculation of the Earth's circumference on several known facts, inventing an ingenious relationship between these facts. He derived the distance between Alexandria and Syrene (now Aswan) from the time a camel train took to travel that distance, which he estimated to be 5,000 stadia. He also knew that at midday on the summer solstice the Sun shone directly down a deep well in Syrene, so he measured the angle of the Sun's rays in Alexandria at that same moment; he found the angle to represent 1/50 of a whole circle, and thus the whole circle must measure 250,000 stadia around. The precision of this number cannot be verified since the modern equivalent of a stadia is not known with any degree of accuracy. In these books Eratosthenes also divided the terrestrial globe into zones—two frigid zones at the poles, then two temperate zones, and a torrid zone between the two tropics, bisected by the equator.

Sources reveal that Eratosthenes wrote many other works and made many other contributions to intellectual life. In *The Platonicus* he discussed proportion and progression and developed from these a theory of musical scales. In *Chronography and Olympic Victors* he established a chronology of his culture's history, setting dates for important events, such as the fall of Troy. He was considered an expert on drama and wrote *On the Old Comedy* in a series that included no fewer than 12 books of literary criticism. He also composed poetry in the books *Hermes, Erigone,* and *Anterinys/Hesiod Poetry.* In mathematics he solved the problem of doubling the cube and described a mechanism for arriving at this measurement. He also invented the sieve of Eratosthenes, a method of finding prime numbers by filtering out composites. He measured the obliquity of the ecliptic, or the tilt of the Earth's axis: 23°51′20″. He became blind late in life, and rather than suffer his inability to read, he committed suicide by starvation.

Erlanger, Joseph
(1874–1965)
American
Physiologist

Joseph Erlanger ushered in the era of modern neurophysiology with his pioneering research on the nervous system. He won the 1944 Nobel Prize in physiology or medicine with HERBERT SPENSER GASSER for their joint discovery of nerve fibers within the same nerve cord that had different functions.

Erlanger was born on January 5, 1874, in San Francisco, California, the sixth of seven children—five boys and two girls. Both of his parents had emigrated from

Joseph Erlanger, whose seminal research on the nervous system ushered in the era of modern neurophysiology. *Bernard Becker Medical Library, Washington University School of Medicine.*

Württemberg, southern Germany, in 1842 at age 16. The Gold Rush drew his father, Herman Erlanger, to California; however, he had more luck in business than in prospecting. In 1849 the elder Erlanger married Sarah Galinger, the sister of his business partner. Erlanger himself married Aimée Hirstel on June 21, 1906, and together the couple had three children—Margaret, Ruth Josephine, and Hermann.

Erlanger attended the College of Chemistry at the University of California at Berkeley from 1891 through 1895, when he earned his bachelor's degree for his study of the development of newt eggs. He graduated second in his class from Johns Hopkins University School of Medicine in 1899. In medical school he spent the summer of 1896 in Lewellys Barker's histology laboratory locating horn cells in the spinal cord of rabbits, and the subsequent summer he studied the digestive system of dogs.

In 1901 Erlanger based his first publication on this canine research in the article "A Study of the Metabolism of Dogs with Shortened Small Intestines." This report caught the attention of William H. Howells, the professor of physiology at Johns Hopkins, who recruited Erlanger as an assistant professor of physiology. He was promoted to associate professor before 1906, when he became the first to hold a chair of physiology at the medical school of the University of Wisconsin at Madison. He left this appointment in 1910 bound for St. Louis, where he became a professor of physiology at the better-funded Washington University School of Medicine; he ascended to the position of department chair and remained there until his retirement.

In 1904 Erlanger hand-constructed a sphygmomanometer, improving on previous designs. He used the apparatus to study blood pressure and published his results in the article "A New Instrument for Determining the Minimum and Maximum Blood-Pressures in Man." Soon after he arrived in St. Louis his former student, Gasser, joined him at Washington University, and the two collaborated on significant research into the neural system. In 1922 the pair succeeded in amplifying the electrical responses of a single nerve fiber. A decade later they discovered different thresholds of excitability in certain nerve fibers, which led to their formulation of the law that the velocity of nervous impulses is directly proportional to the diameter of the nerve fibers. Erlanger and Gasser achieved these results with the help of an amplified cathode-ray oscilloscope, the instrument that became the basis of modern neurophysiological research.

Late in life Erlanger suffered the losses of his wife, his son Hermann, and his son-in-law with dignity. He died of heart failure on December 5, 1965, in St. Louis, Missouri.

Esaki, Leo

(1925–)
Japanese
Physicist

Leo Esaki opened up a new field of scientific inquiry by applying the highly theoretical realm of quantum mechanics to the highly practical realm of semiconductors. He correctly hypothesized that an electrical current could "tunnel" through a semiconductor in a way that defied classical conceptions of physics but obeyed the laws of quantum physics. This work resulted in his discovery of the tunnel diode, otherwise known as the Esaki diode. It also resulted in the 1973 Nobel Prize in physics, which he shared with Ivar Giaevar and Brian Josephson.

Esaki was born on March 12, 1925, in Lsaka, Japan. His father, Soichiro Esaki, was an architect, and his mother was Niyoko Ito. Esaki attended Tokyo University, where he earned his master's degree in physics in 1947 and his Ph.D. in 1959. That year he married Masako Araki, and together the couple had three children—Nina, Anna, and Eugene.

While working on his doctorate Esaki simultaneously worked in the corporate world. He accepted a position at the Kobe Kogyo company in 1947, the year he received his master's degree. The Sony Corporation recruited him in 1956 as the chief physicist of a small team of researchers. In 1960 he joined IBM at the Thomas J. Watson Research Center in Yorktown, New York. Though he intended to stay only for a year, he remained with IBM until 1993, retaining his Japanese citizenship while living in the United States. IBM granted him a fellowship in 1965, and throughout his tenure at IBM he maintained academic appointments at the University of Pennsylvania and Tokyo University. In 1993 he accepted the presidency of Tsukuba University, charged with the goal of fulfilling its mission of becoming an outstanding technological institution.

In 1957 Esaki came up with the idea of "tunneling" by considering the disjunction between classical and quantum physics. Classical physics ruled that an electrical current could not pass through an insulator because of its high resistance. However, quantum physics allowed for indeterminacy, as the exact position of electrons can never be verified, thus allowing for the possibility of passing electricity through an insulator. Esaki experimented with different conditions that might encourage this phenomenon and discovered that by "doping" the semiconductors, or adding impurities to them, he made them passable under quantum conditions. The resulting effect was a backward diode, or the opposite polarity of a standard diode, which became known as an Esaki diode. This technology could operate at very high speeds with little noise and power consumption while maintaining a small size and thus found use in computer and microwave technology.

In addition to the Nobel Prize, Esaki won such prestigious awards as the 1961 Morris N. Liebman Memorial Prize of the Institute of Radio Engineers, the 1961 Stuart Ballantine Medal from the Franklin Institute, the 1991 Medal of Honor from the Institute of Electrical and Electronic Engineers, and the Japanese Nishina Memorial Award. Esaki gained a lasting place in scientific history by bridging the gap between the theoretical and the practical, between quantum physics and hands-on technology.

Euclid

(c. 300 B.C.)
Greek
Mathematician

Euclid played a significant and influential role in the development of Western scientific and mathematical thought. His main text, the 13 books of *The Elements,* reportedly has had more reprints and commentaries published on it than the Bible. In it he laid the foundations of mathematical reasoning and the systematic organization of ideas, a style adopted by many other schools of thought and writings. Ironically Euclid was not a particularly original, strong, or innovative mathematician; he was, however, an excellent teacher, judging from his texts, and his genius resided in his ability to synthesize diverse existing knowledge with new ideas to fill in the gaps of a coherent overall system. This system held sway until the advent of non-Euclidean geometry in the 19th century and Einstein's application of his theory of relativity to spatial concepts in the 20th century, which required a new geometry.

Though his texts have survived more than 2,000 years, biographical details have disappeared, leaving scant information about his life. Only one aspect has been confirmed as fact—that he founded and taught at a school in Alexandria during the reign of Ptolemy I Soter (323 B.C.–285/283 B.C.). He may also have been a pupil of Plato in Athens.

The Elements, a treatise on geometry, consisted of 13 books—6 books on plane geometry, 4 on the theory of numbers (including a proof that there is an infinite number of prime numbers), and 3 on solid geometry. Euclid based his work on that of his predecessors, namely, Eudoxus, whose discoveries fill Books V and XII, and Theaetetus, whose discoveries fill books X and XIII. His most recent predecessor was Theudius, who compiled the mathematics text supposedly used during ARISTOTLE's time while a student at the Academy of Athens. The strength of Euclid's own explication often depended on the quality of his sources. Though he borrowed liberally from sources, the overall design of the work was original. This design relied on the axiomatic method, in which Euclid proposed 10 basic axioms, on which he based all subsequent theorems. This method set the precedent for the organization of many ensuing texts and schools of thought and ensured the enduring strength of his text.

The manuscript of *The Elements* was translated first into Arabic; it was translated into Latin in the 12th century and printed in the 13th century. The 16th century saw the first direct Latin translation from the Greek. By the early 20th century the definitive English edition of *The Elements* was in print. Euclid produced many other, less significant works, including *Data, On Divisions, Optics,* and *Catoptrica.* Several texts ascribed to Euclid have been lost, including *The Pseudaria* (The Fallacies), the three books of *The Porisms* (The Corollaries), the four books of *The Conics,* and the two books of *The Surface-Loci.*

The most famous statement attributed to Euclid is his response when Ptolemy inquired whether there were an easier way to learn geometry than by studying *The Elements,* to which Euclid reportedly replied, "There is no royal road to geometry."

Euler, Leonhard

(1707–1783)
Swiss
Mathematician

Leonhard Euler, one of the most prolific mathematicians in history, is considered one of the founders of pure mathematics and analysis. He utilized his innate calculating abilities on both pure and applied mathematics, and he devoted his arithmetic and analytical skills to practical concerns, such as mechanics and acoustics.

Euler was born on April 15, 1707, in Basel, Switzerland. His father, Paul Euler, a Calvinist minister, and his mother, Margarete Brucker, married the year before his birth. Leonhard married Katharina Gsell, the daughter of the Swiss painter George Gsell, in 1733, and their first son, Johann Albrecht, was born the next year. Their second son, Karl, was born in 1740. The couple had a third son, Christoph, and two daughters; eight of their children died in infancy. In 1735 Euler lost sight in his right eye after observing the Sun.

Euler's father, an amateur mathematician, introduced his young son to Christoff Rudolf's *Algebra,* an exceedingly difficult text for such a young boy. In 1720 Euler entered the University of Basel at the age of 14; there he discussed mathematics with DANIEL BERNOULLI. In 1722 Euler earned a bachelor of arts, and in 1723 he earned a master's degree in philosophy.

Daniel Bernoulli used his influence to secure Euler an associateship in physiology at the newly established St. Petersburg Academy of Sciences, where they both commenced work in 1727. In 1731 Euler moved to the department of physics as a professor, and in 1733 he succeeded Bernoulli in the mathematics chair. Frederick the Great invited Euler to become a member of the Berlin Academy in 1741, and he held joint appointments in Berlin and in St. Petersburg until he was appointed the director of the St. Petersburg Academy of Sciences in 1766.

It would be difficult to catalog all of Euler's writings since he was so productive in his career. During his tenure in Berlin alone he wrote 380 works, about 275 of which were published. His most important publications, however, include his 1736 treatise *Mechanica,* which transformed mechanics through the application of analysis, and his 1748 publication, *Introductio in analysin infinintorum,* which set the precedent for expressing mathematics and physics in arithmetical terms and helped establish modern analytical geometry and trigonometry. Euler investigated the three-body problem of the relationship of the motions of the Sun, Moon, and Earth and published two important theories of lunar motion, the first in 1753 and the second in 1772. Euler also published two highly influential textbooks, *Institutiones calculi differentialis* in 1755 and *Institutiones calculi integralis* in 1768 through 1770.

Euler's name is associated with numerous mathematical concepts, the most prominent of which is *Euler's e,* which refers to Euler's rule regarding polyhedrons. Euler made contributions to the understanding of diverse fields, including number theory, algebra, infinite series, the concept of function, differential equations, the calculus of variations, geometry, hydromechanics, astronomy, and physics. Euler submitted many of his papers to the St. Petersburg Academy, and between 1738 and 1772 he won 12 prizes for them. In 1749 the Royal Society of London elected him a member, as did the Society of Physics and Mathematics in Basel in 1753 and the Académie des Sciences of Paris in 1755. Euler's devotion to his work lasted until the day he died, September 18, 1783, in St. Petersburg, Russia, where he spent time formulating laws of ascent for the recently invented hot-air balloon.

Alice Evans, whose research on bacteria found in fresh cows' milk led to the understanding that humans could contract brucellosis by drinking fresh cows' milk. *National Library of Medicine, National Institutes of Health.*

Evans, Alice Catherine

(1881–1975)
American
Microbiologist

Alice Evans showed that a dangerous disease was transmitted in fresh milk, forcing the dairy industry to begin heat-treating milk to kill bacteria. She was born in Neath, Pennsylvania, on January 29, 1881. Her father, William, was a farmer, surveyor, and teacher. Alice obtained a minimal education at the Susquehanna Institute in Tonawanda, then taught school for 4 years. A Cornell University nature study course for teachers turned her interest toward science, and she enrolled in the Cornell College of Agriculture, from which she earned a B.S. in 1909. She chose bacteriology as her specialty.

Evans won a scholarship to do graduate work at the University of Wisconsin at Madison, from which she earned a master's degree in 1910. She then joined the Dairy Division of the Bureau of Animal Industry of the U.S. Department of Agriculture, working at first at its branch on the Madison campus. When the division gained permanent research laboratories in Washington, D.C., in 1913, Evans transferred there. She was the first woman given permanent employment in the Dairy Division.

Evans worked with a group looking for ways to prevent disease-causing bacteria from contaminating fresh milk. Independently she also studied the bacteria in uncontaminated milk, which was thought to be safe to drink. She was especially interested in two types of supposedly unrelated bacteria. One, *Bacillus abortus,* caused a contagious disease that made pregnant cattle miscarry. The other, *Micrococcus melitensis,* produced a debilitating and sometimes fatal human illness that was called undulant fever because of its pattern of rising and falling body temperature. It had first been identified in British soldiers on Malta who drank milk from infected goats. Evans discovered that *B. abortus* was common in the milk of apparently healthy cows. She also found that *B. abortus, M. melitensis,* and a third microbe from pigs were almost identical. Together these facts suggested to her that a germ often found in fresh cow's milk could cause human disease.

When Evans presented her findings at a meeting of the Society of American Bacteriologists (now the American Society of Microbiology) in 1917, other bacteriologists were skeptical. (Evans commented later that at least one may have opposed her because he "was not accustomed to considering a scientific idea proposed by a woman.") In 1920, however, some other bacteriologists confirmed her work, reclassifying her goat, cow, and pig bacteria into a new genus, *Brucella.* The disease they caused was renamed *brucellosis.* By the end of the decade reports from all over the world proved Evans's claim that humans could contract brucellosis by drinking fresh cow's milk.

Evans showed in the 1930s that brucellosis had a chronic or long-lasting form that had previously been unknown because it mimicked other diseases. This explained why the number of human brucellosis cases had appeared to be small even though infection in cows was common. It turned out that there were 10 times as many cases of brucellosis in the United States as had been thought. Evans herself contracted chronic brucellosis in 1922 and suffered from it for 23 years, until it became treatable with antibiotics.

Evans had pointed out from the start that the threat of brucellosis and other diseases carried in milk could be removed by a heat treatment called pasteurization, invented by French bacteriologist LOUIS PASTEUR in the 1860s. Dairies had resisted pasteurization because it required buying new equipment, but by the 1930s they were finally persuaded to begin using it. Pasteurization is required for all milk sold in the United States today.

Evans's work won recognition and awards, including several honorary degrees. In 1928 she became the first woman to be elected president of the American Society for Microbiology. Around 1939 she turned to research on streptococci, bacteria that infect wounds, and continued this work until her retirement in 1945. She died on September 5, 1975. According to the science historian Elizabeth O'Hern, Evans's work on brucellosis "has been cited as one of the outstanding achievements in medical science in the first quarter of the 20th century."

Ewing, William Maurice
(1906–1974)
American
Oceanographer

Ewing utilized his innovative seismic refraction technique to measure the thickness of the oceanic crust, which turned out to be only 3 to 5 miles thick, much thinner than the continental crust, which averages approximately 25 miles thick. Ewing also investigated the extent of midocean ridges, which were discovered when transatlantic cable was laid earlier in the century but had not been well defined. Ewing and his colleagues discovered a central rift in this ridge.

Ewing was born on May 12, 1906, in Lockney, Texas. He attended Rice University, in Houston, Texas, and received his Ph.D. in 1931, then went on to teach at Lehigh University in Pennsylvania from 1934 to 1944. In 1944 Ewing joined the faculty of Columbia University, where he also worked at the Lamont-Doherty Geographical Observatory in New York, helping to organize it in its early days. In 1949 he became the director of the observatory and transformed it into one of the leading geophysical research facilities in the world. A decade later Columbia promoted him to the position of full professor of geology.

Ewing's oceanographic work set precedents for research procedures, as he used technology inventively to achieve his purposes. For example, in 1935 Ewing recorded the first seismic measurements in open sea, a technique that he used extensively throughout his career. He developed another innovative technique in 1939, taking the first deep-sea photographs.

In 1956 Ewing and his colleagues set out to map the Mid-Atlantic Ridge, whose existence was known but dimensions unknown. This investigation resulted in the discovery that the ridge was similar to a long mountain range, in that it extended some 40,000 miles throughout the underwater world. In 1957 Ewing along with Marie Tharp and Bruce Heezen discovered a central rift that divided the ridge. This rift was, in places, twice as wide and as deep as the Grand Canyon. Ewing also investigated the Mediterranean Sea and the Norwegian seas.

Ewing and his colleagues expected to find deep deposits of oceanic sediment, possibly up to 10,000 feet thick, surrounding the ridge, but to their surprise they found little or no sediment within about 30 miles of the ridge. After this zone the sediment grew in thickness to a mere 130 feet, still far below the projections. Ewing hypothesized that these discoveries could support the theory propounded by Harry H. Hess that the ocean floor was engaged in a spreading process.

Ewing reserved his full validation of this theory until it could be proved. Later, Frederick Vine and Drummond H. Matthews devised a method to prove it, utilizing magnetic reversals discovered by B. Bruhnes in 1909. Ewing also suggested in conjunction with William Donn a theory to explain the periodic ice ages, which was based on the variation of freezing waters in the Arctic.

Ewing's important texts include *Propagation of Sound in the Ocean,* published in 1948; *Elastic Waves in Layered Media,* published in 1957; and *The Floors of the Oceans: I. The North Atlantic,* published in 1959. Ewing died on May 4, 1974, in Galveston, Texas.

F

Faber, Sandra Moore

(1944–)
American
Astronomer

Sandra Faber has provided ground-breaking new information about the material that makes up galaxies and how those galaxies are formed. She has helped to show that the universe is "lumpy," with clusters of galaxies drawing together to form still larger aggregates. She has also played a major role in the repair of the Hubble space telescope and the construction of the Earth's largest optical telescopes, the twin 400-inch Keck telescopes in Hawaii.

Sandra Moore was born in Boston on December 28, 1944, but grew up in Cleveland, Ohio. When she was a child, she told the *Omni* interviewer Paul Bagne, "science was as natural to me as breathing." Her father, a civil engineer, encouraged her interest in astronomy by buying her a pair of binoculars to help her observe the stars.

A favorite teacher at Swarthmore College inspired Moore to make astronomy her career. Moore graduated from Swarthmore with high honors in physics in 1966 and a year later married Andrew Faber, a physicist she had met at college (he later became an attorney). She earned her Ph.D. in astronomy from Harvard in 1972. The Fabers then moved to northern California, where Sandra joined the Lick Observatory of the University of California at Santa Cruz, becoming the observatory's first female member. Soon afterward she gave birth to a daughter, the first of two. In 1975 she and a fellow Lick astronomer, Robert Jackson, discovered the Faber-Jackson relation between the size and brightness of elliptical galaxies and the speeds of stars orbiting within them. This was the first of several major research advances by Faber.

Around 1979 Faber and other astronomers proved that about 90 percent of the matter in the universe is dark, or invisible. She and two Santa Cruz colleagues proposed in 1984 that this dark matter was "cold" and consisted of relatively massive subatomic particles. Their theory also stated that galaxies developed from relatively dense "seeds" that were formed soon after the big bang, the gigantic explosion believed to have given birth to the universe. This was the first comprehensive theory of how galaxies evolved, and although some details of it are being modified, Faber says the theory is still "the current working paradigm for structure formation in the Universe." She now believes that galaxies may consist of a mixture of cold dark matter and ordinary matter.

Ever since the big bang, all objects in space have been streaming away from each other. While measuring the motion of certain galaxies in the late 1980s, Faber and six coworkers, later nicknamed the "Seven Samurai," found other "peculiar" kinds of motion occurring as well, causing nearby galaxies to move faster than expected. They concluded that gravity is pulling our local supercluster of galaxies toward a section of the sky that they called the Great Attractor. This area, about 150 million light-years from Earth and 300 million light-years across, has the mass of tens of thousands of galaxies. Faber now thinks the Great Attractor and its attendant galaxies, in turn, are flowing toward a still larger mass somewhere else. The discovery of the Great Attractor, like the work of VERA COOPER RUBIN and MARGARET JOAN GELLER, suggests that the universe is far "lumpier" than had been thought.

Faber has also improved the technology with which she and other astronomers study the universe. In the late 1980s she was one of three scientists who diagnosed the flaw in the mirror of the Hubble space telescope, and she also helped to design the procedure that repaired it. That project was exhausting but was also, she has written, "the most exhilarating phase of her career" so far. She has also played a major part in managing the construction of the Keck telescopes, built in the 1990s, and she is currently building a new spectrograph for the second Keck, which will increase its power to observe distant galaxies by 13 times.

Faber has received many awards for her work, including the Bok Prize of Harvard University (1978) and the Heineman Prize of the American Astronomical Society (1986). She has been elected to both the National Academy of Sciences (1985) and the American Academy of Arts and Sciences (1986). Since 1996 she has been one of the three University Professors of the University of California.

Faber has used the Hubble space telescope to study the centers of galaxies, which, according to evidence she and others have gathered, often hide massive black holes. She has also observed extremely distant galaxies, which date from early in the formation of the universe. As she expected, they are smaller and less organized than later galaxies, but she does not yet know why they changed in exactly the way they did. She has stated that she hopes her work will help her discover how galaxies developed in the early universe.

Fahrenheit, Gabriel Daniel

(1686–1736)
German
Physicist

Although Gabriel Daniel Fahrenheit received little formal training in science, he developed a standard thermometric scale that revolutionized experimental physics. The Fahrenheit scale—as his innovation came to be called—enabled scientists in different locations to compare temperatures for the first time since the invention of the thermometer. An instrument maker, Fahrenheit was the first to create a thermometer that used mercury. He also discovered that different liquids have distinct boiling points.

Fahrenheit was born on May 24, 1686, in Danzig (now Gdansk), Poland, into a family of wealthy merchants. His parents, Daniel and Concordia Schumann Fahrenheit, had four other children, of whom Gabriel was the oldest. When his parents died unexpectedly in 1701, Fahrenheit's guardian sent him to Amsterdam to acquaint himself with the business that had generated his family's wealth.

After his arrival in Amsterdam Fahrenheit became interested in the small but burgeoning industry of creating scientific instruments. He also worked as a glassblower. In 1707 Fahrenheit embarked on an extensive period of travel, which took him to some of Europe's largest cities. During this time he met and observed other scientists and instrument makers, including Olaus Roemer, a Danish astronomer who had conceived an earlier (and unsuccessful) thermometric scale. In 1714 Fahrenheit built the first mercury thermometer. The original thermometer, invented by GALILEO GALILEI around 1600, had relied on changes in air volume to measure temperature shifts. But when it became apparent that the volume of air changes with atmospheric pressure, scientists substituted many different kinds of liquid in thermometers. Alcohol was the most common substance in use before Fahrenheit introduced mercury.

Fahrenheit returned to Amsterdam in 1717; there he continued his career as an instrument maker and began the work that would be his legacy—the thermometric scale that would later bear his name. At the time no standard scale had been invented that would allow for accurate comparisons of temperatures—each thermometer demarcated temperature in its own way. Building on Roemer's efforts, Fahrenheit chose to base his system on two fixed points. Fahrenheit selected the temperature at which a mixture of water and salt melted as the lower of these two fiducial points (this was, in fact, the lowest temperature he could obtain). The higher point was the temperature of the human body. He then divided this range into 96 equal parts. The freezing temperature of water fell at 32° and the boiling point at 212°.

Fahrenheit made other important discoveries as well. He invented a superior hygrometer, which is a device used to measure the amount of moisture in the air. Moreover he determined that other liquids besides water have fixed boiling points and that these points are affected by atmospheric pressure. In 1724 he published his only scientific work, *Philosophical Transactions*.

Despite his lack of formal training, Fahrenheit was admitted to the Royal Society in 1724 on the strength of his achievements. But the impact of the Fahrenheit scale went far beyond garnering the accolades of his contemporaries. The scale allowed for precise temperature data to be obtained and compared, fundamentally promoting the progress of experimental physics in the 18th century. Fahrenheit's scale soon became standard throughout the Netherlands and Britain. It remains the most common temperature scale used in English-speaking countries.

Faraday, Michael

(1791–1867)
English
Physicist, Chemist

Michael Faraday, a humble and devoutly religious man, rose from extreme poverty to become one of the greatest experimental physicists in history. He educated himself and applied his innate scientific skills first to chemical experimentation, and later to the study of electricity.

Faraday was born on September 22, 1791, in Newington, Surrey, England. He was the third of four children born to James Faraday and Margaret Hastwell. James Faraday was a blacksmith whose ill health often prevented him from working, leaving the support of his family to his

Michael Faraday, who in 1821 discovered the principles of the electric motor. *Photo by John Watkins, courtesy AIP Emilio Segrè Visual Archives.*

wife. These abject circumstances created close connections between Faraday and his siblings, especially with Robert, his older brother, who became a gas fitter, and Margaret, his younger sister, whose education he helped support. Between the time of his father's death in 1809 and his mother's in 1838 Faraday married Sarah Barnard in 1821. The couple never had children.

Faraday quit school in 1804, when he was 13 years old, to help support the family as an apprentice bookseller and binder. His reading of an *Encyclopaedia Britannica* entry on electricity represented a formative educational experience, prompting him to join the City Philosophical Society and to pursue his interests in science. In 1812 he attended talks by the scientist Sir HUMPHRY DAVY, and he later bound the extensive notes he had taken at these lectures. He showed Davy these notes while working briefly for him, prompting Davy to hire Faraday full time as a laboratory assistant in 1813. Faraday traveled with Davy on a continental European tour between 1813 and 1815 and met many of the significant scientists of the day, furthering his education by this direct contact with great minds.

In 1815 the Royal Institution appointed Faraday as the assistant and superintendent of the apparatus of the laboratory and the meteorological collection, commencing a relationship that lasted until 1861. In 1825 the institution promoted him to the position of director of laboratories, and in 1833, to the position of Fullerian Professor of Chemistry.

Faraday focused on chemical experimentation in the early part of his career, and he was the first to liquefy chlorine, in 1823. Two years later he isolated benzene. Faraday had set the foundation for his later work on electricity with his 1821 discovery of the principle of the electric motor. In 1831 he proposed Faraday's law of induction and in 1833 Faraday's laws of electrolysis. In 1845 he described the Faraday effect, or the rotation caused by a magnetic field on a plane of polarization. Faraday's publications reveal his broadening interests, as early texts such as *Chemical Manipulation,* published in 1827, specialized in one topic, whereas later work, such as the posthumously published *On the Forces of Nature* (1873), focused on broader topics. Other significant titles included *Experimental Researches in Electricity,* published between 1839 and 1855; *Experimental Researches in Chemistry and Physics,* published in 1859; and *A Course of Six Lectures on the Chemical History of a Candle,* published in 1861.

The royal offer of knighthood to Faraday demonstrates his importance as a historical figure, but his refusal of this honor reveals his integrity and commitment to his own principles of humility. Faraday died on August 25, 1867, in Hampton Court, Surrey.

Fawcett, Stella Grace Maisie

(1902–1988)
Australian
Botanist

In pioneering fieldwork on Australia's high plains Maisie Fawcett showed how overgrazing affected plant life and led to soil erosion. Stella Grace Maisie Fawcett was born in 1902 in Footscray, a suburb of Melbourne, and grew up there. After several years of teaching school she won a scholarship to the University of Melbourne at age 20. Told that her first career choice, geology, was "not for women," she changed to botany. She earned an M.S. in 1936 and remained at the university as a demonstrator and researcher.

Cattle and sometimes sheep had grazed on the high plains of the Australian state of Victoria since the mid-19th century. By the early 1940s the toll they had taken on the land was becoming clear. Victoria's Soil Conservation Board (SCB) asked John Turner, head of the University of Melbourne's botany department, to find a "suitable man" to study the effect of grazing on the plant life of the plains. World War II made men scarce, so Turner suggested that the best "man" for the job was Maisie Fawcett.

The 29-year-old Fawcett reached the Bogong High Plains in September 1941 and fenced off two study areas on the steep, forest-covered slopes. The fences kept grazing animals out, so she could note which plants sprang up there. Linden Gillbank describes some of the hardships of Fawcett's work: "Thistles ripped through her clothes, it poured [rain] non-stop for days, fences were damaged, and . . . she was absolutely physically exhausted from riding and recording."

Dealing with the ranchers was difficult, too. Fawcett wrote: "The Board will never realize the amount of charm, pasture wisdom, and general knowledge I have expended on the locals to get them to . . . stir themselves." The ranchers at first were amused by Fawcett, nicknaming her "Washaway Woman" (a washaway was an erosion gully) or "Erosion Girl," but in time they became fond of her.

During the early 1940s Fawcett showed that overgrazing was causing erosion in both lower and higher pastures. She recommended allowing fewer cattle on the land, moving them to the high pastures later in the year (to give plants a head start on their growing season), and completely banning sheep, which graze down to the ground, from the area. The government backed her proposals, and the soil situation improved.

Fawcett moved back to Melbourne in 1949 and became a temporary lecturer in ecology at the university. She was made a permanent senior lecturer in 1952. She started a massive book on the plant life of Victoria, which the botany students who helped her dubbed "the Monster." Fawcett married Denis Carr, a fellow professor, in the 1950s, and they moved to Canberra in 1967. Fawcett was a visiting fellow at Australian National University until her death in 1988.

Fermi, Enrico

(1901–1954)
Italian/American
Physicist

Enrico Fermi stands out from other physicists in that he excelled in both theoretical and experimental physics. In the former category Fermi developed a statistical system for quantum mechanics that PAUL ADRIEN MAURICE DIRAC also arrived at independently, hence the name *Fermi-Dirac statistics*. Fermi also proposed a theory for beta decay. In the latter category Fermi discovered a method for inducing radioactivity with slow neutrons, though he did not recognize the process as nuclear fission. Later, working under the auspices of the Manhattan Project, he directed the first controlled nuclear chain reaction.

Enrico Fermi, who served as a member of the Manhattan Project and directed the first controlled nuclear reaction. *NARA, courtesy AIP Emilio Segrè Visual Archives.*

Fermi was born on September 29, 1901, in Rome, Italy, to Alberto Fermi, an administrator for the Italian railroads, and Ida de Gattis, a teacher. Fermi had an older sister, Maria; his older brother, Giulio, died when Fermi was 14 years old. In 1928 Fermi married Laura Capon, the daughter of an admiral in the Italian navy. The couple had two children—Nella in 1931 and Giulio in 1936.

In 1918 the Reale Scuola Normale Superior awarded Fermi a fellowship on the strength of his application essay, which evinced his genius. In 1922 he graduated magna cum laude from the University of Pisa with a doctoral dissertation he wrote on research that he performed by using X-ray experimentation.

Fermi then received a fellowship from the Italian Ministry of Public Instruction to study at the University of Göttingen under MAX BORN and at the University of Leiden under Paul Ehrenfest. In 1925 Fermi returned to Italy as a lecturer at the University of Florence. During his tenure there Fermi wrote the paper that applied WOLFGANG PAULI's exclusion principle on electrons to atoms in a gas, a similar proposition to Dirac's a few months later, which resulted in Fermi-Dirac statistics. On the strength of this important advance Fermi won the new chair in theoretical physics established in 1927 at the University of Rome. The next year he published *Introduction to Nuclear Physics,* the first Italian textbook on the subject. In 1933 Fermi rounded out his work with theoretical physics by proposing his theory for beta decay, in which he proposed the existence of a new kind of force resulting from neutrons transitioning from higher to lower states of energy, later called weak force.

In 1934 FRÉDÉRIC JOLIOT and IRÈNE JOLIOT-CURIE discovered artificial radioactivity by alpha bombardment. Fermi repeated the process, using neutrons instead of alpha particles to bombard stable isotopes as a means of converting them to unstable, radioactive elements. Fermi attempted this process with all of the elements and by accident discovered that bombarding through paraffin wax, which contained hydrocarbon molecules, slowed the neutrons enough to elicit more assured reactions. Uranium proved to be the most interesting reaction, as it created a new element, with an atomic number one higher than that of uranium. In fact, Fermi had discovered nuclear fission, a fact that became apparent when German scientists replicated the process.

Upon acceptance of his 1938 Nobel Prize in physics for this discovery Fermi and his family defected from Italy to the United States, where he became a professor of physics at Columbia University. In 1939 he drafted a letter carried to President Roosevelt by ALBERT EINSTEIN warning of the potential destructive power of nuclear weapons, especially in the hands of the Nazis. After Roosevelt instituted the Manhattan Project to build the first atomic bomb, Fermi played an instrumental role. He was responsible for the activation of an atomic pile on December 2, 1942, at 3:21 P.M., initiating a 28-minute self-sustaining nuclear chain reaction, the first step into the nuclear age.

The Fermis became U.S. citizens on July 11, 1944, and in 1946 the University of Chicago named him the Charles H. Swift Distinguished-Service Professor for Nuclear Studies. That year he received the Congressional Medal of Merit. After battling stomach cancer, he died on November 28, 1954, in Chicago. The U.S. Department of Energy awarded him the first Enrico Fermi Prize, named in his honor, posthumously. Element 100 was named fermium for him as well.

Feynman, Richard Philip

(1918–1988)
American
Physicist

Richard Philip Feynman's scientific career greatly contributed to the development of theoretical physics. He was awarded the 1965 Nobel Prize in physics for his work on the theory of quantum electrodynamics—the branch of physics that accounts theoretically for the interaction of electromagnetic radiation with atoms and their electrons. Feynman reformulated quantum electrodynamics through his path integral approach, which resolved several inconsistencies in the theory of quantum electrodynamics as it was first conceived in the 1920s. Moreover, his Feynman diagrams, which provided pictorial representations of the interactions of subatomic particles, greatly simplified quantum electrodynamics calculations.

Born in New York City on May 11, 1918, Feynman was the oldest surviving child of Melville Arthur and Lucille Phillips Feynman. His father, an emigrant from Minsk, Byelorussia, was employed as a uniform manufacturer.

Feynman entered the Massachusetts Institute of Technology (MIT) in 1936, with plans to major in mathematics. However, he became enchanted by the emerging field of nuclear physics and received a bachelor's degree in physics in 1939. Despite rabid anti-Semitism in Ivy League schools' admissions policies, Feynman was admitted in 1939 to Princeton University, where he studied under John Archibald Wheeler and received his Ph.D. in theoretical physics in 1942.

In 1943 Feynman was recruited by J. ROBERT OPPENHEIMER to join the Manhattan Project, devoted to creating the first atomic bomb. Feynman worked under HANS ALBRECHT BETHE in the theoretical division, where he calculated the speed at which neutrons diffuse through a critical mass of uranium or plutonium approaching the reaction that causes a nuclear explosion. Despite the scientific and moral challenges of the Manhattan Project, Feynman was more concerned with personal matters. In 1942 he had married Arline Greenbaum, who was afflicted with tuberculosis. She died on June 16, 1945, exactly one month before the first atomic bomb was successfully detonated. After his 1952 marriage to Mary Louise Bell ended in divorce in 1956, he wed Gweneth Howarth in 1960. The couple had two children.

After the war ended, Feynman accepted a position as associate professor of physics at Cornell University. There he immersed himself in the theory of quantum electrodynamics, which had been formulated in the 1920s by PAUL ADRIEN MAURICE DIRAC, WERNER KARL HEISENBERG, and WOLFGANG PAULI. The theory accounted for how light could be composed of both waves and particles but remained plagued by several inconsistencies, which Feynman resolved to address. Inspired by a student who was spinning a plate in the Cornell cafeteria, Feynman formulated the path integral approach, which posited that the probability of an event that can occur in a number of different ways (such as an electron's existing in a certain place) is the sum of all the probabilities of all the possible ways. Feynman expounded on this concept in a 1948 article in *Physical Review.* In 1949 he first presented his famous Feynman diagrams in the paper "Space-Time Approach to Quantum

Electrodynamics." The diagrams illustrated how subatomic particles interact through time and space.

In 1950 Feynman moved to the California Institute of Technology, where he would remain for the duration of his career. During a 1951 sabbatical he investigated the role of elemental particles called mesons in the cohesion of the atomic nucleus. Four years later he proposed an innovative model of the structure of liquid helium. In formulating a theory of superfluidity, Feynman was able to explain some of liquid helium's bizarre behavior, such as the way it appeared to defy gravity. He later theorized the existence of partons—the constituent elements of protons—thus helping forge the current conception of quarks (paired elementary particles). In 1986 Feynman was appointed to the presidential commission on the *Challenger* disaster; he explained that the space shuttle crash was the result of cold temperatures' affecting the O-rings that sealed the joints of the rocket boosters.

Feynman was recognized as one of the great physicists of the latter half of the 20th century. He shared the 1965 Nobel Prize with JULIAN SEYMOUR SCHWINGER and SHINICHIRO TOMONAGA and also received the 1964 Albert Einstein Award and the 1973 Niels Bohr International Gold Medal. His books, *Feynman Lectures on Physics* (1963) and *Surely You're Joking Mr. Feynman!* (1985), won him popular acclaim as well. Feynman died in 1988 after a long bout with cancer.

Fibonacci, Leonardo Pisano

(c. 1170–c. 1250)
Italian
Mathematician

Often considered the greatest European mathematician of the Middle Ages, Leonardo Pisano Fibonacci was best known for introducing the Hindu-Arabic numeral system to Europe and for developing a numeric sequence that later came to be known as the Fibonacci sequence. Fibonacci wrote several books on mathematics and was recognized by Frederick II, the Holy Roman Emperor from 1212 to 1250, for his mathematical genius.

Born around 1170 in Pisa, Italy, Fibonacci was the son of Guilielmo Bonacci, who worked as a customs officer of sorts. *Fibonacci,* a shortened form of *filius Bonacci,* meant "the son of Bonacci," and Fibonacci primarily used this name. In about 1192 Fibonacci's father was sent by the Republic of Pisa to Bugia (now Bejaia), located in northeastern Algeria. Bugia was a port town, and Bonacci represented Pisan merchants who were involved in the trading colony there. Some years later Fibonacci accompanied his father to Bugia.

Fibonacci was educated in North Africa, and because his father wished for him to become a merchant, Fibonacci was rigorously schooled in mathematics and calculation techniques. It was in Bugia that Fibonacci was introduced to the Hindu-Arabic number system. Bonacci eventually sent Fibonacci on business trips to meet with various merchants, and Fibonacci observed the mathematical systems used in these countries and learned of the Hindu-Arabic system's many practical uses. Among the places Fibonacci visited were Egypt, Syria, Sicily, Greece, and the region of Provence in France.

Around 1200 Fibonacci returned to Pisa and began writing his mathematical works. *Liber abbaci,* or Book of the abacus, was completed in 1202 and introduced the Hindu-Arabic numbering system to Europe. The system used the Arab numerals 1, 2, 3, 4, 5, 6, 7, 8, 9, and a symbol for 0. The "new" system also utilized the decimal point. Fibonacci's book described mathematical rules for adding, subtracting, dividing, and multiplying numbers by using the system. The book also included mathematical problems relevant to merchants, such as the calculation of profit and conversion of currencies, and introduced a problem that resulted in what came to be known as the Fibonacci sequence: Fibonacci's problem proposed that one pair of rabbits be set aside in an enclosed space; Fibonacci then asked how many pairs of rabbits could be produced from that pair each year if each pair produced a new pair every month, and if each new pair began producing a pair from the second month. In the resultant sequence each number was the sum of the preceding two numbers.

Frederick II learned of Fibonacci through his court scholars, who had been corresponding with Fibonacci. Interested in meeting Fibonacci, Frederick II arranged a gathering in 1225. One member of his court, Johannes of Palermo, assembled some problems for Fibonacci. Fibonacci's solutions were outlined in *Flos;* among the problems Fibonacci solved was one taken from an algebra book by the astronomer and mathematician OMAR KHAYYÁM.

Fibonacci's other works included *Liber quadratorum* in 1225, a book on number theory that explored methods for finding Pythagorean triples, and *Practica geometriae* in 1220, a group of geometry problems. Fibonacci also completed a second edition of *Liber abbaci* in 1228. Little is known of Fibonacci's life after 1228 except that he received a stipend from the Republic of Pisa in 1240. Fibonacci was instrumental in introducing the Hindu-Arabic numeral system to Europe, where the Roman numeral system had long been employed, and the Fibonacci sequence found many useful applications in mathematics and science.

Fischer, Emil Hermann

(1852–1919)
German
Chemist

Acknowledged as the father of biochemistry, Emil Hermann Fischer made a number of important discoveries. Fischer

Emil Fischer, who is considered by many to be the father of biochemistry. *Smith Collection, Rare Book & Manuscript Library, University of Pennsylvania.*

was the first scientist to ascertain the molecular structure of glucose and fructose, and the first to synthesize glucose and other simple sugars. Fischer also determined the formula of many purine derivatives, such as uric acid, xanthine, and caffeine. His work in proteins was ground-breaking as well: He discovered a new type of amino acid and established the type of bond that links amino acids (which are the building blocks of proteins). He was awarded the 1902 Nobel Prize in chemistry for his research into sugars and purines (organic bases containing nitrogen found in many substances, including nucleic acids).

Fischer was born in Euskirchen, Germany, on October 9, 1852, to Laurenz and Julie Poensgen Fischer. A successful businessman, Laurenz wanted nothing more than that his son join him in the family lumber enterprise and was bitterly disappointed when Emil showed little aptitude for this profession. In 1888 Fischer married Agnes Gerlach, who died a mere seven years after their wedding. The couple had three sons, two of whom were killed during World War I. Fischer's only surviving son, Hermann, became an organic chemist.

Although Fischer initially wanted to specialize in physics, a cousin convinced him to study chemistry instead. After entering the University of Bonn in 1871, Fischer followed his mentor, Adolf von Baeyer, to the University of Strasbourg, where he received his Ph.D. in chemistry in 1874. During the course of his dissertation work Fischer discovered the first hydrazine base, phenylhydrazine, a reagent he would later use in his seminal research on sugars.

In 1875 Fischer accepted a position as an assistant professor in organic chemistry at the University of Munich. He moved to the University of Erlangen in 1881; he remained there until 1888, when he took a post at the University of Würzburg. In 1892 he was named chair of the chemistry department at the University of Berlin. Fischer began his famous work into sugars and purines during his tenure at the University of Erlangen, conducting his first experiments in 1882. He determined that various substances—adenine, guanine, caffeine, animal excrete, and uric acid—all belonged to one homogeneous family of compounds consisting in part of a double-ringed, nitrogen-containing molecule. He named this family *purine* in 1884 and succeeded in synthesizing it in his laboratory at the University of Berlin in 1898. During this period he also formulated a number of purine derivatives and began to study sugars. Fischer quickly established the relationships among various monosaccharides (simple sugars), and after discovering mannose (one such monosaccharide) in 1888, he determined the molecular structure of all known sugars in 1890. He was also able to synthesize the monosaccharides glucose, fructose, and mannose.

Fischer contributed to the current understanding of proteins as well. Between 1899 and 1908 he sought to separate and identify the individual amino acids that proteins comprise. In the process he discovered a new category of amino acids—the cyclic amino acids. Fischer also discovered that peptide bonds connect amino acids together.

During World War I Fischer organized German chemical resources and strove to rebuild chemistry departments in the wake of the destruction wrought by the war. The Nobel Prize Fischer received in 1902 acknowledged his work on sugars and purines. In addition, he is now credited with providing the foundation for much later work on proteins. Fischer died of cancer on July 15, 1919.

Fischer, Ernst Otto

(1918–)
German
Inorganic Chemist

Ernst Otto Fischer was a pioneer in the field of organometallic chemistry—the study of organic compounds containing both metal and carbon. He was awarded the Nobel

Prize in chemistry for his X-ray analysis of a molecular complex called ferrocene, composed of iron and carbon. Fischer's research revealed that ferrocene's structure was that of a "sandwich compound," in which an iron atom was sandwiched between two parallel rings of carbon but formed bonds with the electrons in the rings rather than with the individual carbon atoms. After this discovery Fischer shifted his focus to the analysis and synthesis of transition metals, which are substances such as dibenzenechromium that exist in a state between metallic and organic. He also was first to synthesize carbene and carbyne complexes—carbon atoms triply joined to metal atoms.

The third child of Karl and Valentine Danzer Fischer, Ernst Fischer was born on November 10, 1918, in Solln, a suburb of Munich, Germany. Fischer's father was a physics professor at Munich's Technische Hochschule.

After completing high school in 1937, Fischer served two years in compulsory service in the German army. His tour of duty was then extended because of the onset of World War II. Between periods of military service in Poland, France, and Russia, Fischer found the time in the winter of 1941 to 1942 to begin his studies in chemistry at the Technische Hochschule in Munich. His work was interrupted, however, when he was captured by the American army and held until after the end of the war. Undeterred, he resumed his chemistry classes in Munich in 1946 and earned his Ph.D. in 1952 for a dissertation on carbon-to-nickel chemical bonds.

Fischer's first project as an assistant researcher at the Technische Hochschule, a position he assumed after completing his doctorate, was the investigation of an unknown compound consisting of five-carbon rings linked by an iron atom. This compound, initially named *dicyclopentadienyl,* was notable for its unusual thermal and chemical stability. Certain chemists had proposed that this compound's structure was that of an iron atom joined between two consecutive longitudinal rings of carbon. This theory was unsatisfying, however, as it was inconsistent with the compound's stability. Sir GEOFFREY WILKINSON postulated that dicyclopentadienyl (which had by then been dubbed *ferrocene*) had an entirely different sort of chemical structure: He believed that the iron atom was sandwiched between two parallel carbon rings, one on top of the other. In other words, the iron atom formed bonds with all the carbon atoms in both rings and also with the electrons within the carbon rings. Fischer used X-ray crystallography to prove Wilkinson's theory and demonstrate that ferrocene was indeed a "sandwich molecule."

Fischer also pursued research into organometallic compounds. He mapped the structure of various other transition metals and proceeded to synthesize them. He demonstrated that dibenzenechromium was another sandwich molecule, which comprises two rings of benzene with a chromium atom between them. In 1954 he was named an assistant professor at the Technische Hochschule; in 1957 he became a full professor at the University of Munich. He returned to the Technische Hochschule in 1964, to become the director of the Institute for Inorganic Chemistry. His laboratory became the center of organometallic research. In the early 1970s Fischer synthesized a new class of organometallic compounds. He also produced the first carbene and carbyne compounds, which consisted of carbon atoms triply bonded to metal atoms.

The implications of Fischer's work were profound. His findings fueled the creation of special catalysts used in the pharmaceutical and petroleum industries. Moreover, his ground-breaking discoveries concerning ferrocene's structure laid the foundation for the dynamic field of organometallic chemistry. He shared the 1973 Nobel Prize in chemistry with Wilkinson and also won the 1957 Göttingen Academy Prize and the 1959 Alfred Stock Memorial Prize of the Society of German Chemists. He was granted honorary membership in the American Academy of Arts and Sciences, as well as full membership in the German Academy of Scientists. A lifelong bachelor, Fischer continued to explore organometallic compounds in the late 1990s.

Fischer, Hans

(1881–1945)
German
Biochemist

Hans Fischer devoted nearly 25 years to the study and synthesis of pyrroles—molecular compounds that give biological substances such as blood and plant leaves their distinct colors. As director of the Institut für Anoreganische Chemise he oversaw a vast effort to discover the structure of hemin (the pigment that makes hemoglobin red), chlorophyll (the green pigment in plants), and bile. He was awarded the Nobel Prize in chemistry in 1930 for successfully synthesizing hemin and noting its similarities to chlorophyll. During his career he synthesized over 130 porphyrins—pigments such as hemin and bile that are composed of pyrroles joined in a chemical ring—and published 129 papers.

On July 27, 1881, Fischer was born to Eugen and Anna Herdegen Fischer at Höchst am Main, Germany. His father, a dye chemist, exerted a considerable influence over Fischer's future. Fischer developed an early interest in the chemical nature of the pigments his father worked with as the laboratory director at the Kalle Dye works.

Fischer was avidly interested in both chemistry and medicine. After receiving his doctorate in chemistry from the University of Marburg in 1904, Fischer attended the University of Munich's medical school and completed his M.D. in 1908.

After fulfilling various roles at the Physiological Institute in Munich and at the University of Innsbruck, Fischer accepted a position as director of the Institut für Anoreganische Chemise at the Technical University in Munich in 1921. He remained at Munich for the duration of his career and occupied himself almost exclusively with the study of pigments. He organized a number of laboratories to participate in his effort and, by segmenting his work in this manner, achieved an unparalleled degree of efficiency in his attempts to analyze and synthesize pigments. His first breakthrough occurred in 1926, when he synthesized porphyrin, thereby overturning an accepted scientific theory. It had been believed that a single, basic porphyrin structure was the building block of all naturally existing pigments. Fischer determined instead that porphyrins differed from one another. He also identified the specific molecular structures of individual porphyrin groups that compose certain pigments. After he ascertained that porphyrins are composed of pyrrole nuclei bound by methane groups into a ring structure, he was able to synthesize porphyrine in a laboratory setting.

In 1929 Fischer became the first person ever to synthesize hemin. He also demonstrated that hemin and chlorophyll are related. For these accomplishments he received the Nobel Prize. Thereafter he continued to unravel the chemical structures of biological pigments and synthesize them. He married Wiltrud Haufe in 1935.

In 1944 he produced bilirubin (the pigment of bile); he later established that chlorophyll's pyrrole rings had magnesium at their center instead of iron, as in hemin. In addition, he nearly was able to synthesize chlorophyll, but that achievement was beyond his grasp. Chlorophyll was not synthesized until 1960, and then mainly on the basis of his work.

Despite his many scientific triumphs, Fischer's life did not end happily. He was plagued by tuberculosis, an illness to which he had lost a kidney in 1917. After a bombing campaign during World War II destroyed his laboratory, Fischer became convinced that his work, the central passion of his existence, had been obliterated and his legacy lost. Despondent, he committed suicide in 1945. In addition to the Nobel Prize, Fischer had been awarded the Leibig Memorial Medal in 1929, the Davy Medal in 1936, and an honorary degree from Harvard University in 1935. His three-volume textbook, *Die Chemie des Pyrroles,* remains the definitive work on the topic.

Fleming, Sir Alexander

(1881–1955)
Scottish
Bacteriologist

Sir Alexander Fleming's significance rests on his discovery in 1928 of the penicillin antibiotic, which transformed medicine profoundly in the 20th century by combating bacterial infections. Chance played a large part in Fleming's breakthrough, as the spores of *Penicillium notatum* drifted into his laboratory from a downstairs lab that was studying whether this mold triggered asthma. Also, growth of the mold, which must occur first, is triggered by cold, whereas growth of the *Staphylococcus aureus* bacteria that Fleming was culturing requires heat. Fortuitously London experienced a cold wave followed by a heat wave at just that time. Furthermore, Fleming wouldn't have seen this growth except for his predilection to keep sample dishes long after they had cultured. In this case he noticed a small ring around the mold free of bacteria. It took another 12 years until the full significance of this discovery was acknowledged and penicillin was recognized as a kind of wonder drug.

Fleming was born on August 6, 1881, in a farmhouse in Lochfield, Ayrshire, Scotland. He was the third of four children born to Hugh Fleming, a farmer, who died when Fleming was seven, and his second wife, Grace Morton. Fleming had two stepbrothers and two stepsisters from his father's first marriage. In 1915 Fleming married Sarah Marion McElroy, an Irish nurse. In 1924 Sarah gave birth to their only child, Robert, who, like his father, became a physician. In 1949 Sarah died; in 1953 Fleming married Amalia Coutsouris-Voureka, a fellow bacteriologist and former student.

After moving to London, Fleming studied business at the Regent Street Polytechnic for two years. In 1901 he inherited £250 and applied this money to medical studies at St. Mary's Medical School, where he won a scholarship in his first year. In 1906 he graduated; he then pursued an M.B. and B.S. at London University, which he received along with the university's Gold Medal in 1908. The next year he passed the Fellowship of the Royal College of Surgeons exam.

After graduation from medical school in 1906 Fleming served as a junior assistant to Sir Almroth Edward Wright in his laboratory at St. Mary's. Before World War I Fleming was a member of the London Scottish Regiment from 1900 to 1914. During the war Fleming served in the British Royal Army Medical Corps, stationed in Boulogne, France. With Wright, he studied the treatment of infected wounds and found that antiseptics, the treatment of the time, not only killed bacteria but also killed white blood cells, thus inhibiting healing. This experience prompted him to investigate other modes of battling bacterial infection without impinging on the body's own defense mechanisms. After the war the Royal College of Surgeons named Fleming the Hunterian Professor in 1919. In 1921 he became the assistant director of the Inoculation Department at St. Mary's.

That year he discovered antibacterial lysozyme in a culture of his own nasal mucus that he collected while suffering from a cold. Though this antibody proved safe to the human body, it did not inhibit bacterial infection. Then in 1928, the same year he was named the Arris and Gale Lecturer at the Royal College of Surgeons and appointed as a professor of bacteriology at London University, Fleming discovered penicillin. The significance of this discovery was not recognized because Fleming was unable to isolate the antibody. In 1940 two chemists from Oxford University, Sir ERNST BORIS CHAIN and Sir HOWARD WALTER FLOREY, stumbled across the mention of penicillin in a journal and decided to test it as an antibody. Their isolation, purification, and testing of penicillin revealed it to be incredibly effective against bacterial infection.

Fleming won the 1945 Nobel Prize in medicine along with Chain and Florey, though Fleming received most of the recognition. Indeed, he had been knighted the year before. Fleming died of a heart attack at his home in London on March 11, 1955.

Fleming, Williamina Paton Stevens

(1857–1911)
American
Astronomer

Going from housemaid to astronomer, Williamina Fleming helped to devise a way to classify stars according to patterns in their light. She also identified a number of unusual stars. She was born Williamina Paton Stevens on May 15, 1857, in Dundee, Scotland. Her father, Robert,

A group of women computers at the Harvard College Observatory, directed by Williamina Fleming (standing), who became the foremost American woman astronomer of her time. Edward C. Pickering, director of the observatory, is looking on. *Harvard College Observatory.*

carved and gilded picture frames and furniture. He died when Mina, as people called her, was seven.

Mina did so well in school that she became a student teacher at age 14. She married James Fleming in 1877, and they sailed for Boston the next year. There Fleming abandoned his pregnant bride. Mina looked for domestic work and was hired by Edward C. Pickering, head of the Harvard Observatory.

Pickering was starting a huge project that classified stars according to their spectra, the patterns revealed when their light was broken up into a rainbow (by passing it through a crystal called a prism) and photographed. Differences in spectra reflect such characteristics as a star's surface temperature and chemical elements. According to legend, Mina Fleming got her start in astronomy when Pickering lost patience with a young man hired to analyze the spectra and snapped, "My Scottish maid could do better!" He took the Scottish maid—Fleming—to the observatory and found that indeed she could.

Whether or not her introduction happened this way, Pickering did eventually show Fleming how to study spectra with a magnifying glass and found that she had a knack for sorting them according to similarities and making the calculations necessary to determine the stars' position. She was hired as a permanent employee of the observatory in 1881 and put in charge of the star project in 1886. She and Pickering devised the system (later improved by ANNIE JUMP CANNON) used to classify the spectra of 10,351 stars in the massive *Draper Catalogue of Stellar Spectra*, published in 1890, and she did most of the classification. When Pickering began work on an even larger star catalog and hired a "harem" (as people joked) of young women as "computers" on the project, Fleming supervised them with a kindly but stern eye. They included ANNIE JUMP CANNON, HENRIETTA SWAN LEAVITT, ANTONIA CAETANA MAURY, and Cecilia Payne-Gaposchkin.

In addition to her work on the star catalogs, Fleming made some original contributions to astronomy. She was the first to notice, for instance, that variable stars, whose light regularly brightened and dimmed, could be identified by bright lines in their spectra. In 1907 she published a list of 222 variable stars, most of which she had discovered.

Despite her humble beginnings and lack of formal training, Williamina Fleming became the foremost American woman astronomer of her time. In 1898 the Harvard Corporation, which ran the university, made her the observatory's curator of photographs, the first appointment the corporation awarded a woman. In 1906 Fleming became the sixth woman, and the first American woman, to be elected to Britain's prestigious Royal Astronomical Society. Fleming died of pneumonia on May 21, 1911, at the age of 54.

Florey, Howard Walter

(1898–1968)
Australian
Pathologist

Howard Walter Florey played a crucial role in the development of penicillin, an antibiotic that treats bacterial infections such as pneumonia, meningitis, and scarlet fever. With his colleague, Sir ERNST BORIS CHAIN, Florey built upon Sir ALEXANDER FLEMING's 1928 discovery that the *Penicillium* mold produced an antibacterial substance. In their laboratory at Oxford University in 1940 Florey and Chain succeeded in isolating and purifying penicillin for clinical use, a difficult task bacause of penicillin's instability under laboratory conditions. Florey received numerous awards for his pioneering work, including the 1945 Nobel Prize in medicine or physiology, which he shared with Chain and Fleming. Thereafter he continued to study other antibiotics.

Born on September 24, 1898, in Adelaide, Australia, Florey was the only son of Joseph and Bertha Mary Wadham Florey. In addition to his two sisters, young Howard Florey had two half-sisters from his father's first marriage, to Charlotte Aimes. Joseph Florey was a boot manufacturer; Howard chose not to follow his father's career path.

After completing his bachelor's degree at Adelaide University in 1921, Florey was awarded a Rhodes Scholarship to Oxford University. From 1922 until 1924 he studied under the neurophysiologist CHARLES SCOTT SHERRINGTON at Oxford. In 1924 Florey left Oxford for Cambridge University, and he embarked on a Rockefeller Traveling Scholarship at the University of Pennsylvania the following year. He returned to Cambridge in 1926 and was awarded his Ph.D. in 1927. He and his wife, Mary Ethel Hayter Reed, whom he had married in 1926 had two children, Charles and Paquita.

Florey remained at Cambridge from 1927 to 1931 as the Huddersfield Lecturer in Special Pathology. His research explored lysozyme, an enzyme with antibacterial properties found in human tears, nasal secretions, and mucous membranes. He succeeded not only in purifying lysozyme, but also in characterizing the substances upon which it acted. His interest in such naturally occurring antibacterial substances led him to the work of Sir Alexander Fleming, who had discovered in 1928 that the *Penicillium* mold harbored a bacteria-fighting agent. Florey left Cambridge in 1931, for a professorship at the University of Sheffield. In 1935 he returned to Oxford to become the director of the Sir William Dunn School of Pathology, where he collaborated with Chain, a biochemist, to study antibacterials found in molds.

Despite the challenges posed by the instability of the *Penicillium* mold, Florey and Chain succeeded in isolating 100 milligrams of penicillin from broth cultures by 1940.

Florey first tested penicillin's bacteria-fighting potential in experiments on mice. He infected eight mice with lethal doses of streptococcus bacteria, then injected four of them with penicillin. He found that those four lived, whereas the others died the following day. In 1941 Florey conducted his first clinical trial on humans, and penicillin provided a cure for patients suffering from deadly cases of staphyloccal and streptococcal infection. Florey wished to build on this success, but the outbreak of World War II hampered his ability to produce the vast amounts of the drug needed. As a result Florey left for the United States in 1942 and relied on American drug factories to further his research.

Even after this breakthrough Florey remained interested in antibacterial substances. He turned his attention to cephalosporins, a group of drugs that worked similarly to penicillin. In 1944 he was invited by the prime minister of Australia to oversee a review of the country's medical research facilities. Beginning in the late 1940s Florey made annual trips to Australia to implement a program to improve medical laboratories there. After his lengthy tenure at the Oxford school of pathology, he was named provost of Queen's College at Oxford in 1962. In 1965 he was appointed chancellor of the Australian National University.

Florey's achievements were widely recognized. In addition to winning the Nobel Prize in 1945, he was named president of the Royal Society in 1960. He was knighted in 1942 and granted a peerage in 1965 (thus becoming Baron Florey of Adelaide and Marston). The year after his wife's death in 1966 he married Margaret Jennings. He died the following year, leaving a tremendous legacy marked by the countless lives his discovery saved.

Flourens, Pierre

(1794–1867)
French
Physiologist

Pierre Flourens excelled as an experimental physiologist who worked under the assumption that the anatomical division of organs coincided with a division of functions performed by the body, and set out experimentally to isolate which body parts related to which physical functions. His work on the brains of vertebrates established a correspondence between cerebral sections and functions. He devoted his later career to scientific biography, excelling in this discipline as well.

Flourens was born on April 15, 1794, in Maureilhan, France, into a family of humble means. He attended medical school at the University of Montpelier and graduated in 1813. Flourens then traveled to Paris, where, on the recommendation of the botanist Augustin de Candolle, GEORGES LÉOPOLD CHRÊTIEN FRÉDÉRIC DAGOBERT CUVIER took him under his wing, securing positions for him and ensuring that his own positions passed on to Flourens. Between 1814 and 1822 Flourens performed some of his most important work, basing it on the advances made by the Italian anatomist Luigi Rolando on the nervous system. Flourens experimented on the brains of pigeons by systematically removing different cerebral sections and observing the effect this had on their behavior. Flourens noted that when he removed the cerebral hemispheres in the forefront of the brain, the pigeons' judgment and perception disappeared; when he cut out the cerebellum, so too went the pigeons' coordination and equilibrium; when he dissected the medulla oblongata, respiration ceased; and when he cut out the cerebral cortex, the pigeons' vision vanished. Flourens also exposed the entire length of a dog's spinal cord and noticed that stimulation lower on the cord led to muscle movement, whereas the dog's muscles did not react to stimulation higher on the cord. He also established a connection between the semicircular canals in the ears and balance. Flourens felt confident in reporting on the correspondences between certain sections of the brain and the functions associated with them, but he was not confident enough to assert the localization of functions on the grounds of his experiments alone. It was not until 1870 that Gustav Fritsch and Eduard Hitzig experimentally confirmed cerebral localization.

In 1822 Cuvier presented the results of the cerebral experiments to the Academy of Sciences, securing Flourens's fame. Flourens published the results of these experiments in 1824 in the text *Experimental Researches on the Properties and Functions of the Nervous System in Vertebrates.* The academy awarded him the Montyon Prize in both 1824 and 1825. Flourens served as Cuvier's deputy lecturer at the Collège de France from 1828, the year that the Academy of Sciences elected him as a member. Upon Cuvier's death in 1832 Flourens assumed his chair at the Collège de France; in 1833, he replaced Cuvier as the permanent secretary to the academy, according to Cuvier's wishes.

In the latter part of his career, in 1847, Flourens demonstrated the efficacy of tricholoromethane as an anesthetic on animals; that demonstration led a Scottish obstetrician to use it on women in childbirth. In 1864 Flourens attacked Darwinism in the text, *Examination of Mr. Darwin's Book,* in which he labeled CHARLES ROBERT DARWIN's theories "childish and out of date personifications." Flourens died on December 6, 1867, in Montgeron, France.

Fossey, Dian

(1932–1985)
American
Zoologist

Living like a hermit in a mountain rain forest in Africa, Dian Fossey learned more about the endangered mountain gorilla than had ever been known. As more and more gorillas were killed by poachers, she turned from scientist to fierce conservationist. She eventually gave her life for the animals she loved.

Dian Fossey was born and raised in San Francisco, California. From her father, George, she learned to love nature, but her parents divorced when she was six and her mother, Kitty, and stepfather, Richard Price, did not let her have pets. Because of her love for animals she entered the University of California at Davis in 1950 with plans to become a veterinarian. After two years, however, she transferred to San Jose State College, where she trained as an occupational therapist. She obtained her B.A. in 1954. In 1956 she moved to Louisville, Kentucky, where she became head of the occupational therapy department at Kosair Crippled Children's Hospital.

"I had this great urge, this *need* to go to Africa," Fossey once told a *Chicago Tribune* interviewer. In 1963 she borrowed money to finance a seven-week safari to the continent. She met the British anthropologist LOUIS SEYMOUR BAZETT LEAKEY and first saw mountain gorillas during this trip. When she saw Leakey again in Louisville in 1966 and he said he was looking for a woman to do a long-term study of mountain gorillas like the one his protégée JANE GOODALL was doing with chimpanzees, Fossey eagerly volunteered. She even followed Leakey's half-joking suggestion that she have her appendix removed before going to Africa because, camping in the rain forest, she would be far from medical help.

Mountain gorillas are much more rare than lowland gorillas. They live only in the Virunga Mountains, a group of volcanoes in east-central Africa in the countries of Rwanda, Uganda, and the Democratic Republic of Congo. Fossey began her research in Congo in early 1967, but the country was involved in a civil war, and soldiers drove her out of her camp and imprisoned her after she had been there only six months. She escaped and fled to Rwanda, where she set up a new camp on Mount Visoke near another mountain, Karisimbi. She named the camp Karisoke and settled down at last to her research. Local people soon began calling her *Nyirmachabelli,* "the woman who lives alone in the forest."

Contrary to their fearsome "King Kong" image, the gorillas were very shy. Fossey finally learned to soothe the animals' fears by imitating the loud belches and other sounds they made while eating. She observed nine groups, each consisting of 5 to 19 members, and made close contact with four. She gave the animals in these groups whimsical names such as Digit (because of the animal's twisted finger), Uncle Bert, and Beethoven. Her studies uncovered many details of the gorillas' family life, mating, diet, and communication that had never been observed before. For instance, she noted that the "gentle giants," as she called them, were almost never truly aggressive. To gain more scientific validation for her work, she wrote it up as a thesis for a doctorate in zoology, which she obtained from Cambridge University in 1974.

In a census she did in 1970 Fossey found that there were only 375 gorillas left in the Virunga Mountains. She became increasingly determined to protect the animals—no easy task. The people of Rwanda, then the most heavily populated country in Africa, needed more land for themselves, and they often invaded what was supposed to be parkland. Cattle herders and woodcutters damaged the gorillas' forest habitat. Worse still, poachers killed the animals themselves—sometimes accidentally and at other times deliberately—in order to obtain body parts that could be sold as trophies or for magic.

When poachers killed Digit, Fossey's favorite gorilla, in 1977, she felt as if a beloved family member had been murdered. The following year she established a fund in Digit's name to pay Rwandan guards to track and drive off poachers. She also began a personal war against the intruders, using tactics that ranged from scaring them with a Halloween mask to kidnaping their children. Her approach angered local people, Rwandan government officials, and even some wildlife protection groups.

Desperate for funds to continue her work, Fossey went to the United States in 1980 and stayed for three years, teaching and lecturing at Cornell University and writing a popular book about her experiences called *Gorillas in the Mist.* The book, published in 1983, became a best-seller and earned enough money to allow her to return to Rwanda. It was also made into a film in 1988, starring Sigourney Weaver as Dian Fossey.

Fossey now suffered from emphysema and other health problems and had to abandon her gorilla research to assistants. Growing moodiness and her obsession with fighting poachers isolated her from those around her. In mid-1985 she said, "I have no friends. The more you learn about the dignity of the gorilla, the more you want to avoid people."

Unfortunately she was not able to continue her research. On December 27, 1985, one of her Rwandan guards found Fossey in her hut, slashed to death by a machete. The murderer was never identified. Fossey's friends buried her in the graveyard she had set up for the slain gorillas, under a tombstone that reads: "No one loved gorillas more. Rest in peace, dear friend, eternally protected in this sacred ground, for you are home where you belong."

Foucault, Jean Bernard Léon

(1819–1868)
French
Physicist

Although Jean Bernard Léon Foucault originally planned to become a physician, he eventually pursued a career in experimental physics instead. Among other discoveries, Foucault determined that the speed of light in the air is faster than the speed of light in water. This finding was influential, for it definitively proved the wave theory of light. Using a rotating-mirror device, Foucault then accurately measured the speed of light. He is even better known for his experiments with Foucault's pendulum, which he used to demonstrate the Earth's rotation. After noting that a pendulum continues to swing in the same plane, he deduced that the Earth's rotation left the pendulum's path fixed with respect to the stars. He also improved the technology of the telescope.

Foucault was born on September 19, 1819, in Paris. Little is known about his family. His father was a poor bookseller, who encouraged him to pursue medicine. Because he was frail and sickly, Foucault was educated at home.

Despite his ambitions to become a surgeon, Foucault was revolted by the sight of blood and human suffering. He therefore quit his medical studies at the École de Médecine, but not before he had forged a relationship with Alfred Donné, a professor of clinical microscopy at the school. Donné appointed Foucault his assistant lecturer, and the duo collaborated in writing a textbook in 1845.

The same year Foucault succeeded Donné as science reporter for the newspaper *Journal des débats*, for which he wrote a regular column on scientific innovations geared to the paper's lay audience. Simultaneously Foucault began conducting experiments at a laboratory in his home. He first investigated a question concerning the speed of light. D.F.J. Arago had sought to determine whether light traveled faster in air or in water but had been unsuccessful in finding a solution. The issue was not picayune. Physicists had debated for some time whether light traveled as a wave or as particles (as Sir ISAAC NEWTON believed). Determining whether light indeed moved faster in air could resolve this matter, as the wave theory required this to be true. Foucault used a rotating-mirror apparatus to measure light's velocity and in 1850 announced that light did indeed travel faster in air. After providing a full description of his experiments, Foucault was awarded his doctorate in 1853. He continued his research and in 1862 obtained the first accurate value for the speed of light with his apparatus.

But Foucault chose not to confine himself to one field of inquiry. While he was attempting to create an accurate timing device for his work on light, he observed that a pendulum remained swinging in the same plane when he rotated the apparatus. His curiosity piqued, he began experimenting with pendulums in 1850. Two years later he performed his famous experiment, in which he suspended a pendulum more than 200 feet from the ceiling of the Panthéon in Paris. A needle attached to the ball inscribed a mark in the sand, and the mark moved as the Earth rotated around the plane of the pendulum's swing. Foucault had conducted the first physical demonstration of the Earth's rotation. The same year he invented the gyroscope. In recognition of his achievements he was given a post at the Paris observatory in 1853.

Foucault met with success in other endeavors as well. With a colleague, Hippolyte Fizeau, he produced the first daguerreotype of the Sun in 1845. In 1857 he pioneered the technique for silvering glass to make mirrors for reflecting telescopes, and in 1858 he introduced a simple and accurate method for testing and correcting the mirrors and lenses on telescopes.

Foucault's accomplishments were recognized in his lifetime. He received the Cross of the Legion of Honor in 1851 and the Copley Medal of the Royal Society in 1855; he was inducted into the Académie des Sciences in 1865. On February 11, 1868, he died of a brain disease. Foucault's work not only provided an impetus for further investigations into theoretical mechanics, but also rooted the study of the Earth's movement in the realm of physical experiments rather than theoretical hypotheses.

Fowler, William Alfred

(1911–1995)
American
Physicist

William Alfred Fowler conducted ground-breaking theoretical and experimental research into the nuclear reactions within stars and the energy and elements produced by those reactions. Awarded the 1983 Nobel Prize in physics for his work with ELEANOR MARGARET PEACHEY BURBIDGE, Geoffrey Burbidge, and Sir FRED HOYLE, Fowler explained how the chemical elements of the universe had been created and definitively proved that not all of the Earth's elements were formed during the big bang.

Born on August 9, 1911, in Pittsburgh, Pennsylvania, Fowler was the oldest of John and Jennie Watson Fowler's three children. When Fowler was two years old, the family moved to Lima, Ohio, where his father worked as an accountant. Fowler excelled academically and athletically and graduated from high school in 1929 as valedictorian and a varsity football player.

Fowler matriculated at Ohio State University in 1929 and, after receiving his bachelor's degree in engineering physics, entered the graduate program in physics at the California Institute of Technology (Caltech) in 1933. Under the supervision of C. C. Lauritsen, the director of Caltech's W. W. Kellogg Radiation Laboratory, Fowler wrote his doctoral dissertation on radioactive isotopes formed by proton bombardment. He was awarded his Ph.D. in physics in 1936. With Ardiane Olmstead Fowler, whom he wed in 1940, Fowler had two daughters.

Fowler remained at Caltech for almost his entire career. At the invitation of Lauritsen he took a position in 1936 as a research fellow at Kellogg. Three years later Fowler began his ascent of the academic ladder, when he was made an assistant professor at Caltech. In 1942 Fowler was sent to Washington, D.C., to assist the United States' military efforts during World War II. In this capacity he developed proximity fuses for bombs, shells, and rockets. Upon returning to Caltech at the end of the war, Fowler embarked on the project that would eventually earn him the Nobel Prize. Fowler was interested in how the various chemical elements of the universe had been formed. HANS ALBRECHT BETHE had proposed that the conditions of thermonuclear reactions within stars were sufficient to account for the creation of at least one element—the conversion of hydrogen into helium. (These reactions also generated the energy stars emitted.) But although Bethe's theory could explain the creation of helium, it could not account for the genesis of heavier elements. GEORGE GAMOW's hot big bang theory attempted to address this puzzle. Gamow postulated that heavier elements were produced when atoms captured additional neutrons, thereby forming new atoms with heavier atomic masses. But Gamow's hypothesis could not be confirmed experimentally.

Fowler, however, believed that the heavier elements were formed after the big bang rather than during it. He teamed up in the 1950s with the astrophysicists Hoyle, Margaret Burbidge, and Geoffrey Burbidge to investigate how helium could have been converted into heavier elements. In 1957 the quartet published an important paper that articulated the idea that the process of neutron capture can account for the conversion of helium into carbon, carbon into iron, and finally iron into the heavier elements. In 1965 Fowler and Hoyle produced a more complete description of their work, "Nucleosynthesis in Massive Stars and Supernovae," and in 1975 the four scientists published one of the seminal papers of modern physics, "Synthesis of the Elements of the Stars." Fowler devoted the rest of his career to the details of stellar nucleosynthesis. In addition to investigating stellar neutrino flux, Fowler explored how to determine the amount of helium and deuterium in the universe. A full professor at Caltech since 1946, he was made an emeritus professor there in 1982.

Fowler's research is recognized as a cornerstone of modern astrophysics. In addition to the 1983 Nobel Prize in physics (which he shared with SUBRAHMANYAN CHANDRASEKHAR), Fowler received a bevy of honorary degrees from institutions including the University of Chicago and the Observatoire de Paris. The year after his wife's death in 1988, Fowler married Mary Dutcher. He died in 1995.

Franck, James

(1882–1964)
German/American
Physicist

James Franck's most lasting scientific contribution arose from his collaboration with a fellow physicist, GUSTAV HERTZ. Their experiments revealed that atoms absorb energy only in discrete and specific amounts. These results provided the first empirical confirmation of an atomic theory postulated by NIELS HENDRIK DAVID BOHR and proved to be a significant step in the development of 20th-century physics. Later in his career, Franck turned his attention to photochemical phenomena and was one of the scientists involved in the Manhattan Project, which developed the first atomic bomb.

Franck, the child of observant Jews, was born on August 26, 1882, in Hamburg, Germany. His father, Jacob, was a banker. His mother was named Rebecka Nahum Drucker. James's first wife, Ingrid Josephson, died in 1942, and he married a physics professor, Hertha Sponer, in 1946. The couple had two daughters.

Franck studied chemistry at the University of Heidelberg but transferred after two terms to the University of Berlin, where he became interested in the field of physics. He graduated with a D.Phil. in 1906. Upon completing his studies, he took a position at the University of Frankfurt on Main; he became an assistant to a physics professor in Berlin in 1908. It was there that he met Gustav Hertz, with whom he would produce his most notable work bearing out Bohr's insight. Bohr had previously theorized that the center of an atom, its nucleus, was surrounded by groups of electrons that inhabited defined "orbits" or shells. Bohr believed that these electrons had defined positions within their orbits, but that when a quantum of energy was applied to the atom, electrons would jump to different shells, the distance of the jump depending on the amount of energy applied. Furthermore Bohr hypothesized that electrons would absorb only the exact amount of energy necessary to enable them to jump. It was this last idea that Franck and Hertz proved.

Curiously, Franck and Hertz were unaware of Bohr's work at the time they conducted their experiments. They were simply hoping to track the amount of energy required to ionize mercury atoms. To do so, they subjected atoms of mercury vapor to a bombardment of electrons moving at specific speeds. They discovered that electrons traveling below a certain speed bounced off the atoms, whereas those moving faster than that speed allowed a transfer of energy from the electrons to the atoms, causing the mercury gas to glow (a phenomenon called resonance). The duo realized that the energy value of the light given off by the glowing substance was identical to that transferred by the electrons. This finding not only confirmed Bohr's theory but also led to the recognition that the atomic structure of an atom can be analyzed by examining its light spectrum. The particular energy levels of that spectrum could reveal the range of jumps an electron could make within that atom.

After service in World War I Franck was appointed to chair the physics section of the Kaiser Wilhelm Institute for Physical Chemistry in 1918. He moved to the University of Göttingen in 1920; there he was named director of the university's Second Physical Institute, continuing his study of atomic collisions all the while. This research kindled his interest in photochemical processes, and in 1925 he authored a paper on the basic properties of photochemical reactions. Franck formulated a general rule for the distribution of vibrational energy as part of this project. The physicist Edward Condon subsequently described this rule in terms of quantum mechanics, and it came to be known as the Franck-Condon principle.

After Hitler's ascension to power in 1933, Franck publicly resigned from his Göttingen post; he eventually settled in the United States, working first at Johns Hopkins University and later at the University of Chicago, where he directed a laboratory dedicated to researching photosynthesis. He also participated in the Manhattan Project but criticized the decision to drop atomic bombs on Japanese population centers.

Franck received much acclaim for his achievements. On top of the 1925 Nobel Prize in physics, which he shared with Hertz, Franck was awarded the Max Planck Medal in 1953 and the Rumford Medal in 1955. He became a member of the English Royal Society and the American National Academy of Sciences. He died in 1964.

Frank, Ilya Mikhailovich

(1908–1990)
Russian
Physicist

Ilya Frank's most significant scientific contribution was his work on devising a theoretical explanation for the Cherenkov effect, the blue glow created by radiation when gamma rays pass through water. Discovered by PAVEL ALEKSEYEVICH CHERENKOV in 1934, the phenomenon had defied explanation. But in 1937, in conjunction with his colleague IGOR EVGENIEVICH TAMM, Frank posited that the effect was caused by charged particles traveling faster than the speed of light. Although a basic law of physics holds that nothing can move faster than the speed of light in a vacuum, an object may exceed that speed in another medium. The glow is the product of the extreme velocity, not unlike a sonic boom created by a plane traveling faster than the speed of sound. Frank and Tamm's theory proved instrumental in developing measuring devices that remain important for particle accelerators.

Frank, the second son of Mikhail Luydvigovich and Yelizaveta Mikhailovna Gratsianova Frank, was born in St. Petersburg on October 23, 1908. Mikhail was a professor of mathematics, and Yelizaveta was a physician. Little is known about Ilya's childhood.

Frank attended Moscow State University, where he specialized in the photo-luminescence of solutions. Upon receiving his bachelor's degree in physics in 1930, Frank returned to Leningrad (the renamed St. Petersburg) to begin study at the State Optical Institute. There he conducted meticulous experiments concerning light-induced chemical reactions. He received his doctorate in physical and mathematical sciences from the institute in 1935.

After his graduation Frank accepted an invitation from a former teacher at Moscow State University to return to Moscow to work at the recently founded P. N. Lebedev Institute. He remained at this institution for the duration of his academic career. One of Frank's new colleagues was Cherenkov, who had discovered his eponymous effect in 1934. In 1936 Frank teamed up with another institute scientist, Tamm, in an effort to explain the Cherenkov effect. The duo met with success the following year. In 1937 Frank married Ella Abramova Beillikhis, who was a historian.

The practical significance of Frank and Tamm's theory was quickly recognized, as it led to the creation of the Cherenkov detector. This device, made from glass or some other transparent substance through which the high-velocity particles could pass, makes possible photoelectric analysis of the Cherenkov radiation. Cherenkov detectors are now widely used to study particles produced in cyclotrons and other types of acceleration.

During World War II Frank conducted research into the science behind the development of nuclear weapons and power plants. He continued this vein of study after he became a professor of physics at Moscow State University in 1944 (a position he held jointly with his post at the Lebedev Institute). He was appointed the director of the school's Laboratory of Radioactive Radiation as well. After the war Frank was also named director of the Lebedev Institute's Laboratory of the Atomic Nucleus. He held this latter post until his death. Frank continued to work with Tamm, and the two collaborated with Cherenkov to examine the phenomenon of electron radiation. In 1956 Frank left Moscow State to supervise the creation of the Laboratory of Neutron Physics in Dubna. Frank investigated radiation into the late 1980s, concentrating particularly on gamma rays and neutron beams. His son, Alexander, became a physicist and continued his father's work at Dubna.

Frank was highly decorated for his work. He shared the 1958 Nobel Prize in physics with Tamm and Cherenkov, the first Russian so honored. Moreover, Frank was elected a member of the Soviet Academy of Sciences, and received the Stalin Prize. Ilya Frank died in 1990.

Franklin, Benjamin
(1706–1790)
American
Physics, Oceanography

Although Benjamin Franklin is remembered mainly for his role in the American Revolution, he was also one of the preeminent scientists of his era. Self-taught, he investigated the then-mysterious phenomenon of electricity. Franklin proposed a "one-fluid" theory of electricity, which was underpinned by the concept that charges lost by one body must be gained simultaneously by another in equal amounts. Now known as the law of conservation of charge, it remains a fundamental scientific tenet. His famous kite experiment proved that lightning is an electric charge and spurred him to invent the lightning rod.

The 15th child of Josiah Franklin, Benjamin Franklin was born in Boston, Massachusetts, on January 17, 1706. Franklin's mother, Abiah Folger, was his father's second wife. In 1730 Franklin would marry Deborah Read.

After learning the printing trade in Boston and Philadelphia, Franklin worked as a printer in London from 1724 until 1726; there he made the acquaintance of several notable British scientists. In 1726 he returned to Philadelphia, where he published the *Pennsylvania Gazette* in 1729 and the wildly popular *Poor Richard's Almanac* in 1733. He entered public service in 1736, when he was appointed clerk of the State Assembly.

Benjamin Franklin, whose one-fluid theory of electricity remains a fundamental scientific tenet. *Smith Collection, Rare Book & Manuscript Library, University of Pennsylvania.*

Franklin was 40 when he embarked on his scientific career. Although not formally educated, he devoted tremendous effort to learning the theories of the day. He read the works of ROBERT BOYLE and Sir ISAAC NEWTON and continued to correspond with the scientists he had met abroad. A Leyden jar (a type of condenser—or capacitor—used for storing an electrical charge) donated to a Philadelphia library ignited Franklin's interest in electricity. From 1743 he made several important discoveries in the field. Franklin postulated that electrical effects are caused by the movement or transfer of an electric "fluid" that comprised particles of electricity, which could permeate other materials. In addition, he theorized that whereas these particles repel each other, they are attracted to the particles of other matter. Franklin dubbed his concept the *one fluid* theory of electricity and introduced the terms *positive* and *negative* into scientific parlance. In Franklin's framework a charged body is one that has either lost or gained electrical fluid, thus becoming negative or positive. Implicit in Franklin's formulation was a concept that would later be named the law of conservation of charge, which holds that when one body loses a charge, another body must neutralize the effect by gaining an equal charge at the same time. He published his findings in his 1751 book, *Experiments and Observations on Electricity, Made at Philadelphia in America.* In 1752, Franklin undertook his famous (and dangerous) experiment to test whether lightning was an electric charge. By flying a kite in a thunderstorm, Franklin concluded that it was. The experiment also convinced him of the efficacy of lightning rods—which he invented and sold.

Franklin conducted other scientific endeavors as well. In the 1770s he pioneered the study of the Gulf Stream. In addition to measuring the ocean's temperature at different locations and depths, he collected data from sea captains and produced the first printed chart of that meteorological phenomenon. He also invented bifocal spectacles, the rocking chair, and the Franklin stove.

It is a testament to Franklin's many talents that his exceptional scientific career was only one facet of his life. After serving as the clerk of the State Assembly until 1751, he became deputy postmaster for the colonies from 1753 until 1774. He was one of the signers of the Declaration of Independence in 1776 and was sent by the revolutionary government to France to seek military aid, which he singlehandedly secured. Before he retired from public life in 1788, he was a delegate to the 1787 Constitutional Convention.

Franklin is credited with forging the branch of science dealing with electricity. After winning the Copley

Medal in 1753, he was elected to the Royal Society in 1756. He was the chief founder of the American Philosophical Society, the colonies' first permanent scientific society, and played an important role in the creation of the University of Pennsylvania. Franklin died in 1790. His "one-fluid" theory influenced Count ALESSANDRO GIUSEPPE ANASTASIO VOLTA, who would later apply many of Franklin's principles in producing the first battery.

Franklin, Rosalind Elsie

(1920–1962)
British
Chemist

Deoxyribonucleic acid, or DNA, carries the inherited information in the genes of most living things. In the early 1950s scientists realized that the key to finding out how this information was stored and reproduced lay in the structure of DNA's complex molecules. Rosalind Franklin took X-ray photographs that gave two rival scientists, JAMES DEWEY WATSON and FRANCIS HARRY COMPTON CRICK, the clues they needed to work out the structure of DNA.

Rosalind Franklin was born on July 25, 1920, in London. Her father, Ellis, was a well-to-do banker, and her mother, Muriel, did volunteer social work, while raising five children. Rosalind decided at age 15 that she wanted to be a scientist. Her father objected, believing like many people of the time that higher education and a career made women unhappy, but she finally overcame his resistance. She studied chemistry at Newnham, a women's college at Cambridge University, and graduated in 1941.

As a way of helping her country during World War II, Franklin became assistant research officer at the Coal Utilization Research Association (CURA). She studied the structure of carbon molecules, introducing, according to one professor, "order into a field which had previously been in chaos." She turned some of this work into the thesis for her Ph.D., which she earned from Cambridge in 1945.

Seeking new challenges, Franklin went to work for the French government's central chemical research laboratory in 1947. Friends later said that her three years there were the happiest of her life. She enjoyed an easy camaraderie with her coworkers, chatting at cafés and on picnics. She also learned the technique of X-ray crystallography, to which she would devote the rest of her career.

Many solid materials form crystals, in which molecules are arranged in regular patterns. In 1912 a German scientist named Max von Laue found that if a beam of X rays is shone through a crystal, some of the rays bounce off the crystal's atoms, whereas others pass straight through. When photographic film, sensitive to X rays, is placed on the far side of the crystal, the resulting photograph shows a pattern of black dots that can reveal important facts about the three-dimensional structure of the molecules in the crystal.

Chemists eventually also found ways to use X-ray crystallography on amorphous compounds, which did not form obvious crystals. Most of the complex chemicals in the bodies of living things are amorphous compounds. Molecular biologists were beginning to realize that the structure of these compounds revealed much about their function, and crystallography was a promising tool for revealing that structure. One of the molecules about whose structure scientists were most curious was DNA.

Scientists knew that the DNA molecule consisted of several smaller molecules. It had a long chain, or "backbone," made of alternating molecules of sugar and phosphate (a phosphorus-containing compound). Four different kinds of other molecules called bases were attached to the backbone. No one knew, however, whether the chain was straight or twisted, how the bases were arranged on it, or how many chains were in each molecule. Franklin and Maurice Wilkins, the researcher with whom she worked at King's College, hoped that Franklin's X-ray photographs would provide this information.

Franklin photographed two forms of DNA, a "dry," or crystalline, form and a "wet" form that contained extra water molecules. No one had photographed the wet form before. At the time Franklin was not sure which type gave the more useful information. She took an excellent photograph of the wet form in May 1952, but she put it aside in a drawer and continued working with the dry form.

Two Cambridge scientists, a brash young American named James Watson and a somewhat older Briton, Francis Crick, were also trying to work out the structure of DNA. Although Watson saw himself and Crick as competitors of the King's College group, he and Wilkins became friends, and on January 30, 1953, he visited Wilkins at King's College. Without asking Franklin's permission, Wilkins showed Watson the photograph of "wet" DNA that she had made in May 1952. When he saw the photo, Watson wrote later, "My mouth fell open and my pulse began to race." He hurried back to Cambridge to describe the photo to Crick.

To Watson the "GillSans Light"–shaped pattern of dots in Franklin's photo showed clearly that the DNA molecule had the shape of a helix. On the basis of this and other evidence, he and Crick concluded, as by this time Franklin also had, that the molecule consisted of two helices twined around each other. The backbones were on the outside and the bases stretched across the center. In other words, the molecule was shaped like a spiral staircase or a twisted ladder with the bases as steps or rungs.

Watson and Crick published a ground-breaking paper on the structure of DNA in Britain's chief science journal, *Nature,* on April 25, 1953. Neither then nor later did they fully credit Franklin for the important part her photograph had played in their discovery, and Franklin herself probably never realized its role. By the time the Cambridge scientists' paper appeared, she was no longer working on DNA. She had moved from King's College to Birkbeck, a college of the University of London, and was beginning an X-ray study of a common plant virus, the tobacco mosaic virus. Almost nothing was known about the structure of viruses at that time. Franklin drew on her crystallography studies to make a model of the tobacco mosaic virus, which was exhibited at the 1957 World's Fair in Brussels. The virus's inherited information was carried in ribonucleic acid (RNA), a chemical similar to DNA. Franklin showed that the RNA molecule was also a helix.

In 1956 Rosalind Franklin discovered that she had ovarian cancer. The cancer proved untreatable, and she died of it on April 16, 1958. Four years later Watson, Crick, and Wilkins shared the 1962 Nobel Prize in physiology or medicine for their work on DNA. Supporters and critics still debate whether she would or should have been included if she had lived.

Fraunhofer, Joseph von

(1787–1826)
German
Physicist, Optician

A dedicated and skilled optician and glassmaker, Joseph von Fraunhofer was the first to study the dark lines of the spectra of the Sun and stars, which later came to be called Fraunhofer lines. Fraunhofer was also the first to use the diffraction grating, an instrument used to produce optical spectra by the diffraction of reflected light. Fraunhofer's achievements not only helped Germany become the center of advanced optics in the late 18th century, but also led to the development of spectroscopy.

Born on March 6, 1787, in Straubing, Germany, Fraunhofer grew up in a poor family. The youngest of 11 children, Fraunhofer was the son of Franz Xaver Fraunhofer, a glazier, and the former Maria Anna Fröhlich. He was raised in an environment in which many of his acquaintances were involved in the optical and glass industries.

Fraunhofer had little formal education, entering his father's trade at the young age of 10. When his father died in 1798, Fraunhofer was apprenticed to a mirror maker and glass cutter. Though the apprenticeship was unpleasant, Fraunhofer never swayed from his desire to become an optician. In 1801 Fraunhofer was involved in an accident at work, which resulted in his receiving a significant amount of money. The money permitted Fraunhofer to buy books on optics, a glass-working machine, and release from part of his six-year apprenticeship.

In 1806 Fraunhofer began working at the Munich Philosophical Instrument Company, a manufacturer of scientific instruments. At the time the quality of glass was rather poor, and opticians were forced to construct lenses by trial-and-error methods. To improve the quality and homogeneity of optical glass, Fraunhofer began to learn glassmaking skills from a master glassmaker, Pierre Louis Guinand, in 1809. In addition, Fraunhofer, interested in optical theory and mathematics, endeavored to approach optics from a scientific standpoint and abandon the trial-and-error technique. Fraunhofer thus began to study the dispersion powers and refractive indices of assorted types of optical glass. In 1814 he compared light from the Sun with light from a flame and noticed dark lines in the Sun's spectrum. Fraunhofer eventually observed 574 spectral lines and succeeded in accurately measuring the dispersion and refractive properties of many types of glass.

Continuing with his observations of spectra, Fraunhofer in 1821 developed a grating consisting of 260 wires and studied the spectra produced by diffraction gratings. Fraunhofer was able to determine that the dispersion of the spectra was greater with the diffraction grating than with the prism because of the presence of dark solar lines. Fraunhofer then studied the relationship between the dispersion and the separation of wires in the grating, concluding that the dispersion was inversely related to the distance between successive slits in the grating. This research also enabled Fraunhofer to establish the wavelengths of certain colors of light. Fraunhofer later developed reflection gratings, which allowed him to study the effect of diffraction on oblique rays.

By 1811 Fraunhofer had become a partner and director of glassmaking at the Munich Philosophical Instrument Company. In 1823 he became the director of the Physics Museum of the Bavarian Academy of Sciences in Munich and was given the honorary title of Royal Bavarian Professor. Although Fraunhofer did not publish his studies of glassmaking, he did publish works on diffraction gratings and spectral lines. An explanation of the dark lines was not offered, however, until 1859, when the physicist Gustav Robert Kirchhoff provided one. Never married, Fraunhofer died in 1826 at the age of 39 after contracting tuberculosis.

Freud, Sigmund

(1856–1939)
Austrian
Psychoanalyst, Neuropathologist

Sigmund Freud founded the revolutionary theory of psychoanalysis, a method devoted to studying and analyzing psychological phenomena to diagnose and treat neuroses and mental conditions. Freud developed the psychoanalytic technique of free association—expressing, without censorship, the first thought that enters the mind. Freud used free association to study the unconscious mind and gain a better understanding of psychological disorders. He was also instrumental in promoting the analysis of dreams and formed pioneering and controversial theories regarding sexuality; he believed infantile psychosexual development played a critical role in adult psychological development.

Freud was born on May 6, 1856, in Freiberg, Moravia (now Pribor, Czech Republic). Freud was the eldest child of his wool merchant father's second family. Freud's mother, Amalie Nathanson, was 20 years younger than Jakob, Freud's father. Freud's older half-brother, who was about the age of Freud's mother, had a child close to Freud's age. Making sense of this confusing family situation heightened Freud's intellect and curiosity. The family moved to Vienna in 1860 after the wool trade in Freiberg deteriorated, and Jakob was frequently unemployed.

Despite the family's lack of wealth, Freud's parents encouraged him in his studies and made financial sacrifices to further his education. Freud entered the University of Vienna in 1873 and graduated with a M.D. in 1881. Freud hoped for a career in biological research but chose instead to practice medicine at the Vienna General Hospital to support his new wife, Martha Bernays. Freud worked in a number of departments at the hospital, remaining the longest in the nervous diseases department because of his interest in neuropathology.

After leaving the hospital in 1885, Freud went to Paris to study with the neurologist Jean-Martin Charcot. It was during this four-month period, in which Freud worked with patients labeled as hysterics, that he began theorizing that neuroses may be caused not by organic disease but by psychological factors.

Freud returned to Vienna and started a private practice as a neuropathologist. Though he published several works on neuropathology, it was the publication in 1895 of *Studies in Hysteria* that marked the early stages of psychoanalytical theory. The publication, the result of a decade-long collaboration with a physician, Josef Breuer, introduced Freud's method of free association. An unpublished work, "Project for a Scientific Psychology," written in 1895, presented a neurological approach to normal and abnormal psychology and also introduced Freud's definitions of the ego, the conscious, and the id, the unconscious.

Freud published what was considered to be his most important and original work—*The Interpretation of Dreams*—in 1901. The publication theorized that the formation of dreams was influenced by unconscious experiences and desires. Another important work was *Three Essays on the Theory of Sexuality*, published in 1905, which introduced Freud's theories on the stages of psychosexual development and infantile sexuality. Freud believed that many disorders were a result of suppressed sexual wants. In 1923 Freud published a third work on psychoanalytic theory, *The Ego and the Id*. In this volume he further developed his concepts of id, ego, and superego, or conscience.

In 1902 Freud began holding weekly meetings at his home with colleagues to discuss psychoanalytic theory. The group evolved into the International Psychoanalytic Association in 1910, and participants included such notable members as CARL GUSTAV JUNG and Alfred Adler. The association soon disbanded, however, as a result of disagreements and differences of opinion. By this time Freud was already well known in Europe and discovering a highly receptive audience in the United States.

Freud and his wife had six children in the first 10 years of their marriage, creating a considerable demand on Freud's time and energy, but he was undeterred in his pursuit to unveil the workings of the human mind. Though diagnosed with cancer in 1923, Freud continued to work and write. He was compelled to leave Vienna for London in 1938 when the Nazis gained control of Austria, but he proceeded to treat patients and work on his final book, *Moses and Monotheism*.

Freud had jaw cancer for many years and underwent 33 operations to treat the disease during the last 20 years of his life. He died in London in September 1939 but continues to be remembered as the father of psychoanalysis. He believed that psychoanalysis was more significant as theory than as treatment, but his work strongly influenced modern-day psychotherapy.

Frith, Uta Auernhammer

(1941–)
German/British
Psychologist, Brain Researcher

Uta Frith has provided a new understanding of the brain and of several brain disorders. She was born Uta Auernhammer in Rockenhausen, Germany, on May 25, 1941, to an artist father and a writer mother. She started her elementary education in a girls' school, but at age 12 she insisted on attending the more demanding boys' school instead.

At college in Saarbrücken, Auernhammer decided on psychology as a career. She traveled to Britain in 1964 and earned a Ph.D. in psychology from the Institute of Psychiatry at London University in 1968. She then took a job at the Medical Research Council's Developmental Psychology Unit. Until 1998 she was a senior scientist with its successor, the Cognitive Development Unit. Frith moved to the Institute of Cognitive Neuroscience at University College, London (UCL), when the Cognitive Development Unit closed, and became an honorary professor at UCL in 1996. She was married to Christopher Frith.

One of Uta Frith's specialties is a brain disorder called autism, which isolates people emotionally. It is thought to be caused by brain damage or failure of the brain to develop normally before birth. Frith believes that the main kind of thinking defect in autism is a lack of ability to form what psychologists have called a theory of mind. Autistic people cannot understand that other people's perceptions, thoughts, beliefs, and feelings differ from their own. "Autism [is] a kind of . . . mindblindness," Frith told the writer Karen Gold in 1996.

Frith and her coworkers Alan Leslie and Simon Baron-Cohen demonstrated the theory-of-mind defect in 1986 by telling autistic and normal children a story about two girls, Sally and Anne, who were playing with a marble. Frith explained that Sally put the marble into a basket and left the room. While she was gone, Anne put the marble into a box. Frith then asked the children where Sally would look for the marble when she returned. Normal children as young as four years old said that Sally would look for the marble in the basket because she would not know that Anne had moved it. Autistic children, however—even teenagers of normal or superior intelligence—said that Sally would look for the marble in the box. They could not grasp that even though *they* knew the marble had been moved, Sally did not. In more recent years Frith and her coworkers have used imaging techniques to tie this inability to "mind-read" to lack of activity in a particular small area of the brain.

Frith has also studied dyslexia, a brain disorder best known for its ability to cause trouble in reading and spelling by making people see groups of letters in words as reversed. Frith believes, however, that it is a basic disorder of speech processing that can reveal itself long before a child attempts to read. It may be inherited. In 1995 Frith and her husband used an imaging technique called positron emission tomography (PET) to show that when normal people took language tests, two brain areas just above the ear and a third part called the insula (island), which connects them, were all active. The insula, however, was not active in dyslexics. "Each of the language areas deals with a specific aspect of word processing and in normal people the insula synchronizes this work," Frith told the magazine *New Scientist*. "In dyslexics the areas are disconnected."

Frith has received several honors for her work on autism and dyslexia, including the British Psychological Society's President's Award in 1990. She was made a fellow of the British Psychological Society in 1991 and a member of the Academia Europa in 1992. She edited a book in the field of reading development, *Cognitive Processes in Spelling,* and is the author of *Autism: Explaining the Enigma.* She and a collaborator, John Morton, are working on a general framework within which to explain developmental disorders such as autism and dyslexia that will combine the biological, cognitive (mental), and behavioral aspects of the defects. In recent years, making links between these levels has continued to be her main objective.

Fukui, Kenichi

(1918–1998)
Japanese
Chemist

Kenichi Fukui was best known for his Nobel Prize–winning research into the role of electron orbitals in chemical reactions. He applied complex mathematics and quantum theory to practical chemistry in order to explain the nature of chemical reactions. His work made it possible to predict the outcomes of many of these reactions. Although Fukui was not immediately given the credit he was due—mainly because the high-level mathematical formulas he employed to test his theory about molecular interactions and chemical bonds were incomprehensible to other chemists—the significance of his concept of frontier orbitals was eventually recognized, and Fukui gained international acclaim.

The eldest of the three sons of Chie and Ryokichi Fukui, Kenichi Fukui was born October 4, 1918, in Nara, Japan. His father, a merchant and factory manager, exerted a considerable influence over his son's future by persuading the younger Fukui to study chemistry, a subject in which Kenichi initially had little interest.

Fukui acceded to his father's wishes and soon discovered that he liked the discipline of chemistry. He enrolled in the Department of Industrial Chemistry at Kyoto Imperial University; he graduated in 1941. He was then called upon to apply his skills to the Japanese war effort and spent much of World War II studying synthetic fuels. After the war ended, Fukui returned to Kyoto University; he received his Ph.D. in engineering in 1948. In 1947 he married Tomoe Horie, with whom he had two children, Tetsuya and Miyako.

Fukui was a member of the Kyoto University faculty for his entire career. After being appointed a full professor of chemistry in 1951, he embarked on the project that would give him renown when he investigated the effect of electron orbitals on molecular reactions. At the time it was understood that electrons encircle the nucleus of an atom in orbitals of different energy levels and that the atoms in a molecule are held together by the force of electron bonds. When molecules react with one another, at least one of these electron bonds is broken and changed. As a result of such reactions molecules form new bonds, which alter the molecular structure of the substance. Beyond these generalities, however, little was actually known about how this process occurred.

Applying sophisticated mathematical analysis, Fukui proposed that this molecular interaction occurred in the highest occupied molecular orbital (HOMO) of one molecule and the lowest unoccupied molecular orbital (LUMO) of another molecule. He theorized that because the HOMO has a high energy level, it is willing to "give up" an electron to the LUMO, which has a low energy level and is therefore willing to "accept" the HOMO's electron. Fukui named these orbitals *frontier orbitals*. He asserted that the new bond—formed between the HOMO and the LUMO—was at an intermediate energy level between those of the HOMO and LUMO.

The importance of Fukui's insight was not recognized until two other scientists, ROALD HOFFMAN and ROBERT BURNS WOODWARD, working independently of each other and Fukui, reached conclusions similar to Fukui's. But the scientific community came to grasp the work of these three men and realized that it offered a more effective way to predict the outcomes of chemical reactions and to understand the complexity of these processes. In 1981 Fukui shared the Nobel Prize with Hoffman, the first Japanese to win this award in the field of chemistry. Fukui's accomplishments have been recognized in other fields as well. He was elected a member of a number of scientific organizations, including the Chemical Society in Japan, the U.S. National Academy of Sciences (he was named a foreign member), and the International Academy of Quantum Molecular Science. Fukui died on January 9, 1998.

G

Gabor, Dennis

(1900–1979)
Hungarian/British
Physicist

Dennis Gabor is best known for inventing holography, a system of three-dimensional photography without lenses. Gabor devised this process in the course of trying to overcome the most significant limitation of early electron microscopes—that past a certain magnification the image would become distorted, thereby hindering thorough observation. Gabor hit upon the idea of recording on a photographic plate the phase patterns of the object under observation. When placed in a beam whose rays are all in the same phase, the plate produces a three-dimensional image of the object, which shifts as the observer alters perspective. Early versions of this technique were only modestly effective, as there was no reliable method by which to generate the requisite beam. With the invention of the laser in 1960, however, myriad applications for Gabor's discovery were found.

Gabor was born in Budapest, Hungary, on June 5, 1900, the eldest of his parents' three sons. His father, Berthold, was the grandson of Russian-Jewish immigrants and the director of the Hungarian General Coal Mines. His mother, Ady Jacobvits, had been an actress prior to giving birth.

After serving briefly in the Austro-Hungarian army at the end of World War I, Gabor enrolled in the Budapest Technical University, where he studied mechanical engineering. Rather than reenlist after being called up again during his third year of study, Gabor decided to leave the country, unwilling to serve what he viewed as an authoritarian government. He then moved to Berlin and entered the Technische Hochschule, from which he earned a diploma in 1924 and a doctorate in engineering in 1927.

Upon graduation Gabor took a position in the Siemens and Halske physics lab in Siemensstadt, Germany, where he invented the quartz mercury lamp. However, Gabor's contract was terminated shortly after Adolph Hitler assumed power in 1933, despite the fact that his family had converted to Lutheranism in 1918. Gabor went back to Hungary but moved to England in 1934 and began a long association with British Thompson-Houston Company (BTH). At BTH he first worked on developing a plasma lamp (a new sort of fluorescent lamp) that he had tinkered with in Hungary; he then shifted to electron optics. He also married a fellow BTH employee, Marjorie Louise Butler, in 1936. The couple had no children.

World War II limited Gabor's work, as he was restricted from access to much of BTH's military research by virtue of his foreign birth. After the war, however, he focused his attention on the electron microscope. In 1947 he had the insight that led to his developing the technique of holography. He published his theory the same year and coined the term *hologram* (from the Greek meaning

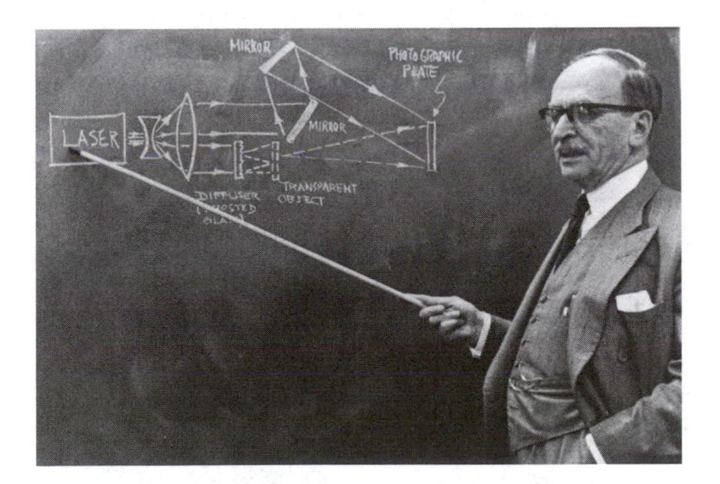

Dennis Gabor, who invented holography in 1947. *AIP Emilio Segrè Visual Archives, Physics Today Collection.*

"completely written") to describe it. The practical implementation of his insight was beset by technical problems, however, including a persistent double image produced by the photographic plate. Gabor left BTH in 1949 to become a reader (equivalent to an associate professor in the United States) at the Imperial College of Science and Technology at the University of London. In 1958 he was appointed professor of applied electron physics. With his students he built a flat television tube, an analog computer, and a Wilson cloud chamber.

The development of the laser in 1960 led to renewed interest in Gabor's work. The ability to concentrate a narrow light source with all waves in the same phase eliminated the problems that had nagged Gabor's earlier efforts. After retiring from Imperial College in 1967, Gabor continued his own research and demonstrated significant potential applications of holography in computer data processing and a number of other fields.

The importance of Gabor's work was widely recognized. He was awarded the Nobel Prize in physics in 1971 and became a member of several scientific societies. He also held more than 100 patents. Moreover, he became a popular speaker, in part as a result of two books he wrote on the importance of scientists using their knowledge to benefit society. Gabor died in 1979, but his legacy lives on. Holography remains an important tool in many aspects of modern life, including photography, mapmaking, computing, medicine, and even supermarket checkout scanners.

Galdikas, Biruté M. F.

(1946–)
German/Canadian
Zoologist

Biruté M. F. Galdikas has studied and protected one of humanity's closest cousins, the red-haired Asian ape called the orangutan. She discovered much of what is known about this solitary animal. It was her belief that she inherited her love of nature from her ancestors in Lithuania, the heavily forested central European country where her parents, Antanas and Filomena Galdikas, grew up. They fled the country separately during World War II and met in a refugee camp. They were married in 1945, and Biruté, their oldest child, was born in Wiesbaden, Germany, on May 10, 1946.

After the war the Galdikases immigrated to Canada, where Antanas Galdikas worked as a miner. Biruté grew up in Toronto with two brothers and a sister. In high school she first read about orangutans, whose name means "people of the forest" in the language of their homeland, Indonesia.

Biruté's family moved to the United States in 1965, and she attended college at the University of California at Los Angeles (UCLA). She stayed there to do graduate work in anthropology, the study of humankind.

While at UCLA Galdikas heard about two young women, JANE GOODALL and DIAN FOSSEY, who were doing ground-breaking studies of chimpanzees and gorillas by observing the animals in their natural habitat for years at a time. Galdikas became "obsessed" with the idea of doing similar work with orangutans. "I was born to study orangutans," she wrote.

Galdikas knew that the great apes—chimpanzees, gorillas, and orangutans—are humans' closest living relatives. All shared an ancestor millions of years ago. Orangutans are the only apes that still live in trees, as human ancestors did 15 million years ago. Galdikas thought that learning more about these apes might help people understand their own origins. In her autobiography she writes: "Orangutans reflect, to some degree, the innocence we humans left behind in Eden."

Galdikas found a path to her dream in 1969, when she attended a lecture given by the British anthropologist LOUIS SEYMOUR BAZETT LEAKEY. Leakey had sponsored the work of Goodall and Fossey. Galdikas introduced herself to him and urged him to help her as well. After some discussion he agreed.

Galdikas finally reached Indonesia, a string of islands off the southeast Asian coast, in September 1971. Rod Brindamour, a fellow Canadian whom she had met at UCLA and married in 1969, went with her. Officials sent them to Tanjung Puting, a forest reserve on the southern coast of the country's largest island, Borneo.

Galdikas and Brindamour set up camp in a bug-infested hut with a grass roof and bark walls, an old hunters' shelter. Honoring Galdikas's mentor, they named it Camp Leakey. Temperatures averaged 90°F, humidity hovered around 100 percent, and bloodsucking leeches "dropped out of our socks and off our necks and fell out of our underwear" when the couple undressed each night. Nonetheless, Galdikas loved the rain forest, which she called "a great cathedral."

At first the couple seldom saw the orangutans, which moved silently through the trees far above their heads. With time and patience, however, Galdikas learned to spot the orange apes and follow their progress. The orangutans, in turn, slowly came to ignore the intruders. Galdikas took notes as the apes searched for fruit, their chief food, during the day and made sleeping nests in the trees at night. She saw adult males, built like wrestlers, fight over a female. She watched females care for their babies, who clung constantly to their mothers during their first years. She verified that, unlike gorillas and chimpanzees, adult orangutans usually lived alone.

Galdikas and Brindamour remained in the forest for four years without a break, spending 6,804 hours observing 58 individual orangutans. After an additional three years Galdikas wrote up her research as her Ph.D. thesis for UCLA. It described details of orangutans' daily lives that had never been reported before. Galdikas has said, "My main contribution [to science] is . . . following one population longer than anyone."

Almost as soon as she began observing wild orangutans, Galdikas also started rehabilitating young orangutans that had been seized from people who captured them illegally to sell to zoos or keep as pets. Forestry officials asked Galdikas and Brindamour to provide a haven for these repossessed babies and help them return to life in the wild.

Galdikas did most of her research with orphan orangutans clinging to her body. As they grew older, the "unruly children in orange suits," as she called them, tore up or tried to eat almost everything in camp. Seeing them slowly go back to forest life was worth the struggle, however. Galdikas has returned more than 80 formerly captive orangutans to the wilderness.

Meanwhile Galdikas was also raising her own son, Binti Paul, born in 1976. Binti's life in the forest ended in 1979, when Rod Brindamour divorced Galdikas, remarried, and returned to Canada, taking the boy with him. Galdikas herself married again in 1981. Her second husband, Pak Bohap bin Jalan, is a Dayak, one of the aboriginal people of Borneo. He and Galdikas have two children, Frederick and Jane.

In the 1990s Galdikas divided her time between overseeing research and rehabilitation work at her camp—now much enlarged—and raising funds for Orangutan Foundation International (OFI), a group she founded in 1986. OFI works to protect the world's 10,000 to 20,000 remaining orangutans, most of which live on Borneo or another Indonesian island, Sumatra. Galdikas has tried to educate people both locally and worldwide about the need to preserve these gentle, intelligent animals and their forest home, "a world which is in grave danger of vanishing forever." In 1997 her work earned the Tyler Prize for Environmental Achievement.

Galen

(c. A.D. 129–c. A.D. 216)
Greek
Physician

The Greek physician Galen's medical discoveries in anatomy and pathology influenced the theory, practice, and teaching of medicine in Europe for 15 centuries. Galen was renowned for his methodical experimentation techniques and his progressive theories regarding human anatomy and physiology. Galen successfully concluded that arteries carried blood rather than air, as was commonly believed in his time. Although some of Galen's ideas regarding physiology were later found to be false, his theories were original and influential, and many were supported by sufficient experimental evidence to be credible. Galen was also a prolific writer and documented his medical observations in about 300 titles, of which about 150 survive today, either in full or in part.

Galen's exact date of birth is unknown, but most likely he was born in A.D. 129 or 130 in Pergamum (now Bergama, Turkey). Galen's decision to study medicine was influenced by his father, Nikon, a wealthy and successful architect, with whom Galen shared a close relationship. A shrine to Asclepius, the god of medicine, was located in Pergamum, and Nikon supposedly had a dream in which Asclepius appeared, convincing him that Galen should pursue medical studies.

At the age of 16, Galen began studying medicine in Pergamum. Five years later he went to Smyrna to learn anatomy, and from there he traveled to Asia Minor to study drugs. Galen's final destination was Alexandria, Egypt, the most important center of medicine during Galen's times. There he was able to study the human skeleton. After more than a decade of training, Galen returned to his hometown and worked as a physician to a group of gladiators.

In about A.D. 162, Galen moved to Rome, where he gained respect through his large practice, public lectures on anatomy, and treatment of several Roman emperors, including Marcus Aurelius and Commodus. Galen considered anatomy to be the foundation of medical knowledge and conducted experiments to feed his appetite for medical learning. He was a proponent of frequent dissection for honing of surgical and research skills, and he dissected animals regularly to study anatomy and physiology. Through his experiments, Galen was able to describe heart valves, cranial nerves, the differences between veins and arteries, the digestive system, the function of spinal nerves, and more. Dissection also led Galen to the accurate and important conclusions that arteries carried blood and that urine traveled from the kidneys to the bladder down the ureters.

Galen was heavily influenced by the works of the physician HIPPOCRATES OF COS and believed in the Hippocratic theory that disease was the result of an imbalance in the four body humors—blood, yellow bile, black bile, and phlegm. Galen modified the theory and claimed the imbalances could be found in specific organs as well as throughout the body. Though humoral theory was later found to be untrue, Galen's contribution allowed physicians to diagnose and treat patients better.

Galen's works enjoyed wide circulation during his lifetime, and by 500 his theories were widely accepted and were part of the medical curriculum in Alexandria. His titles were translated into a number of languages, and the Latin versions provided the foundation of medical training at Western European universities during the medieval era. Though Galen died in about A.D. 216, his observations of physiological processes prevailed for centuries, inspiring many to continue the study of anatomy and physiology.

Galilei, Galileo

(1564–1642)
Italian
Astronomer, Physicist

Galileo Galilei made significant contributions in the fields of astronomy and physics, particularly in mechanics. His discoveries in astronomy, accomplished with the telescope, a relatively new invention, changed the way the universe was viewed. Galileo was the pioneering force behind the application of mathematics to the analysis of mechanics, and he demonstrated that falling bodies accelerate uniformly and independently of their weight. Galileo's support of NICOLAUS COPERNICUS's theory that the planets revolve around the Sun was contrary to the teachings of the Roman Catholic Church and caused years of conflict and unrest for Galileo.

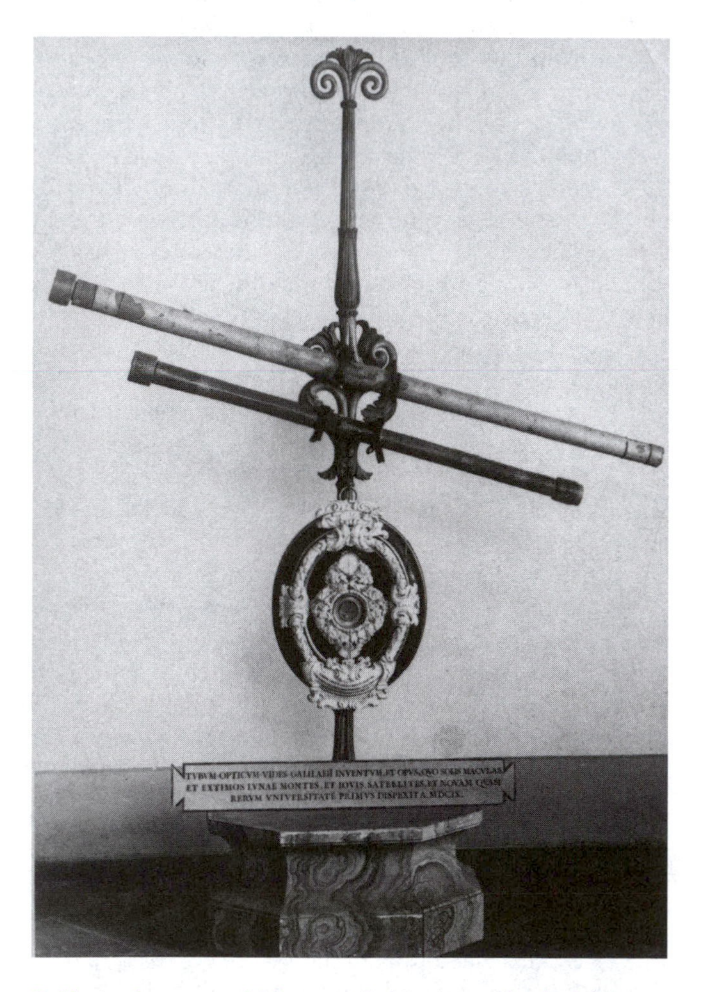

Galileo's telescope and "discoverer" objective, with which he observed in 1610 that Jupiter had four satellites. *AIP Emilio Segrè Visual Archives, E. Scott Barr Collection.*

Galileo was born February 15, 1564, in Pisa, Italy, the oldest of seven children. His mother was Giulia Ammannati of Pescia, and his father was Vincenzio Galilei, a musician and descendant of a Florentine patrician family. The family moved to Florence in 1575, and Galileo was educated at a monastery near Florence. When he entered the order as a novice in 1578, his father, who did not approve, removed him to Florence.

Galileo entered the University of Pisa in 1581 to study medicine but became fascinated with mathematics. He studied mathematics with a private tutor and left the university in 1585 without a degree. Galileo returned to Florence to teach and continued his studies privately. He secured a teaching position at the University of Pisa in 1589 and moved to Padua in 1591; there he gained a professorship in mathematics. Galileo returned to Pisa in late 1610 and became the chief mathematician at the University of Pisa.

Galileo made one of his first discoveries in 1583 while he was still a university student. Observing that lamps swinging in the wind took the same amount of time to swing regardless of the degree of movement, Galileo proposed the use of pendulums for clocks. In 1586 he created a hydrostatic balance to calculate relative densities. He also formulated the law of uniform acceleration for falling bodies through a series of experiments in which he rolled balls down sloped planes. (Galileo did not drop weights from the leaning tower of Pisa, as was widely reported.) He later documented many of his observations on mechanics in *Discourses Concerning Two New Sciences* (1638).

Galileo is perhaps best known for his contributions in astronomy. Though he did not invent the telescope, he improved upon the basic design to create an astronomical instrument and was the first to study the stars and planets. Galileo was amazed by the abundance of stars, invisible to the naked eye, and observed that the Moon had mountains and craters, not a smooth surface, as was traditionally thought. He also discovered sunspots. In 1610 Galileo viewed Jupiter and discovered that it had four satellites—a major finding. He quickly published his observations in *Starry Messenger* and enjoyed immediate acclaim.

Galileo's studies of the universe convinced him to adopt the Copernican principle that the Earth revolved around the Sun. This posed problems between Galileo and the Roman Catholic Church, and Galileo had no choice but to retract his support of the theory. He tried to avoid controversial issues in subsequent years, but his thinly veiled support of the Copernican theory in *Dialogue on the Two Chief World Systems, Ptolemaic and Copernican,* in 1632 was the last straw. In 1633 Galileo was condemned by the Inquisition and placed under house arrest until his death in 1642.

Galileo is regarded as the father of mathematical physics for his extensive studies on mechanics, and his contributions to physics and astronomy render him one of the greatest scientists of all time. Galileo never married, but at the age of 35 he bore two daughters and a son with a Venetian mistress, Marina Gamba. The older daughter, Virginia, became a cherished companion to Galileo in his later years. Galileo spent his final years under house arrest in Arcetri, near Florence, and for the last four he was totally blind. He died on January 8, 1642, a few weeks before his 78th birthday.

Gamow, George

(1904–1968)
Russian/American
Physicist

In addition to his numerous advances in nuclear physics and molecular biology, the Russian-born George Gamow made discoveries in the field of radioactivity through his research of nuclear decay. He was also known for his work in astrophysics. A proponent of the big bang theory of the origin of the universe, he speculated about the atmospheric changes that may have taken place in the universe shortly after its creation. Gamow's research on deoxyribonucleic acid (DNA) influenced modern genetic theory, and he succeeded in popularizing physics and astronomy through his many writings.

George Gamow's attraction to learning may have been influenced by his parents, Anthony M. Gamow and Alexandra Lebedinzeva, who were both educators. Born on March 4, 1904, in Odessa (now Ukraine), Russia, Gamow showed an early fascination with mathematics and science. While his fellow classmates studied conventional subjects, Gamow taught himself advanced math, such as differential equations, and the theory of relativity. Gamow's interest in astronomy and optics was sparked in 1917, when he received a telescope from his father for his birthday.

Gamow entered Novorossysky University in Odessa in 1922 to study mathematics but transferred to the University of Petrograd (now St. Petersburg) after one year. Gamow was not particularly interested in his coursework, however, and though he passed classes, he rarely attended. It is believed the university never granted him a doctoral degree.

It was during a summer at the University of Göttingen in 1928 that Gamow achieved his first major scientific discovery. In an effort to understand why Lord ERNEST RUTHERFORD's attempts to bombard an atomic nucleus with positively charged particles had failed, Gamow developed his quantum theory of radioactivity, which effectively explained the behavior of radioactive elements.

George Gamow, who made significant discoveries in the fields of nuclear physics, molecular biology, and radioactivity. *AIP Emilio Segrè Visual Archives.*

Recognized for his work, Gamow was awarded a fellowship at the Institute for Theoretical Physics in Copenhagen from 1928 to 1929. He then traveled to the Cavendish Laboratories at Cambridge, where he worked with the nuclear scholar ERNEST RUTHERFORD. There Gamow succeeded in calculating the energy required to split an atom's nucleus using protons. Gamow's findings were used as the foundation for modern theories of nuclear fission and fusion.

Gamow returned to Russia in 1931; he married Loubov Wochminzewa and received a professorship in physics at the University of Leningrad (now St. Petersburg). Political unrest was extreme, however, and Gamow and Wochminzewa defected in 1933. Gamow traveled to the United States in 1934, and in 1935 he became a professor of physics at George Washington University in Washington, D.C.; he held that post until 1956. There Gamow collaborated with the physicist EDWARD TELLER to devise the theory of beta decay, a nuclear decay process in which a high-speed electron, or beta particle, is emitted.

Gamow's interest in astrophysics steered him to study the nuclear, physical, and chemical changes that might have taken place during the creation of the universe. Gamow was able to develop a theory explaining the formation of hydrogen and helium, but it was not until later that he devised a theory regarding the formation of heavier elements. He concluded in 1956 that these elements must have formed in the center of stars. Gamow's initial findings were published in a paper called "Alpha-Beta-Gamma," a joint effort with Ralph Alpher and HANS ALBRECHT BETHE.

Molecular biology was the next field to interest Gamow, and after 1953 he investigated the structure of DNA molecules. Gamow theorized that DNA molecules held a genetic "code," composed of the four types of nucleic acid bases found in a DNA molecule.

Gamow was a member of a number of professional societies, including the American Physical Society and the U.S. National Academy of Sciences. Though Gamow made contributions in a number of scientific disciplines, he was also known for his many writings, including books featuring the character "Mr. Tompkins." The works explained science in layperson's terms and earned Gamow the Kalinga Prize from the United Nations Educational, Scientific, and Cultural Organization (UNESCO) in 1956. Also in 1956 Gamow left George Washington University and separated from his wife. He moved to Boulder, Colorado, and taught physics at the University of Colorado. He married Barbara Perkins in 1958 and died in Boulder on August 19, 1968.

Gardner, Julia Anna

(1882–1960)
American
Geologist, Paleontologist

Julia Gardner used the fossils of snail-like ocean animals to identify rocks that contained oil. She was born on January 26, 1882, in Chamberlain, South Dakota. Her father, Charles, a physician, died when she was just four months old. She and her schoolteacher mother, also named Julia, moved several times as she grew up.

Gardner attended college at Bryn Mawr, where the pioneer geologist Florence Bascom interested her in the subject of geology and became her mentor and friend. Gardner also studied paleontology, the science of fossils. She graduated in 1905 and, after a year of teaching school, returned to Bryn Mawr to earn her master's degree in 1907. Bascom helped her obtain a scholarship to Johns Hopkins University, where she earned her Ph.D. in paleontology in 1911. She stayed on to teach at Hopkins, sometimes without pay, until 1917. She also began to work as a contractor for the U.S. Geological Survey (USGS). Her specialty was identifying certain kinds of rocks, formed when ancient seabeds turned to stone, by identifying the fossils of mollusks (shelled sea creatures) embedded in them.

Gardner joined the USGS as a full-time assistant geologist in 1920. She worked in the coastal plain section on a survey of part of Texas and then, starting in 1936, in the paleontology and stratigraphy section. She became an associate geologist in 1924 and a geologist in 1928. The kinds of rocks she mapped were important because many of them contained oil. Writing in *Notable American Women: The Modern Period,* Clifford M. Nelson and Mary Ellen Williams say that "by the 1940s, Gardner's work . . . was of national and international importance."

During World War II Gardner worked for the Military Geology Unit of the USGS, which analyzed maps, aerial photographs, and other sources for information that might be helpful in the war. When Japan sent firebombs to America's northwest coast, Gardner used shells in the recovered bombs' sand ballast to identify the beaches where the sand and the bombs originated.

Gardner's war work aroused her interest in the geology of the Pacific, and after the war she began work on geological maps of western Pacific islands. She retired from the USGS in 1952 but was immediately rehired as a contractor to continue her study of fossil mollusks on the islands. In the year she retired she received the Interior Department's Distinguished Service Award and served as president of the Paleontological Society. She was also a vice president of the Geological Society of America in 1953. A stroke ended her career in 1954, and she died on November 15, 1960.

Gasser, Herbert Spencer

(1888–1963)
American
Neurophysiologist

With his collaborator JOSEPH ERLANGER, Herbert Spencer Gasser conducted important research into the composition and workings of nerve fibers. Together the duo demonstrated that every nerve fiber could be categorized as one of three types, based on thickness, and that each type performed specific functions. Their work, which earned them a Nobel Prize, did much to advance the understanding of the human nervous system.

Born on July 5, 1888, in Platteville, Wisconsin, Gasser was one of three children of Herman and Jane Griswold Gasser. His mother was a teacher, and his father, an immigrant from the Tyrol region of Austria, was a country doctor. Herman Gasser's interest in the evolutionary debates of the time led him to name Herbert after the British philosopher Herbert Spencer.

Herbert Gasser, whose research on the human nervous system earned him the Nobel Prize for physiology or medicine in 1944. *Bernard Becker Medical Library, Washington University School of Medicine.*

Gasser matriculated at the University of Wisconsin, where he earned a bachelor's degree in zoology in 1910 and a master's degree in anatomy in 1911. While at the university he took a physiology class taught by Erlanger, whose lectures on medical research inspired the young Gasser to attend medical school. Gasser received his medical degree in 1915 from Johns Hopkins University.

Gasser spent most of his early career at Washington University in St. Louis, serving as a professor of physiology from 1916 to 1921 and a professor of pharmacology from 1921 to 1931. During this period he teamed up with Erlanger, his undergraduate mentor, in a series of experiments that would forever alter the study of the nervous system. Physiologists had already discovered that impulses (also known as action potentials) travel along nerves to convey sensations and to control muscle movement. Although these impulses could be recorded by electrical instruments, there were no devices precise enough to measure changes in their duration and strength. It had been hypothesized that thicker nerve fibers carried impulses more rapidly than thinner fibers, but this theory remained unverifiable—until Gasser and Erlanger applied a new instrument to the problem in 1922.

The pair successfully modified a cathode-ray oscilloscope, which had been invented by the Western Electric Company and ran at a low voltage, enabling them to chart the diminutive currents of nerve impulses. Moreover, their cathode-ray oscilloscope allowed them to record the time elapsed between impulses and to measure changes in nerve reactions. Their experiments proved that there were different types of nerves and that thicker nerve fibers conveyed impulses faster than thinner ones. They also demonstrated that different types of fibers serve discrete functions. For example, impulses conveying pain sensations travel along thinner and slower fibers, whereas muscle commands move on the thicker, faster pathways. Gasser and Erlanger further discovered that nerve impulse transmission occurs in two phases—with an initial surge in electric potential (which they called the spike), followed by a sequence of small, slow potential changes (which they described as the after-potential).

In 1931 Gasser left Washington University for a professorship in physiology at Cornell University Medical College, where he remained until 1935. That year he was named the scientific director of the Rockefeller Institute for Medical Research, a post he held until 1953. Upon his retirement from the institute Gasser returned to the laboratory, where he methodically took up his work on nerve physiology. He studied differentiation of the thinnest nerve fibers.

Gasser's many accomplishments were recognized during his lifetime. In 1944 he and Erlanger were awarded the Nobel Prize in physiology or medicine for their work on nerve fibers. Gasser was elected to the National Academy of Sciences in 1934 and received the Kober Medal from the American Association of Physicians in 1954. Gasser's research with Erlanger fueled further advances in neurophysiology. Their conclusions about the different types of nerve fibers and functions led to a better understanding of the mechanisms of the human nervous system. Their technical innovation in applying the cathode-ray oscilloscope to measuring nerve impulses paved the way for the use of ever more precise machinery. Gasser died on May 11, 1963. He never married or had children.

Gauss, Karl Friedrich

(1777–1855)
German
Mathematician

Karl Friedrich Gauss, considered one of the greatest mathematicians of all time, made numerous and important contributions in algebra, probability theory, number theory (higher arithmetic), and differential geometry, as well as in mathematical astronomy. Gauss's findings influenced nearly every facet of mathematics and mathematical physics. Gauss failed to publish many of his discoveries, however, and the depth of his knowledge, as well as the extent to which he influenced a century of mathematics, were largely unknown until after his death.

Gauss was somewhat of a child prodigy. Born on April 30, 1777, in Brunswick, Germany, to a peasant family, Gauss taught himself to read and count at an early age. He succeeded in finding and correcting an error in his father's calculations at the age of 3, and at 10 he figured out the formula for the sum of an arithmetic series. Gauss possessed an uncanny ability to perform mental calculations, an ability that persisted throughout his life. By the age of 17 Gauss had already conceived a substantial number of his mathematical discoveries. His mother encouraged Gauss to pursue a profession rather than a trade, and at 14 he obtained the patronage of the duke of Brunswick, who funded Gauss's education.

Gauss attended Caroline College in Brunswick and the University of Göttingen from 1795 to 1798. He earned his doctorate at the age of 22 from the University of Helmstedt. In his thesis Gauss proved the fundamental theorem of algebra and developed the theory of complex numbers. Gauss's proof was the first genuine, error-free proof and demonstrated the rigorous standards he placed on himself. Two years after receiving his doctorate, Gauss published *Examinations of Arithmetic,* which discussed his theory of numbers. Considered his most important achievement, the book essentially founded the discipline of number theory. Among the problems Gauss successfully proved in the work was the impossibility of creating a regular heptagon with a compass and straightedge, a problem that had challenged mathematicians since ancient times.

In 1807, a year after the death of the duke, Gauss became director of the observatory at Göttingen, a position that required little teaching and mainly emphasized research, which he preferred. Gauss continued his work in mathematics, but his interests led him to study other disciplines, such as mathematical astronomy, as well. In 1801 Gauss calculated the orbits of small planets; when the asteroid Ceres was discovered and subsequently lost by the Italian astronomer GIUSEPPE PIAZZI, Gauss successfully predicted the position in which it would reappear. Gauss published a work on planetary motion in 1808 and made design modifications and improvements to astronomical instruments used in his observatory.

From 1820 to 1840 Gauss tackled geodesy, terrestrial mapping, electromagnetism, and optics. He introduced the Gaussian error curve and invented the heliotrope, which was used to study the size and shape of the Earth. In 1833 Gauss collaborated with Wilhelm Weber to create the electromagnetic telegraph.

In 1805 Gauss married Johanne Osthof, who died in 1809 after the birth of their third child. Though Gauss remarried and had three more children, the loss of his first wife affected him profoundly, and he threw himself into reclusive mathematical research. Gauss received acclaim during his lifetime for his studies in mathematical astronomy, but the depth of his genius was unknown until after his death, when personal notebooks of his discoveries were unearthed. For example, Gauss had predicted non-Euclidean geometry 30 years before it was created by Janos Bolyai and Nikolai Lobachevsky. Had Gauss published more of his discoveries, mathematics would have been advanced by several decades. Gauss died in his sleep on February 23, 1855, in Göttingen.

Geller, Margaret Joan

(1947–)
American
Astronomer

Astronomers used to believe that the universe was as smooth as pudding, with galaxies scattered through it evenly like raisins. Margaret Geller, however, has shown that it is more like a dishpan full of soap bubbles. She has helped to make the most extensive existing maps of the nearby universe. The largest of the structures shown in Geller's maps, the Great Wall, contains thousands of galaxies and is at least a half-billion light-years across.

Margaret Geller was born on December 8, 1947, in Ithaca, New York. Her father was a crystallographer at Bell Laboratories and sometimes took her to work with him. "I got the idea that science was an exciting thing to do," she says. He also gave her toys that helped her visualize in three dimensions, a skill that proved essential in her later work. Her mother encouraged her interest in language and art; Geller has said that if she had not become a scientist, she probably would have been an artist. These two interests combine in Geller's fondness for patterns in nature.

At the University of California at Berkeley, Geller changed her major from mathematics to physics and graduated in 1970. Then, excited by new findings in astronomy, she specialized in astrophysics in graduate school. She obtained her master's degree in 1972 and her Ph.D. in 1975 from Princeton University. She was not comfortable with the scientific atmosphere there, however, and often considered dropping out.

After several years of postdoctoral work at the Harvard-Smithsonian Center for Astrophysics, Geller spent a year and a half at Britain's Cambridge University, mostly thinking over her career. "I realized that if I was going to stay in science, I was going to have to make some changes and do problems because I was interested in them," she says. What interested her, she decided, was the large-scale structure of the universe—a subject that she came to realize was largely unexplored. She returned to the Harvard-Smithsonian Center in 1980, continuing her affiliation with both it and Harvard University, where she became a professor in 1988.

In 1985 Geller and a fellow Harvard-Smithsonian astronomer, John Huchra, decided to map the distribution of galaxies in a given volume of space. They examined a pie-slice-shaped segment of the sky with a 60-inch telescope on Mount Hopkins, Arizona. Huchra and a French graduate student, Valerie de Lapparent, gathered data at the telescope, scanning the sky in a strip and measuring about 20 galaxies a night. Geller worked to interpret the results.

The researchers expected the galaxies to be more or less evenly distributed. When de Lapparent plotted a computer-generated map showing the location of the galaxies in the pie slice, however, it revealed a remarkable pattern. "A lot of science is really . . . mapping," Geller says. "You have to make a map before you understand."

When Geller, Huchra, and de Lapparent looked at the map of about 1,000 galaxies in the fall of 1986, they were astounded. Instead of an even distribution of galaxies, the map showed clusters of star systems curving around dark voids almost bare of visible matter. Geller said that their slice of the universe looked like a "kitchen sink full of soapsuds." Astronomers had seen clusters of galaxies before, but, Geller says, "nobody had seen *sharp* structures. . . . The pattern is so striking." Most striking of all was the strange shape in the center of the map, which looked like a child's sketch of a person and has been nicknamed "the stickman." In 1989, as the group's survey continued, they discovered another structure, now called the Great Wall.

Geller and Huchra's discoveries have "changed our understanding of the universe," the *Discover* writer Gary Taubes wrote in 1997. They showed that the universe did not have the smooth structure that everyone had expected. Geller believed that the patterns she and her coworkers saw were explained by the action of gravity. She agreed that the distribution of matter in the universe was uniform immediately after the big bang, 10 to 15 billion years ago, but said that the details of the universe's formation and evolution to its present structure remain to be discovered. In the late 1990s Geller and another Smithsonian astrophysicist, Dan Fabricant, began working on an even deeper map that they hoped would allow them to see the evolution of the distribution of galaxies directly. They expect to survey more than 50,000 galaxies, reaching out to a distance of 5 billion light-years from Earth.

Geller, in 1998 a professor of astronomy at Harvard and a senior scientist at the Harvard-Smithsonian Center for Astrophysics, had won several awards, including the MacArthur Genius Award (1990) and the Newcomb-Cleveland Award of the American Association for the Advancement of Science (1991). Even more than awards, however, she has loved the process of discovery and admits that her long-range scientific goals have been "to discover what the universe looks like and to understand how it came to have the rich patterns we observe today."

Germain, Marie Sophie

(1776–1831)
French
Mathematician

Because Sophie Germain taught herself, she saw mathematical problems in a fresh way and solved some of them better than any man of her time. She is considered one of the founders of mathematical physics.

Marie Sophie Germain was born in Paris on April 1, 1776, to Ambrose Germain, a silk merchant, and his wife. Although the family was financially comfortable, the bloody chaos of the French Revolution surrounded Sophie's childhood, and she retreated to her father's library.

Sophie's parents, like most at the time, feared that mental effort would harm their daughter's health, so they tried to stop her from studying. When she was 13, they took away her clothing, candles, and the fire that warmed her room to prevent her from getting out of bed at night. Sophie got up anyway, wrapped quilts around herself, lit candles that she had hidden, and went on with her math, even though the room was so cold that her ink froze.

In 1798 Germain wanted to hear Joseph Lagrange's mathematics lectures at the École Polytechnique in Paris, but the school did not admit women. Undaunted, Germain borrowed the notes of men friends who attended the lectures. To escape "the ridicule attached to a woman devoted to science," she submitted an end-of-term paper under the name Monsieur Leblanc. Lagrange praised the paper and was astounded when Germain revealed her identity.

Lagrange became Germain's mentor and helped her meet or correspond with other mathematicians, such as the German KARL FRIEDRICH GAUSS. Gauss exchanged letters with Germain under her Leblanc pseudonym for several years before learning that she was a woman. When he found out, he wrote to her that since she had learned mathematics in spite of all the obstacles that society put in the way of female education, she must possess "the most noble courage, extraordinary talent, and superior genius."

Germain made contributions in both pure and applied mathematics. Her studies in pure mathematics involved number theory, Gauss's field. At the age of 25 she offered a proof of part of an equation called Fermat's last theorem. No one else of her time was able to prove even a part of the theorem, and no one proved the whole theorem until 1993.

The physicist Ernest Chladni inspired Germain's contribution to applied mathematics. Chladni caused a sensation in 1808 by placing sand on a metal or glass plate on a stand and making the plate vibrate by drawing a violin bow along its edge. Lagrange, for one, believed that mathematics was not advanced enough to produce equations that could describe the curving patterns the sand formed on this flexible, or elastic, surface. When the French Academy of Sciences announced a competition to attempt that task in 1809, Sophie Germain was the only person brave—or naive—enough to enter. Lagrange and two other mathematicians reviewed her paper, which she submitted anonymously. They found serious flaws in it, thanks mostly to gaps in her home-grown education, but praised her effort.

The academy held the contest again in 1813 and 1815, and each time Germain was the best entrant, although her work was still far from perfect. She finally was awarded a prize in a public ceremony on January 8, 1816. When her paper was published in 1821, the French mathematician Claude Navier said of it, "It is a work which few men are able to read and which only one woman was able to write." In 1822 the Academy of Sciences formally admitted Germain to its meetings, an unusual honor for a woman. Germain died of breast cancer in 1831.

Gilbert, Grove Karl

(1843–1914)
American
Geologist

Grove Karl Gilbert carried out significant work in the area of geomorphological theory. Known for his careful application of quantitative techniques to physiographic problems, he researched sediment transport in streams and rivers, evaluated the effects of mining on California river systems, and studied the ancient lakes of the Great Basin (the large inland region stretching from the Sierra Nevada to the Wasatch Mountains). He also theorized about Earth's climate during the Pleistocene era (the period of geological time when humans first appeared, and when the polar ice caps expanded and then later retracted) and postulated a system of laws pertaining to landform development after he recognized a topographic form caused by faults in the Earth's crust.

Gilbert was born on May 6, 1843, in Rochester, New York. His parents, Grove Sheldon and Eliza Stanley Gilbert, presided over a closely knit family. Gilbert's father was a self-taught painter, and the family was poor. After finishing high school at the age of 15, Gilbert enrolled at the University of Rochester, from which he graduated in 1862. While at Rochester, Gilbert took a geology class taught by Henry Ward. In addition to his professorial duties, Ward operated Ward's Natural Science Establishment (NSE), a business that provided geological and zoological specimens—particularly fossils—to schools and museums.

At Ward's invitation, Gilbert worked at NSE from 1863 to 1868. Gilbert's interest in geology arose during an excavation of a mastodon, when he became more fascinated by potholes in the riverbed where the skeleton was located than by the excavation itself. In 1869 Gilbert volunteered for the second geological survey of Ohio. His supervisor was impressed by Gilbert's work and offered him a salaried job the following year.

A major breakthrough in Gilbert's career occurred in 1871, when he was chosen for one of the four survey teams exploring the United States west of the 100th meridian. By 1874 these distinct surveys were incorporated into one unit, the U.S. Geological Survey (USGS), which was headed by John Wesley Powell. Gilbert and Powell developed a collaborative relationship that spurred both men to interpret the unique geological features they encountered in the American West. One of Gilbert's first important discoveries arose from his investigation of the Henry Mountains in Utah during 1875 and 1876. He recognized that the rock formations there were laccoliths, volcanic rocks intruding in layers of sedimentary rock. Gilbert was the first to conclude that rock could bulge upward and be deformed in this manner.

Gilbert's time in the Great Basin led to his major contribution to geomorphological theory. After surveying the ancient lakes of the region, he was able to make determinations about past climatic conditions. His study of Lake Bonneville's displaced shorelines prompted him to draw conclusions about the forces at work on Earth's crust. He also deduced that the topographic features of the Great Basin were caused by faults in the crust, and he helped formulate the concept of isostasy—that an equilibrium exists in the Earth's crust, in which elevating forces are balanced by depressing ones. His 1890 publication, *Monograph on Lake Bonneville,* amplified this point. Gilbert served as chief geologist of the USGS from 1889 until 1892.

After a stint heading the USGS's Appalachian division, Gilbert returned westward and conducted innovative research on sediment transport. Between 1905 and 1908 he measured the effect of hydraulic-mining debris on the Sacramento River System in California. This early "environmental impact report" cataloged the monumental effect humanity can have on the planet's geologic features. He continued to pursue this research in Berkeley, California, and in 1909 proposed that streams tend to make channels and slopes that transport exactly the amount of sediment that is deposited in them.

Despite his professional modesty, Gilbert's accomplishments were well recognized by his peers. He was selected to lead the Geological Society of America in 1892 and 1909—the only person ever to hold this position twice. In addition to his theories on isostasy and sediment transport, Gilbert is acknowledged for his study of glacial geology, his cogent history of the Niagara River, and his participation on the 1899 Harriman expedition to Alaska. Gilbert died on May 1, 1918.

Gilbert, Walter

(1932–)
American
Molecular Biologist

Although Walter Gilbert was trained as a physicist and mathematician, he has devoted almost his entire career to molecular biology. In the late 1960s Gilbert proved the existence of repressor molecules, which modify or inhibit the activity of certain genes. He then turned his attention to sequencing deoxyribonucleic acid (DNA). In 1980 he shared the Nobel Prize in chemistry for his efforts in this area. After a stint as the CEO of Biogen, a technology company he helped found, Gilbert returned to his research and contributed his expertise to the Human Genome Project—a vast undertaking to map human DNA.

Gilbert was born on March 21, 1932, in Boston, Massachusetts, to Richard and Emma Cohen Gilbert. His father was an economist at Harvard University, and his mother, a child psychologist, taught the children at home when they were young. In 1953 Gilbert married Celia Stone, and the couple had two children.

At Harvard University, Gilbert majored in chemistry and physics, earning his B.A. summa cum laude in 1953, and his master's degree in physics in 1954. He then pursued a doctoral degree at Cambridge University, where he studied theoretical physics under Abdus Salam. At Cambridge, Gilbert met JAMES DEWEY WATSON and FRANCIS HARRY COMPTON CRICK, who had identified the structure of DNA. In 1957 Gilbert was awarded his Ph.D. in mathematics; he thereupon returned to Harvard as a National Science postdoctoral fellow in physics.

Despite the considerable time Gilbert had dedicated to physics, his interests soon changed. In 1959 he accepted a position as an assistant physics professor at Harvard, but in 1960 Gilbert struck up a friendship with Watson, who was also at Harvard. On Watson's invitation Gilbert collaborated on a project to isolate messenger ribonucleic acid (RNA), the substance thought to transmit genetic information from DNA to ribosomes (the cellular structure where protein synthesis takes place). Despite his lack of formal training, Gilbert was convinced by his role in the project to devote his future research to molecular biology.

In the 1960s Gilbert began his first substantial project in this field, when he explored how genes are activated within a cell. François Jacob and Jacques Lucien Monod had speculated that cells must contain an element that represses some genes, preventing them from acting at all times. Their hypothesis explained how cells could perform different functions, even though all cells contain identical genetic components. Gilbert sought to determine whether an actual "repressor" substance existed. To do so he used an innovative experimental technique called equilibrium dialysis on studies of *Escherichia coli* bacterium. When lactose (milk sugar) is present, *E. coli* produces an enzyme named betagalactosidase, which enables the bacterium to split lactose into its constituent sugars. Without lactose *E. coli* does not manufacture the enzyme. In order to see whether or not a repressor molecule was prohibiting the production of betagalactosidase when lactose was not available, Gilbert added radioactive molecules that were similar to lactose, making it possible to trace any potential repressor activity. In 1966 he definitively proved the existence of a repressor molecule. Two years later Gilbert was named a full professor of biochemistry at Harvard.

Gilbert next investigated the topic of sequencing, or chemically describing, DNA. Using the technique of gel electrophoresis—in which an electric current pushed DNA fragments that were tagged with radioactive markers through a gel substance—Gilbert was able to "read" the chemical code of the DNA fragment when he exposed it to X-ray film. Although he did take a sojourn into private enterprise during the late 1970s and early 1980s, he has been an American Cancer Society Professor of Molecular Biology since 1974. In 1987 he was appointed chairman of the developmental biology department and became involved in the Human Genome Project, which sought to map human DNA.

Gilbert's efforts to increase the understanding of genetic functions and processes helped advance the fields of molecular biology and biochemistry. In 1980 his work in DNA sequencing was recognized when he was coawarded the Nobel Prize in chemistry with Frederick Sanger and PAUL BERG. He also shared the 1979 Albert Lasker Basic Medical Research Award with Sanger. He is a member of the American Academy of Arts and Sciences, the National Academy of Sciences, and the British Royal Society.

Gilbreth, Lillian Evelyn Moller

(1878–1972)
American
Psychologist, Engineer

Working with her husband, Frank Gilbreth, Lillian Gilbreth studied workers' movements and redesigned workplaces and homes to make labor both more efficient and easier.

She also made managers see the importance of workers' psychological needs. In an essay in *Women of Science,* the author, Martha Trescott, calls Gilbreth "probably the best-known woman engineer in history."

Lillian Evelyn Moller was born into a prosperous family in Oakland, California, on May 24, 1878. Her father, William, owned a large hardware business. As the oldest of eight children, Lillian often cared for the younger ones when her mother, Annie, was sick or pregnant. Although a shy girl, she insisted on the unusual step of going to college. She earned a B.A. in literature from the nearby University of California at Berkeley in 1900 and became the first woman to speak at the university's commencement.

Moller obtained a master's degree from UC–Berkeley and planned to go on to a doctorate, but her plans changed after she met Frank Bunker Gilbreth while visiting friends in Boston in 1903. The energetic Gilbreth had taught himself construction skills and engineering and become one of the country's leading building contractors. He specialized in "speed building," using techniques and devices that helped workers do their jobs more efficiently. He and Moller married at her home on October 19, 1904.

The Gilbreth marriage was a partnership in every sense. Lillian edited Frank's writing, learned the construction business, and even devised new building techniques. The couple had 12 children, and Frank, for his part, took a full share in raising them. Applying their efficiency techniques to the household, the Gilbreths taught their children to take responsibility for themselves and each other. Two of the brood described the family's happy if eccentric life in a humorous memoir called *Cheaper by the Dozen,* which was made into a movie in 1950. The family lived first in New York City, then in Providence, Rhode Island, and finally in Montclair, New Jersey.

Frank Gilbreth admired the ideas of Frederick Taylor, who had developed a concept called time and motion study. By analyzing workers' movements, Taylor said, managers could eliminate unnecessary motion and thus increase the amount of work done in a given length of time. Gilbreth insisted that workers' comfort as well as their efficiency be considered. After filming workers on the job, he redesigned workplaces so that, for instance, tools could be reached without stretching or bending. In both his own business and in others that hired him as a consultant, he looked for the "one best way"—the method that required the least effort—to do each task. He became a pioneer in what was called scientific management.

Lillian Gilbreth, in turn, went beyond Frank's ideas, pointing out that changes that improved physical efficiency but psychologically stressed or isolated workers would fail to improve production in the long run. The Society of Industrial Engineers said in 1921 that Lillian was "the first to recognize that management is a problem of psychology and . . . to show this fact to both the managers and the psychologists." She helped managers and workers find ways to cooperate rather than oppose each other.

The Gilbreths began full-time management consulting in 1914, and their company, Gilbreth, Inc., soon earned an international reputation. Besides lecturing and visiting factories, they held summer classes in their home. They wrote many articles and books about their ideas, including *A Primer of Scientific Management* (1912) and *Applied Motion Study* (1917).

At Frank's urging, Lillian studied for a Ph.D. in psychology at Brown University; she obtained it in 1915. Her doctoral thesis grew into a book, *Psychology of Management,* which was published in 1914. Martha Trescott says that this book "open[ed] whole new areas to scientific management. . . . [It] formed a basis for much modern management theory."

Frank Gilbreth died of a heart attack on June 14, 1924. As recounted in *Belles on Their Toes,* the sequel to *Cheaper by the Dozen,* life was hard for his family for a while. Managers who had hired the Gilbreths doubted that Lillian Gilbreth could be a useful consultant on her own. Gilbreth, however, continued teaching and speaking as well as running Gilbreth, Inc., and in time she persuaded clients that she was more than competent.

Gilbreth also took over her husband's position as visiting lecturer at Purdue University in Indiana after his death. From 1935 to 1948 she was a professor of management in Purdue's School of Mechanical Engineering, the first woman to be a full professor in an engineering school. She was also Purdue's consultant on careers for women from 1939 until her death. In addition to teaching at Purdue, she became head of the new Department of Personnel Relations at the Newark School of Engineering in 1941.

Beginning in the late 1920s, Gilbreth extended her ideas about efficient work into the home, taking industrial engineering "through the kitchen door." She pointed out that homemakers lost time and effort to badly designed room layouts and appliances, just as factory workers did. She designed an "efficiency kitchen" for the Brooklyn Gas Company that was featured in women's magazines. She also worked out kitchen and household arrangements for disabled people that helped them achieve greater independence. Her books on these subjects include *Normal Lives for the Disabled* (1944, with Edna Yost) and *Management in the Home* (1954).

Lillian Gilbreth continued writing, teaching, and lecturing well into her 80s. She received many awards, including more than 20 honorary degrees and commendations. She was made an honorary member of the Society of Industrial Engineers in 1920. In 1966 she became the first woman to win the Hoover Medal for distinguished public service by an engineer. She also was awarded the National Institute of Social Science's medal "for distinguished service to humanity." She died on January 2, 1972, at the age of 93.

Glashow, Sheldon Lee

(1932–)
American
Physicist

Sheldon Lee Glashow's work in the highly theoretical field of particle physics led him to postulate a means of extending what had come to be known as the Weinberg-Salam theory to two other classes of subatomic particles, for which he shared the 1979 Nobel Prize. He also assigned a new property to subatomic particles, which he called "charm," that had important applications to the study of quarks. Both these insights were significant developments in the field.

Glashow's parents, Lewis and Bella (née Rubin) emigrated from Russia to New York City to escape czarist anti-Semitism. Upon arrival in America Lewis founded a successful plumbing business. Sheldon, born on December 5, 1932, in New York City, was the couple's third and youngest son.

Glashow attended the Bronx High School of Science, where his classmates included the future Nobel laureates (both in physics) STEVEN WEINBERG and Gerald Feinberg. He entered Cornell University in 1950. After graduation in 1954 he enrolled at Harvard University; he earned a master's degree in 1955 and a Ph.D. in 1959. Under the tutelage of the Nobel laureate JULIAN SEYMOUR SCHWINGER Glashow's dissertation represented an early effort to devise a theory combining two of the basic natural forces—the weak and the electromagnetic.

After completing his studies at Harvard, Glashow conducted postdoctoral research at the Bohr Institute, the European Organization for Nuclear Research, and the California Institute of Technology. In 1960 he took a faculty position at Stanford University; he moved to the University of California at Berkeley the following year. In 1967 he was appointed professor of physics at Harvard, where he has remained except for brief leaves of absence to other institutions. He married Joan Shirley Alexander in 1972, and the couple had four children.

Physicists had long held that four fundamental forces operate in nature—strong, weak, electromagnetic, and gravitational. Underpinning this belief, however, was the sense that these forces might actually represent varying manifestations of the same basic force. ALBERT EINSTEIN had been one of the first scientists who sought to propound a theory to unify these forces, but this project proved to be exceptionally complex and highly theoretical in nature. During the 1960s the physicists Steven Weinberg and Abdus Salam had independently devised a theory that explained the weak and electromagnetic forces as aspects of the same force, which they termed the *electroweak* force. However, this theory was limited in scope, as it was applicable to only one class of subatomic particles—leptons (which include electrons and neutrinos).

Glashow's innovation was to devise a method to expand this theory to two other classes of particles—mesons and baryons (which include protons and neutrons). For his theory to work Glashow had to create a new property, "charm," for these subatomic particles. The basic building blocks of mesons and baryons are called quarks, and Glashow theorized that a fourth type of quark, the charmed quark, had to exist. This "charm" explanation proved useful in expanding on the quark theory propounded by Murray Gell-Mann and also helped explain the behavior of the J/psi particle, which was discovered in 1974 by Burton Richter and SAMUEL CHAO CHUNG TING. During the 1970s experimental data gathered primarily from particle accelerators supported the theories of Weinberg, Salam, and Glashow, and the three shared the Nobel Prize in physics in 1979. In 1983 Glashow accepted a joint professorship at Texas A&M University.

Glashow's work has exerted significant influence on the field of particle physics. Gell-Mann's insight has been broadened to include another type of quark, the colored quark, and Glashow's work underpins much of the current thought in the field. Along with the Nobel Prize, Glashow received the J. Robert Oppenheim Memorial Medal in 1977, the George Ledlie Prize in 1978, and the Castiglione di Sicilia Prize in 1983.

Goddard, Robert Hutchings

(1882–1945)
American
Physicist

The physicist Robert Hutchings Goddard was a pioneer in rocketry and space flight theory. Interested in space travel from an early age, Goddard thought of rockets as a means to such flight. He developed and launched the first liquid-fueled rocket in 1926 on his aunt's farm in Massachusetts. Many of Goddard's innovations in rocket technology were incorporated in the development of rockets and missiles for both weaponry and space flight programs. Goddard's contribution to modern rocketry was great, but it may have been greater still had the publicity-shy Goddard published more and communicated effectively his findings to other scientists and engineers.

Goddard was born on October 5, 1882, in Worcester, Massachusetts. His father, Nahum Danford Goddard, and his mother, Fannie Louise Hoyt Goddard, were of modest means. His father worked as a machine-shop owner, book-

Robert Hutchings Goddard, for whom NASA's Goddard Space Flight Center is named. *NASA, courtesy AIP Emilio Segrè Visual Archives.*

keeper, and salesman, but his interest in invention influenced him to steer his son toward experimentation and invention. Goddard was a sickly boy and was frequently absent from school. Science fiction proved to be a solace, and Goddard often dreamed of inventions and space-flight devices.

Because of ill health Goddard did not graduate from high school until 1904, at the age of 21. He immediately enrolled at Worcester Polytechnic Institute and majored in general science with a concentration in physics. He then studied physics at Clark University in Worcester and earned his master's degree in 1910 and his doctorate in 1911. Goddard was conducting research in physics at Princeton University when he fell dangerously ill with tuberculosis in 1913. After he recovered, he returned to Clark University in 1914. He became a full professor by 1919 and remained at Clark for the duration of his career.

Though Goddard had long been fascinated by rocketry, it was not until 1909, when he was a student at Clark, that he began to research the subject seriously. In 1914 Goddard secured his first two patents—one for a liquid-fuel gun rocket and one for a two-stage powder rocket. Goddard made a number of additional discoveries in his laboratory at Clark. He was the first to prove that rockets could be fired in a vacuum and did not have to react against air, as was commonly believed at the time. Goddard's experimentation also allowed him to accomplish higher degrees of energy efficiency and exhaust velocities with his rockets than had previously been achieved. The Smithsonian Institution recognized his successes and provided Goddard with funding from 1917 to help him to advance his research.

During World War I Goddard helped create tube-launched rockets, which were later used in the design of bazookas, prevalent in World War II. Goddard wrote about his findings on solid-propellant rockets in the paper "A Method of Reaching Extreme Altitudes." Published in 1919, the seminal work detailed many of the fundamental ideas of modern rocketry.

The 1920s were significant years for Goddard, in both his professional and personal lives. In 1921 Goddard began to explore liquid propellants after becoming frustrated with problems presented by solid propellants, and in 1924 he married his wife, Esther. In 1926 he launched the first liquid-fueled rocket, and though it climbed only 41 feet, it was a significant event in the history of rocket flight and led to a generous grant from Daniel Guggenheim. Goddard then set up a research facility in Roswell, New Mexico, in the early 1930s to continue developing rockets. Among his creations were gyroscopic steering, fuel-injection systems, and the automatic launch sequential system. In 1935 he successfully shot a liquid-propelled rocket faster than the speed of sound.

During the later years of his life Goddard turned to defense-related research and worked on jet-assisted takeoff devices and variable-thrust rockets, used for aircraft. Goddard died on August 10, 1945, in Baltimore, Maryland. He received significant posthumous recognition for his achievements. In 1960 his widow, Mrs. Esther C. Goddard, was awarded $1 million by the U.S. government for rights to use Goddard's patents, which numbered more than 200. Also in 1960 Goddard posthumously received the Smithsonian Institution's Langley Gold Medal for excellence in aviation. The following year Goddard was awarded a Congressional Gold Medal and a building, NASA's Goddard Space Flight Center, was named in his honor.

Golgi, Camillo

(1843–1926)
Italian
Histologist, Cytologist

Although Camillo Golgi was awarded the Nobel Prize for his ground-breaking discoveries about the nervous system, his extensive contributions to science ranged across several disciplines. Early in his career he invented a method for staining nerve tissue, which made it possible to examine nerve cells in great detail under the microscope. Employing this method, he made a number of important discoveries, including the identification of a specialized type of nerve cell. This finding laid the foundation for modern neurology. He later discovered and described a small organ, a network of fibers and cavities, located within a cell's cytoplasm. Golgi also spent several years studying malaria. His findings provided for the diagnosis of the different types of malaria and, eventually, the treatment of the disease.

Golgi was born on July 7, 1843, in Corteno (later renamed Corteno Golgi in his honor), Brescia, Italy. His father, Alessandro, was a doctor. Later in his life Golgi married Donna Lina Aletti, and the couple adopted Golgi's nephew, Aldo Perroncito.

Golgi received his medical degree from the University of Pavia in 1865. While continuing to work at the university, he joined the Ospedale di San Matteo in Pavia, where he developed an interest in the field of histology. In 1868 Golgi became an assistant in the Psychiatric Clinic of Cesare Lombrose. Although his tenure there was brief, Lombrose kindled in Golgi a lifelong interest in the workings of the central nervous system.

In 1871 Golgi left the Ospedale di San Matteo to become the chief resident physician at a home for incurable patients in Abbiategrasso. After setting up a makeshift laboratory in the home's kitchen, Golgi made his first important discovery in 1873, with little more than his microscope and assorted kitchen utensils. He developed a staining method, treating thinly sliced nerve tissue with silver nitrate, that made possible the first real examination of nerve tissue; previously nerve tissue had appeared to be an impenetrable knot of neurons when viewed under a microscope. Golgi's technique enabled observers to differentiate clearly the features of the nerves. While still at Abbiategrasso Golgi began to scrutinize nerve tissue samples treated with his stain. In 1874 he identified a type of nerve cell, later named the *Golgi body,* that had many short, branching dendrites connecting it to other nerve cells. His discovery was profound. SANTIAGO RAMÓN Y CAJAL later established the nerve cell as the basic structural unit of the nervous system. Much of modern neurology is built upon this premise.

Golgi accepted a position as a lecturer in histology at the University of Pavia in 1875. After a brief stint as chair of the anatomy department at the University of Siena in 1879, Golgi returned to Pavia as the chair of general pathology in 1880. Between 1885 and 1893 he conducted his influential research on malaria. Building upon the work of CHARLES-LOUIS-ALPHONSE LAVERAN, who discovered the disease-causing parasite *Plasmodium* in 1880, Golgi found that the two types of malarial fevers—tertian (occurring every other day) and quartan (occurring every third day)—were caused by different species of *Plasmodium.* He also correlated stages of the malarial fever with the parasite dividing in the bloodstream. These discoveries, along with his observation that quinine helped treat the disease, provided a sound basis for diagnosing and treating malaria.

In 1898 Golgi used his staining method to make an essential cytological discovery. He noted a hitherto unobserved organelle within the nerve cell's cytoplasm, the cellular substance between the membrane and the nucleus. This internal reticular apparatus, now known as the Golgi apparatus, appeared to be bundles of stacked membranes. It was not until recent advances in electron microscopy that scientists ascertained that the Golgi apparatus is involved in the synthesis and secretion of proteins.

Golgi remained in Pavia until his death in 1926. He served in the Royal Senate in 1900, and later in the Superior Council of Public Instruction and Sanitation. He was appointed president of the faculty of medicine at the University of Pavia. His numerous awards included the 1906 Nobel Prize in physiology or medicine (which he shared with Cajal) for his work on the central nervous system. His discoveries are considered central to the development of the modern science of neurology.

Goodall, Jane

(1934–)
British
Zoologist

Jane Goodall's research on chimpanzees in Africa is the longest continuous study of animals in the wild and, according to the naturalist Stephen Jay Gould, "one of the Western world's great scientific achievements." For it she was named a Commander of the British Empire, given the U.S. National Geographic Society's Hubbard Medal, and awarded the Kyoto Prize, Japan's equivalent of the Nobel Prize. She has educated people about chimpanzees, worked to rescue captive chimpanzees and save the habitat of wild ones, and sponsored a program that teaches children to care about animals, the environment, and their home communities.

Jane, the older of Mortimer and Vanne Morris-Goodall's two daughters, was born in London on April 3, 1934. Her father was an engineer, her mother a housewife and writer. Her favorite toy as a baby was a stuffed chimpanzee named Jubilee, which she still owns.

An incident that happened when Jane was just five years old showed her patience and determination as well as her interest in animals. One day, while on a farm, she vanished for more then five hours. Vanne Goodall called the police, but before a search could be launched, Jane reappeared. She explained that she had been sitting in the henhouse, waiting for a hen to lay an egg. "I had always wondered where on a hen was an opening big enough for an egg to come out," Goodall recalled later. "I hid in the straw at the back of the stuffy little hen house. And I waited and waited."

Jane's family moved to the seaside town of Bournemouth at about that time, and she stayed there with her mother and sister after her parents divorced several years later. They had no money for college, so she went to work as a secretary. Then in 1957 a former school friend invited Jane to visit her in Kenya. The invitation revived a childhood dream of going to Africa that Jane had formed after reading books such as Hugh Lofting's fantasies about Dr. Doolittle, who lived in Africa and could talk with animals. Jane began saving her money and left as soon as she could pay the fare.

While in Kenya, Jane Goodall met the famed British anthropologist LOUIS SEYMOUR BAZETT LEAKEY, who had made pioneering discoveries about early humans. Leakey, then the head of Kenya's National Museum of Natural History, liked Goodall and hired her as an assistant secretary. In time he told her about his belief that the best way to learn how human ancestors might have lived was to study the natural behavior of their closest cousins, the great apes—chimpanzees, gorillas, and orangutans—over long periods. He wanted to start with chimpanzees, and he asked Goodall whether she would like to do the research. "Of course I accepted," she says.

Following Leakey's recommendation, Goodall decided to work at Gombe Stream, a protected area on the shores of Lake Tanganyika in the neighboring country of Tanzania. When British officials informed her that they could not let her live alone in the wilderness, her mother agreed to stay with her for a few months. The Goodalls and their African assistants set up camp at Gombe in July 1960. Goodall could not get anywhere near the chimpanzees at first, but as the months passed, they grew used to the "peculiar, white-skinned ape." She, in turn, learned to recognize them as individuals. She gave them names such as Flo and David Graybeard.

In the first year of her research Goodall made several observations that overturned long-held beliefs about chimpanzees. Scientists had thought that the animals ate only plants, insects, and perhaps an occasional rodent, but Goodall saw them eating meat from larger animals. Later she saw groups of them hunt young baboons. Goodall one day watched something even more amazing as David Graybeard lowered a grass stem into the hard-packed open tower of a termite mound. After a few moments he pulled it out with several of the antlike insects clinging to it, then ate them. She later saw other chimps do the same thing. The animals were clearly using the grass blades as tools to get the insects, whereas the common belief was that only humans used tools. Goodall even saw the chimps make tools by selecting twigs and stripping the leaves from them. She observed other kinds of tool use as well. "Chimpanzees are so inventive," Goodall told the writer Peter Miller.

In 1961 Goodall enrolled in the Ph.D. program in ethology, the study of animal behavior, at Britain's prestigious Cambridge University. She was only the eighth student in the university's long history to be admitted to a Ph.D. program without having first obtained a bachelor's degree. She wrote up her chimpanzee studies as her thesis and obtained her degree in 1965.

In 1962 the National Geographic Society sent a Dutch photographer, Baron Hugo van Lawick, to take pictures of Goodall at work. Van Lawick and Goodall fell in love and married on March 28, 1964. Their wedding cake was topped by a statue of a chimpanzee. In 1967 they had a son, whom they named Hugo after his father, but everyone called the blond youngster *Grub*, Swahili for "bush baby." Imitating the behavior of good chimpanzee mothers, Goodall kept the child close to her and showered him with affection. Goodall and van Lawick divorced in 1974. Grub later attended boarding school in England, but he returned to Africa as an adult.

As Goodall's observations continued, she discovered dark sides to chimpanzee behavior. For instance, she saw the animals wage war. One group staged repeated sneak attacks on a neighboring group over a period of four years, eventually wiping them out. "When I first started at Gombe, I thought the chimps were nicer than we are," she said in a 1995 *National Geographic* article. "But time has revealed that they are not."

In 1975 Goodall married Derek Bryceson, who was in charge of Kenya's national parks. He was the only white official in the country's cabinet. Unfortunately Bryceson developed cancer in 1980 and died within a few months.

Another tragedy struck Goodall soon after her marriage. Her camp had grown to include students from the United States and Europe, and on May 19, 1975, rebel soldiers from nearby Zaire invaded the camp and seized four of the students. The soldiers demanded ransom for the students' return. After two months of negotiations, money was paid and the students were released unharmed, but the Eden-like peace of Gombe had been permanently shattered. About this time Goodall decided that "I had to use the knowledge the chimps gave me in the fight to save them." She left the continuing observation of the Gombe chimps to her students and began to travel the world as a spokesperson for the animals. In 1977 a nonprofit organization, the Jane Goodall Institute, was formed to help with her work. In 1998 its headquarters were in Silver Spring, Maryland, and from that base she has worked to educate people about chimpanzees, save the chimps' habitat, and limit the use of chimpanzees and other animals in medical research.

In 1984 Goodall received the J. Paul Getty Wildlife Conservation Prize for "helping millions of people understand the importance of wildlife conservation to life on this planet." Her scientific articles have appeared in *National Geographic*, and her books include *Wild Chimpanzees* and *In the Shadow of Man*. In 1998 she published a book for children, *With Love: Ten Heartwarming Stories of Chimpanzees in the Wild*.

Gould, Stephen Jay
(1941–)
American
Evolutionary Biologist

Stephen Jay Gould, a professor of geology and biology, has not only influenced the scientific community but also attracted a mass audience of readers with his clear and elucidating prose. In addition to a monthly column published in the magazine *Natural History*, Gould has written several award-winning and critically acclaimed books. His theory of punctuated equilibrium, which he formulated in the early 1970s with Niles Eldredge, revised a tenet of CHARLES ROBERT DARWIN's theory of evolution. Gould's research on the Cerion snail has also made him an expert on the subject.

Gould was born on September 10, 1941, in New York City. With his brother, Peter, he was raised in a lower-middle-class neighborhood in Queens. His father, Leonard Gould, was a court stenographer, a self-educated man with an interest in natural history. Stephen absorbed his father's hobby. On a visit to the American Museum of Natural History when he was five years old, Stephen was so impressed by a fossil exhibit about dinosaurs that he vowed to spend his life studying geological periods.

After graduating from high school, Gould enrolled at Antioch College in Pennsylvania, where he received his B.A. in 1963. He pursued graduate studies at Columbia University. At Columbia he investigated fossil land snails and was awarded his Ph.D. in paleontology in 1967.

Upon receiving his doctorate, Gould immediately accepted a position at Harvard University, where he became an associate professor of geology in 1971 and a full professor in 1973. He also served as curator of invertebrate paleontology at Harvard's Museum of Comparative Zoology. Gould made an important contribution to evolutionary biology in 1972, when he and Eldredge published "Punctuated Equilibria: An Alternative to Phyletic Gradualism," in the journal *Models of Paleobiology*. The premise set forth in the paper was bold. Gould and Eldredge refuted Darwin's theory of phyletic gradualism, which held that evolutionary adaptations are caused by a slow, continuous process of development. Gould and Eldredge noted that the fossil record provided no confirmation of Darwin's postulate—few examples of transitional species had been found. In place of phyletic gradualism, Gould and Eldredge proposed the model of punctuated equilibrium, which hypothesized that long stable periods were punctuated by shorter periods of rapid change. It was during these evolutionary moments of change that new adaptations suddenly appeared.

Gould began his career as a popular writer in 1974 when *Natural History* published his monthly column, "This View of Life." He was lauded by readers and critics alike for his clear prose and his ability to distill difficult concepts into understandable language without unduly simplifying the material. Each column made a cogent argument and was relevant to a contemporary issue. His columns ranged across such topics as Adam's navel, the black widow spider, and the development of Mickey Mouse. He eventually published eight volumes of his collected essays, which include *Ever Since Darwin: Reflections in Natural History* (1977), *The Panda's Thumb* (1980), and *The Flamingo's Smile* (1985). He also wrote several books on a variety of subjects. *The Mismeasure of Man* (1981) criticized intelligence tests as a means of serving political agendas and furthering cultural prejudices. His 1989 monograph *Wonderful Life: The Burgess Shale and the Nature of History* elaborated on his theory of punctuated equilibrium. Using the Burgess shale (an unusual fossil bed in western Canada) as an example, Gould once more asserted that evolution is not progressive. He passionately argued that some of the strange fossils embedded in the Burgess Shale revealed that evolution is random and disorderly.

Gould's writing has received numerous awards. *The Panda's Thumb* won the 1981 American Book Award in Science, and *The Mismeasure of Man* garnered the 1982 National Critics Circle Award for Essays and Criticism. His eloquent arguments about evolution and natural history have done much to shape the beliefs of his academic colleagues as well as countless readers. He married the artist and writer Deborah Lee in 1965, and the couple had two sons. In the late 1990s he was still teaching at Harvard University.

Gourdine, Meredith Charles

(1929–1998)
American
Engineer

Meredith Charles Gourdine conducted ground-breaking research in electrogasdynamics (EGD) technology and applied his discoveries in this field to the creation of energy conversion systems. Electrogasdynamics involves the interaction between an electrical field and charged particles suspended in a gas, resulting in the production of high-voltage electricity. Gourdine's chief innovation was to employ the principle of electrogasmagnetics to produce enough electricity for practical applications. As president and CEO of his own company, Energy Innovations, Gourdine holds over 70 patents for his electrogasmagnetic-related inventions.

Born on September 26, 1929, in Newark, New Jersey, Gourdine was one of four children. His father, a maintenance worker, had once won a scholarship to Temple University but had opted not to attend. His mother, a devout Catholic, was a teletype operator with an interest in mathematics. When Gourdine was seven, the family moved to Harlem, New York. As a high school student he juggled his schoolwork with athletics (he was a star track athlete and swimmer) and a job to help pay for college.

After graduation from Brooklyn Technical High School, Gourdine enrolled at Cornell University in 1948. There he became interested in the rigorous field of engineering physics, which involved the practical application of mathematics. In addition to completing his course work, Gourdine spent his time at Cornell training for the 1952 United States Olympic track team. He won a silver medal in the broad jump at the Olympic Games in Helsinki, Finland, and received his bachelor's degree in engineering physics in 1953. Gourdine married June Cave in 1953, and the couple had four children. After he and his first wife divorced, Gourdine wed Carolina Baling. After a stint in the navy he received a Guggenheim Fellowship and entered the California Institute of Technology (Caltech) to pursue graduate studies in 1955. During his last year at Caltech he worked at the Jet Propulsion Lab, where he developed the formula for electrogasdynamics. In 1960 he was awarded his Ph.D. in engineering science, an interdisciplinary field that integrated branches of physics.

In his early career Gourdine served in a research capacity at scientific companies. From 1960 to 1962 he worked as the laboratory director at Plasmadyne Corporation, where he explored magnetohydrodynamics (MHD), a conversion method similar to EGD that generates power through an interaction between magnetic fields and gases. The company did not encourage his work, however. In 1962 he accepted a position as chief scientist at Curtiss-Wright Corporation's Aero Division. He remained there for two years, then founded his own company, Gourdine Systems, in 1964. At the helm of his own business Gourdine was free to return to the topic that had captivated him in graduate school—conversion methods. He devoted himself to the quest for practical applications of EGD and MHD. During the next nine years Gourdine obtained 70 patents for his work. To prevent overtaxing of the resources of his fledgling company, Gourdine chose not to manufacture his inventions, but rather to license his patents to other companies. His inventions from this period include the Electradyne Spray Gun, which used EGD to atomize and electrify paint, thereby making it an easy task to spray-paint any object.

In the early 1970s, however, Gourdine's company began to manufacture and market some of its own products. As he had originally feared, the strategy backfired and the company was stretched too thin. In 1974 Gourdine launched a new company, Energy Innovations, that was committed to inventing and licensing practical applications of EGD, MHD, and other direct conversion methods.

In 1986 diabetes-induced blindness limited some of Gourdine's technical involvement in Energy Innovations. Nevertheless, he continued to produce new applications for his technologies with one of his sons, who assisted him in his work after he lost his vision. Gourdine's peers recognized his many achievements. He was elected to the National Academy of Engineering and is a member of the Black Inventors Hall of Fame. Many of Gourdine's applications provide cost-saving and energy-conserving alternatives. For instance, he invented a battery for electric cars, a method for repairing potholes by using rubber from old car tires, and a device to clear fog from airports. Gourdine died in Houston, Texas, on November 20, 1998, of complications of multiple strokes.

Grignard, François Auguste Victor

(1871–1935)
French
Organic Chemist

François Auguste Victor Grignard is best known for his work in organic synthesis. Prior to Grignard's efforts, combining different organic chemical species was a difficult and time-consuming task. Grignard developed a process that made joining types of compounds simple and inexpensive. While still a graduate student Grignard discovered that when magnesium was treated with methyl iodide in ether, the result was an organomagnesium compound, the first of a class of organic reagents, later named *Grignard's reagents* in his honor, that could synthesize a variety of organic compounds. In reactions with carbonyl and other compounds Grignard's reagents created organic alcohols, organic metals, and other organic compounds. The importance of Grignard's findings was immediately recognized, and he shared the Nobel Prize in chemistry in 1912.

On May 6, 1871, Grignard was born in Cherbourg, France, to Théophile Henri and Marie Hérbert Grignard. His father, a sailmaker and a foreman at the local marine arsenal, encouraged his son to pursue a career in education. The young Grignard strove to do so, but after two years at the École Normale Spécial (which routinely produced a great many secondary school educators), the school closed. Grignard transferred to the University of Lyons but ultimately failed the mathematics exams required for teaching.

After completing his compulsory military service in 1893, Grignard returned to Lyons and followed the advice of a friend to study chemistry. He discovered that the topic fascinated him, and he resolved to earn a graduate degree. In 1898 Grignard began the research that would ultimately lead to his Nobel Prize. Under the guidance of his professor, Philippe Barbier, Grignard tackled the problem of organic synthesis. In his own work Barbier had explored organometallic compounds, which are formed by combining two organic radicals (chemical groups that are unchanged by a reaction) with a metal atom. However, using these organometallic compounds was difficult. Some were highly reactive (organosodium, for instance), whereas others were entirely unreactive (such as organomercury). Organozinc compounds often spontaneously ignited at room temperature. Although Barbier had used magnesium to join organic radicals, he had not obtained consistent results.

Grignard set out to investigate organomagnesium compounds. Wary of the spontaneous flammability of organozinc compounds, he took precautions with magnesium. Grignard knew that performing organozinc reactions in anhydrous ether prevented the spontaneous fires and decided to apply the method to his magnesium experiments. He added magnesium shavings in ether to methyl iodide. The resulting solution, the first of Grignard's reagents, reacted with an aldehyde or ketone to produce a number of organic compounds. He first presented his discovery in a paper at a meeting of the Academy of Sciences in 1900. The following year Grignard submitted his thesis on organomagnesium compounds and their role in synthesis. He was awarded his Ph.D. in July 1901.

Grignard accepted a position at the University of Nancy in 1909; there he continued to study organomagnesium compounds and their reactions with various chemical compounds. During World War I Grignard was commissioned to research the detection of mustard gas and to synthesize toluene, a solvent. In 1919 he was offered a professorship at the University of Lyon, where he succeeded his mentor, Barbier. He remained at Lyon until his death. In 1935 he published the first two volumes of his *Treatise on Organic Chemistry,* which he planned as a 15-volume work. His sudden death prevented its completion.

In 1912 Grignard was awarded the Nobel Prize in chemistry for his pioneering discoveries about organomagnesium compounds and their application in synthesis. (He shared the prize that year with PAUL SABATIER.) However, this was not the only official recognition Grignard received. He received honorary doctoral degrees from the universities of Louvain and Brussels and was named Commander of the Legion of Honor. After his death he was survived by his two children from his 1910 marriage to Augustine Marie Boulant. His son, Roger, became an organic chemist and supervised the posthumous publication of two additional volumes of Grignard's *Treatise.*

Grimaldi, Francesco Maria

(1618–1663)
Italian
Physicist

Francesco Maria Grimaldi became known for his work in astronomy and optics. Grimaldi was responsible for discovering the diffraction of light, which he outlined in the treatise *Physicomathematical Studies of Light, Colors, and the Rainbow,* published posthumously in 1665. Grimaldi was also known for his astronomical work conducted in collaboration with Giovanni Battista Riccioli and for his studies of the physical features of the Moon.

Grimaldi was born on April 2, 1618, in Bologna, Italy. Paride Grimaldi, his father, was a silk merchant from a wealthy noble family. In 1614 Paride married Grimaldi's mother, Anna Cattani, after his first wife died. Anna and Paride had six sons, of whom five survived. After Paride's death Anna managed her grandfather's chemist shop and Grimaldi entered the Jesuit order, the Society of Jesus, in 1632 with his brother, Vincenzo Maria.

Through the 1630s Grimaldi studied philosophy at various Jesuit colleges, including those at Novellara, Parma, and Ferrara. After teaching rhetoric and humanities at the College of Santa Lucia at Bologna from 1638 to 1642, he studied theology for several years, most likely at the College of Santa Lucia. He earned his doctorate in theology in 1647 and was to teach philosophy, but an illness prompted him to enter a less time-consuming discipline—mathematics. Grimaldi took full vows for the priesthood in 1651.

In the 1640s and early 1650s Grimaldi focused his efforts on the field of astronomy. His findings in astronomy were connected to the work of Giovanni Battista Riccioli, the prefect of studies at Bologna and a Jesuit since 1614. Grimaldi conducted numerous experiments and studies to assist Riccioli in preparing his *Almagestum novum.* By dropping weights from a tower and timing the fall with a pendulum, Grimaldi explored the theory of free-fall. Grimaldi's ability to develop and build new instruments was notable, and he made a map of the Moon and measured the heights of its clouds and mountains. Grimaldi initiated the custom of naming planetary regions after scientists as well. When Riccioli completed *Almagestum novum* in 1651, he gave Grimaldi much credit.

Beginning in the mid-1650s, Grimaldi turned his attention to optics and dedicated himself to the preparation of his treatise on light. The paper detailed new theories on its diffusion and discussed reflection, refraction, and diffraction. Grimaldi did not support any corpuscular theories of light, believing instead that light was fluid and had a wave nature. He determined this through an experiment in which he passed a beam of light through two successive narrow holes. The resulting pattern of light was slightly bigger than it should have been if the light had actually traveled in a straight line. The diffraction thus provided evidence supporting a fluid nature of light. Grimaldi also proposed that color was not the addition of something to light but an alteration of light.

Though Grimaldi received recognition from Riccioli for his assistance with Riccioli's astronomical research, he did not live long enough to learn of the influence of his contributions in optics. Grimaldi died of a sudden illness on December 28, 1663, in Bologna, soon after completing the treatise.

Guillaume, Charles Édouard

(1861–1938)
Swiss
Physicist

For his discovery of a nickel-steel alloy and his work in metrology, the scientific study of measurements and weights, Charles Édouard Guillaume was awarded the 1920 Nobel Prize in physics. Guillaume researched thermometry; set standards for the meter, kilogram, and liter; and made a more precise measurement of the volume of the liter. Guillaume's investigations of assorted alloys, and his resultant discovery of the nickel-steel alloy called invar, advanced the field of metrology and facilitated the production of measuring instruments, as well as other products, including clocks and watches.

Guillaume was born on February 15, 1861, in Fleurier, Switzerland. Guillaume's father, Édouard Guillaume, had moved to Switzerland from London, where he had managed a clockmaking business. After moving to Switzerland, his father established a watchmaking company in Fleurier. Guillaume's grandfather was also a watchmaker, who had originally moved from France to England to set up a watchmaking firm. Guillaume's father was learned in the sciences and shared his knowledge with his son.

After gaining his high school education at the gymnasium in Neuchâtel, Guillaume entered the Zurich Polytechnic (later known as the Federal Institute of Technology) to study science. He also studied German and French literature and became increasingly interested in physics. In 1882 Guillaume received his doctorate; his dissertation concerned electrolytic capacitors. Guillaume then entered the military to serve his compulsory year. Working as an artillery officer, Guillaume had the opportunity to study ballistics and mechanics.

In 1883 Guillaume began work at the newly established International Bureau of Weights and Measures, located near Paris in Sèvres. It was at the bureau that Guillaume made his many contributions in metrology. His early research was on mercury thermometers, and in 1889 he published a treatise on thermometry that was widely used by metrologists. Guillaume also worked on the international standardization of the meter, kilogram, and liter, and his investigations led him to a more precise measurement of the volume of the liter. He spent time calibrating thermometers and making duplicates of the standard meter bar to send to countries all over the world. It was while making copies of the bar that Guillaume began to investigate metal alloys. The standard bar consisted of a platinum-iridium alloy that staved off corrosion. The high cost of the metals inspired Guillaume to search for other materials to use for the duplication of the standard meter bar. He prepared a number of ferronickel alloys, including one that was seven parts iron to three parts nickel. Guillaume observed that this particular alloy expanded 10 times less than iron and had a coefficient of expansion of 0. This alloy came to be known as invar, and its contribution to the field of metrology was great—measuring instruments composed of invar were resistant to temperature changes. Guillaume also discovered another alloy, known as elinvar, whose elasticity was resistant to changes in temperature. Elinvar and invar found practical applications in the manufacture of such products as watches and clocks. Balances, a part of a watch, containing one of the two alloys later came to be known as Guillaume balances.

Guillaume stayed at the International Bureau of Weights and Measures for the duration of his career, becoming its director in 1905 and honorary director when he retired in 1936. In addition to the Nobel Prize, Guillaume received numerous honors for his contributions to metrology. He was a Grand Officer of the Legion of Honor and president of the French Physical Society, and received honorary degrees from several universities. Guillaume married A. M. Taufflich in 1888 and had three children.

Gullstrand, Allvar

(1862–1930)
Swedish
Ophthalmologist

For his breakthrough work on the human eye and its light-refracting abilities, known as dioptrics, Allvar Gullstrand was awarded the 1911 Nobel Prize in medicine or physiology. Gullstrand extended the understanding of the human eye and advanced the field of ophthalmologic science. In addition to his research on the human eye as an optical system, Gullstrand developed a number of useful devices for the study of the eye, including the slit lamp and the reflector ophthalmoscope. These instruments proved useful in eye examinations and in the diagnosis of optical disorders.

Born on June 5, 1862, in Landskrona, Sweden, Gullstrand was the son of a well-known city physician, Pehr Alfred Gullstrand, and Sophia Korsell. His father encouraged him to pursue medicine rather than engineering, a field for which Gullstrand demonstrated a natural talent. In 1885 Gullstrand married Signe Christina Breitholtz. They had a daughter, who died at an early age.

Gullstrand attended a number of universities, including those in Uppsala, Stockholm, and Vienna. He received his medical degree from the Royal Caroline Institute in Stockholm in 1888. In 1890 he was granted his doctorate. His dissertation topic dealt with astigmatism, a visual defect that leads to blurred vision. He worked as a chief physician at the Stockholm Eye Clinic and as a lecturer at the Royal Caroline Institute. In 1894 he was offered a position as professor of eye diseases at the University of Uppsala.

At the University of Uppsala, Gullstrand researched the dioptrics of the eye, expanding upon the findings of Hermann Ludwig Ferdinand von Helmholtz, a German physicist and physiologist, who had invented an ophthalmoscope in 1851. Helmholtz investigated physiological optics and concluded that the eye carries out accommodation, the ability to adjust to objects near and far, by undergoing a change in the curvature of the lens surface. Helmholtz proposed that the lens became convex when viewing nearby objects and concave when viewing those in the distance. Gullstrand showed that Helmholtz's theory could account only for two-thirds of a normal eye's accommodation. The remainder could be accounted for by what Gullstrand termed *extracapsular accommodation,* whereby needed adjustments were carried out by fibers behind the optical lens.

Gullstrand's other achievements in ophthalmology included the development of aspheric lenses designed for surgical patients whose lenses had been removed for the treatment of cataracts and the creation of optical instruments such as the slit lamp and the ophthalmoscope. The slit lamp, a diagnostic tool now commonly used in eye examinations, was a combination of a light and a microscope. The lamp allows for the detailed study of the eye and permits physicians to locate foreign objects or tumors in it. The Gullstrand ophthalmoscope consisted of a light and a magnifying lens and was designed to enable physicians to study the optic disk and the retina. The instrument was useful in the diagnosis of eye disorders and defects.

In addition to the Nobel Prize, Gullstrand received numerous honors before his death in 1930, including the Uppsala Faculty of Medicine's Björken Prize, the Swedish Medical Association's Centenary Gold Medal, and the German Ophthalmological Society's Graefe Medal. Gullstrand also served on the Nobel Academy's Physics Committee and received honorary degrees from several universities.

Gutierrez, Orlando A.

(1928–)
Cuban/American
Engineer

Besides his support of the education and training of Hispanics in the sciences, Orlando A. Gutierrez is known for his work in the areas of thermodynamics and aeroacoustics. A well-published engineer, Gutierrez has been active in such organizations as the National Aeronautics and Space Administration (NASA) Headquarters' Hispanic Employment and Minority University programs, as well as the Society of Hispanic Professional Engineers, an organization for which Gutierrez served as president beginning in 1993.

Born on July 23, 1928, in Havana, Cuba, Gutierrez was the son of Antonio Maria Gutierrez, the manager of a newspaper, and Flora Maria Izaguirre. Though his mother hoped he would pursue a career in medicine, Gutierrez displayed an interest in engineering at an early age.

Because Cuba lacked high-quality engineering programs, Gutierrez, with his parents' agreement, journeyed to the United States to enter college after graduating from Academia Baldor, a private high school in Havana. Gutierrez began attending Rensselaer Polytechnic Institute in Troy, New York, in 1945 and received his bachelor's degree in mechanical engineering in 1949. He married Helen LaBarge in 1947, and the couple had five children.

Gutierrez's first job after graduation from Rensselaer was with the IBM World Trade Corporation. The position allowed him to work in the United States and in Cuba. Gutierrez began working as a design and test engineer with the American Locomotive Company in Schenectady, New York, in 1951. He later became the manager of the company's heat transfer laboratory. After 10 years with the American Locomotive Company, Gutierrez was offered a job at NASA's Lewis Research Center in Cleveland, Ohio. There he conducted engineering research in thermodynamics, studying heat transfer and other subjects related to the creation of power plants designed for spacecraft. Gutierrez also explored jet acoustic studies, also known as aeroacoustics, and was particularly interested in the problem of noise reduction in aircraft engines. In 1975 he helped write and publish a NASA report, *Forced-Flow*

Once-Through Boilers: NASA Research. Gutierrez became increasingly involved in the human resource aspect of his job, first acting as the director of the Lewis Research Center's Hispanic program. Gutierrez's career interest shifted entirely to human resources in 1982.

In 1982 Gutierrez moved to Washington, D.C., to become the manager of NASA Headquarters' Hispanic Employment Program. His objective was to increase the number of Hispanic engineers employed at all branches of NASA. In 1990 he became the head of NASA's Minority University Program and remained in the position for two years until his retirement. During the time he worked in Washington, D.C., Gutierrez was also the national treasurer for the Society of Hispanic Professional Engineers. He became the society's national president in 1993.

Gutierrez was involved in the preparation and publication of more than 25 papers on heat transfer and aeroacoustics. A dedicated and skilled engineer, Gutierrez found equal satisfaction in the realm of human resources, particularly in the advancement of the engineering and scientific careers of Hispanics. Among the honors he has received for his work are NASA's Equal Opportunity Medal, NASA's Exceptional Service Medal, and the Mexican American Engineering Society's Medalla de Oro.

Haber, Fritz
(1868–1934)
German
Chemist

Fritz Haber's achievements ranged across disciplines: physical chemistry, organic chemistry, physics, and engineering. His best-known accomplishment may have been his development of an economical process for synthesizing large amounts of ammonia. For this work Haber was awarded the Nobel Prize in chemistry in 1918. Haber considered the scientific community to be global and made significant advances in expanding communication among countries; after World War I the Kaiser Wilhelm Institute for Chemistry in Berlin, Germany, which Haber managed, became a leading research center where scientists from around the world gathered.

Haber's father, Siegfried, was a wealthy merchant and manufacturer of chemical dyes and pigments, and Haber was encouraged to study chemistry to enable him to take over the business. An only child, Haber was born on December 9, 1868, in Breslau (now Wroclaw, Poland). His mother, Paula, died in childbirth, and his father remarried in 1877 and had three daughters. Though Haber did not have a close relationship with his father, he was treated kindly by his stepmother, Hedwig Hamburger.

Haber enrolled at the University of Berlin in 1886, intending to study chemistry. After one semester he transferred to the University of Heidelberg, where he studied physical chemistry, physics, and mathematics under the direction of ROBERT WILHELM BUNSEN. After earning his doctorate in 1891, Haber entered the Federal Institute of Technology in Zurich, Switzerland, to study the latest advances in chemical engineering.

After a failed attempt to run the family business, Haber took a teaching position at the Technical Institute, Karlsruhe, in 1894. There he researched electrochemistry and thermodynamics and published several works, including the original *Thermodynamics of Technical Gas Reactions* in 1905. In this book Haber discussed the behavior of gases in relation to thermodynamic theory in order to set industrial requirements for creating reactions. The work influenced subsequent teaching and research in thermodynamics. His successes led to a position as professor of physical chemistry in 1898.

Fritz Haber, whose team of scientists invented chemical weapons and poison gas to aid the German war effort during World War I. *Smith Collection, Rare Book & Manuscript Library, University of Pennsylvania.*

In 1905 Haber also began his work on the synthesis of ammonia. The production of large quantities of ammonia was becoming more critical as the growing population in Europe created greater demands on agricultural production and, in turn, industrial fertilizer. Nitrates, used in fertilizer, required ammonia for their production. Haber developed an inexpensive and practical method for the large-scale synthesis of ammonia from nitrogen and hydrogen. The engineer CARL BOSCH then developed Haber's process for industrial production, creating the Haber-Bosch ammonia process. Industrial output of the ammonia began in 1910.

Haber began his reign as director of the Kaiser Wilhelm Institute for Chemistry, located just outside Berlin, in 1912. When World War I broke out in 1914, Haber volunteered the institute's services to help the German war effort, and he and his team invented chemical weapons and poison gas. In 1916 Haber headed the Chemical Warfare Service, and his ammonia process was used to create explosives. After the war Haber continued as director of the Kaiser Wilhelm Institute; in 1919 he started a continuing research seminar, the Haber Colloquium. The seminars and the institute gathered together notable chemists, physicists, and other scientists. Haber traveled extensively to promote communication and cooperation among scientific communities, and in 1930 he helped found the Japan Institute, designed to foster mutual understanding and cultural interests.

Haber married Clara Immerwahr, a fellow chemist, in 1901, and together they had one son. Clara disapproved of Haber's work in poison gases, and after a fierce argument in 1915 she committed suicide. Haber married Charlotte Nathan in 1917 and had two children, but the couple divorced in 1927.

The rise of the Nazis altered Haber's life irreversibly. Haber, a Jew, resigned from his post in 1933 and traveled to England, where he worked for several months at the Cavendish Laboratory in Cambridge. Recovering from a heart attack, the ailing Haber was en route to a position in what is now Israel when he died on January 29, 1934, in Basel, Switzerland.

Though some scientists had protested Haber's Nobel Prize because of his involvement in chemical weaponry, his contributions to chemistry were widely recognized, well respected, and influential in the discipline for many years.

Hadley, George

(1685–1768)
English
Meteorologist

A half-century after EDMOND HALLEY proposed his theory partially explaining the nature of the trade winds, George Hadley completed it by explaining the reason behind the hemispheric differential in the prevailing direction of these winds. Hadley is thus credited with formulating the modern theory of the trade winds, and his contribution is known as the Hadley cell.

Hadley was born on February 12, 1685, in London, England. His parents were Katherine FitzJames and George Hadley. His older brother, John Hadley, invented the reflecting telescope as well as the reflecting quadrant, also known as the octant, the precursor of the sextant. The younger Hadley studied law at Pembroke College of Oxford University, and he became a barrister in 1709.

Though Hadley practiced law as his vocation, his avocation was meteorology, and he was charged with the duty of recording all official meteorological observations for the Royal Society of London. Hadley published some of this work under the title *Account and Abstract of the Meteorological Diaries Communicated for 1729 and 1730*. However, Hadley's name would have resided in relative obscurity if he had not built upon Halley's attempt to explain the nature of the trade winds.

In 1686 Halley put forth the explanation that the trade winds resulted from the fact that the Sun heats the air at the equator more intensely than anywhere else on Earth, causing equatorial air to rise. This effect pulls air from the Tropics, which is colder than equatorial air, creating an effect of convection. However, Halley could not explain why the trade winds blew from the northeast in the Northern Hemisphere, but from the southeast in the Southern Hemisphere.

This mystery persisted for almost 50 years, until 1735, when Hadley published *Concerning the Cause of the General Trade Winds*. In this text he explained that the rotation of the Earth from west to east would affect airflow from the equator. Specifically, this effect would cause the direction of airflow in the Northern Hemisphere to be the opposite of that in the Southern Hemisphere. This effect became known as the Hadley cell, or the cycle of airflow between the Intertropical Convergence Zone at the equator and the horse latitudes.

Subsequent theorizing accounted for airflow beyond the equatorial and tropical zones. Atmospheric circulation was broken down into three cells—the Hadley cell, the Ferrel cell, and the polar cell. The Ferrel cell represented the westerlies between the subsiding air currents at the horse latitudes and the rising air on the polar front; the polar cell represented the polar easterlies at both ice caps.

The Royal Society elected Hadley as a fellow in 1754. Hadley died on June 28, 1768, in London. Hadley and his older brother exerted a profound influence on maritime navigation in the 18th century. The elder Hadley produced a means of charting the position of a ship at sea based on its relation to a star or even the Sun, and the younger Hadley brought about a greater understanding of the most commercially important wind system, the trade winds.

Hale, George Ellery

(1868–1938)
American
Astrophysicist

George Ellery Hale, a pioneering figure, had a substantial impact on 20th-century astronomy. Not only did he invent new instruments, such as the spectroheliograph, the spectrohelioscope, and various telescopes, but he also detected the existence of magnetic fields in sunspots. A proponent of large-scale powerful telescopes, Hale was a tireless fund-raiser and was instrumental to the founding of several of the leading observatories in the United States.

Hale's inventiveness and interest in instrument design were influenced by his father, William Hale, a prosperous engineer, and by his mother, Mary Scranton Browne. Hale was born on June 29, 1868, in Chicago, Illinois, and was a sickly child. Nevertheless he spent his childhood conducting experiments and building instruments, including a telescope. He was introduced to astronomy and spectroscopy by a neighbor, Sherburne W. Burnham, who also encouraged him to obtain a Clark telescope, which was the world's largest telescope from 1897 to 1909 and is still renowned as the world's largest refractor.

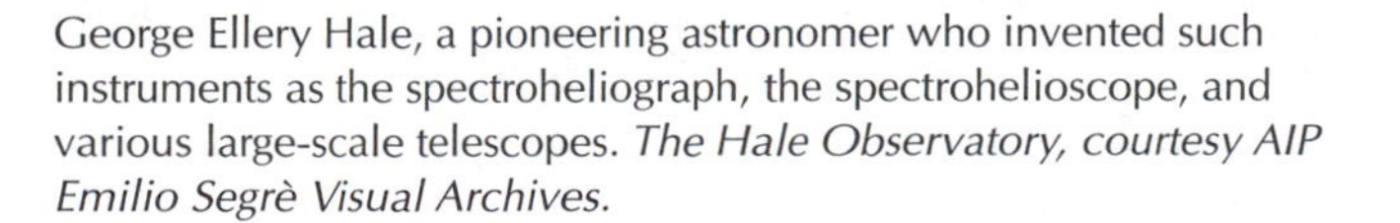

George Ellery Hale, a pioneering astronomer who invented such instruments as the spectroheliograph, the spectrohelioscope, and various large-scale telescopes. *The Hale Observatory, courtesy AIP Emilio Segrè Visual Archives.*

Hale enrolled at the Massachusetts Institute of Technology in 1886 to study physics but spent considerable time studying astronomy and spectroscopy independently. He also volunteered at the Harvard College Observatory and built a solar observatory behind his house. Hale made one of his first significant discoveries while he was still in college; in 1880 he invented the spectroheliograph, an instrument designed to photograph solar prominences, or clouds of flaming gas. Hale detailed his observations about the spectroheliograph in his senior thesis, and the instrument became a basic tool in solar astronomy. Hale graduated from college in 1890 and married Evelina Conklin the day after graduation. The couple had two children.

In 1892 Hale began a job teaching astrophysics at the University of Chicago. The same year he obtained funding from a businessman, Charles Yerkes, to build a 40-inch refracting telescope. From 1895 to 1905 Hale acted as the director of the university's Yerkes Observatory in Williams Bay, Wisconsin, which opened in 1897. Hale continued to research sunspots and designed a spectroheliograph to attach to the telescope, still the largest refractor ever built.

With the Yerkes Observatory established, Hale turned his efforts to building another observatory. In 1904 he received financial support from the Carnegie Institute in Washington, D.C., and began planning the Mount Wilson Observatory near Los Angeles, California, which he managed from 1904 to 1923. In 1905 Hale took the first photographs of a sunspot spectrum, and in 1908 he discovered evidence of magnetic fields within sunspots. Hale built a 60-inch reflecting telescope in 1908 and the 100-inch Hooker telescope in 1918. The Hooker telescope was financed in large part by the businessman John D. Hooker and remained the biggest reflector for 30 years.

Hale was known as the father of solar astronomy and was considered to be a tireless visionary in pursuit of astronomical advances. Included among his many achievements are his participation in founding the American Astronomical and Astrophysical Society in 1899 (renamed the American Astronomical Society in 1914), his reformation of the National Academy of Sciences starting in 1902, and his assistance in organizing the National Research Council in 1916. Hale also helped to establish several journals, including *Astronomy and Astrophysics* and *Astrophysical Journal*, which became the leading journal in its field soon after it was founded in 1895.

Retirement in 1923 failed to slow Hale. In 1928 he undertook plans for yet another observatory, this one with a 200-inch telescope. The Rockefeller Foundation donated funding to the California Institute of Technology in Pasadena, California, to build the telescope, and construction of the Palomar Observatory began in 1930. The project took nearly 20 years, and Hale unfortunately did not live to see the finished telescope, which was named the Hale telescope in 1948.

Hall, James

(1761–1832)
Scottish
Geologist, Physicist

Sir James Hall was a determined and skilled experimenter known for founding experimental geology. A geologist and physicist, Hall produced various types of rock through artificial means in his laboratory. He succeeded in proving that igneous rocks form crystalline rocks rather than glassy masses when allowed to cool slowly. Hall's findings refuted many geological beliefs of his time.

Hall grew up in a well-to-do family as the son and heir of Sir John Hall of Dunglass, Scotland. Hall was born on January 17, 1761, in Dunglass and succeeded to his father's baronetcy and significant fortune in 1776, when he was only 15 years old. His guardian and granduncle, Sir John Pringle, who was president of the Royal Society, then directed Hall's education.

Hall studied for two years at Christ's College in Cambridge and left in 1779 without a degree. He spent the next few years traveling and studying in Europe before enrolling in 1781 at the university in Edinburgh, where he became interested in chemistry. After completing school in 1783, Hall again set off for Europe. He traveled extensively for three years and met with many scientists and scholars. Hall's interest in geology and chemistry intensified; he studied volcanic activity in Italy and spent time in Paris learning about the new ideas in chemistry proposed by ANTOINE-LAURENT LAVOISIER. Hall married Lady Helen Douglas, the earl of Selkirk's second daughter, in 1786.

After his European travels Hall returned to Scotland and spent his years in Edinburgh and at his estate in Dunglass. In Edinburgh he began conducting numerous geological experiments. Hall's first experiments were stimulated by JAMES HUTTON, a close friend and fellow geologist. Hutton, in his *Theory of the Earth,* argued that rocks were formed by intense heat and thus had an igneous origin. Hall did not entirely support Hutton's ideas but grew to believe they were worthy of investigation. In 1798 he carried out his first significant experiment, in which he melted minerals and cooled them at a slow, controlled rate. Hall discovered that the slow cooling process, which he theorized would have been most likely under natural conditions, caused crystalline rocks, and not glass, to form.

After making his first discovery in support of Hutton's theories, Hall tackled the second major criticism lodged against Hutton's ideas—that the Earth could not have been formed through intense, subterranean heat as Hutton proposed because of the abundance of limestone on the surface of the planet. Limestone, opponents argued, decomposed when subjected to heat. Hutton claimed that limestone would not decompose if heated under great pressure, and Hall attempted to prove this through his experiments. Between 1798 and 1805 Hall conducted more than 500 experiments using a trial-and-error method; finally he was able to prove that limestone, when heated under pressure, does not decompose during cooling but instead turns to marble.

Hall published a number of papers describing his geological observations. In 1815 he wrote *On the Vertical Position and Convolutions of Certain Strata,* in which he suggested that folds in the Earth's surface were caused by lateral pressure. In *On the Revolutions of the Earth's Surface* Hall suggested that tsunamis, or tidal waves, affected the features of the Earth's surface. Hall's final paper, *On the Consolidation of the Strata of the Earth,* published in 1826, outlined his experimental attempts to prove that loose sand would turn into a firm sandstone when heated in concentrated brine.

Active in the scientific community, Hall was elected fellow of the Royal Society of Edinburgh in 1784 and in 1812 became its president. He was elected fellow of the Royal Society of London in 1806. Some of Hall's experimental instruments are on display in the British Museum and in the Geological Museum in London, attesting to the significance of his contributions in experimental geology.

Hall, Lloyd Augustus

(1894–1971)
American
Chemist

Lloyd Augustus Hall was a pioneer in the field of food technology. Although he experienced considerable discrimination because of his race, Hall made several crucial discoveries that revolutionized the food industry, particularly by developing better techniques to preserve meat and sterilize spices. After 35 years in industry Hall retired and applied his considerable expertise to humanitarian concerns.

On June 20, 1894, Hall was born in Elgin, Illinois, to Augustus and Isabel Hall. Hall received a scholarship to Northwestern University, where he received his B.S. in chemistry in 1916. He then began a graduate program in chemistry at the University of Chicago but a lack of funds forced him to leave school and seek full-time work. Lloyd married Myrrhene E. Newsome in 1919.

Rampant discrimination made finding a job difficult. For instance, after being hired over the telephone for a position at the Western Electric Company, he was fired when he reported in person for his first day of work. By the end of 1916, though, he found a position testing food products for purity at the Chicago Department of Health Laboratories. Within a year he was promoted to chief chemist. Hall worked at several industrial laboratories during the next six years. After a year as chief chemist at Boyer Chemical Laboratory in Chicago, Hall struck out on his own in 1922 when he launched Chemical Products Corporation, a consulting laboratory.

One of his clients was Griffith Laboratories, owned by Carroll Griffith, an old college friend of Hall's. In 1929 Hall abandoned his consulting enterprise and joined Griffith Laboratories, which sold flavorings to meat packers, as chief chemist and director of research. Hall would remain there until 1959 and would make his most important discoveries during this period. His first breakthrough was to invent a better way to preserve meat. Typically meat was stored in one of two problematic ways: preserved in brine (a solution of salt and water) or treated with nitrogen-based compounds. Although the brine allowed meat to last for months, it adversely affected the meat's appearance and taste. Nitrogen compounds, by contrast, maintained the color and taste of the meat but caused it to turn soft and gelatinous. Hall formulated a new technique, *flash-drying,* which incorporated the positive qualities of both methods. He dissolved plain salt and nitrogen compounds in water, then sprayed the solution on hot rollers. As the water evaporated, the dried salt mixture bound enough nitrogen compounds to ensure that the meat stayed firm, but not so many that the meat lost its proper color and taste.

Hall next turned his attention to difficulties with the spices used to flavor preserved meat. These spices often ruined the cured meat, because of contamination by bacteria, molds, or yeasts. Sterilization could prevent this problem, but the spices lost their flavor when heated above 240 degrees. Hall solved the dilemma by using a vacuum chamber to remove all moisture from the spices, then applying ethylene oxide—a gas previously used to kill insects—to the spices to eradicate the biota. This discovery not only benefited the meat packing industry, but proved useful to other businesses as well. For instance, hospital supply companies used Hall's technique to sterilize bandages and gauze.

By the time he retired from Griffith in 1959 Hall held over 105 patents for his inventions. In addition to the important scientific contributions he made to food chemistry, he played an essential role in establishing the field as a legitimate scientific discipline. In 1939 he cofounded the Institute of Food Technology, and he edited the organization's journal, *The Vitalizer,* for the next four years. His achievements were widely recognized. He was elected chairman of the Chicago chapter of the American Institute of Chemists in 1954, and the following year became the first African American appointed to its national board. He was awarded an honorary doctorate from Virginia State College in 1944, and President Kennedy nominated him to the American Food Peace Council in 1962. In his later years Hall served as a consultant to a number of organizations, including the United Nations. He remained involved in community affairs until his death on January 2, 1971.

Halley, Edmond

(c. 1656–1742)
English
Astronomer, Physicist

Astronomy, mathematics, and geophysics were among the numerous disciplines studied and advanced by Edmond Halley. Halley's most significant and best-known contribution was his study of cometary orbits. Halley successfully identified the comet of 1531 as the same comet that appeared in 1607 and 1682 and correctly predicted its return in 1758. The comet was later named *Halley's comet* in recognition of his achievement. Halley was also instrumental in the publication of Sir ISAAC NEWTON's treatise on gravitation, offering editorial services as well as financing.

Halley grew up during a significant period of scientific advance and enjoyed a wealthy upbringing as the son of Edmond Halley, a landowner and merchant. Though there is doubt concerning his precise date of birth, it is generally regarded to be October 29, 1656, in London. Halley's father encouraged his studies and paid for a private tutor before sending him to St. Paul's School. Halley became interested in astronomy at an early age and acquired a substantial collection of astronomical instruments, paid for by his father. Halley's mother died in 1672, a year before Halley entered college.

During Halley's years at Queen's College, Oxford, which began in 1673, he was encouraged to study astronomy by the astronomer John Flamsteed of the Royal Greenwich Observatory. Halley was inspired by Flamsteed's work and left school in 1676 to study the southern skies. Halley spent two years on St. Helena, a British territory in the South Atlantic, and he cataloged the stars of the Southern Hemisphere and published his results in 1678. The *Catalog of the Southern Stars,* the first work detailing telescopically determined star locations, established his reputation as an astronomer. He was elected a fellow of the Royal Society in 1678 and was granted a master's degree from Oxford University with the intervention of King Charles II. In 1682 he married Mary Tooke and the couple had three children.

Halley first visited Sir Isaac Newton in 1684, marking what was to become an important relationship. At that time both Newton and Robert Hooke were attempting to understand and explain planetary motion. Halley discovered that Newton held the correct answer and encouraged him to continue with his studies on celestial mechanics, despite a priority dispute with Hooke. Halley edited the text of Newton's *Principia Mathematica* and paid the printing costs.

After the publication of Newton's seminal work Halley began his study of cometary movement. Once he had calculated the parabolic orbits of 24 known comets, he determined that the comets of 1531, 1607, and 1682 were the same comet. He also calculated its next return, the first correct prediction of its kind. Halley published his observations in 1705 in *A Synopsis of the Astronomy of Comets*.

Halley made numerous other contributions in astronomy and in other fields as well. He was responsible for publishing in 1686 the first meteorological chart of the world, and he published the first magnetic charts of the Atlantic and Pacific areas in 1701. These charts were of great use in navigation. Halley also discovered stellar motion and founded scientific geophysics, completing studies on the trade winds and tidal phenomena. Halley's most significant contribution to geophysics was his theory of terrestrial magnetism, in which he proposed that the Earth had four magnetic poles. Halley also created the first mortality tables, studied the optics of rainbows, and drew conclusions as to the age of the Earth. During the later years of his life Halley worked as Astronomer Royal at Greenwich, a position he received in 1720. In his 60s, he commenced an observational study of the 19-year lunar cycle and was able to confirm the secular acceleration of the Moon.

Halley's work affected nearly all facets of astronomy and physics. Though his study of cometary orbits interested astronomers of his time, it was not until the return of the comet in 1758, as predicted, that Halley achieved wide recognition. Halley, unfortunately, had already been dead for nearly 15 years.

Hamilton, Alice

(1869–1970)
American
Physician

When Alice Hamilton began investigating industry in the early years of this century, no laws protected workers on the job. Many lost their health or even their lives to poisoning or accidents. Hamilton almost singlehandedly established industrial medicine in the United States, and her research helped persuade employers and legislators to make workplaces safer.

Hamilton was born in New York City on February 27, 1869, the second of Montgomery and Gertrude Hamilton's four children. She grew up with other members of the large Hamilton clan at her grandmother's home in Fort Wayne, Indiana. Montgomery was part owner of a wholesale grocery firm, though he had little love for business. Gertrude schooled her children at home, teaching them not only languages, literature, and history, but also responsibility and independence. Alice said later that she had learned from her mother that "personal liberty was the most precious thing in life."

Following family tradition, Alice was sent to Miss Porter's School for Girls in Farmington, Connecticut, when she was 17. To the surprise of those who knew her Alice decided to become a doctor. She had never felt any interest in science, but she chose medicine, she said later, "because as a doctor I could go anywhere I pleased and be of use."

After taking classes at the Fort Wayne College of Medicine, Alice Hamilton entered the University of Michigan medical school in 1892 and earned her M.D. degree in 1893. She then took practical training at two hospitals, plus a year of graduate study in Germany and another at Johns Hopkins University. Beginning in 1897, she taught at the Women's Medical School of Northwestern University, near Chicago, but lost that job when the school closed in 1902. She then began research at the Memorial Institute of Infectious Diseases, also near Chicago.

The move to Chicago allowed Hamilton to fulfill a longtime dream by becoming a member of Hull House, the famous settlement house that Jane Addams had founded in 1889. The house, located in one of the city's worst slums, was staffed by young women who worked without pay to help the immigrant families who crowded into the area. Hamilton took on several health-related jobs at Hull House, including teaching mothers about nutrition and health care and holding a well-baby clinic in the house's basement washroom. She later said that washing those babies was the most satisfying work she ever did. Nonetheless, after 10 years at Hull House, Hamilton, by then almost 40, was dissatisfied with her life. Her laboratory research seemed "remote and useless," and her settlement work left her feeling "pulled about and tired and yet never doing anything definite." She felt that neither made much difference in the world.

In the course of her work at Hull House, Hamilton often met laborers who were severely ill. A British book called *The Dangerous Trades* opened her eyes to the fact that many of these illnesses could be caused by lead or other poisons that the workers were exposed to on the job. She learned that Britain and Germany had tried to reduce poisoning in workplaces, but no such attempts had been made in the United States.

Hamilton was not the only American concerned about workplace safety. In 1908 the governor of Illinois appointed a commission to make a survey of work-related illness, specifically, poisoning by lead, something no state had done before. Hamilton applied for a place on the commission and was put in charge of the survey, which began in 1910.

Although workers themselves often refused to talk to her out of fear that they or family members would lose their jobs, Hamilton's survey finally documented 578 cases of work-related lead poisoning. Her commission published its report, *A Survey of Occupational Diseases*, late in 1910, and a year later Illinois passed a law requiring safety measures, regular medical checkups, and payments for workers who became ill or injured on the job.

While still working on the Illinois survey, Hamilton attended an international meeting on occupational accidents and diseases in Belgium. There she met Charles O'Neill, commissioner of labor in the U.S. Commerce Department, who asked her to conduct a similar survey of the United States. She said later that this meeting "resulted in . . . the taking up of this new specialty as my life's work."

Hamilton began the national survey in 1911 and continued it through the years of World War I. She had become one of the world's chief authorities on industrial diseases and virtually the only expert in the United States when, in 1919, Harvard University—which did not yet accept women students and had never had a woman on its faculty—asked her to teach industrial medicine in its School of Public Health. Hamilton took the position (part-time, so she could continue her survey work), but Harvard hardly went out of its way to make her welcome. She was kept at the lowly rank of assistant professor during all 16 years she taught there. She was also slighted in small ways, such as being forbidden to enter the faculty club or march in the commencement procession.

In the wider world Hamilton could see that her work was making a difference. In the 1920s, after she finished her national survey, one state after another passed laws requiring safety measures in workplaces and payments to workers who suffered illness or injury on the job. All states had such laws by the end of the 1930s. In 1925 she summarized her work in *Industrial Poisons in the United States*, the first American textbook on the subject. She published another textbook, *Industrial Toxicology*, in 1934.

Hamilton worked for many social causes besides industrial safety. She joined Jane Addams and other women on an international mission to try to stop World War I. After the war she raised money for starving German children. She served two terms on the Health Committee of the League of Nations, the forerunner of the United Nations, in the 1920s. On the national front she supported birth control and opposed capital punishment. In 1963 at the age of 94 she wrote an open letter demanding withdrawal of U.S. troops from Vietnam. She died on September 22, 1970, at the age of 101, knowing that the "dangerous trades" were no longer as dangerous because of her. Near the end of her life she said, "For me the satisfaction is that things are better now, and I had some part in it."

Harvey, William

(1578–1657)
English
Physician

The physician and anatomist William Harvey discovered the nature of the circulation of blood in the human body, thereby providing a viable alternative to the theory that had been embraced for many centuries. Harvey also conducted studies in embryology and served as the personal physician to two kings. He relied on observation in the study of nature and was an original and independent thinker. His findings transformed physiological thought and influenced a new generation of anatomists.

Harvey was the oldest of the seven sons of Thomas Harvey, a yeoman farmer, and Joan Halke. Harvey's father later became a successful merchant and rose to the gentry. Harvey's brothers grew up to become wealthy London merchants. Born on April 1, 1578, in Folkestone, a coastal town, Harvey attended King's School in Canterbury.

Harvey attended Gonville and Caius College in Cambridge from 1593 to 1599 before enrolling in medical school at the University of Padua, the top medical school in Europe at that time. There he studied under the anatomist Girolamo Fabrici and, most likely, Cesare Cremonini, an Aristotelian philosopher. Harvey earned his M.D. in 1602. In 1604 he married Elizabeth Browne, the daughter of a prominent physician; the couple had no children.

After returning to England to begin his practice, Harvey was elected a fellow of the Royal College of Physicians in 1607. Two years later he became a physician at St. Bartholomew's Hospital, a position he held until 1643. Harvey became the personal physician to King James I in 1618 and to Charles I in 1625. Harvey traveled with Charles I on his many excursions and campaigns during the English Civil War and remained loyal to Charles I until his surrender in 1646.

Despite his success as a physician, Harvey's primary interest was in research. By 1615 Harvey had formed an idea regarding blood circulation, and he continued to research his ideas through dissection and experimentation. The traditional view regarding circulation was that developed by the Greek physician and anatomist GALEN in the second century. According to Galenic theory, blood originated in the liver, and arteries and veins carried different substances. Harvey supposed that blood traveled through the arteries and returned through the veins and proposed a circulatory theory. To prove this, Harvey dissected animals and cold-blooded creatures, including frogs and fish. Harvey found that blood moved through arteries, veins, and heart valves in one direction and that the heart acted as a pump. Harvey published his observations in 1628 in *On the Motions of the Heart and Blood.* The work was well received but did not find universal support at the time. Later it received recognition as announcing one of the most important discoveries made in anatomy and physiology during the modern era.

Harvey's interests were not restricted to anatomy—he was also drawn to embryology. Harvey believed that all life originated in an egg and rejected the idea of spontaneous generation. Harvey created the concept of epigenesis—the theory that an egg develops independently. He published his ideas in *On the Generation of Animals* in 1651.

Harvey remained active in the Royal College of Physicians throughout his career and donated funds for a library. Harvey left his research library to the college after his death in 1657. After retiring in 1646 Harvey, a widower, lived with his brothers in various homes around London. Harvey's discovery of blood circulation was one of the major findings of early modern science and encouraged scientists of his day to experiment, think independently, and question the authority of ancient theories.

Hawes, Harriet Ann Boyd

(1871–1945)
American
Archaeologist

When she led a group excavating a 3,000-year-old Cretan town in 1901, Harriet Hawes became the first woman to head a large archaeological dig. She was born in Boston, Massachusetts, on October 11, 1871, to Alexander Boyd, who owned a leather business, and his wife, also named Harriet. Harriet's mother died when she was a baby, so she grew up in a household of men—her father and four older brothers. Her second brother, Alexander, passed on to her his interest in ancient civilizations.

Boyd studied classics at Smith College, from which she graduated in 1892. She did four years of graduate work at the American School of Classical Studies in Athens, Greece. Her professors told her to stick to library research, but, as she wrote to her family, "I was not cut out for a library student."

In 1900, using money from a fellowship plus her own savings, Boyd and a woman friend traveled to the island of Crete, home of a spectacular ancient civilization whose remains archaeologists were just beginning to study. Boyd was the first American archaeologist to go there. An account by the Archaeological Institute of America pictured her "daring to travel—as few women had yet done—through rugged mountains . . . riding on muleback in Victorian attire. . . . Here we have all the elements of romance and danger." At a spot called Kavousi, on the eastern side of the island, Boyd uncovered several Iron Age tombs. She discussed this discovery in the thesis for her master's degree, which she received from Smith in 1901.

To Boyd's delight a grant from the American Exploration Society let her return to Kavousi that same year. This time at a nearby site named Gournia she discovered the remains of an entire town. Dating to the early Bronze Age, it was even older than the tombs she had already found. Because Gournia had been occupied chiefly by workers, it offered valuable glimpses of the era's ordinary people. Boyd led further expeditions to Gournia in 1903 and 1904, supervising over a hundred villagers. It was the first Bronze Age Cretan city to be fully excavated. She described the excavation in a report published by the exploration society in 1908.

Boyd married Charles Henry Hawes, a British anthropologist, whom she had met on Crete, on March 3, 1906. They collaborated on a popular book, *Crete: The Forerunner of Greece,* which was published in 1909. They later had a son and a daughter, and Harriet spent most of her time caring for them. Eventually the family moved to Boston, where Charles Hawes became assistant director of the Museum of Fine Arts. In 1920 Harriet resumed her career, teaching pre-Christian art at Wellesley College until 1936. She died on March 31, 1945.

Hawking, Stephen William

(1942–)
English
Theoretical Physicist

The theoretical physicist and cosmologist Stephen Hawking has been a celebrity not only within the scientific community but among the general public as well. His wide-ranging fame is due in large part to his book *A Brief History of Time: From the Big Bang to Black Holes,* published in 1988. The publication was written for a general audience and received significant worldwide attention. Hawking is known for his pioneering work in the study of black holes, and his interests have included such disciplines as physics, supersymmetry, and quantum gravity. Hawking, considered by many to be the most talented theoretical physicist since ALBERT EINSTEIN, has also been concerned with the origin and fate of the universe.

Hawking was born in Oxford, England, on January 8, 1942, coincidentally on the 300th anniversary of GALILEO GALILEI's death. Hawking grew up in a scholarly and accomplished family—his father, Frank Hawking, was a

Stephen Hawking, the celebrated physicist, whose book *A Brief History of Time: From the Big Bang to Black Holes* won him popular acclaim. *AIP Emilio Segrè Visual Archives, Physics Today Collection.*

research biologist and physician, and his mother, Isobel Hawking, was active in Britain's Liberal Party. An eccentric child, Hawking knew at a young age that he wished to pursue science. By his teen years, he had narrowed his choices to physics or mathematics, and in high school his math skills flourished.

Receiving a scholarship in natural sciences, Hawking entered University College in Oxford in 1959. There his talents and abilities began to surface more clearly. Hawking graduated in 1962 and continued on to Cambridge University for his doctorate in cosmology, which he received in 1965. There he began his studies of black holes and space-time singularities, applying formulas developed much earlier by Einstein. After Cambridge he obtained a fellowship in theoretical physics at Gonville and Caius College, Cambridge, where he continued his research on black holes. He collaborated often with the mathematician Roger Penrose and began to build a reputation as an outspoken scientist. Hawking married Jane Wilde in 1965, and the couple had three children before separating in 1990.

Hawking received a position at the Institute of Astronomy in Cambridge in 1968; there he, with Penrose, began to use mathematics to apply the rules of thermodynamics to black holes. In 1973 Hawking published the highly technical *The Large Scale Structure of Space-Time.* Soon he made the significant discovery that black holes were capable of emitting thermal radiation under certain conditions. The accepted thinking had been that black holes could not emit anything. The subatomic particles that were emitted were thereafter called Hawking radiation.

After being named a fellow of the Royal Society in 1974, Hawking continued his efforts to create a theory on the universe's origin. Toward this end, he began to link the theory of relativity, which concerned gravity, with quantum mechanics, which dealt with minute events inside the atom. Through the 1980s Hawking began to question the big bang theory, wondering whether there had actually been a beginning to space-time—the big bang—or whether one universe or state of affairs had given birth to another. Hawking theorized that it was possible that the birth of new universes was common as a result of anomalies in space-time.

It was the publication of *A Brief History of Time: From the Big Bang to Black Holes* in 1988 that made Hawking famous. The book explained, in layman's language, Hawking's ideas regarding the cosmos. It received significant media attention, and Hawking was featured on a number of television programs and films. The book climbed onto best-seller lists in the United States and Britain.

Among the numerous awards granted to Hawking were the Eddington Medal of the Royal Astronomical Society (1975), the Maxwell Medal of the Institute of Physics (1976), the highly prestigious Albert Einstein Award of the Lewis and Rose Strauss Memorial Fund (1978), and the Gold Medal of the Royal Society (1985). Hawking was named the Lucasian Professor of Mathematics at Cambridge in 1979. During his years as a student Hawking developed amyotrophic lateral sclerosis, a debilitating disease. Though wheelchair-bound and required to use a voice synthesizer to speak, he continued his studies and his rounds on the lecture circuit through the 1990s.

Hazen, Elizabeth Lee

(1885–1975)
American
Medical Researcher

Elizabeth Lee Hazen improved methods for identifying human diseases caused by fungi and, with the chemist Rachel Brown, discovered a drug that provided the first successful treatment for many of them. She was born to William and Maggie Hazen on a cotton farm in Rich, Mississippi, on August 24, 1885. Her parents died when she was a child, and an aunt and uncle in nearby Lula raised her and her sister.

Hazen graduated from the State College for Women (later Mississippi University for Women) in Columbus in 1910. After several years of teaching high school science, she entered Columbia University; she obtained a master's degree in bacteriology in 1917. She worked for a West Virginia hospital for years before returning to Columbia for a Ph.D., which she received in 1927 at the age of 42. After four years of teaching at the university she joined the New York City branch laboratory of the New York State Department of Health's Division of Laboratories and Research. She became an expert in identifying disease-causing fungi, which few other researchers had studied.

Penicillin and other antibiotics developed in the 1940s had no effect on fungi, so Hazen decided to search for a drug that would kill these stubborn microbes. Soil, the source of several antibiotics, seemed a likely place to look. Hazen traveled around the country gathering samples of soil, which she took back to her laboratory and sprinkled in nutrient-filled dishes so that microorganisms in them would form colonies. She sent these cultures to Rachel Brown at the Department of Health's laboratory in Albany, and Brown extracted substances that the microbes made and sent them back to Hazen. Hazen tested them to see which ones killed disease-causing fungi.

In 1948 Hazen collected a soil sample near Warrenton, Virginia, which proved to contain a microbe that produced two fungus-killing substances. One was new to science, and Hazen and Brown announced its discovery late in 1950. The drug made from this substance, named *nystatin* in honor of the New York State Health Department, was first marketed in 1954. The women patented the drug but gave the $13 million profit from it to a foundation that endowed medical research. Hazen and Brown received several awards for their discovery, including the Squibb Award (1955) and Chemical Pioneer Awards from the American Institute of Chemists (1975). The New York Department of Health honored Hazen with a Distinguished Service Award in 1968.

Nystatin proved to have a surprising range of uses. In addition to attacking fungi that caused human illnesses, it destroyed the fungus that caused Dutch elm disease, a widespread killer of trees. It killed molds that spoiled stored fruit, livestock feed, and meat. It saved priceless paintings and manuscripts from mold damage after a flood in Florence, Italy, in 1966.

After the New York City branch laboratory closed in 1954, Hazen worked at the central laboratory in Albany. In 1955 she wrote the textbook *Identification of Pathogenic Fungi Simplified*. She was an associate professor at Albany Medical College from 1958 until her retirement from both this position and the one at the Department of Health in 1960. After that she became a guest investigator at Columbia's Medical Mycology [study of fungi] Laboratory. She died on June 24, 1975.

Heisenberg, Werner Karl

(1901–1976)
German
Physicist

The physicist Werner Karl Heisenberg founded quantum mechanics and influenced greatly the progress of nuclear and atomic physics. In 1932 Heisenberg won a Nobel Prize in physics for his formulation of the uncertainty principle, which declared the impossibility of specifying both the position and the momentum of a particle at the same time. The discovery of the principle had a significant

Werner Karl Heisenberg, acknowledged by many as the founder of quantum mechanics. *AIP Emilio Segrè Visual Archives, Segrè Collection.*

effect on the field of physics and refuted the fundamental scientific theory of cause and effect. Heisenberg also developed matrix mechanics, a new version of quantum theory used to solve problems regarding atomic structure.

Born in Würzburg, Germany, on December 5, 1901, Heisenberg was raised in a scholarly family. His father, August Heisenberg, was a professor of Greek, and his mother, Annie Wecklein, was the daughter of a headmaster. Heisenberg was nine years old when the family moved to Munich for his father's new position at the university.

In 1920 Heisenberg enrolled at the University of Munich to study physics. Under the supervision of Professor Arnold Sommerfeld, Heisenberg quickly became immersed in theoretical physics and discovered an explanation for the Zeeman effect, the splitting of spectral lines, for which none had existed. Heisenberg's name began to spread in the scientific community of Europe for his work on the Zeeman effect, and in 1923 he earned his doctorate. Soon afterward he went to Göttingen to study under the physicist MAX BORN.

Heisenberg began formulating his theory of matrix mechanics in 1925. He studied the unusual results emerging from quantum theory and concluded that the common approach of considering the atom in visual terms was a mistake. Heisenberg claimed that our knowledge of the atom consisted of its observable aspects—its frequency, its intensity, and the light it emits. What was thus needed, Heisenberg proposed, was a set of formulas or equations that would accurately predict such atomic phenomena. Heisenberg worked with Born and Born's assistant Pascual Jordan and published a polished version of the theory, marking the start of a new age in atomic physics.

Heisenberg traveled to Cophenhagen in 1926 to become the assistant to the physicist NIELS HENDRIK DAVID BOHR. There he developed the uncertainty principle in 1927. He reached his conclusion about the impossibility of determining both the position and momentum of particles, such as the electron, through "thought experiments." Heisenberg noted that trying to pinpoint the exact position of an electron required the use of rays with very short wavelengths, such as gamma rays. Using such rays, however, would alter the electron's momentum because the electron would interact with the rays. Using waves with lower energy would not alter the electron's momentum, but the radiation of lower-energy waves also would not provide an exact location.

In 1927 Heisenberg became professor of theoretical physics at the university in Leipzig and continued researching such fields as cosmic radiation, nuclear physics, and quantum electrodynamics. After Sir JAMES CHADWICK discovered the neutron in 1932, Heisenberg developed a theory that the nucleus consisted of protons and neutrons and not, as was believed by some physicists, protons and electrons.

As political unrest began to grow in Germany through the 1930s, many scientists fled to other countries. Heisenberg chose to stay in Germany with the intention of preserving the nation's scientific traditions and institutions. He married Elisabeth Schumacher in 1937 and had seven children. The Nazi government appointed Heisenberg director of the German atomic bomb project at the beginning of World War II, and though Heisenberg later claimed he took the position to ensure that an atomic bomb would not be produced, he met with much criticism. After the war Heisenberg helped found the German Research Council and worked to restore the values and esteem of German science.

He died in 1976, seven years after his official retirement. Throughout his career Heisenberg published a number of papers and works on quantum mechanics, cosmic rays, and atomic physics, including *The Physical Principles of Quantum Theory* in 1930. His development of quantum theory influenced the progress of modern atomic and nuclear physics.

Hero of Alexandria

(c. A.D. 62)
Greek
Mathematician, Inventor

A number of written works on mathematics, geometry in particular, and physics have been attributed to Hero of Alexandria. Hero developed the formula for calculating the area of a triangle from its three sides, now known as Hero's formula, and invented the aeolipile, the first steam-powered engine. *Metrica,* Hero's most significant work on geometry, was lost until 1896.

Besides his works, which were written in Greek, nothing is known about the life of Hero. The earliest written mention of Hero appeared in a work by PAPPUS OF ALEXANDRIA around A.D. 300, which quoted from Hero's book *Mechanics.* The tentative date of A.D. 62 for Hero's life was derived from the description of an eclipse in one of Hero's works, found to correspond undeniably to the eclipse of A.D. 62. Because of the nature of the works written by Hero, it was believed that he taught at the University of Alexandria in such disciplines as mathematics, physics, pneumatics, and mechanics.

The treatise *Metrica,* which dealt with geometry, comprised three books. *Metrica* introduced his formula for finding the area of a triangle. The first book encompassed plane figures and surfaces of common solids. Hero explained how to calculate the area of quadrilaterals, circles, regular polygons, and ellipses. He also described how to find the surface area of cylinders, cones, spheres, and segments of spheres. Also in Book I was a method for estimating the square root of a number, a method later used in computers; Book II provided methods for determining the volume of such figures as the cone, pyramid, cylinder, prism, and sphere; Book III discussed procedures for dividing certain areas and volumes into parts of given ratios.

Hero's longest work was *Pneumatics,* considered to be a collection of notes for a textbook. The treatise tackled a number of mechanical problems and included discussion of the aeolipile, the steam-powered engine designed and built by Hero. The engine was made of a sphere with two nozzles positioned so that steam jets produced from the inside would cause the device to turn. *Pneumatics* also discussed the pressure of air and water, as well as the occurrence of a vacuum in nature. Discussion of siphons, coin-operated machines, a fire pump, and a water organ also appeared in the work. Hero also described a number of toys and playthings, such as puppet shows and trick jars that released wine and water separately.

Another important work by Hero was *Mechanics,* a textbook for engineers, builders, and architects that discussed the mechanical problems of everyday life. The theory of the wheel and the theory of motion were covered in the work, and Hero also explained how to construct plane and solid figures in a given proportion to a given figure. Hero presented the theory of the center of gravity and equilibrium and the theory of the balance, using the ideas of ARCHIMEDES. *Mechanics* also discussed the five simple machines—the lever, the pulley, the winch, the wedge, and the screw. The work described cranes, devices for transport, wine presses, and screw cutters, believed to have been invented by Hero.

Dioptra was about land surveying and contained a description of a diopter, a surveying instrument most likely developed by Hero. The work also described an odometer, another of his inventions, and dealt with astronomy, offering a method for calculating the distance between Alexandria and Rome through the simultaneous observation of a lunar eclipse in both cities. Yet another work, *Catoptrica,* offered the theory of reflection and provided instructions on how to make mirrors.

Hero's works have been widely studied, and evidence exists to suggest his texts enjoyed a large audience—*Pneumatics* was read by many during the Middle Ages and the Renaissance, and more than 100 copies have surfaced. His writings indicate that Hero was a man of great scientific knowledge and creativity.

Herophilus of Chalcedon

(c. 335 B.C.–c. 280 B.C.)
Greek
Physician

Often called the father of anatomy, Herophilus of Chalcedon was a well-respected physician and teacher of medicine in Alexandria. Herophilus lived during a rare period in Greek medical history when human dissection was allowed, and he greatly advanced the study of scientific anatomy through his extensive dissections of human cadavers. In conjunction with his studies, Herophilus coined many medical terms, some of which are still in use.

Little is known about the life of Herophilus. He was from Chalcedon, located on the Asiatic side of the Bosphorus, and studied under Praxagoras of Cos. Later he lived in Alexandria and practiced medicine. Herophilus was an extremely popular teacher, inspiring many students to gather in Alexandria to attend his lectures. He embraced a dialectical approach to medicine rather than an empirical method, which relied on practical experience rather than scientific theory.

An important aspect of Herophilus' research was his work on the anatomy of the brain. He uncovered the correct view of the functions of the brain and the heart and was able to differentiate between the cerebrum and the cerebellum. His dissections also allowed him to theorize that the nerves originated in the brain and were responsible for motion. Praxagoras had accurately distinguished between arteries and veins but believed that certain parts of some arteries controlled movement. Herophilus adapted Praxagoras' assumption and attributed the function of movement to the nerves. Herophilus also classified the nerves as sensory or motor and accurately traced the sensory nerves leading from the brain to the eyes. He traced the sinuses of the dura mater to their junction, which he called *torcular Herophili.*

Herophilus was also very interested in the liver and studied it by applying the methods of comparative anatomy. By dissecting numerous specimens, he discovered that the liver differed in size and structure in different individuals of the same species. He also observed that the liver sometimes appeared on the left side instead of the right. Herophilus attached the name *duodenum* to the beginning section of the small intestine.

Herophilus was the first to measure the pulse, which he computed by using a clepsydra, or water glass. He studied its size, strength, rate, and rhythm and was believed to have considered particular cardiac rhythms as typical of various periods of life. Pulsation, Herophilus claimed, was involuntary and caused by the contraction and enlarging of the arteries, which reacted to impulses received from the heart. Another significant finding regarding the pulse was Herophilus' observation of the rhythmic contraction and expansion of the heart to convey blood to the lungs, referred to as the pulmonary systole and diastole.

Though Herophilus' interests were primarily in anatomy and dissection, his studies were wide in range. He made significant discoveries in gynecology, completing a treatise on midwifery in which he successfully described the uterus, ovaries, and cervix. He also completed a work on dietetics and promoted exercise and diet for the preservation of health. A supporter of HIPPOCRATES OF COS's doctrine of the four humors, Herophilus also believed in the healing powers of drugs. Though none of Herophilus' written works survived, he wrote at least 11, covering such topics as anatomy, ophthalmology, therapeutics, the pulse, and the causes of sudden death.

Herschel, Caroline Lucretia

(1750–1848)
British
Astronomer

Caroline Herschel not only assisted her famous astronomer brother Sir WILLIAM HERSCHEL but also "swept the skies" on her own, spotting at least five comets for the first time. The youngest of Isaac and Anna Herschel's 10 children, she was born in Hanover, later a part of Germany, on March 16, 1750. She wrote that her father, a musician, was also a "great admirer" of astronomy and once took her "on a clear and frosty night into the street, to make me acquainted with several of the beautiful constellations." Isaac wanted to see all his children educated, but Anna opposed him, believing that girls should be taught only household skills.

Caroline's brothers, William and Alexander, moved to England in 1758, and William became a music teacher and organist at Bath, a popular resort. Caroline remained behind as a sort of household servant until William rescued her in 1772. In Bath he helped her train as a singer and taught her English, mathematics, and, above all, astronomy, which he loved more than music.

Unlike most astronomers, William made his own telescopes. His devotion—and Caroline's—to this painstaking task produced instruments much better than the average. William polished his telescope mirrors by hand, once for 16 hours straight, while Caroline read to him and fed him. When he built a 40-foot telescope in the 1780s, Caroline pounded up horse manure and forced it through a sieve to make molds for the device's giant mirrors.

In 1779 the Herschels began the first round of a project to map all the objects in the night sky. They made several "sweeps," each more extensive than the last, between that year and 1802. Night after night William looked through his telescope and called out his observations to Caroline, who recorded them. Next day she made a clean copy of her notes and calculated each object's position. It is not clear when she found time to sleep.

William Herschel's first sky sweep uncovered nothing less than a new planet, now known as Uranus, in 1781. King George III was so impressed that he named Herschel the Royal Astronomer the following year with an annual salary of £200. That meant that he could quit music and work full-time on astronomy. William's quitting meant that Caroline quit too, since she refused to sing under the direction of anyone else. She felt "anything but cheerful" about it, but she did not protest.

In 1786 the Herschels settled in Slough, where they found a house with a yard big enough for their telescopes. William had taught Caroline how to use a telescope by this time and had even built small telescopes for her, but she could do concentrated work or observe through his big telescopes only when he was away on trips. On August 1, 1786, during one such absence, she found her first new comet. She sent a letter about it to Britain's chief scientific body, the Royal Society, and it was published in their journal.

Caroline Herschel's find made her famous in the scientific world. Soon afterward, when George III gave William extra money to help pay for the 40-foot telescope, he also granted a yearly allowance of £50 to Caroline, officially acknowledging her as William's assistant. The amount was small, but the recognition was unusual. Caroline proudly wrote in her diary that this was "the first money I ever in all my lifetime thought myself to be at liberty to spend to my liking."

William married Mary Pitt in 1788, and his bride insisted that Caroline live separately from the couple. Caroline was grieved at first, but eventually she realized that the situation left her more time to "mind the heavens." Between 1788 and 1797, in addition to helping William, she found seven more comets, as well as other heavenly objects such as star clusters.

Caroline's most important works were the *Index to the Catalogue of 860 Stars Observed by Flamsteed but Not Included in the British Catalogue* and *Index to Every Observation of Every Star in the British Catalogue,* published by the Royal Society in 1798. These books resolved discrepancies between unpublished notes left by Flamsteed, the first Royal Astronomer, and a catalog he had published in 1725.

William Herschel died in 1822. Desolate, Caroline moved back to Hanover. She became a family drudge again, but she also continued William's work, assembling a catalog of 2,500 nebulae (cloudlike star formations) and star clusters that he (and she) had observed. In 1825 she sent the finished catalog to William's son, John, who had also become a famous astronomer. When John presented it to the Royal Astronomical Society in 1828, the group awarded Caroline a gold medal, calling the book "the completion of a series of exertions probably unparalleled either in magnitude or importance, in the annals [history] of astronomical labour." The society made Caroline an honorary member in 1835.

In 1846 on her 96th birthday the king of Prussia awarded Caroline another gold medal. Awards meant little to her, however; she always insisted that she had done no more for William than "a well-trained puppy-dog" could have done. She died on January 9, 1848, never guessing that she would later be called "the most famous and admired woman astronomer in history."

Herschel, Sir John Frederick William

(1792–1871)
English
Astronomer

The son of a famous astronomer, Sir John Frederick William Herschel advanced his father's studies in stellar and nebular astronomy considerably. Herschel discovered more than 500 nebulae and conducted a thorough survey of the Southern Hemisphere. A scientist with many interests, Herschel made significant contributions to such varied fields as mathematics, photography, and astrophysics.

John Frederick Herschel was the only child of the astronomer Sir WILLIAM HERSCHEL and Mary Pitt. His father was intensely devoted to astronomical observation and spent much time working. Herschel developed a close relationship with his aunt, CAROLINE LUCRETIA HERSCHEL, his father's assistant and a respected astronomer, and received much encouragement from her. Herschel attended Eton briefly and was educated privately after the age of eight.

In 1809 Herschel entered Cambridge University to study mathematics. There he befriended CHARLES BABBAGE, who later invented the computer, and George Peacock. Together they founded the Analytical Society of Cambridge in 1812 to promote continental methods of mathematical calculus. Herschel submitted his first mathematical paper to the Royal Society in 1812 and was elected a fellow in 1813. He completed his undergraduate education at Cambridge in 1813, earning first place in the school's mathematical examinations; he began to study law but was diverted by his interest in physics, when he began conducting home experiments to study polarized light. Herschel earned his master's degree in 1816 at Cambridge.

Herschel did not pursue astronomy seriously until 1816, when he somewhat unwillingly returned home to assist his father with astronomical observations. After his father's death in 1822 Herschel began the task of continuing his father's work, reobserving the double stars and systems cataloged by him. Herschel's father had confirmed that double stars orbited about a common center, and Herschel hoped to investigate the gravitational forces of the universe by studying the movements of double stars. With James South, who owned two refracting telescopes, Herschel published a catalog of 380 double stars in 1824. For the work the two were awarded the Lalande Prize of the French Academy in 1825 and the following year received the Gold Medal of the Royal Astronomical Society, which Herschel helped found in 1820. Herschel married Margaret Brodie Stewart in 1829 and had 12 children.

To extend his father's survey of the universe Herschel decided to travel to the Southern Hemisphere to view the skies not visible in England. In 1834 he arrived at the Cape of Good Hope, armed with a large reflecting telescope for studying nebulae and a refracting telescope for observing double stars. There he surveyed the entire southern sky, cataloguing double stars and nebulae. He also invented an instrument designed to measure solar radiation, described sunspot activities, and observed Halley's comet and Saturn's satellites. After four years of observation Herschel journeyed home with the locations of 68,948 stars in 3,000 sky areas and a listing of 2,102 pairs of double stars and 1,707 nebulae. He published *Results of Astronomical Observations Made during the Years 1834–38 at the Cape of Good Hope* in 1847.

An accomplished chemist, Herschel contributed greatly to the field of photography after his return from South Africa. He created the first photograph on a glass plate in 1839, introduced the terms *negative* and *positive* in regard to photographic images, and invented sensitized paper, the blueprint paper used in developing film, that is made sensitive to light by a solution of iron salts. Herschel greatly advanced astronomical photography and was the first to prepare colored photos of the Sun's spectrum.

In 1831 Herschel was knighted, and in 1838 he was made a baronet. He won the Royal Society's Copley Medal in 1821 and 1847 and its Royal Medal on three occasions. He was elected president of the British Association for the Advancement of Science in 1845. In 1850 Herschel became Master of the Mint, a post that proved exhausting and unenjoyable. He resigned in 1856 and spent his last years working on catalogs of double stars and nebulae. He died on May 11, 1871, in Collingwood, Hawkhurst, Kent, England.

Herschel, Sir William

(1738–1822)
German/English
Astronomer

Though Sir William Herschel became a professional astronomer late in his adulthood, his contributions to astronomy were vast. Herschel discovered the planet Uranus, the first discovery of a new planet since prehistoric times, and made thorough and extensive observations of the universe. Herschel was unique among astronomers of his time because of his interest in studying distant celestial bodies rather than nearby planets. Through his observations and studies Herschel exposed the true nature of the Milky Way as a galaxy and studied nebulae, theorizing that they were composed of stars.

Sir William Herschel, who discovered the planet Uranus. *Smith Collection, Rare Book & Manuscript Library, University of Pennsylvania.*

Herschel was born in Hanover, Germany, on November 15, 1738, the third of six children. Like his father, Isaac Herschel, he became a musician in a regimental band. He fled to England in 1757 after the French occupation of Hanover and worked as a freelance musician. After acquiring a job in 1766 as an organist at a chapel, he moved to Bath. His sister, CAROLINE LUCRETIA HERSCHEL, who became his devoted assistant and an astronomer in her own right, joined him in 1772.

Through the 1770s Herschel's interest in astronomy grew, and he began constructing his own reflecting telescopes. Herschel soon realized that to investigate the sidereal universe successfully he needed more powerful telescopes with large mirrors to gather adequate light. Because of the prohibitive costs of such large mirrors, he began to grind his own and embarked on his surveys of the night sky.

Herschel's first major finding, and the one that made him famous, was his discovery of the planet Uranus in 1781. Herschel had been focusing on studying double stars when he came upon an object that was not an ordinary star. Initially Herschel believed it to be a comet, but further examination revealed it to be a primary planet of the solar system. For his discovery Herschel was awarded the Copley Medal of the Royal Society, and in 1782 he was appointed Court Astronomer to George III, allowing him at the age of 43 to give up music and turn his full attention to astronomy.

Developing an understanding of the construction of the heavens was Herschel's goal, and he worked tirelessly, studying the skies night after night. Herschel's interest in nebulae developed in the early 1780s after the release of Charles Messier's catalog, which raised the number of known nebulae to just over 100. Herschel found that his telescopes were well suited for studying nebulae, which appeared as luminous, milky patches in the sky. Several theories regarding the nature of nebulae existed; one supposed that they were composed of a luminous fluid. Through his big telescopes Herschel was able to separate several nebulae into distinct stars, thereby encouraging him to theorize that all nebulae were formed of stars.

Herschel was the first to recognize the structure of the galaxy. By conducting a large number of star counts, Herschel found that stars were more numerous in the galactic plane and in the Milky Way and less abundant toward the celestial poles. Herschel was also responsible for establishing the motion of the Sun and trying to calculate its speed. In 1787 he discovered two satellites of Uranus and two of Saturn.

From 1786 Herschel lived in Slough in what became known as Observatory House. With financing from George III he built a 40-foot reflecting telescope with which he made many of his astronomical discoveries. Herschel married Mary Pitt, the widow of a neighbor, in 1788 when he was nearly 50 years of age, and in 1792 they had a son, Sir JOHN FREDERICK WILLIAM HERSCHEL, who grew up to become a notable astronomer and to continue his father's cataloging of double stars. The elder Herschel published more than 70 papers, cataloged 800 double stars, and in 1820 published a catalog of more than 5,000 nebulae. Herschel was knighted in 1816 and died six years later, on August 25, 1822.

Hershey, Alfred Day

(1908–1997)
American
Microbiologist

Hershey gained his fame from the 1952 "blender experiment" he performed with Martha Chase to demonstrate that deoxyribonucleic acid (DNA) and not protein was the genetic foundation of life. MAX DELBRÜCK and SALVADOR EDWARD LURIA conducted experiments simultaneously but independently that yielded the same results, confirming the discovery. This trio shared the 1969 Nobel Prize in physiology or medicine for their work.

Hershey was born on December 4, 1908, in Owosso, Michigan. His mother was Alma Wilbur and his father was Robert Day, an autoworker. On November 15, 1945, Hershey married Harriet Davidson, a former research assistant who went on to edit the journal *Cold Spring Harbor Symposia on Quantitative Biology.* Together the couple had one son, Peter, who was born on August 7, 1956.

Hershey studied bacteriology at Michigan State College, which granted him a bachelor of science degree in 1930. He remained at what later became Michigan State University, where he earned the Ph.D. in chemistry in 1934 with his dissertation on the chemical characteristics of *Brucella,* the bacterial agent that caused undulant fever.

The Department of Bacteriology at Washington University School of Medicine hired Hershey as a research assistant under Jacques Jacob Bronfenbrenner. Washington University retained Hershey's services for the next 16 years, promoting him to the position of instructor in 1936, to assistant professor in 1938, and to associate professor in 1942. In 1950 the department of genetics of Carnegie Institute at Cold Spring Harbor, New York, hired Hershey as a staff scientist. In 1962 he took on the directorship of his department, renamed the Genetics Research Institute, a position he held until his 1974 retirement.

Hershey studied viral replication in the 1940s, hypothesizing in 1945 that bacteriophages, or simply "phages," mutated spontaneously, contradicting the contemporaneous belief that mutation occurred gradually. In 1946 he discovered what he called genetic recombination, whereby viruses and bacteria could exchange genetic material, mimicking the genetic transfer that occurred sexually in more advanced species, such as humans. Hershey, Delbrück, and Luria collaborated to form the Phage Group, a loosely knit collective of scientists studying bacteriophages who shared their results in order to advance the progression of understanding of phages more efficiently. In Hershey's famous 1952 experiment he and Chase "tagged" DNA with radioactive phosphorus and protein with radioactive sulfur, then traced them after blending them to separate one from the other. Hershey found the tagged DNA invading the host bacteria, proving his hypothesis.

Besides the Nobel Prize, Hershey received many awards, including the 1958 Albert Lasker Award from the American Public Health Association. That year the National Academy of Sciences elected him as a member. He then received the 1965 Kimber Genetics Award from the National Academy of Sciences. Hershey died on May 22, 1997, in Syosset, New York. Besides contributing their discoveries to science, the three 1969 Nobel laureates also contributed a methodology for advancing science collaboratively—in the hope that the results would benefit all of science—instead of competitively, which served individual egos rather than the common good.

Hertz, Gustav

(1887–1975)
German
Physicist

Gustav Hertz received the 1925 Nobel Prize in physics with his collaborator, JAMES FRANCK, in recognition of their experimental verification of NIELS HENDRIK DAVID BOHR's quantum theory of the atom.

Hertz was born on July 22, 1887, in Hamburg, Germany. His mother was Auguste Arning; his father, Gustav, was an attorney. Hertz's uncle was HEINRICH RUDOLF HERTZ, the scientist after whom the unit of electrical frequency was named. In October 1919 Hertz married Ellen Dihlmann, and together the couple had two sons, Hellmuth and Johannes. Dihlmann died in 1941, and in 1943 Hertz married Charlotte Jollasse.

In 1906 Hertz began the study of mathematics and physics at the Universities of Göttingen, Munich, and Berlin. In 1911 Hertz earned his Ph.D. from the University of Berlin with a dissertation on the infrared absorption spectrum of carbon dioxide. After receiving his doctorate, he remained at the university's physical institute as a research assistant.

Almost immediately Hertz fell in with Franck, collaborating on an experiment whereby they accelerated positively charged electrons through a field of mercury vapor, then detected the electrons on the other side, tracking the loss of energy in the process. They discovered almost no loss of energy until the potential difference of 4.9 volts was reached, when the detector capturing the electrons on the far side of the mercury fell practically to zero. The two could not account theoretically for the significance of this specific voltage reading until they applied Bohr's new theory advancing a quantum model of the atom, which predicted their results almost precisely. Hertz and Franck reported the results in a 1913 edition of a German physics journal under the title "Impacts between Gas Molecules and Slowly Moving Electrons," and in another German physics journal under the title "Collisions between Electrons and Molecules of Mercury Vapor and the Ionizing Voltage for the Same."

Soon after Hertz completed this experiment, the German army drafted him to serve in World War I, in which he was severely wounded. After the war Hertz returned to the University of Berlin as a *Privatdozent,* or an unsalaried lecturer, for three years. In 1920 he transferred from academia to the private sector, working for Philips Incandescent Lamp Works in Eindoven, Netherlands, one of the first companies to support its own research laboratories. Hertz then returned to academia in October 1925 as a physics professor and the director of the physics institute at the University of Halle. In 1928 he transferred to the Charlottenberg Technical Institute in Berlin as a professor of physics, but the rise of Nazi rule forced him from this position because of his Jewish heritage.

Luckily he found a job in the private sector, working for the Siemens and Halske Company in Berlin in 1934. After World War II, Hertz immigrated to the Soviet Union as a researcher on atomic energy, radar, and supersonics based in Sukhumi on the Black Sea. After another decade, the Karl Marx University in Leipzig appointed him director of its physics institute in 1955, a position he retained until his 1961 retirement. Hertz died on October 30, 1975, in Berlin.

Hertz, Heinrich Rudolf

(1857–1894)
German
Physicist

Though he died at the young age of 36, Heinrich Rudolf Hertz made significant advances in electrodynamics. Hertz was the first to produce radio waves artificially and succeeded in proving that radio and light waves were electromagnetic waves. A persistent and highly self-motivated researcher, Hertz studied various aspects of electricity, including cathode rays and dielectrics.

Born into a wealthy family, Hertz was encouraged by his parents, particularly his mother, to excel in his studies. He was born on February 22, 1857, in Hamburg, Germany, the son of Gustav F. Hertz, a barrister, and Anna Elisabeth Pfefferkorn. The eldest of five children, he was educated at a strict private school. At an early age he showed skill at woodworking, as well as a gift for languages. As a teenager he studied Greek and Arabic. In 1875 Hertz went to Frankfurt to gain hands-on experience in engineering but soon decided to pursue an academic and scientific career.

In 1877 he entered the University of Munich to study mathematics; he transferred a year later to the University of Berlin, where he studied under the physicist Hermann von Helmholtz. Helmholtz proved to be a profound influence on Hertz, who flourished in Berlin's rigorous research environment. Hertz received his doctorate in 1880 and spent the next three years as Helmholtz's assistant. During that period Hertz completed the research for 15 publications, the majority of which dealt with electricity. Hertz discussed electromagnetic induction, the inertia of electricity, dielectrics, and cathode rays. It was also during this time that he, with Helmholtz's prompting, became interested in the electromagnetic theories of JAMES CLERK MAXWELL.

From 1885 to 1889 Hertz worked as a professor of physics at Karlsruhe Technical College in Karlsruhe, Germany. There Hertz began the experimental trials that would dramatically impact the field of electrodynamics. Hertz married Elisabeth Doll in 1886 and had two children. By late 1888 Hertz had successfully produced electromagnetic waves using an electric circuit. A metal rod in the circuit possessed a small gap in the middle, and when sparks traversed this gap, high-frequency oscillations resulted. To prove that these waves were transmitted through the air, Hertz set up a similar circuit a short distance from the first and used the circuit to detect the waves. Hertz was also able to demonstrate that the waves were reflected and refracted in the same way as light waves, and that, though they traveled at the same speed as light waves, they had a longer wavelength. These waves were called Hertzian waves but later came to be known as radio waves. Hertz published his findings in nine papers, which drew significant attention.

Hertz's discoveries secured him a position in 1889 at the University of Bonn, where he continued his theoretical study of Maxwell's theory. He also returned to the topic of cathode rays and in 1891 turned his attention to a theoretical study of the principles of mechanics as discussed by Helmholtz in his work on the principle of least action.

Among the numerous awards presented to Hertz for his findings on electric waves were the Matteucci Medal of the Italian Scientific Study (1888), the Baumgartner Prize of the Vienna Academy of Sciences (1889), the La Caze Prize of the Paris Academy of Sciences (1889), the Rumford Medal of the Royal Society (1890), and the Bressa Prize of the Turin Royal Academy (1891). A unit of frequency, the hertz, was named in his honor. After several years of poor health, Hertz died in 1894 of blood poisoning. Had he lived another decade he would have learned of GUGLIELMO MARCONI's discovery that the transmission of radio waves could be used as a means of communication.

Herzberg, Gerhard

(1904–)
German/Canadian
Physicist, Physical Chemist

Gerhard Herzberg is considered the founder of molecular spectroscopy, the science of recording spectral waves emitted by atoms and molecules as a means of measuring energy radiance as well as measuring the light transmitted and absorbed by molecules. Herzberg received the 1971 Nobel Prize in chemistry, the first Canadian laureate in chemistry.

Herzberg was born on December 25, 1904, in Hamburg, Germany. His mother was Ella Biber and his father was Albin Herzberg. In 1929 Herzberg married Luise H. Oettinger, and together the couple had two children, Paul and Agnes, both of whom became teachers. In 1945 Herzberg became a naturalized citizen of Canada.

Herzberg attended the Darmstadt Institute of Technology and in 1927 earned his B.S. in engineering. The next year the institute bestowed a doctorate on him for his dissertation on the reaction of matter with electromagnetic radiation. He continued on with postdoctoral study at the University of Göttingen and the University of Bristol in England for two years.

Herzberg returned to Darmstadt as an instructor in 1930; he remained there until Nazi purges ousted him in 1935 because of his Jewish heritage. Herzberg emigrated from Germany to Canada, where he filled a Carnegie guest professorship at the University of Saskatchewan. A decade later he worked as an astrophysicist for three years at the Yerkes Observatory of the University of Chicago. Then in 1948 he returned to Canada to head the physics division of the Canadian National Research Council as its first Distinguished Research Scientist. There he founded a laboratory that the Swedish Royal Academy of Science called "the foremost center for molecular spectroscopy in the world."

Herzberg focused his spectroscopic research on the basic elements of life—diatomic molecules such as oxygen and carbon dioxide. His research cleared a path through the wilderness of spectral understanding. In fact, Herzberg discovered new bands in the oxygen spectrum, now known as Herzberg bands. Herzberg also published prolifically, intending to write a 200-page overview of molecular spectroscopy but ultimately composing a three-volume, 2,000-plus-page tome, *Molecular Spectra and Molecular Structure: Spectra of Diatomic Molecules* (volume I, published in 1939), *Infrared and Raman Spectra of Polyatomic Molecules* (volume II, published in 1945), and *Electronic Spectra and Electronic Structure of Polygamic Molecules* (volume III, published in 1966). The same year he received the Nobel Prize he published *The Spectra and Structure of Simple Free Radicals: An Introduction to Molecular Spectroscopy.*

Besides the Nobel Prize, Herzberg received numerous awards: The American Chemical Society awarded him the Willard Gibbs Medal and the Linus Pauling Medal, the Canadian Association of Physicists granted him its Gold Medal, and the Royal Society of London bestowed on him its Royal Medal. Interestingly, Herzberg always considered himself a physicist, though his Nobel was in chemistry and he made some of his most significant contributions to the understanding of astrophysics with his interstellar spectroscopic observations of specific molecular activity, such as the existence of elemental hydrogen in some planetary atmospheres.

Hess, Victor Francis Franz

(1883–1964)
Austrian
Physicist

Victor Hess shared the 1936 Nobel Prize in physics with CARL DAVID ANDERSON, whose discovery of the positron and muon could not have occurred without Hess's prior discovery that the source of a commonly known but little-understood form of radiation that was pervasively present on Earth was not terrestrial but a region in the cosmos beside the Sun. This so-called Hess radiation was eventually renamed *cosmic radiation.*

Hess was born on June 24, 1883, in Schloss Waldstein, Austria. His mother was Serafine Edle von Grossbauer-Waldstätt; his father, Vincens, was the chief forester for Prince Oettinger-Wallerstein. Hess married Mary Berta Waermer Breisky on September 6, 1920. When she died in 1955, Hess married Elizabeth Hoencke on December 13 of that year. Neither of his marriages produced children.

Hess graduated from the Humanitisches Gymnasium in Graz in 1901, then studied at the University of Graz, where he earned his Ph.D. in physics summa cum laude in 1906. In 1910 he accepted a position as a research assistant with the Institute for Radium Research at the University of Vienna, a post he retained for the next decade. During this time he accompanied the Austrian Air Club on 10 balloon ascensions as high as six miles, where he measured radiation with readings up to eight times as great as terrestrial readings. He conducted several of these ascensions at night to ascertain the influence of direct sunlight on the readings. He also ascended in a balloon during the April 12, 1912, solar eclipse. As in his nighttime readings he measured no significant difference in radiation readings when the Sun didn't shine, confirming that it also was not the source.

In 1920 Hess returned to the University of Graz as an associate professor of experimental physics. In 1928 he published *The Electrical Conductivity of the Atmosphere and Its Causes.* In 1931 the University of Innsbruck recruited him to head its institute for radiation research as a professor of experimental physics. The next year he published the article "The Cosmic Ray Observatory in the Hafelekar" in the journal *Terrestrial Magnetism and Atmospheric Electricity,* chronicling his establishment of this facility. He fled the Nazi regime in 1937 to protect his first wife, whose Jewish heritage endangered her. Fordham University appointed him as a professor of physics that year. In 1940 he published his account "The Discovery of Cosmic Radiation" in *Thought* magazine. In 1944 he became a citizen of the United States. Hess remained at Fordham until his 1956 retirement.

After World War II Hess switched his attention from extraterrestrial radiation to radiation produced by nuclear explosives testing, which he opposed vehemently. Radiation readings he took atop the Empire State Building confirmed the increase in natural background radiation levels after nuclear testing. Besides the Nobel Prize, Hess received numerous awards, including the 1919 Lieben Prize from the Austrian Academy of Sciences and the 1932 Ernst Abbé Prize from the Carl Zeiss Foundation. Hess died on December 17, 1964, in Mount Vernon, New York.

Hess, Walter Rudolf

(1881–1973)
Swiss
Physiologist

Walter Rudolf Hess received the 1949 Nobel Prize in physiology or medicine for his research into the diencephalon, the mechanism in the interbrain that controls internal organ functions. Hess shared the award with ANTÓNIO EGAS MONIZ, a Portuguese neurosurgeon who worked with white brain matter.

Hess was born on March 17, 1881, in Frauenfeld, Switzerland. His mother was Gertrud Fischer Saxon and his father was Clemens Hess, a physics teacher who passed on his love for science to his son. Hess married Louise Sandmeyer in 1908, and together the couple had two children, Rudolf and Gertrud.

Hess attended several universities throughout his undergraduate and graduate careers, and at the University of Zurich he received his M.D. in 1905. During his first residency under Dr. Konrad Brunner, Hess designed an improved blood viscometer, an instrument that measured the blood's consistency or relative thickness. Hess served his second residency in ophthalmology, mistakenly believing that it would allow him time to conduct research on the mechanisms of the eyes. Hess did not find ample time for research during this residency, but the specialization did provide him with a lucrative practice after his residency.

In 1912 Hess abandoned his successful practice in favor of pure research, when he began work at the Institute of Physiology. In 1917 the University of Zurich promoted him to the status of professor of physiology, and he later chaired the Department of Physiology until his 1951 retirement. Hess focused his physiological research on the inner parts of the brain that control both physical and emotional responses in the body. He conducted his experiments on cats, inserting tiny electronic probes into their brains to stimulate, sometimes to the point of destruction, the hypothalamus and the medulla oblongata, observing the reactions in the subject's breathing, blood pressure, temperature, and digestion, as well as its emotions such as fear and hunger. Hess found that he could simulate a cat's frightened response to the sight of a dog, even with no dog present, simply through interbrain stimulation.

Hess published throughout his career. In 1930 and 1931, respectively, he published two companion books, *The Regulation of the Circulatory System* and *The Regulation of Respiration*. In 1957 he published a more detailed account of his research, *The Functional Organization of the Diencephalon*. He contributed an autobiographical sketch, "From Medical Practice to Theoretical Medicine," to the 1963 collection *A Dozen Doctors*. His later publications included *The Biology of the Mind*, published in 1964, and *Biological Order and Brain Organization*, a selection of his writings published posthumously in 1981.

Besides the Nobel, Hess won the 1933 Marcel Benorst Prize from the Swiss government and the 1938 Ludwig Medal of the German Society for Circulation Research. Hess died on August 12, 1973, in Locarno, Switzerland. Interestingly, the success of Hess's research rested to a large degree on its sheer mass, as no single experiment convinced the scientific community of the validity of his results. Rather, it was his impressively large body of evidence that made it impossible to ignore his findings.

Hewish, Antony

(1924–)
English
Astrophysicist

Antony Hewish's greatest achievement was the 1967 discovery of pulsars. For this accomplishment Hewish and his colleague Martin Ryle, a fellow astronomer, were awarded the 1974 Nobel Prize in physics, the first time the prize had been awarded for astrophysics. Hewish also investigated the scintillation of quasars, which are similar to twinkling stars but related to radio waves. His work furthered the field of radio astronomy considerably.

The youngest of three sons, Hewish was born on May 11, 1924, in Fowey, Cambridge, England. His father was Ernest William Hewish, a banker, and his mother was Frances Grace Lanyon. Not interested in following his father into banking, Hewish decided at an early age to pursue a career in science. He showed a proclivity for physics at the age of 11, when he was a student at King's College in Taunton.

Hewish entered Cambridge University in 1942, but his college education was interrupted by World War II. From 1943 to 1946 Hewish worked for the British government designing airborne radar countermeasure instruments. It was at the government's Telecommunications Research Establishment that he met Martin Ryle. After the war Hewish returned to Cambridge University and completed his degree in 1948. He then joined Ryle's research team at Cambridge's well-known Cavendish Laboratory, where he worked with leading physicists. Hewish married Marjorie Richards in the early 1950s and had two children. He earned his doctorate in 1952 and in the early 1960s became a lecturer in physics at Cambridge.

It was during his early years at the Cavendish Laboratory that Hewish became very interested in radiation and the up-and-coming field of radio astronomy. For some years Hewish studied the twinkling of radio stars, hoping to determine its cause. In 1964 he made a significant advance; he was able to demonstrate that when radio waves from a source with a small diameter, such as a quasar (a starlike object that emits radio waves), passed through a group of charged particles, the waves were diffracted. Hewish observed various quasars and successfully confirmed the scintillation outcome.

In 1967 Hewish completed the building of a large high-resolution radio telescope. The telescope, which was sensitive to weak radio sources and operated at a wavelength of 3.7 meters, was used by Hewish and his research students to investigate radio galaxies. In the summer of 1967 a graduate student, SUSAN JOCELYN BELL BURNELL, happened to notice an unusually regular but weak signal. Hewish and his team then started investigating the nature of the signal, first eliminating human causes, such as electrical machinery, satellites, and radar echoes. Hewish found, however, that the signal was being emitted outside the solar system. Working secretively, the team explored various possibilities, theorizing that the scintillations might be originating from a flare star or perhaps from alien beings. After additional observations Hewish concluded that the regularly patterned radio pulses were generated by certain condensed stars, which later were classified as pulsars. Hewish and his colleagues published their findings in 1968 and touched off a flurry of excitement in the astrophysics community.

Hewish's discovery of pulsars inspired astrophysicists around the globe to search for the source. The American Thomas Gold's proposal became the commonly accepted theory—Gold had determined that pulsars were rapidly spinning neutron stars. Hewish continued his studies in radio astronomy, working as professor of radio astronomy at the Cavendish Laboratory from 1971 to 1989. In 1989 he became professor emeritus. His ground-breaking work greatly advanced the fields of radio astronomy and astrophysics.

Jaroslav Heyrovsky, who won the 1959 Nobel Prize in chemistry for his invention of the polarograph. *Smith Collection, Rare Book & Manuscript Library, University of Pennsylvania.*

Heyrovsky, Jaroslav

(1890–1967)
Czech
Physical Chemist

Jaroslav Heyrovsky invented the polarograph, an instrument used for analyzing the composition of solutions electrochemically by means of a polarogram, which plots a curve of voltage against current output. Heyrovsky's new method reduced the time needed to determine the chemical composition of solutions from one hour to 15 minutes; eventually he improved the technology to reduce the analysis time to a matter of milliseconds. This work earned him the 1959 Nobel Prize in chemistry.

Heyrovsky was born on December 20, 1890, in Prague, in what was then Austria-Hungary. His mother was Klára Hanl; his father, Leopold, was a professor of Roman law at Charles University in Prague and later the rector of the college. In 1926 Heyrovsky married Marie Koranova, and together the couple had two children—a daughter named Jitka who became a biochemist, and a son named Michael who succeeded his father at the Institute of Polarography.

In 1909 Heyrovsky commenced studying physics, chemistry, and mathematics at Charles University, where he was influenced by Frantisek Zaviska and Bohumil Kucera. The next year he transferred to University College of Cambridge, enticed there by the opportunity to study under Bohumil Brauner. Heyrovsky earned his B.S. in 1913 and returned to his homeland; the outbreak of World War I prevented his return to England, so he served as a chemist and radiologist in a military hospital from 1914 through 1918. He managed to continue his doctoral research throughout the war, earning a Ph.D. in 1918 from Charles University, which retained him as an assistant professor in 1919 and promoted him to the position of appointed lecturer in 1920. That year he published his dissertation "The Electroaffinity of Aluminum" in *Journal of the Chemical Society (Transactions)*. On the strength of this publication University College granted him a second doctorate in 1921.

Charles University again promoted Heyrovsky in 1922, appointing him as an associate professor and the head of the chemistry department. Two years later the university established a new Institute of Physical Chemistry, naming Heyrovsky as its director and as an Extraordinary Professor. Heyrovsky lectured at the University of Paris under a Rockefeller Fellowship in 1926, and that same year he became a full professor at Charles University, where he remained for the next quarter-century. He interrupted this appointment in 1933 to accept a Carnegie Visiting Professorship at the University of California at Berkeley, and he ended his tenure at Charles University in 1950 to accept the directorship of the Polarographic Institute, newly established by the Czechoslovak Academy of Sciences.

In addition to his establishment of the discipline of polarography, Heyrovsky championed the advancement of the sciences in his homeland by founding the Czech-language chemical journal *Collection of Czechoslovak Chemical Communications* with the help of the Royal Bohemian Society of Sciences. In recognition of his work for his native country, Heyrovsky received the 1951 Czech State Prize and the 1955 Order of the Czechoslovak Republic. The most appropriate distinction bestowed upon him was the renaming of the institute he directed to J. Heyrovsky Institute of Polarography in 1964. Heyrovsky died on March 27, 1967, in Prague.

Hill, Sir Archibald Vivian

(1886–1977)
English
Physiology

Sir Archibald Vivian Hill's pioneering research into the physiology of muscles earned him the 1922 Nobel Prize in physiology or medicine, which he shared with OTTO FRITZ MEYERHOF. By painstakingly measuring the minute amounts of heat produced by muscles, Hill concluded that muscles emit heat in two distinct phases. He also found that oxygen is not essential for the act of muscle contraction, although it does play a critical role in muscle recovery.

Hill was born on September 26, 1886, in Bristol, England, to Jonathon and Ada Priscilla Rumney Hill. After his father deserted the family in 1889, Archibald and his sister, Muriel, were raised by their mother. Hill married Margaret Neville Keynes, the sister of the British economist John Maynard Keynes, in 1913. The couple had four children.

In 1905 Hill accepted a scholarship to study mathematics at Trinity College, Cambridge, but soon abandoned math in favor of physiology. He graduated from Trinity in 1907 with a medical degree and in 1909 completed his examinations with honors.

Hill remained at Cambridge from 1909 until 1914 as a research fellow at the Cambridge Physiological Laboratory. There he began to study the production of heat in the process of muscle contraction, an inquiry that the Cambridge physiologist Frederick Gowland Hopkins had initiated. Using a precise measuring device, the thermocouple recorder, Hill demonstrated in 1913 that, contrary to the accepted scientific wisdom of the time, heat production occurs in two distinct phases during muscle contraction. He observed that when frog muscle fibers begin to contract, a small amount of heat is generated; after this initial phase of contraction a recovery phase occurs, during which a greater amount of heat is produced. By observing muscle tissue at work in an atmosphere of pure nitrogen, Hill also determined that heat was produced at the onset of contraction even in the absence of oxygen.

Although service as an officer in the British army during World War I interrupted his work, Hill returned to his research in 1920, when he accepted a position as chair of the Physiology Department at Manchester University. There he worked with Meyerhof, a German-American biochemist, to apply chemical dynamics to his hypotheses about lactic acid breakdown in muscles. Previous researchers had found that lactic acid is created during muscle contraction and is eliminated in the presence of oxygen. Because of his conviction that heat was produced during two phases, Hill believed that the initial output of heat he had recorded was caused by the formation of lactic acid from a precursor substance and that the release of heat during recovery was caused by the removal of lactic acid through oxidation. (Lactic acid is generated by muscles only in the absence of oxygen.) Hill also proposed that lactic acid was converted back into its precursor form. Meyerhof's experiments bolstered Hill's theories by establishing the existence of a precursor substance, glycogen, and proving that the energy levels created by oxidizing lactic acid were sufficient to reconvert it back into glycogen.

Hill left Manchester in 1923, when he was named the Foulerton Research Professor of the Royal Society at University College, London. After analyzing lactic acid accumulation in human muscles, he concluded that intense exercise can create an oxygen debt (as he termed it), causing an excessive buildup of lactic acid that can be dissipated only through reabsorption over time. But war once again impeded Hill's work. Before and during World War II Hill vocally opposed German anti-Semitism toward Jewish scientists and formed networks to help Jewish researchers escape. While retaining his academic position, Hill also served as a member of Parliament (1940 to 1946) and on the War Cabinet Scientific Advisory Committee. He devoted himself fully to his muscular research after his retirement in 1952.

Hills is credited with stimulating the growth of biophysics. His techniques of measuring temperature and chemical changes were later used to study the mechanisms of the passage of nerve impulses. Along with the Nobel Prize, Hill was honored with election into the Royal Society in 1918 and was knighted for his World War I military service. He died on June 3, 1977.

Hill, Henry Aaron

(1915–1979)
American
Chemist

Henry Aaron Hill overcame the discrimination facing African-American scientists in the 1930s and 1940s to become an expert on polymers. He founded his own company, held a number of management positions at various chemical corporations, and was appointed to the National Commission on Product Safety by President Lyndon Johnson. Hill also served as the president of the American Chemical Society.

On May 30, 1915, Hill was born in the town of St. Joseph, Missouri. Little is known about his childhood and family. Hill attended John C. Smith College, a small liberal arts school in North Carolina. After receiving his B.S. in chemistry in 1936, he began graduate studies at the University of Chicago. He transferred to the Massachusetts Institute of Technology (MIT) in 1937; there he developed a strong relationship with James Flack Norris. Unlike many of his other professors and classmates, Norris found the color of Hill's skin irrelevant and was impressed by the youth's considerable talents. In 1942 Hill was awarded his Ph.D. in organic chemistry.

Upon graduating Hill worked at a variety of chemical companies. In addition to learning more about the industrial, rather than the academic facets of his field, Hill gained management experience. From 1942 to 1943 he headed the chemistry research division at Atlantic Research Associates in Massachusetts. He was elevated to the position of research director in 1943 and promoted again in 1945 to vice president of research. (By the time he assumed this latter position, the company was officially known as the National Atlantic Research Company.) During his tenure at National Atlantic, Hill formulated water-based paints, synthetic rubber, and rubber adhesives. Although he hoped some day to launch his own laboratory, Hill continued to garner experience in the chemical industry. In 1946 he accepted as position as group leader at the Dewey & Almy Chemical Company, where he remained until 1952. In this new capacity Hill focused on polymers—large molecules, such as protein, plastics, and silk, that are composed of similar or identical small linked molecules. Seeking a more collegial environment, Hill moved to the Wilmington, Massachusetts–based National Polychemicals Inc. in 1952. National Polychemicals produced the chemical intermediaries necessary to synthesize polymers, and Hill ascended the corporate ranks. During his nine years there he rose from assistant manager to vice president, and the company's fortunes waxed as well. By 1961 its annual sales topped $10 million.

Hill achieved his longtime goal of operating his own research company when he founded Riverside Research Laboratory in 1961. Although Riverside's overarching mission was to offer consulting and research and development services across the broad field of organic chemistry, Hill was particularly interested in resin, rubber, and plastics. He quickly became a recognized authority in polymer chemistry.

At the same time that Hill developed his research credentials he participated in the professional aspect of the field. In 1968 he was appointed to serve on President Johnson's National Committee on Product Safety, which involved him in matters of product liability. That same year Hill was selected to chair the American Chemical Society's committee on professional relations. A member of the American Chemical Society for 38 years, he served on its board of directors from 1971 to 1978. In 1977 his colleagues bestowed upon him the ultimate honor of electing him president of the society. He was also a fellow of both the American Association for the Advancement of Science and the American Institute of Chemists. Hill died of a heart attack on March 17, 1979, and was survived by one child.

Hipparchus

(c. 170 B.C.–c. 120 B.C.)
Greek
Astronomer

Known for his methodical observational skills and innovative mind, Hipparchus did much to advance the field of astronomy and to transform it into a quantitative science. Hipparchus was responsible for constructing the first known star catalog, was the first to assign a scale of magnitude to classify the brightness of stars, and developed what became trigonometry. Hipparchus was best known, however, for his discovery of the precession of the equinoxes. This finding allowed him to calculate the length of the year.

Little is known about Hipparchus' life except his birthplace of Nicaea (now in Turkey). It is believed he spent much of his career in Rhodes, where he built an observatory, and also, though there is some doubt, in Alexandria. Only one of his written works survived, and much of what is known about Hipparchus is through the writings of other scientists. The astronomer CLAUDIUS PTOLEMAEUS, who followed Hipparchus by about three centuries, cited Hipparchus' works frequently in his writings and used them as a basis for his own theories.

Hipparchus' most significant achievement was his discovery of the precession of the equinoxes. Precession is the slow movement of the Earth's axis around the ecliptic pole. The movement is caused primarily by the gravitational pull of the Sun, Moon, and planets. Hipparchus' finding resulted from observation and comparison of the positions of the stars to the positions recorded 150 years earlier by Timocharis of Alexandria. He also studied observations made even earlier in Babylonia. Through these comparisons Hipparchus discovered that Spica, one of the brightest stars in the sky, had been eight degrees from the autumn equinox, in contrast to Hipparchus' own observations, which indicated that it was six degrees from the equinox. Hipparchus used this information to determine the east-to-west movement, or precession, of the equinoctial point and estimated that the rate of the precession as about 45 seconds of arc per year. With these data Hipparchus was also able to theorize on the length of the sidereal year, the time required for the Earth to revolve completely around the Sun, relative to the fixed stars, and on the length of the solar year, the time required for the Earth to make one complete revolution around the Sun relative to the equinoxes. His calculation of the solar year was accurate to within six and one-half minutes.

Hipparchus was also responsible for creating the first known star catalog. Inspired by the discovery of a new star in 134 B.C., he constructed a catalog of about 850 stars. He also introduced the scale of magnitudes to classify the luminosity of stars. The brightest stars were classified as first magnitude and those faintest to the naked eye as sixth magnitude. This scale remains in use, though in a much more refined form. Hipparchus also contributed to geography, improving upon the work of ERATOSTHENES to suggest better methods for determining latitude and longitude. A table of chords constructed by Hipparchus was a precursor of the sine and laid the foundation for the development of trigonometry.

Hipparchus enjoyed a distinguished reputation during his lifetime, and his likeness appeared on coins of his hometown. His belief that astronomy was an evolving science was progressive for his time, as was his willingness to disregard traditional ideas. Hipparchus greatly advanced astronomy with his numerous original discoveries.

Hippocrates of Cos

(c. 460 B.C.–c. 370 B.C.)
Greek
Physician

Traditionally regarded as the father of medicine, Hippocrates was a Greek physician who helped transform medicine from a field ruled by magic, religion, and superstition to a scientific discipline. Hippocrates based his medical practice on the study of the human body and on objective observations. He believed a person's well-being depended on the balance of the four humors—phlegm, black bile, yellow bile, and blood—and supported the doctrine of treating the body as a whole. His opinion that ill health should be treated not with drugs but with rest, good diet, fresh air, and exercise was progressive.

It is believed Hippocrates was born around B.C. 460 on the Greek island of Cos, the son of Heraclides and Phenaretes. His father was a physician, and Hippocrates reportedly studied under him, as well as under the atomist Democritus and the sophist Gorgias. Plato referred to Hippocrates in some of his writings, as did Aristotle. Hippocrates seems to have traveled extensively in Greece and Asia Minor throughout his career, teaching and practicing medicine. It is believed he spent considerable time teaching at the medical school at Cos.

Dismissing the common view that illness was caused by disfavor of the gods or possession by evil spirits, Hippocrates believed that sickness had a rational, physical explanation. He treated the body as a whole rather than as a series of parts, and in his practice he gathered information about parts of the body to form an overall concept before breaking down the whole into parts. Hippocrates also embraced the commonly held belief that disease was caused by an imbalance of the four bodily humors, but he expanded on the theory and maintained that humors were glandular secretions and their imbalance was caused by outside forces.

Hippocrates was an advocate of natural healing processes. The goal of medicine, he held, should be to avoid the conditions known to cause disease and to build the strength of the ill through proper diet, cleanliness, exercise, and rest. His emphasis on diet stemmed from his belief that diseases were often caused by undigested residues that were a result of a poor diet. These residues excreted vapors that passed into the body and caused illnesses. Drugs and drastic treatment, Hippocrates argued, should be used only as last resorts. Hippocrates' progressive and prevention-oriented beliefs contrasted with popular beliefs of the time, which promoted the detailed diagnosis and classification of diseases.

A substantial collection of writings, numbering about 60 and known as the Hippocratic Collection, detailed many of Hippocrates' medical theories and case histories, but few were believed to have been written by Hippocrates himself. Hippocrates' influence, however, is apparent throughout the works. The collection most likely constituted the library of a medical school. The works detailed a number of subjects, including prognosis, clinical subjects, anatomy, surgery, medical ethics, and diseases of women and children. Significant among these were *Epidemics,* which described the course of many illnesses; *Aphorisms*, which included a collection of observations regarding diagnosis, prognosis, and treatment; and *Airs, Waters, and Places*, which recognized a link between environment and disease.

Hippocrates accurately described symptoms of diseases and was the first to detail the symptoms of pneumonia and epilepsy in children. He was also the first to support the idea that thoughts and feelings originated in the brain rather than the heart, as was commonly believed. Hippocrates was an acclaimed physician during his lifetime, and his theories grew to dominate the medical field, persisting for centuries. Hippocrates' contribution to modern clinical medicine is unrivaled.

Hodgkin, Alan Lloyd

(1914–)
English
Neurophysiologist

Alan Lloyd Hodgkin collaborated with Sir ANDREW FIELDING HUXLEY to determine the mechanics of nerve impulses, confirming earlier theories of neural resting and action potentials. Hodgkin and Huxley identified the chemical function of the "sodium pump," whereby sodium flooded inward and potassium flooded outward across the neural membrane, creating momentary neural action before returning to equilibrium in a resting state. Hodgkin and Huxley shared the 1963 Nobel Prize in physiology or medicine with Sir JOHN ECCLES, who furthered their work on neural impulses by focusing on the synaptic level.

Hodgkin was born on February 5, 1914, in Banbury, Oxfordshire. His mother was Mary Wilson; his father, George L. Hodgkin, died in Baghdad during World War I soon after Hodgkin's birth. In 1944 Hodgkin married Marion Rous, daughter of Peyton Rous, who won a Nobel Prize two years after his son-in-law. Hodgkin had four children.

In 1932 Hodgkin matriculated at Trinity College of Cambridge University. In 1936 he was named a fellow of Trinity College, filling the positions of lecturer and assistant director of research at the physiological laboratory. Hodgkin spent the year 1937–1938 at the Rockefeller Institute in New York City. While he had been experimenting on the nervous systems of shore crabs at Cambridge, American researchers were conducting experiments on the *Loligo* genus of squid, which had a giant axon nerve that allowed for more exacting observations. Upon returning to Cambridge, Hodgkin inserted microelectrodes into the axon nerves of squids as well as on the outside of these nerves to measure chemical activity, a study that led to the identification of the sodium pump. Essentially, a nervous impulse elicits the reversal of sodium and potassium concentrations inside and outside the nerve, effectively reversing the electrical charge inside the nerve from negative in its resting potential to positive in its action potential.

World War II interrupted these researches, as Hodgkin served the Air Ministry by working on radar systems for aircraft. Hodgkin returned to Cambridge in 1945 to continue work on nerve impulses and published his results with his colleagues in 1951. It was the 1952 publication of the mathematical formulae describing the nerve impulses that caught the eyes of the Nobel Prize committee. That year the Royal Society of London named Hodgkin the Foulerton Research Professor, and in 1958 it awarded him its Royal Medal. In 1970 the Royal Society appointed Hodgkin as its president. Meanwhile Cambridge promoted him that same year by naming him the John Humphrey Plummer Professor of Biophysics, a position he retained until 1981. Hodgkin served as Master of Trinity College from 1978 through 1984.

Hodgkin's publications covered a wide range. His 1963 publication *Conduction of Nerve Impulses* reported on his specialty; his collaborative publication *The Pursuit of Nature* ventured further afield. He waxed more philosophical in his 1992 publication *Chance and Design: Reminiscences of Science in Peace and War.* However, Hodgkin's lasting influence on science resulted from his collaborative investigations into the biochemical mechanics of nervous impulses.

Hodgkin, Dorothy Crowfoot

(1910–1994)
British
Crystallographer

By interpreting X-ray photographs of crystals, Dorothy Hodgkin worked out the three-dimensional structure of the vital and complex molecules of penicillin, vitamin B_{12}, and insulin. Her work won a Nobel Prize in 1964.

Dorothy Mary Crowfoot was a globe-trotter from birth. She was born on May 12, 1910, in Cairo, Egypt, where her father, John Crowfoot, worked for the Ministry of Education, part of the British government that controlled Egypt at the time. Dorothy's mother, Molly, was an expert on ancient cloth and a keen student of plants.

World War I broke out when four-year-old Dorothy was visiting England with her family. Her parents felt they had to go back to Egypt but thought it too dangerous to take their three daughters with them, so they left them in England under the care of a maid, Katie, and their grandmother. The girls saw their mother only once during the next four years.

As a teenager Dorothy became interested in chemistry, especially the study of crystals—solids whose molecules are arranged in regular patterns. For her 16th birthday her mother gave her a book by William Henry Bragg, a pioneer in the new field of X-ray crystallography. Bragg wrote, "The discovery of X-rays has increased the keenness of our vision over ten thousand times. We can now 'see' the individual atoms and molecules." Dorothy decided that she, too, wanted to see molecules with X-ray crystallography.

Dorothy Crowfoot entered Somerville, a women's college at Oxford University, in 1928. She graduated with a bachelor's degree in chemistry in 1932, then began working at Cambridge, Britain's other most famous university, with J. D. Bernal, who was among the first to use X-ray crystallography to study the complex molecules made in the bodies of living things. She made a habit of "clearing Bernal's desk," analyzing the crystals that scientists from all over the world sent to him for study. These included vitamin D, vitamin B_1, sex hormones, and the digestive enzyme pepsin. "My research with Bernal formed the foundation for the work I was to do during the rest of my career," she said later.

Somerville persuaded Crowfoot to return as a researcher and teacher in 1934, but it failed to provide her with a decent laboratory. She had to work in the basement of the Oxford Museum, surrounded by dinosaur bones and cases of dead beetles. A far more serious difficulty was the rheumatoid arthritis that developed around 1934. This painful disease twisted and deformed the joints of her hands and feet.

Throughout her career Dorothy Crowfoot chose tasks that everyone else believed could not be done, then developed techniques to make them possible. One of the first molecules for which she did this was cholesterol, a fat best known today for its role in heart disease. To discover cholesterol's structure, Crowfoot first applied a method that used thousands of calculations to produce a diagram something like a topographical map. Instead of showing the elevation of hills and valleys, the map showed the density of electrons struck by the X-ray beam, which in turn revealed where atoms were located. She supplemented this technique with another that she developed herself, which involved making crystals that were just like natural ones except that they contained an extra atom of a heavy element such as mercury. She filled in missing parts of a crystal's structure by comparing X-ray photos of natural and artificial crystals.

Crowfoot was so excited when she finally figured out the structure of cholesterol that she literally danced around her lab. She used this work as her thesis and received her Ph.D. in 1937. Cholesterol was the most complex molecule yet analyzed by crystallography and the first to have its structure worked out by this technique alone.

In that same year Crowfoot met Thomas L. Hodgkin, a cheerful, outgoing man who seemed to balance her own quiet nature. They married on December 16, 1937. For eight years they could spend only weekends together because Thomas taught in the north of England and Dorothy taught at Oxford. In 1945, however, he too obtained a post at Oxford. They had three children, raised partly by family members and live-in helpers.

As World War II began, several scientists were trying to find ways to make the antibiotic penicillin in large quantities, and Dorothy Hodgkin met one of them, Sir ERNST BORIS CHAIN. Penicillin, discovered by the Scottish bacteriologist Sir ALEXANDER FLEMING in 1928, was made naturally by a mold, but no one then knew how to grow the mold in large amounts. Chain hoped to create the drug synthetically. He told Hodgkin that knowing the exact structure of its molecules would help in this task.

Hodgkin and a graduate student, Barbara Rogers-Low, solved the penicillin puzzle in 1946 after four years of hard work. Making the drug naturally in vats of mold turned out to be the best way to mass produce it after all, but Hodgkin's work helped chemists create synthetic penicillins that were better than the natural form at attacking certain kinds of bacteria.

Hodgkin's work on penicillin made her internationally famous. Britain's top scientific body, the Royal Society, elected her to membership in 1947; she was only the third woman to receive this honor. By that time she had a larger laboratory and an official post and salary as a lecturer and demonstrator, but Oxford did not make her a reader, the equivalent of an American full professor, until 1957 or give her a fully modern laboratory until 1958.

In 1948 a scientist from the drug company Glaxo gave Hodgkin a vial of wine-colored crystals that he had just made and asked her to work out the structure of the molecules in them. The crystals were vitamin B_{12}, a compound essential for healthy blood. Some people were not able to extract the vitamin from their food and needed to take it as a drug. As with penicillin Glaxo wanted to learn the structure of B_{12}, so it could be produced it in large quantities.

Even less was known about B_{12} than about penicillin, and it was a much larger molecule. Hodgkin's research team gathered data about the molecule for six years before even trying to analyze it. When they did start the analysis, they got some welcome help. In 1953 Hodgkin met a scientist named Kenneth Trueblood, who had programmed an early computer at the University of California at Los Angeles (UCLA) to do crystallography calculations. Hodgkin said later that she deciphered the B_{12} molecule "by post and cable," sending her data to be analyzed by Trueblood's computer. J. D. Bernal called this work, which she finished in 1956, "the greatest triumph of crystallographic technique that has yet occurred."

In October 1964 Dorothy won the Nobel Prize in chemistry, making her the first British woman to receive a science Nobel. The following year she also received the Order of

Merit, one of Britain's highest awards. Only one other British woman, Florence Nightingale, had received this honor.

Hodgkin's next big project was on insulin. She had wanted to determine the structure of the insulin molecule since the mid-1930s, but its 777 atoms defeated even her until 1969. Working out insulin's structure required analysis of 70,000 X-ray spots—no mean feat even with a computer. Her work helped scientists understand how insulin functions. Dorothy Hodgkin retired in 1977. She died on July 30, 1994, at the age of 84.

Hoffmann, Roald

(1937–)
Polish/American
Chemist

Roald Hoffmann is best known for developing a method to predict the course of organic chemical reactions. With ROBERT BURNS WOODWARD Hoffmann explored how the orbital symmetry of molecules determines the outcome of different reactions. The duo conceived a mathematical method, now termed the Woodward-Hoffmann rules, for predicting reactions without tedious and complex equations. In later years Hoffmann successfully applied his method to inorganic compounds as well as organic ones, transcending the traditional divide between the two areas. Hoffmann has also sought to explain the principles of molecular chemistry to a lay audience and is a published poet.

Although his adult life has been extraordinarily successful, Hoffmann's childhood was marked by tragedy and loss. Born Roald Safran in Zloczow, Poland (now Zolochez, Ukraine), on July 18, 1937, Hoffmann lived his first years at the onset of one of the darkest periods in European history. In 1941 German troops occupied his town. Like most European Jews during World War II, Hoffmann's family was interned at a labor camp. His father, Hillel Safran, arranged for the escape of his wife, Clara Rosen Safran, and his son. Roald and his mother then spent a year hiding in the attic of a Ukrainian schoolhouse; his father was executed by Nazi troops when he attempted to join his family. After the Soviet army liberated Zloczow in 1944, Hoffmann and his mother lived in a number of displaced persons camps, where Clara met and married Paul Hoffmann, whose own wife had been killed in the death camps. The family immigrated to the United States in 1949 and settled in New York City. Only 80 of the 12,000 Zloczow Jews survived the war.

Hoffmann enrolled at Columbia University in 1955 and obtained his bachelor's degree in chemistry in 1958. In 1960 he married Eva Börjesson, with whom he had two children. He pursued graduate studies in chemical physics at Harvard University, where he earned his Ph.D. in 1962, and remained as a fellow until 1965. In 1964 he began his fruitful collaboration with Woodward on the unusual class of chemical reactions called the pericyclics.

After accepting a position as an associate professor at Cornell University in 1965, Hoffmann continued his project with Woodward. It was understood that the release of energy that occurs during a chemical reaction is caused by changes in the motion of electrons moving from one energy level (or orbital) to another. Less clear was how to predict the course of the reaction to determine how energy would be released. Hoffmann discovered that the course of the reaction was determined by the symmetry of the electron orbitals. Using quantum mechanics, Hoffman and Woodward promulgated what came to be known as the Woodward-Hoffmann rules to predict these symmetries and ascertain whether certain combinations of chemicals would result in a reaction. In 1969 they published their ground-breaking paper, "Conservation of Orbital Symmetry," detailing this work.

Hoffmann continued to explore the significance of his findings; with his research team at Cornell he painstakingly analyzed both inorganic and organometallic (organic compounds with a metal) molecules and found that they conformed to the Woodward-Hoffman rules as well. After being named a full professor at Cornell in 1968, Hoffmann became the John A. Newman Professor of Physical Science at Cornell in 1974.

Most recently Hoffmann has investigated the connections between art and science. In addition to publishing two volumes of poetry, *The Metamict State* and *Gaps and Verges,* he has collaborated with several artists in joint projects. He has also worked to introduce his field of expertise to a wider audience. He wrote and hosted several episodes of the television program *The World of Chemistry* and in 1955 published *The Same and Not the Same,* which sought to explain to a nonscientific audience how the world behaves at a molecular level.

The impact of Hoffmann's work was acknowledged when he received the 1981 Nobel Prize in chemistry (which he shared with KENICHI FUKUI). Among many other honors, Hoffmann won the 1974 Pauling Award and the 1990 Priestley Medal. The Woodward-Hoffmann rules revolutionized the study of reactions, and the method is widely considered to be the most important contribution to organic chemistry since World War II.

Hollerith, Herman

(1860–1929)
American
Engineer

Regarded as the father of information processing, Herman Hollerith created a system for recording and retrieving information on punched cards. The system was used for data processing and gained widespread usage after successfully handling the computation of the 1890 U.S. census. Hollerith went on to found the company that eventually became International Business Machines Corporation (IBM).

Hollerith was born on February 29, 1860, in Buffalo, New York, to parents who were German immigrants. His father was a teacher of classics. Equal emphasis was placed on coursework and practical work at the Columbia School of Mines in New York, where Hollerith was an engineering student. He studied physics and chemistry, as well as surveying, geometry, drawing, and assaying. The school's requirement that engineering students visit places of industry to observe practical methods likely drew Hollerith to machine and metallurgy shops, which would later play a significant role in his career.

After graduation in 1879 Hollerith worked with his instructor, W. P. Trowbridge, on the 1880 U.S. Census. By then, the census had become a labor-intensive, time-consuming project. Though the census itself took only a few months to carry out, tabulation and analysis of the 1880 data were projected to take nearly a decade. A mechanical solution was needed. Dr. John Shaw Billings, the head of the division of vital statistics, recognized potential in Hollerith and discussed with him the possibility of a tabulating system that would solve the problem the Census Bureau faced. Thus, after Hollerith left the bureau to teach at the Massachusetts Institute of Technology (1882–1884) and work at the U.S. Patent Office in Washington, D.C., he continued to investigate the census problem.

By 1884 Hollerith had developed a first design and applied for a patent. The system used punched tape that ran over a metal drum and under brushes. Whenever the brushes encountered a hole, a circuit was completed and the data recorded. The use of electricity increased the speed and efficiency of the system. Hollerith continued to make design modifications to the machine, including the use of punched cards rather than paper. The machine-readable cards, which came to be known as the Hollerith cards, were more easily replaced and corrected than tape. By 1887 he had developed a machine designed for the census that could handle up to 80 cards per minute. The code used to record the alphanumeric information on the punch cards was called the Hollerith code.

When the bureau began preparing for the 1890 Census, it held a competition to find the best system. Competitors were to input and tabulate data from one city, and each method was timed. The systems of Hollerith's competitors William Hunt and Charles Pidgin required more than 100 hours to input the data, whereas Hollerith's system took just over 72 hours. For the computation Pidgin's system took more than 44 hours and Hunt's more than 55. Hollerith's machine tabulated the data in 5 hours, 28 minutes, and won the competition. Using Hollerith's system, the 1890 Census took three years to process, compared to the seven years required to complete the 1880 census.

Hollerith adapted his machine for commercial use after the 1890 census, and in 1896 he established his own company, the Tabulating Machine Company, which manufactured and marketed his systems. His business was successful, and his machines, because of their versatility and ability to work with data of almost any kind, were used in a variety of trades. In 1911 Hollerith sold his share of the company, and the company in 1924 became IBM. Hollerith's census methods persisted into the 1960s and helped move the field of data processing into the computer age.

Hopper, Grace Brewster Murray

(1906–1992)
American
Computer Scientist

Grace Hopper knew computers almost from their birth, when the hulking machines filled a room and could do only three calculations a second. She helped make computers practical for businesses and individuals to use by devising ways for them to do some of their own programming and helping to develop a programming language that used English words.

Grace Brewster Murray was born in New York City on December 9, 1906, to Walter Murray, an insurance broker, and his wife, Mary. She was inspired both by her great-grandfather, an admiral in the navy ("a very impressive gentleman," she recalled), and her grandfather, a civil engineer, who sometimes took her with him on surveying trips. She tried her own first engineering project at age seven, when she took apart seven of the family's clocks but failed to put them back together.

Murray graduated from Vassar College with a B.A. in mathematics and physics in 1928. She earned a master's degree at Yale in 1930 and shortly afterward married Vincent Foster Hopper, a teacher of English and literature. The next year she began teaching mathematics at Vassar. She continued with her graduate studies as well, earning a Ph.D. from Yale in 1934.

In December 1943, the height of World War II, Grace Hopper left her post as an associate professor at Vassar and enlisted in the Naval Reserve. By that time she had separated from her husband (they divorced in 1945). She was assigned to the Bureau of Ordnance computing project at Harvard University. She reported to Howard Aiken, the head of the project, on July 2, 1944, and got her first look at the Mark I, 51 feet long and 8 feet high—the country's first modern computer. Aiken's instructions for using this monster consisted of telling Hopper, "That's a computing engine." Hopper said later that she was "scared to death" by the machine, but she also thought it was "the prettiest gadget I ever saw."

During the war the Mark I and its successors, Marks II and III, performed the calculations needed to aim complex navy guns and rockets accurately. The machines worked night and day, and so did Hopper and the rest of the crew who ran them. The computer operators sometimes slept on their desks so they could spring into instant action when one of the machine's thousands of mechanical switches failed, as occurred often. The most unusual cause of a failure was a moth that was caught in a switch and beaten to death. After the moth incident Hopper and the others began to call finding and correcting failures "debugging." Computer programmers still use this term.

While Hopper and her cohorts were debugging the Marks, a pair of inventors named John Mauchly and J. Prosper Eckert built ENIAC, the world's first general-purpose electronic computer. They were among the few people at the time who believed that computers would eventually be useful to ordinary businesses, and Hopper came to share their enthusiasm. She joined their fledgling company in 1949.

The company built UNIVAC, the first mass-produced commercial computer, in 1951. UNIVAC could calculate 1,000 times faster than the Mark I, but it was still too big, too expensive, and, above all, too difficult to use to appeal to any but the largest businesses. For instance, each new program had to be entered into the machine, even though parts of many programs were the same. In 1952 Hopper devised a new type of program called a compiler, which allowed a computer to assemble its own programs from shorter routines stored in its memory. This not only saved time but also eliminated errors introduced during retyping.

Another problem was that the "languages" in which computer instructions had to be written were complex and their use required special training. In 1957 Hopper designed the language Flowmatic, which used English words in both its data and its instructions. Flowmatic became one of the ancestors of Common Business-Oriented Language (COBOL), which Hopper and other computer experts designed in 1959. COBOL used English words in structures that resembled ordinary sentences and aided greatly in making computers acceptable to business.

Hopper received many awards during her career, including the Naval Ordnance Development Award (1946), the Legion of Merit (1973), induction into the Engineering and Science Hall of Fame (1984), and the National Medal of Technology (1991). The award she treasured most, however, she received on November 8, 1985, when the navy raised her to the rank of rear admiral—the only woman admiral in the country's history.

Hopper retired from the Sperry Corporation (the descendant of Eckert and Mauchly's firm) in 1971, and she finally retired from the navy for good in 1979. At that time she received the Distinguished Service Medal, the Department of Defense's highest honor. "Amazing Grace," as she was lovingly called, died at the age of 85 on January 1, 1992.

Horney, Karen Danielsen

(1885–1952)
German/American
Psychologist

Karen Horney was what one biographer called a "gentle rebel." As a child she rebelled against her father's authoritarian discipline. As an adult she rebelled against Sigmund Freud's equally authoritarian ideas about how the mind works.

Karen Danielsen was born in Blankenese, near Hamburg, Germany, on September 16, 1885. Her father, the Norwegian-born Berndt Danielsen, was a ship captain. She admired him in some ways but disliked his sternness and his belief that women should confine themselves to the home. When her parents separated in 1904, Karen remained with her mother, Clotilde.

While a teenager Karen decided to be a doctor. She persuaded her mother, who in turn persuaded her father, to help her get the education she needed. She studied medicine at the universities of Freiburg, Göttingen, and Berlin and obtained her M.D. from the University of Berlin in 1911. In 1909, while still a student, she married Oskar Horney, a lawyer and businessperson. They had three daughters, then separated in 1926 and later divorced.

Karen Horney decided to specialize in psychoanalysis, the form of psychiatry created by SIGMUND FREUD. She began seeing patients around 1912, obtained a Ph.D. in 1915, and worked at a psychiatric hospital for several years. In 1919 she joined the highly respected Berlin Psychoanalytic Clinic and Institute. She was a lecturer, analyst, and trainer of other analysts there until 1932.

From the beginning Horney's ideas caused controversy. Freud had claimed that women envied and felt inferior to men, but Horney claimed that the idea "that one half of the human race is discontented with the sex assigned to it . . . is decidedly unsatisfying, not only to feminine narcissism but also to biological science." She said it was just as likely that men envied women their ability to give birth and nurture children. If women envied anything about men, it was their social and economic power. These views pleased feminists but shocked many of Horney's Berlin colleagues.

In 1932 Franz Alexander, a former student of Horney's who had become the director of the Chicago Institute for Psychoanalysis, invited her to be the institute's assistant director. Horney accepted, glad to leave the increasing power of the Nazis, who had labeled psychoanalysis a "Jewish science" and therefore liable to persecution. Once in Chicago, however, she found that she and Alexander had different ideas, and they parted in 1934.

Horney moved to New York City and began lecturing at the New School for Social Research, teaching at the New York Psychoanalytic Institute, and carrying on a thriving private practice. She then once again rebelled against her colleagues' ideas. Freud and his followers blamed most mental illness on instinctive sexual conflicts with parents during infancy, but Horney believed that social and cultural factors were more important. She presented her views in lectures and in books such as *The Neurotic Personality of Our Time* (1937) and *New Ways in Psychoanalysis* (1939).

According to Horney's theory, social factors often put a strain on parents, who respond by becoming less affectionate or more controlling toward their children. The children, in turn, develop behavior that they hope will protect them from their parents. These actions usually continue into adulthood. If they fail to provide protection or are in conflict, the form of mental illness called neurosis develops. "The genesis [cause] of a neurosis," Horney wrote, "is . . . all those adverse influences which make a child feel helpless . . . and which make him conceive the world as potentially menacing."

The traditional psychoanalysts who dominated the New York Institute resented Horney's departure from Freud's theories and her introduction of ideas from sociology and anthropology into psychoanalysis. In 1941 they voted to bar her from training other analysts. Horney, four supporters, and 14 students responded by resigning. "Reverence for dogma has replaced free inquiry," they complained. The group formed its own professional organization, the Association for the Advancement of Psychoanalysis, and training center, the American Institute for Psychoanalysis. Horney was dean of the institute and editor of its journal for the rest of her life.

In the 1940s Horney presented other new ideas in her lectures and in several books, including *Self-Analysis* (1942), *Our Inner Conflicts* (1945), and *Neurosis and Human Growth* (1950). She said that people have a natural tendency to improve themselves and develop their full potential. Unlike Freud and his followers, who felt that most neurotic people could improve only after a lengthy series of sessions with an analyst, Horney thought that people often could learn to analyze themselves. She "encourage[d] people to make the attempt to do something with their own problems." When people did undertake psychoanalysis with a professional, Horney recommended that the analyst take a nonjudgmental approach and focus on present problems and solutions rather than dwelling on the patient's early childhood, as traditional Freudians did.

Horney died of cancer on December 4, 1952, but many of her ideas live on as accepted parts of psychiatry. The Karen Horney Foundation in New York, founded in 1955, carries on her work.

Hoyle, Sir Fred

(1915–)
English
Astronomer

A controversial and outspoken character, Sir Fred Hoyle has been best known for his advocacy of the steady-state theory of the creation of the universe. He also played an important role in the development of radar and proposed a theory of stellar evolution that explained the nuclear processes that fueled the stars. Hoyle is also a skilled writer, producing many technical works and popular science fiction works.

The son of Benjamin Hoyle, a textile merchant, and Mabel Picard Hoyle, Fred Hoyle was born on June 24, 1915, in Bingley, a village in Yorkshire, England. He attended Bingley Grammar School; throughout his childhood Hoyle was a bit of a rebel and a long-standing truant. Hoyle married Barbara Clark in late 1939, and together they had two children.

Hoyle attended Emmanuel College at Cambridge to study mathematics and astronomy. After earning his master's degree in 1939, Hoyle earned a fellowship at St. John's College. Hoyle then spent six years during World War II developing radar technology with the British Admiralty. In 1945 he returned to Cambridge to teach astronomy and mathematics.

During the early 1940s, Hoyle worked on his theory of stellar evolution, expanding on work by the physicist HANS ALBRECHT BETHE on the energy production of stars. In 1938 Bethe had proposed that stars were powered by a sequence of nuclear reactions, but his theory did not explain the production of heavy elements within some stars. Hoyle provided a workable theory that involved the nuclear fusion of stars and accounted for the existence of heavy elements in the solar system.

Hoyle's most significant contribution to astronomy was his support of the steady-state theory of the universe. In 1948 the physicist GEORGE GAMOW proposed the big bang theory of the universe, which stated that the universe was created by a huge explosion that occurred billions of years ago. According to the theory galaxies were still being formed and moving as a result of the explosion. In reaction to this theory the astronomer Thomas Gold and the mathematician Hermann Bondi formulated the steady-state theory, which suggested that the universe expanded perpetually and matter was created continuously to maintain the mean density of matter at a constant value. Unlike in the big bang theory, there was no beginning or end of the universe in the steady-state theory. Hoyle became a strong supporter of the steady-state theory and developed it further; he modified and applied the equations of ALBERT EINSTEIN to produce a mathematical model that supported the theory. His contributions provided the theory with respectability and made it plausible. Hoyle subsequently became the official spokesperson for the steady-state theory, publishing a number of works explaining and supporting it.

The theory was not without controversy, however, and for the next 15 years proponents of each theory searched for ways to support the favored idea. In 1964 Arno Penzias and Robert W. Wilson detected microwave radiation in space by using radio telescopes. They claimed that the radiation indicated remnants of the big bang explosion, and the big bang theory emerged as the preferred theory.

Hoyle received many honors in recognition of his numerous contributions. He was elected to the Royal Society of London in 1957, became the director of the Institute of Theoretical Astronomy in 1962, and was knighted in 1972. A prolific writer, Hoyle has produced textbooks, technical treatises, and popular science fiction works in the form of short stories, plays, and novels. Since leaving Cambridge University in 1972, after disputes with the administration, Hoyle has worked as a visiting professor at a number of institutions.

Hrdy, Sarah Blaffer

(1946–)
American
Anthropologist

Sarah Hrdy has shown that the needs, strategies, and behavior of female animals are just as important as those of males in shaping the way a species evolves. The daughter of an oil-rich Texas family, she was born Sarah Blaffer on July 11, 1946, in Dallas, but grew up in Houston. Although her mother encouraged her desire to get an education and have a career, society did not. When she attended a girls' boarding school in Maryland at age 16, "It was the first time in my life when the things I loved were valued and vindicated," she told an interviewer, Lucy Hodges.

At Wellesley College, Blaffer majored in philosophy but also took creative writing courses. She decided to write a novel about the Maya culture of South America and began researching the folklore of the Maya as background. Finding the research more interesting than the novel, she transferred to Radcliffe and changed her major to anthropology. For her undergraduate thesis Blaffer wrote about the demon H'ik'al, who took the form of either a bat or a black man and punished women who violated sexual taboos. Her research earned her a B.A. in 1969 and was published as *The Black-Man of Zinacantan* in 1972.

Sarah Hrdy later said, "This is a lot of fun, but I want to do something relevant to the world." Deciding to learn how to make films that could teach people from developing nations about subjects such as health care, she signed up for filmmaking courses at Stanford University. The courses were disappointing, but while at Stanford she attended a class by PAUL EHRLICH, who taught about problems caused by overpopulation. This suggested an idea for her Ph.D. thesis. She had heard that black-faced Indian monkeys called langurs sometimes lived in overcrowded colonies and that male langurs often killed infants in their groups. Blaffer decided to test the hypothesis that overcrowding caused the infanticide.

In 1972 Blaffer married Daniel Hrdy, then a fellow anthropologist, of Czech ancestry. Daniel Hrdy traveled with Sarah to study the langurs living on Mount Abu in Rajasthan. The couple later had three children.

To her surprise Sarah Hrdy found that under certain conditions, langur males killed infants whether they were overcrowded or not. Each langur troop consists of one male and several females, plus a number of other males that swarm around the central group. Every 28 months or so one of these outside males ousts a troop leader and takes over his harem. The new leader then usually kills all the infants in the troop. Hrdy concluded that, far from being the "sick" response to environmental stress that other observers had thought it was, this behavior made evolutionary sense from the killer's point of view. Killing the infants made the females receptive to mating, so the new male could sire offspring that would pass on his own genes instead of wasting his energy raising babies sired by another male.

Hrdy summarized her langur research for her Ph.D., which she received from Harvard in 1975. Two years later she published her findings in the book *The Langurs of Abu*. When her work first appeared, Hrdy told the interviewer Thomas Bass, "I was attacked by some of the most eminent anthropologists in the country. . . . [They] couldn't believe [that what she reported] was happening." Her discoveries challenged the common belief that primates (monkeys and apes) acted for the good of their group. Instead, it fitted with a new evolutionary doctrine called sociobiology, which stated that animals act in ways that maximize the chances of passing on their genes. The pattern of infanticide Hrdy described was later found not only in other primates but also in animals ranging from hippos to wolves.

Like most researchers of the time, Hrdy began by watching the males in the langur troops. After a while, though, she began to pay more attention to the females. She noticed that they mated not only with their troop leader, old or new, but with as many of the outside males as possible, even when they were already pregnant. Hrdy concluded that the females did this as a strategy to protect their babies, since a male would not kill babies that might be his own. Her evidence that in female primates important sexual strategies had evolved helped to change the way anthropologists thought.

Hrdy's interest in the behavior and strategies of female primates, including human females, has expanded over the years. She published a book on the subject, *The Woman That Never Evolved,* in 1981. Much behavior of female primates, she believes, evolved because of competition between females, which she calls an "evolutionary trap." Human women, however, can resist evolutionary pressures and cooperate. "The female with 'equal rights' never evolved," Hrdy wrote. "She was invented and fought for consciously with intelligence, stubbornness, and courage."

Hrdy has been a professor of anthropology at the University of California at Davis since 1984. In the late 1990s she was investigating animal motherhood. She wanted to know whether the amounts of resources that mothers invest in male and female offspring are equal and, if not, which conditions produce more investment in either sons or daughters.

Hubble, Edwin Powell

(1889–1953)
American
Astronomer

Recognized as the founder of extragalactic astronomy for his discovery of the existence of galaxies outside the Milky Way, Edwin Powell Hubble significantly influenced the field of astronomy. His impact on science has even been compared to that of such scientists as the physicist Sir ISAAC NEWTON and the astronomer GALILEO GALILEI. Hubble also showed that the universe, which was believed to be static, was in reality expanding. He developed a mathematical model of the expansion, which is now known as Hubble's law.

Born on November 20, 1889, in Marshfield, Missouri, Hubble was one of seven children born to John P. Hubble, a fire insurance agent, and Virginia Lee James Hubble. The family moved to the Chicago area in 1898, when Hubble's father received a job transfer, and Hubble attended Wheaton High School outside Chicago. Hubble was an outstanding student and a strong athlete as well, and he graduated at the age of 16 in 1906.

Hubble earned an academic scholarship to the University of Chicago for his achievements and academic promise. Studying mathematics, physics, chemistry, and astronomy, he tutored and worked during the summers to earn enough to cover his college expenses. After graduation in 1910 with bachelor's degrees in both mathematics and astronomy, Hubble was awarded a Rhodes Scholarship and traveled to the University of Oxford to study law. Hubble returned to the United States and started practicing law in Louisville, Kentucky, where his family had moved, but within a year he became bored and returned to the University of Chicago to study astronomy. While studying at the university's Yerkes Observatory in Wisconsin, Hubble met the astronomer GEORGE ELLERY HALE, who founded the Yerkes Observatory and was then the director of the Mount Wilson Observatory in Pasadena, California. In 1916 Hale offered Hubble a future position on the Mount Wilson staff. Hubble earned his doctorate in astronomy in 1917.

After service in World War I Hubble accepted Hale's invitation, and he began work at Mount Wilson in 1919. Hubble's first major accomplishment there was his discovery of galaxies outside the Milky Way. Through his astronomical observations Hubble identified a Cepheid, a certain type of variable star, in the Andromeda nebula (now Andromeda galaxy). Using information about luminosity, magnitude, and distances of Cepheid stars in our galaxy, Hubble calculated the distance to the Cepheid in the Andromeda nebula to be about 1 million light years, well beyond the Milky Way system. His findings were announced in 1924 and forever changed the way astronomers thought of the cosmos. Hubble married Grace Burke Leib in 1924.

A second important achievement involved Hubble's studies in the mid-1920s concerning the expansion of the universe. Through continued studies of the distances of galaxies, Hubble found that a number of galaxies were moving away from the Milky Way at high speeds. In addition, he observed that the more distant galaxies moved at a higher velocity than those closer to the Milky Way. This latter became known as Hubble's law. The universe had long been considered fixed and static, and Hubble's discovery again revolutionized the field of astronomy.

Hubble identified two extragalactic nebulae, M31 and M33, between 1923 and 1925; and then introduced a classification system for galaxies in 1925, the same year he helped develop the Mount Palomar Observatory in California. In 1929 he published his findings about the velocity-distance relation, which forms the basis of modern cosmology.

He received numerous awards for his contributions to astronomy—he was named an honorary fellow of Queen's College, Oxford, and the National Aeronautics and Space Administration (NASA) named the Hubble space telescope in his honor. Hubble was a promising amateur heavyweight boxer during his college years and had a number of interests outside astronomy, including fly fishing and collecting antique books, which he enjoyed during his later years. He died in San Marino, California, on September 28, 1953.

Hubel, David Hunter

(1926–)
Canadian/American
Neurobiologist

David Hunter Hubel collaborated for years with TORSTEN NILS WIESEL, and together they discovered how neurons in the brain interacted with the visual system to create what is perceived as sight. Specifically they investigated the architecture of cellular structure as it is related to the visual process. For this work the pair shared half of the 1981 Nobel Prize in physiology or medicine; the other half was awarded to ROGER WOLCOTT SPERRY, who researched split-brain physiology.

Hubel was born on February 27, 1926, in Windsor, Ontario, to American parents. His mother was Elsie M. Hunter; his father, Jesse H. Hubel, was a chemical engineer. In 1953 Hubel married Shirley Ruth Izzard, and together the couple had three sons.

Hubel entered McGill University in Montreal in 1944 and earned his bachelor of science degree with honors in mathematics and physics in 1947. That year he decided at the last moment to attend McGill Medical School, without any background in biology. He overcame this handicap immediately and spent his summers studying the nervous system at the Montreal Neurological Institute. Hubel earned his M.D. in 1951, and he proceeded to spend the next four years studying clinical neurology at the Montreal Neurological Institute and then at Johns Hopkins University in Baltimore, Maryland. The United States Army drafted Hubel in 1955, assigning him to the Neurophysiology Division of the Walter Reed Army Institute of Research.

It was at Walter Reed that Hubel developed the tungsten microelectrode, an instrument that allowed him to track and record neural activity at the cellular level in the cerebrum while the subject slept. Electrical activity between nerve cells registered as pops when fed through amplifiers and loudspeakers, and intense activity sounded like the firing of a machine gun, Hubel reported. He continued this line of research after the army discharged him in 1958 and he took up a position at the Wilmer Institute of Johns Hopkins. Researching in the laboratory of Stephen Kuffler, Hubel collaborated with Wiesel to apply the former's techniques on the relationship between vision and the brain. In 1959 they published some of their results in the article "Receptive Fields of Single Neurones in the Cat's Striate Cortex" in the *Journal of Physiology.*

That year Kuffler moved his entire team, including Hubel and Wiesel, to the Harvard School of Medicine, where he became the chairperson of the newly formed department of neurobiology in 1964. Hubel took over as head of the department in 1967, and in 1968 Harvard named him the George Packer Berry Professor of Physiology. Meanwhile he and Wiesel continued to research and publish their results—for example in the 1962 *Journal of Physiology* article "Receptive Fields, Binolular Interaction and Functional Architecture in the Cat's Visual Cortex."

In 1988 Hubel published his comprehensive book on his topic of specialization, *Eye Brain, and Vision.* In it he explained his reasons for focusing on what was known as "area 17" of the brain, since it contained the visual mechanism. Hubel's discoveries led to the development of treatments for congenital cataracts and strabismus, and they have revolutionized the understanding of how vision functions in concert with the brain.

Hutton, James

(1726–1797)
Scottish
Geologist

The geologist James Hutton is regarded as the founder of geology as a modern science. The uniformitarian principle, which he developed, proposed that the Earth was very old and that the features of the Earth's surface were caused by natural processes that took place over a long period. This theory was contrary to the beliefs of the time, which corresponded to the biblical Creation and considered the Earth to be only a few thousand years old. Hutton's ideas thus advanced the field of geology and permanently altered the way the Earth was viewed.

James Hutton (right), founder of modern geology. *Smith Collection, Rare Book & Manuscript Library, University of Pennsylvania.*

Born in Edinburgh, Scotland, on June 3, 1726, Hutton was one of four children born to William Hutton and Sarah Balfour Hutton. Hutton's father, a merchant and former city treasurer, died when Hutton was only three years old. The will left the family with considerable sums of money, and it is believed Hutton felt little pressure to earn a living.

In 1740 Hutton entered the University of Edinburgh. There he developed an interest in chemistry, but he decided to become a lawyer instead and left the university in 1743 to work as a lawyer's apprentice. Hutton lasted less than a year, choosing to study medicine instead of law. He returned to the University of Edinburgh and studied there from 1744 to 1747. Hutton then spent two years in Paris, studying anatomy and chemistry. He received his M.D. degree in 1749 in Leiden, Holland.

Hutton went to London after getting his degree, and there he and his friend James Davie agreed to go into business manufacturing ammonia salts, using an inexpensive method the two had previously developed. The undertaking proved successful and supplied Hutton with a good income. Hutton then returned to Edinburgh in 1750, and though he had a medical degree, he decided to take up farming. This endeavor prompted him to travel and learn about agriculture, and around 1753 he began to study geology. By 1765 both the sal ammoniac business and the farm were thriving, and in 1768 Hutton gave up farming in order to spend all his time pursuing his scientific interests. He then moved to Edinburgh and lived with his unmarried sisters. Hutton never married but did have a son who was most likely born while Hutton was still a student.

The uniformitarian principle was Hutton's main contribution to science. Geological observations regarding rocks, fossils, and strata had been made and documented by the late 18th century, but a general scientific theory regarding the formation and features of the Earth had yet to be postulated. Many subscribed to the belief stated in the biblical book of Genesis that the Earth had been created 6,000 years earlier. Hutton proposed, however, that the Earth was immeasurably old and that observable processes could explain geologic phenomena. For instance, Hutton believed that continental topographic features were caused in large part by the erosive action of rivers, and that sedimentation in the ocean collected and formed new rocks through geothermal heat. These processes of erosion, sedimentation, deposition, and upthrusting, Hutton claimed, were cyclical and had been taking place with uniformity over long periods. These tremendously long cycles suggested that the Earth must be ancient. Hutton also stated that these geologic processes scientifically accounted for landforms all over the world; biblical explanations were not needed to justify their existence. Hutton announced his beliefs in papers he presented to the Royal Society of Edinburgh in 1785.

The uniformitarian principle that Hutton proposed provided the foundation for the science of geology and provided a scientific explanation for the features of the Earth's crust. Though Hutton's theories gained popularity, his 1795 *Theory of the Earth,* which explained his views and provided evidence for his conclusions, was not widely read because Hutton's writing style was hard to understand. In 1802 Hutton's friend JOHN PLAYFAIR provided a more easily understandable version with *Illustrations of the Huttonian Theory of the Earth.*

Huxley, Sir Andrew Fielding

(1917–)
English
Physiologist

Sir Andrew Fielding Huxley shared the 1963 Nobel Prize in physiology or medicine with his collaborator, ALAN LLOYD HODGKIN, and with the Australian scientist who built upon their discoveries, Sir JOHN ECCLES. Huxley and Hodgkin identified what they called the *sodium pump,* or the mechanism whereby impulses are transmitted across the nervous system.

Huxley was born on November 22, 1917, in Hampstead, London, England. His mother was Rosalind Bruce; his father, Leonard Huxley, was a writer, as was his half-brother, Aldous Huxley. Both his grandfather, Thomas Henry Huxley, and his other half-brother, Julian Sorel Huxley, were prominent biologists. In 1947 Huxley married Jocelyn Richenda Gammell Pease, and together the couple had six children—Janet Rachel, Stewart Leonard, Camilla Rosalind, Eleanor Bruce, Henrietta Catherine, and Clare Marjory Pease.

Huxley studied physical sciences at Trinity College of Cambridge University and earned his B.A. in 1938. While pursuing his master's degree, which he received in 1941, he switched to studying physiology. In 1939 he had started conducting the physiological experiments at the Plymouth Marine Biological Laboratory with Hodgkin that would lead to their Nobel Prize. However, World War II interrupted their efforts, as Huxley conducted operational research for the Anti-Aircraft Command from 1940 through 1942, when he transferred to working for the Admiralty from 1942 through 1945.

Huxley returned to Cambridge after the war in 1946, serving as a demonstrator until 1950, as an assistant director of research from 1951 through 1959, as the director of studies from 1952 until 1960, and as a reader in experimental biophysics from 1959 to 1960. It was early in his tenure at Cambridge that he conducted his famous experiments with Hodgkin on squid axons.

In 1952 Huxley published a series of articles coauthored by Hodgkin chronicling these experiments. The first of this series, "Properties of Nerve Axons, I. Movement of Sodium and Potassium Ions during Nervous Activity," appeared in the journal *Cold Spring Harbor Symposia on Quantitative Biology.* Though Huxley and Hodgkin's research built upon the earlier work of Julius Bernstein, JOSEPH ERLANGER, and HERBERT SPENSER GASSER, the pair discovered a new aspect of neural impulses: They observed that the impulse is elicited by the exchange across the neural membrane of potassium and sodium, with the former flooding out of the axon and the latter flooding into the axon, causing its charge to change from negative to positive; the charge is retained until the axon reaches a threshold of positivity, when sodium is repelled back out of the membrane. All this transpires in a matter of milliseconds, Huxley and Hodgkin discovered, and represents the nerve impulse.

In 1960 University College in London named Huxley the Jodrell Professor of Physiology, and from 1967 through 1973 he also served as the Fullerian Professor of the Royal Institution. In 1969 University College renamed him as the Royal Society Research Professor of Physiology, a position he retained until 1983. In 1980 he published his book *Reflections on Muscle.* Huxley served as the editor of the *Journal of Physiology* from 1950 through 1957. The Royal Society elected him a fellow in 1955 and awarded him its Copley Medal in 1973. Huxley was knighted the next year.

Hyde, Ida Henrietta

(1857–1945)
American
Physiologist

Ida Hyde invented the microelectrode, an essential tool for research on nerve cells. She was born on September 8, 1857, in Davenport, Iowa, to Meyer and Babette Heidenheimer, emigrants from Germany. Meyer Heidenheimer, a merchant, changed the family name to Hyde soon after his arrival in the United States.

Young Ida's future seemed to hold nothing more ambitious than her job in a hat shop, but she attended night classes and at age 24 won a scholarship to the University of Illinois. She was there only a year before she had to go to work as a teacher to help her family. She finally obtained a B.S. from Cornell University in 1891, when she was 34 years old. Her specialty was physiology, the study of the functioning of the body and its parts.

Hyde's first job as a researcher was at Bryn Mawr College. An eminent German professor was so impressed by her published work that he invited her to study in that country, but her experience there was best described in the title she later chose for an article about it: "Before Women Were Human Beings." She finally obtained a Ph.D. from the University of Heidelberg—the first woman to do so—in 1896.

Back in the United States, Hyde worked and taught at several schools and universities, including Harvard Medical School, at which she was the first woman researcher. In 1898 she moved to the University of Kansas, where she spent most of the rest of her career. The university created a separate department of physiology in 1905 and chose Hyde to head it, making her a full professor. She wrote two textbooks on physiology, which were published in 1905 and 1910.

Hyde investigated the circulatory, respiratory, and nervous systems of a variety of animals, but her greatest contribution to science was the microelectrode. The science historian G. Kass-Simon calls this device the "most useful and powerful tool in electrophysiology." To understand how nerve and muscle cells work, scientists need to be able to stimulate individual cells electrically or chemically and record the resulting changes in the tiny electric current that the cells produce. Devices that added chemicals to a cell or recorded its current had been created earlier, but Hyde in 1920 was the first to make a single tool that could do both at once. Unfortunately her work was not well known, and other scientists later unwittingly repeated it and received credit for her invention.

Hyde was elected to the American Physiological Society in 1902, its first woman member. She endowed several scholarships for women science students but insisted that women meet the highest academic standards. She retired in 1920 and then moved to California. She died on August 22, 1945.

Hyman, Libbie Henrietta

(1888–1969)
American
Zoologist

Libbie Hyman produced a multivolume study of invertebrates, or animals without backbones, that is still a standard reference. She was born in Des Moines, Iowa, on December 6, 1888, but grew up in Fort Dodge. Both her parents were immigrants; her father, Joseph, from Poland and her mother, Sabina, from Russia. Libbie escaped from her father's poverty and her mother's bossiness by exploring the woods and fields near her home.

Hyman graduated in 1905 as both the youngest member of her high school class and its valedictorian. A teacher helped her get a scholarship to the University of Chicago, and she earned a B.S. in zoology in 1910 and a Ph.D. in 1915. One of her professors, Charles Manning Child, then invited her to stay on as his research assistant. She did experiments to provide support for his theories about the way certain animals regenerated lost limbs or other tissues. Her knowledge of chemistry proved especially helpful. She also taught and wrote two books, *Laboratory Manual for Elementary Zoology* (1919) and *Laboratory Manual for Comparative Vertebrate Anatomy* (1922), which became standard works.

Though one of her books dealt with vertebrates, Hyman once said, "I just can't get excited about [vertebrates]. . . . I like invertebrates . . . [especially] the soft delicate ones, the jellyfishes and corals and the beautiful microscopic organisms." She became an expert on the taxonomy, or scientific classification, of invertebrates and grew used to receiving parcels from all over the world containing odd creatures for her to identify. Her specialty was worms, which few other biologists studied.

In 1931, when she was over 40, Hyman found herself truly on her own for the first time. Her mother had recently died, Child was on leave, and royalties from her books generated enough money to live on. She left the University of Chicago and eventually settled in New York City, where, working first in her apartment and after 1937 at the American Museum of Natural History, she began a massive reference work on the biology and classification of all million known invertebrates. The first volume of *The Invertebrates* appeared in 1940 and the fourth, the last on which she worked, in 1967, two years before her death on August 3, 1969. A fifth volume was added later.

Hyman received several awards for her work, including gold medals from the Linnean Society in 1960 and from the American Museum of Natural History in the year that she died, 1969. She was elected to the National Academy of Science in 1954. Her books are still the primary references for scientists who study invertebrates.

Hypatia

(370–415)
Egyptian
Mathematician, Astronomer, Physicist

Hypatia is the earliest woman scientist about whom much is known and is one of the most famous women scientists of all time. She was born around A.D. 370 in Alexandria, Egypt. Her father, Theon, a mathematician and astronomer, headed the famous Museum of Alexandria, which was the equivalent of a large university. Wanting his daughter to be a "perfect human being," Theon taught Hypatia mathematics, philosophy, and astronomy. He then sent her for further study to Italy and Athens, where she impressed everyone with her beauty as well as her intelligence.

After returning to Alexandria, Hypatia lectured and wrote about mathematics, astronomy, philosophy, and mechanics. Her students called her "The Muse" and "The Philosopher." Her writings have been lost, but historians know about her from the writings of some of these students, such as Synesius of Cyrene, later bishop of Ptolemais in Libya. According to one account, Hypatia took over her father's leadership of the museum at age 31.

Hypatia is believed to have written a 13-volume commentary on the *Arithmetica* of Diophantus, an Alexandrian mathematician who had recently invented algebra. She also wrote an eight-book popularization of a work on conic sections, the geometric figures formed when a plane passes through a cone, by another Alexandrian, Apollonius. Hypatia worked in astronomy as well, compiling the *Astronomical Canon*, a set of tables describing the movements of heavenly bodies.

According to Synesius, Hypatia invented a device for removing salt from seawater as well as a plane astrolabe, which determined the positions of the Sun, stars, and planets and was useful for navigation and telling time. She also invented a planisphere for identifying stars and their movements, a device for measuring the level of water, and a hydrometer for determining the density or specific gravity of liquids.

Hypatia was a close friend of Orestes, the Roman prefect of Egypt, and, as one of her students wrote: "The magistrates were wont to consult her first in their administration of the affairs of the city." Cyril, the head of the powerful Christian church in Alexandria, denounced Orestes, Hypatia, and other non-Christians as evil. In March 415, inspired by Cyril and possibly following his orders, a mob attacked Hypatia's chariot, dragged her into a nearby church, cut her to pieces with sharpened oyster shells, and publicly burned her remains. Many historians have seen her brutal murder as the start of an eclipse of both science and women's rights that lasted more than a thousand years.

I

Ildstad, Suzanne
(1952–)
American
Immunologist

The discoveries of Suzanne Ildstad, a surgeon turned researcher, may lead to breakthroughs in organ transplantation and new treatments for acquired immunodeficiency syndrome (AIDS), diabetes, and other diseases. She was born in Minneapolis on May 20, 1952. Both her mother, Jane, and her grandmother were nurses; her grandmother was a scrub nurse for the famous Mayo brothers, who founded the Mayo Clinic and Medical School in Rochester, Minnesota. After receiving a B.S. in biology, summa cum laude, from the University of Minnesota in 1974, Suzanne earned her M.D. from the Mayo Medical School in 1978.

Ildstad became a surgeon, specializing in transplanting organs in children. She found surgery exciting, but she knew that it was the easy part of the transplantation process. The hard parts were waiting for an organ to be found—far more people need organs than donate them—and then fighting a lifelong battle to prevent the recipient's immune system from destroying the transplant. This can be done only with the help of drugs that partly suppress the immune system, leaving the patient vulnerable to cancer and infections. In transplants of bone marrow, from which almost all blood and immune cells arise, an opposite but equally deadly problem occurs unless donor and recipient are carefully chosen to be genetically very similar: Immune cells from the grafted marrow attack the recipient.

To fight these problems, Ildstad turned from the operating room to the laboratory, beginning at the National Institutes of Health in the early 1980s and continuing at the University of Pittsburgh, which she joined in 1988. She found that removal of a certain type of immune cell, called the T cell, prevented marrow grafts from attacking the body, but very similar but much rarer cells (she calls them facilitating cells), which she was the first to see, had to remain if the graft were to survive. After years of work Ildstad succeeded in removing T cells from grafts while preserving the facilitating cells. Marrow treated in this way survived in mice without attacking them, even marrow from rats. The result was a chimera, a mouse with an immune system that was partly mouse and partly rat. Such an animal could accept organ grafts from rats without needing drugs.

Ildstad also designed a procedure for transplanting bone marrow from baboons, which cannot contract AIDS, into humans. Surgeons used this procedure on Jeff Getty in December 1995 after destroying Getty's own AIDS-weakened marrow with radiation. Although Getty had been expected to die in a few months without treatment, his health improved considerably in the year after he received the baboon marrow. The baboon cells apparently survived for only two weeks, however, so it is unclear whether or how the transplantation helped him. In 1998 he was still in relatively good health three years after transplantation.

Ildstad, an elected member of the prestigious Institute of Medicine, moved to the Allegheny University of the Health Sciences in Philadelphia in September 1997 and became head of its new Institute for Cellular Therapeutics. Her husband, the public health physician David J. Tollerud, directs another center there. (They married in 1972 and have two children.) She continues to refine her human bone marrow transfer. In an initial test in which five people with advanced leukemia were given marrow from unmatched donors—minus T cells but with facilitating cells—the marrow transplants survived and did not attack the recipients. If Ildstad's treatment proves dependable, it could be a transplant breakthrough. Marrow transplants from people who are naturally resistant to AIDS, she believes, could treat those who have the disease. Marrow transplants could also help people who suffer from diabetes, arthritis, or other conditions in which the immune system attacks the body's own tissues. Blending marrow from potential donors and recipients, as she did with the rats and mice, could make subsequent organ transplants—even, perhaps, between animals and humans—possible without immunity-suppressing drugs.

Imes, Elmer Samuel

(1883–1941)
American
Physicist

Interested in the field of infrared spectroscopy, Elmer Samuel Imes had the distinction of being only the second African American to earn a doctorate in physics. His research provided early spectroscopic evidence for the existence of nuclear isotopes, and his findings have made their way into modern physics textbooks. Imes was the first chair of Fisk University's department of physics. A dedicated and stimulating teacher, Imes also worked in schools operated by the American Missionary Association.

Born on October 12, 1883, in Memphis, Tennessee, Imes was one of three sons born to Benjamin A. and Elizabeth W. Imes. Both of his parents worked as missionaries. Imes's father graduated from Oberlin College and the Theological Seminary and was active in promoting the importance of the church and education as part of the American Missionary Association.

Imes attended Fisk University, which was one of the historically black colleges and universities and founded in 1866, and graduated with a bachelor's degree in 1903. Before continuing into graduate studies in physics, however, Imes spent several years teaching in schools run by the American Missionary Association. Imes taught at the Albany Normal School and at facilities in the Albany, Georgia, area. Returning to Nashville, Tennessee, for graduate work at Fisk University, Imes earned his master's degree in 1910. Imes again took some time off prior to pursuing his doctorate, working as a physicist and research and consulting engineer in New York City. Imes then entered the University of Michigan's doctoral program in physics. From 1916 to 1918 he was a fellow at the University of Michigan. He received his doctorate in physics in 1918, the second African American to achieve that accomplishment. The title of his dissertation was "Measurements on the Near Infra-Red Absorption of Some Diatomic Cases."

After receiving his doctorate, Imes again worked in private industry as a consulting engineer. He was offered a position as research physicist in 1922 at the Federal Engineer's Development Corporation. After two years Imes accepted a job at the Burrows Magnetic Equipment Corporation. In 1927 he moved again, this time to E. A. Everett Railway Signal Supplies to work as a research engineer. Imes returned to academia in 1930 after being offered the position of professor and director of the newly established physics department at Fisk University. Imes remained at Fisk until his death at the age of 58. While at Fisk, Imes developed the department of physics considerably, creating a successful and respected division. He was also instrumental in the establishment of an infrared spectroscopy laboratory at the university.

Not only was Imes a skilled experimenter in infrared spectroscopy, he was also a tireless advocate for education, instrumental in building the reputation of Fisk University's physics department. Imes was a member of such scientific and professional societies as the Physical Society and the Society of Testing Materials.

Itakura, Keiichi

(1942–)
Japanese/American
Molecular Biologist

Best known for his research on the synthesis of genes for peptides, Keiichi Itakura has worked in the field of molecular biology, particularly in the study of deoxyribonucleic acid (DNA). The peptide synthesis work was an important achievement because it provided scientists with the tools needed to develop hormones such as insulin and somatostatin in a laboratory setting. The synthetic process allowed for greater production of these medically significant hormones.

Born in Tokyo, Japan, on February 18, 1942, Itakura is the son of Tsuneo Itakura and Nobuko Orimoto. He married Yasuko Shimada in 1970. They have two children.

Itakura studied at Tokyo Pharmaceutical College and graduated in 1965. He continued with his studies there and received a doctorate in organic chemistry in 1970. After completing his doctoral work, Itakura received a fellowship to study DNA synthesis under Dr. S. Narang at the National Council of Canada, located in Ottawa. Itakura found DNA synthesis to be a compelling field of research and was interested in learning more about the recently discovered synthesized gene of transfer ribonucleic acid (t-RNA), which had been found by the Indian-born American biochemist HAR GOBIND KHORANA. After his fellowship ended, Itakura went to the California Institute of Technology in Pomona to work as a senior research fellow.

In 1975 Itakura joined the staff of the City of Hope National Medical Center in Los Angeles, California, as an associate research scientist. There he continued his DNA research but also began work on the synthesis of peptides, compounds made up of two or more linked amino acids.

Itakura and his colleagues soon synthesized a gene for somatostatin, a peptide produced mainly by the hypothalamus that hinders the secretion of particular hormones, including insulin, glucagon, and somatotropin. Using *Escherichia coli* bacteria, Itakura and his team artificially developed and cloned a somatostatin gene. Somatostatin was an ideal choice for their experiment because it is not very toxic or difficult to handle. The hormone was frequently employed to treat diseases such as diabetes, pancreatitis, and acromegaly. Their method made the production of large quantities of synthetic somatostatin not only possible but also commercially viable. Additionally, Itakura's work allowed medical researchers artificially to develop other peptides, such as insulin, quickly and economically. Itakura's method, in fact, was 10 times quicker than earlier systems. The work of Itakura and his colleagues enabled scientists to develop gene mutations and guide DNA sequencing more effectively.

Itakura has continued to work at the City of Hope National Medical Center. In 1980 he became senior research scientist, and in 1989 he was named the director of the genetics laboratory. Itakura received the David Rumbough Scientific Award in 1979 from the Juvenile Diabetes Foundation for his contributions in the development of synthetic peptides. He joined the New York Academy of Sciences in 1991. An avid athlete, Itakura has participated in such activities as running, soccer, skiing, and tennis in his spare time.

J

Jackson, Shirley Ann

(1946–)
American
Physicist

Shirley Jackson, the first black woman to earn a Ph.D. from the Massachusetts Institute of Technology, has researched solid-state physics at AT&T Bell Laboratories and has headed the federal Nuclear Regulatory Commission (NRC) since 1995. She was born on August 5, 1946, in Washington, D.C., the second daughter of George and Beatrice Jackson. Her father encouraged her interest in science by helping her with school science projects.

Jackson attended college at the Massachusetts Institute of Technology, earning a B.S. in 1968 and a Ph.D. in 1973 in theoretical elementary particle physics. After research on strongly interacting subatomic particles at the Fermi National Accelerator Laboratory and the European Center for Nuclear Research, she joined AT&T Bell Laboratories in Murray Hill, New Jersey, in 1976 and remained there until 1991. There she did research in theoretical, solid-state, quantum, and optical physics, especially on the behavior of subatomic particles in solid material. From 1991 to 1995, in addition to serving Bell Labs as a consultant in semiconductor theory, she was a professor of physics at New Jersey's Rutgers University. Jackson has won honors including the New Jersey Governor's Award in Science and MIT's Karl Taylor Compton Award, and she has been elected a fellow of the American Physical Society and of the American Academy of Arts and Sciences. She is married to the physicist Morris A. Washington and has one son, Alan.

Jackson was a member of the New Jersey Commission on Science and Technology in the late 1980s and early 1990s. In 1995 President Bill Clinton appointed her to head the Nuclear Regulatory Commission (NRC). She was the first woman and first African American to chair this federal agency. The NRC oversees safety in the nuclear industry, including preventing accidents at nuclear power plants and ensuring safe disposal of nuclear waste. "My job is public service at a very high level," Jackson says.

Ten months after Jackson took over the NRC a whistle-blower revealed many violations of NRC rules at the nuclear division of Northeast Utilities, which operates nuclear power plants in New England, and accused the NRC of failing to enforce safety codes. Jackson admitted that "there is truth in . . . those charges" and called the scandal "a wake-up call." She vowed to toughen up her agency, stating, "We must demonstrate vigilance, objectivity and consistency." Since then, in addition to shutting down several Northeast Utilities plants, Jackson has added eight other plants to the NRC's "watch list" of places with possible problems. Bill Megavern, director of the Critical Mass Energy Project at Ralph Nader's Public Citizen group, in 1997 characterized Jackson as "the toughest [NRC] chairman we've seen."

Jacquard, Joseph-Marie

(1752–1834)
French
Inventor

Joseph-Marie Jacquard was an inventor whose innovation had much wider application than he had originally intended. He invented the Jacquard loom as a means of expediting the arduous task of hand-looming intricate patterns; instead, Jacquard programmed the pattern into a series of punch cards that symbolized the pattern through holes punched in the card. This hole functioned as a type of on/off switch, telling the loom whether or not to weave. This system of communicating with machines grew into the binary system of 0s and 1s, indicating to the machine whether to switch on or off. This system lent itself perfectly to the use of transistors, which function as on/off switches. The binary system evolved into the most basic language of computers.

Jacquard was born in 1752 in Lyon, France. He apprenticed in several different trades—bookbinding, cutlery making, and typefounding—before committing himself to the trade of weaving. Jacquard based this decision mostly on the fact that he had just inherited a small weaving business from his family. He entered the profession wholeheartedly, devoting himself to producing the most intricate and aesthetically pleasing designs. However, weaving these complex designs proved to be time-consuming, making the business a losing proposition. Jacquard's weaving business failed, and he moved back to cutlery making, but his mind remained fixed on weaving.

By 1801 Jacquard had devised a means of increasing the efficiency of the weaving process exponentially: by replacing the weaver with an automated system. Jacquard's system relied on punch cards whose holes corresponded to weaving instructions. However, this system reduced the role of weavers in the weaving process from hand manufacturers to mere designers of a product manufactured by machine. Weavers of the day revolted against their potential obsolescence by destroying Jacquard looms, but once the technology became available, it was impossible to stay its growth into the position previously occupied by human workers. Not only did this invention help incite the Industrial Revolution, but it also presaged the technological revolution of more than a century later.

Between 1801 and 1804 Jacquard improved upon the design and construction of his looms as well as perfecting his punch card system at the Paris Conservatoire des Artes et Métiers. His demonstration of the Jacquard loom and its attendant punch card system in Paris in 1804 so impressed Napoleon that the dictator awarded Jacquard a medal, a patent, and a pension. By 1812 there were 11,000 Jacquard looms in France alone, and they were spreading into new countries.

Jacquard died in 1834, but the significance of his work lives on. His invention of the punch-card system acted as a harbinger of the advent of computers under the binary system. In fact, this application had already been realized in the 1830s by CHARLES BABBAGE, who theorized and built a calculating machine utilizing punch cards. Though Jacquard's invention was extremely progressive theoretically, its immediate practical implications were less positive, as the machine put many weavers out of work and foretold the future mechanization of the labor process.

Jansky, Karl Guthe

(1905–1950)
American
Radio Engineer

Though little celebrated during his lifetime, Karl Guthe Jansky was responsible for founding the field of radio astronomy through his discovery of radio waves from an extraterrestrial source. Radio astronomy allowed astronomers to study the universe to a greater extent than had been possible before the detection of astronomical radio sources. Jansky unfortunately did not have the opportunity to further his studies of radio waves because of a lack of financial and scientific support.

Born on October 22, 1905, in Norman, Oklahoma, two years before it became a state, Jansky grew up in an academic household. Jansky's father, Cyril Jansky, was a professor of electrical engineering who later became the head of the School of Applied Science at the University of Wisconsin. One of six children, the younger Jansky was named after one of his father's college professors from the University of Michigan.

Jansky attended the University of Wisconsin and studied physics. He also played on the ice hockey team and wished to join the Reserve Officer's Training Program. Jansky was deterred when he was diagnosed with a chronic kidney disease known as Bright's disease. After earning his degree in 1927, Jansky remained at the University of Wisconsin for an additional year with plans to earn his master's degree. He left in 1928 without completing his thesis and secured a job with Bell Communications Laboratories. Being hired proved a challenge, as the company believed his disease might interfere with his work. With help from his older brother, who had contacts at the company, Jansky was hired and placed in New Jersey, considered to be a less stressful location than the headquarters in New York. It was in New Jersey that Jansky made his breakthrough discovery of radio waves.

In 1931 Jansky was assigned the task of determining the various forms of static and interference that were disrupting telephone communications. The interference was in the form of hissing, crackling, clicking, and banging noises. Jansky first built an antenna to help identify the sources of disturbance. The antenna was directional, enabling it to receive a greater range of wavelengths than standard antennas of the time. The receiver Jansky made produced very little static, making it easier for Jansky to measure static from outer sources. Additionally he developed a device designed to record the variations in static. Jansky set up the equipment in the rural area of Holmdel, New Jersey. The antenna was mounted in such a way that it could rotate and thus scan the entire sky. Jansky soon determined that thunderstorms were the cause of much of the static, but a steady hiss-type static could not be identified.

Jansky recorded and tracked the intensity of the hiss-type static and observed that it was most intense every 23 hours, 56 minutes, which corresponded with the amount of time required for a complete rotation of the Earth in relation to the stars. Jansky was able to show that the static was coming from the direction of the center of the Milky Way and theorized that it was caused by interstellar ionized gas. Jansky's discovery of radio waves from an astronomical source was a major advance during a time when radio waves generated by a source beyond the Earth had not even been considered. In late 1932 Jansky announced his discovery in a report presented to the Institute of Radio Engineers. His discovery created little commotion, largely because astronomy was an optical endeavor and knowledge of radio measurements was minimal.

Though Jansky attempted to continue his research of radio waves, Bell assigned him to a different project, and later attempts to study the phenomenon were disrupted by various circumstances. Jansky died of complications associated with Bright's disease in 1950 at the young age of 44. Surviving Jansky were his wife, Alice, whom he married in 1929, and two children. Jansky never received recognition for his pioneering contributions to radio astronomy while he was alive, but the significance of radio waves came to light after World War II. The jansky, the unit of radio emission strength, was named in his honor in 1973.

Jeans, Sir James Hopwood

(1877–1946)
English
Astronomer

Though trained in mathematics and physics, Sir James Hopwood Jeans was best known for his work in astronomy. Jeans was the first to suggest that matter is continually created throughout the universe and pioneered a number of ideas furthering the disciplines of astronomy and astrophysics. He also introduced the tidal theory of the origin of the solar system. A prolific writer, Jeans was author of a number of widely read, popular books and textbooks.

The son of William Tullock Jeans, a London journalist who wrote two books about scientists, Jeans was born on September 11, 1877, in Ormskirk, located in Lancashire, England. When Jeans was three years old, the family moved to London. Jeans did not have a particularly happy childhood and had a shy, standoffish personality. A bright and skilled boy, Jeans spent much time in his early years taking apart and reassembling clocks; at the age of nine Jeans wrote a pamphlet about clocks.

In 1896 Jeans entered Cambridge University to study mathematics. He did extremely well on the university mathematics exams, earning a Smith's Prize in 1900 for his top score, and was elected a fellow in 1901. Jeans received his master's degree in 1903. A year later he published his first treatise, *Dynamical Theory of Gases*, which later became a standard textbook because of its precision and style. Much of Jeans's own research was incorporated into the publication.

From 1905 to 1909 Jeans lived in the United States and taught mathematics at Princeton University. There he completed two textbooks, one on theoretical mechanics and another on the mathematical theory of electricity and magnetism. Jeans returned to Cambridge in 1910 and was Stokes Lecturer in applied mathematics for two years. After 1912 Jeans dedicated himself to conducting research and publishing his findings; *Report on Radiation and the Quantum Theory* advanced early quantum theory. Jeans later became more interested in astronomy; in 1923 he became a research associate at the Mount Wilson Observatory in California, where he remained until 1944. His first wife, the American Charlotte Tiffany Mitchell, died in 1934, and in 1935 he married Suzanne Hock, a concert organist. Jeans had four children.

The development of the tidal theory resulted from his investigation of the fragmentation of rapidly spinning bodies succumbing to centrifugal force. Jeans's research led him to reject the theory of Marquis PIERRE-SIMON DE LAPLACE, who proposed that the Sun and planets condensed from a single cloud of gas and dust. With Harold Jeffreys, Jeans introduced the tidal theory, which had first been proposed by an American geologist, Thomas C. Chamberlin. The updated theory suggested that a star had narrowly avoided collision with the Sun. As the star passed by the Sun, it drew stellar debris away from it. This debris then condensed to form the planets.

Jeans also studied other astronomical phenomena, including spiral nebulae, double stars, giant and dwarf stars, and the source of stellar energy, and he applied mathematics to thermodynamic and radiant heat problems. The tidal theory was eventually discarded in the 1940s, when a revised version of Laplace's nebular theory came into favor. Jeans received a number of honors, including the Royal Medal of the Royal Society in 1919, the Gold Medal of the Royal Astronomical Society in 1922, and the Franklin Medal of the Franklin Institute in 1931. Jeans was knighted in 1928.

Jemison, Mae Carol

(1956–)
American
Physician

Mae Jemison has taken her skills as a physician around the world and was the first African-American woman astronaut. She was born in Decatur, Alabama, on October 17, 1956, to Charlie Jemison, a roofer, carpenter, and maintenance supervisor, and Dorothy, a teacher, but grew up in Chicago. From childhood Mae planned to be a scientist. To her, as she wrote in *Odyssey* magazine, being a scientist "meant that I wondered about the universe around me and wanted to devote a significant part of my life to exploring it." She also was determined to be an astronaut, even though—until she was almost through college—all astronauts, like most scientists, were white and male.

Jemison entered California's prestigious Stanford University at age 16 and graduated in 1977 with a B.S. in chemical engineering and a B.A. in African and Afro-American Studies. When asked why she studied such different fields, she told the interviewer, Maria Johnson, "Someone interested in science is interested in understanding what's going on in the world. That means you have to find out about social science, art, and politics."

In 1981 Jemison earned her M.D. from Cornell University Medical College in New York City. While still in medical school she began working in Cuba, Kenya, and a Cambodian refugee camp in Thailand, and in 1983 she joined the Peace Corps to continue her overseas work. She was the medical officer for the West African nations of Sierra Leone and Liberia for two and a half years.

When Jemison returned to the United States in 1985, she began working for CIGNA, a health maintenance organization in Los Angeles. She also applied to the National Aeronautics and Space Administration (NASA) to become an astronaut candidate. In June 1987 she learned that she was 1 of 15 people accepted from among 2,000 applicants. After a year of grueling training she was qualified as a "mission specialist" (scientist) astronaut.

Jemison achieved her dream of going into space on September 12, 1992, as one of a seven-person crew aboard the shuttle *Endeavour.* During her eight days in space as part of a U.S./Japan joint mission called Spacelab J, she fertilized frog eggs and found that the resulting embryos developed into normal tadpoles under weightless conditions. She also designed and carried out an experiment to study calcium loss from astronauts' bones during their time in space. Jemison has emphasized that space travel "is a birthright of everyone who is on this planet" and that space and its resources "belong to all of us." She told Constance Green, "This is one area where we [African Americans and women] can get in on the ground floor and . . . help to direct where space exploration will go."

Mae Jemison resigned from the astronaut program in March 1993 and formed a Houston company called the Jemison Group, which develops advanced technology for export to developing nations. In 1998 she was still the company's president as well as serving as a professor in environmental studies and as the director of the Jemison Institute for Advancing Technology in Developing Countries at Dartmouth College. The institute researches, designs, implements, and evaluates "cutting-edge" technology to ensure that it works to the benefit of people in developing countries.

Jemison's honors include the Kilby Science Award and induction into the Women's Hall of Fame. The Mae Jemison Academy, a Detroit school that focuses on science and technology, is named after her. In 1998 Jemison wrote that her aim was to "focus on the beneficial integration of science and technology into our everyday lives—culture, health, environment and education—for all on this planet."

Jenner, Edward

(1749–1823)
English
Physician

Though Edward Jenner was elected a fellow of the Royal Society in 1789 for his studies of the cuckoo, he is best known for his discovery of the smallpox vaccine. Considered the founder of immunology and a pioneer of virology, Jenner was responsible not only for introducing the practice of vaccination but also for coining the term *virus.* A tireless advocate of the cause of vaccination, Jenner saved innumerable human lives from the deadly smallpox virus.

Born on May 17, 1749, in Berkeley, a small town in Gloucestershire, England, Jenner was the youngest of six children. He grew up in a family of clergy. His father, Stephen Jenner, was rector of Rockhampton and vicar of Berkeley. Jenner's mother was the daughter of a former vicar of Berkeley. When Jenner's parents died, he was only five years old and was raised by an older brother, who had become rector of Rockhampton. As a young boy Jenner gained a love of nature and spent hours exploring the countryside for fossils.

It was standard practice for surgeons to acquire training and education through apprenticeships, and thus Jenner, at the age of 13, was apprenticed to a nearby surgeon. After his eight-year apprenticeship Jenner went to London in 1770 to study anatomy and surgery under John Hunter, who later became a prominent surgeon. Hunter instilled in Jenner an interest in biological phenomena, keen observational skills, and a trust in experimental investigation. The two developed a deep and long-lasting friendship.

In 1773 Jenner returned to Berkeley and opened a medical practice. Popular and successful, Jenner found time to participate in medical organizations, play music, and, an ardent naturalist, make observations in natural history. In particular, he made significant discoveries by studying the migration of birds and observing the nesting habits of cuckoos. In 1788 he married Katherine Kingscote; the couple had four children.

Jenner was equally busy and observant in his medical practice, and he soon began to uncover the key to smallpox prevention. Smallpox was common in the 18th century, and sporadic intense outbreaks resulted in a high fatality rate. Inoculation was the only preventative measure available at the time, and it consisted of infecting a healthy person with matter from smallpox pustules taken from a patient suffering from a mild case of the disease. In this manner the healthy individual would acquire a mild case of smallpox and build up an immunity against subsequent infection. The problem, unfortunately, was that inoculation was not foolproof—all cases were not mild and occasionally the inoculated patient died. Jenner began to inoculate patients against smallpox and observed that some patients appeared to be immune to the disease. Upon further investigation he learned that these patients had previously been afflicted with cowpox, a disease contracted from cattle. Jenner then concluded that cowpox could prevent smallpox.

To test his new theory, Jenner inoculated a healthy young boy with matter taken from a patient's fresh cowpox lesions. The boy developed a slight fever and a small lesion. Two months later Jenner inoculated the boy with smallpox matter, to which the boy had no reaction. Jenner outlined his findings in a paper and submitted it to the Royal Society in 1797, but it was rejected. The following year Jenner printed at his own expense a small publication describing his findings.

The practice of vaccination spread rapidly after the publication of Jenner's work, and the death rate from smallpox decreased. Jenner devoted much of his life to supporting the cause of vaccination. He was awarded an honorary M.D. degree from the University of Oxford in 1813, and in the early 1800s the British government granted him £30,000 for his discovery. After his wife's death, in 1815, Jenner remained in Berkeley and returned to his studies in natural history.

Joliot-Curie, Frédéric

(1900–1958)
French
Nuclear Physicist

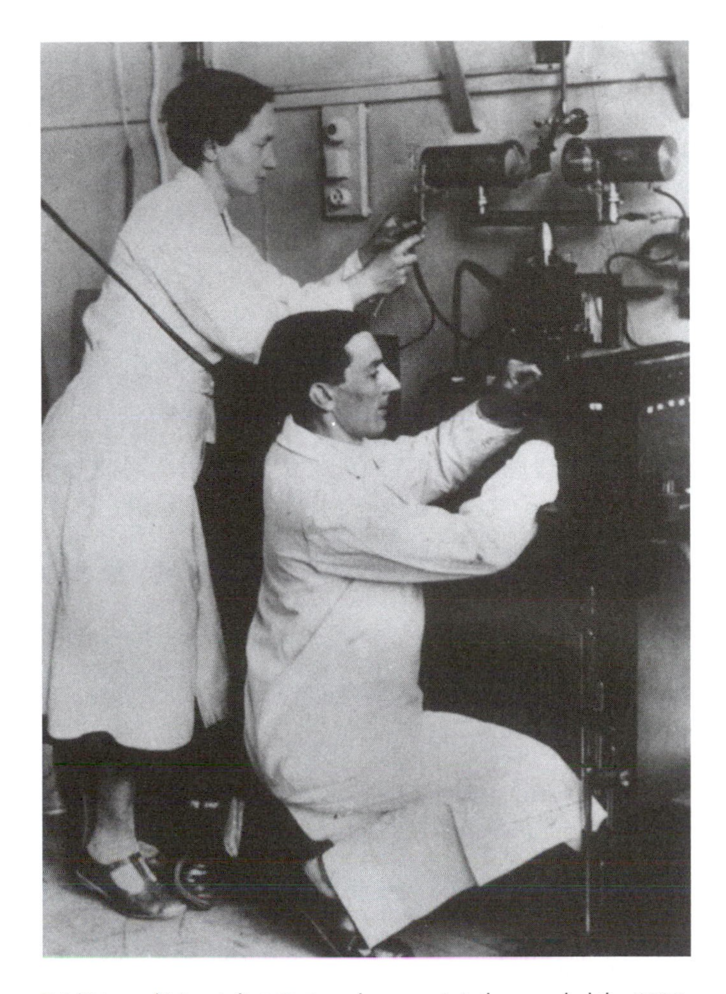

Frédéric and Irène Joliot-Curie, who were jointly awarded the 1935 Nobel Prize in chemistry for their discovery of artificial radioactivity. *Société Française de Physique, Paris, courtesy AIP Emilio Segrè Visual Archives.*

Frédéric Joliot-Curie was a member of a prominent family of scientists—his mother-in-law was MARIE SKLODOWSKA CURIE and his father-in-law was PIERRE CURIE. In 1935 he and his wife, IRÈNE JOLIOT-CURIE, were awarded the 1935 Nobel Prize in chemistry for their pioneering discovery of artificial radioactivity. Joliot-Curie made advances in nuclear fission studies and was the director of France's Atomic Energy Commission, which was established after World War II.

Heavily influenced by his liberal and active family, Joliot-Curie was born Jean-Frédéric Joliot on March 19, 1900, in Paris, France, the youngest of six children. His father, Henri Joliot, was a dry goods merchant who held leftist political views. He also loved the outdoors, a passion that would greatly influence his son. Joliot-Curie's mother, Emilie Roederer, was from a liberal family as well, so he was introduced to progressive social ideas from an early age. The family did not attend church, and Joliot-Curie remained a lifelong atheist and political leftist.

In 1920 Joliot-Curie entered the School of Industrial Physics and Chemistry to study engineering. He was influenced strongly by the school's director, the physicist Paul Langevin, who recognized in Joliot-Curie an aptitude for scientific research and encouraged him to take a position as a research assistant at the Radium Institute at the University of Paris after graduation. Joliot-Curie became assistant to the Nobel Prize winner MARIE SKLODOWSKA CURIE in 1925. Madame Curie instructed Joliot-Curie to learn the basics of radioactivity under Irène Curie, the Curies' oldest daughter, who worked at the institute. Though their personalities were quite different, the two married in 1926 and combined their last names to carry on the scientific Curie family line.

Frédéric and Irène did not begin joint research until 1931, a year after Frédéric earned his doctorate. The pair began conducting research on alpha particles to explore further the findings of the German physicists Hans Becker and WALTHER WILHELM GEORG BOTHE, who discovered that strong radiation was sent out from some of the lighter elements when they were bombarded with alpha rays. Frédéric and Irène started bombarding nuclei of various elements, including aluminum, boron, and magnesium, with alpha particles and analyzed the results. After a period of irradiation the pair would remove the source of alpha rays. What resulted were radioactive isotopes of elements that were not typically radioactive. Frédéric and Irène had thus artificially produced radioactive isotopes. They announced their discovery to the Academy of Science in 1934 and a year later won the Nobel Prize in chemistry.

In 1937 Joliot-Curie became a professor at the Collège de France and continued his work in radiation. Joliot-Curie found that a nucleus of a radioactive element could be split into two smaller nuclei of comparable mass with a discharge of energy emissions. It was believed that this phenomenon, known as nuclear fission, was capable of occurring rapidly in a chain reaction. If the prediction was correct, nuclear fission would have to be controlled in some way. This finding ultimately led to the development of atomic energy as well as the creation of the atomic bomb.

During World War II Joliot-Curie became more politically active and joined the French Communist Party in 1942. Arrested a few times, he was compelled to send his wife and two children to Switzerland for safety as he went underground. After the war, in 1946, Joliot-Curie became involved in the establishment of the Atomic Energy Commission. As commissioner, he was instrumental in building a major nuclear research center. Joliot-Curie lost his position in 1950, largely because of his leftist activities.

Joliot-Curie and his wife began to suffer from poor health due to their extended exposure to radiation, and his wife died in 1956. Joliot-Curie took over as head of the Radium Institute and carried on his wife's labors to establish a new physics laboratory south of Paris. Joliot-Curie died in 1958 at the age of 58.

Joliot-Curie, Irène

(1897–1956)
French
Physicist, Chemist

Daughter of the Nobel Prize winners PIERRE CURIE and MARIE SKLODOWSKA CURIE, Irène Joliot-Curie won her own Nobel Prize (with her husband, FRÉDÉRIC JOLIOT-CURIE) for discovering artificial radioactivity. She was born in Paris on September 12, 1897, just as her parents were beginning their ground-breaking research on radioactivity. She was their first child. Marie Curie doted on the baby, calling her "my little Queen," but she was often busy in the laboratory, so Irène's grandfather, Eugène Curie, took over most of the child's care.

World War I broke out in 1914, when Irène was just 17 years old. Marie organized a fleet of wagons (nicknamed "little curies") to carry portable X-ray equipment to battlefields, and Irène traveled to hospitals near the battlefront to teach surgeons how to use the machines. Neither Marie nor Irène questioned the young woman's fitness for this mission. "My mother had no more doubts about me than she doubted herself," Irène said later.

When Marie Curie's new Radium Institute opened in Paris after the war, Irène became her mother's laboratory assistant. Studying chemistry, physics, and mathematics, she published her first research paper in 1921. She did her Ph.D. project on the alpha particles (nuclei of helium atoms) given off by the nuclei of polonium atoms as they broke down. She received her degree from the Sorbonne in 1925.

That same year Irène Curie met another laboratory assistant at the Radium Institute, a young army officer named Frédéric Joliot. The charming, outgoing Joliot was very different from Curie in personality, but they shared many interests. Fred, as everyone called him, later said of Irène, "I discovered in this girl, whom other people regarded . . . as a block of ice, an extraordinary person, sensitive and poetic." They married on October 9, 1926, and thereafter both used the hyphenated last name *Joliot-Curie.* They had a daughter, Hélène, born in 1927, and a son, Pierre, born in 1932.

Much of the Joliot-Curies' research at the Radium Institute resulted in frustrating near-misses of great discoveries. In 1931, for instance, they showed that when beryllium, a lightweight metal, was bombarded with alpha particles from polonium, it gave off powerful rays that could make protons burst at high speed from the atomic nuclei in paraffin wax. They concluded that the rays were a new type of gamma ray, the most powerful form of atomic radiation then known. When the British physicist Sir JAMES CHADWICK repeated their experiment, however, he realized that the rays included a new kind of massive subatomic particle, which he called a neutron. He later won a Nobel Prize for this insight.

Finally, however, it was the Joliot-Curies' turn for success. Early in 1933 they found that when they placed polonium beside aluminum foil, neutrons and positrons (electrons with a positive charge—another particle they had just missed discovering) flew out of the foil instead of the protons they expected. Furthermore, when they moved the polonium away, the positrons kept streaming from the aluminum. A Geiger counter, which detects radioactive particles, ticked for several minutes when placed beside the aluminum. Somehow the alpha particles from the polonium had made the aluminum radioactive.

The Joliot-Curies concluded that the nuclei of the aluminum atoms had absorbed alpha particles from the polonium, ejecting a neutron in the process and changing to a radioactive form of phosphorus that did not exist in nature. The phosphorus nuclei broke down in a few minutes, emitting a positron and changing to a stable, nonradioactive form of silicon. The couple had created the first artificial radioactive isotope, or variant form of an element. Irène used her chemical expertise to devise a test that proved that the short-lived phosphorus actually existed.

The aging Marie Curie was thrilled when her daughter and son-in-law showed her their experiment early in 1934, about the time they published their results. She was sure they would win a Nobel Prize for it, and she was right, though she did not live to see their triumph. The Joliot-Curies received the chemistry prize in 1935, when Irène was 37 years old. Their work led to the creation of radioactive isotopes of many other elements, which proved useful in physics, chemistry, industry, and medicine. The Joliot-Curies also won other prizes for their discovery, including America's Bernard Gold Medal for Meritorious Service to Science (1940), the Henri Wilde Prize, and the Marquet Prize of the French Academy of Sciences.

Although they remained as close as ever in private life, the Joliot-Curies stopped doing research together after they won the Nobel Prize. Fred became a professor at the Collège de France, the country's foremost research institution. Irène, meanwhile, continued to direct research at the Radium Institute, as she had since 1932, and also became a professor at the University of Paris.

When the Popular Front was elected to political power in France in 1936, it asked Irène to become the undersecretary of state for scientific research. She thus became one of France's first woman cabinet ministers. By prearrangement she gave up the post after three months and returned to her beloved laboratory, where she continued to work in spite of ill health caused by tuberculosis and radiation exposure.

Irène Joliot-Curie's research included one more important near-miss, which, if it had succeeded, might have allowed her to duplicate her mother's record of two Nobel Prizes. Like several other eminent scientists in the late 1930s, she studied what happened when neutrons bombarded uranium, the heaviest natural element. All expected that when the neutrons penetrated the uranium atoms' nuclei, they would create artificial elements, probably short-lived and radioactive, that were heavier than uranium. In 1938, therefore, when Joliot-Curie detected what appeared to be lanthanum, an element lighter than uranium, in the wake of such an experiment, she assumed she had made a mistake. The German scientists Otto Hahn and Fritz Strassmann also doubted their results when soon afterward they obtained another lightweight element, barium, from a similar experiment. They, however, had an advantage that Joliot-Curie did not have: the imagination of LISE MEITNER, who guessed after hearing of their experiment that they had split the nuclei into two parts.

Irène became the head of the Radium Institute in 1949. Her health had improved after the war when an antibiotic cured her tuberculosis, but in the mid-1950s she became ill again. Early in 1956 she learned that she had leukemia, the same cancer that had killed her mother. "I am not afraid of death," she told a longtime friend. "I have had such a thrilling life!" Irène Joliot-Curie died on March 17, 1956, at the age of 58.

Joule, James Prescott

(1818–1889)
English
Physicist

One of the leading experimentalists of his time, James Prescott Joule made his main contributions to science with the discovery of the first law of thermodynamics, the law of the conservation of energy, and his findings concerning the mechanical equivalent of heat. Joule also collaborated with Lord WILLIAM THOMSON KELVIN in the

James Prescott Joule, who discovered the first law of thermodynamics. *AIP Emilio Segrè Visual Archives, Physics Today Collection.*

formulation of the Joule-Thomson effect, which stated that the temperature of an expanding gas cooled if the gas did not perform external work.

The second of five children, Joule grew up in a wealthy brewing family. Born on December 24, 1818, in Salford, near Manchester, England, to Benjamin and Alice Prescott Joule, Joule was a shy and frail child. Joule and a brother were tutored at home, and from 1834 to 1837 the brothers learned mathematics and science from the chemist JOHN DALTON, known for his work on atomic theory. Joule developed an interest in physics at an early age and set up a laboratory near the brewery to conduct experiments.

Though Joule did not receive a formal education or a college degree, he made significant discoveries, most of them before the age of 30. At age 19 he began independent research under the guidance of William Sturgeon, an amateur scientist. Joule was influenced by Sturgeon's interest in electromagnetic theories. At about the same time Joule began to investigate the problems of heat, particularly the heat developed by an electric current. He found that the heat produced in a wire by an electric current was related to the current and resistance of the wire. Joule announced his findings in 1840 in the paper "On the Production of Heat by Voltaic Electricity."

From 1837 to 1847 Joule studied the mechanical equivalent of heat and other forms of energy and established the principle of conservation of energy. He systematically studied the thermal effects caused by the production and passage of current in an electric current, and in 1843 he determined the amount of mechanical work needed to produce a given amount of heat; his discovery was guided by precise experiments in which Joule measured the degree of heat produced by rotating paddle wheels, powered by an electromagnetic engine, in water. Joule presented his observations in 1849 to the Royal Society in "On the Mechanical Equivalent of Heat." A year earlier he had also written a paper on the kinetic theory of gases. The paper included the first estimation of the speed of gas molecules. During this productive decade Joule also married Amelia Grimes of Liverpool, in 1847. When she died in 1854, Joule was left to raise their two children.

Joule worked with Thomson from 1852 to 1859 on experiments in thermodynamics. Their most significant discovery was that an expanding gas's temperature cooled under certain conditions. This became known as the Joule-Thomson effect and provided the basis for the development of a large refrigeration industry in the 19th century.

In 1850 Joule was elected to the Royal Society and enjoyed broad recognition and a strong reputation. Joule continued to carry out experimental investigations, but his findings failed to match the accomplishments of his early years. Though Joule never received an academic appointment, his work on thermodynamics and the mechanical equivalent of heat was widely accepted and helped advance the sciences. The joule, a unit of energy, was named in his honor.

Jung, Carl Gustav

(1875–1961)
Swiss
Psychiatrist

The founder of analytical psychology, Carl Gustav Jung did work that greatly impacted the field of psychiatry. Jung developed the concepts of the extroverted and introverted personality types and proposed the theory of the collective unconscious, a part of the unconscious mind shared by all humans, which holds archetypes, including those of religion, morality, and science. Jung collaborated closely with the psychoanalyst SIGMUND FREUD for several years until disagreements over psychological theories forced them to part ways.

The religious atmosphere in which Jung was raised had a significant impact on his adult life. Born on July 26, 1875, in Kesswil, Switzerland, Jung was an imaginative and solitary child. His father, a pastor and philologist, had enjoyed a successful university career. Though his father began to have intellectual doubts regarding religion, he was adamant on the need for belief. This attitude eventually led Jung to reject formal religion. Jung's mother was traditional on the one hand and superstitious on the other and communicated her beliefs to Jung. An extremely bright child, Jung was able to read at a young age and began learning Latin at the age of six. His family expected him to join the clergy, but Jung chose to seek another career.

In 1895 Jung entered the University of Basel. Interested in a number of subjects, including zoology, paleontology, and geology, he decided to study medicine. In 1900 Jung joined the staff of the Burghölzli Asylum of the University of Zurich to work toward his medical degree, which he earned in 1902. One year later he married Emma Rauschenbach and the couple eventually had five children.

While at the Burghölzli Asylum Jung began to apply association tests to both normal and mentally ill individuals. Jung investigated the illogical responses to stimulus words and connected these responses to repression. Jung coined the term *complex* to describe these mental conditions. These studies earned Jung a strong reputation and led to a close collaboration with Freud, whose theories were confirmed by Jung's findings. Jung was considered the most likely candidate to succeed Freud as the most influential proponent of psychoanalysis, but in 1912 the relationship ended, in large part because Jung disagreed with Freud's emphasis on the sexual bases of neuroses.

An early achievement of Jung's was his classification of personality types as either extroverted or introverted. Later, Jung categorized four functions of the mind as thinking, feeling, sensing, and intuiting. Jung's theory of the collective unconscious arose after years of self-analysis in which he studied his dreams, visions, and fantasies. He believed that his strange experiences arose from a part of the unconscious mind he termed the *collective unconscious,* which was the consequence of ancestral experience. The collective unconscious was made up of archetypes, which were instinctive patterns with a universal character.

Jung continued to develop his ideas and form psychotherapeutic methods. He introduced a process he called *individuation,* which involved working out layers of psychological conflict to unify the self. In treating patients who felt their lives had lost meaning, for example, Jung prompted the patients to examine their dreams and imagination to discover their own myths, which in turn would provide meaning to their existence.

From 1933 to 1941 Jung was a psychology professor at the Federal Polytechnical University in Zurich, and in 1943 he became professor of medical psychology at the University of Basel. Until World War II Jung traveled extensively, delivering lectures and studying primitive cultures. Jung was a great pioneer in furthering psychoanalysis and the study of the human mind.

K

Kastler, Alfred

(1902–1984)
French
Physicist

Alfred Kastler developed two different methods of observing Hertzian resonances within atoms by atomic excitation—double resonance in spectroscopy, and optical pumping—which subsequently proved important to the invention of the maser and the laser. Kastler received the 1966 Nobel Prize in physics for these innovations.

Kastler was born on May 3, 1902, in the Alsatian village of Guebwiller, Germany (now France). His parents were Anna Frey and Frederic Kastler. Kastler married Élise Cosset on December 24, 1924, and together the couple had three children: two boys, Daniel and Claude-Yves, who became teachers, and one girl, Mireille, who became a physician.

In 1920 Kastler started study at the École Normale Supérieure of the University of Paris thanks to a dispensation extended to residents of the newly annexed province of Alsace, though he had failed the entrance examination. He graduated in 1926 with a teaching certificate, which he used to secure teaching jobs at lycées in Mulhouse, Colmar, and Bordeaux. In 1931 he returned to university as a research assistant at the University of Bordeaux. While there he wrote his dissertation on the excitation of mercury atoms to earn the degree of docteur des sciences physiques in 1936.

Between 1936 and 1938 Kastler served as a lecturer in physics at Clermont-Ferrand University. He returned to the University of Bordeaux as a professor of physics in 1938. In 1941 he returned to the École Normale Supérieure as the director of the Hertzian spectroscopy group and as an assistant professor. In 1945 the University of Paris promoted him to the position of full professor. Kastler stayed there for more than a quarter of a century, until he took up the director of research position at the National Center of Scientific Research.

On May 30, 1950, Kastler and his collaborator (and former student) Jean Brossel announced their development of double resonance to a meeting of la Société Français de Physique. The method was named for the dual application of a light beam to excite atoms to a higher energy level, and then a radiofrequency field to return these atoms to their original energy level, emitting the energy they'd absorbed in the process as light, which Kastler could then analyze. Kastler developed a related technique, called optical pumping, within a few months. With this method he shined polarized light on groups of atoms, again to excite them into a higher energy level. However, some sublevels of atoms do not absorb polarized light and thus remain stable. Once the excited atoms return to equilibrium, they return to sublevels that absorb as well as to the levels that do not absorb the light. This method of atom excitation was instrumental in the development of the maser, the laser, the atomic clock, and the magnetometer. In France, Kastler was called the "grandfather of the laser."

Besides contributing to scientific understanding, Kastler contributed to the peace movement by actively opposing the war in Vietnam and the French occupation of Algeria. Kastler died on January 7, 1984, in Bandol, France.

Kato, Tosio

(1917–)
Japanese/American
Mathematical Physicist

Tosio Kato specialized in the perturbation theory, or the study of systems that deviate slightly from the ideal, as do many of the real-life systems that physicists study. Kato specifically focused on the relation of the theory to linear operators, or functions. Kato distinguished himself as a prolific writer, contributing hundreds of articles to professional journals.

Kato was born on August 25, 1917, in Tochigiken, Japan. His mother was Shin Sakamoto and his father was Shoji Kato. In 1941 he earned his bachelor's degree from the University of Tokyo, commencing a relationship that would last two more decades. After a year away Kato rejoined the University of Tokyo in 1943 to start his teaching career while also pursuing a doctorate there. In 1944 he married Mizue Suzuki. In 1951 he became a doctor of science, and the same year he became a full professor. In 1962 he immigrated to the United States to accept a professorship at the University of California at Berkeley from which he retired in 1989.

In 1949, while still working on his doctorate, Kato published one of his early important papers, "On the Convergence of the Perturbation Method, I, II" in *Progressive Theories of Physics*. In 1966 he gathered together the knowledge he had accumulated thus far into a comprehensive survey of his focal topic, entitled *Perturbation Theory for Linear Operators*. He had started this text while still in Japan, but it wasn't until after he had moved to the United States that he could finalize the book for publication. Kato later condensed this text into a more concise statement, incorporating the development of the theory in the meantime in *A Short Introduction to Perturbation Theory for Linear Operators*.

The perturbation theory had its roots in the work of John Rayleigh and ERWIN SCHRÖDINGER, conducted in the 1920s. In advancing the theory Kato demonstrated that Schrödinger operators are symmetric, thus simplifying calculations in quantum mechanics. Earlier Kato had laid the mathematical foundation for the theory through analysis and function theory; Kato later studied these operators by researching their spectral properties, that is, how the sets can be split just as light can be diffused into separate colors of the spectrum. Kato's work in the United States focused on applying functional analysis to solve problems in hydrodynamics and to figure out evolution equations.

The significance of Kato's work was recognized early in his career in Japan by the 1960 Asahi Award. However, the United States was slower to recognize it: He won the Norbert Wiener Prize for applied mathematics jointly from the American Mathematical Society and the Society for Industrial and Applied Mathematics some two decades later, in 1980. Kato received the highest honor for his work in mathematics in the year of his retirement, 1989, when the University of Tokyo held the International Conference on Functional Analysis in Honor of Professor Tosio Kato.

Kelsey, Frances Oldham

(1914–)
Canadian/American
Physician

Only the "stubbornness" of the U.S. Food and Drug Administration medical officer Frances Oldham Kelsey prevented the epidemic of birth defects that swept Europe in the early 1960s from striking in the United States as well. Frances Oldham was born on July 14, 1914, to Katherine Oldham and Frank Oldham, a retired British army officer. She grew up in Cobble Hill on Vancouver Island, in the Canadian province of British Columbia. She received a bachelor's degree in 1934 and a master of science degree in 1935 from McGill University in Montreal.

Oldham studied pharmacology at the University of Chicago, obtaining a Ph.D. in 1938. She then joined the university's faculty. In 1943 she and another pharmacologist, Fremont Ellis Kelsey, discovered that the fetuses in pregnant rabbits' wombs could not break down a common drug that adult rabbits' bodies were able to detoxify. This was one of the first explanations for the fact that the effects of drugs on the unborn could be very different from those on adults.

Oldham and Kelsey married near the end of 1943. That decision cost Frances Kelsey her job, since the university, like many others, did not permit a husband and wife to work in the same department. Kelsey entered medical school instead, meanwhile giving birth to two daughters. She obtained her M.D. from the University of Chicago in 1950.

After eight years at the University of South Dakota, during which (in 1956) Frances became a U.S. citizen, the Kelseys moved to Washington, D.C. Frances Kelsey became a medical officer at the Food and Drug Administration (FDA), reviewing applications from drug companies that wished to market new medicines.

Kelsey's first application, a seemingly routine request from the respected William S. Merrell Company of Cincinnati, Ohio, arrived on September 8, 1960. Merrell asked permission to sell in the United States a drug already widely used in Europe to help people sleep, relieve anxiety, and ease the nauseating "morning sickness" that often plagued pregnant women. The drug's scientific name was *thalidomide*. Merrell claimed that thalidomide had had no demonstrated major side effects in either animals or the people who used it in Europe, but Kelsey rejected Merrell's application on November 10, on the grounds that the company had not proved the drug's safety. When she rejected Merrill's resubmission in January 1961, Merrell's representative complained to her supervisor, Ralph Smith, but Smith refused to overrule her.

Kelsey's reservations were justified a month later when she read a British medical report that thalidomide sometimes apparently caused polyneuritis, a nerve inflammation that could produce lasting damage. Concern about this side effect made her reject Merrell's application a third time at the end of March. Furthermore, remembering her earlier rabbit research, she asked the company to provide evidence that the drug was safe for mothers and fetuses if used during pregnancy.

Kelsey's worst fears were confirmed when she began to hear disturbing reports about unusual numbers of severely deformed babies born in Europe. The babies' hands and feet were often attached directly to their shoulders and hips like flippers, giving the defect the name of *phocomelia,* or "seal limbs." Many of the babies had severe internal defects as well. Doctors in Germany and Australia reported that most of the women who gave birth to these babies had taken thalidomide during their first three months of pregnancy. Faced with these reports, thalidomide's German manufacturer stopped making the drug in November, and Merrell, too, withdrew its FDA application. By then Kelsey had singlehandedly kept the drug off the American market for $14^1/_2$ months.

On August 7, 1962, President John F. Kennedy awarded Kelsey the Distinguished Federal Civilian Service Medal, praising "her high ability and steadfast confidence in her professional decision." More important in Kelsey's eyes, late in 1962, Congress passed a law barring companies from distributing new drugs for testing purposes, as Merrell had thalidomide, without FDA approval.

In that same year a new branch of the FDA was formed to oversee the distribution of "experimental" knowledge or drugs, and Frances Kelsey was put in charge of it. In 1967 she became director of a newly formed division that monitored the performance of animal and human studies conducted to determine the safety and effectiveness of drugs. The division also tracked the activities of institutional review boards and local committees that help to ensure the rights and welfare of human research subjects in drug trials. She continued this work until 1995, when she became the deputy for science and medicine at the FDA's Office of Compliance, a position she still held in 1998.

Kelvin, William Thomson, Lord

(1824–1907)
Irish
Physicist

William Thomson, Lord Kelvin, who introduced the absolute scale of temperature, which uses the kelvin as the unit of measure. *AIP Emilio Segrè Visual Archives.*

An engineer, mathematician, and physicist, William Thomson, Lord Kelvin was a progressive scientist with diverse interests who greatly influenced the scientific community of his day. Kelvin made numerous discoveries during his career, including the development of the conservation law of energy, the first law of thermodynamics. Kelvin also introduced the absolute scale of temperature, which used the kelvin as the unit of measure, and supervised the laying of the first transatlantic cable. A prolific writer, Kelvin published more than 600 scientific works and patented 70 inventions. His enthusiasm and zest for knowledge did much to advance scientific thought.

An unusually bright child, Kelvin was born William Thomson on June 26, 1824, in Belfast, located in the county of Antrim in Ireland (now Northern Ireland). His mother died when he was six years old, leaving behind a large family. Kelvin's father educated the children in mathematics, and they advanced to a high level. In 1834 Kelvin's father became a mathematics instructor at the University of Glasgow.

At the amazingly young age of 10 Kelvin entered the University of Glasgow. He published his first papers on mathematics at the ages of 16 and 17. A paper he wrote at 15, "An Essay on the Figure of the Earth," won an award from the university. Kelvin entered Cambridge University in 1841, and four years later he graduated with a bachelor's degree. He then went to Paris to gain practical experience in laboratory and experimental work. He returned to Glasgow and at the age of 22 became a professor of physics at the University of Glasgow, a position he would hold for 53 years. Although he married twice, Kelvin never had any children.

Kelvin's studies on electromagnetism advanced the field considerably. He, together with the physicist MICHAEL FARADAY, was responsible for introducing the concept of an electromagnetic field. Kelvin's mathematical analysis of electricity and magnetism laid the foundation for JAMES CLERK MAXWELL's electromagnetic theories. Also of importance were Kelvin's investigations in thermodynamics: In 1848 he developed the absolute scale of temperature and the concept of absolute zero, the temperature at which substances hold no thermal energy. Kelvin was one of the first to appreciate the findings of JAMES PRESCOTT JOULE regarding the mechanical equivalent of heat, and the two collaborated from 1852 to 1859. The outcome of their association was their postulation of the Joule-Thomson effect, which stated that the temperature of an expanding gas cooled under certain conditions. Through their many experiments Kelvin also developed one of the first versions of the second law of thermodynamics. He described his findings in the paper "On the Dynamical Theory of Heat," which was published in 1851.

In 1854 Kelvin began working on the project of laying transatlantic cables. He determined that the flow of electric current was equivalent to the flow of heat. The problems of sending electrical signals over long distances were solved by applying ideas of heat flow. In addition to supervising the project, Kelvin invented the mirror galvanometer, patented in 1858 as a long-distance telegraph receiver. His involvement with the transatlantic cable project made him wealthy.

Kelvin had numerous other interests, which he pursued with vigor throughout his long career. Interested in the ages of the Sun and the Earth, he calculated values for them. Kelvin also contributed to the theory of elasticity and the field of hydrodynamics and invented the flexible wire conductor and a gyrocompass. He was knighted in 1866 and in 1892 became Baron Kelvin of Largs. He was elected a fellow of the Royal Society in 1851 and its president in 1890. In 1902 he received the Order of Merit.

Kepler, Johannes

(1571–1630)
German
Astronomer

The German mathematician and astronomer Johannes Kepler is widely regarded as the founder of modern astronomy for his discovery of the laws of planetary motion. A strong advocate of the views of NICOLAUS COPERNICUS, who proposed that the planets revolved around the Sun, Kepler, through his laws of planetary motion, greatly furthered Copernicus's theories. Kepler worked for a number of years with the astronomer TYCHO BRAHE and completed the *Rudolphine Tables*, a large volume charting Brahe's findings. Throughout his life Kepler was concerned with the harmony of the universe.

Johannes Kepler, whose discovery of the laws of planetary motion greatly advanced modern astronomy. *AIP Emilio Segrè Visual Archives.*

Born on December 27, 1571, in Weil der Stadt, Germany, Kepler had a childhood that was rather turbulent. At the age of three he suffered from smallpox, which permanently damaged his eyesight and the use of his hands. That same year Kepler's father, Heinrich, joined a legion of mercenary soldiers and went to Holland to fight the Protestant uprising. This act disgraced the family. Kepler's father returned in 1576 only to leave soon thereafter to join the Belgian military service. In 1588 Kepler's father abandoned the family for good. Kepler's mother, Katharina Guldenmann, generally ill-tempered, had shown Kepler the comet of 1577, an event that was influential in his life. Years later she was accused of witchcraft.

Though destined for the clergy, Kepler entered the University of Tübingen in 1587 and studied science and mathematics. At the university he was introduced to Copernican astronomy by Michael Maestlin, an astronomy professor who was well versed in Copernican thought. Kepler was an outstanding student; he received his bachelor's degree in 1588 and his master's degree in 1591. Subsequently he commenced the theological program but left in his third and final year to become the instructor of mathematics at the Protestant seminary in Graz, Austria.

In Graz, Kepler taught mathematics and began issuing calendars that contained predictions for the coming years. These calendars enhanced Kepler's local status and supplemented his income considerably. He also issued horoscopes, although his feelings regarding astrology were mixed. In 1596 Kepler published his first book, *Mystery of Cosmography.* In it he detailed his belief in a mathematical harmony of the universe. The following year Kepler married the twice-widowed Barbara Müller. After her death in 1611 he married Susanna Reuttinger in 1613. Kepler had 12 children, only 4 of whom survived to adulthood.

In 1600 Kepler was invited to work with Tycho Brahe, the best observational astronomer of the time, in Prague. Brahe instructed Kepler to work out the orbit of Mars. Brahe died a year later, and Kepler succeeded Brahe as imperial mathematician. Twenty years' worth of Brahe's observations were left with Kepler, and he used these accurate observations to formulate his laws of planetary motion. Kepler was unable to fit Brahe's observations of Mars into the circular orbits prescribed by Copernican theory, and in 1609, he developed his first two laws: that the planets moved about the Sun in elliptical orbits, and that the line joining a planet to the Sun sweeps out equal areas in equal periods. In 1618 Kepler formulated the third law of planetary motion: that the squares of the sidereal periods are proportional to the cubes of their mean distances from the Sun.

Also of significance was Kepler's completion in 1627 of the *Rudolphine Tables,* a task handed down from Brahe. The work tabulated Brahe's results and contained, in addition to tables for calculating planetary positions and tables of refraction and logarithms, a catalog of more than 1,000 stars. In 1604 Kepler had published *Optics,* which included an explanation of the pinhole camera and improved refraction tables. Kepler also assisted modern optics by proposing the ray theory of light to explain vision. A book on the problems of measuring volumes of liquids in wine casks proved to be influential in the development of infinitesimal calculus.

Kepler's laws of planetary motion persuaded many to abandon the Ptolemaic view that the planets revolved around the Earth. His work also enabled Sir ISAAC NEWTON to develop his theory of gravitational force.

Khayyám, Omar

(c. 1050–c. 1123)
Persian
Astronomer, Mathematician

Perhaps best known in the 20th century for *The Rubáiyat of Omar Khayyám,* a collection of poetry, Omar Khayyám made contributions to mathematics, particularly in algebra and analytical geometry, that were revolutionary. Khayyám completed numerous treatises on mathematics and tackled EUCLID's theory of ratios and theory of parallel lines. Khayyám also published a number of philosophical works, proposed a theory of music, and studied astronomy. One of the leading Arab scholars of his day, Khayyám was quite famous in his lifetime as a result of his studies in mathematics and astronomy.

Little is known about Khayyám's personal life. It is believed he was born around 1050 in Nishapur, the provincial capital of Khurasan (now Iran). His name suggests that his father was a maker of tents. What is known, however, is that Khayyám grew up in a time of astronomical development. When he was a child, the explosion of a supernova, now know as the Crab Nebula, occurred; the event would have been observed by Khayyám. Twelve years later a luminous comet streaked across the sky, encouraging Khayyám to pursue astronomy.

Much of Khayyám's education took place in Nishapur, but it was likely he also traveled to several renowned educational institutions, including those at Samarkand, Ispahan, and Bukhara. Though it is possible Khayyám became a tutor, teaching would have allowed him little time to pursue scientific studies.

Khayyám's reputation began to build quickly, and the Seljuk sultan Malik Shah commissioned Khayyám to reform the solar calendar using astronomical observations. Khayyám then moved to Ispahan and built an astronomical observatory. He stayed at Ispahan for nearly 18 years, working with the best astronomers of the time. Under Khayyám's supervision a work of astronomical tables was produced. Khayyám also, as was his official task, proposed a plan for calendar reform in about 1079. Khayyám's calendar was based on a cycle of 33 years in honor of the sultan. The calendar proposed the average length of the year to be 365.2424 days, which deviated from the true solar calendar by 0.0002 day. A difference of 1 day thus accumulated over 5,000 years. With the Gregorian calendar the 1-day difference occurred in 3,333 years.

An accomplished mathematician, Khayyám made major contributions to the field. He developed a geometrical method for solving equations and solved cubic equations by intersecting a parabola with a circle. Khayyám also classified numerous algebraic equations according to their complexity and identified 13 different forms of cubic equations. He was the first to develop the binomial theorem and establish binomial coefficients. Especially noteworthy were Khayyám's commentaries on the Greek mathematician Euclid's theory of ratios and theory of parallel lines. Khayyám developed a new concept of number that included a divisible unit; the Greeks had thought of number as a collection of indivisible units. Khayyám was thus able to extend Euclid's work to include the multiplication of ratios.

Khayyám contributed to various other fields of science as well—he introduced a method for the accurate calculation of specific gravity and completed two books in metaphysics. Not all of his written works survived, but those that did covered such topics as physics, geometry, and algebra. A monument to Khayyám was constructed at his tomb in Nishapur in 1934.

Khorana, Har Gobind

(1922–)
Indian/American
Biochemist

Har Gobind Khorana won the 1968 Nobel Prize in physiology or medicine with MARSHALL NIRENBERG and Robert W. Holley for work showing how the synthesis of protein is controlled by genetic components of the cell nucleus. Khorana is widely recognized as the "author" of the "genetic code word dictionary," as he discovered the genetic codes of A, C, G, and T, which combine to create all codes passed through the genes. Khorana also developed an inexpensive method for synthesizing acetyl coenzyme A, which expedited biochemical research relying on this enzyme that had previously been very expensive.

Khorana was born on January 9, 1922, in Raipur, India (now Pakistan). He was the youngest of five children born to Krishna Devi and Ganpat Rai Khorana, a tax collector for the British government. In 1952 Khorana married Esther Elizabeth Sibler, and together the couple had three children—Julia Elizabeth, Emily Anne, and Dave Roy. Khorana became a citizen of the United States in 1966.

Khorana attended Punjab University in Lahore on a government scholarship, studying chemistry; he graduated with honors in 1943. Two years later Khorana graduated with honors again, this time earning his master of science degree in 1945. Khorana traveled to the University of Liverpool under a Government of India Fellowship to pursue his doctorate in organic chemistry. He received the Ph.D. in 1948, having researched the structure of the bacterial pigment violacein. Khorana then traveled to Zurich, where he conducted postdoctoral work on alkaloids, or organic bases, under Vladimir Prelog at the Swiss Federal Institute of Technology.

Between 1950 and 1952 Khorana conducted research on nucleic acids under Baron ALEXANDER ROBERTUS TODD on the Nuffield Fellowship at Cambridge University. In 1952 he continued his travels westward to take up the directorship of the British Columbia Research Council's Organic Chemistry Section at the University of British Columbia in Victoria. In 1960 he migrated south to become codirector of the Institute for Enzyme Research at the University of Wisconsin at Madison. There he became a professor of biochemistry in 1962, and in 1964 the university named him the Conrad A. Elvelijem Professor of Life Sciences. Khorana then moved east to take up the Alfred P. Sloan chair in biology and chemistry at the Massachusetts Institute of Technology.

In 1959 Khorana and his colleague John G. Moffat announced their synthesis of acetyl coenzyme A, which could previously be obtained only from yeast through a difficult and costly process. Thus Khorana and Moffat's artificial synthesis process revolutionized biochemical research and earned them international recognition in the scientific community. In 1964 Khorana followed up on his earlier work on nucleic acids by synthesizing parts of the nucleic acid molecule. He then demonstrated that genetic codes are transmitted by these nucleic acids in combinations of three "letters" and thereby duplicated all 64 possible combinations of the code molecules of nucleic acid. In 1970 Khorana synthesized a deoxyribonucleic acid (DNA) gene from yeast, the first artificial genetic synthesis. Khorana was also the first to synthesize *Escherichia coli*, or *E. coli*.

Khorana received the Laskar award and the American Chemical Society award. He was known for his disciplined work ethic, working as long as 12 years between vacations.

King, Mary-Claire

(1946–)
American
Geneticist

Some geneticists merely analyze deoxyribonucleic acid (DNA) in test tubes, but Mary-Claire King saw genetics as closely tied to politics. "I've never believed our way of thinking about science is separate from thinking about life," she told an interviewer, David Noonan, in 1990. She has used genetics to study human origins, discover why some people are more susceptible to breast cancer or acquired immunodeficiency syndrome (AIDS) than others, and identify the lost children of people murdered by repressive governments. As the reporter Thomas Bass has noted, "Any one of her accomplishments could make another scientist's full-time career."

Mary-Claire King was born in Wilmette, Illinois, a suburb of Chicago, on February 27, 1946. Her father, Harvey, was head of personnel for Standard Oil of Indiana. Clarice, her mother, was a homemaker. A childhood love of solving puzzles drew Mary-Claire to mathematics, which she studied at Carleton College in Minnesota; she graduated cum laude in 1966.

King then went to the University of California (UC) at Berkeley to learn biostatistics, or statistics related to living things. Eventually King went to work for Allan C. Wilson, who was trying to trace human evolution through genetics and molecular biology, in his molecular biology laboratory. Wilson assigned King to compare the genes of humans and chimpanzees. At first she thought she must be doing something wrong because "I couldn't seem to find any differences," but her results proved to be accurate. She proved that more than 99 percent of human genes are identical to those of chimpanzees. This startling research not only became the thesis that earned her a Ph.D. in genetics from Berkeley in 1973 but also was featured on the cover of *Science* magazine.

In 1975 King turned to a quite different aspect of genetics: the possibility that women in certain families inherit a susceptibility to breast cancer. Such women have the disease more often and at a much earlier age than average. Scientists at the time were discovering that all cancer grows out of damaging changes in genes, but usually those changes occur during an individual's lifetime and are caused either by chance or by factors in the environment, such as chemicals or radiation. The damaged genes involved in only a few rare cancers were known to be inherited. King eventually proved that about 5 percent of breast cancers are inherited.

In 1990 King localized the breast cancer gene, which she called BRCA1, at halfway down the lower arm of the 17th of the human cell's 23 chromosomes. By then she had become a professor of epidemiology at UC–Berkeley's School of Public Health (in 1984) and a professor of genetics in the university's Department of Molecular and Cell Biology (in 1989).

King's most unusual genetic project, tied to her lifelong concern for human rights, was helping to reunite families torn apart during the "dirty war" waged in Argentina between 1976 and 1983. During that time the country's military government kidnaped, tortured, or murdered an estimated 12,000 to 20,000 citizens. Babies born in prison or captured with their mothers were sold or given away, thus becoming lost to their birth families.

In 1977, while the military government was still in control, a group of courageous older women began gathering every Thursday on the Plaza de Mayo in Buenos Aires, opposite the government's headquarters, to protest the loss of their sons and daughters and demand the return of their grandchildren. They called themselves *Abuelas de Plaza de Mayo* (Grandmothers of the Plaza of May). When a more liberal government took power in 1983, the group stepped up its campaign to locate the missing children. Even when the children were found, however, the families who had them usually refused to admit that they were adopted or give them up.

Knowing that genetic tests could show, for instance, whether a man was the father of a certain child, two representatives of the grandmothers' group traveled to the United States and asked for a geneticist to help them prove their relationship to the disputed children. They were sent to Luca Cavalli-Sforza, a renowned Stanford geneticist with whom King had worked; he in turn referred them to King. King began working with the group in 1984; she now calls Argentina her "second home."

At first King used marker genes to test for relationships in the Argentinian families, but later she adapted a better technique that Allan Wilson developed in 1985. Most human genes are carried on DNA in the nucleus of each cell, but small bodies called mitochondria, which help cells use energy, also contain DNA. Unlike the genes in the nucleus, which are from both parents, mitochondrial DNA is passed on only through the mother and therefore is especially useful in showing the relationship between a child and its female relatives. King says that mitochondrial DNA examination "has proved to be a highly specific, invaluable tool for reuniting the grandmothers with their grandchildren." As a result of King's work some 50 Argentinian children have been reunited with their birth families to date. The same technique has since been used to identify the remains of people killed in wars or murdered by criminals.

In 1995 King moved to Seattle (which she called "the Athens of genetics") to head a laboratory at the University of Washington. Her laboratory has pursued a number of projects, including continuation of the human rights work of identifying military murder victims and children separated from their families by war; this work, under the direction of Michele Harvey, now encompasses Bosnia, Rwanda, and Ethiopia as well as Argentina. The lab is also investigating the genetic characteristics of inherited deafness and genetic variations that may determine why some people are more readily infected with the human immunodeficiency virus (HIV) after exposure and develop full-blown AIDS more quickly after infection than others. The laboratory's chief focus, however, continues to be the study of BRCA1 and other genes involved in breast and ovarian cancer, including noninherited forms of the disease. "Our goal is to eliminate breast cancer as a cause of death," King says.

King's work has earned awards such as the Susan G. Komen Foundation Award for Distinguished Achievement in Breast Cancer Research (1992) and the Clowes Award for Basic Research from the American Association for Cancer Research (1994).

Mary-Claire King believes that women contribute a special gift to science. "Women tend to tackle questions in science that bridge gaps," she said. "We're more inclined to pull together threads from different areas, to be more integrative in our thinking."

Kirch, Maria Margaretha Winkelmann

(1670–1720)
German
Astronomer

Maria Kirch, the first woman credited with discovering a comet, not only was denied a formal post as an astronomer because of her gender but was even reprimanded for appearing in public in an observatory. She was born Maria Margaretha Winkelmann in 1670 at Panitsch, near Leipzig, in what is now Germany. Her father, a Lutheran minister, educated his daughter at home. When she showed an interest in astronomy, he sent her to study under Christoph Arnold, a self-taught astronomer nicknamed the "astronomical peasant."

Gottfried Kirch, Germany's foremost astronomer, went to consult the peasant master one day, met Maria, and fell in love. The two married in 1692 and moved to Berlin in 1700. By then Maria had become Gottfried's working partner. They either observed different parts of the sky on the same night or took turns on alternate nights, one watching while the other slept. During one of these solo nights in 1702 Maria spotted a new comet, then considered a major astronomical find. Gottfried took credit for the discovery in the first published account of it, perhaps because he feared ridicule if people learned that the comet had been found by a woman, but in a 1710 revision he admitted that his wife had been the discoverer.

The Kirches also worked together to make calendars sold by the Royal Academy of Sciences in Berlin, and Maria made many of the astronomical calculations the calendars required. (She also published three astrological pamphlets under her own name between 1709 and 1711; at the time astronomy and astrology overlapped.) After Gottfried died in 1710, Maria asked the academy to let her continue making the calendars, but in spite of the support of the academy's president, the renowned mathematician and philosopher Gottfried von Leibniz, its executive counsel, fearing that "what we concede to her could serve as an example in the future," refused.

Kirch joined the private observatory of Baron Frederick von Krosigk in 1712. When the Academy of Sciences appointed her son, Christfried, to his father's old position in 1716, however, she returned to Berlin to act as his "assistant." A year later the council complained that Kirch was "too visible at the observatory when strangers visit" and ordered her to "retire to the background and leave the talking to . . . her son." When she refused to comply, they forced her to leave, even making her give up her house on the observatory grounds. She died of a fever in 1720. Her biographer, Vignole, wrote, "She merited a fate better than the one she received."

Klein, Christian Felix

(1849–1925)
German
Mathematician

In his first series of lectures for his first professorship at the age of 23, Felix Klein proposed the Erlanger programm, which unified increasingly divergent geometries by considering geometry as a generalized theory of invariants of a particular group of transformations. Klein's system placed traditional Euclidean and non-Euclidean geometries under the same umbrella, a categorization that lasted until 1916, when new geometries were discovered that defied the definition of the Erlanger Programm, which nevertheless remained important. Klein also conceived of the Klein bottle, a topological space that, as with the Möbius strip, had no inside and no outside but was one continuous surface.

Klein was born on April 25, 1849, in Düsseldorf, Germany. In August 1875 he married Anne Hegel, the granddaughter of the famous philosopher, and together the couple had four children—one son and three daughters. In 1865 he commenced study of mathematics and physics at the University of Bonn, where he received his doctorate in 1868. In 1869 he conducted self-directed studies in Göttingen, Berlin, and Paris; this last course was interrupted by his service as a medical orderly in the Franco-Prussian War.

Klein worked for one year as a lecturer at the University of Göttingen in 1871 before his appointment as a full professor at the University of Erlangen in 1872. The title of his famous unification of geometry derived from the name of the institution under which he developed the concept. Klein moved on to the University of Leipzig, where he served from 1880 until 1886, when he returned to the University of Göttingen to commence a long-standing relationship that lasted until his retirement in 1913, though he continued to lecture from his home throughout World War I and afterward despite his failing health. Thanks largely to Klein, the University of Göttingen developed an international reputation as a center for research and innovation in the exact sciences.

From 1872 Klein served as the editor of the journal *Mathematische Annalen* of Göttingen. In 1884 he published his influential "Lectures on the Icosahedron," and in 1895 he founded the *Encyklopädie,* which he supervised until his death. In 1897 and 1902 he published "Lectures on the Theory of Automorphous Functions." His later publication of *Elementary Mathematics from an Advanced Standpoint* introduced a less specialized audience to his complex ideas.

In 1870 Klein collaborated with Sophus Lie of Norway to posit the theory of groups of geometrical transformations, setting the foundation for Klein's subsequent Erlanger Programm. This more important theory grew out of Klein's fascination with the unifying power of group theory. Besides his work with group theory, Klein contributed to number theory, differential equation theory, and function theory; in fact, he recategorized Riemannian geometry under function theory. In the 1890s Klein even developed a theory of the gyroscope. Though Klein's mind worked well with wide-ranging concepts, it did not function as well with details, and thus Klein avoided mathematical calculations, preferring to leave this work to his students. Klein died on June 22, 1925, in Göttingen, Germany.

Klein, Melanie Reizes

(1882–1960)
German/British
Psychologist

Melanie Klein extended the concepts of Freudian psychoanalysis to young children and people with severe mental illness. She was born Melanie Reizes on March 30, 1882, in Vienna, Austria, the youngest of the four children of Moriz Reizes and his wife, Libussa. The family was poor.

Reizes married her cousin, the engineer Arthur Klein, in 1903 and had a daughter and two sons, but the marriage was not happy. Around 1912, while the Kleins lived in Budapest, Melanie entered psychoanalysis as a result. She became interested in SIGMUND FREUD's ideas, and her analyst, Sandor Ferenczi, encouraged her to become an analyst herself.

Freud had doubted that young children could be psychoanalyzed, but Klein disagreed. Her shy five-year-old son, Erich, became her first patient. A paper about Erich's treatment earned her membership in the Hungarian Psychoanalytic Society in 1919. It stressed that exploring the unconscious roots of anxiety was vital in treating children as well as adults.

In 1921 Klein separated from her husband and moved to Berlin. She became a member of the Berlin Psychoanalytic Society in 1922 and began analyzing other children. Some Berlin analysts criticized her work because she lacked academic credentials, but her views proved more acceptable to British psychoanalysts. She moved to England in 1926 and joined the British Psycho-Analytic Society a year later.

In contrast to Freud, who thought that the childhood conflicts that sometimes produced mental illness grew out of instinct and focused on the father, Klein felt that the most important conflicts grew out of children's relationships with others, especially their mothers. When children felt that their mothers were denying them, they fantasized hurting the mothers and then feared punishment for this. Klein believed that children—or adults—could overcome such fears, and resulting illness, by realizing that they had the goodness and power to repair imagined damage done to their mothers.

Klein analyzed children as young as three years old. Her techniques, which she described in *The Psychoanalysis of Children* (1960), are still widely used. She gave her young patients toys such as miniature houses, cars, and dolls and encouraged them to use the toys to act out stories about themselves and their families. From symbols and actions in this play she learned what the children were thinking and feeling. She also studied how they reacted to her as a mother figure.

Freud had concentrated on the relatively mild mental illness called neurosis, but Klein claimed that psychoanalytic techniques could be modified to treat more serious disturbances, including depression and schizophrenia. Her ideas and personality caused great controversy among psychoanalysts. One called her "the most impressive human being I have known," but others complained of her "overweening self-righteousness" and "adamantine dogmatism." Klein died of cancer on September 22, 1960, at the age of 78.

Klug, Aaron

(1926–)
South African/British
Chemist, Molecular Biologist

Sir Aaron Klug won the 1982 Nobel Prize in chemistry for his research on the three-dimensional structures of viruses. Most importantly Klug created a technique that combined electron microscopy with X-ray crystallography to create crystallographic electron microscopy, whereby electron micrographs are taken from different angles, then X-ray diffractions are taken of these micrographs and combined into a three-dimensional image of the molecule being studied.

Klug was born on August 11, 1926, in Zelvas, Lithuania. His parents, Bella Silin and the cattle dealer Lazar Klug, moved with their two-year-old boy to Durban, South Africa, to join his mother's family. Klug married Liebe Bobrow in 1948, and together the couple had two sons—Adam and David.

An early experience that influenced Klug toward biology was his reading of Paul De Kruif's *Microbe Hunters*, first published the year of Klug's birth. Klug entered the University of Witwatersrand at Johannesburg in 1942 as a premedical student, but he graduated in 1945 with a degree in science. The University of Cape Town offered Klug a scholarship to study crystallography in its doctoral physics program; however, Klug left after receiving his master's degree in 1949 to pursue a fellowship at Trinity College in Cambridge, England. Klug conducted research in the renowned Cavendish Laboratory on the structural changes at the molecular level when molten steel solidifies. Klug received his Ph.D. in 1952, after which he commenced a research fellowship at Birkbeck College of the University of London, where he studied the structure of the tobacco mosaic virus with ROSALIND ELSIE FRANKLIN, who took the X-ray crystallography photographs used by

FRANCIS HARRY COMPTON CRICK and JAMES DEWEY WATSON to develop their Nobel Prize–winning double-helical model of deoxyribonucleic acid (DNA).

After Franklin's untimely death in 1958 Klug succeeded him as the director of the Virus Research Group at Birkbeck College. Then in 1962 he joined Crick and a host of renowned scientists at the British Medical Research Council's Laboratory of Molecular Biology. Klug worked there for a number of years, becoming the joint head of the division of structural studies in 1978 and the director of the entire laboratory in 1986.

Klug commenced his work with both viruses and electron microscopy early in his career. At Birkbeck he studied the polio virus with J. D. Bernal and small viruses with Donald Caspar. There he also pioneered improvements of electron microscopy by shining laser light onto the micrograph to illuminate more detailed information about diffraction patterns. This innovation led him to consider further improvements on electron microscopy, which retained many flawed details as a result of the constraints of two dimensions. However, the more precise X-ray crystallography could not provide the magnification of electron microscopy. Klug simply decided to combine both methods into one, taking a series of electron micrographs from different angles and then performing X-ray diffraction on these micrograph plates, producing clearer, three-dimensional pictures.

Besides winning the Nobel Prize, Klug received the H. P. Heineken Prize from the Royal Netherlands Academy of Arts and Sciences and the Louisa Gross Horwitz Prize from Columbia University. Klug was knighted in 1988.

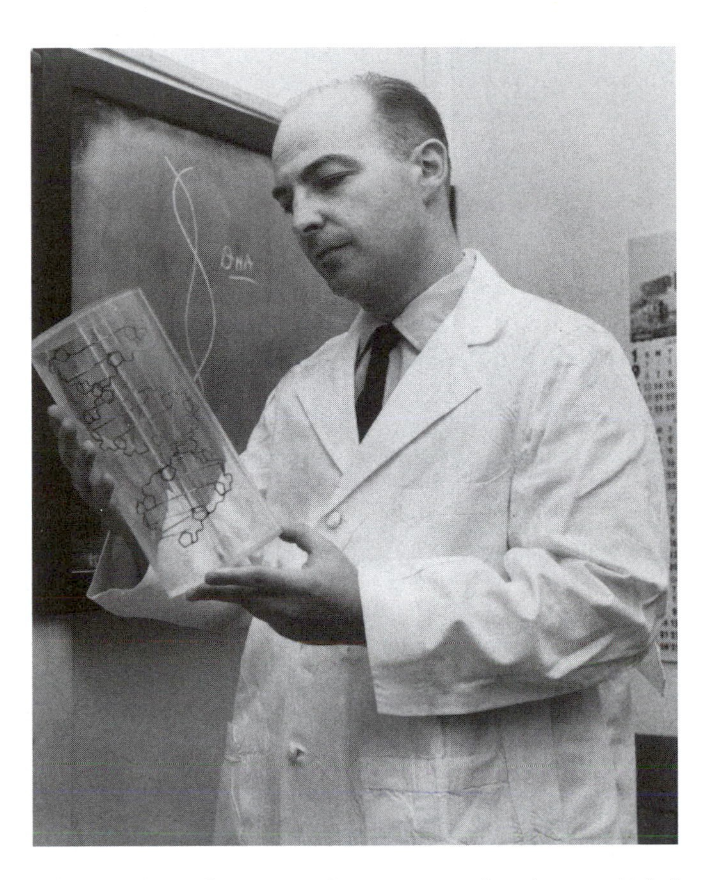

Arthur Kornberg, whose research on DNA won him the 1959 Nobel Prize in physiology or medicine. *Bernard Becker Medical Library, Washington University School of Medicine.*

Kornberg, Arthur

(1918–)
American
Biochemist

Arthur Kornberg received the 1959 Nobel Prize in physiology or medicine with SEVERO OCHOA. The award recognized his discovery of the process by which deoxyribonucleic acid (DNA) molecules replicated in bacterial cells. Later in his career Kornberg succeeded in synthesizing DNA polymerase, facilitating genetic research.

Kornberg was born on March 3, 1918, in Brooklyn, New York. His parents were Lena Katz and Joseph Kornberg. In 1943 Kornberg married Sylvy Ruth Levy, a fellow biochemist, who collaborated with Kornberg often. Together the couple had three sons.

Kornberg attended the City College of New York on scholarship as a premedical student and graduated in 1937 with a B.S. in biology and chemistry. He then went upstate to the University of Rochester School of Medicine, where he earned his M.D. in 1941. That year he commenced his internship at the university's facility, the Strong Memorial Hospital.

With the advent of World War II, Kornberg served with the United States Public Health Service, and from 1942 until 1952, he worked at the National Institutes of Health (NIH) in Bethesda, Maryland, first in the nutrition section of the division of physiology until 1945, and then as the chief of that division's enzymes and metabolism section. Throughout his tenure with the NIH, Kornberg researched at other facilities as well. In 1946 he worked at the New York University School of Medicine with Ochoa; in 1947 he conducted investigations with Carl Cori and GERTY THERESA RADNITZ CORI at Washington University in St. Louis, Missouri; in 1951 he worked with H. A. Barker at the University of California at Berkeley. In 1953 he returned to the Washington University School of Medicine as a professor of microbiology and as the chief of that department. In 1959 he moved on to Stanford University, where he served as the chairman of the department of biochemistry.

When he contracted hepatitis, which often causes jaundice as a side effect, and used his own case as the basis for the paper "The Occurrence of Jaundice in an

Otherwise Normal Medical Student," Kornberg commenced his publishing career while still in medical school. More important, however, was his 1961 publication *Enzymatic Synthesis of DNA*, as well as his 1980 publication *DNA Replication*.

While at the NIH Kornberg developed methods for synthesizing the coenzymes diphosphopyridene nucleotide (DPN) and flavin adenine dinucleotide (FAD) by means of the condensation reaction, in which phosphate is eliminated in much the same way, Kornberg hypothesized, as in bodily DNA synthesis. At Washington University, Kornberg discovered what he called DNA polymerase, or the enzyme responsible for bodily DNA synthesis, in the intestinal bacterium *Escherichia coli*. At Stanford he and his collaborator Mehran Goulian announced on December 14, 1967, their synthesis of DNA polymerase.

Besides the Nobel Prize, Kornberg won the 1951 Paul-Lewis Laboratories Award in Enzyme Chemistry from the American Chemical Society. In 1968 he won several major awards, including the Scientific Achievement Award from the American Medical Association and the Borden Award in Medical Sciences from the Association of American Medical Colleges. In 1980 he won the National Medal of Science, an honor that recognized his long and influential career.

Kovalevskaia, Sofia Vasilyevna ("Sonya")
(1850–1891)
Russian
Mathematician

Sofia Kovalevskaia overcame many barriers to become the first tenured woman professor in modern Europe and won the prestigious Bordin Prize from the French Academy of Sciences. One of her biographers, Ann Hibner Koblitz, writes, "During her lifetime, Kovalevskaia was regarded as one of the most eminent mathematical analysts in the world."

Sofia, usually called Sonya, was the middle of Vasily and Yelizaveta Korvin-Krukovsky's three daughters (they also had a son). She was born in Moscow on January 15, 1850. Her father was a nobleman and a general in the Russian artillery. When Sonya was six, he retired from the army and moved to an isolated country estate in Palabino, near the Lithuanian border. The family started to put new wallpaper in the house there, but the supply ran out before one room was covered, so they finished the job with printed pages of calculus lectures from the general's student days. "These sheets, spotted over with strange incomprehensible formulae, soon attracted my attention," Sonya wrote in her *Recollections of Childhood*. "I passed whole hours before that mysterious wall, trying to decipher even a single phrase, and to discover the order in which the sheets ought to follow each other."

Wallpaper was just the beginning of Sonya's self-education. When a family friend gave her a physics text, she discovered that she needed to know trigonometry in order to understand it and worked out the basic ideas of that branch of mathematics on her own. The friend's amazed report of this, coupled with the girl's own pleas, persuaded Korvin-Krukovsky to let her obtain private tutoring in St. Petersburg.

That seemed as far as Sonya could hope to go, since Russian universities did not admit women. Some universities elsewhere in Europe did, but unmarried Russian women could not leave the country without their parents' permission. Young women in Sonya's forward-looking set of friends sometimes got around this by making marriages of convenience with men who accompanied them to a university and then left them. Sonya and her sister explored this idea with Vladimir Kovalevsky, a 26-year-old geology student. Kovalevsky agreed—so long as the bride was Sonya. He wrote to his brother that in addition to being "extremely well-educated," she was "lively, sweet, and very pretty." They married in October 1868.

In 1869 the Kovalevskys went to Germany's University of Heidelberg. Sonya was not allowed to enroll there, but she attended lectures. According to Julia Lermontova, a friend who moved in with the Kovalevskys shortly after their arrival, professors who met Kovalevskaia (female form of Kovalevsky) "spoke of her as something extraordinary." Sonya and her new husband, too, appeared to be truly in love. "This was the only time I have known Sonya to be really happy," Lermontova wrote later. Soon, however, more female friends arrived; the Kovalevskys' apartment became overcrowded, and Vladimir Kovalevsky moved out.

After two years at Heidelberg Kovalevskaia went to Berlin in the hope of studying under Karl T. Weierstrass, the "father of mathematical analysis." Weierstrass tried to get rid of her by giving her problems that he normally assigned only to his most advanced students, but when she solved them, he became one of her staunchest supporters. He tutored her without charge for the next four years and began to see her almost as a daughter.

Weierstrass helped Kovalevskaia gain permission to apply for a doctorate from the University of Göttingen, and she earned her degree, summa cum laude, in 1874, one of the first women to obtain a degree from a German university.

Difficult as it had been, obtaining a degree proved far easier than obtaining a job. Kovalevskaia finally gave up and returned to Russia late in 1874. There she became reunited with her husband, who now taught paleontology in Moscow. For the first time they truly lived as man and wife, and they had a daughter in 1878. The girl, Sofia, was nicknamed Fufa.

For six years Kovalevskaia abandoned her mathematical career and tried to be a conventional wife and mother, but she later referred to this life as "the soft slime of bourgeois existence" and said that Fufa was the only good thing that came out of it. She and Vladimir Kovalevsky did not get along, and their problems were worsened by his growing mental instability and tendency to make disastrous financial investments. They separated in 1880, and, leaving Fufa in Moscow, Kovalevskaia returned to Berlin and then to Paris. There she was elected to the Mathematical Society and carried out research on the way light travels through crystals. Back in Russia, Kovalevsky killed himself on April 15, 1883.

Weierstrass had asked Gösta Mittag-Leffler, a former student who was now a professor of mathematics at the University of Stockholm, to help Kovalevskaia find work. Mittag-Leffler finally persuaded the Swedish university to hire her as a lecturer in autumn 1883. On her arrival some Stockholm newspapers hailed her as "a princess of science," but the playwright August Strindberg claimed that "a female professor of mathematics is . . . a monstrosity." After a year, apparently agreeing with Mittag-Leffler's view of Kovalevskaia rather than Strindberg's, the university offered her a five-year salaried professorship. In 1885 she began teaching mechanics as well.

Kovalevskaia was delighted to be lecturing, teaching, and doing research. In 1886 she sent for Fufa. The tradition-minded townspeople criticized her, however, and she in turn began to find Swedish society boring and longed for Paris or St. Petersburg. She even became somewhat tired of mathematics. She joined Anna Charlotte Leffler-Edgren, Mittag-Leffler's sister, in writing a play, and on her own she wrote short stories, articles, poetry, novels, and her autobiography.

In 1886 the French Academy of Sciences offered its highest award, the Bordin Prize, to the best paper describing the rotation of a solid body around a fixed point. Kovalevskaia had studied this subject when she wrote about the rings of Saturn, so she decided to enter the contest. To prevent favoritism, all entries were submitted anonymously, with a short phrase in place of the author's name. The judges agreed that the best of the 15 entries was the one signed with the saying, "Say what you know, do what you must, come what may," noting, "The author's method . . . allows him to give a complete solution [of the problem] in the most precise and elegant form." When the authors' identities were revealed, the astounded judges learned they had used the wrong pronoun: The paper was Kovalevskaia's.

The French science academy was so impressed with Kovalevskaia's work that it not only awarded her the Bordin Prize in December 1888 but also added another 2,000 francs to it. When she refined the paper in 1889, it won an additional 1,500 kroner from the Swedish Academy of Science and earned her a permanent professorship at the university—the first granted to a woman in Europe in modern times. Her paper was later called "one of the most famous works of mathematical physics in the nineteenth century."

In spite of these successes Kovalevskaia was not happy. She had fallen in love with a Russian sociologist and historian, Maxim Kovalevsky (a distant relative of her late husband's family), and their relationship was stormy. They spent the holidays in France at the end of 1890 and may have been planning to marry the following year, but Kovalevskaia caught a chill on the journey back to Stockholm that developed into pneumonia. She died of it on February 10, 1891, just three weeks after her 41st birthday.

Krebs, Sir Hans Adolf

(1900–1981)
German/British
Biochemist

Sir Hans Adolf Krebs received the 1953 Nobel Prize in physiology or medicine for his identification of the tricarboxylic cycle, also known as the citric acid cycle but best known by the nomenclature that acknowledged his role in its discovery, the Krebs cycle. Krebs shared the prize with FRANCIS ALBERT LIPMANN, who discovered coenzyme A, an important component of the Krebs cycle.

Krebs was born on August 25, 1900, in Hildesheim, Germany. His parents were Alma Davidson and Georg Krebs, an ear, nose, and throat surgeon. Krebs had two siblings, a sister named Elisabeth and a brother named Wolfgang. The German military drafted Krebs in 1918, but a month later World War I ended, thereby releasing him from his duties and enabling him to pursue his education. Krebs matriculated at the University of Göttingen as a premed student for one year, then transferred to the University of Freiburg. In 1921 he again transferred, this time to the University of Munich, after which he spent a mandatory year in clinical practice at the Third Medical Clinic of the University of Berlin. In 1925 Krebs received his M.D. from the University of Hamburg.

From 1926 Krebs worked at the Kaiser Wilhelm Institute for Cell Physiology in Berlin under Otto Warburg, conducting research on amino-acid degradation and excretion of nitrogen in urea, which led to his identification of the urea cycle. In 1930 Krebs returned to the University of Freiburg, but his research and clinical work there lasted only until 1933, when the advent of Hitler's Nazi regime forced Krebs to emigrate. Krebs received a Rockefeller Research studentship at Cambridge University in England, and in 1934 Cambridge appointed him as a demonstrator in biochemistry. In 1935 he moved to Sheffield University, where he served as a lecturer in pharmacology until 1938, when he was appointed as a lecturer in charge of the Department of Biochemistry. In 1945 the university named him a professor of biochemistry.

During his early tenure at Sheffield, Krebs conducted the research that led to his identification of the Krebs cycle. Gerty Cori and Otto Meyerhof had already identified that glycogen was broken down in the liver anaerobically into lactic acid. Krebs went a step beyond this to realize that lactic acid as well as carbohydrates aerobically metabolized into carbon dioxide, water, and energy that fueled the body. In 1937 he published his results in the journal *Enzymologia* in an article cowritten with his graduate student William Arthur Johnson, "The Role of Citric Acid in Intermediate Metabolism in Animal Tissues." In 1938 Krebs married Margaret Fieldhouse, a teacher of domestic science at Sheffield University, and together the couple had three children. In September 1939, on the eve of World War II, Krebs became a naturalized British citizen.

In 1947 the Royal Society elected Krebs as a fellow; the society awarded him its Royal Medal in 1954 and its Copley Medal in 1961. Krebs spent the end of his career, from 1954 through 1967, at Oxford University as the Whitley Professor of Biochemistry and as a fellow of Trinity College. His identification of the self-regenerating cycle that created energy for the body to use marked a vital step in the understanding of biochemical processes. Krebs died in Oxford on November 22, 1981.

Kuhn, Richard

(1900–1967)
Austrian/German
Chemist

Richard Kuhn was awarded the Nobel Prize in chemistry in 1938, though Nazi opposition to the prize, when it was awarded to the concentration camp prisoner Carl von Ossietzky in 1934, prevented him from accepting the medal until 1949, after World War II. He received the honor for his work on vitamins and carotenoids, naturally occurring fat-soluble yellow coloring agents, of which Kuhn discovered at least eight.

Kuhn was born on December 3, 1900, in Vienna, Austria. His father, Hofrat Richard Clemens Kuhn, was a hydraulics engineer, and his mother, Angelika Rodler, was an elementary school teacher who taught Kuhn until the age of nine. In 1928 Kuhn married Daisy Hartmann, who studied under him in Zurich. Together the couple had four daughters and two sons.

Kuhn entered the University of Vienna on November 22, 1918, four days after his discharge from the Austro-Hungarian army. Three semesters later he transferred to the University of Munich, where he studied chemistry under Richard Willstätter. Kuhn earned his Ph.D. in 1922 with the dissertation "On the Specificity of Enzymes in Carbohydrate Metabolism," which demonstrated both discipline and creativity.

After serving as Willstätter's assistant, Kuhn traveled to Zurich in 1926 to fill the professorship of general and analytical chemistry at the Federal Institute of Technology. In 1929 Kuhn moved permanently to the Kaiser Wilhelm Institute for Medical Research in Heidelberg as a professor of organic chemistry and the director of the chemistry department. In 1937 he was named the director of the institute, and in 1950 his professorship transferred into the department of biochemistry. That year he helped rename the institute as the Max Planck Institute. From 1946 to 1948 he had similarly helped transform the Kaiser Wilhelm Society for Scientific Research into the Max Planck Society for the Advancement of Science and later served as its vice president. From 1948 on he edited the journal *Justus Liebig's Annals of Chemistry.*

Kuhn's research focused on the intersection between chemistry and biology, or how chemical compounds interacted with the body. This interest led him to consider carotenoids such as carotene, the pigment of carrots. Kuhn discovered that carotene was related to vitamin A, and he also discovered three distinct kinds of carotene: beta-carotene, which bends light; alpha-carotene, which does not bend light; and gamma-carotene, which has the same chemical formula as the others but a different molecular structure; thus the three are isomers. Kuhn determined the structure of vitamin A and, simultaneously with his rival Paul Karrer, isolated and crystallized a gram of vitamin B_2, or riboflavin, from skim milk. Kuhn showed the important role this vitamin played in respiratory enzyme function. In 1938 he synthesized vitamin B_6.

In addition to the Nobel Prize, Kuhn won the Pasteur, Paterno, and Goethe Prizes. He was a prolific researcher and writer, publishing more than 700 scientific papers. The University of Heidelberg honored him with its first commemorative medal given to a scientist shortly before Kuhn's death on July 31, 1967, in Heidelberg, Germany.

Kuiper, Gerard Peter

(1905–1973)
Dutch/American
Astronomer

A celebrated astronomer, Gerard Peter Kuiper made a number of advances in astronomy, including discoveries of the moons of Uranus and Neptune, the calculation of the diameter of Pluto, and a prediction of what the surface of the Moon would be like. Kuiper also played a significant role in the advancement of the United States' space program and formulated a theory of the origin of the solar system.

Born on December 7, 1905, in Harencarspel in the Netherlands, Kuiper was one of the four children of Gerard Kuiper and Anna de Vries. He attended the Gymnasium in Haarlem and graduated in 1924. Kuiper married Sarah Parker Fuller in 1936 and had two children.

Kuiper entered the University of Leiden to study astronomy and physics. After receiving his bachelor's degree in 1927, he was an assistant observer at Leiden and had the opportunity to travel to Sumatra in 1929 with the Dutch Solar Eclipse Expedition. In 1933 Kuiper earned his doctorate with a thesis on binary stars. Immediately after receiving his degree he moved to the United States, serving as a fellow at the University of California's Lick Observatory until 1935 and then working at Harvard University for a year. In 1936 Kuiper began working at the University of Chicago's Yerkes Observatory, where he remained until 1960. He also worked at the McDonald Observatory in Texas, which was also run by the University of Chicago.

After studying stellar astronomy, Kuiper turned to planetary astronomy in the 1940s. In 1944 he was the first to verify the existence of an atmosphere around a moon of the outer planets when he confirmed the presence of a methane atmosphere around Titan, Saturn's moon. In 1948 Kuiper made several breakthrough achievements: He successfully predicted that carbon dioxide was a primary part of the atmosphere of Mars, theorized that the rings of Saturn were made up of ice particles, and discovered Miranda, Uranus's fifth moon. Kuiper also suggested that higher life forms did not exist on Mars. The following year he made another discovery—Nereid, Neptune's second moon. He also introduced his theory on the origin of the solar system, which suggested that the planets had formed by the condensation of a cloud of gas around the Sun. Billions of years ago, Kuiper proposed, the cloud of gas collapsed, and when its density reached a grave value, the cloud scattered into baby planets.

In 1950 Kuiper used the 200-inch reflecting telescope at Mount Palomar to become the first to calculate Pluto's diameter dependably. A year later he introduced a theory regarding the source of comets and proposed that a disk-shaped ring of comets orbited around the Sun. In 1956 Kuiper demonstrated that the polar icecaps on Mars were not composed of carbon dioxide, as had been previously thought, but of thinly spread ice crystals. Kuiper also found bands of methane on Uranus and Neptune, which became known as Kuiper bands.

Kuiper left the University of Chicago in 1960 and moved to the University of Arizona to found the Lunar and Planetary Laboratory. Kuiper's 1964 prediction that the surface of the Moon would resemble crunchy snow would be verified in 1969 when humans first walked on the Moon. Kuiper promoted the use of jet airplanes as high-altitude observatories; mounting telescopes on aircraft allowed Kuiper to make infrared observations of Venus.

Kuiper was a very active and highly admired astronomer who received numerous honors. He won the French Astronomical Society's Janssen Medal for the discovery of Nereid and Miranda and was named to the Order of the Orange Nassau in the Netherlands. Craters on the Moon, Mercury, and Mars were named after him, as was the Kuiper Airborne Observatory, a jet aircraft housing a telescope. Kuiper's theory regarding the ring of comets was confirmed in the 1990s, and it was named the Kuiper belt.

L

Ladd-Franklin, Christine

(1847–1930)
American
Mathematician, Psychologist

Christine Ladd-Franklin made important contributions in two very different scientific areas: symbolic logic and the theory of color vision. She was born to Eliphalet and Augusta Ladd in Windsor, Connecticut, on December 1, 1847. She spent her childhood in New York City, where her father was a well-to-do merchant, and then in Windsor, to which he moved his family after he retired in 1853. After Christine's mother died when she was 13, she lived with her grandmother in Portsmouth, New Hampshire.

Ladd studied at Vassar College for two years, majoring in mathematics (she would have preferred physics but could not get access to laboratory equipment), and obtained her B.A. in 1869. She then taught high school science for nine years. In 1878 she applied to Johns Hopkins University for admission as a graduate student in mathematics. The university did not accept women at the time, but the Johns Hopkins mathematics professor James J. Sylvester had read her papers in a British mathematical journal and asked that she be allowed to attend as a "special status" student. The university agreed, but in order to avoid setting a precedent, it did not put her name in its official list of students.

Ladd fulfilled all the requirements for a Ph.D. by 1882, but the university refused to grant the degree. (It finally did in 1926, when she was 78 years old.) She married Fabian Franklin, a fellow Johns Hopkins mathematician, in August of that year. They had two children, of whom one, Margaret, survived into adulthood.

Ladd-Franklin, as Christine now called herself, made her first contributions in the field of symbolic logic. They concerned syllogisms, three-part statements such as "If all *a*'s are *b*'s, and all *b*'s are *c*'s, then all *a*'s are *c*'s." In the early 1880s, she reduced all such statements to a single formula and showed how to use it to determine whether a syllogism was valid.

Around 1886 Ladd-Franklin moved from mathematics to psychology, specifically the theory of vision. The connection, for her, lay in the way two eyes combine separate images into one, the study of which requires both mathematics and psychology. Later she concentrated on color vision. She studied with two authorities in the field, G. E. Müller of Göttingen University and Hermann von Helmholtz of the University of Berlin, and combined parts of their opposing theories into her own hypothesis about the way color vision had evolved. She first presented it at the International Congress of Psychology in 1892. She believed that the most primitive form of vision, which did not include color vision, evolved into vision of blue and yellow. The chemical associated with yellow then broke down into products that allowed vision of reds and greens, producing advanced color vision.

Johns Hopkins made Ladd-Franklin a lecturer in psychology and logic in 1904. She continued to teach there until 1909, when Fabian Franklin, now a journalist, moved to New York City. Ladd-Franklin lectured and did research at Columbia University from 1914 until her retirement in 1927. Her ideas, some of which are still accepted, were collected in the book *Colour and Colour Theories* in 1929. She died in New York on March 5, 1930, at the age of 82.

Lamarck, Jean-Baptiste-Pierre-Antoine de Monet, chevalier de

(1744–1829)
French
Biologist

Though celebrated for his findings in botany and invertebrate zoology, Jean-Baptiste-Pierre-Antoine de Monet,

Jean-Baptiste-Pierre-Antoine de Monet, chevalier de Lamarck, developer of Lamarckism, one of the first comprehensive theories of evolution. *Smith Collection, Rare Book & Manuscript Library, University of Pennsylvania.*

chevalier de Lamarck was generally scorned for his ideas on evolution, for which he is best known. Despite the widespread rejection of his theory, known as Lamarckism, his was one of the first comprehensive theories of evolution and influenced CHARLES ROBERT DARWIN to formulate his theory of evolution based on natural selection—a theory that invalidated Lamarckism. A celebrated naturalist, Lamarck made important contributions in botany, paleontology, and zoology and was the first to use the word *biology*.

The youngest of 11 children, Lamarck was born on August 1, 1744, in Bazantin, France. His father, Philippe Jacques de Monet de La Marck, was a military officer, and the family was rather poor despite aristocratic blood. His mother was Marie-Françoise de Fontaines de Chuignolles. Lamarck's parents intended that he enter the priesthood, and at the age of 11 Lamarck entered a Jesuit school. When his father died in 1759, however, Lamarck left the school and joined the military, which had interested him considerably more than a religious career. He married three or four times during the course of his life and had a total of eight children, but his personal life was full of tragedy.

Lamarck served several years in an infantry regiment and fought in the Seven Years' War. After the war he was stationed at a number of French forts on the Mediterranean and eastern French borders, where he developed his interest in botany. In 1768 Lamarck resigned from the military service because of ill health. Lamarck studied medicine for four years and grew interested in meteorology, chemistry, and shell collecting. He then immersed himself in botany, studying at the Jardin du Roi, the royal botanical gardens, in Paris.

In 1778 Lamarck published *Flowers of France,* a three-volume work that was extremely well received. In it he introduced a key for plant classification that was far easier to use than other identification methods of the time. Because of this work Lamarck was elected to the Academy of Sciences. Lamarck continued to study botany until the revolution of 1789, when the Jardin du Roi was reorganized. It became the Jardin des Plantes (National Museum of Natural History) in 1793, with Lamarck in charge of animals without backbones, which he later named *invertebrates.* Lamarck not only introduced the concept of museum collection, in which objects were arranged in a systematic order, but also in 1800 revised the classification of invertebrates. He systematically analyzed the functions and complexities of essential organs, thereby discerning the differences between invertebrates with outward similarities. Lamarck described his findings in his breakthrough work *Natural History of Invertebrate Animals,* published between 1815 and 1822.

Lamarck's evolutionary theory was first introduced in *Zoological Philosophy,* published in 1809. He proposed that there was a natural tendency to increasing complexity. Lamarck held that organs were strengthened or weakened through use and disuse and that acquired characteristics are inherited. His most famous example concerned the giraffe; Lamarck stated that the giraffe's neck had grown longer because of the need to stretch for high leaves, and that this change was passed on to offspring.

Though Lamarck was well respected for his achievements in botany and zoology, by the early 1800s he had severed many ties with the scientific community; science was becoming more detail-driven, and Lamarck was concerned that it would no longer be a logical system enhancing general understanding of the world. He began to suffer from poor health in 1809, and by 1818 he was completely blind. Only one of his eight children was financially successful, so when Lamarck died in 1829, his family had to solicit funds to pay for the funeral.

Langmuir, Irving

(1881–1957)
American
Chemist, Physicist

Ambitious and successful, Irving Langmuir was a chemist, physicist, and industrial researcher. Langmuir won a Nobel Prize in chemistry in 1932 for his work in surface chemistry; he was the first scientist employed in industry and the second American to win a Nobel Prize. Langmuir worked at the General Electric Company (GE) for much of his career and developed modifications of a number of devices and objects, including the light bulb and vacuum pump. Later in life he became interested in meteorology and atmospheric science.

Born on January 31, 1881, in Brooklyn, New York, Langmuir was the third of four sons born to Charles Langmuir, an insurance executive, and Sadie Comings, the daughter of an anatomy professor whose ancestors immigrated to America on the *Mayflower.* Because of the travel involved with his father's work Langmuir was educated for a short time in France, but his father died when Langmuir was 17 years old.

Irving Langmuir, the first scientist employed in industry and the second American to win a Nobel Prize in science, which he was awarded for chemistry in 1932. *Smith Collection, Rare Book & Manuscript Library, University of Pennsylvania.*

After graduation in 1903 from the Columbia University School of Mines, where he studied chemistry, physics, and mathematics, Langmuir traveled to Germany to study at the University of Göttingen. Under the guidance of the physical chemist and Nobel laureate Walther Nernst, Langmuir completed his doctoral dissertation on the chemical reactions that occur between hot gases at low pressure and radiant platinum wire. After receiving his doctorate in 1906, Langmuir returned to the United States, where he accepted a position teaching chemistry at the Stevens Institute of Technology in Hoboken, New Jersey. Langmuir married Marion Mersereau in 1912, and the couple adopted two children.

After a few years at the Stevens Institute, Langmuir applied for a job at GE's research laboratory in Schenectady, New York, hoping for research opportunities. Though he had applied for a summer position, it grew into a permanent career. Langmuir's first achievement at GE was his development of a better light bulb. His assignment was to improve a new type of light bulb that used a tungsten metal wire as a light source rather than the commonly used carbon filament. Langmuir filled the glass bulb with a combination of inert gases and used a coiled filament to produce an energy-efficient, longer-lasting light bulb. He received a patent for his design in 1916. Another early success was Langmuir's invention of a vacuum pump that was considerably more powerful than existing models. Vacuum tubes were significant because of their use in radio broadcasting and in electrical power. Langmuir's pump enhanced the making of vacuum tubes and was widely adopted by industry.

Langmuir's most celebrated work was his research in the chemistry of surfaces. He investigated and introduced the principles of adsorption, which involved the inclination of gas molecules to accumulate in a single layer on the surfaces of liquids or solids. Langmuir also observed adsorption in the actions of thin films of oil on the surfaces of liquids. For this work Langmuir was awarded the Nobel Prize in 1932.

Langmuir made numerous other contributions to science. He researched the chemical reactions of gases at high temperature and low pressure, provided explanations on the structure of the atom, and studied *plasmas,* a term Langmuir coined to refer to electrically neutral, highly ionized gases. Langmuir also designed an improved electric stove and a hydrogen welding torch. His interests in meteorology led him to investigate aircraft deicing methods and to create rain and snow by seeding clouds with dry ice and silver iodide. Langmuir received 63 patents and wrote more than 200 papers.

Laplace, Pierre-Simon, marquis de

(1749–1827)
French
Mathematician, Astronomer

Best known for his theories regarding the stability and origin of the solar system, Marquis Pierre-Simon de Laplace made numerous contributions in mathematics, astronomy, and physics. Laplace developed celestial mechanics, the science of the movement of celestial bodies under the influence of gravitational forces, and established probability theory, a division of mathematics that investigates the likelihood of occurrence of random events in order to predict the behavior of defined systems.

Born in Beaumont-en-Auge in Normandy, France, on March 23, 1749, Laplace had a childhood that is somewhat of a mystery. His father, Pierre, was a farmer and small estate owner and probably in the cider business. His mother, Marie-Anne Sochon, was from a family of successful farmers. Laplace attended a school run by the Benedictines from the age of 7 to 16. Students of the school generally entered the clergy or the army; Laplace's father intended for him to join the clergy.

In 1766 Laplace entered the University of Caen, and it was there he realized his mathematical talents. Rather than continue on in the department of theology, he left for Paris with a letter of recommendation from a mathematics professor, Pierre La Canu, to the mathematician Jean d'Alembert. Laplace also presented a letter on principles of mechanics to d'Alembert, who was impressed enough to arrange for him to become a professor of mathematics in the École Militaire in Paris.

Laplace began his studies of the solar system in 1773. He applied Sir ISAAC NEWTON's theories of gravitation to the solar system and attempted to solve the puzzle of why Jupiter's orbit seemed to be continually shrinking whereas Saturn's continuously grew. Newton acknowledged the existence of irregularities in planetary movement and believed they would lead to the end of the universe without divine intervention. Laplace, however, demonstrated that these gravitational perturbations, variations in orbits resulting from the influence of external bodies, would not lead to instabilities in planetary orbits. This discovery was a crucial advance in confirming the stability of the solar system. Laplace continued to investigate planetary perturbations and in 1786 showed that the inclinations of orbits to each other would forever remain constant, minor, and self-correcting. Laplace thus concluded that the results of these perturbations were not cumulative but periodic and harmless. In 1787 he announced that the acceleration of the Moon depended on the eccentricity of the Earth's orbit and provided further evidence as to the stability of the solar system.

Marquis Pierre-Simon de Laplace, who developed the science of celestial mechanics. *Smith Collection, Rare Book & Manuscript Library, University of Pennsylvania.*

In 1796 Laplace published *The System of the World,* in which he detailed his findings in celestial mechanics. In this work Laplace introduced his nebular hypothesis, which stated that the solar system was created by the cooling and contracting of a rotating mass of gas—a gaseous nebula—which condensed to form the Sun. Laplace's greatest work was probably *Celestial Mechanics,* published between 1799 and 1825 in five volumes, which detailed his observations concerning the stability of the solar system. In 1812 Laplace completed the *Analytic Theory of Probability,* which described the methods he developed for mathematically predicting the probabilities that certain events will occur in nature. Two years later Laplace published a similar work geared to a general audience.

Laplace also studied spheroids and worked with the chemist ANTOINE-LAURENT LAVOISIER to invent an ice calorimeter and prove that respiration was a form of combustion. Laplace helped found the Society of Arcueil, helped organize the metric system, and was made a marquis and a count. Laplace expanded on Newton's theories and was an extremely influential scientist.

Laveran, Charles-Louis-Alphonse

(1845–1922)
French
Physician, Pathologist, Parasitologist

In Algeria in 1880, Alphonse Laveran discovered the causative agent of human malaria, a parasite, though he only hypothesized that mosquitoes acted as the carriers. In the 1890s Sir Ronald Ross confirmed Laveran's hunch, proving that *Anopheles* mosquitoes carried malaria. Laveran continued to work on pathogenic protozoans throughout his career, and in 1907 the Swedish Academy of Arts and Sciences recognized this work by awarding him the Nobel Prize in physiology or medicine.

Laveran was born on June 18, 1845, in Paris, France. His mother was Marie-Louise Anselme Guénard de la Tour, and his father, in whose footsteps he walked throughout much of his life, was Louis-Theodore Laveran. In 1885 Laveran married Sophie Marie Pidancet. The couple had no children, but his older sister was a companion to Laveran and his wife.

After attending the College Sainte-Barbe and the Lycée Louis-le-Grand, he began to study medicine in 1863 in Strasbourg at the École Imperiale du Service and with the Faculté de Médecine simultaneously, as his father had. In 1867 he defended his doctoral dissertation on experimental research on the regeneration of nerves, which retained its currency long after its composition.

Laveran served as an army surgeon in the Franco-Prussian War from 1870 to 1871. In 1874 he won a competitive examination to fill the chair his father formerly held at the École du Val-de-Grace military medical school in Paris. Between 1878 and 1883 Laveran served in a series of military hospitals in Algeria, where he conducted research on the numerous patients he treated for malaria. On November 6, 1880, in Constantinople, he discovered the living agent of malaria, a parasite that he found in 148 of 192 autopsies he performed. He presented his findings to the Academy of Medicine in Paris on November 23, 1880, and then on December 24, 1880, the Société Médicale des Hospitaux published a paper documenting these findings along with further research. Laveran presented his combined research on the malarial parasite, later named *Plasmodium* instead of his suggestion of *Oscillaria malariae,* in the 1881 monograph *Nature parasitaire des accidents de l'impaludisme* and more completely in the 1884 text *Traité des fièvres palustres avec la description des microbes du paludisme.*

The scientific community and the military did not recognize the significance of Laveran's discovery, and the latter institution relegated him to the relatively lowly position of professor of military hygiene at the École du Val-de-Grace. Laveran finally resigned from the military in frustration to take up a post at the Pasteur Institute in Paris, where he remained for the rest of his career.

In 1901 the French Academy of Sciences elected Laveran as a member. He used half of the prize money from his 1907 Nobel Prize to establish the Laboratory of Tropical Diseases at the Pasteur Institute, and in 1908 he founded the Société de Pathologie Exotique. He was named a Commander of the Legion of Honor in 1912 and president of the Academy of Medicine in 1920. Laveran died after a short illness on May 18, 1922, in Paris.

Lavoisier, Antoine-Laurent

(1743–1794)
French
Chemist

Widely regarded as the founder of modern chemistry, Antoine-Laurent Lavoisier was a versatile and talented scientist who made numerous contributions not only to chemistry but also to geology, physiology, economics, and social reform. Through experimental observations Lavoisier discovered oxygen and its role in combustion. He also isolated the major components of air and introduced the

Antoine-Laurent Lavoisier, who is regarded by many as the founder of modern chemistry. *Smith Collection, Rare Book & Manuscript Library, University of Pennsylvania.*

method of classifying chemical compounds. Lavoisier was also instrumental in the development of thermochemistry.

The father of Lavoisier, Jean-Antoine, was a wealthy Paris lawyer who inherited his family's estate in 1741. He married Émilie Punctis, the daughter of an attorney, in 1742; on August 26, 1743, she gave birth to Lavoisier. Lavoisier's mother died when he was five years old. He was educated at the Collège des Quatre Nations, commonly known as the Collège Mazarin, beginning in 1754, when he was 11. The school had an outstanding reputation and Lavoisier gained a thorough classical and literary education and the best scientific training available in Paris.

In 1761 Lavoisier began to study law, intending to follow the family tradition. He earned his degree in 1763 and received his license to practice law the following year. The pull of science was strong, however, and his friend Jean-Étienne Guettard, a geologist, encouraged him to study geology, mineralogy, and chemistry, which Guettard considered to be imperative for the analysis of rocks and minerals. From 1762 to 1763 Lavoisier took courses in chemistry given by the popular Guillaume Francois Rouelle. He also expanded his knowledge of geology under Guettard's guidance. Lavoisier completed the first geological map of France, which earned him an invitation to the Royal Academy of Sciences in 1768. Also that year he invested in a tax-collecting company, which was prosperous enough to allow him to build his own laboratory.

Lavoisier is best known for his studies of combustion. During that time it was believed that combustible matter contained a substance known as phlogiston, which was emitted when combustion occurred. The primary flaw of this theory was that substances sometimes increased in weight as a result of combustion. Lavoisier sought to solve this problem and carried out a range of experiments. He burned phosphorus, lead, and a number of other elements in closed containers and observed that the weight of the solid increased, but the weight of the container and its contents did not. In 1772 he found that burning phosphorus and sulfur caused a gain in weight when combined with air. Lavoisier then heard of JOSEPH PRIESTLEY's discovery that mercury oxide released a gas when heated and left behind mercury; Priestley referred to the gas as dephlogisticated air. Expanding on Priestley's findings, Lavoisier determined in 1778 that the gas that mixed with substances during combustion was the same gas that was emitted when mercury oxide was heated. Lavoisier called the gas *oxygen*. He also recognized the existence of a second gas, which later was named *nitrogen*.

In 1776 Lavoisier worked at the Royal Arsenal, in charge of gunpowder production. It was there that Lavoisier collaborated with Marquis PIERRE-SIMON DE LAPLACE to create an ice calorimeter and measure the heats of combustion and respiration. This work marked the beginning of thermochemistry. In 1787 Lavoisier worked with three other French chemists to propose a method of classifying chemical compounds. This system is still used today. Lavoisier published an important and influential work in 1789 called *Elementary Treatise on Chemistry*. It summarized his observations, discussed the law of conservation of mass, and provided a list of the known elements.

In addition to his combustion studies, Lavoisier explained the formation of water from hydrogen and sought to discover a method for improving the water supply to Paris. He also worked toward social reform, endorsed scientific agriculture, and was a member of the commission that promoted the metric system in France. In 1771 Lavoisier, at age 28, married Marie Anne Pierrette Paulze, who was 14 years old. The couple had no children, but Lavoisier's wife became his close collaborator. Lavoisier's life was cut short in 1794, when he was executed during the Reign of Terror.

Lawrence, Ernest Orlando

(1901–1958)
American
Physicist

Ernest Orlando Lawrence invented the cyclotron, a nuclear particle accelerator that improved upon the existing model by transforming the path of particles from a straight line to a circular motion, allowing for much greater speeds in much smaller space. With this device he and his colleagues discovered the radioactive isotope of plutonium. He won the 1939 Nobel Prize in physics for this work.

Lawrence was born on August 8, 1901, in Canton, South Dakota. He was the eldest son of Gunda Jacobson, a mathematics teacher, and Carl Gustav Lawrence, the state superintendent of public education and later the president of a teacher's college. His younger brother, John, worked with Lawrence at Berkeley. In 1932 Lawrence married Mary Kimberly Blumer, better known as Molly, the daughter of the dean of the Yale University School of Medicine. Together the couple had six children—four daughters and two sons.

Lawrence commenced his undergraduate career at St. Olaf's College in Minnesota before transferring to the University of South Dakota, where he received his B.S. in 1922. A year later he earned his M.S. from the University of Minnesota. There he studied under W.F.G. Swann, and Lawrence pursued his doctorate by following his mentor first to the University of Chicago and then to Yale University, where he wrote his dissertation on the photoelectric effect to earn his Ph.D. in 1925. That year he published

Ernest Orlando Lawrence, who discovered with his colleagues the radioactive isotope of plutonium. *Department of Special Collections, Stanford University Libraries.*

"The Photoelectric Effect in Potassium Vapour as a Function of the Frequency of Light" in *Philosophical Magazine*.

Lawrence worked briefly at Yale before moving to the University of California at Berkeley in 1928. In 1930 at the age of 29 he became the youngest full professor of physics at the university. In 1936 the university appointed him as the director of its Radiation Laboratory, later dubbed the Lawrence Berkeley Laboratory.

Lawrence conceived of the idea for the cyclotron in 1929, and by 1932 he and his colleagues, including his graduate students Niels E. Edlefson and M. Stanley Livingston, succeeded in disintegrating a lithium nucleus in the cyclotron. Along with Livingston and M. G. White, Lawrence reported these results in the journal *Physical Review* in the article "Disintegration of Lithium by Swiftly Moving Photons." Lawrence subsequently improved upon the cyclotron by inventing the bevatron, a larger and more powerful version. During World War II Lawrence worked on the Manhattan Project from his own lab while his Berkeley colleague J. ROBERT OPPENHEIMER directed the laboratories at the Los Alamos, New Mexico, facility where the atomic bombs were made. After the war Lawrence and Oppenheimer divided over the future of nuclear technology, Lawrence in favor of armament and Oppenheimer favoring disarmament.

Before Lawrence received the Nobel Prize, he had won the Comstock Prize of the National Academy and the Hughes Medal of the Royal Society, as well as the 1957 Fermi Award. Lawrence died of recurrent ulcerative colitis on August 27, 1958, in Berkeley, California. The transuranium element 103, discovered in 1961 at Berkeley, was named *lawrencium* in his honor.

Leakey, Louis Seymour Bazett

(1903–1972)
British
Anthropologist, Archaeologist

Persistent, controversial, and tirelessly enthusiastic, Louis Seymour Bazett Leakey made breakthrough discoveries concerning the evolution of human beings and was a pioneer in paleoanthropology. The African-born Leakey, through fossil findings and ancient cultural artifacts, was able to prove that human life was older than had been previously believed and also demonstrated that its evolution was concentrated in Africa and not Asia as was commonly accepted. Leakey and his team of researchers found numerous remains of apes, humans, and plant and animal species through their many archaeological excavations.

Born on August 7, 1903, in Kabete, Kenya, Leakey was the son of Mary Bazett and Harry Leakey, missionaries of the Church of England. Leakey spent his childhood at the Church Missionary Society, where his parents worked, and learned the language and customs of the Kikuyu. While pursuing his childhood interest in birds, Leakey frequently came upon stone tools that had been washed out of the ground by rains. Leakey believed these tools to have prehistoric roots, but scientists of the age considered Asia to be the center of human evolution, and there was little interest in East Africa as a possible site for locating evidence of early humans. Leakey left Africa for England soon after World War I ended and attended a public school there.

Leakey attended Cambridge University, studying anthropology and archaeology. He took a year off from school in 1924 and traveled to Tanganyika (now Tanzania) with the British Museum East African Expedition on a fossil-hunting trip. This expedition, coupled with his academic studies, encouraged Leakey to seek out the origins of humanity, which he supposed could be found in Africa. In 1926 Leakey received his doctorate in African prehistory.

From 1926 through 1935 Leakey organized a number of expeditions to East Africa. On his earliest trip Leakey unearthed a 200,000-year-old hand ax in addition to materials from the late Stone Age. In 1931, despite attempts to dissuade him by the paleontologist Hans Reck, Leakey made his first journey to Olduvai Gorge, a ravine in Tanzania, in search of remains. He found animal fossils and flint tools, which provided conclusive evidence of the presence of hominids. The following year Leakey discovered remains of modern humans, *Homo sapiens,* near Lake Victoria in the form of skulls and a jaw. It was also in the mid-1930s that Leakey married Mary Douglas Nicol, his second wife. MARY DOUGLAS NICOL LEAKEY became an integral member of Leakey's the research team and made some of their most important discoveries.

Toward the end of the 1930s Leakey shifted his focus from searching for stone tools, which provided important evidence of human habitation, to the excavation of human and prehuman remains. Regular excavations at Olduvai Gorge commenced in 1952, and in 1959 the team discovered fragments of a hominid skull that resembled that of the australopithecines, hominids with small brains and near-human dentition. Leakey believed the discovery to represent a new genus, and he named it *Zinjanthropus boisei.* In the early 1960s Leakey's team found more fragments. Upon studying the bones, Leakey discovered that they must have belonged to a creature more similar to modern humans than *Zinjanthropus,* with a larger brain and the ability to walk fully upright. He suggested that it existed simultaneously with *Zinjanthropus.* Dating of the bed in which the bones were found indicated an age of 1.75 million years. Leakey named the new creature *Homo habilis* and viewed it as a direct ancestor of modern humans. This discovery suggested that human life was much older than scientists had believed, and Leakey's ideas were widely criticized.

Leakey made other significant finds, including the discovery of remains of *Proconsul africanus* and *Kenyapithecus,* both links between apes and early humans. In the 1960s Leakey devoted more time to the Centre for Prehistory and Paleontology in Nairobi and was influential in the careers of JANE GOODALL and DIAN FOSSEY, who studied primates. One of Leakey's sons, RICHARD ERSKINE FRERE LEAKEY, became a noted paleontologist and continued his father's work.

Leakey, Mary Douglas Nicol

(1913–1996)
British
Paleontologist, Anthropologist

Mary Leakey made many of the discoveries of bones of human ancestors for which her better-known husband, LOUIS SEYMOUR BAZETT LEAKEY, became famous. After his death she was celebrated in her own right for such finds as the earliest known fossil footprints of humans walking upright. "It was Mary who really gave that team scientific validity," Gilbert Grosvenor, chairman of the National Geographic Society, once said.

Leakey was born Mary Douglas Nicol on February 6, 1913, in London. Her father, Erskine Nicol, was a landscape painter. One of his favorite places to paint was southwestern France, where beautiful paintings by Stone Age artists had been discovered in caves. Mary loved exploring the caves and decided that she wanted to study early humans.

Leakey met the archaeologists Alexander Keiller and Dorothy Liddell in the late 1920s and began assisting Liddell on summer digs, drawing artifacts found at the sites. She also increased her knowledge of archaeology by attending lectures at the London Museum and London University. Another woman archaeologist, Gertrude Caton-Thompson, soon asked her to illustrate one of her books as well.

In 1933 Caton-Thompson invited Leakey to a dinner party she was giving for a visiting scientist, Louis S.B. Leakey. Leakey, born of British parents and raised in Kenya, was starting to become known for his studies of early humans and their ancestors. He asked Mary to make some drawings for him, and the two fell in love, even though Leakey was 10 years older than Nicol, married, and the father of two children.

Mary traveled to Africa with Leakey in 1935 and for the first time gazed "spellbound" at his favorite site, Olduvai Gorge in Tanzania, near the Kenyan border, a view that she once said "has . . . come to mean more to me than any other in the world." The two married on December 24, 1936, as soon as Leakey's divorce became final, and were, in Mary's words, "blissfully happy." They later had three sons, Jonathan, Richard, and Philip. RICHARD ERSKINE FRERE LEAKEY, like his parents, would become famous for studies of early humans. The Leakeys became close professional as well as personal partners. They excavated several Stone Age sites in Kenya and Tanzania during the 1930s, and Mary continued during World War II while Louis worked for British intelligence. After the war Louis spent part of his time at the Coryndon Museum in Nairobi, of which he became curator in 1946, while Mary went on digging at Olduvai and elsewhere. In 1948 she made the couple's first big find, the skull of an ape called *Proconsul africanus.* The finding of this 16-million-year-old distant human ancestor in East Africa gave weight to the idea that humans had originated in Africa, rather than in Asia, as had been thought. It also made the Leakeys world famous.

Mary continued her work at Olduvai during the 1950s and also explored a site in Tanzania where people of the late Stone Age, when the Sahara was a fertile valley, had painted thousands of humans and animals on rocks, revealing such details as clothing and hairstyles. She copied some 1,600 of these paintings, a task she later called "one of the highlights" of her career. The best of her drawings were published in a book in 1983.

Mary Leakey made another major discovery on July 17, 1959. She and Louis had been excavating together at Olduvai, but on that day Louis was sick and stayed in camp. Walking in the oldest part of the site, Mary spotted a piece of bone in the ground. It proved to be part of an upper jaw, complete with two large, humanlike teeth. She dashed back to camp and burst into Louis's tent, shouting, "I've got him! I've got him!" What she eventually had were about 400 bits of bone, which she painstakingly assembled into an almost complete skull of a 1.75-million-year-old humanlike creature that the Leakeys named *Zinjanthropus,* or "Zinj" for short.

The discovery of *Zinjanthropus* extended the timeline of human evolution back by a million years. Louis Leakey at first believed that Zinj was a "missing link" between humans and apes, but later research showed that it was an australopithecine, a member of a family of humanlike beings that developed alongside the earliest true humans. A few years later the Leakeys found a skull of a new human species as well. They named it *Homo habilis,* or "handy man," because of the many tools found nearby.

Louis and Mary Leakey drifted apart during the 1960s, and their marriage was over in all but name by the time Louis died of a heart attack in 1972. Mary continued working, and in 1978 she made what she felt was her most important discovery: three sets of fossil footprints crossing a patch of hardened volcanic ash at Laetoli, Kenya, about 30 miles south of Olduvai. The footprints looked so fresh that, she said, "they could have been left this morning," but tests showed that they had been made about 3.6 million years ago, much earlier than when humans had been thought to be walking upright. Leakey noted that no tools were found in the area, a finding that suggested that humans had begun to walk upright before they started to make tools. "This new freedom of forelimbs posed a challenge," Leakey wrote in *National Geographic.* "The brain expanded to meet it. And mankind was formed."

Mary Leakey received many awards for her work, including the Hubbard Medal of the National Geographic Society, which she shared with Louis in 1962; the Boston Museum of Science's Bradford Washburn Award; and the Gold Medal of the Society of Women Geographers. Age and failing eyesight forced her to give up fieldwork in the early 1980s. She died in Nairobi on December 9, 1996, at age 83.

Leakey, Richard Erskine Frere

(1944–)
Kenyan
Anthropologist, Paleontologist

A member of the famous Leakey family of paleoanthropologists, Richard Erskine Frere Leakey is known for his extensive discoveries of ancient human remains in East Africa. Continuing his parents' work, Leakey proposed that humans were much older than had been formerly proposed; Leakey's findings led him to conclude that humans existed about 3 million years ago: nearly double the time span of previous estimates. Leakey also adopted his father's controversial idea that there were at least two parallel lines of human evolution, with one leading to modern humans.

Richard Leakey was the son of the Kenyan-born British anthropologist LOUIS SEYMOUR BAZETT LEAKEY and the British anthropologist MARY DOUGLAS NICOL LEAKEY. Born on December 19, 1944, in Nairobi, Kenya, Leakey made his first fossil discovery at the age of six, when he accompanied his parents on an excavation near Lake Victoria. The remains he found belonged to an extinct giant pig. Leakey was still determined, however, not to follow in his parents' footsteps. In the early 1960s, when Leakey was a teenager, he dropped out of high school to lead photographic safaris in Kenya. At this time he also found a talent for trapping animals. During a safari in 1963 Leakey took a group of paleontologists to an area in Tanganyika (now Tanzania) that had fossil beds. The discovery of an australopithecine jaw, the first of a complete lower jaw and the only skull fragment found since his mother's discovery in 1959, spurred him to pursue a career in anthropology. In 1964 Leakey married Margaret Cropper, who had worked on his father's research team. The couple divorced in the late 1960s, and in 1970 Leakey married Meave Gillian Epps.

Intending to pursue studies in anthropology, Leakey traveled to London and entered school. Though Leakey completed a two-year program in just six months, his money and his interest in classroom work ran out, and he returned to Kenya without a degree. In 1967 Leakey led an expedition to the Omo Delta region in southern Ethiopia, where the excavation site was estimated to have fossils from about 150,000 years ago. The remains of two human skulls, believed to be of modern humans, were found, providing evidence contradictory to the commonly accepted view that modern humans came into being about 60,000 years ago.

Though his parents spent a great deal of time excavating in the Olduvai Gorge in Tanzania, Leakey focused on regions in northern Kenya and southern Ethiopia, most notably the Koobi Fora area, located on the border of Ethiopia and Kenya. There his team unearthed more than 400 hominid fossils, at one point at the rate of nearly 2 fossils per week, and a large amount of stone tools. In 1972 the team discovered a skull that seemed to resemble one identified by Leakey's father, which his father had named *Homo habilis.* The skull, called "Skull 1470," had about double the cranial capacity of *Australopithecus boisei* (then *Zinjanthropus*) and more than half that of modern humans. Leakey believed the skull was 2.9 million years old; that idea led him to theorize that early hominids may have been present about 2.5 to 3.5 million years ago. This supported the theory that *Homo habilis* was not a descendant of the australopithecines but coexisted with them. In 1975 the skull of a *Homo erectus,* the ancestor of modern humans, was found at Koobi Fora. The skull dated back about 1.5 million years and represented the earliest known remains of *Homo erectus* in Africa.

Leakey continued to make significant anthropological findings, including the discovery in 1984 of a nearly complete *Homo erectus* skeleton. Leakey's findings did much to corroborate the theories of his parents and significantly advanced the field of physical anthropology. In 1979 Leakey suffered kidney failure and survived with the donation of a kidney by his brother Philip. Leakey lost both legs as a result of an airplane crash in 1993 but continued his activities.

Leavitt, Henrietta Swan
(1868–1921)
American
Astronomer

Henrietta Leavitt, who discovered the first method of measuring large-scale distances in the universe. *Harvard College Observatory.*

In addition to determining the brightness of thousands of stars, Henrietta Swan Leavitt discovered the first method of measuring large-scale distances in the universe. She was born on July 4, 1868, in Lancaster, Massachusetts, one of George and Henrietta Leavitt's seven children. Her father was a minister. She grew up in Cambridge and later in Cleveland, Ohio.

Leavitt studied for two years at Oberlin College in Ohio, then transferred to Radcliffe in 1888 and graduated in 1892. A class she took in her senior year interested her in astronomy. She began doing volunteer work for the Harvard Observatory in 1895 and joined the paid staff in 1902. She was one of a number of women "computers," including WILLIAMINA PATON STEVENS FLEMING and ANNIE JUMP CANNON, whom Edward Pickering, the head of the observatory, hired to work on large projects.

Astronomers had first measured stars' brightness by comparing stars by eye. As more astronomical work came to be done through photographs, the scale of brightness had to be recalibrated because film registers slightly different wavelengths of light than the eye. Pickering began a large project to do this in 1907 and put Leavitt in charge of it, making her the chief of the photographic photometry department. She first calculated the brightness of 46 stars near the North (Pole) Star, selected to represent all degrees of brightness, then extended the work to other parts of the sky. In 1913 the International Committee on Photographic Magnitudes adopted Leavitt's "North Polar Sequence" as its brightness standard.

Leavitt's true interest, however, was in variable stars, which brighten and dim on a regular schedule. She discovered some 2,400 of these stars, most of them in the two Magellanic Clouds, which were later shown to be small galaxies. Many variables in the Small Magellanic Cloud were a type called Cepheids, and Leavitt showed in 1912 that the longer a Cepheid's period (the time of one cycle of brightening and dimming), the brighter the star was at its brightest. Stars' apparent brightness is affected by their distance from Earth, but since all Cepheids in the cloud were about equally far away, she concluded that their periods were related to their true brightness.

Leavitt would have liked to follow up on her finding, but Pickering did not encourage his women "computers" to do independent or theoretical work. Cecilia Payne-Gaposchkin, another of the Harvard women astronomers, wrote that restriction of Leavitt to photometry "was a harsh decision, which condemned a brilliant scientist to uncongenial work, and probably set back the study of variable stars for several decades."

Other astronomers turned Leavitt's discovery into what the astronomer VERA COOPER RUBIN calls "the most fundamental method of calculating distances in the universe." If a Cepheid's period were known, its true brightness could be calculated, and if its true brightness were known, its distance from Earth could be calculated by comparing the true brightness with the apparent brightness. By using Cepheids as yardsticks, astronomers could determine the distance of any group of stars in which they were embedded.

Leavitt never achieved the fame that Cannon or Fleming did. Some of her fellow women astronomers, however, thought she was the brightest of the Harvard group. She died of cancer on December 12, 1921, at the age of 52.

Lee, Tsung-Dao

(1926–)
Chinese/American
Physicist

Tsung-Dao Lee won the 1957 Nobel Prize in physics in tandem with CHEN NING YANG for their explanation of the seemingly inexplicable behavior of the K-meson. The two scientists hypothesized that the K-meson violated the law of the conservation of parity, a supposedly inviolable natural law defining symmetry in subatomic physics. Lee and Yang even suggested experiments to prove their theory correct, which CHIEN-SHIUNG WU performed successfully within months, verifying the validity of Lee and Yang's revolutionary theory.

Tsung-Dao Lee, who won the 1957 Nobel Prize in physics for work on the unusual behavior of the K-meson. *AIP Emilio Segrè Visual Archives, Segrè Collection.*

Lee was born on November 24, 1926, in Shanghai, China. He was the third of six children born to Ming-Chang Chang and Tsing-Kong Lee, a businessman. Lee met Hui-Chung Chin, also known as Jeanette, while both were studying in Chicago, and the couple married on June 3, 1950. Together they had two sons, James and Stephen.

Lee commenced his undergraduate education at the National Chekian University in Kweichow, though the Japanese invasion of 1945 interrupted his course of study, which he continued after fleeing southward to the National Southwest Associated University in Kunming. In 1946, before Lee had completed his bachelor's degree, he was invited to accompany his professor, Ta-You Wu, back to the United States. However, only the University of Chicago would accept Lee for graduate study without an undergraduate degree. There Lee studied under ENRICO FERMI and received his Ph.D. in 1950 after writing his dissertation on the hydrogen content of white dwarf stars.

In 1950 Lee went to work for the Yerkes Astronomical Observatory at Lake Geneva, Wisconsin, and later that year he accepted a research associateship at the University of California at Berkeley. From there he moved on in 1951 to the Institute for Advanced Studies at Princeton University, where he reunited with Yang, an acquaintance from Kunming and Chicago. Even after 1953, when Lee accepted an assistant professorship of physics at Columbia University, he and Yang continued to collaborate by meeting weekly, alternately in New York City and in Princeton, New Jersey. Columbia University promoted Lee to the position of associate professor in 1955 and full professor in 1956.

That same year Lee and Yang focused their attention on the question of the K-meson, which seemed to be one particle in all ways except for one: It did not abide by the law of conservation of parity, or the symmetry of subatomic particle interactions. Physicists solved this problem by labeling it as two different particles, tau mesons and theta mesons, which exhibited even and odd parity. However, the only factor supporting this solution was the sanctity of the law of conservation of parity. Lee and Yang put aside the weight of scientific precedent to suggest a more logical explanation: that the two were in fact one particle that did not abide by the law. The June 22, 1956, issue of *Physical Review* printed their paper, "Question of Parity Conservation in Weak Interactions." Their simple yet elegant theory withstood the test of experimental substantiation, earning them international recognition immediately.

Though some scientists wait years to receive a Nobel Prize for their ground-breaking work, Lee and Yang received theirs within one year. In 1963 Columbia University named Lee its first Enrico Fermi Professor of Physics, wooing him back from his three-year stint at the Institute for Advanced Studies in Princeton. In 1984 Columbia further honored Lee by naming him a University Professor.

Lee, Yuan Tseh
(1936–)
Chinese/American
Chemist

Yuan Tseh Lee won the 1986 Nobel Prize in chemistry with Dudley Herschbach and John C. Polanyi for their work on chemical kinetics or reaction dynamics—the study of the chemical transformations that occur when molecules collide at supersonic speeds. Lee's primary role consisted of improving the instrumentation to achieve the degree of sensitivity necessary to conduct the experiments.

Lee was born on November 29, 1936, in Hsinchu, Taiwan. His father was an artist and an art teacher, and his mother was an elementary school teacher. In 1963 Lee married Bernice Chinli Wu, a childhood friend, and together the couple had three children—a daughter and two sons. In 1974 Lee became a citizen of the United States.

Reading a biography of MARIE SKLODOWSKA CURIE influenced Lee to study science. He was admitted to the National Taiwan University without taking an entrance examination as a result of his outstanding performance at Hsinchu High School. He graduated with a B.S. in 1959, then moved on to the National Tsinghua University, where he earned his M.S. in 1961. Lee immigrated to the United States in 1962 to attend the University of California at Berkeley, where he earned his Ph.D. in 1965. Lee performed his doctoral research on the collision of neutral hydrogen atoms with positively charged nitrogen ions with the help of an instrument that he designed and constructed himself, a precursor to his later building of an experimental apparatus that won him the Nobel.

Lee stayed on at Berkeley for postdoctoral study until Herschbach invited him to Harvard in 1967 to continue this work. Specifically Herschbach was seeking Lee's expertise in instrumentation, as his experiments required more technical sophistication than was available to him at the time. Within a year Lee had devised and built an improved mass spectrometer to measure with much more accuracy the results of Herschbach's colliding molecular beam experiments. Herschbach's experiments differed from those previously performed in the field by focusing not simply on the physical results of molecular collisions, as physicists tended to measure the energy released in this process, but also on the chemical changes that occurred in the process. Without the technological innovations introduced by Lee, Herschbach's experiments never would have achieved Nobel status. Furthermore Lee's apparatus subsequently aided in diverse researches, earning it the title of a "universal" mechanism.

The University of Chicago hired Lee as an assistant professor in 1968 and promoted him to associate professor in 1971 and then to full professor in 1973. In 1974 Lee returned to the University of California at Berkeley as a professor of chemistry.

In 1981 the United States Energy Research and Development Agency awarded Lee its Ernest Orlando Lawrence Memorial Award. In addition to the Nobel Prize, Lee won the Peter Debye Award from the American Chemical Society and the National Medal of Science from the National Science Foundation in 1986. Herschbach once described Lee as "the Mozart of physical chemistry," who combined innate brilliance with technical precision.

Leeuwenhoek, Antoni van
(1632–1723)
Dutch
Microscopist

Although Antoni van Leeuwenhoek had no formal scientific education, he contributed immensely to scientific understanding through his empirical observations at the microscopic level. Isolated from the scientific community

Antoni van Leeuwenhoek, whose microscopic observations of the mechanism of sexual reproduction contributed significantly to scientific understanding of reproduction in plants and animals. *Smith Collection, Rare Book & Manuscript Library, University of Pennsylvania.*

by his lack of training and his understanding only of his native language, Leeuwenhoek relied on his own faculties, such as his keen eyesight, his excellent hand-to-eye coordination, and his innate mathematical abilities, to guide his scientific experimentation.

Leeuwenhoek was born on October 24, 1632, in Delft, Holland. His parents were Margriet Jacobsdochter van den Berch and Philips Thoniszoon, a basket maker. On July 29, 1654, Leeuwenhoek married Barbara de Mey, daughter of Elias de Mey, a British serge merchant. Together the couple had five children, but only their daughter Maria survived her father. Leeuwenhoek spent his early life as a draper and haberdasher; incidentally, he ground and polished his own magnification lenses to inspect cloth quality. He also served as the chamberlain to the sheriffs of Delft, as an alderman, and as a wine gauger (or inspector of weights and measures). Leeuwenhoek's first wife died in 1666, and on January 25, 1671, he married Cornelia Swalmius, daughter of a Calvinist minister.

That year Leeuwenhoek's scientific pursuits commenced in earnest, as his financial situation was secure and his second wife's social position put him into contact with the Dutch intelligentsia. In his microscopic observations he focused his attention on the mechanisms of sexual reproduction and the nourishing systems in plants and animals. In 1674 he became the first person to observe blood corpuscles; that same year he discovered what he called animalcules, otherwise known as protozoa. Two years later in a letter dated October 9, 1676, he reported to the Royal Society of London this discovery of microorganisms with life force, a revelation that startled the fellows of the society. That same year he observed bacteria in tooth scrapings, and in 1677 he observed spermatozoa.

Though Leeuwenhoek was more given to observation than to interpretation, he hypothesized that spermatozoa penetrate the egg in mammals and thus introduce the life force to sexual reproduction. This theory became known as animalculism, against which the ovists argued, as they believed that the egg contributed the life force. This controversy remained unsolved for the next two centuries, until the fusion of sperm and egg was demonstrated in 1875. In 1680 the Royal Society elected Leeuwenhoek as a fellow, and he continued with his microscopic observations: In 1682 he recognized striations in skeletal muscle; in 1683 he described blood capillaries; and in 1684 he observed red blood cells. By that year Leeuwenhoek had written 110 letters reporting his findings to the Royal Society, which he gathered together for publication.

Throughout his lifetime Leeuwenhoek ground 419 lenses, which magnified samples 30 times at the weakest and 266 times at the strongest magnification. In 1716 he received a silver medal from the Louvain College of Professors, and throughout his late life he received eminent visitors, including Czar Peter the Great and Frederick the Great of Prussia, a testament to his significance as a prominent scientist of his time. Leeuwenhoek died on August 26, 1723, in Delft.

Lehmann, Inge

(1888–1993)
Danish
Geologist

Inge Lehmann used the effects of waves sent through the Earth by earthquakes to prove that the planet has a solid core. She was born in Copenhagen, Denmark, in 1888. Her father was a professor of psychology at the University of Copenhagen. Lehmann studied mathematics and physical science there from 1907 to 1910 and also spent a year at Britain's Cambridge University. She then worked in the insurance industry for eight years before returning to the University of Copenhagen, where she earned master of science degrees in mathematics in 1920 and in geodesy in 1928.

Lehmann began her professional life in 1925 at the new Royal Danish Geodetic Institute and became the head of the institute's seismology department in 1928. Indeed, for two decades she was Denmark's only seismologist. Geologists knew that a layer of molten rock lay below the Earth's surface, and at the time they thought this layer extended all the way to the planet's center. Lehmann, however, noticed that one type of earthquake wave, called P prime, did not act as it would be expected to do if the Earth's entire core were molten. She concluded in 1936 that these waves were being reflected off a dense core of solid material that began 869 miles from the Earth's center. Other geologists refused to accept this idea at first, but in time the evidence supporting it grew too great to deny.

Lehmann retired in 1953 and died 40 years later at the amazing age of 105. Among the honors she received were the Gold Medal of the Royal Danish Society of Science (1965), the Bowie Medal of the American Geophysical Union (1971), and the Medal of the Seismological Society of America (1977). In 1997 the American Geophysical Union named a medal after her—the first that the group has named for a woman—to be awarded in recognition of outstanding research on the structure, composition, and dynamics of the Earth's core.

Leloir, Luis Federico

(1906–1987)
French/Argentine
Biochemist

Luis Federico Leloir discovered sugar nucleotides that proved to be what he identified as the "missing link" in

Luis Leloir, the first Argentinean to win a Nobel Prize in chemistry, which he was awarded in 1970. *Bernard Becker Medical Library, Washington University School of Medicine.*

the mystery of how the body converts carbohydrates into energy. For this work he won the 1970 Nobel Prize in chemistry, the third Argentinian to win the prize and the first in the field of chemistry.

Leloir was born on September 6, 1906, while his parents, Federico and Hortensia Aguirre Leloir, were vacationing in Paris, France, from their home in Buenos Aires, Argentina. Leloir's grandparents had emigrated from France to Argentina and become landowners; that move secured future wealth for the family when cattle and land became important to the Argentine economy. Leloir married Amelie Zuherbuhler in 1943, and together the couple had one daughter.

Leloir attended the University of Buenos Aires, earning his M.D. there in 1932. His internship at the university hospital convinced him of his unsuitability for practicing medicine, so he decided to pursue medical research instead. Toward this end he transferred to the university's Institute of Physiology, where he conducted his doctoral research under Bernardo Houssay, who exerted a lasting influence on Leloir's career and later won the 1947 Nobel Prize for his work on the adrenal gland. The faculty voted Leloir's dissertation on the influence of adrenal glands on carbohydrate metabolism the best thesis of the year. Leloir spent a year conducting postdoctoral enzyme research at the Biochemical Laboratory of Cambridge University in England, after which he returned to Buenos Aires.

In 1943 Leloir sought exile in the United States because of political differences with the tumultuous Argentinian government, which had dismissed Houssay for supposedly opposing the government. Leloir ended up at Washington University in St. Louis as a research assistant in the biochemical laboratory. From there he moved to the Enzyme Research Laboratory of the College of Physicians and Surgeons at Columbia University in New York City before returning to Buenos Aires.

In 1945 he founded the private Institute for Biochemical Investigations with the backing of Jaime Campomar, the owner of a textile business, who died in 1957, leaving the question of future funding up in the air. Leloir's appeal to the National Institutes of Health in the United States yielded a surprising response of funding until 1958, when the Argentinian government stepped in to offer the institute the use of a former girls' school and financial support from the newly formed Argentine National Research Council. The institute then officially became affiliated with the University of Buenos Aires.

In 1946 Leloir and his colleagues published *Renal Hypertension,* which described their discovery of antihypertensives. More significantly they announced in 1957 their discovery of a new coenzyme, uridine triphosphate (UTP), which formed a new sugar nucleotide, uridine diphosphate glucose (UDPG). This model replaced the model suggested by Carl Cori and GERTY THERESA RADNITZ CORI in the 1930s, which worked in vitro but did not replicate what happens in the body as did Leloir's model.

Leloir went on to help establish the Argentine Society for Biochemical Research and the Panamerican Association of Biochemical Sciences. Leloir died on December 2, 1987, in Buenos Aires.

Lenard, Philipp Eduard Anton von

(1862–1947)
German
Physicist

An experimental physicist, Philipp Eduard Anton von Lenard received a Nobel Prize in physics in 1905 for his work on cathode rays. He produced beams of cathode rays, or electrons, in air, allowing for the study of electrons. In this manner Lenard was able to conclude that an atom is primarily empty space. He made numerous other discoveries, many of which advanced the understanding of atomic physics. Lenard also studied phosphorescence for more than four decades and conducted breakthrough research on the photoelectric effect.

Born on June 7, 1862, at Pressburg, Austria-Hungary (present-day Bratislava, Slovak Republic), Lenard was an only child and spent most of his life in Germany. Lenard's father was a successful wine maker. His mother, Antonie Baumann, died when Lenard was a child, and he was raised by an aunt who later married his father. Interested in mathematics and physics from an early age, Lenard conducted experiments and studied these subjects on his own, using college textbooks as a guide.

Though Lenard wished to study science, his father planned for him to enter the family wine business. Only if Lenard would study the chemistry of wine at the technical college in Vienna and Budapest would his father allow him to pursue a higher education. Uninspired by the schooling, Lenard soon dropped out to work for his father. In 1883 he had managed to save enough money for a trip to Germany. There he met the chemist ROBERT WILHELM BUNSEN, whom he had long admired. Bunsen's lectures inspired Lenard to pursue science, and that same year he enrolled at the University of Heidelberg. After studying physics at Heidelberg and at the University of Berlin, Lenard received his doctorate in 1886.

In 1891 Lenard went to the University of Bonn to become the assistant of the physicist HEINRICH RUDOLPH HERTZ, who had discovered electromagnetic waves. Continuing Hertz's experiments with cathode rays, a topic that had interested him since 1880, when he had read a paper by Sir WILLIAM CROOKES discussing the movement of cathode rays in discharge tubes, Lenard in 1892 planned an experiment to observe the rays outside discharge tubes. Lenard developed a tube that used aluminum foil as a window and was able to demonstrate that cathode rays could pass through the window and move about in the air, where they could be studied independently of the discharge process. The window came to be known as the Lenard window. Lenard theorized that because the rays were able to move through the metal, the atoms in it must be made up of a large portion of empty space.

After Hertz died in 1894, Lenard became head of Hertz's laboratory; he left to accept a professorship at Heidelberg in 1896. In 1898 he was offered a full professorship and command of the physics laboratory at the University of Kiel. In 1902 Lenard began to focus on photoelectricity. Lenard discovered numerous significant properties of the photoelectric effect, including the fact that the velocity of electrons ejected was dependent not on intensity of light but on the wavelength. ALBERT EINSTEIN was able to provide an explanation for this phenomenon in 1905 with the theory of light quanta, but Lenard rejected that theory.

After receiving the Nobel Prize in 1905, Lenard became professor and director of the physics and radiology laboratory at the University of Heidelberg in 1907. He remained at Heidelberg until 1931, and the laboratory was renamed the Philipp Lenard Laboratory in 1935. Lenard received honorary doctorates from the Universities of Christiania (1911), Dresden (1922), and Pressburg (1942). He also received the Franklin Medal in 1905. Lenard was an early and avid supporter of Adolf Hitler and the National Socialist Party. He was married to Katharina Schlehner.

Leopold, Aldo

(1886–1948)
American
Conservationist

Aldo Leopold exerted an inestimable influence on American public policy regarding wilderness use and game management. His breakthrough book, *Game Management,* shaped the direction of national laws immediately after its publication in 1933; it continued to drive wilderness and wildlife legislation long after Leopold's death. Leopold's writing lived on in the public imagination long after his death as well, as the posthumously published *A Sand County Almanac, and Sketches Here and There* collected his personal observations and essays displaying his encyclopedic knowledge of and sensitivity to nature.

Leopold was born on January 11, 1886, in Burlington, Iowa. His parents, Clara Starker and Carl Leopold, sent him east for college to Yale University, where he studied at the Sheffield Scientific School for his degree in 1908. Leopold remained at Yale for graduate work at its newly established school of forestry. After graduation in 1909 he took up a job as a Forest Service ranger in New Mexico and Arizona, where he quickly climbed the ranks to deputy supervisor and then supervisor of Carson National Forest. While in Santa Fe, New Mexico, he met Estella Bergere, whom he married on October 9, 1912. Together the couple had five children, all of whom became scientists.

In 1913 the United States Forest Service appointed him assistant district forester in Albuquerque, New Mexico. During this period Leopold's philosophies began to diverge from the prevailing conservation theory of the "wise use" of natural resources, which in practice amounted to the exploitation of nature for economic gain. He expressed his more radically conservationist views in *The Pine Cone,* the newsletter of the New Mexico Game Protection Association, a group he helped organize in 1915 and served as secretary in 1916. The group's persistent lobbying placed the issue of big game protection at the heart of the gubernatorial race, leading to the eventual appointment of a state game warden sympathetic to the group's platform.

In 1924 Leopold accepted a position as associate director of the U.S. Forest Products Laboratory in Madison, Wisconsin. In 1928 he left the Forest Service to become a private consultant on forestry and wildlife issues. The University of Wisconsin created a professorship in wildlife management expressly for Leopold in 1933, and he inhabited the chair the rest of his career. That year he published *Game Management*, which influenced a host of national laws: the 1934 Fish and Wildlife Coordination Act, the 1969 National Environmental Policy Act requiring environmental impact statements, the 1973 Endangered Species Act, the 1974 Forest and Rangelands Renewable Resources Planning Act, the 1976 National Forest Management Act, and the 1976 Federal Land Policy and Management Act, among others.

Leopold served on numerous committees and organizations promoting his conservationist theories. President Franklin D. Roosevelt appointed him to his Special Committee on Wildlife Restoration. The next year Leopold helped found the Wilderness Society and became the director of the Audubon Society. In 1947 three important appointments came his way: honorary vice president of the American Forestry Association, president of the Ecological Society of American, and member of the advisory council for the Conservation Foundation. At the height of his influence Leopold died of a heart attack while fighting a fire at a neighboring farm on the Wisconsin River on April 21, 1948. His burial plot overlooks the Mississippi River from his birthplace in Burlington, Iowa.

Lester, William Alexander, Jr.

(1937–)
American
Chemist

William Lester distinguished himself by excelling in all of the roles of a modern scientist—researcher, mathematician, teacher, and administrator. Lester worked in both the academic and commercial sectors, in both pure and applied science. Lester was a frontrunner in the application of computer technology to further the interests of science. Lester rested his laurels not on a single theory or discovery, but rather on a diverse career of service to the field of science.

Lester was born on April 24, 1937, in Chicago, Illinois, the only boy in a family of four children. His mother was Elizabeth Clark and his father was William Lester, a mailman and later a postal supervisor. In 1959 Lester married Rochelle Diane Reed, and together the couple had two children.

Lester began high school in a recently desegregated school at the age of 12, having skipped two grades previously. He entered the University of Chicago on a scholarship to study history, though he gravitated toward the sciences, as he had worked in the university's physics department between his junior and senior years of high school. Lester earned his B.S. in 1958 and stayed on at the university to earn his M.S. in 1959. He started to pursue his doctorate at Washington University in St. Louis, though he stayed there only a year before moving to the Catholic University of America in Washington, D.C., where he earned his Ph.D. in 1964.

In order to support his family while pursuing his doctorate, Lester started working as a theoretical physical chemist in 1961 at the National Bureau of Standards. After receiving his doctorate, he moved to the Theoretical Chemical Institute of the University of Wisconsin at Madison, first as a research associate and then as the assistant director. In 1968 he joined the IBM Research Laboratory in San Jose, California, where he immersed himself in the study of quantum chemistry and molecular collisions, his specialties. His research helped explain the chemical reactions behind such phenomena as combustion, sound propagation, and atmospheric reactions. He left to serve on the technical planning staff for the director of research at IBM's T. J. Watson Research Center in Yorktown Heights, New York, in 1975. IBM then called him back to the San Jose facility to manage the molecular interactions group.

In 1978 Lester accepted the directorship of the National Resource for Computation in Chemistry at the Lawrence Berkeley Laboratory of the University of California at Berkeley. This position charged Lester with the responsibility of coordinating efforts throughout the scientific communities to facilitate their research by computerization. Simultaneously Lester served as the associate director and senior staff scientist at the Lawrence facility itself. In 1981 he shifted his focus to teaching as a professor of chemistry at the University of California at Berkeley while serving as the faculty senior scientist at the Materials and Molecular Research Division of the Lawrence lab. A decade later he returned to administration as the associate dean of the College of Chemistry of University of California at Berkeley in 1991.

Lester served as an editor of multiple journals and conference proceedings. He also received many awards, including the 1974 IBM Outstanding Contribution Award. The National Organization of Black Chemists and Chemical Engineers awarded him the 1979 Percy L. Julian Award in Pure and Applied Research and the 1986 Outstanding Teacher Award. Lester was a modern "renaissance man" of science.

Levi-Montalcini, Rita

(1909–)
Italian/American
Medical Researcher

Drawing on research begun in her bedroom during World War II, Rita Levi-Montalcini discovered a substance that makes nerves grow. It plays a vital part in the development of humans and animals and in the future may be used in medical treatment. For its discovery Levi-Montalcini and

Rita Levi-Montalcini, who discovered nerve growth factor (NGF), a substance that makes nerves grow and may someday be used in treatment of such conditions as Alzheimer's disease. *Bernard Becker Medical Library, Washington University School of Medicine.*

her coworker, STANLEY H. COHEN, shared the Nobel Prize in physiology or medicine in 1986.

Rita Levi-Montalcini was born in Turin, Italy, on April 22, 1909. She had a twin sister, Paola, and an older brother and sister. Her father, Adamo Levi, was an engineer and factory owner. Rita admired him but feared his temper and felt closer to her mother, Adele, whose maiden name was Montalcini. Rita later combined her father's and mother's last names in her own.

Rita felt she was "drifting along in the dark," with no idea of what to do with her life, until she was 20, when a family friend's painful death of cancer made her decide to become a physician. Having received only the minimal education allotted to girls, she had to be tutored before she could qualify for medical school. She entered the Turin School of Medicine in 1930 and earned her M.D. with honors in 1936. While still a student she became an assistant to one of her professors, Giuseppe Levi (no relation), and she continued in that capacity after graduation. From him she learned how to study spiderlike nerve cells in the nervous systems of embryo chickens.

The Fascist leader Benito Mussolini had taken control of Italy in 1925. His regime did not persecute Jews as intensely as that of Nazi Germany, but in 1938 it passed a law that deprived all Jews of academic jobs. Both Levi and Levi-Montalcini thus became unemployed. Levi-Montalcini was discouraged until a friend from medical school reminded her of all the good scientific work that had been done with limited laboratory equipment. "One doesn't lose heart in the face of the first difficulties," he told her. He suggested that she try to continue her chick embryo research at home.

The attempt seemed "a voyage of adventure" to Levi-Montalcini, so she set up what she called a "private laboratory *a la* Robinson Crusoe" in her bedroom. She made a small heater into an incubator for her eggs and sharpened sewing needles to use as scalpels for cutting up the tiny embryos. She continued even after World War II began and her family had to move to a small house in the hills to escape the Allied bombing of Turin. There her laboratory shrank to a corner of the living room, and the eggs she experimented on later became part of the family's meals.

In her makeshift laboratory Levi-Montalcini set out to duplicate the experiments of a researcher named Viktor Hamburger, who had shown that when a limb was cut off a developing embryo, the nerves starting to grow into that limb died. He thought this happened because the cells were no longer receiving some substance from the limb that they needed in order to mature. Levi-Montalcini, however, found that the cells did mature before they died. She suspected that the unknown substance kept the cells alive and attracted them to the limb rather than helping them mature. She published the results of her research in a Belgian science journal.

Mussolini fell from power in July 1943, to the relief of many Italians, but German troops took control of the country a month and a half later. Now for the first time Jewish families in Italy were in real danger. Using assumed names, Rita and most of her family hid for a year with a friend of Paola's in Florence. They returned to Turin after the Allies freed Italy in the spring of 1945.

In 1946 Levi-Montalcini received a surprise letter from Viktor Hamburger, who had read the report of her findings about his research. He was now a professor at Washington University in St. Louis, and he invited her to go to the United States and work with him for a semester. She agreed, little knowing that the "one-semester" visit that began in 1947 would last 30 years.

Levi-Montalcini's work took a new direction in 1950, when Hamburger told her about research done by a former student, Elmer Bücker. Bücker had grafted tissue from a mouse cancer onto a chick embryo and found that the tumor made nerve fibers from the embryo multiply and grow toward it, just as a grafted limb would have. Levi-Montalcini could not see why cancer tissue should have this effect, so she set out to repeat Bücker's experiments. She found that some mouse cancers produced an amount of nerve growth "so extraordinary that I thought I might be hallucinating" when she viewed it under the microscope. Nerves grew not only into the tumor but into nearby organs of the embryo. Levi-Montalcini noted that the nerve cells did not make contact with other cells, however, as they would have if stimulated by an extra limb.

Levi-Montalcini decided that she would have a better chance of identifying this "nerve-growth promoting agent" if she studied its effects on nerve tissue in laboratory dishes rather than on whole embryos, which contained many substances that affected growth and development. She therefore spent several months in Rio de Janeiro, Brazil, learning the techniques of tissue culture from a former student of Giuseppe Levi's. At first she had trouble making her experiments work, but she finally found one tumor that made nerves grow out from a chick embryo ganglion, or nerve bundle, "like rays from the sun."

In January 1953 a new coworker, the biochemist Stanley H. Cohen, joined Hamburger's group. He set out to learn the chemical nature of Levi-Montalcini's mystery substance, which those in the lab were beginning to call nerve growth factor (NGF). His skill in chemistry perfectly complemented Levi-Montalcini's in working with embryos and tissues. "You and I [separately] are good," Cohen once told her, "but together we are wonderful." Levi-Montalcini, for her part, later called the six years she worked with Cohen "the most intense and productive years of my life." Among other things, they learned that NGF was a protein, one of a large family of chemicals that carry out most activities in living cells.

Levi-Montalcini had been made a full professor at Washington University in 1958, but Hamburger was unable to gain a similar appointment for Cohen. Deciding that he needed more security, Cohen reluctantly gave up his research with Levi-Montalcini and moved to Vanderbilt University in Tennessee in 1959. Levi-Montalcini heard the news of his departure "like the tolling of a funeral bell."

Levi-Montalcini had become a United States citizen in 1956, but she also maintained Italian citizenship and stayed in close touch with her family. In 1959 she decided that she wanted to spend more time with them, so she arranged with Washington University to establish a research outpost in Rome. This research unit was greatly enlarged in 1969 and became part of the new Laboratory of Cell Biology, run by Italy's National Council of Research. Levi-Montalcini directed the cell biology laboratory for the next 10 years, and in 1977 she moved back to Italy for good. She shared an apartment in Rome with Paola, an artist, whom she called "part of myself."

Levi-Montalcini received many awards for her work on NGF, including the prestigious Lasker Award for medical research in 1986. The Lasker is often considered a prelude to a Nobel Prize, and so it proved to be for her. In October 1986 she learned that she was to share that year's prize in physiology or medicine with Cohen.

Levi-Montalcini officially retired in 1979 but continued to do research. In 1986 she found that NGF can spur the growth of brain cells as well as those of the spinal cord; that finding suggests that it might someday be used in a treatment for brain-damaging conditions such as Alzheimer's disease and strokes. She has also found that some cells in the immune system both produce and respond to NGF, suggesting a link between this disease-fighting system and the nervous system.

Levy, Jerre

(1938–)

American

Brain Researcher, Psychologist

Jerre Levy has helped to show that the two halves, or hemispheres, of the brain process information in different ways. She was born on April 7, 1938, in Birmingham, Alabama, and grew up in the small Alabama town of Demopolis. Her father, Jerome, owned a clothing store.

Even more than most children, young Jerre asked her parents constant questions. "When you grow up, you can spend your entire life asking and answering questions," her mother told her. "People who get paid for asking questions are called *scientists*." When she reached high school, however, an aptitude counselor told her that American universities did not allow women to become scientists—except possibly in the field of psychology, and even then only with difficulty.

Levy experienced some of that difficulty firsthand. After receiving an undergraduate degree in psychology from the University of Miami in 1962, she entered the university's graduate program in biological psychology. When she applied for a research assistantship to support her studies during her third year, however, she was informed that such assistantships were now restricted to male students. The chairman of the department explained that it was a waste of money to support female students, since "no university would ever hire a woman." Levy, divorced and the mother of two young children, finally obtained an assistantship in the university's School of Marine Science.

After a year and a half Levy transferred to the graduate program in biology at the California Institute of Technology (Caltech) and joined the laboratory of ROGER WOLCOT SPERRY. Sperry was studying people whose corpus callosum, a thick bundle of nerve fibers that normally connects the two hemispheres of the brain, had been cut surgically as a treatment for uncontrollable epilepsy. Sperry showed that each isolated hemisphere had its own mental world that could not be communicated to the other hemisphere. It was therefore possible to study the functions of each side of the split brain without the influence of the other side.

Studies of patients with damage to one side of the brain had already established that the left hemisphere processes most verbal information, whereas the right hemisphere is the specialist in processing faces and spatial information. On the basis of her studies of split-brain patients Levy proposed that the two hemispheres differ in their strategy of information processing. For instance, the left hemisphere, she suggested, analyzes specific details of pictures, whereas the right hemisphere combines the details into a whole.

After receiving her Ph.D. from Caltech in 1970, Levy did postdoctoral work at the University of Colorado and Oregon State University. She then went to the University of Pennsylvania, where she was an assistant and then associate professor from 1972 to 1977. She joined the University of Chicago in 1977 and became a full professor in that university's psychology department.

Levy has changed her thinking about the brain many times. For example, she found that the conclusion that the left hemisphere is analytical and the right hemisphere holistic—a distinction still widely repeated in the popular press—may be true only of nonverbal pictorial information, in which the right hemisphere specializes, whereas for verbal information, the specialty of the left hemisphere, the reverse is true. Each hemisphere can do higher-order, holistic thinking about the kind of information in which it specializes.

Most of Levy's research has concentrated on the links between brain organization and behavior in normal people, including studies of differences in brain function between men and women and between left- and right-handers. She has also studied the dominant role of the right hemisphere in the perception and expression of emotional information, communication between the two brain hemispheres in children and adults, and the different ways that the left and right hemispheres process printed words.

Clearly, Levy has achieved the life her mother had promised her. "I want to find out how the brain operates," she said. "And to do that, I have to ask all sorts of questions."

Li, Ching Chun

(1912–)
Chinese/American
Biostatistician, Population Geneticist

Ching Chun Li, or C.C. as he was known, was responsible for introducing the practice of random and "blind" testing in clinical trials while working on a chemotherapy study for the National Institutes of Health in the 1950s. This method has been instituted as a standard practice in medical research ever since. He, along with his colleagues, also introduced the "singles method" in segregation analysis, thus expediting the tracing of genetic defects in family lineages. Li considered the most important influence on his biostatistical work to be his fight against Lysenkoism, the belief in influential genetic traits, which had been adopted by Soviet and Chinese Communists.

Li was born on October 27, 1912, in Tientsin, China. His father was a businessman and a biblical scholar. The younger Li married Clara Lem in September 1941. The couple's first son, Jeff, died of dysentery in China in 1943. They subsequently had two more children, Carol, born in 1945, and Steven Muller, named after the Nobel Prize–winning scientist HERMAN JOSEPH MULLER, who helped Li and his family immigrate to the United States from China.

Li attended the University of Nanking from 1932 to 1936. In 1937 he traveled to the United States to attend the College of Agriculture at Cornell University to study plant breeding. That same year Theodosius Dobzhansky published the book that inspired Li to devote his life to population genetics, *Genetics and the Origin of Species.* Cornell awarded Li his Ph.D. in 1940, and he pursued postdoctoral studies in mathematics at the University of Chicago, Columbia University, and North Carolina State University.

In 1941 Li returned to China, though the ravages of World War II prevented him from finding his father in Chungking and his mother in Shanghai. He taught at the Agricultural College of National Kwangsi University in 1942, then at the University of Nanking at Chengtu from 1943 to 1946. From 1946 until 1950 he served as a professor and the chairman of the Department of Agronomy at Peking University. While he was there, the Maoist government tried to persecute him for his belief in Mendelian genetics, in opposition to the Communist Party's doctrine of Lysenkoism. Li was eventually forced to resign from Peking University, at which point he immigrated to the United States.

Li spent the remainder of his career, from 1951 on, at the University of Pittsburgh. The university promoted him to the status of professor of biometry in 1960, and from 1969 through 1975 he served as the head of the Department of Biostatistics and Human Genetics. In 1975 he was honored with the title of University Professor.

Li published extensively throughout his career. He started writing his most important book, *Population Genetics,* in 1946 while still in China; it was not published until 1955. His other books included *Human Genetics, Principles and Methods,* in 1961; *Introduction to Experimental Statistics,* in 1964; *Path Analysis, A Primer,* in 1975; and *Analysis of Unbalanced Data: A Pre-Program Introduction,* in 1982. Li also authored more than 150 scientific articles. Most importantly, he promoted the cause of scientific integrity, as evidenced by his fight against Lysenkoism and his refusal to link intelligence to race.

Lim, Robert Kho-seng

(1897–1969)
Singaporean/American
Physician, Physiologist

Robert Lim conducted physiological studies on the gastrointestinal tract and the central nervous system. He especially interested himself in the relationship between gastric juice and blood. He conducted research on the neurophysiology of pain, applying this study specifically to the case of aspirin. More important was his influence in improving medical care in the Chinese army during the Sino-Japanese War and World War II and in introducing the study of physiology by Western techniques into China.

Lim was born on October 15, 1897, in Singapore. He was the eldest son of Lim Boon-Kang, a renowned physician, who graduated from the University of Edinburgh. Lim received his bachelor's degrees in medicine and chemistry from the Edinburgh University Medical School in 1919. He continued on at the university, studying under the physiologist Edward A. Sharpey-Schafer in pursuit of his doctorate, which he received in 1920. That year he married Margaret Torrance in Scotland. Together the couple had two children—a son, named James, and a daughter, named Effie. Lim survived his first wife and later married Tsing-Ying Tsang.

Lim taught physiology at the University of Edinburgh for four years after receiving his doctorate there. In 1923 he was elected a fellow of the Royal Society of Edinburgh. Soon thereafter he continued west to the University of Chicago for postgraduate study under Julius Carlson and Andrew Ivy. From there he traveled to China, where he was the youngest person, as well as the first Chinese person, appointed as a professor and chairman of the physiology department at the Peking Union Medical College in 1924. In 1926 he helped establish the Chinese Physiological Society, acting as its first chairman. He served for many years as the managing editor of the *Chinese Journal of Physiology*. He also served as the president of the Chinese Medical Association.

At the onset of the Sino-Japanese War in 1931 Lim again applied his organizational skills to help found the Chinese Red Cross Medical Relief Corps. The Chinese government subsequently appointed him as the surgeon general of medical services for the Chinese army, a post that commenced during World War II but continued after its end. Lim also aided the American effort in the war, supporting the U.S. general Joseph Stilwell's Chinese Expeditionary Force in 1942 with medical services. The United States government recognized Lim's contributions to the Allied war efforts with a Medal of Freedom and the Legion of Merit.

After the war Lim immigrated to the United States; he became an American citizen in 1955. He worked at the University of Illinois and at Creighton University before settling down at the Miles Medical Science Research Laboratories in Indiana. Lim had been elected as a foreign associate member of the National Academy of Sciences in 1942, and upon his naturalization in 1955 he became a regular member of the academy. Lim died of esophageal cancer on July 8, 1969, at his son's home in Kingston, Jamaica.

Linnaeus, Carl

(1707–1778)
Swedish
Botanist

An ambitious and tireless worker, Carl Linnaeus revolutionized botany through his introduction of the modern classification system for plants and animals. Linnaeus's system of binomial nomenclature designated a generic

Carl Linnaeus, whose classification system for plants and animals revolutionized botany. *Smith Collection, Rare Book & Manuscript Library, University of Pennsylvania.*

name and a specific name for each plant. Linnaeus completed many works on botany and was a popular and inspirational university professor. A confident self-promoter, Linnaeus was also successful in gaining widespread acceptance among the scientific community for his classification system.

Born Carl von Linné on May 23, 1707, in Råshult, a town in the province of Småland, Sweden, Linnaeus developed an interest in plants at an early age. His father, Nils Linnaeus, was a country pastor and avid gardener who introduced the wonders of botany to Linnaeus. At the age of eight Linnaeus was nicknamed "the little botanist." He entered the Latin School in a nearby city in 1716; there natural history became his favorite subject. Though Linnaeus was an average student, his enthusiasm for learning about plants was unsurpassed.

In 1727 Linnaeus entered the University of Lund and began medical studies. A year later he transferred to the University of Uppsala to continue his studies. Influenced by the botanist Olof Celsius to pursue botanical studies, Linnaeus began to investigate the recently proposed theory that plants showed evidence of sexuality. By 1730 Linnaeus had started forming a categorization system based on the stamens and pistils of plants. Also that year Linnaeus became a lecturer in botany. Two years later he traveled to Lapland to conduct explorations for the Uppsala Academy of Sciences. There he found 100 new species of plants.

After traveling through parts of Europe from 1733 to 1735, Linnaeus settled in Holland and published *The System of Nature,* his first major work. In it he presented a systematic arrangement of the animal, plant, and mineral kingdoms. He categorized whales as mammals and acknowledged humans' similarity to apes. Linnaeus also introduced a basic system of plant nomenclature. Flowering plants were separated into classes according to the number of stamens, then subdivided into orders based on the number of pistils. Because this system was based primarily on sexual characteristics of plants, it did not fully express the natural relationships between plants. The system was successful, however, in its simplicity and ease of application, and thus it was widely used. As a result of this work and a number of other publications, Linnaeus had gained a solid reputation.

Linnaeus continued his travels from 1736 to 1738, visiting physicians and botanists around Europe. In 1738 he opened a medical practice in Stockholm. In 1741 he was appointed to the chair of medicine at the University of Uppsala, only to trade it for the chair of botany a year later. Linnaeus devoted himself to teaching and writing, and in 1753 he published his seminal work, *Species of Plants,* which detailed the system of binomial nomenclature that he had introduced in 1749. The system assigned each plant a Latin or latinized name of two sections: The first name represented the genus, and the second indicated a specific name. For instance, *Juglans regia* was the name of the English walnut, *Juglans* referring to walnut and *regia* specifying that the walnut was the English walnut. *Juglans nigra* was the black walnut, as *nigra* means "black." Until the advent of Linnaeus's system, scientific plant names consisted solely of a Latin description of the plant's distinctive features.

Linnaeus's innovative classification system provided the foundation for modern nomenclature. A popular and captivating instructor, he inspired many of his students to travel around the world in search of plant specimens. Linnaeus married Sara Moraea in 1739. His manuscripts and collections are preserved by London's Linnaean Society.

Lipmann, Fritz Albert

(1899–1986)
Prussian/American
Biochemist

In 1941 Fritz Lipmann published his landmark article "Metabolic Generation and Utilization of Phosphate Bond Energy," which laid the foundation for most of the biochemical research to follow. In it he explained group potential and the role of group transfer in biosynthesis, defining elements in biochemistry. Four years later Lipmann discovered what he called coenzyme A (CoA), a key to metabolism and the transformation of carbohydrates into energy. After refining the understanding of CoA for the next eight years, Lipmann received the 1953 Nobel Prize in physiology or medicine with Sir HANS ADOLPH KREBS for this work.

Lipmann was born on June 12, 1899, in Königsberg, the capital of East Prussia (now Kaliningrad, Russia). His parents were Leopold, a lawyer, and Gertrud Lachmanski, and he had one brother, Heinz. Lipmann married Freda Hall on June 21, 1931. Together the couple had one son, Stephen Hall.

Lipmann's undergraduate study, which he commenced in 1917 at the University of Königsberg, was interrupted by World War I, as he served in medical service to the army in 1918. After the war in 1919 he continued his study of medicine in Königsberg as well as in Munich and Berlin. He earned his M.D. in 1922 from the University of Berlin, then worked in the pathology department in Berlin while working on his doctorate, which he earned in 1927. Lipmann worked on his doctorate at the University of Amsterdam, where he studied pharmacology, and back at the University of Königsberg, where he studied chemistry.

After receiving his Ph.D. Lipmann conducted research in the laboratory of OTTO FRITZ MEYERHOF, at both the University of Heidelberg and the Kaiser Wilhelm Institute in

Berlin, between 1927 and 1931. He fled Germany in 1932 to work at the Biological Institute of the Carlsburg Foundation in Copenhagen. In 1939 he immigrated to the United States to work at the Cornell Medical School in New York City until 1941. Between 1941 and 1957 he directed the biochemistry research department at the Massachusetts General Hospital in Boston, while also serving on the faculty at Harvard University as a professor of biological chemistry from 1949 to 1957. From 1957 until his retirement in 1970 he served as a professor of biochemistry at the Rockefeller Institute for Medical Research in New York City.

After his 1945 discovery of CoA, Lipmann worked to isolate it in 1947; he had named it and determined its molecular structure by 1953. Although Lipmann received the Nobel Prize for this work, his work furthering the biochemical field more generally holds equal significance. He produced 516 publications between 1924 and 1985, and one paper alone, written on acetyl phosphate in 1944, has been cited in other papers more than 700 times, attesting to the primacy of Lipmann's ideas. Besides the Nobel Prize, Lipmann won the 1966 National Medal of Science. One of his most important attributes was his ability to see the bigger picture in science, thus keeping in mind the larger implications of his detailed work. Lipmann died on July 24, 1986, in Poughkeepsie, New York.

Lipscomb, William Nunn, Jr.

(1919–)
American
Physical Chemist

William Lipscomb won the 1976 Nobel Prize in chemistry for his research on the chemical bonding of boron compounds, especially boron hydrides. His investigations on boranes, or artificially synthesized combinations of boron with other elements, proved important because they revealed how these compounds broke certain laws of chemistry, thus necessitating a rethinking and formulation of chemical properties, especially those involving the bonding of atoms. Had Lipscomb accepted the contemporary wisdom concerning boron, he might not have made his discoveries, but one of his main strengths was his ability to look past the limitations imposed by scientific dogma, using instead his research and his imagination as his guides.

Lipscomb was born on December 9, 1919, in Cleveland, Ohio; his family moved south within a year to Lexington, Kentucky. His mother was Edna Porter and his father was William Lipscomb. Lipscomb married his first wife, Mary Adele, in 1944, and together the couple had one son and one daughter. They divorced in 1983, and Lipscomb subsequently married Jean Evans.

Lipscomb earned his bachelor's degree in chemistry from the University of Kentucky in 1941. He then applied his scientific expertise to the war effort, serving as a physical chemist with the United States Office of Scientific Research and Development during and after World War II, from 1942 to 1946. He earned his Ph.D. in physical chemistry from the California Institute of Technology, having done his doctoral research under the Nobel laureate LINUS CARL PAULING.

Lipscomb commenced his professional career as an assistant professor at the University of Minnesota in 1946. The university promoted him to the status of professor of chemistry in 1954, when he also took on the role of chief of the physical chemistry division. In 1959 Lipscomb moved to Harvard University, where he remained for the rest of his career; he retired in 1990. From 1962 until 1965 he served as the chairman of the department of chemistry there.

In 1954 Lipscomb and his colleagues Bryce Crawford and W. H. Eberhardt published an important article in the *Journal of Physical Chemistry* expanding on the discussion of the unusual "three-center" bond involving one atom of hydrogen and two atoms of boron. Lipscomb's research on boranes utilized innovative techniques, including low-temperature X-ray crystallography. These techniques not only aided the specific research performed at the time, but also had much broader applications to chemical research in general. Lipscomb gathered his thoughts on this research in his 1963 text, *Boron Hydrides.* Lipscomb subsequently applied nuclear magnetic resonance studies to his work on chemical bonding, which he reported in his 1969 collaboration with G. R. Eaton, *Nuclear Magnetic Resonance Studies of Boron and Related Compounds.*

Lipscomb devoted his later career to the study of proteins and enzymes. His enthusiasm for scientific research earned him the title of "Colonel" from his colleagues and students. Besides the Nobel Prize, Lipscomb received the Distinguished Service Award from the American Chemical Society and the George Ledlie Prize from Harvard University.

Lonsdale, Kathleen Yardley

(1903–1971)
British
Crystallographer

Kathleen Lonsdale used X-ray crystallography, in which X-ray beams are shone through the regularly arranged atoms and molecules in crystals, to determine the shapes of key parts of the molecules of carbon-containing compounds. She was born Kathleen Yardley, the youngest of 10 children, in Newbridge, Ireland, on January 28, 1903. Her postmaster father, Henry, was an alcoholic, and her mother left him when Kathleen was five, taking her younger children to Seven Kings, a London suburb.

Yardley studied mathematics and physics at the Bedford College for Women and earned a B.S. in 1922. William H. Bragg then invited her to join his X-ray crystallography laboratory at University College, London. She accompanied him to the Royal Institution the following year. She earned an M.S. in 1924 from University College.

In 1927 Yardley married Thomas Lonsdale, another crystallographer. They had three children, Jane, Nan, and Stephen, between 1929 and 1934. Until 1930 they lived in Leeds, and Kathleen began some of her most important research at Leeds University. It concerned a group of six atoms called a benzene ring, an important part of many organic, or carbon-containing, compounds, including explosives, dyes, and drugs. Lonsdale showed that, contrary to what many organic chemists had believed, the ring was flat and hexagonal. This ground-breaking crystallography study, published in 1929, became her doctor of science thesis and, says the UCLA professor K. N. Trueblood, "had an enormous impact on organic chemistry."

Even before her work on the benzene ring, Kathleen Lonsdale and a coworker, William Astbury, began calculating tables of X-ray patterns in common crystals. This painstaking work, which Lonsdale continued at home after her family's return to London in 1930, made her the first woman crystallographer to earn a worldwide reputation. She continued to revise and edit expanded versions of these tables for much of her life, and they are still an essential reference for crystallographers.

Lonsdale rejoined Bragg's laboratory at the Royal Institution in 1934. There she developed the technique of shining multiple X-ray beams through a crystal from different angles and used it to determine the spacing of carbon atoms in natural and artificial diamonds. A rare form of diamond found in meteorites was named *lonsdaleite* in her honor in 1966.

In 1946 after Bragg's death Lonsdale returned to University College, where she became the institution's first woman professor and established her own crystallography department. She became head of the department in 1949. In her later years she studied the mineral deposits, or "stones," that can form in the kidney, bladder, and gallbladder. She retired in 1968.

In 1945 Lonsdale became one of the first two women admitted as full members to Britain's top science organization, the Royal Society. She was made a Dame Commander of the British Empire, the equivalent of a knighthood, in 1956, and received the Davy Medal of the Royal Society in 1957. In 1966 she became the first woman president of the International Union of Crystallography and in 1967 the first woman president of the British Association for the Advancement of Science. She died of cancer on April 1, 1971.

Love, Susan

(1948–)
American
Physician, Medical Researcher

Susan Love is one of the country's most respected—and most outspoken—authorities on women's health, particularly breast cancer. She was born in Little Silver, New Jersey, on February 9, 1948, but grew up partly in Puerto Rico, because her father, James Love, a machinery salesman, was transferred there when Susan was 13 years old. After another move she went to high school in Mexico City.

An order of nuns, the Sisters of Notre Dame, ran the school Susan attended in Puerto Rico, and one of the nuns sparked her interest in medicine. She entered a small women's college run by the order, Notre Dame of Maryland, in 1966. After two years at the college she decided to become a nun herself, but she found the life too restrictive and quit after six months.

Love took premedical training at Fordham University in New York, then applied to medical schools on the East Coast, only to find that most had already filled their tiny quota of women. She was finally admitted to the State University of New York's Downstate Medical College in Brooklyn, from which she graduated near the top of her class in 1974. She decided to become a surgeon. "In surgery, you figure out what's wrong and then you go in and fix it. . . . That's what I like," she told *Boston Magazine*'s Anita Diamant.

When Love began private practice in 1980, she found that most of the cases other surgeons referred to her were breast cancer patients. "I saw I could make a contribution" as a breast cancer surgeon, she told the *Advocate* interviewer Tzivia Gover. She became director of the breast clinic at Beth Israel Hospital and in 1982 a surgical oncologist (cancer specialist) for the Dana Farber Cancer Institute Breast Evaluation Center. In 1988 she helped to found the Faulkner Breast Center, part of Boston's Faulkner Hospital; this center was the first in the country to have a multidisciplinary all-woman staff. Love became an assistant professor of surgery at Harvard Medical School as well.

Love began to feel that the standard cancer treatments—surgery, radiation, and drugs, which she refers to as "slash, burn, and poison"—were woefully inadequate. So were the education and psychological support that most women with breast cancer received from their (usually male) doctors. She summarized her opinions and advice in *Dr. Susan Love's Breast Book* (1990), which *Ms.* magazine called "one of the most important books in women's health in the last decade." In 1990 she cofounded the National Breast Cancer Coalition, a lobbying organization that has helped to raise the national budget for breast cancer research from $90 million to $430 million.

Love has challenged several widely held medical opinions. For instance, she has questioned the value of monthly breast self-examination and of mammograms (breast X rays for early detection of cancer) for most women under 50. She also complains that many women with early breast cancer are encouraged to have a mastectomy, or complete removal of the breast, whereas simple removal of the cancerous tissue and its immediate surroundings (lumpectomy) followed by radiation would give them an equally good chance of survival. More recently she has questioned the widespread use of hormone replacement for women who have reached menopause.

In 1992 Love became the director of the new Revlon/UCLA Breast Center in Los Angeles. The center's aim was to integrate all the specialists involved in each woman's treatment, but this approach proved too difficult to carry out. Love left the Breast Center and retired from surgery in 1996. "I . . . want to do research and . . . the political stuff, and I can't do it all," she says. She is now attending business school to become what she calls "bilingual"—speaking the languages of both business and medicine. She also raises her daughter, Katie, with her longtime partner, Helen Cooksey, and directs the Santa Barbara Breast Cancer Institute.

At the UCLA School of Medicine, where she is an adjunct professor of surgery, Love is researching a new method of breast cancer detection, which involves threading a tiny fiber-optic tube, or endoscope, through the milk ducts in a woman's nipple to look for cancerous tissue. She hopes that the work of other scientists, such as her friend MARY-CLAIRE KING, will reveal ways to prevent or cure breast cancer through gene or hormone therapy.

Whether or not Susan Love proves correct in her criticism of particular medical treatments, she is surely right to encourage women to take a more educated and assertive role in deciding about their health care. As she told Anita Diamant, "My job is . . . to be an educator. I teach a woman what she needs to know to make a valid decision [about her treatment] for herself."

Lovelace, Augusta Ada Byron

(1815–1852)
British
Mathematician, Computer Scientist

Even though they were written 100 years before electronic computers were invented, Ada Lovelace's instructions for CHARLES BABBAGE'S "analytical engine" have been called the world's first computer programs. She was born Augusta Ada Byron in London on December 10, 1815. The short, stormy marriage between her father, George Gordon, Lord Byron, the famous British poet, and her mother, the former Anne Isabella (Annabella) Milbanke, ended a month after Ada's birth, when Byron departed, never to see his daughter again. Byron's flamboyant life-style had put him completely at odds with the quiet but equally strong-willed Annabella, whose love of mathematics made Byron call her the "Princess of Parallelograms."

Ada, tutored extensively at home, shared her mother's fondness for mathematics. At a party in 1833 she met the mathematician Charles Babbage, who had invented a machine that he called a "difference engine." Navigation, insurance, and other fields were coming to depend on tables of figures that required repeated calculations, but these calculations, done by hand, often contained errors. Babbage's machine solved polynomial equations (then called "difference equations") automatically, making the preparation of tables based on them faster and more accurate.

Babbage had built a small model of his device and obtained a government grant to produce a full-scale version. Unfortunately the technology of the time could supply neither the power nor the thousands of precisely machined parts that the machine needed, and by the time he met Ada, Babbage had lost the government funding and much of his own money without ever building the machine.

Ada also met a young nobleman, William King, who later became the earl of Lovelace, and they married in 1835 and later had three children. King encouraged his wife to pursue her intellectual interests, including her friendship with Babbage.

In addition to the difference engine, Babbage had designed a more general-purpose machine that he called the analytical engine. It would have performed any kind of calculation once figures and instructions were programmed into it. A Frenchman, JOSEPH-MARIE JACQUARD, had invented an automatic loom that wove patterned cloth through instructions fed into it on punched cards, and Babbage thought that his machine, too, might use punched cards. After the failure of the difference engine, however, no one wanted to invest money in something even more complex, so this second machine existed only on paper. Conceptually it is the ancestor of modern computers.

The Italian mathematician Luigi F. Menebrea in 1842 wrote a paper in French that described the workings of the analytical engine and the theory behind it. Babbage wanted the paper translated into English, and Ada Lovelace offered to do the job. Her finished work included notes of her own that were three times as long as Menebrea's manuscript, including sample sets of instructions for the machine. Babbage wrote that these notes "entered fully into almost all the very difficult and abstract questions connected with the subject." It was not considered proper for a woman of Lovelace's class to put her name on a public document, so her work, published in a collection of scientific papers in 1843, was signed only with the initials A.A.L. For 30 years no one knew that she was the author.

In her notes to Menebrea's paper Lovelace offered a warning against overestimating the powers of computing machines that is still timely: "The Analytical Engine has no pretensions whatever to originate anything. . . . The machines . . . must be programmed to think and cannot do so for themselves." On the other hand, she noted, also correctly, that in the process of reducing operations into forms that could be used by the machine, "the relations and the nature of many subjects . . . are necessarily thrown into new lights, and more profoundly investigated." She wrote to Babbage in 1845, "No one knows what . . . awful energy and power lie yet undeveloped in that wiry little system of mind."

Differences in their working styles (Babbage tended to be sloppy, whereas Lovelace, like her mother, was meticulous and bossy) put a strain on the relationship between Lovelace and Babbage, and the two never worked together on a major project again, although they remained friends. Meanwhile Lovelace pursued a variety of interests, including ideas about the functioning of the brain. She hoped to work out "a law or laws for the mutual action of the molecules of the brain . . . a *Calculus of the Nervous System,*" an idea that, like her computer programs, was far ahead of its time.

Another interest, unfortunately, was horse racing, for which Lovelace believed that she had developed an infallible mathematical betting system. Lord Lovelace joined his wife in placing bets at first, but he stopped when he lost money. Ada, however, became addicted to gambling and fell deep into debt, having to pawn the Lovelace family jewels twice (her mother redeemed them). She also became addicted to laudanum, or morphine.

Ada Lovelace died of cancer of the uterus on November 27, 1852, when she was only 36 years old, but she was not forgotten. In the late 1970s the U.S. Department of Defense created a computer language for programming missiles, planes, and submarines and named it Ada in her honor.

Lovelock, James Ephraim

(1919–)
English
Climatologist

An extremely versatile scientist, James Ephraim Lovelock has worked in biology, medicine, physics, and chemistry. A prolific inventor, he has patented about 60 of his devices. Lovelock is best known, however, for his Gaia hypothesis, which proposes that the Earth is a living organism that regulates and maintains itself to create an environment favorable for organisms. Though the theory is dismissed by the majority of scientists, Lovelock found supporters among environmentalists and conservationists.

Born in the village of Letchworth in Hertfordshire, England, on July 26, 1919, Lovelock spent his childhood exploring the country fields and rolling hills. Lovelock's father, Thomas A. Lovelock, was an art dealer, and his mother, Nellie A.E. March Lovelock, was a local official. Lovelock's father stimulated Lovelock's interest in nature by identifying various plants, birds, and insects during long hikes. Little interested in school, Lovelock believed he was better suited to teach himself.

Lovelock's college career began with night school courses at the University of London's Birbeck College. After a few years Lovelock entered the University of Manchester to study chemistry as a full-time student. He graduated with a bachelor's degree in 1941 and the following year married Helen M. Hysop, with whom he had four children. After his graduation Lovelock took a job as a staff scientist with the National Institute for Medical Research (NIMR) in London. Though he remained with NIMR for 20 years, Lovelock managed to earn his M.D. in 1949 and a doctorate in biophysics in 1959 from the University of London.

It was at NIMR that Lovelock's inventiveness first surfaced. Faced with numerous technical problems that arose during the course of projects, Lovelock invented solutions. For example, while studying the effects of thawing and freezing on living tissue, Lovelock was hampered by the lack of a device for generating heat in tissue and living cells. To solve the problem, Lovelock invented such a device. He also developed numerous measuring instruments, as well as a gauge designed to monitor blood pressure in scuba divers and a procedure for freezing the sperm of bulls. Lovelock's 1957 invention of an electron capture detector allowed for the measurement of chemical contamination levels in the environment. The detector was also used to confirm the existence of chlorofluorocarbons in the atmosphere, thereby providing the basis for a clearer understanding of global warming and ozone layer depletion.

After two decades with NIMR Lovelock moved to Houston, Texas, to work with the National Aeronautics and Space Administration (NASA). NASA employed him to assist in the development of devices and carrying out of experiments to determine the existence of life beyond Earth. Lovelock also took a teaching position in chemistry at the Baylor College of Medicine. Lovelock commuted from Houston to NASA's Jet Propulsion Laboratory in California, where he first worked on the Gaia theory. Lovelock began to consider the unstable blend of gases that made up the Earth's atmosphere, which are not in equilibrium but remain relatively constant over long periods. That effect led Lovelock to wonder whether the Earth regulated itself to create an environment that was welcoming to life. When Lovelock introduced his theory in a 1968 publication of the American Astronautical Society, there was little response. Lovelock publicized his hypothesis the following year at a meeting in Princeton, New Jersey, and faced negative reactions. In 1979 he published *Gaia: A New Look at Life on Earth,* which provided an expanded version of his hypothesis. Scientists remained skeptical about the theory, but Lovelock found supporters among environmentalists. Lovelock continued to work on his Gaia hypothesis and in 1991 published another book, *Healing Gaia.*

Despite the controversy surrounding the Gaia theory, Lovelock's ideas were enthusiastically embraced by many and presented a unique view of the Earth. His wife died in 1989; he married Sandra J. Orchard in 1991. Lovelock is a fellow of the Royal Society, an associate of the Royal Institute of Chemistry, and a Commander of the British Empire.

Lowell, Percival

(1855–1916)
American
Astronomer

A member of a prominent Boston family, Percival Lowell devoted his energies to astronomical observation after extensive travels in the Far East. Lowell predicted the existence and position of the planet Pluto and spent the later years of his life searching for it. His extensive observations of Mars led him to propose that it was inhabited. Lowell's realization that astronomical observations were best made where optimal atmospheric conditions prevailed was a breakthrough discovery. This recognition led him to build an observatory in a then-remote area of Arizona.

Born on March 13, 1855, in Boston, Massachusetts, Lowell was a descendant of some of the most distinguished families in New England. Lowell's mother, Katharine Bigelow Lawrence Lowell, was the daughter of Abbott Lawrence, who served as the United States minister to England. His father, Augustus Lowell, was associated with the cultural life of Boston. Lowell's grandfather instilled in Lowell a lifelong interest in botany and horticulture. Lowell's younger brother, Abbott Lawrence Lowell, later became president of Harvard University, and his sister Amy became a renowned poet and critic. Lowell's early education filled him with many interests, including astronomy.

After graduation from Harvard University in 1876 with a degree in mathematics Lowell traveled extensively. In 1883 he visited Japan and studied the people, customs, and language. While there he was asked to serve as foreign secretary and counselor to the first diplomatic mission from Korea to the United States. This oppointment led Lowell to visit Korea. Until 1893 Lowell traveled throughout the Far East, soaking up the customs and detailing his observations in a number of books.

It was the Italian astronomer GIOVANNI VIRGINIO SCHIAPARELLI's observation of lines, or "canals," on the surface of Mars that motivated Lowell to devote his full attention to astronomy. Lowell decided to study Mars and built a private observatory at Flagstaff, Arizona, using his family's fortune to finance the endeavor. The observatory was built at a height of 7,200 feet, and in 1894 he commenced his observations. Lowell was able to see Schiaparelli's canals and also reported seeing signs of vegetation and oases. For the following 15 years Lowell studied Mars, and his research led him to conclude that it was inhabited. He proposed that the inhabitants built a system of irrigation that utilized water from the polar ice caps, which melted annually. The canals, he believed, were gardens cultivated by the inhabitants and were dependent on the irrigation. Lowell published a number of books presenting his theories of Mars.

Lowell began his pioneering work on the existence of Pluto in the early 1900s. While studying the orbit of Uranus, Lowell discovered irregularities that could not be accounted for by the effects of Neptune. He theorized that there was a ninth planet beyond Neptune that was causing the anomalies. Lowell calculated its orbit and position and began a search for the planet from 1905 to 1914. He published his negative results in *Memoir on a Trans-Neptunian Planet* in 1915. Pluto was finally discovered in 1930, 14 years after Lowell's death, by Clyde William Tombaugh.

Lowell's astronomical discoveries influenced not only the scientific community but the general public as well. His beliefs regarding life on Mars fascinated the public until the theory was finally dispelled in the 1960s by information gathered by U.S. spacecraft voyages. Lowell was a popular and captivating speaker who gave frequent lectures. His interest in botany led to his discovery around Flagstaff of plants and a new species of ash tree, which was named in his honor. Lowell received medals from several astronomical societies and was an honorary member of the Royal Astronomical Society of Canada. He married Constance Savage Keith in 1908.

Luria, Salvador Edward

(1912–1991)
Italian/American
Molecular Biologist

Salvador Luria won the 1969 Nobel Prize in physiology or medicine along with the other founding members of the so-called American Phage Group, MAX DELBRÜCK and ALFRED DAY HERSHEY, for their work on bacteriophages, or viruses that infect bacteria.

Luria was born on August 13, 1912, in Turin, Italy. His mother was Ester Sacerdote; his father, David Luria, was an accountant and manager of a small printing business who hailed from a prominent Jewish family in Turin. In 1945 Luria married Zella Hurwitz, a psychologist, and in 1948 the couple had a son, Daniel, who became a political economist. The previous year Luria had become a naturalized citizen of the United States.

In 1929 Luria commenced study at the University of Turin School of Medicine under the histologist and nerve-tissue expert Giuseppe Levi. Luria earned his M.D. summa cum laude in 1935. Between 1935 and 1937 he served as a medical officer in the Italian army, after which he studied radiology in Rome and conducted research at the Physics Institute of the University of Rome under ENRICO FERMI. However, the rise of fascism soon required Luria to flee Italy for Paris, where he learned bacteriophage research techniques. The Nazi invasion of France prompted Luria to flee even farther, to the United States.

Luria, who landed a position with Columbia University in 1940, first collaborated with Delbrück at the university's Biological Laboratory at Cold Spring Harbor on Long Island in 1941, and subsequently at Vanderbilt University on a Guggenheim Fellowship from 1942 through 1943. Luria taught bacteriology and virology at Indiana University in Bloomington from 1943 to 1950, when he accepted a professorship in bacteriology from the University of Illinois at Champaign-Urbana. In 1959 Luria commenced a long relationship with the Massachusetts Institute of Technology as a professor and the chairman of the Department of Microbiology. In 1964 MIT named Luria the Sedgwick Professor of Biology, and after winning the Nobel Prize, he became an Institute Professor. In 1974 he took over the directorship of MIT's Center for Cancer Research.

In 1943, while attending a faculty dance in Bloomington, Luria was inspired by the sight of a slot machine to apply the same theory of probability to bacterial mutation, a theory immediately substantiated by mathematical analysis by Delbrück. The pair collaborated on a paper introducing their notion of a fluctuation test for spontaneous mutation. In 1953 Luria published an important textbook, *General Virology,* and in 1973 he published *Life: The Unfinished Experiment* to a more general readership. The book won the 1974 National Book Award in the sciences.

Besides the Nobel Prize, Luria won the 1965 Lenghi Prize of the National Academy of Science of Italy and the 1965 Louisa Hurwitz Prize from Columbia University. Luria defined himself politically as a socialist and a pacifist, and he worked actively for these causes. He also loved fiction and poetry and taught a world literature course later in his career at MIT. Luria died of a heart attack on February 6, 1991, in Lexington, Massachusetts.

Lyell, Sir Charles

(1797–1875)
English
Geologist

The geologist Sir Charles Lyell was responsible for the general acceptance of the uniformitarian principle, which proposed that the Earth was ancient and that the features of the Earth's surface were caused by physical, chemical, and biological processes that occurred over a long period. The concept was first set forth by the geologist James Hutton, but Lyell was instrumental in its recognition. Lyell's accomplishments provided the basis for evolutionary biology and advanced the understanding of the Earth's development.

The eldest of 10 children, Lyell was born on November 14, 1797, at the family's Kinnordy estate, located in what was then the county of Forfarshire in eastern Scotland. Lyell spent most of his childhood at the Bartley Lodge at Lyndhurst, located on the border of the New Forest near Southampton, England, where his family moved in 1798. Lyell's mother was Frances Smith and his father was Charles Lyell, Esquire. Though he had studied law, Lyell's father was interested in botany and spent much time collecting rare plants in the New Forest region. His father also had an extensive library, which included books on geology. Lyell was not an attentive student, preferring to explore the outdoors. In 1808 Lyell pursued his first scientific hobby—collecting insects and studying their habits.

In 1816 Lyell entered Oxford University, where he became a more enthusiastic student and studied classics, mathematics, and geology. Lyell earned his bachelor's degree in late 1819 and moved to London with the intent of studying law. As a result of his law studies Lyell suffered from severe eye strain. During his recovery periods he traveled widely, and on one journey he observed evidence of vertical movements of the Earth's crust in Sussex. On a trip to Paris in 1823 Lyell met the naturalists Baron ALEXANDER VON HUMBOLDT and Baron Georges-Leopold-Chretien-Frédéric-Dagobert Cuvier and studied the Paris Basin with the geologist Louis-Constant Prévost. Lyell completed his law studies and was admitted to the bar in 1825, but his interests lay solidly in geology, and with his father's financial assistance Lyell was able to pursue geology more heartily.

After publishing his first scientific papers in 1825, Lyell began planning a book that included his ideas that all geologic phenomena had natural explanations. He also planned to propose that the Earth was extremely old because the natural processes that occurred daily were slow and took place over long periods. Lyell journeyed to France and Italy, investigating the geological characteristics of the land, and compiled his results in the three-volume *Principles of Geology,* the first volume of which was published in 1830. The work convincingly supported the uniformitarian principle, thereby refuting the traditional view that corresponded to the biblical Creation, which suggested the Earth was only a few thousand years old. Lyell's book gave him considerable fame, and he was recognized as a leader in geology. Lyell completed 11 editions of the book in his lifetime. In 1832 he married Mary Horner, who participated in his geological work for 40 years.

In the 1840s and 1850s Lyell traveled throughout the United States. He also investigated the prevention of mine accidents with the physicist MICHAEL FARADAY in 1844 and helped set in motion education reform at Oxford University in the early 1850s. Lyell's theories influenced CHARLES ROBERT DARWIN in developing his evolutionary theory. After initially rejecting Darwin's theory, Lyell accepted it in 1865. Lyell received the Royal Society of London's Copley Medal and a number of honorary degrees. He was knighted in 1848 and made a baron in 1864.

M

Maathai, Wangari Muta

(1940–)
Kenyan
Ecologist

Wangari Maathai, the first woman in East and Central Africa to obtain a Ph.D., has used her scientific background to design grass-roots programs that preserve the environment. The former U.S. assistant secretary of state for African affairs Chester Crocker has called her "the leading environmentalist on the African continent."

Wangari Muta Maathai was born on April 1, 1940, and raised in Nyeri, Kenya, a rural area that she told the writer Aubrey Wallace was "very green, very productive." Her Kikuyu family taught her to respect nature. In 1960 she won an American government scholarship and attended Mount St. Scholastica College in Atchison, Kansas, from which she earned a B.S. in biology in 1964. She received an M.S. from the University of Pittsburgh in 1965 in anatomy and tissue culture, growing cells in laboratory dishes.

Maathai returned to Kenya in 1966 and earned a Ph.D. from the University of Nairobi in 1971. She worked her way up through the university, becoming a senior lecturer, associate professor, and in 1976 head of the anatomy department. She was the first woman in the university to do each of these things. She also married in 1969 and went on to have three children.

When Maathai's marriage ended in a bitter divorce in the mid–1970s, she decided to run for parliament. This process required quitting her job at the university. After her political campaign was barred on a technicality, the university refused to take her back, ending her academic career. This move, on top of the divorce, made Maathai painfully aware for the first time of the jealousy her successes had aroused in her male colleagues. "Sometimes we don't quite realize that not everybody's clapping when we're succeeding," she told an interviewer, Mary Anne French.

Maathai began to focus on the human and environmental degradation she had seen in the countryside. She had become aware that, as she said later, "poverty and need have a very close relationship with a degraded environment. It's a vicious circle." Women and trees, she concluded, were at the heart of the problem. Up to 97 percent of Kenya's forests had been cut down, either to clear land for cash crops or to provide wood for cooking fires. Women therefore had to spend more time looking for firewood, and that left less for cooking. They began to feed their children low-nutrient processed foods that did not require cooking, which often led to malnutrition. Cutting down trees also removed the roots that held the continent's thin topsoil in place, causing much of it to be eroded by wind and water. This reduced soil fertility and made a further contribution to hunger.

Maathai decided that women and trees might also be a solution. Most of Kenya's farmers were women, and if each planted a "green belt" of trees around her farm, it would make a huge difference. With a little help anyone could plant and tend a tree, and native trees could provide food, firewood, shade, and beauty as well as reducing soil erosion and air pollution. "Trees are miracles," she has said.

Typically Maathai started her project with herself. On June 5—World Environment Day—1977 she planted seven trees in a park in Nairobi, and the Green Belt movement was born. Maathai's new organization gained the backing of the National Council of Women of Kenya, which helped it spread the word in cities and the countryside. "Soon, people from all over the country were asking where they could find seedlings," Maathai reported to *UNESCO Courier.* By 1997 some 80,000 Green Belt movement members in Kenya, most of them women, had planted 15 million trees, and the movement had spread to over 30 other African countries. Maathai described this work in *The Green Belt Movement* (1985) and *The Green Belt Movement: Sharing the Approach and the Experience* (1988).

One reason for the movement's success, Maathai has explained, is that it is kept firmly under local control. The organization enters an area only when it is invited. Green Belt movement employees then provide training and equipment to help local women establish a nursery for raising seedling trees. Village women receive free trees from the nursery, plant them, and appoint "rangers," usually children or disabled people, to follow up on the trees' care. The women are paid a small sum for each tree that survives more than three months. Maathai also emphasized in a 1992 speech that the movement does much more than plant trees. It "is about hope. . . . It tells people that they are responsible for their own lives. . . . It raises an awareness that people can take control of their environment."

In the late 1980s, concluding that planting trees was not enough, Maathai began taking a more direct role in politics in the hope of improving Kenyans' social and economic conditions. This put her in conflict with the country's government, which has frequently harassed and arrested her and once clubbed her into unconsciousness. She nonetheless continues her political as well as environmental activities. Awards she has received for those activities include the Woman of the Year Award (1983), the Right Livelihood Award (1984), the Windstar Award for the Environment (1988), the Woman of the World Award (1989), the Goldman Environmental Prize (1991), and the Jane Addams International Women's Leadership Award (1993).

Maathai sometimes has missed her scientific career. "I would love to go back into an academic institution," she wrote in *Ms.* magazine in 1991. Nonetheless, she said, "I have . . . felt . . . like I was being useful, so I have no regrets." Above all, she says, she is proud to have shown that "one person *can* make the difference."

Mach, Ernst

(1838–1916)
Austrian
Physicist

Ernst Mach made a number of contributions in an array of fields, but his true influence was as a bold critic of science. His conclusions inspired future generations of scientists, including such notable thinkers as ALBERT EINSTEIN. Mach vehemently adhered to the notion that science should restrict itself to the description of phenomena that could be perceived by the senses, rather than delving into unquantifiable metaphysical concepts. He investigated shock waves and published the first photographs of projectiles in motion that revealed the accompanying shock waves. The Mach number—the ratio of the speed of an object to the speed of sound in the atmosphere—was named in his honor. In what is now known as Mach's principle, he denied the existence of absolute space and thus refuted Sir ISAAC NEWTON'S concept of inertia. For Mach inertia depended on the relationship of one body to another, and he therefore concluded that a body would have no inertial mass in a universe in which no other body was present.

Born on February 18, 1838, in Turas, Austria (now in the Czech Republic), Mach was the first of three children of Johann Mach and Josephine Lanhaus Mach. He was raised on a farm in Untersiebenbrunn (near Vienna), where his family grew an orchard and raised silkworms. His father was an educated man who instilled a knowledge of the classics and science in young Ernst, as well as the practical skills of farming and carpentry. Mach's mother, whose own family practiced law and medicine, taught her children to love music and poetry.

Until he was 14, Mach was educated at home by his father. He was tutored in Greek and Latin grammar, literature, algebra, and geometry and devoted his afternoons to manual labor on the farm. In 1853 he briefly attended the Gymnasium in Kremsier, before he proceeded to the University of Vienna at the age of 16. He obtained his doctorate in physics in 1860; his dissertation was on electrical discharge and induction.

He then taught mechanics and physics at the University of Vienna until 1864, when he became a professor of mathematics at the University of Graz. During this period he discovered Mach's bands, the physiological phenomenon in which the human eye tends to see bright or dark bands between objects of different illuminations. In 1867 he married Ludovica Marussig in Graz. The couple moved together to Prague, where Mach worked at the Charles University as professor of experimental physics for the next 28 years, producing his most influential work. Between 1873 and 1893, in addition to studies on kinesthetic sensation, he developed optical and photographic techniques for measuring sound waves. He established the principle of supersonics in 1887; ultimately that principle led to the Mach number's being named for him in 1929, in recognition of his pioneering work on the relationship between speed and sound.

Mach's diverse projects were informed by his rigorous adherence to observable phenomena. He forcefully argued that science dispense with metaphysical constructs that could not be empirically verified, a position he laid out in his 1897 book *Contributions to the Analysis of Sensations.* This theoretical framework led him to reject Isaac Newton's formulation of absolute space and time in the course of his own studies of inertia. Mach posited that inertia—the tendency of a body at rest to stay at rest and a body in motion to stay in motion—was not an intrinsic property of matter but rather the result of its relationship with other matter in the universe. This radical theory came to be known as Mach's principle.

Although Mach returned to the University of Vienna as a professor of inductive philosophy in 1895, in that year he suffered a stroke that left the right side of his body paralyzed. He retired from his position in 1901, whereupon he was elected to the Austrian parliament. He continued to research and write, and in 1913 he moved to the country home of his son. He died in 1916, after his work on inertia inspired Einstein to devise his theory of relativity.

Macleod, John James Rickard

(1876–1935)
Scottish
Physiologist

J. J. R. Macleod worked on carbohydrate metabolism throughout his career, focusing in the early 1920s on the mystery of the pancreatic hormone that controlled blood-sugar levels. The team of Macleod, Sir FREDERICK GRANT BANTING, Charles Herbert Best, and James Bertram Collip discovered and prepared the hormone, named insulin, that helped control the disease diabetes. Banting and Macleod shared the 1923 Nobel Prize in physiology or medicine for this work, and they in turn recognized the contributions of their colleagues by sharing their prize money with Best and Collip.

Macleod was born on September 6, 1876, in Cluny, near Dunkeld, Scotland. His father was the Reverend Robert Macleod. Macleod attended the Aberdeen Grammar School and Aberdeen University before entering medical school at Marischal College, where he earned his M.D. and graduated with honors in 1898. There he received the Matthews Duncan and Fife Jamieson Medals, and subsequently the Anderson Traveling Scholarship, which he used to study biochemistry for one year at Leipzig's Physiology Institute. After serving at the London Hospital Medical College as a demonstrator in physiology in 1900 and as a biochemistry lecturer in 1901, Macleod was named the Mackinnon Research Scholar by the Royal Society in 1901. He then earned a diploma in public health from Cambridge University in 1902.

In 1903 Macleod traveled to the United States to take up the post of professor of physiology at Western Reserve University in Cleveland, Ohio. After serving there for 15 years he moved on to a similar appointment at the University of Toronto as a professor of physiology, adding to that post duties as the associate dean of the medical faculty. In 1928 he returned to Scotland to head the department of physiology at the University of Aberdeen and conduct research at the Rowett Institute.

Macleod published *Practical Physiology* the year he arrived at Western Reserve. While there he published an influential series of papers, *Studies in Experimental Glycosuria,* in 1907. In 1913 he published *Diabetes: Its Physiological Pathology.* In 1918 he published his masterwork, the 1,000-page text *Physiology and Biochemistry in Modern Medicine,* which went through seven editions in his lifetime alone.

In 1901 Macleod studied intracranial circulation and caisson's disease, which resulted from moving between differently pressurized atmospheres. He then proceeded to study salt and urea metabolism. In 1921 Banting approached Macleod for permission to search for the pancreatic hormone that controlled blood-sugar levels, and Macleod was dubious, as physiologists had not yet been able to identify this agent. Nevertheless he gave the go-ahead, then traveled from Toronto to Scotland. Upon his return Banting and Best announced that they had isolated the hormone from the pancreas's islet of Langerhans. Macleod named it *insulin,* and charged Collip with obtaining as much as possible. They then proceeded to conduct trials on humans.

Besides the Nobel Prize, Macleod received the 1923 Cameron Prize from Edinburgh University. That year he also became a member of the Royal Society. In 1925 he became the president of the Royal Canadian Institute. Macleod died on March 16, 1935, in Aberdeen, Scotland.

Marconi, Guglielmo

(1874–1937)
Italian
Physicist, Engineer

Although Guglielmo Marconi did not discover radio waves, he harnessed them to practical use by inventing and patenting devices for the transmission and reception of "wireless" communications, as they were then called. Marconi received the 1909 Nobel Prize, the first Italian so honored in physics, with KARL F. BRAUN, who increased the range of radio transmission fivefold.

Marconi was born on April 25, 1874, in Bologna, Italy. His father, Giuseppe Marconi, a wealthy landowner, had two older sons, Luigi from his first marriage and Alfonso from his second marriage to Annie Jameson, the daughter of the Irish whiskey distiller. On March 16, 1905, Marconi married Beatrice O'Brien of the Irish aristocracy, and together the couple had three daughters—Lucia, who died in infancy; Degna; and Goia Jolanda—and one son, Giulio. Their marriage ended in a 1927 annulment, and on June 12 of that year Marconi married Maria Bezzi-Scali of papal nobility. They had one daughter, Elettra.

Marconi's formal education was limited to study at the Instituto Cavallero in Florence and at the Technical Institute of Leghorn. More important was his private tutoring in physics by Professor Vincenzo Rosa. On his own Marconi began experimenting with HEINRICH RUDOLF HERTZ's "invisible rays" or "electric waves." In 1895 he succeeded in transmitting a signal the one-and-a-half-mile length of his father's estate, Villa Grifone. Unable to interest the Italian government in his innovation, Marconi traveled in

February 1896 to London, where his cousin, Henry Jameson Davis, aided him in preparing a patent application that was accepted on July 2, 1897.

That year Marconi formed the Wireless Telegraph and Signal Co., Ltd., which he renamed Marconi's Wireless Telegraph Co., Ltd., in 1900. The previous year he had achieved the first international radio transmission of 85 miles across the English Channel from Chelmsford, England, to Wimereux, France. Scientific belief at the time held that radio waves traveled in straight lines, thus hampering long-distance transmission. Marconi believed otherwise, that the waves would bend with the Earth's curvature, a hypothesis he set out to prove by setting up a transmitter at Chelmsford, England, and a receiver on Cape Cod, Massachusetts. When a storm destroyed the Cape Cod antenna, Marconi traveled to St. John's, Newfoundland, to fly a kite 400 feet high as an antenna. On December 12, 1901, the kite worked, receiving the transmission over the 1,800 miles from Chelmsford of the Morse code for the letter *S*.

Marconi subsequently attempted to increase the distance he could transmit radio signals by investing interest first in short wave, setting up a global short-wave system through the British government by 1927, and later in microwaves, which he began experiments with in 1932. Besides the Nobel Prize, Marconi was honored with the title of *marquis* in 1929 and with the presidency of the Royal Italian Academy in 1930. His most treasured honor, though, was a gold tablet given to him by 600 survivors of the *Titanic,* who felt they were saved by his wireless technology. Marconi died on July 20, 1937, in Rome. In Britain telegraph and wireless offices marked the time of his funeral with two minutes of dead air in his honor.

Margulis, Lynn Alexander

(1938–)
American
Microbiologist, Geneticist

Lynn Margulis has proposed or supported several ideas that, although rejected at first, eventually resulted in major changes in biologists' thinking. A colleague calls her "one of the most outspoken people in biology."

Lynn Alexander was born on March 5, 1938, in Chicago, the oldest of the four daughters of Morris and Leona Alexander. Her father headed a company that made marker stripes for roads and was also a lawyer and politician. Her mother was, she says, a "glamorous housewife." In an autobiographical essay Lynn described her child self as "passionate, hungry for knowledge, grabby of the leading roles, . . . and nature loving."

Lynn entered the University of Chicago when she was 15. Inspired by an innovative program that featured works by great scientists instead of textbooks, she decided to become a scientist. She met CARL EDWARD SAGAN, a physics graduate student, who "shared with me his keen understanding of the vastness of time and space." They married in 1957, just after Lynn earned her B.A., and later had two sons, Dorion and Jeremy. Lynn Sagan earned an M.A. in zoology and genetics at the University of Wisconsin at Madison in 1960, then studied cells and their evolution at the University of California at Berkeley and completed a Ph.D. in 1965.

Scientists at the time believed that in most living things all genetic information was carried in a part of the cell called the nucleus; the only exceptions were simple microorganisms that lacked a nucleus. Lynn Sagan, however, learned that some geneticists in the early part of the century had proposed that certain other bodies within the cell, called organelles, might also contain genetic material. These scientists had suggested that the organelles were once free-living bacteria that had come to reside inside other bacteria early in the evolution of living things. Eventually the bacteria formed a mutually beneficial relationship, or symbiosis, that became so close that they could not survive without each other; indeed, they became a single organism. This microorganism was the ancestor of cells with nuclei.

Most other geneticists regarded this "serial endosymbiosis theory" as ridiculous, but Sagan disagreed. From her own research and that of others she gathered a wealth of material that supported it. For instance, in the early 1960s she and other researchers found that chloroplasts, the organelles that make food in green plants, and mitochondria, organelles that help cells use energy, both contain deoxyribonucleic acid (DNA), the chief chemical carrier of genetic information. This DNA was similar to that found in bacteria, just as the endosymbiosis theory predicted. Furthermore, both chloroplasts and mitochondria proved to resemble certain types of free-living microorganisms.

Sagan assembled her ideas into a long paper and began submitting it to scientific journals. Fifteen rejected or lost it before the *Journal of Theoretical Biology* finally printed it in 1966. By then her "turbulent" marriage to Carl Sagan had ended in divorce, and she had just begun teaching and research at Boston University.

In greatly expanded form this paper became Lynn Sagan's first book, *Origin of Eukaryotic Cells* (cells with nuclei), published in 1970 in spite of a letter from the National Science Foundation saying that the book's ideas were "totally unacceptable to important molecular biologists." Eleven years later, when it was issued in a revised edition as *Symbiosis in Cell Evolution,* the ideas in it had become widely accepted. William Culberson, professor of botany at Duke University, has said, "The reason that the symbiotic theory is taken seriously [today] is Margulis. She's changed the way we look at the cell." More recently Margulis has maintained that bacteria called spirochetes were the ancestors of cell organelles that provide movement and even of sensory and nerve cells. She has provided evidence to support this, but it remains more debatable than the case for mitochondria and chloroplasts.

Meanwhile, Lynn married Thomas N. ("Nick") Margulis, a crystallographer, in 1967 and with him had two more children, Zachary and Jennifer. Although she has described this marriage as "healthier and happier" than the one with Sagan, it, too, ended in divorce in 1980.

After Lynn Margulis became a professor at Boston University in 1977, she continued to espouse unpopular ideas and see them proved right. For instance, she supported a classification scheme first proposed by the late Robert H. Whittaker of Cornell University. Instead of dividing all living things into the traditional two kingdoms of plants and animals, Whittaker listed five kingdoms: animals, plants, fungi, protists (organisms with cell nuclei that do not belong to the first three groups; Margulis prefers the term *protoctists*), and monera (bacteria and other microorganisms without cell nuclei). This classification is widely used today.

Perhaps the most fiercely debated of Margulis's stands is her support of JAMES E. LOVELOCK's Gaia theory. Lovelock, a British chemist, proposed in the early 1970s that, as Margulis puts it, "life does not randomly 'adapt' to an inert environment; rather, the nonliving environment of the Earth is actively made, modulated and altered by the . . . sum of the life on the surface of the planet." For instance, processes in the bodies of living things, chiefly microorganisms, maintain the planet's surface temperature within the narrow limits that can sustain life. The theory, named after the ancient Greek goddess of the Earth, has been interpreted by some to mean that the Earth is, in essence, a single living organism.

Lynn Margulis is no more discouraged by others' doubts now than she ever was—and she now has the position and honors to make herself heard. She was elected to the National Academy of Sciences in 1983 and became a Distinguished University Professor at the University of Massachusetts in Amherst in 1988. Her writings, both scientific and popular, about evolution and microbes (some coauthored with her oldest son, Dorion Sagan) have been prolific. The paleontologist Niles Eldredge has called her "one of the most original and creative biologists of our time."

Maria the Jewess (Mary, Miriam)

(first century A.D.)
Egyptian
Chemist

In the first centuries after Christ alchemists blended religion, art, and science into a mixture that contained the seeds of modern chemistry. Some early alchemists were women, and the best known of these wrote under the name of Maria (or Mary or Miriam) the Jewess. None of her complete works survives, but fragments and references to her appear in the writings of other alchemists. Historians believe that Maria lived in the first century A.D. in Alexandria, Egypt, then one of the world's chief seats of learning.

The work of Maria's about which the most is known is called the *Maria Practica,* which combined the mystical theories of alchemy with practical descriptions of laboratory devices and processes. Among these were several pieces of equipment that Maria probably invented or at least perfected.

The simplest and best known of Maria's devices is the water bath, still found in many kitchens as a double boiler. It consists of two containers, one suspended inside the other. Water is placed into the outer container and heated. In turn it slowly heats the material in the inner container.

Maria also provided the oldest known description of a still, a device used to separate substances in a liquid through the process of distillation. In distillation a liquid mixture is heated in a closed container until part of the liquid evaporates into a gas. The gas is then piped into a different container and cooled until it condenses back into liquid. Maria's was a *tribikos,* or three-armed still.

Maria's most complex device was the *kerotakis,* an apparatus for allowing gases to color or otherwise act on metals. The kerotakis was shaped like a globe or a cylinder, covered by a domed top. It was placed over a fire, and sulfur, mercury, or arsenic solution was heated in a pan near its bottom. A piece of metal was put on a plate suspended from the cover. As gas from the heated solution rose past the plate, it reacted with the metal. At the top of the dome, as in the still, the gas cooled back into a liquid. The liquid was not carried off, however, but rather ran back down the side of the kerotakis to be evaporated again.

Massey, Walter Eugene

(1938–)
American
Theoretical Physicist

Walter E. Massey focused his early career on the research of cryogenics, or the study of materials at very low temperatures, publishing articles on liquid helium. As his career progressed he shifted his focus to administration, starting with an initiative to improve science education in inner cities and progressing to positions deciding policies for science at the national level and for university education at the statewide level.

Massey was born on April 5, 1938, in Hattiesburg, Mississippi. His parents were Essie Nelson and Almar Massey. Massey excelled in school, leaving the Royal Street High School in 10th grade to accept a scholarship to attend Morehouse College in Atlanta, where he studied physics with Sabinus H. Christensen, often in one-on-one tutorials because of a lack of other physics majors his year. After receiving his B.S. in 1958, Massey continued on at Morehouse as an instructor in physics. He then proceeded to Washington University in St. Louis, where he focused his doctoral research on cryogenics under Eugene Feenberg and earned his Ph.D. in 1966. In 1969 he married Shirley Streeter, and together the couple had two sons, Keith Anthony and Eric Eugene.

In 1966 Massey accepted a postdoctoral research fellowship at the Argonne National Laboratory in Batavia, Illinois, a facility funded by the United States Department of Energy and run by the University of Chicago. In 1967 Massey published a paper documenting his research, "Variational Calculations of Liquid Helium 4 and Helium 3," coauthored by C. Woo, in *Physical Review.* That same year he accepted the position of associate professor of physics at Brown University, where he organized the Inner City Teachers of Science program to improve urban science instruction, an initiative that earned him the 1975 Distinguished Service Citation from the American Association for the Advancement of Science. That year Brown promoted him to the rank of full professor as well as dean of the college. The next year Massey and his cowriter H. Eschenbacher published a paper in *The Physics Teacher,* "Training Science Teachers for the Inner City."

In 1979 the University of Chicago invited Massey to return to the Argonne National Laboratory as its director as well as a professor of physics at the university. In response to increasing criticism of the public funding of labs such as Argonne, Massey organized the Argonne National Laboratory/University of Chicago Development Corporation to promote research that would benefit the public as well as private corporations. In 1983 he assumed more responsibility as the vice president of research for the University of Chicago.

After serving on the board of directors of the American Association for the Advancement of Science from 1981 to 1985, Massey became the organization's president in 1988, the first African American to fill this position. In 1990 President George Bush tapped Massey to head the National Science Foundation, a position he intended to hold for six years. However, in 1993 the University of California offered him the position of provost and senior vice president for academic affairs, an opportunity he could not pass up, and he returned to academic life.

Massie, Samuel Proctor

(1919–)
American
Chemist

Samuel Massie focused his early research on the chemical phenothiazine, a three-ring organic molecule that became important in the production of the psychiatric drug thorazine. His later research focused on human health and environmental protection. Massie's most significant contributions to the sciences, however, were made as an educator and an administrator, and the end of his active career was punctuated with recognitions of his importance in educating future scientists, especially those of African-American heritage.

Massie was born on July 3, 1919, in North Little Rock, Arkansas. Both of his parents were teachers, and he received his early education in his mother's grade-school classrooms and his secondary education in his father's biology classrooms. In 1947 he married Gloria Tompkins, and together the couple had three sons, who all became attorneys.

Massie graduated from high school at the age of 13, and in 1934 he entered Dunbar Junior College. Massie moved on to the Agricultural, Mechanical, and Normal College of Arkansas (Arkansas A, M & N, now the University of Arkansas), where he graduated at the top of his class in 1938 at the age of 18. He attained a National Youth Administration Scholarship for graduate work at Fisk University, where he worked as a lab assistant in chemistry. He earned his master's degree in 1940, after which he returned to Arkansas A, M & N as an associate professor of mathematics. During World War II Massie worked as a research associate in chemistry at Iowa State University under Henry Gilman, who was a member of the Manhattan Project working on the atomic bomb. Massie received his Ph.D. in organic chemistry from Iowa State University in 1946.

Langston University offered Massie not only a full professorship, but also the chair of its chemistry department, in 1947. Massie then returned to Fisk University in 1953 to head its chemistry department. During his tenure there he also served as the Sigma Xi Lecturer at Swarthmore College. In 1960 he took on a more administration-oriented role as the associate program director of the National Science Foundation, but the next year he returned to academia, in the administrative role of head of the pharmaceutical chemistry department at Howard University. He returned to administration to become the president of North Carolina College in Durham in 1963. In 1966 his career settled as he became a chemistry professor at the United States Naval Academy, a position that allowed him to continue his research. He chaired the department there between 1977 and 1981.

While at the Naval Academy, Massie served for 21 years on the Maryland State Board of Community Colleges, 10 of those as its chair. In 1989 the board named the Massie Science Prize, awarded annually to the outstanding science scholar in the system, in his honor. In 1992 the National Naval Officers Association similarly named the Samuel P. Massie Educational Endowment Fund in his honor, offering Naval Academy scholarships to local minorities and women. In 1994 a coalition of 100 of America's largest companies joined with nine colleges and universities and the U.S. Department of Energy to endow the Samuel P. Massie Professorial Chair of Excellence at each of the nine member institutions, which receive $1.6 million apiece to fund environmental research. In 1988 Massie had received the National Lifetime Achievement Award at the White House for his advancement of minority education, and in 1989 he was inducted into the National Black College Alumni Hall of Fame.

Maury, Antonia Caetana

(1866–1952)
American
Astronomer

Antonia Maury improved a star classification system and studied pairs of stars that orbit each other. She was born on March 21, 1866, in Cold Spring, New York, to parents who both had ties to science. Her father, Mytton Maury, was an amateur naturalist as well as an Episcopal minister. Her mother, Virginia, was the sister of Henry Draper, the first person to photograph stars' spectra, the patterns of rainbow colors and dark lines made by passing the stars' light through a prism.

Antonia Maury was educated at home, mainly by her father, until she went to Vassar College, where MARIA MITCHELL interested her in astronomy. Maury graduated in 1887 with honors in astronomy, physics, and philosophy. By that time her aunt, Henry Draper's wealthy widow, had endowed a large project at the Harvard Observatory to classify stars according to differences in their photographed spectra, and Maury's father asked Edward Pickering, the observatory's director, to employ Antonia on this work. In 1888 Pickering added her to his staff of women "computers."

Pickering and Maury clashed constantly. He wanted her to apply the classification system he and WILLIAMINA PATON STEVENS FLEMING had worked out, but Maury devised a new system. He demanded that she work quickly, whereas she wanted to dwell on minute details such as differences in the thickness and sharpness of dark lines in the spectra. Maury worked on a catalogue of bright northern stars until the early 1890s, then resigned from the observatory. The catalogue was finally published in 1897 with Maury's name on the title page, the first time an observatory publication had so credited a woman.

For almost 20 years Maury pursued a varied career of lecturing, tutoring, and environmental work. She also sometimes visited the observatory to continue research on her special interest, pairs of stars called spectroscopic binaries. The stars in these pairs were so close together that the eye could not distinguish them, even with the best telescopes. Only the doubling of lines in their spectra revealed that there were two stars. She and Pickering had first identified such pairs in 1889. Maury observed changes in the spectra of one particularly intriguing binary, Beta Lyrae, for years and published a book about them in 1933.

Meanwhile other astronomers vindicated the approaches that Pickering had derided. The Danish astronomer Ejnar Hertzsprung told Pickering in 1905 that in some ways Maury's star classification system was better than his. Hertzsprung used the differences in the appearance of spectral lines that Maury had pointed out to confirm that stars of the same color could differ in size and brightness. The diagram of star development that he and Henry Russell constructed on the basis of these differences became a cornerstone of astrophysics.

Maury again joined the Harvard Observatory staff in 1918; she remained there until her retirement in 1935 and continued to visit yearly thereafter. She got along well with Pickering's successor, HARLOW SHAPLEY, who encouraged her research on binaries. The American Astronomical Society awarded her the Annie Jump Cannon Prize in 1943 for her star catalog and classification system. She died in New York on January 8, 1952.

Maury, Matthew Fontaine

(1806–1873)
American
Oceanographer

Matthew Fontaine Maury, who is recognized as the father of physical oceanography, organized the first systematic collection of data about winds and ocean currents. This information about oceanic and atmospheric circulation allowed ship pilots to shorten the length of their voyages dramatically. His work inspired the first international marine conference, which in turn enabled Maury to produce detailed charts of the Atlantic, Pacific, and Indian Oceans. He also published the first modern oceanographic text.

Maury, his parent's seventh child, was born near Fredericksburg, Virginia, on January 14, 1806. His father, Richard Maury, was a small planter whose Huguenot ancestors were early settlers in the area. His mother, the former Diana Minor, also had roots dating back to colonial times. Maury's family moved from Virginia to Williamson County, Tennessee, in 1810. Maury went on to have a large family of his own. In 1834 he married Ann Herndon, and together they had five daughters and three sons.

After graduation from Harpeth Academy in 1825 Maury joined the United States Navy as a midshipman. He accumulated a great deal of his scientific background in the service, especially through such experiences as circumnavigating the globe. In 1836 he was promoted to the rank of lieutenant and published his *Treatise on Navigation*. His promising naval career was abruptly ended, however, when an 1839 stagecoach injury permanently damaged his leg and made him unfit for active duty.

His expertise in navigation nevertheless led to his appointment in 1842 to head the Navy's Depot of Charts and Instruments (DCI), which later became the U.S. Naval Observatory and Hydrographic Office. As director of DCI Maury gave logbooks to ship captains in order to obtain from them data about winds and currents. From these records Maury began to publish his *Wind and Current Charts* (also known as "pilot charts"). The information contained in these charts gave captains the ability to take into account the powerful forces of winds and currents when plotting their routes. This capacity allowed them to cut almost a month from the time required to travel by boat from New York to California. Maury gave his pilot charts to captains at no charge in exchange for additional logs of the winds and currents they encountered on their trips.

His efforts provided the impetus for the first international marine conference, which was held in Brussels in 1853. The Brussels meeting led to the foundation of the International Hydrographic Bureau. Maury used the flood of worldwide data loosed by this organization to produce charts of the Atlantic, Pacific, and Indian Oceans. He also conducted the first bathymetric survey of the Atlantic, which proved that it was possible to lay transatlantic telegraph cable on the floor of the ocean. In 1855 he published the first modern oceanographic textbook, *The Physical Geometry of the Sea*. Maury resigned his commission in 1861 to join the Confederate Navy. In 1865 he traveled to Mexico, where he spent a year serving Emperor Maximilian. He then wrote textbooks in England and did not return to the United States until 1868; he was a professor of physics at the Virginia Military Institute until his death in 1873.

Maury's scientific accomplishments were due to his organizational and empirical abilities. His overarching goal was to advance the cause of maritime commerce, especially in an era increasingly dominated by rail transport. His weakness lay in his sometimes inaccurate interpretations of the data he collected. Nevertheless he was honored for his achievements on many occasions. He was decorated by rulers in Denmark, Portugal, and Russia. He received honorary degrees from Columbia College in 1853 and from the University of North Carolina in 1852. His birthday is celebrated as a school holiday in Virginia.

Maxwell, James Clerk

(1831–1879)
Scottish
Physicist

James Clerk Maxwell, who is known for his unified theory of electromagnetism and his kinetic theory of gases. *AIP Emilio Segrè Visual Archives.*

Regarded as one of the preeminent physicists of the 19th century, James Clerk Maxwell explored an array of scientific topics. His most renowned works were his unified theory of electromagnetism and his kinetic theory of gases. Building upon the ground-breaking research of MICHAEL FARADAY, Maxwell showed the connection between electricity and magnetism. Maxwell wrote four differential equations (known simply as Maxwell's equations) describing the transmission of electromagnetic waves. When taken together, these equations provided a complete description of how electric and magnetic fields were produced and how they were related. These equations influenced ALBERT EINSTEIN in his formulation of the theory of relativity. Maxwell's studies of gases and his theory that gases were composed of molecules in random motion led to the postulation of the Maxwell-Boltzmann distribution law, a statistical formula for determining a gas molecule's probable energy level.

Maxwell was born in Edinburgh on June 13, 1831, to John Clerk Maxwell and Frances Kay Maxwell. After his mother died when he was eight, Maxwell was raised by his father, a lawyer with a profound interest in technical matters. Indeed, the elder Maxwell published a scientific paper and was a member of the Royal Society of Edinburgh. It was from his father that Maxwell inherited both his methodical manner and the family estate in Scotland, to which he retired to undertake the bulk of his scientific writing. He married Katherine Mary Dewar in 1858; the couple had no children.

Maxwell's talents were recognized early. At the age of 14 he published his first paper, an examination of the optical and geometric properties of ovals. He entered the University of Edinburgh in 1847, when he was 16, and obtained a mathematics degree from Trinity College, Cambridge, in 1854. At Cambridge he was influenced by William Hopkins.

Upon his graduation Maxwell held professorships of natural philosophy at Marischal College, Aberdeen (1856–1860), and King's College, London (1860–1865). He resigned in 1865 and retired to research and write at his family estate. In 1871 colleagues persuaded him to return to the world of academia, and he accepted an appointment as the first Cavendish Professor of Experimental Physics at Cambridge. Over the course of his career Maxwell published four books and 100 papers on a variety of topics. His early work was devoted to such divergent problems as the mechanism of color vision and the composition of Saturn's rings. He concluded in 1849 that all colors were derived from the three primary colors—red, yellow, and blue—and in 1861 produced the first color photograph. From 1855 to 1859 he showed that Saturn's rings were neither fluid nor solid, as had heretofore been supposed, but rather were composed of small bodies in orbit.

Maxwell's most significant work was conducted during the 1860s. It was then that he formulated an electromagnetic theory and a kinetic theory of gases. He illustrated that oscillating electric charges (that is, charges fluctuating at regular intervals between the highest and lowest values) would produce waves transmitted through the electromagnetic field. He concluded that light was an electromagnetic wave, as were infrared and ultraviolet radiation, and he postulated the existence of additional kinds of electromagnetic radiation. This theory was supported in 1888 when HEINRICH RUDOLF HERTZ discovered radio waves. In 1864 Maxwell published *Dynamical Theory of the Electric Field*, which contained his famous differential equations. One of the more influential aspects of these equations was that Maxwell established that the speed of a wave was independent of the velocity of its source.

Maxwell's research into gases built upon the existing idea that gas consisted of constantly moving molecules colliding with one another. In 1860 Maxwell (and LUDWIG EDUARD BOLTZMANN independently) employed statistical methods in the Maxwell-Boltzmann distribution to account for the wide variation in the velocities of the various molecules in gas. Maxwell divined that this variation resulted from temperature. He also proposed that heat from the motion of the gas molecules is stored in a gas, a theory that led to explanations of the viscosity and diffusion of gases.

Maxwell's untimely death of abdominal cancer in 1879 cut short his further investigation into electromagnetism. His innovations, nevertheless, were considered to be on a par with those of Sir ISAAC NEWTON and Albert Einstein.

Mayer, Maria Gertrude Goeppert

(1906–1972)
American
Physicist

Maria Mayer worked out a theory that explains how particles are arranged in the atomic nucleus. It won her a share of the Nobel Prize in physics in 1963. Nonetheless she was denied a paying academic position for most of her life.

Maria Gertrude Goeppert was born in Kattowitz, then part of Germany (it is now Katowice, Poland), on June 28, 1906. When she was four, her physician father moved his family to Göttingen and became a professor of pediatrics at the famous university there. (Maria liked to brag, "On my father's side, I am the seventh straight generation of university professor.") He fostered Maria's interest in science, and her mother, also named Maria, shared with her a love of music and social life.

Maria began studying mathematics at Göttingen University in 1924. After meeting professors and students in the "young and exciting" field of atomic physics, however, she changed her major in 1927. Physics, like mathematics, involved "puzzle solving," but in physics, she said later, the "puzzles [were] created by nature, not by the mind of man."

Maria's father died in 1927, and her mother began taking in boarders to support the family. One boarder was a lanky American chemistry student named Joseph E. (Joe) Mayer, who looked for a room in the Goeppert house in 1928 after a German friend told him that "the prettiest girl in Göttingen" lived there. Mayer apparently agreed with that evaluation, and Maria Goeppert was equally pleased with him. They married on January 19, 1930. Shortly thereafter Maria finished her Ph.D. Her thesis is considered a fundamental contribution to quantum mechanics.

The newlywed Mayers moved to the United States. Joe became an assistant professor of chemistry at Johns Hopkins University, but there was no job for Maria because antinepotism rules, common at the time, forbade the hiring of both husband and wife—and it was always the wife who was left out. Maria did important research on dye chemistry during this time, but the only money she received was a few hundred dollars a year for helping a professor with his German correspondence.

Maria became a naturalized U.S. citizen in 1933, and her daughter, Maria Anne (Marianne), was born shortly thereafter. In 1937 she and Joe began work on a textbook on statistical mechanics, which described the behavior of molecules. (It was published in 1940.) Maria gave birth to her second child, Peter, just before Joe moved to Columbia University in 1938. Like Johns Hopkins, Columbia refused to give Maria a job.

As World War II loomed, the Mayers were among the European expatriate scientists who urged the United States to sponsor a research program to develop an atomic bomb. When the program, code-named the Manhattan Project, began, the head of the Columbia chemistry department asked Maria to help search for ways to separate the bomb's potential fuel, the radioactive form of uranium, from the more common, nonradioactive form. He even offered to pay her. Her work proved to have little direct effect on the bomb project, but, she said later, "It was the beginning of myself standing on my own two feet as a scientist, not leaning on Joe."

After the war ENRICO FERMI, EDWARD TELLER, and several other of the Mayers' friends from their Göttingen days worked at the University of Chicago's new Institute for Nuclear Studies (later the Enrico Fermi Institute), and the Mayers were invited to join them early in 1946. This time Maria was made an associate professor—she said that Chicago was "the first place where I was not considered a nuisance, but greeted with open arms"—though she still did not receive a salary. She did, however, earn part-time pay as a senior researcher at the nearby Argonne National Laboratory. The lab's head, Robert G. Sachs, had been one of her first graduate students and was glad to hire her.

The group of physicists at the University of Chicago in the late 1940s and 1950s was one of the premier gatherings of scientists in the 20th century. One observer called their weekly seminars "conversation[s] of the angels." One subject they often discussed was the arrangement of particles in the atomic nucleus. Physicists knew that electrons orbited the nucleus in distinct layers called shells, and some had suggested that particles in the nucleus might also be arranged in shells, but little physical or mathematical evidence had been found to support this idea. The most commonly accepted model of the nucleus pictured it as something like a drop of water, in which protons and neutrons moved randomly.

Beginning in 1947, Maria Mayer worked with Edward Teller on a theory of the origin of the chemical elements. In the course of researching this project Mayer noticed that elements whose nuclei contained certain numbers of protons or neutrons—what came to be called her "magic numbers"—were unusually abundant and stable, almost never undergoing radioactive decay. Teller and others shrugged off this finding, but Mayer "kept thinking why, why, *why* do they exist?" She suspected that they had some relation to nuclear particles' being arranged in shells. The numbers might represent filled shells in the nucleus that prevented atoms from breaking down radioactively, just as filled electron shells prevented elements from reacting chemically. Still, she could not prove her idea.

Mayer frequently talked over her ideas with Enrico Fermi. One day in 1948, as Fermi was leaving her to answer a telephone call, he asked offhandedly, "Is there any indication of spin-orbit coupling?" Mayer suddenly realized that this was the missing piece of her puzzle. Spin-orbit coupling is the phenomenon that the direction in which a particle spins helps to determine which orbit or shell it will occupy. Spin-orbit coupling among electrons is weak, but Mayer realized that if it was powerful in the nucleus, requiring much more energy for particles to spin in one direction than in the other, it could explain her magic numbers and prove that nuclear particles are arranged in shells. "I got so excited it wiped everything out," she said later. She worked out the calculations to confirm her idea in 10 minutes.

Maria published a description of her theory in April 1950. A German researcher, Hans D. Jensen, thought of the same idea independently and published his account of it at about the same time. The two met later in 1950 and worked together on a book, *Elementary Theory of Nuclear Shell Structure,* which was published in 1955.

In 1959 the newly formed University of California at San Diego invited both Mayers to join its faculty—and offered to hire Maria at the same salary and rank (full professor) as Joe. The Mayers were glad to accept. Unfortunately, a few months after their arrival in 1960 Maria suffered a stroke. She continued her research, however.

Maria Mayer's shell theory won the 1963 Nobel Prize in physics. She shared the prize with Jensen and another nuclear physicist, Eugene Wigner. She was only the second woman to win the physics prize. She was pleased, of course, but she said later that "winning the prize wasn't half as exciting as doing the work itself." She continued to refine her theory in the years that followed, meanwhile receiving other awards, including election to the National Academy of Sciences. Mayer died of heart disease on February 20, 1972, at the age of 65.

McClintock, Barbara

(1902–1992)
American
Geneticist

Working in her cornfields while other geneticists investigated molecules, Barbara McClintock was ignored for decades because her discoveries ran as counter to the mainstream as her methods. In the end, however, she proved that, contrary to what almost everyone had thought, genes could move and control other genes. Organisms thus could partly shape their own evolution.

Barbara McClintock, the first woman to win an unshared Nobel Prize in physiology or medicine. *Cold Spring Harbor Laboratory Archives.*

Barbara preferred her own company almost from her birth in Hartford, Connecticut, on June 16, 1902. As a baby she played happily by herself; as a teenager she liked to spend time simply "thinking about things." She grew up in Brooklyn, then a somewhat rural suburb of New York City, to which her father, a physician for Standard Oil, moved the family when she was six. Her parents, Thomas and Sara McClintock, encouraged independence in their four children by allowing them to skip school.

Barbara became determined to attend college, even though, as she said later, her mother feared that a college education would make her "a strange person, a person that didn't belong to society. . . . She was even afraid I might become a college professor." Thomas McClintock took Barbara's side, however, and in 1919 she enrolled in the College of Agriculture at Cornell University, which offered free tuition to New York residents.

McClintock took a course in the relatively new science of genetics in her junior year, and by the time she graduated in 1923 she had decided to make genetics her career. As a graduate student in the university's botany department she studied Indian corn, a type of maize (corn) in which the kernels on each ear have different colors. The color pattern is inherited.

McClintock made her first important discovery while still a graduate student. Geneticists were realizing that inherited traits were determined by information contained in microscopic wormlike bodies called chromosomes. Each cell has a number of pairs of chromosomes, and each pair appears only slightly different from the others. McClintock was the first to work out a way to tell the 10 pairs of maize chromosomes apart.

McClintock earned her M.A. in botany in 1925 and her Ph.D. in 1927, after which Cornell hired her as an instructor. In 1931 she and another scientist, Harriet Creighton, carried out an experiment that firmly linked changes in chromosomes to changes in whole organisms, a link that some geneticists had still doubted. This experiment has been called "one of the truly great experiments of modern biology."

McClintock was earning a national reputation in genetics, but Cornell refused to promote her because she was a woman. Rather than remain an instructor forever, she resigned shortly after her landmark paper was published. For the next several years she led an academic gypsy life, living on grants and dividing her research time among three universities in different parts of the country. She did her commuting in an old Model A Ford, which she repaired herself whenever it broke down.

In 1936 the University of Missouri at Columbia, one of the institutions at which McClintock had done part-time research, gave her a full-time position as an assistant professor. While there she studied changes in chromosomes and inherited characteristics made by X rays, which damaged genetic material and greatly increased the number of mutations, or random changes, that occurred in it. This university, too, refused to treat her with the respect she felt she deserved, and she resigned in 1941.

McClintock was unsure what to do next until a friend told her about the genetics laboratory at Cold Spring Harbor, on Long Island. Run by the Carnegie Institution of Washington, which the steel magnate and philanthropist Andrew Carnegie had founded, it had been the first genetics laboratory in the United States. McClintock moved to Cold Spring Harbor in 1942 and remained there the rest of her life.

The discovery in the 1940s that the complex chemical deoxyribonucleic acid (DNA) was the carrier of most genetic information and the working out of DNA's chemical structure and method of reproduction in 1953 revolutionized genetics, turning attention away from whole organisms or even cells and toward molecules. Geneticists saw genes, now shown to be parts of DNA molecules, as unalterable except by chance or the sort of damage that X rays produced. FRANCIS HARRY COMPTON CRICK, the codiscoverer of DNA's structure, expressed what he called the "central dogma" of the new genetics: "Once 'information' has passed into protein [chemicals that carry out cell activities and express characteristics] *it cannot get out again.*"

Barbara McClintock meanwhile went her own way, working with her unfashionable corn and "letting the material tell" her what was happening in its genes. Contrary to Crick's central dogma, she found genes that apparently could change both their own position on a chromosome and that of certain other genes, even moving from one chromosome to another. This movement, which she called transposition, appeared to be a controlled rather than a random process. Furthermore, if a transposed gene landed next to another gene, it could turn on that gene (make it active, or capable of expressing the characteristic for which it carried the coded information) if it had been off, or vice versa. Genes that could control their own activity and that of other genes had not been recognized before. McClintock suspected that such genes and their movement played a vital part in organisms' development before birth.

Even more remarkable, some controlling genes appeared able to increase the rate at which mutations occurred in the cell. McClintock theorized that these genes might become active when an organism found itself in a stressful environment. Increasing the mutation rate increased the chances of a mutation that would help the organism's offspring survive. If a gene that increased its mutation rate could be turned on by something in the environment, then organisms and their environment could affect their own evolution, something no one had thought possible.

McClintock attempted to explain her findings at genetics meetings in the early 1950s, but her presentations were met with blank stares or even laughter. She offered ample evidence for her claims, but her conclusions were too different from the prevailing view to be accepted. The chilly reception "really knocked" McClintock, as she later told her biographer, Evelyn Fox Keller, and after a while she stopped trying to communicate her research. Most geneticists forgot, or never learned, who she was; one referred to her as "just an old bag who'd been hanging around Cold Spring Harbor for years." She did not let rejection stop her work, however. "If you know you're right, you don't care," she said later.

McClintock received a certain measure of recognition in the late 1960s. In 1967, the same year in which she officially retired (in fact, her work schedule continued unchanged), she received the Kimber Genetics Award from the National Academy of Sciences. She was awarded the National Medal of Science in 1970. Only in the late 1970s, however, did other geneticists' work begin to support hers in a major way. Researchers found transposable elements, or "jumping genes" as they became popularly known, in fruit flies and other organisms, including humans. The idea that some genes could control others was also proved.

The trickle of honors became a flood in the late 1970s and early 1980s. McClintock won eight awards in 1981 alone, the three most important of which—the MacArthur Laureate Award, the Lasker Award, and Israel's Wolf Prize—occurred in a single week. Then in October 1983, when she was 81 years old, she learned that she had won the greatest scientific award of all, a Nobel Prize. She was the first woman to win an unshared Nobel Prize in physiology or medicine.

These honors and their attendant publicity irritated McClintock more than they pleased her. She complained, "At my age I should be allowed to . . . have my fun," which meant doing her research in peace. McClintock continued to have her scientific "fun" among the corn plants almost until her death on September 2, 1992, just a few months after her 90th birthday.

Some people saw Barbara McClintock's solitary life as lonely and perhaps even sad, but she never viewed it that way. As she said shortly before she died, "I've had such a good time. . . . I've had a very, very satisfying and interesting life."

McMillan, Edwin Mattison

(1907–1991)
American
Physicist

Edwin Mattison McMillan revolutionized physics when he created a new element by bombarding uranium with neutrons. This 1940 discovery of neptunium sparked a flurry of attempts to create other transuranic elements—those literally "beyond" (or heavier than) uranium in the periodic table. In 1941 McMillan and his colleague GLEN THEODORE SEABORG synthesized the 94th element, plutonium, which became an essential component of the nuclear bombs dropped on Japan. In addition to this Nobel Prize–winning work, McMillan conceived the principle of phase stability, which led him to invent the synchrocyclotron, an improved version of the cyclotron particle accelerator.

McMillan was born in Redondo Beach, California, to Edwin Harbaugh McMillan, a physician, and the former Anna Marie Mattison, on September 18, 1907. In 1941 McMillan wed Elsie Walford Blumer, with whom he had three children.

McMillan received his B.S. in physics from the California Institute of Technology in 1928, and his Ph.D. in physics from Princeton University in 1932. He was then awarded a National Research Fellowship and elected to attend the University of California, Berkeley.

McMillan spent two years as a research fellow and a third as a research associate at Berkeley before becoming an instructor on its physics faculty in 1935. At Edward E.O. Lawrence's Radiation Laboratory, McMillan built upon the work of ENRICO FERMI. Fermi believed that bombarding an element with neutrons could convert it into a new one (with the next-highest atomic number) and sought to create the 93rd element by bombarding uranium. He failed, but his work opened the door for McMillan, who was able to transmute uranium into what he dubbed neptunium by bombarding it with neutrons in the cyclotron particle accelerator. He conducted extensive chemical tests to verify his discovery. (McMillan had long been interested in chemistry as well as physics.)

McMillan immediately aimed to create the 94th element. He believed he had succeeded but was unable to verify his work before being directed by the War Department to report to the Massachusetts Institute of Technology in 1940 to develop radar for the American World War II effort. McMillan provided his colleague Glen Seaborg with his notes. In McMillan's absence Seaborg obtained definitive proof that McMillan had produced element 94, named *plutonium,* before he departed for Boston. For the duration of the war McMillan investigated sonar at the navy's Radio and Sound Laboratory and added his expertise to the Manhattan Project's effort to develop an atomic bomb.

Upon returning to Berkeley after the war, McMillan turned his attention to the limitations of cyclotron particle accelerators, which could not accelerate particles to optimal velocities. Applying his newly minted principle of phase stability, McMillan vastly improved the capabilities of the cyclotron in 1945. The shortcomings of prewar cyclotrons were explained by the special theory of relativity, which held that as a particle increases in velocity, it also increases in mass, eventually "falling out" of the radio frequency field that propelled it to high speeds. McMillan proposed "stabilizing" the particle either by adjusting the magnetic field that held it or by changing the frequency of the accelerating particle so that it kept pace with the frequency field. This synchrocyclotron, as he named his invention, proved to be the model for all future particle accelerators.

After attaining the rank of full professor at Berkeley in 1946, McMillan was appointed director of the Lawrence Radiation Laboratory in 1958. Under his leadership the laboratory prospered, as its new cyclotron facilitated the discovery of elements 103, 104, and 105. He retired in 1973. The recognition he achieved during his lifetime was considerable. In addition to the 1951 Nobel Prize in chemistry (which he shared with Seaborg), McMillan was jointly awarded the 1963 Atoms for Peace Prize (with Vladimir Veksler) for his invention of the synchrocyclotron. McMillan also chaired the U.S. National Academy of Sciences from 1968 to 1971. McMillan's impact on the development of physics was astounding. He pioneered the research of transuranic elements, and his ground-breaking discoveries directly resulted in the synthesis of plutonium, which dramatically altered the political landscape of the second half of the 20th century. His phase stability theory made possible all modern particle accelerators and thus engendered later discoveries of antiprotons and antineutrons.

McNally, Karen Cook

(1940–)
American
Geologist

Karen McNally has predicted earthquakes by identifying "seismic gaps," areas where expected Earth movement has not taken place and sudden large movements are thus more likely. Born in 1940, she felt earthquakes while growing up in Clovis, California, but she was more interested in learning ranch work from her father and music from her mother.

McNally's parents persuaded her to attend nearby Fresno State College, but she sought independence in an early marriage. She studied part-time while raising two daughters, which she calls a "superhuman task." Magazine articles interested her in geology, and in 1966 she divorced her husband and moved, daughters and all, to the University of California at Berkeley to learn more about it. She completed her bachelor's degree in 1971, a master's degree in 1973, and a Ph.D. in geophysics in 1976.

While doing postdoctoral work at the California Institute of Technology (Caltech) in Pasadena, McNally learned about the seismic gap theory, a new approach to earthquake prediction. Normally the immense plates that make up the Earth's crust slide past each other smoothly, a bit at a time, causing "microquakes" detectable only with a seismograph. Friction, however, can lock plates together in a certain area for decades or even centuries. Because the plates as a whole continue to move, pressure builds up in that area. When the pressure finally fractures the rock and releases the plates, a large amount of motion usually occurs all at once, causing a major earthquake. Some geologists believed that if seismographs were placed in areas identified as having "seismic gaps"—long preceding periods of little or no Earth movement—they might detect fracturing that presaged a big quake.

One seismic gap lay on the western coast of Mexico south of Oaxaca. In 1977 geophysicists from Japan and Texas warned that this area was ripe for a large quake. McNally and geologists from Mexico's National University had seven portable seismographs installed there by November 1. For weeks their recordings showed a buildup of tremors; then, on November 29, McNally says, "there was absolute silence" for part of the day—followed by a Richter 7.8 quake within 31 miles of where her group had predicted it. Their seismographs captured a complete picture of this major earthquake, which McNally compares to finding a live dinosaur.

McNally and others predicted five more Mexican quakes, including a Richter 8.1 quake that devastated Mexico City on September 19, 1985. She has used information from these and other quakes to refine the seismic gap theory. She believes that as pressure in a gap area builds up, weak spots at the ends give way first, producing small quakes. Cracks then start spreading throughout the area's rock, perhaps producing the phenomena sometimes seen just before earthquakes, such as changes in groundwater levels. Cracking and shaking increase until, finally, the strongest rocks break and a major quake occurs.

McNally remained at Caltech until 1986, eventually rising to the rank of associate professor, and then moved to the University of California at Santa Cruz, where in 1998 she was a professor of geophysics. She has been the director of the university's Institute of Tectonics and its Charles F. Richter Seismology Laboratory.

Mead, Margaret
(1901–1978)
American
Anthropologist

Margaret Mead, perhaps the best-known anthropologist of her time, introduced basic ideas from the study of humankind into mainstream culture. Her writings suggested that people's personalities and expectations about such phenomena as gender roles are shaped more by their culture than by their genes.

Margaret was born on December 16, 1901, in Philadelphia to Edward Mead, who taught economics at the University of Pennsylvania's Wharton School of Business, and Emily, a sociologist. The oldest of five children, Margaret was educated mostly at home by her mother and grandmother, who taught her to observe the people around her (starting with her brother and sisters) and even to take notes.

In 1919 Margaret entered DePauw University in Indiana, planning to major in English, but she felt out of place there and transferred to Barnard College, a women's college affiliated with Columbia University. After meeting Franz Boas and RUTH FULTON BENEDICT, the leading anthropologists of their day, at Columbia, she changed her major to anthropology. She graduated from Barnard in 1923.

Mead married Luther Cressman, who was then planning to become a minister, and began work for a master's degree in psychology at Columbia, which she earned in 1924. She continued to use her maiden name professionally. For her Ph.D. fieldwork she decided to find out whether teenage girls on the Pacific islands of Samoa felt the anxieties and frustrations that American adolescents did. Beginning in 1925, when she was 23 years old, she spent nine months on the island of Tau, where she became friends with the native women and children and lived as they did. The island girls called the diminutive Mead "Malekita" and, as far as she could tell, spoke freely to her about their lives.

After she returned from Samoa in 1926, Mead became assistant curator of ethnology (cultural anthropology) at the American Museum of Natural History in New York. She maintained an office there all her life and became associate curator in 1942 and curator in 1956.

Meanwhile Mead's Samoan research earned her a Ph.D. from Columbia in 1929 and also became the basis of her first book, *Coming of Age in Samoa,* which was published in 1928. Aimed at a popular audience, the book painted a charming picture of an easygoing people who saw nothing wrong with sexual activity before marriage and seemed to suffer few of the anxieties that American teenagers felt. It became a best-seller and made Mead famous.

Mead set out to study the Manus people of the Admiralty Islands in New Guinea, north of Australia, in 1928. This time she did not do her fieldwork alone. On the ship to Samoa three years before she had met a young New Zealand anthropologist, Reo Fortune, and the two had fallen in love. On her way to New Guinea, Mead, now divorced from Cressman, stopped in New Zealand to marry Fortune, and the two went on to the islands together.

Mead and Fortune studied four peoples of New Guinea over the next several years. Their field trips became the basis for scientific papers and Mead's next two popular books, *Growing Up in New Guinea* (1930) and *Sex and Temperament in Three Primitive Societies* (1935). Both books presented the idea that personality and behavior patterns, including those assigned on the basis of gender, were determined largely by culture and that each culture encouraged a certain type of temperament and discouraged others, giving the culture itself a kind of personality. Mead had learned these ideas from Boas and Benedict.

Mead met Gregory Bateson, another anthropologist doing research in New Guinea, in 1932 and again fell in love. She divorced Fortune and then married Bateson in March 1936. They had a daughter, Mary Catherine, in December 1939. Mead and Bateson did further research in New Guinea and also worked together on a study of the Balinese people of Indonesia, *Balinese Character* (1941), which featured many photographs. They were among the first anthropologists to use photography extensively in their work. After 15 years together, to Mead's grief, Bateson ended their marriage.

Mead partly put aside her research to do government-related work during World War II. She also turned her anthropologist's eye on the culture of the United States and presented the results in another popular book, *And Keep Your Powder Dry: An Anthropologist Looks at America* (1942). The book compared American culture with seven others Mead had studied. According to Roger Revelle of the American Association for the Advancement of Science, Mead "became a kind of modern oracle because of her sensitivity to what was significant in American life."

Mead thereafter constantly expanded the range of her prolific writings and lectures, both academic and popular. In addition to making additional research trips to the Pacific islands, New Guinea, and Bali, during which she noted changes in groups she had visited earlier, she wrote and spoke on everything from family life and education to ecology and nutrition. She once commented, "The anthropologist's one special area of competence is the ability to think about a whole society and everything in it."

Mead received many honorary degrees and other awards for her work, including a gold medal from the Society for Women Geographers in 1942 and election to the National Academy of Sciences in 1974, the latter by one of the highest votes recorded in an academy election. She was president of the American Association for the Advancement of Science in 1974.

Mead officially retired from the Museum of Natural History in 1964, but she continued to work there for many years afterward. "Sooner or later I'm going to die, but I'm not going to retire," she told reporters on her 75th birthday. She died in New York on November 15, 1978, at the age of 76.

In 1983, several years after her death, Mead's reputation suffered a blow when the Australian anthropologist Derek Freeman published a book claiming that the conclusions in her famous early book on Samoa had been incorrect. Drawing on his own field research and early documents, Freeman maintained that Samoans suffered as much anxiety as any other people and were rather possessive sexually. He said that Mead's informants had told her false stories about their free sexual life both because they thought that was what she wanted to hear and because they liked to tease her gently and she had been young and naive enough to believe them.

Most of Mead's admirers say that even if Freeman's allegations are correct, they have little impact on her overall importance, which is based on a lifetime of research and thought. The biographer Jane Howard notes, "Her early fieldwork may have been hurried and imperfect, but her generous view of human nature endures and the energy and volume of her later achievements are staggering."

Meitner, Lise

(1878–1968)
German/Swedish
Physicist

Walking through the snowy Swedish countryside, Lise Meitner and her nephew, Otto Frisch, concluded that the nucleus of an atom could be split. This realization not only transformed scientists' understanding of atoms but also led to the creation of nuclear power and the atomic bomb.

Lise Meitner, who, with her nephew Otto Frisch, announced in 1939 that the nucleus of an atom could be split. *AIP Emilio Segrè Visual Archives.*

Lise Meitner was born on November 7, 1878, in Vienna, Austria, to Philipp, a wealthy lawyer, and Hedwig. Lise, the third of the Meitners' eight children, learned a love of music from her mother. From childhood on she also displayed what she later called a "marked bent" for physics and mathematics. Inspired by reading about MARIE SKLODOWSKA CURIE, she decided that she, too, wanted to become a physicist and study radioactivity.

Standard education for girls of the time did not prepare them for a university, but Lise had extra tutoring and entered the University of Vienna in 1901. She earned a doctorate in physics in 1905, the first physics doctorate that the university had ever granted to a woman.

Wanting to study under MAX PLANCK, the founder of quantum physics, Meitner went to the University of Berlin in 1907. There she persuaded Otto Hahn, a chemist about her own age who was also studying radioactivity, to hire her as an assistant. The only problem was that Emil Fischer, the head of the institute in which Hahn worked, refused to allow a woman in its classrooms or laboratories. Hahn and Meitner had to work in a basement room that had been a carpentry workshop. The years they spent there, however, were happy and productive. When work was going well, Meitner recalled later, "we sang together in two-part harmony, mostly songs by Brahms." The pair's working conditions improved in 1912, when they moved to the new Kaiser Wilhelm Institute in Dahlem, a Berlin suburb.

World War I interrupted their work—Hahn remained at the university to do war research, and Meitner served as an X-ray technician in a field hospital—but they continued it when they could. In 1918, soon after the war ended, they announced their discovery of a new element, the second heaviest then known. They named this radioactive element *proto-actinium*, "before actinium" in Latin, because it slowly broke down into another element, actinium. The name was later shortened to *protactinium*.

The discovery of protactinium made Hahn and Meitner famous. Meitner won the Leibniz Medal from the Berlin Academy of Science and the Leiben Prize from the Austrian Academy of Science. She was made head of a new department of radioactivity physics at the Kaiser Wilhelm Institute in 1918 and became Germany's first woman full physics professor (at the University of Berlin) in 1926. The Swiss physicist ALBERT EINSTEIN called her "the German MARIE [SKLODOWSKA] CURIE." She and Hahn no longer worked together as often as before, but they remained good friends. Meitner's chief research subject in the 1920s was the behavior of beta particles, negatively charged particles given off during the breakdown of radioactive atoms. Meitner believed, correctly, that these particles were electrons from the nucleus.

Meitner and her friends knew that she must leave the country immediately when in March 1938 Germany seized control of Austria. They decided that the best way for her to escape was to pretend to take a vacation to Holland, which did not require a visa. The train was stopped at the German border, however, and a Nazi military patrol confiscated her passport, an Austrian one that had expired 10 years before. "I got so frightened, my heart almost stopped beating," Meitner wrote later. "For ten minutes I sat there and waited, minutes that seemed like so many hours." Fortunately the patrol then returned her passport and allowed the train to proceed.

Meitner went on to Sweden and began working at the Nobel Institute of Theoretical Physics in Stockholm, but anxiety and homesickness made research difficult. In Germany meanwhile Otto Hahn and a new partner, Fritz Strassmann, like a number of other European nuclear physicists, were bombarding heavy elements with subatomic particles called neutrons in the hope of creating artificial elements that were heavier than uranium, the heaviest natural element. (Physicists had learned that an atom under the right conditions could capture a neutron and emit a beta particle, thus becoming an atom of the next heaviest element.) To their amazement when Hahn and Strassmann exposed uranium to neutrons, they got what appeared to be barium, a much lighter element. Hahn described this puzzle to Meitner in a letter in December 1938 and asked whether she could explain it.

When she received Hahn's letter, Meitner was spending Christmas vacation in the Swedish village of Kungälv with her nephew, Otto Frisch, another physicist. The two discussed the letter as they walked in the snow, and it suddenly occurred to them that Hahn's results could be explained if, rather than simply adding or subtracting a particle or two from the uranium nucleus, he had split it almost in half. Doing so would produce barium and krypton, a gaseous element that was hard to detect.

Physicists had thought splitting atomic nuclei was not possible because very powerful forces held the nucleus together. Most physicists thought the nucleus was like a drop of water, held together by the equivalent of surface tension. The electric charge of a heavy nucleus partly offsets this force, however, and Meitner and Frisch concluded that under some conditions, as Frisch wrote later, the nucleus might become "a very wobbly drop—like a large thin-walled balloon filled with water." A neutron might then split it.

Scrawling calculations in the snow, Meitner and Frisch figured out that, according to the theories of Albert Einstein, splitting a uranium nucleus should release 200 million electron volts, 20 million times more than an equivalent amount of trinitrotoluene (TNT). This energy would not be obvious in a sample of uranium as small as those Hahn had used, but it could be detected with the right instruments. Frisch hurried back to his laboratory to repeat Hahn's experiments and look for the energy. He found it, and he and Meitner began drafting a paper about their discovery. Borrowing the term biologists use to describe the process by which a cell splits in half to reproduce, Frisch called the atomic splitting *fission*. Meitner and Frisch's announcement that the atom could be split rocked the world of physics in 1939.

Six years later it rocked the wider world as well. Living in semiretirement in neutral Sweden, Meitner knew nothing of the development of the atomic bomb until August 6, 1945, when headlines announced that the United States had dropped such a bomb on the Japanese city of Hiroshima. She called the news "a terrible surprise, like a bolt of lightning out of the blue." In an interview shortly afterward she said, "You must not blame us scientists for the uses to which war technicians have put our discoveries. . . . My hope is that the atomic bomb will make humanity realize that we must, once and for all, finish with war."

Otto Hahn received a Nobel Prize in 1944 for his part in the discovery of nuclear fission. For unknown reasons Meitner was not so honored. She did receive other awards, however, including Germany's Max Planck Medal (1949) and the Enrico Fermi Award from the U.S. Atomic Energy Commission in 1966 (she shared this prize with Hahn and Fritz Strassmann). Meitner was the first woman to receive the latter prize.

Lise Meitner continued to live and do research in Sweden, even after she officially retired in 1947. In 1960 she moved to England to be near Otto Frisch, then a professor at Cambridge University. She died on October 27, 1968, a few days before her 90th birthday. In 1982, when element 109 was created, its inventors named it *meitnerium* in honor of Lise Meitner. The leader of the research team called Meitner "the most significant woman scientist of this century."

Mendel, Johann Gregor

(1822–1884)
Austrian
Geneticist

Johann Gregor Mendel's pioneering plant hybridization experiments laid the foundation for the modern field of genetics. His careful analysis of research concerning the transmission of plant characteristics across generations provided the first mathematical basis for that science. His work led to the formulation of Mendel's laws of segregation and independent assortment and represented the first application of statistical analysis to the explanation of a biological phenomenon.

Mendel's intense interest in horticulture and plant hybridization was rooted in his childhood. Born on July 22, 1822, in Heinzendorf, Austria (now Hyncice in the Czech Republic), the young Mendel was raised helping his peasant farmer father, Anton, graft trees in the family orchard. Anton, who had served in the Napoleonic army, was dedicated to improving his crops by applying farming methods he had observed during his travels. Since the family lived on the border between German- and Czech-speaking regions, Mendel became fluent in both languages at an early age.

Mendel's parents and teachers noted his abilities and arranged for him to be educated at the Gymnasium in Troppau, where he studied from 1834 to 1840. After suffering a nervous breakdown brought on by stress, Mendel attended Olmütz University. In 1843 at the age of 21 he entered the Augustinian monastery in Brno. Although he felt he lacked a religious vocation, the monastery afforded him an excellent environment for his studies and freed him of financial worries. After a brief stint as a substitute teacher he failed to pass the teachers' qualifying exam. His order sent him to the University of Vienna from 1851 to 1853. In addition to learning botany and chemistry there, Mendel studied under Franz Unger, a plant physiologist who taught him how to organize botanical experiments. Mendel then returned to Brno but again failed the teachers' exam.

Although he continued to serve as a substitute teacher until 1868, Mendel's true work took place in the monastery's garden. From 1856 to 1863 he conducted a massive study involving 28,000 edible pea plants. He had noted several constant plant traits, including stem length (whether tall or short); flower position (whether axial or terminal); pod color (green or yellow); pod form (smooth or wrinkled); flower seed coat color (purple or white); and cotyledon color (yellow or green). He then crossed plants bearing these distinct attributes. The plants were self-pollinated and then individually wrapped to prevent pollination by insects. After collecting the seeds, Mendel scrupulously studied the offspring.

His findings overturned the accepted heredity theory of the day when he proved that the seven characters did not amalgamate on crossing: that is, rather than offspring bearing a blend of these characteristics, the progeny manifested only one such trait (for example, either a tall stem or short stem rather than some combination). However, when two of these "hybrid" progeny were crossed, the product of that union could manifest the traits of either grandparent, indicating that those elements had remained distinct, even when one was latent. This insight became known after 1900 as Mendel's first law, or the law of segregation. Another discovery was dubbed Mendel's second law, or the law of independent assortment. This principle holds that the various characters transmitted from parent to offspring combine randomly, rather than in any grouped bunch. Mendel's experiments revealed that a given pea plant was statistically no more likely to have a tall stem and smooth pod form, for example, than a tall stem and a wrinkled pod form. Moreover, all of the various permutations of the seven pea traits occurred with equal frequency.

Although Mendel presented his conclusions in 1865 at a meeting of the Natural Sciences Society and published his results in 1866, the significance of his experiments was lost on his contemporaries. After he was appointed abbot of his monastery, his bureaucratic duties expanded, leaving him less time for research. It was not until the 20th century that his work became widely recognized. Then not only did Mendel's experiments help forge the science of genetics but they also bolstered CHARLES ROBERT DARWIN's theory of evolution. Mendel's demonstration of the mechanism of variability in plant species buttressed Darwin's overarching theory of natural selection by supplanting Darwin's erroneous notion of pangenesis (the hypothetical theory that the reproductive egg or bud contains particles from all parts of the parent) as an explanatory principle.

Mendeleev, Dmitri Ivanovich

(1834–1907)
Russian
Chemist

Dmitri Mendeleev's greatest achievement was his formulation of the periodic law. In the process of writing a chemistry textbook Mendeleev arranged the elements according to their increasing atomic weight. He recognized that the resulting table revealed a periodicity of properties among groups of elements. This empirical method of ordering the elements shaped future developments in inorganic chemistry. In 1955 a new element was discovered (with the atomic number 101) and named *mendelevium* (Md) in his honor.

Mendeleev, the youngest of 17 children, was born on February 8, 1834, in Tobolsk, Siberia. His father, Ivan Pavlovich Mendeleev, a teacher who was the director of the Tobolsk Gymnasium, became blind the year Dmitri was born. In order to support the family, his mother, Maria Dmitrievna Kornileva Mendeleev, bought and operated a glass factory nearby. Shortly after his father's death in 1847 a massive fire destroyed his mother's business. She nevertheless enrolled him in the Pedagogic Institute in St. Petersburg at the age of 15.

Dmitri Mendeleev, who formulated the periodic table of elements. *Smith Collection, Rare Book & Manuscript Library, University of Pennsylvania.*

After graduation from the institute Mendeleev qualified to be a teacher in 1855. He continued his studies in 1859 at the University of Heidelberg, where he studied under the German chemist ROBERT WILHELM BUNSEN. In September 1860 Mendeleev attended the first International Chemical Congress at Karlsruhe, where the concept of atomic weights was explored.

Although he returned to St. Petersburg in 1861, Mendeleev did not find a permanent position until 1864, when he became a professor of chemistry at the Technical Institute. Unable to find a suitable textbook for his classes, he wrote his own, *The Principles of Chemistry,* in 1868. As he pondered the properties of the elements, he conceived the idea of arranging them in order of increasing atomic weight. Upon realizing that his table indicated a periodicity of properties, he championed his method as an ideal classification scheme. His 1869 paper "On the Relation of the Properties to the Atomic Weights of the Elements" posited that other elemental properties varied from moment to moment or yielded conflicting results. He was so sure of his periodic table that when the atomic weights of some elements did not conform to his law, he declared that they had been assigned incorrect atomic weights. He also opted to leave blank spaces in the table for elements he believed would be discovered in the future. He went so far as to predict the valences and other physical properties of these elements. He was ultimately proved correct on all these counts.

Mendeleev was only 34 when he devised his periodic table. The remainder of his career took him in different directions. He organized the Russian pavilion at the Paris Exposition in 1867 and later improved the Russian soda industry through chemical experimentation. But the czarist regime did not approve of his progressive political views, and he was "retired" from the university in 1890. In 1891 he was commissioned to establish a new system of import duties on heavy chemicals and was named head of the Bureau of Weights and Measures two years later.

Mendeleev married Feozva Nikitichna Leshchevaya in 1862, and they had a son and a daughter. In 1882 he abandoned this family and married Anna Ivanova Popova, a young artist with whom he had two sons and two daughters. Because of his political troubles Mendeleev's accomplishment was never fully recognized within his own nation—he was even denied admission to the Russian Academy of Sciences. Within his lifetime, however, his periodic table provided a framework for chemical research, as it continues to do. His contribution is ranked with those of Sir ISAAC NEWTON and CHARLES ROBERT DARWIN in its scope. In 1894 he was awarded an honorary doctorate by both Oxford and Cambridge, and he was presented with the Copley Medal in 1905. After his death of heart failure in 1907 students carrying the periodic table above their heads followed his funeral procession.

Merian, Maria Sibylla

(1647–1717)
German/Dutch
Zoologist

Maria Sibylla Merian united science and art to produce some of the first detailed drawings of insect life cycles. She was born on April 2, 1647, in Frankfurt, Germany, the daughter of a Swiss artist and engraver, Matthaüs Merian, and his second wife, Johanna. Merian died when Maria was three, but her stepfather, the painter Jacob Marell, trained her in his studio. From the start her favorite artistic subject was insects. At age 13 she was putting descriptions and drawings of the life cycle of silkworms in her journal. She wrote, "I collected all the caterpillars that I could find, in order to observe their metamorphosis [change of form] . . . I withdrew from human society and engaged exclusively in these investigations . . . I learned . . . art . . . so that I could draw . . . them as they were in nature."

In 1665 Maria Merian married a fellow apprentice, Johann Graff, and they moved to Nuremberg. They had two daughters, Johanna and Dorothea. During her married years Maria Graff taught painting and embroidery and sold paints and cloth on which she had painted flowers. She also continued to study caterpillars and in 1679 issued a book, *Wonderful Metamorphosis and Special Nourishment of Caterpillars,* that contained 50 meticulous drawings of the life cycles of caterpillars and the plants they fed on. She was one of the first scientific illustrators to make her drawings from living rather than preserved specimens. She also published several books of flower designs for artists and embroiderers.

Graff moved back to Frankfurt in 1682 to care for her widowed mother. She issued a second volume of caterpillar drawings a year later. Around 1685 she joined a religious sect called the Labadists and moved with her mother and daughters to the group's settlement in Walta Castle in Friesland. She separated from her husband and resumed using her maiden name.

In 1691, after her mother died, Maria Merian and her daughters went to Amsterdam, where she once again began selling paints and painted cloth. When she saw collections of insects and other animals from the Dutch colony of Surinam, on the northern coast of South America, she "resolved . . . to undertake a great and expensive trip to Surinam . . . so that I could continue my observations." Such a trip was difficult, even dangerous, and Merian was now 52 years old. Nonetheless, she and Dorothea traveled to Surinam in July 1699 and remained there for two years.

The colony's Indian natives and African slaves told Merian about the local plants and insects and gave her samples. She drew these and other small animals, such as lizards and snakes, and even a few larger ones, including a crocodile. Illness forced her to return to Amsterdam in September 1701, but by then she had a large portfolio. She published 60 of her drawings in her masterwork, *Metamorphosis Insectorum Surinamensium* (Metamorphosis of the insects of Surinam), in 1705. The book's text included her observations, native lore, and even recipes related to the pictured plants and animals.

Merian's beautiful books remained popular long after her death in 1717. The Russian czar Peter the Great was so impressed with them that he made a collection of her original plates and other works, including her diary. Biologists named six plants, two beetles, and nine butterflies after her.

Metchnikoff, Élie

(1845–1916)
Russian/French
Zoologist, Pathologist

Élie Metchnikoff won the 1908 Nobel Prize in physiology or medicine with PAUL EHRLICH for their work on immunology. Metchnikoff discovered phagocytes, amoebalike cells that engulf foreign bodies such as bacteria and thus form the first line of defense for the body. Phagocytes in humans are called leukocytes, or white blood cells.

Metchnikoff was born on May 16, 1845, in Ivanovka, Ukraine, the youngest of five children. His mother, Emilia Nevahovna, was the daughter of the wealthy Jewish writer Leo Nevahovna, and his father, Ilya Ivanovich Metchnikoff, was an officer in the Imperial Guard at St. Petersburg. In 1869 Metchnikoff married Ludmilla Federovna, who died four years later of tuberculosis in April 1873. Despondent, Metchnikoff attempted suicide unsuccessfully by overdosing on morphine. In 1875 he married the 15-year-old Olga Belokopitova, who contracted typhoid fever. Nevertheless the couple was happy together, and though they never had children, they became the guardians of her two brothers and three sisters when her parents died.

Metchnikoff completed the four-year course of study at the University of Kharkov in two years, graduating in 1864. He then proceeded to study and conduct research in Naples, Göttingen, Giessen, and Munich. While in Naples he collaborated with the Russian zoologist Alexander Kovalevsky to discover the homology between the germ layers of different multicellular animals, earning the Karl Ernst von Baer Prize when Metchnikoff was merely 22 years old. In 1867 Metchnikoff received his doctorate from the University of St. Petersburg for his thesis on the embryonic development of fish and crustaceans.

After working briefly at the University of Odessa, Metchnikoff taught at the University of St. Petersburg. He transferred back to the University of Odessa as a professor of zoology and comparative anatomy in 1870, and he remained there until 1882, when he traveled to Messina, Italy. There he performed his ground-breaking work on phagocytes, transforming his scientific focus from zoology to pathology. He briefly returned to Odessa in 1886 to direct the Bacteriological Institute there, but in 1888 he left for Paris, where he chanced to meet LOUIS PASTEUR, whose career had followed a similar path. Metchnikoff was offered a position at the Pasteur Institute, and he succeeded Pasteur as the institute's director from 1895 until 1916.

Metchnikoff published his findings and beliefs in several important texts, including *Intra-Cellular Digestion* in 1882, *The Comparative Pathology of Digestion* in 1892, *Immunity in Infectious Diseases* in 1901, *The Nature of Man: Studies in Optimistic Philosophy* in 1903, and *The Prolongation of Human Life* in 1910. Other achievements included his demonstration with Émile Roux in 1903 that syphilis could be transmitted to apes. As part of his devotion during the later part of his career to the notion of human longevity, he championed the ingestion of the bacteria produced by lactic acid.

Besides the Nobel Prize, Metchnikoff received the Copley Medal from the Royal Society of London. After a series of heart attacks Metchnikoff died of cardiac failure on July 16, 1916, in Paris.

Meyerhof, Otto Fritz

(1884–1951)
German/American
Biochemist

Otto Fritz Meyerhof believed in the biochemical unity of all life-forms, and his work created the foundation for the field of bioenergetics, or the combination of thermodynamic physics with the study of the chemical functions of living cells. He received the 1922 Nobel Prize in physiology or medicine with Sir ARCHIBALD VIVIAN HILL for their work tracing the relationship between glycogen consumption and lactic acid metabolism as an aerobic and an anaerobic process in muscles. The cycle of glycogen depletion and lactic acid production in muscles became known as the Embden-Meyerhof pathway in honor of the early work on this process by Meyerhof, later built upon by Gustav Embden.

Meyerhof was born on April 12, 1884, in Hannover, Germany. His parents were Bettina May and Felix Meyerhof, a Jewish merchant. Meyerhof received his M.D. from the University of Heidelberg in 1909 with his dissertation on the psychology of mental disturbances. He married Hedwig Schallenberg in 1914, and together the couple had three children—George Geoffrey, Bettina Ida, and Walter Ernst. In 1946 Meyerhof became a citizen of the United States.

Between 1909 and 1912 Meyerhof served as an assistant in internal medicine at the Heidelberg Clinic, where Otto Warburg influenced him to investigate biochemistry. From 1913 to 1924 Meyerhof served in the physiology department at the University of Kiel, where he commenced his application of thermodynamics to the study of the chemical processes in biological systems. In 1919, while studying frog muscle contractions, he established a relationship between the depletion of glycogen and the production of lactic acid. In 1924 he served as the director of physiology at the Kaiser Wilhelm Institute for Biology in Berlin, and that year he also published his most important book, *Chemical Dynamics of Life Phenomena.* Then in 1929 he was appointed director of the Kaiser Wilhelm Institute for Medical Research in Heidelberg. That year adenosine triphosphate (ATP) was discovered in his laboratory, and the molecule proved to be the most important in energy transformations at the cellular level.

The rise of Nazism made it increasingly difficult for Meyerhof to remain in Germany, so in 1938 he, his wife, and his youngest son followed his older children, who had been sent away earlier, to Paris, where Meyerhof served as the director of research at the Institute of Physiochemical Biology for two years. In 1940 the Nazis invaded France, forcing Meyerhof and his family to flee again, first to neutral Portugal and then to the United States, where the Rockefeller Foundation established a chair of physiological chemistry specifically for him at the University of Pennsylvania School of Medicine. There he joined Hill, the man with whom he shared the Nobel Prize. Meyerhof spent his summers researching at the Woods Hole Marine Laboratory on Cape Cod in Massachusetts.

The National Academy of Sciences elected Meyerhof as a member in 1949 in recognition of his lifelong contributions. Meyerhof suffered a heart attack in 1944 but survived for another seven years until he died of a second heart attack on October 6, 1951, in Philadelphia, Pennsylvania.

Michelson, Albert Abraham

(1852–1931)
Prussian/American
Physicist

Albert Michelson's scientific passions revolved around precise measurement, specifically of light. He invented instruments and developed experiments that measured the speed of light with increasing accuracy. The Michelson-Morley experiment, named after him and Edward Morley, which attempted to calculate the speed of light traveling with the rotation of the Earth, as opposed to the speed of light traveling through the supposed "luminiferous ether" surrounding the Earth, was one of the most famous scientific experiments that negated its hypothesis. In 1907 Michelson became the first American to win the Nobel Prize in physics for his optical studies.

Michelson was born on December 19, 1852, in Strelno, Prussia (now Poland). His parents were Rosalie Przlubska and Samuel Michelson. On April 10, 1877, Michelson married Margaret McLean Heminway, the daughter of the New York City banker and lawyer Albert Gallatin Heminway. Their 20-year marriage produced three children—Albert Heminway, Truman, and Elsa.

Michelson entered the United States Naval Academy in 1869 as a "special eleventh appointment." On the entrance exam he had tied two other young men, one of whom was chosen over Michelson. Undeterred, Michelson traveled to the White House to petition President Ulysses S. Grant to grant him one of 10 special appointments reserved each year. When Grant answered that these had already been filled, Michelson appealed to the commandant of the academy, who made a special exception. Michelson's perseverance paid off, and he graduated in 1873.

After serving his mandatory two years at sea, Michelson returned to the academy as a chemistry and physics instructor. He then spent two years traveling in Europe, conducting postgraduate studies in Berlin, Heidelberg, and Paris. Upon his return the Case School of Applied Science in Cleveland appointed him professor of physics. In 1889 Clark University in Worcester, Massachusetts, gave him a similar appointment. In 1893 he moved to the University of Chicago to head the physics department there.

Michelson first measured the speed of light in 1878, improving somewhat on the method used by JEAN-BERNARD-LÉON FOUCAULT. In 1882 he performed an even more exacting measurement of the speed of light as 186,329 miles per second, a calculation that stood for a decade, until Michelson himself made an even more accurate measurement. While in Europe in 1881 Michelson invented the interferometer, an instrument that split a single light beam into two, which traveled separate directions until reunited, when they could be compared to see whether there were any differences between them. Michelson and Morley utilized this instrument in their famous 1887 experiment, which still failed to confirm the assumed existence of ether through which the Earth passed. Not until 1905 was this failure explained, when ALBERT EINSTEIN's theory of relativity hypothesized that the speed of light was a constant, though neither Einstein nor Michelson made the connection at the time.

Michelson published two important texts—*Light Waves and Their Uses* in 1903 and *Studies in Optics* in 1927—as well as other books and articles. The Royal Society of London awarded him the 1907 Copley Medal, and between 1923 and 1927 he served as the president of the National Academy of Sciences. Michelson died of a cerebral hemorrhage on May 9, 1931, while conducting an experiment to measure the speed of light over the 22 miles separating Mount Wilson from Mount San Antonio near Pasadena, California.

Miller, Stanley Lloyd
(1930–)
American
Chemist

Stanley Lloyd Miller conducted his most influential research when he was a graduate student. His experiments showed how amino acids, the building blocks of life, could have first been formed on the primitive Earth. In his laboratory Miller simulated the conditions of the prebiotic world. After running an electrical charge through this "primordial soup," he discovered the presence of amino acids such as alanine and glycine. His work received considerable attention in major newspapers, as well as in scientific journals.

Miller, the younger of two children, was born in Oakland, California, on March 7, 1930. His father, Nathan Harry Miller, was a lawyer, and his mother, Edith Levy Miller, was a homemaker. Miller, himself, never married.

Miller earned his B.S. degree from the University of California at Berkeley in 1951. He then began graduate studies at the University of Chicago. Soon after he arrived in the Midwest, he took a class offered by the Nobel laureate HAROLD CLAYTON UREY on the origin of the solar system. Miller was so intrigued by the theories Urey presented that he abandoned his thesis topic and approached Urey about testing the professor's hypothesis for his dissertation project. Urey had proposed that when the Earth was first formed, atmospheric conditions were ideal for synthesizing organic compounds because of the presence of strong mixtures of methane and ammonia and an excess of molecular hydrogen. Urey tried to dissuade Miller from undertaking this research, fearing that the experiment was too complicated and risky for a dissertation topic. But Miller was persistent, and Urey eventually acceded, under the condition that Miller would devote only one year to the work. If he had not obtained results by that time, Miller would have to begin a new project.

Miller designed an apparatus to simulate primordial conditions. He constructed a model ocean, atmosphere, and rain system within a glass unit and filled it with methane, hydrogen, water, and ammonia—the elements that Urey had posited that the primitive atmosphere comprised. After deciding that the most likely energy source would have been electrical discharges and ultraviolet light, he ran an electric current through his mixture and examined the results. An analysis with paper chromatography revealed the presence of glycine, alpha-alanine, beta-alanine, *a*-amino-*n*-butyric acid, and aspartic acid. Miller's findings were profound: Amino acids are the building blocks of proteins, which make up all living organisms. As proteins grow more complex, they form nucleic acids, which can replicate. On the advice of Urey, Miller quickly published the results of his experiment. His article, "A Production of Amino Acids under Possible Primitive Earth Conditions," appeared in May 1953.

After receiving his doctorate, Miller pursued his research as an F. B. Jewett Fellow at the California Institute of Technology from 1954 to 1955. He performed additional tests to rule out the possibility of contamination and to identify other amino acids present in his experiment. From 1955 to 1960 Miller worked in the department of biochemistry at the College of Physicians and Surgeons at Columbia University. In 1960 he became a professor of chemistry at the University of California at San Diego. He has continued throughout his career to develop his original project through experiments such as one that sought to uncover precursors to ribonucleic acid (RNA).

When a meteorite landed on Earth in 1969, Miller received confirmation that his representation of the primitive atmosphere was accurate. Upon analyzing the meteorite, scientists found that it contained most of the same amino acids that Miller had created. Despite speculation in the popular press that Miller had spontaneously synthesized life in a test tube, he emphasized that forming organic compounds was not synonymous with creating a living organism. In recognition of his achievements he was awarded the Oparin Medal in 1983 from the International Society for the Study of the Origin of Life. He was also elected to the National Academy of Sciences and became a member of Sigma Xi and Phi Beta Kappa.

Minkowski, Hermann

(1864–1909)
German
Mathematician, Physicist

Hermann Minkowski's work laid the mathematical foundation for ALBERT EINSTEIN's general theory of relativity. Minkowski pioneered a new concept of time and space, which had previously been considered independent entities. Minkowski conceptualized time and space as inherently fused into a single fourth-dimensional continuum of "time-space." His work bore out this notion and proved that it was necessary to consider a fourth dimension along with the traditional spatial dimensions of length, width, and depth. His research into quadratic forms became the basis of modern functional analysis; he also introduced a mathematical theory known as the geometry of numbers.

Minkowski was born in Alexotas, Russia (now in Lithuania), on June 22, 1864. His German parents returned to their homeland in 1872, when Minkowski was eight years old. The family settled in the royal Prussian city of Königsberg, where they remained for the duration of his childhood. His brother, Oskar, became a renowned pathologist.

With the exception of three semesters at the University of Berlin, Minkowski pursued his higher education at the University of Königsberg. In 1883, when he was only 19, he was awarded the Grand Prix des Sciences Mathématiques from the Paris Academy of Sciences. His 140-page proof described the number of representations of an integer as a sum of five squares of integers. The prize was also awarded to J. H. Smith, who independently solved the same problem, but Minkowski's proof was considered a better formulation. Minkowski earned his doctorate from the University of Königsberg in 1885.

Minkowski taught at the University of Bonn from 1885 to 1894, then returned to Königsberg for two years. He proceeded to the University of Zurich, where he taught until 1902. During this time Minkowski explored the arithmetic of quadratic forms, especially concerning *n* variables. In 1896 he presented his geometry of numbers, a geometrical method that solved problems in number theory. He applied these conclusions to his studies of convex bodies in 1889, when he analyzed the geometric properties of convex sets. His observations are credited with establishing the underpinnings of modern functional analysis.

In 1902 a friend and fellow mathematician urged the University of Göttingen to create a new professorship for Minkowski, and it was here that he eventually carried out his influential work investigating relativity theory. By the end of the 19th century scientists had begun to realize that Sir ISAAC NEWTON's physical laws did not adequately explain the behavior of subatomic particles. In 1905 Minkowski's pupil, Albert Einstein, proposed a relativity theory that sought to solve these contradictions. Einstein's insight piqued Minkowski's interest and led Minkowski to focus his own work on questions of relativity. Minkowski postulated a unified "time-space" continuum existing in a fourth dimension, a concept that shattered the bounds of conventional Euclidean geometry and Newtonian physics. Minkowski theorized that time and space were not separate entities, but were rather a single presence beyond the three spatial dimensions. Minkowski recognized the import of his own findings: "From now on," he said, "space by itself and time by itself are mere shadows and only a blend of the two exists in its own right." His seminal book *Space and Time*, published in 1907, presented geometric interpretations of relativity. All future mathematical work concerning relativity followed the pathways Minkowski constructed.

Minkowski remained at Göttingen until his sudden death of a ruptured appendix in January 1909. Although he lived to be only 44, he continued to exert a profound influence over Einstein. In 1915 the erstwhile pupil expanded on Minkowski's insight into the time-space continuum and propounded the general theory of relativity, which generated an entirely new approach to the concept of gravity.

Minot, George Richards

(1885–1950)
American
Hematologist, Physician

George Minot received the 1934 Nobel Prize in physiology or medicine with WILLIAM PARRY MURPHY and GEORGE HOYT WHIPPLE for their roles in the discovery that a diet high in raw liver could cure the previously incurable, fatal disease of pernicious anemia. Subsequent research discovered a means of extracting the active agent from liver, vitamin B_{12}.

Minot was born on December 2, 1885, in Boston, Massachusetts. His mother was Elizabeth Whitney, and his father was James Jackson Minot, a prominent physician. On June 29, 1915, he married Marian Linzee Weld. Together the couple had two daughters and one son. In October 1921 Minot was diagnosed with diabetes. In January 1922 insulin was discovered, and by January 1923 insulin became available, and Minot began treatment that managed his diabetes, though he remained on a strict diet the rest of his life.

Minot entered Harvard College in 1904 and graduated in 1908. He deferred his decision about his next step until late September of that year, when he decided five days before the beginning of the academic year to attend Harvard Medical School. Minot received his M.D. in 1912 and commenced his internship at the Massachusetts General Hospital (MGH).

In 1912 Minot became an assistant at the Johns Hopkins University Medical School in Baltimore, Maryland. In January 1915 he returned to the MGH. In 1917 he conducted leukemia research at the Collis P. Huntington Memorial Hospital in Boston, and in 1918 the Harvard Medical School appointed him as an assistant professor. Minot became the chief of medical services at Huntington in 1923, when he was also appointed as an associate in medicine at the Peter Bent Brigham Hospital in Boston. In 1928 Harvard promoted him to the status of full professor, and he also took over the directorship of the Thorndike Memorial Laboratory at the Boston City Hospital that year.

Whipple's experiments acted as a catalyst for Minot's research into pernicious anemia, which he had commenced earlier in his career before abandoning this line of investigation for other hematological studies. Whipple had been inducing anemia in dogs by bleeding them, then nursing them back to health on a diet high in red meat, especially liver. Minot simply applied the same logic to human patients suffering from pernicious anemia, feeding them as much as a half-pound of raw liver a day. Minot and his colleague Murphy observed vast improvements in their patients, with very little remission.

They reported on the improvements for 45 patients at the May 4, 1926, meeting of the American Medical Association in Atlantic City, New Jersey. At the next year's meeting in Washington, D.C., Minot and Murphy reported on the improvement of 105 patients. Minot next enlisted his Harvard colleague Edward J. Cohn to prepare a liver extract, and Cohn identified fraction G as the active ingredient of pure liver. In 1948 other researchers identified vitamin B_{12} as the active ingredient in liver that alleviated pernicious anemia. On April 16, 1947, Minot suffered a severe stroke, and on February 25, 1950, he died in Brookline, Massachusetts.

Mitchell, Maria

(1818–1889)
American
Astronomer

Maria Mitchell won a gold medal for discovering a comet and was the first American woman to become internationally known as an astronomer. She was born on August 1, 1818, in Nantucket, Massachusetts, the third of 10 children. Her father, William Mitchell, was an avid amateur astronomer and shared his interest with her. Her mother, Lydia, encouraged her to read.

Mitchell's formal schooling, some of it in a school run by her father, ended when she was 16. In 1836 she became the librarian of the Nantucket Atheneum, or subscription library, a job she held for 20 years. She read the library's books in the daytime and continued stargazing with her father in the evenings. Their observatory was now on top of the Pacific Bank building, to which the family had moved when her father became the bank's principal officer. There she and her father made measurements for the United States Coast Survey that helped in determination of time, latitude, and longitude.

The 29-year-old Mitchell was scanning the heavens alone on October 1, 1847, when she spotted a blurry object in a part of the sky where none had appeared before. She concluded that it was a comet and ran downstairs to tell her father about it. Recognizing that it was a new one, he informed his friend, William Cranch Bond, the head of the Harvard Observatory. A new comet was considered a major astronomical discovery, and in 1831 the king of Denmark had offered a gold medal to the next person to find one through a telescope. Mitchell received the gold medal a year after she had spotted the comet, and the comet was named after her.

Maria Mitchell, who, at the age of 29, discovered a new comet. She is shown here with her students at Vassar College. *Special Collections, Vassar College Libraries, Poughkeepsie, New York.*

Mitchell's discovery—and the unusual fact that it was made by a woman—made her famous. In 1848 she became the first woman elected to the American Academy of Arts and Sciences. She was also elected to the newly established American Association for the Advancement of Science in 1850; indeed, she was the only woman member for almost 100 years. The director of the Smithsonian Institution sent her a $100 prize.

Mitchell's fame garnered her a new part-time job in 1849 computing information about the movements of Venus (considered a suitable subject to assign to a woman) for the *American Ephemeris and Nautical Almanac*, a national publication that provided tables of data about the movements of the heavenly bodies for the use of sailors and others. On a trip to Europe in 1857 Mitchell's reputation also helped her gain access to observatories and meet such noted scientists as Sir JOHN FREDERICK WILLIAM HERSCHEL (CAROLINE LUCRETIA HERSCHEL's nephew) and MARY FAIRFAX SOMERVILLE.

After Mitchell's mother died in 1861, she and her father moved to Lynn, Massachusetts, where one of her married sisters lived. Four years later her life changed dramatically when she received an offer from a representative of a wealthy brewer and philanthropist, Matthew Vassar, to become a professor of astronomy and director of the observatory at the women's college that Vassar was founding in Poughkeepsie, New York. Mitchell greeted her first students when the Vassar Female College opened on September 20, 1865, and taught there for 23 years. She was the world's first woman professor of astronomy.

Mitchell did some original research and made many astronomical observations at Vassar, including taking daily photographs of sunspots and other changes on the surface of the Sun and studying Jupiter, Saturn, and their moons. She felt, however, that she had to choose between research and teaching, and she believed she had more to give to teaching. Her classes emphasized mathematical training and direct experience rather than lectures or memorization, and she encouraged her students to ask questions. She refused to give grades, saying, "You cannot mark a human mind because there is no intellectual unit." Several of her students, including ANTONIA CAETANA MAURY and ELLEN HENRIETTA SWALLOW RICHARDS, became well-known scientists. (Richards traded astronomy for chemistry.)

Mitchell also advanced the education of women by helping to found the Association for the Advancement of Women in 1873. She was the group's president in 1875 and 1876 and headed its committee on science from 1876 until her death. "I believe in women even more than I do in astronomy," she once said.

Mitchell retired from Vassar on Christmas Day 1888 and returned to Lynn, where some of her family still lived. She died there on June 28, 1889. During her lifetime she was awarded three honorary degrees, and after her death a group dedicated to preserving her memory and carrying on her work, the Maria Mitchell Association, was formed on Nantucket. A crater on the Moon has also been named after her.

Möbius, August Ferdinand

(1790–1868)
German
Mathematician, Astronomer

Although August Ferdinand Möbius was a professor of astronomy, his most important work was in mathematics. In the field of topology—the study of the properties of geometric figures that remain unchanged even when under distortion—Möbius was renowned for conceiving the Möbius strip, a two-dimensional surface with only one side. He introduced the use of homogenous coordinates to analytic geometry and conducted pioneering research into projective geometry, which included inventing the Möbius net.

Born on November 17, 1790, in Schulpforta, Germany, Möbius was the only child of Johann Heinrich Möbius and Johanna Catharine Christiane Keil Möbius, a descendant of Martin Luther. His father, who was the town's dancing instructor, died when Möbius was three, and the young family was subsequently supported by a paternal uncle. Later in his life Möbius lived with his mother, but after she died in 1820 Möbius wed Dorothea Christiane Johanna Rothe. Although his wife became blind, the couple had one daughter and two sons, both of whom became respected literary critics.

Möbius was educated at home until he was 13, at which point he received a formal education at the College in Schulpforta. He expressed an early interest in mathematics, but upon entering Leipzig University in 1809 he initially followed his family's wishes and studied law. He soon abandoned his law classes in favor of mathematics, physics, and astronomy. At Leipzig he worked closely with Karl Mollweide, an astronomer with mathematical interests. In 1813 he embarked on two semesters of theoretical astronomy study at Göttingen under KARL FRIEDRICH GAUSS, another astronomer well versed in mathematics. After completing his thesis, "The Occultation of Fixed Stars," he was awarded his doctorate from Leipzig in 1814.

Möbius became a professor of astronomy at Leipzig University in 1816 and held this position for 50 years. Though he later received better offers from other universities, he remained at Leipzig for his entire career because of his fondness for that region of Saxony and because of the good reputation the university maintained. Like his mentors Möbius worked in the fields of astronomy and mathematics and was quite adept at both. After publishing work on Halley's comet in 1835, and on the fundamental laws of astronomy in 1836, he wrote "The Elements of Celestial Mechanics," in which he thoroughly explored celestial mechanics mathematically without using higher mathematics, in 1843. He also oversaw the creation of the observatory at the university.

Möbius produced ground-breaking mathematical work at Leipzig. His most influential paper, "The Calculus of Centers of Gravity" (1827), introduced homogenous coordinates to analytic geometry, the branch of geometry in which position is indicated by algebraic symbols and solutions are obtained by algebraic operations. This same paper presented insights into projective geometry, including the famous concept of the Möbius net, a configuration that fueled future developments in projective geometry. His 1837 treatise, "Handbook on Statics," applied geometry to the study of statics. His research into topology resulted in his creation of the Möbius strip. This invention, which was discovered in his unpublished papers after his death, was formed by affixing the ends of a rectangular strip of paper after first giving one of the ends a 180-degree twist. The majority of his mathematical findings were published in *Crelle's Journal,* the first journal devoted exclusively to mathematics.

After his death in 1868 Möbius was acknowledged for the thoroughness of his work. His greatest strength was not in finding new results but in inventing simpler and more effective ways of solving existing problems.

Mohs, Friedrich

(1773–1839)
German
Mineralogist

Friedrich Mohs sought to establish a new system for classifying minerals. His experience as curator of several important mineral collections led him to propose a hardness scale for categorizing various minerals, which was later named the Mohs scale in his honor and is still used by mineralogists today. His later work devised a taxonomic scheme of four crystal systems based on external symmetry.

Born in Gernrode, in what is now Germany, Mohs expressed an early interest in science. Although no specific information is known about his family, he was educated at home until 1797, at which point he entered the University of Halle. In 1798 he enrolled at the mining academy in Freiberg and studied mineralogy under ABRAHAM GOTTLOB WERNER. Werner had previously established geology and mineralogy as distinct sciences and was the first to classify minerals systematically. In addition to his work with Werner, Mohs took classes at Freiberg in physics and mathematics. In 1802 he traveled to Great Britain with his classmates George Mitchell and Robert Jameson. The group expected to help plan a mining academy in Dublin. Although the academy was never built, Mohs made important connections with Scottish geologists during his trip.

Upon completing his formal education, Mohs was commissioned in 1802 by a Viennese banker to create a systematic description of the banker's extensive mineral collection. After two years of work Mohs published a two-volume description of the project in 1804. In 1811 Mohs was appointed curator of the mineral collection in Archduke Johann's Johanneum in Graz. The following year Mohs was named professor of mineralogy at the Johanneum, a position he held until 1817. But Mohs made his chief contribution to the science of mineralogy in 1812, in his proposal of a hardness scale to classify minerals. He ordered 10 minerals on a scale ranging from 1 to 10, under the general rule that a higher-numbered mineral would scratch a lower-numbered one. The scale was as follows: talc, 1; gypsum, 2; calcite, 3; fluorite, 4; apatite, 5; feldspar, 6; quartz, 7; topaz, 8; corundum, 9; diamond, 10. Following on from this insight, other mineralogists subsequently added intermediate levels of hardness to describe other minerals. After devising this scale, Mohs spent the remainder of his tenure at Graz exploring the external symmetry of crystal systems as a possible means of classification.

Upon Werner's death in 1817 Mohs was summoned to the mining academy at Freiberg, and in 1818 he assumed Werner's former post as a professor of mineralogy. In 1822 and 1824 Mohs published a two-volume work that introduced a systematic description of minerals based on four crystal systems: rhombohedral (hexagonal), pyramidal (tetragonal), prismatic (orthorhombic), and tessular (cubic). His work created controversy because another mineralogist, Christian Samuel Weiss, accused him of plagiarizing his concepts. Mohs adamantly refuted these claims, and indeed, his analysis had far surpassed Weiss's. Mohs's *Treatise on Mineralogy* was published in 1825.

Mohs resigned his post at Freiberg in 1826 and found work in Vienna, first reorganizing the imperial mineral collection, then working as a professor. In 1835 he left behind the university to become imperial counselor of the exchequer in charge of mining and monetary affairs. In this capacity he traveled frequently about the Austro-Hungarian Empire. He died in Italy in 1839. His Mohs scale of mineral hardness is commonly used today.

Moissan, Ferdinand-Frédéric-Henri

(1852–1907)
French
Chemist

Henri Moissan was the first, after many scientists before him had failed, to isolate the dangerous element fluorine. After this achievement Moissan attempted to synthesize diamonds by inventing the high-temperature electric arc furnace. Though this mechanism was capable of reaching temperatures as high as 3,500 degrees Celsius, diamond synthesis required five times as much pressure as Moissan could generate. Nevertheless, his furnace proved instrumental in a multitude of other scientific experiments. He received the 1906 Nobel Prize in chemistry for these advancements.

Moissan was born on September 28, 1852, in Paris, France. His mother was Josephine Mitel, a seamstress, and his father was François-Ferdinand Moissan, a clerk and railroad worker. In 1882 Moissan married Marie Leonie Lugan, daughter of a pharmacist who supported Moissan's scientific work philosophically and financially. The couple had one son, Louis-Ferdinand-Henri, who was killed in 1915 near the beginning of World War I.

After abandoning school in 1870 for apprenticeships with a watchmaker and a pharmacist as well as a stint in the army, Moissan finally passed the baccalaureate in 1874 while researching in Edmond Frémy's laboratory at the School of Experimental Chemistry at the Paris Museum of Natural History. He proceeded to earn his license, the French equivalent of a B.S., in 1877. In 1879 he attained the rank of pharmacist first-class, and in 1880 he earned his doctorate with a dissertation on pyrophoric iron and various of its oxides.

That year Moissan became an associate professor at the École Superieure de Pharmacie. In 1886 he was promoted to the title of professor of toxicology, and in 1899 he was named a professor of inorganic chemistry. This appointment lasted only one year, as he accepted the offer to fill the Chair of Inorganic Chemistry with the Faculty of Sciences at the Sorbonne, or the University of Paris, in 1900. Moissan's important publications in France included *The Electric Furnace* in 1897, *Fluorine and Its Compounds* in 1900, and *Treatise on Inorganic Chemistry* in 1904 through 1906.

On June 26, 1886, Moissan succeeded in isolating fluorine gas by electrolyzing a solution of potassium fluoride in anhydrous hydrofluoric acid, all encased in an apparatus made of platinum. He subsequently made liquid flourine and, in 1903, solid fluorine. In December 1892 he demonstrated the original design of his furnace, which later went through several improvements, to the Academy of Sciences. The actual synthesis of diamonds was shrouded in controversy, as some scientists charged that Moissan's assistant smuggled real diamonds into the lab and planted them after the experiment.

Besides the Nobel Prize, Moissan received the 1887 Prix Lacaze from the French Academy of Sciences, which elected him a member in 1891. He also received the 1896 Davy Medal from the Royal Society of London, which appointed him as a foreign member in 1905. Moissan's health suffered from his years of working with fluorine, and he died in Paris on February 20, 1907, after surgery on his appendix.

Moore, Stanford

(1913–1982)
American
Biochemist

Stanford Moore received the 1972 Nobel Prize in chemistry with his longtime collaborator WILLIAM HOWARD STEIN, as well as CHRISTIAN BOEHMER ANFINSEN of the National Institutes of Health, for the first complete mapping of the amino acid sequence of pancreatic ribonuclease (RNase), the enzyme that recycles ribonucleic acid (RNA). Moore and Stein subsequently studied deoxyribonuclease, the enzyme that recycles deoxyribonucleic acid (DNA). Moore and Stein pioneered the use of chromatography to separate amino acids, peptides, and proteins.

Moore was born on September 4, 1913, in Chicago, Illinois, and his family soon moved to Nashville, Tennessee. His mother was Ruth Fowler; his father was John Howard Moore, a professor at Vanderbilt University's School of Law. Moore attended Vanderbilt and earned his B.A. in chemistry in 1935. He then proceeded to the University of Wisconsin, where he earned his Ph.D. in 1938.

In 1939 Moore became a research assistant to Max Bergmann at the Rockefeller Institute for Medical Research (RIMR), where he remained for the rest of his career, with the exception of several stints during World War II as a visiting professor in Europe and back at Vanderbilt. Between 1942 and 1945 he served as a technical aide to the National Defense Committee at the Office of Scientific Research and Development, then returned to RIMR. In 1950 he held the Franqui Chair at the University of Brussels, and in 1951 he served as a visiting professor at the University of Cambridge in England. Upon his return in 1952 he was promoted to the position of professor of biochemistry at RIMR. In 1968 Moore spent one last year away from New York City as a visiting professor of health sciences at the Vanderbilt University School of Medicine.

Moore's important publications included "The Chemical Structure of Proteins" in the February 1961 edition of *Scientific American* and "Chemical Structures of Pancreatic Ribonuclease and Deoxyribonuclease" in the May 1973 edition of *Science,* both of which were coauthored by Stein. Moore also served as an editor of *The Journal of Biological Chemistry.*

In 1949 Moore succeeded in separating amino acids from blood and urine, an important first step toward his later achievements. In 1958 he helped develop the automated amino acid analyzer, which reduced the time required to conduct his experiments from one week to one 24-hour day. Moore and Stein made their breakthrough in 1959 when they completely decoded the chemical composition of ribonuclease, identifying the sequence of 124 amino acids in the enzyme. Frederick Sanger had identified the amino acid sequence of a protein in 1955, but Moore and Stein were the first to sequence an enzyme.

Besides the Nobel Prize, Moore received the Richards Medal from the American Chemical Society and the Linderstrøm-Lang Medal. He served as the president of the Federation of American Societies for Experimental Biology and as the treasurer and president of the American Society of Biological Chemistry. Late in life Moore developed Lou Gehrig's disease, or amyotrophic lateral sclerosis, and on August 23, 1982, he committed suicide in his New York City home. Moore, who never married and thus had no direct heirs, left his estate to Rockefeller University.

Morawetz, Cathleen Synge

(1923–)
Canadian/American
Mathematician

An expert on waves, an area of applied mathematics that affects everything from airplane design to medical imaging, Cathleen Morawetz has been one of the few women truly prominent in 20th-century mathematics. She was born in Toronto, Ontario, Canada, on May 5, 1923, to J. L. Synge, an applied mathematician, and Elizabeth Allen Synge, who also studied mathematics at Trinity College, Dublin, but was persuaded not to pursue a career. Both parents were Irish.

Cathleen Synge was not good at arithmetic as a child and began studying mathematics only because a high school teacher urged her to do so in order to win a scholarship to the University of Toronto. There she studied physics and chemistry as well as mathematics. Cecilia Krieger, a mathematician who taught at the university and was also a family friend, encouraged Synge to pursue mathematics as a career. During her senior year Synge met an engineering student, Herbert Morawetz, the son of a Jewish manufacturer who had fled Czechoslovakia after the Germans took over the country. The two became engaged soon after Cathleen graduated in 1945 and married while she was studying for her master's degree in applied mathematics at the Massachusetts Institute of Technology (MIT), which she completed in 1946. They later had three daughters and a son. Synge became a naturalized citizen of the United States in 1950.

As a result of a chance meeting with her father, another famous mathematician, Richard Courant, offered Cathleen Morawetz a job soldering computer connections in his lab at New York University (NYU). When she arrived there in the spring of 1946, however, she found that Courant's assistant had already hired a man to do the work. Not wanting to leave her stranded, Courant then asked her to edit a book he and another mathematician had written, *Supersonic Flow and Shock Waves.* "I learned the subject that way," Morawetz said in 1979. The application of partial differential equations to the behavior of waves became her specialty, and what was later the Courant Institute of Mathematical Sciences became her home. She earned a Ph.D. from NYU in 1951, doing her thesis on differential equations of imploding shock waves in fluids.

Morawetz's studies of waves have been used to improve the design of airplane wings, especially those of planes that fly at about the speed of sound. At these speeds shock waves change the airflow over the wings, increasing drag and slowing the plane. Understanding the wave patterns of the airflow enables engineers to create wing designs that prevent formation of most shock waves. The mathematics of waves is also important in X-ray diffraction studies of crystals and in computed tomography (CT) scans, ultrasound, and other medical imaging technology.

Cathleen Morawetz's honors include the Krieger-Nelson Award of the Canadian Mathematical Society (1997), the Lester R. Ford Award of the Mathematical Association of America (1980), and election to the National Academy of Sciences and the American Academy of Arts and Sciences. She was named Outstanding Woman Scientist by the Association for Women in Science in 1993 and was also recognized by the National Organization of Women for combining a successful career and a family. On receiving the latter honor, she joked, "Maybe I became a mathematician because I was so crummy at housework." In 1995 she became the second woman president of the American Mathematical Society.

Morawetz was a full professor at NYU by 1979 and was director of the Courant Institute from 1984 to 1988—the first woman in the United States to head a mathematical institute. Morawetz was NYU's Samuel F. B. Morse Professor of Arts and Sciences until 1993, when she retired and became a professor emerita. She has also been a trustee of Princeton University and the Sloan Foundation and a director of NCR, a large computer company. Administration "suits me," she said. "I like the relationship with human beings." She believes that her strong point in administration, as in mathematics, is common sense.

Morgan, Thomas Hunt

(1866–1945)
American
Geneticist, Embryologist

Thomas Hunt Morgan's experiments on fruit flies led to important discoveries in what was then the fledgling science of genetics. Morgan established the chromosome theory of heredity and proved that genes (the basic units of heredity) are located in a linear array on chromosomes. Through his research on a mutant variety of fruit fly he also discovered sex-linked genes, and with his research team he devised the first chromosome map. His understanding of chromosome mutations allowed him to offer a genetic explanation for CHARLES ROBERT DARWIN's theory of natural selection. The students he influenced at both Columbia University and California Institute of Technology (Caltech) included many future pioneers in the field of genetics.

Morgan was born in Lexington, Kentucky, on September 25, 1866, to a prominent family. His father, Charlton Hunt Morgan, had been the American consul in Italy during the 1860s, where he had assisted Giuseppe Garibaldi in his movement to unify Italy. His uncle, John Hunt Morgan, had been a general in the Confederate Army, and his maternal great-grandfather was Francis Scott Key, the composer of the American national anthem.

Morgan graduated summa cum laude from the State College of Kentucky (now the University of Kentucky) in 1880 with a B.S. in zoology, then pursued graduate studies in embryology and morphology at the Johns Hopkins University in Baltimore, Maryland. Under the tutelage of W. K. Brooks, Morgan investigated the embryology of sea spiders in an attempt to classify them as either arachnids or crustaceans. He completed his doctoral work and was awarded a Ph.D. in 1890. He remained at Hopkins for an additional year on a Bruce Fellowship.

From 1891 until 1904 Morgan taught at Bryn Mawr College, outside Philadelphia. During his tenure there he applied experimental methods to embryology. His research concerning sea urchin eggs piqued his interest about the hereditary information carried by each cell and the chemical or physical processes that controlled the inheritance of this information. In 1903 he published *Evolution and Adaptation,* which posited that mutation (a concept first proposed by HUGO MARIE DE VRIES) introduced variation, an essential aspect of evolutionary theory, to a population.

In 1904 Morgan accepted a position as head of experimental zoology at Columbia University, where he investigated how an organism's sex was determined. The topic was hotly contested, as some scientists cited environmental factors as the cause of sex determination, and others concluded that inheritance was the key. In 1908 Morgan began to breed *Drosophila melanogaster* (the fruit fly) for laboratory work. The flies were ideal for research because their breeding cycle was a mere three days. In 1910 Morgan showed that the X chromosome was involved with sex inheritance. That same year he postulated that certain genes were connected to the X chromosome and were thus sex-linked. He reached this conclusion by studying a mutant variety of fly—one with white eyes instead of the typical red. After mating these white-eyed males with their red-eyed sisters, Hopkins found that all the offspring had red eyes. However, when he mated these red-eyed offspring, they in turn produced some white-eyed offspring. All of the white-eyed offspring were male, indicating that the genes for eye color were connected to sex-specific chromosomes. By 1911 he had identified five sex-linked genes. Over the next 11 years he devised a map indicating the relative positions of 2,000 genes on the fruit fly chromosomes.

Morgan left Columbia in 1927 to create a biology department at Caltech, in Pasadena, California. Instead of focusing solely on his fruit fly work, he strove to explore larger questions of development and evolution. He remained at Caltech until his death in 1945.

His research garnered many honors. In 1933 Morgan won the Nobel Prize in physiology or medicine. His treatise supplying genetic explanations for Darwin's theories, *A Critique of the Theory of Evolution,* received the Darwin Medal in 1924, and in 1939 Morgan was awarded the Copley Medal of the Royal Society. His findings expanded the field of genetics and helped influence the next wave of genetic research.

Moss, Cynthia

(1940–)
American
Zoologist

Cynthia Moss's study of wild African elephants, lasting more than 25 years, has been compared to JANE GOODALL's long-term study of chimpanzees. Other scientists have called Moss's work "invaluable" and "irreplaceable."

Cynthia was born in Ossining, New York, on July 24, 1940, the younger of the two daughters of a newspaper publisher, Julian Moss, and his wife, Lillian. As a child she loved the outdoors. She attended Smith College, earned a B.A. in philosophy in 1962, then became a news researcher and reporter for *Newsweek*. In 1967, after receiving "beautiful" letters from a friend who had moved to Africa, she took a leave of absence to see the continent for herself. When she arrived, she says, "I had this overwhelming sense that I'd come home."

Moss's trip included a visit to the camp of the British elephant researcher Iain Douglas-Hamilton in Tanzania, where she says she "became completely hooked on elephants." She worked with Douglas-Hamilton off and on until 1970, when his project ended. She then supported herself by writing for *Life* and *Time* and helping other scientists while looking for funding and a site for her own elephant study. The ideal site proved to be Amboseli National Park in Kenya, home to one of the last undisturbed elephant herds in Africa.

What has come to be known as the Amboseli Elephant Research Project began in 1972. Most of its funding has come from Washington, D.C.'s African Wildlife Foundation (AWF). Moss, the project's director, has studied more than 1,600 elephants, which she identifies by differences in their ears. She told *Current Biography* that observing generations of the animals is "like reading a very good . . . family saga. You get so involved you don't want to put it down." By studying a single population over a long period, she has revealed much about elephant behavior and provided information that helps conservationists protect the animals. For instance, she has found that, as she told an interviewer, Marguerite Holloway, "Elephants have a really complex problem-solving intelligence, like a primate might have."

Douglas-Hamilton had discovered that the leader of an elephant family group is the oldest female—the matriarch. The group, whose members eat, rest, and play together, usually consists of several related females, or cows, and their calves. Adult males stay in separate, looser groups and associate with the family groups only during mating.

Moss has found that the family group is only the smallest unit of a many-tiered society. Several such groups make up a larger unit, the bond group. When elephants of different family groups but the same bond group meet, they stage an elaborate greeting ceremony. "I have no doubt even in my most scientifically rigorous moments that the elephants are experiencing joy when they find each other again," Moss has written. Bond groups, in turn, unite in still larger groups called clans, which share the same territory but do not have greeting ceremonies. The biggest subgroup in elephant society is the subpopulation, of which Amboseli has two.

Elephant communication is also complex. Moss and her coworkers have documented 25 different vocalizations, some of which, as KATHARINE BOYNTON PAYNE discovered, involve infrasound—sound too low for human ears to hear, which carries well over long distances. These sounds help different groups stay in contact.

Joyce Poole, formerly of the Kenya Wildlife Service, joined Moss's project in 1976 and has made important contributions to it. For instance, Poole has found that African male elephants often enter a hyperaggressive state called musth, which had previously been reported only in Asian elephants. Moss and Poole have concluded that musth helps males compete for females in heat.

Moss has found that elephant females can change their reproductive patterns in response to the environment. Females normally become sexually mature when they are about 11 years old, but if conditions are harsh, puberty may be delayed until age 15 or later. Similarly, a cow usually has a calf about once every four years, but during a food shortage the interval may be lengthened to seven years or more.

In the late 1980s, after learning that poaching for ivory and loss of habitat had halved Africa's elephant population during a single decade, Moss turned her focus to conservation. She and Poole went to Washington in 1988 to warn the AWF of the growing threat to the animals. Thanks to the efforts of AWF and other conservation groups, the Convention on International Trade in Endangered Species declared the African elephant an endangered species in October 1989 and banned the sale of ivory in January 1990. Since then demand for ivory and loss of elephants to poaching have both dropped dramatically. The species has partly recovered, though Moss has warned that it is still threatened.

Cynthia Moss has received several awards for her ground-breaking studies, including the Smith College Medal for Alumnae Achievement (1985) and a conservation award from the Friends of the National Zoo and the Audubon Society. However, she has said that her greatest reward is continuing to share the lives of the elephant families she knows so well in "their good times and bad times through the seasons and the years."

Mössbauer, Rudolph Ludwig

(1929–)
German
Physicist

During his research into the properties of gamma rays Rudolph Ludwig Mössbauer developed a new measurement technique that enabled scientists in an array of fields to make highly precise measurements. Mössbauer observed the way an atomic nucleus recoiled after emitting a gamma ray (much as a gun does after firing). This recoil action greatly affected the gamma wavelength, making accurate measurement exceptionally difficult. Mössbauer's chief innovation was to discover that if the nucleus were fixed within a crystal lattice, the lattice itself, not just the nucleus, would recoil. As a result the recoil would exert little effect on the gamma wavelength. This method—later named the Mössbauer effect in his honor—allowed for the first testing and verification of ALBERT EINSTEIN's general theory of relativity.

Born on January 31, 1929, in Munich, Germany, Mössbauer was the only son of Ludwig and Erna Mössbauer. The elder Mössbauer was a photo technician who printed color postcards and reproduced photographs. Mössbauer's childhood was shadowed by the political and social events sweeping his nation as Adolph Hitler came to power.

Despite these severe upheavals, Mössbauer was able to receive his primary and secondary education, but new enrollments at universities were curtailed after Germany's defeat in World War II and Mössbauer was unable to attend a university after his graduation from the Munich-Pasing Oberschule in 1948. His father, however, found him a job as an optical assistant at a company in Munich, and he later went on to work for the U.S. Army. With the money he earned, Mössbauer was able to attend the Munich Institute for Technical Physics, where he earned his diploma in 1955. That same year he began his doctoral studies at the Munich Institute, under the supervision of Heinz Maier-Leibnitz. After carrying out some of his research in nuclear resonance fluorescence at the Max Planck Institute for Medical Research in Heidelberg, he was awarded his doctorate in 1958. He married Elizabeth Pritz, and the couple eventually had three children.

It was during the course of his doctoral research that Mössbauer found a solution to a problem that had been perplexing physicists for nearly two decades. W. Kuhn had predicted in 1929 that gamma rays, a form of electromagnetic radiation with a high frequency, would display resonance in the same way that light (another form of electromagnetic radiation) did. When light was shone on certain materials, the atoms constituting those materials absorbed the electromagnetic energy of the light and reemitted it. This reemitted energy had the same frequency as the original light because of resonance within the atoms of the material. This phenomenon—fluorescence—can be witnessed when some materials glow in the dark after exposure to light. No one could prove Kuhn's hypothesis of nuclear resonance fluorescence, however, because the recoil action of the emitting nucleus made measuring gamma ray energy an inexact science at best.

During his doctoral studies Mössbauer conceived the notion of embedding the gamma ray within a crystal lattice. The process allowed the recoil energy to be absorbed and dissipated throughout the entire crystal lattice, thereby rendering negligible the distortion of the gamma ray. Early in 1958 Mössbauer published the results of his work with gamma ray emitters fixed in iridium-191, a radioactive isotope, in the German scientific journals *Naturwissenschaften* and *Zeitschrift für Physic*. After this discovery of what came to be known as the Mössbauer effect he took a position as a research fellow at the Technical University in Munich. In 1960 he left Germany for the California Institute of Technology, where he was appointed professor of physics in 1961. Three years later he returned to Munich to become a professor at the Munich Technical University.

The impact of Mössbauer's findings was profound. In 1959 the Mössbauer effect was used to test and confirm ALBERT EINSTEIN's 1905 theory of relativity, which postulated that the frequency of electromagnetic radiation was influenced by gravity. Thereafter the Mössbauer effect was utilized in many applications, most notably for making exact measurements in the field of solid-state physics. In 1961 Mössbauer was awarded the Nobel Prize in physics. He won numerous additional awards, including the Einstein Medal in 1986.

Muir, John

(1838–1914)
American
Naturalist

Although he was born in Scotland, John Muir had as his greatest legacy the preservation of the American wilderness. A fervent writer, explorer, and naturalist, he was actively involved in the creation of several national parks in the United States. Largely because of Muir's efforts the U.S. Congress created the Yosemite and Sequoia National Parks. With his followers Muir founded the Sierra Club, a conservation organization that exists to this day. His books influenced many politicians and policymakers, including President Theodore Roosevelt, in the development of conservation policies. But Muir was not only an authority on forestry and forest management; he was also the first to postulate that the Yosemite Valley was formed by glacial erosion.

Muir was born on April 21, 1838, in Dunbar, Scotland, to Daniel and Ann Gilrye Muir. The family immigrated to the United States in 1849, when Muir was 11. Although Daniel Muir was a shopkeeper in Scotland, the family eventually settled at Hickory Hill Farm, near the small town of Portage, Wisconsin. Daniel Muir was a strict disciplinarian, who insisted that the family labor on the farm from dawn until dusk. In his free time John Muir roamed the Wisconsin woods and developed a love of nature.

Muir enrolled in the University of Wisconsin in 1860 but left after three years. He became an inventor, conceiving such unique devices as one that tipped a person out of bed before sunrise. But Muir's career path took a turn after he nearly lost an eye in an industrial accident in 1867. Upon his recovery he resolved to devote himself to nature. He walked from the Midwest to the Gulf of Mexico and eventually arrived in California in 1868. Although Muir continued his travels throughout his life, California became his home base thereafter.

He first visited California's high country in 1868, as he explored the spectacular topographical characteristics of the Yosemite Valley. He proposed his theory of the glaciation of the valley in 1871, and in 1879 he traveled to Alaska and discovered Glacier Bay. After Muir married Louise Wanda Strentzel in 1880, the couple had two daughters, Wanda and Helen. The family settled for a time in Martinez, California, on a fruit ranch owned by Muir's father-in-law.

The time Muir had spent in the Sierras convinced him of the devastation caused to the fragile land by ranch animals' grazing in alpine meadows. He began to publish articles urging the creation of a national park to preserve the area. In response to Muir's eloquent lobbying (and the public opinion he galvanized), Congress established the Yosemite and Sequoia National Parks in 1890. In 1892 he formed the Sierra Club, which was dedicated to protecting these new parks. Muir served as president of the organization until his death. Two impassioned Muir articles also impelled Congress to pass President Grover Cleveland's measure to create 13 national forests in 1897.

Muir was a prolific writer, publishing more than 300 articles and 10 books on a range of topics. After publication of *The Mountains of California* in 1894, his influential book, *Our National Parks* (1901), caught the eye of President Theodore Roosevelt. Roosevelt met with Muir, and the two went camping in Muir's beloved Yosemite. Muir is credited with helping shape Roosevelt's massive conservation program. He died of pneumonia in Los Angeles on December 24, 1914.

Muller, Hermann Joseph

(1890–1967)
American
Geneticist

Hermann Joseph Muller made pioneering discoveries in the field of genetics. Early in his career Muller realized that genes are the origin of life in that only genes replicate and thus allow for the continuation of life after death, as one generation passes on its life force and traits to subsequent generations. Muller focused his research on genetic mutations, which he discovered to be essentially chemical transformations. Muller received the 1946 Nobel Prize in physiology or medicine for his discovery that X rays induce genetic mutation.

Muller was born on December 21, 1890, in New York City. His mother was Frances Lyons. His father, after whom he was named, entered the family bronze artwork business but died when Muller was nine years old. In 1923 Muller married Jessie Mary Jacob, a mathematician; the couple's later divorce, combined with work stress, led to Muller's nervous breakdown in 1932. In 1939 he married Dorothea Kantorowitz.

After attending Morris High School in Harlem, Muller received a scholarship to attend Columbia University in 1907; there he studied genetics under Edmund B. Wilson for his bachelor's degree in 1910. While continuing at Columbia with a teaching fellowship in the department of physiology, Muller concurrently enrolled in Cornell Medical School, where he conducted research on the transmission of nerve impulses for his master's degree in 1912. Muller continued with doctoral work at Columbia, collaborating with the zoology professor THOMAS HUNT MORGAN and the doctoral students A. H. Sturtevant and Calvin B. Bridges. The scientists conducted research on the chromosomal genetics of the fruit fly *Drosophila melanogaster,* which they chose for its three-week reproductive cycle. Muller wrote his dissertation on crossing over, a phenomenon that allowed for genetic mapping, which the group reported on in their 1915 collaborative work *The Mechanism of Mendelian Heredity,* which became a classic genetics text. Muller received his Ph.D. in 1916.

Muller taught at Rice Institute in Houston, Texas, from 1916 to 1918, when he returned to Columbia. In 1919 he discovered a correlation between increased temperature and increased genetic mutation. He also theorized that life originated in what he called "naked genes," or viruslike, self-replicating molecules. In 1921 the University of Texas at Austin appointed Muller assistant professor; he was promoted to a professorship in 1925. The next year Muller discovered that X rays induce mutations and reported his findings in the 1927 *Science* article "Artificial Transmutation of the Gene," which earned him international acclaim.

In 1933 Muller received a Guggenheim fellowship for research at the Kaiser Wilhelm Institute for Brain Research in Berlin, but the rise of Nazism led him to seek refuge in the Soviet Union. Between 1933 and 1937 he conducted research on radiation genetics and cytogenetics at the Academy of Sciences in Moscow and Leningrad, but his opposition to Lysenkoist genetics, a flawed theory adopted by Russian Marxists that he argued against in his 1935 text *Out of the Night,* forced him to migrate to Edinburgh, Scotland, where he worked at the Institute of Animal Genetics from 1937 to 1940. He taught at Amherst College during World War II before receiving the appointment of professor of zoology at Indiana University, which he held for the remainder of his career.

Muller used his status as a Nobel laureate to voice his opposition to unnecessary practices of modern medicine, especially unnecessary X rays, as well as to nuclear energy and nuclear weapons testing. Muller favored eugenics in the form of sperm banks housing an endowment from healthy men to counteract the accumulation of genetic mutations inherent in modern life. Muller died on April 5, 1967, in Indianapolis, Indiana.

Müller, Paul Hermann

(1899–1965)
Swiss
Chemist

Paul Müller won the 1948 Nobel Prize in physiology or medicine for his discovery of the use of dichlorodiphenyltrichloroethane (DDT) as an agricultural insecticide, and more specifically as an agent to reduce the spread of tropical diseases such as typhus and malaria that use insects as carriers. Though the agricultural overuse of DDT wreaked environmental havoc, leading to its ban in many countries in the 1970s, it remained important in the fight against tropical diseases.

Müller was born on January 12, 1899, in Olten, Switzerland. His mother was Fanny Leypoldt; his father was Gottlieb Müller, an official with the Swiss Federal Railway. Müller married Friedel Rügsegger in 1927, and together the couple had two sons and one daughter.

Müller forged his path in chemistry by working in industry before attending university. In 1916 he worked as a laboratory assistant at the Dreyfus and Co. chemical factory, and in 1917 he worked as a chemical assistant in the research laboratories of Lonza A.G. Then in 1919 he entered the University of Basel to study chemistry. He conducted his doctoral research under F. Fichter and Hans Rupe and wrote his dissertation on the chemical and electrochemical oxidation of *m*-xylidine. He received his Ph.D. in 1925.

That year he commenced work as a research assistant for J. R. Geigy A.G., where he remained for the rest of his career. By the time of his retirement in 1961 he was the deputy head of pest control research. When he began work there he worked on synthesizing dyes and tanning agents for leather. In 1935 he was assigned to investigate insecticides, in the search for a chemical compound that was cheap and easy to manufacture, that was toxic to insects but not to plants and warm-blooded mammals, and that had enough chemical stability to have long-lasting effects.

In September 1939 Müller identified just such a compound in DDT, first prepared by the German chemist Othmar Zeidler in 1873 with no knowledge of its efficacy as an insecticide. DDT was tested on the Colorado potato beetle that year, and in 1940 Müller obtained a Swiss patent on DDT. The chemical was made available commercially in 1942. In January 1944 it was used in Naples, Italy, to kill typhus-carrying lice, effectively controlling a winter outbreak of typhus for the first time.

Müller published articles on DDT in 1944 and in 1946 that hinted at his foresight about the potential adverse effects to nature of a toxic chemical as stable as DDT. Nevertheless DDT was used extensively for 20 years, and it was estimated in 1968 that 1 billion pounds of DDT remained in the environment, disrupting the natural order of the food chain. DDT's adverse environmental effects were well publicized, undermining its significance as an effective weapon against tropical diseases.

Müller retired in 1961 to set up a personal laboratory in his home. He died on October 13, 1965, in Basel, Switzerland.

Murphy, William Parry

(1892–1987)
American
Physician, Pathologist

William Parry Murphy received the 1934 Nobel Prize in physiology or medicine in conjunction with GEORGE HOYT WHIPPLE and GEORGE RICHARDS MINOT for their roles in the discovery of a diet high in liver as a cure for pernicious anemia. Murphy performed perhaps the most difficult work in the studies, painstakingly counting the microscopic reticulocytes, or red blood cells, that indicated the relative health of the blood samples of the patients participating in the trials. A liver lover, he also coaxed the suffering patients into eating a half-pound of liver every day.

Murphy was born on February 6, 1892, in Stoughton, Wisconsin. His mother was Rose Anna Parry; his father was Thomas Francis Murphy, a Congregational minister. In 1919 Murphy married Pearl Harriet Adams, and together the couple had two children—one son, William P. Murphy, Jr., and one daughter, Priscilla Adams.

Murphy attended the University of Oregon, where he received a bachelor's degree in 1914. After teaching high school math and physics for two years, Murphy returned to the University of Oregon, where he entered the Medical School in 1916 and worked as a laboratory assistant in the anatomy department. Murphy spent the summer of 1918 studying at Rush Medical School and subsequently received the William Stanislaus Murphy Fellowship at Harvard Medical School, where he received his M.D. in 1922.

Murphy then conducted his internship at the Rhode Island Hospital in Providence, followed by residency at the Peter Bent Brigham Hospital, where he remained for the rest of his career. In 1924 Harvard University appointed him an assistant in medicine, and in 1935 Brigham appointed him as an associate in medicine and later promoted him to senior associate in medicine. In 1958 Brigham appointed him as a consultant in hematology.

Following up on Whipple's report of the regenerative effects of a liver-rich diet on dogs with induced anemia, Minot enlisted Murphy to conduct a hushed study on patients suffering from pernicious anemia, a fatal disease with no known cure or relief, by feeding them copious amounts of liver. After noting almost miraculous recovery in these patients, Murphy devoted himself to identifying a more efficient means of introducing the liver into the patients' bodies. Murphy and Minot asked their colleague, the physical chemist Edwin J. Cohn, to prepare a liver extract; Cohn's solution concentrated liver 50 to 100 times. Guy W. Clark of Lederle Laboratories further concentrated liver into an extract that could be injected into the muscle of anemic patients once a month.

Murphy and Minot presented their findings on 45 anemic patients at the May 4, 1926, meeting of the American Medical Association in Atlantic City, New Jersey. An incredulous audience dismissed the treatment as too simple, but continued results eventually convinced the skeptics. Murphy reported their findings in a 1926 *Journal of the American Medical Association* article, "Treatment of Pernicious Anemia by Special Diet," and in a 1927 *Journal of Biological Chemistry* article, "The Nature of the Material in Liver Effective in Pernicious Anemia." In 1934 Murphy and Minot reported on 42 of the original 45 pernicious anemia patients from 1926, 31 of whom had regained their health and 11 of whom had died of other complications.

Besides the Nobel, Murphy received the Cameron Prize and Lectureship from the University of Edinburgh, as well as a Bronze Medal from the American Medical Association and a Gold Medal from the Massachusetts Humane Society. Murphy died on October 9, 1987, in Brookline, Massachusetts.

N

Nambu, Yoichiro

(1921–)
Japanese/American
Physicist

Recognized as one of the leaders in the development of modern particle physics, the theoretical physicist Yoichiro Nambu made significant contributions to the standard model, which explains the fundamental forces and interactions between elementary particles. Nambu developed the theory of spontaneous symmetry breaking in superconductivity, which paved the way for other scientists to formulate the electroweak theory, which successfully unified the theories of electromagnetic and weak interactions. Nambu also worked on quantum chromodynamics, the physics of the relationship between a group of elementary particles known as quarks, and the quantum "string theory."

Born in Tokyo, Japan, on January 18, 1921, Nambu was only two years of age when his family moved to the rural town of Fukui. A destructive earthquake prompted the family to leave Tokyo, and Nambu's father attained a position as an English instructor at a girls' school in Fukui. Nambu spent a considerable amount of his childhood in his father's vast library and demonstrated an early interest in science and a curiosity for the way things work.

After attending a junior college, Nambu entered the esteemed University of Tokyo, where he earned his degree in physics in 1942. Nambu was then drafted into the military and served until the end of World War II. He then pursued graduate studies at the University of Tokyo. Conditions were difficult and provisions were scarce after the war, and Nambu resorted to living in his office and searching for food in the countryside. In 1949 Nambu was fortunate enough to secure a professorship at the new Osaka City University. Nambu received his doctorate from the University of Tokyo in 1952.

Nambu's early work was in solid-state physics, and his first paper was published in *Journal of Theoretical Physics,* an English-language journal established by the Japanese physicist HIDEKI YUKAWA. Nambu also collaborated with SHINICHIRO TOMONAGA, a pioneer in the field of quantum electrodynamics (QED), and published numerous papers on QED before departing for the United States in 1952. Nambu worked at the Institute for Advanced Study in Princeton, New Jersey, until 1954, when he accepted a position at the University of Chicago, where he would remain throughout his career.

The work for which Nambu is best known began in 1959. In a paper he outlined his concept of spontaneous symmetry breaking (SSB) to explain an inconsistency in the theory of superconductivity that had been proposed in 1956 by JOHN BARDEEN, LEON NEIL COOPER, and JOHN ROBERT SCHRIEFFER, the BCS theory. Though the theory was well accepted and resulted in a Nobel Prize in physics for the three creators, it seemed to violate a fundamental rule of the theory of electricity and magnetism. Nambu succeeded in explaining this apparent paradox by proposing the SSB theory, which maintained that the fundamental rule, that of the conservation of charge, was not violated. Symmetry existed, Nambu proposed, but it was hidden.

In the 1960s Nambu turned to the study of color quarks and developed a theory to explain the forces that held quarks together. It was known that three quarks joined to form the proton and the neutron, and Nambu suggested that these quarks could be of three kinds, or colors, as they came to be called, that functioned as sources of eight massless fundamental particles, known as gluons, which supplied the interactive force holding the quarks together. This gauge theory of strong interactions came to be known as quantum chromodynamics.

Nambu began studying string theory in the 1970s; the theory continued to generate interest among physicists for several decades. For his outstanding achievements in physics Nambu has received numerous awards and honors, including the American Physical Society's Dannie Heineman Prize in 1970, the University of Miami's J. Robert Oppenheimer Prize in 1976, the National Medal of Science in 1982, the German Physical Society's Max Planck Medal in 1985, and the Wolf Prize in Physics in 1994. Nambu was elected to the National Academy of Sciences and the American Academy of Arts and Sciences in 1971; was named to the Order of Culture by the Japanese government in 1978; and became an honorary member of the Japan Academy in 1984. Nambu became a U.S. citizen in 1970.

Natta, Giulio

(1903–1979)
Italian
Polymer Chemist

Giulio Natta was the first Italian to receive the Nobel Prize in chemistry, which he shared with KARL ZEIGLER in 1963 for their collaborative work on developing high polymers. Natta developed stereospecific polymers by identifying catalysts that created structures with recurring patterns in the chain at regularly spaced intervals, a phenomenon dubbed *tacticity* by Natta's wife, thus increasing the strength and heat resistance of these plastics.

Natta was born on February 26, 1903, in Imperia, Italy, a resort town on the Ligurian Sea. His mother was Elena Crespi; his father was Francesco Natta, a lawyer and a judge. In 1935 Natta married Rosita Beati, a professor of literature at the University of Milan. Together the couple had two children—one daughter, named Franca, and one son, named Giuseppe.

Natta first attended the University of Genoa to study mathematics but transferred to the Milan Polytechnic Institute to study chemical engineering. He earned his bachelor's degree in 1921 and continued on there with doctoral research. In 1924 at the age of 21 he earned his Ph.D. in chemical engineering.

In 1925 the Milan Polytechnic Institute appointed Natta as an assistant professor of general chemistry and promoted him to the status of full professor in 1927. Natta spent the majority of the 1930s filling professorships at other institutions before returning to Milan in the late 1930s. In 1933 the University of Pavia hired him to direct its chemical institute, and in 1935 he moved to the University of Rome as the chairman of the department of physical chemistry. In 1937 he directed the institute of industrial chemistry at the Turin Polytechnic Institute. Then in 1938 he returned to the Milan Polytechnic Institute to direct the Industrial Chemical Research Center, where he remained for the rest of his career. Natta aided the Italian government run by Benito Mussolini before and during World War II by working to produce synthetic rubber. After the war Natta shifted his focus from synthetic rubber to synthetic plastics.

In 1952 Natta attended a lecture in Frankfurt, Germany, in which Zeigler, the director of the Max Planck Institute for Coal Research in Mulheim, explained his discovery of the Aufbau reaction, or growth reaction, which involved the creation of large molecules from the petroleum by-product ethylene. Realizing the implications of this discovery, Natta teamed up with the Montecatini Company to reach an agreement over research sharing by Natta and Ziegler, with Montecatini controlling the commercial rights in Italy for their discoveries.

In 1953 Zeigler's laboratories synthesized linear polyethylene from ethylene and patented it before informing Natta. Natta's labs in turn synthesized linear polypropylene from propylene, a cheaper petroleum derivative than ethylene, on May 11, 1954, and patented it before informing Zeigler. This mutual disrespect for their research-sharing agreement caused a rift in their friendship, though they reconciled sufficiently to appear together in Stockholm to receive their Nobel Prize in 1963.

In 1959 Natta had contracted Parkinson's disease, which severely crippled him by the time of the Nobel awards ceremony. Two decades after contracting Parkinson's, on May 2, 1979, he died in Bergamo, Italy, of complications of surgery to repair his broken femur bone.

Néel, Louis Eugène Félix

(1904–)
French
Physicist

Louis Néel received the 1970 Nobel Prize in physics for his studies on the magnetic properties of solids. Specifically he discovered antiferromagnetism, a phenomenon that reversed the normal properties of magnetism in certain solids below the threshold of a certain temperature, known as the Néel point or the Néel temperature. Néel shared the Nobel Prize in physics that year with the Swedish astrophysicist HANS ALFVÉN, who conducted research on magnetohydrodynamics.

Néel was born on November 22, 1904, in Lyon, France, to Antoinette Hartmayer and Louis Néel, a director in the civil service. Néel attended secondary school at the Lycée du Parc in Lyon and then at the Lycée St.-Louis in Paris. He entered the École Normale Supérieure in Paris and graduated in 1928. In 1932 he received his Ph.D. from the University of Strasbourg, after conducting his doctoral research under Pierre Weiss. In 1931 he married Hélène Hourticq, and together the couple had one son and two daughters.

The University of Strasbourg retained Néel as a teacher after he earned his doctorate and promoted him to the status of professor in 1937. In 1945 the University of Grenoble appointed him as a professor, and in 1956 he established the Laboratory of Electrostatics and the Physics of Metals there. Concurrently he served as the director of the Center for Nuclear Studies from 1956 until 1971, when he took on the presidency of the Institut National Polytechnique de Grenoble. He retired from these diverse positions in 1976.

In 1930 Néel predicted the phenomenon of antiferromagnetism, whereby the normal magnetic field of certain substances becomes negated below the Néel temperature as a result of spinning of electrons in opposite directions. Néel experimentally confirmed this effect in 1938. This discovery had profound implications for practical applications. Perhaps the most important early application of antiferromagnetism was its use aboard French warships in World War II to neutralize the magnetic field of mines, effectively disabling them.

In 1947 Néel identified the phenomenon of paramagnetism, which resulted from a rise in temperature above the Néel point in an antiferromagnetic substance, when the negated magnetism became spontaneously magnetized in a slightly deformed state of unequal magnetic charges, retaining some of the properties of antiferromagnetism and some of the properties of ferromagnetism. These feromagnetic substances, as they were called, became instrumental in the core memory banks of computers.

Later in Néel's career he identified the phenomenon of magnetic creep, whereby the passage of time affected ferromagnetic substances. Néel subdivided this creep effect into two categories, one affected by temperature changes and the other by atomic redistribution in the crystal that accounts for spontaneous magnetization. Néel studied this phenomenon as a means of determining the past history of the direction and strength of the Earth's magnetic field. Besides the Nobel Prize, Néel received the Gold Medal of the National Center for Scientific Research as well as the Holweck Medal of the Institute of Physics in London.

Newlands, John Alexander Reina

(1837–1898)
British
Chemist

Although John Alexander Reina Newlands was employed as a chemist in the sugar industry, his independent research contributed to the development of the periodic law. He published papers in 1864 that asserted that if the elements were arranged according to their atomic weights, a pattern emerged in which every eighth element shared certain properties. He termed this phenomenon the *law of octaves*. His work was not accepted until after DMITRI IVANOVICH MENDELEEV's periodic table was published in 1869.

Newlands was born on November 27, 1837, in London, England. His father, William Newlands, was a Presbyterian minister, who educated his son at home. His mother, Mary Sarah Reina Newlands, was of Italian descent.

In 1856 Newlands enrolled in the Royal College of Chemistry. For a year he studied under August von Hofmann; he later became the assistant to J. T. Way, a chemist in the Royal Agricultural Society. In 1860, cognizant of his Italian ancestry, Newlands joined the army of Giuseppe Garibaldi in its quest to unify Italy. After participating in Garibaldi's invasion of Naples, Newlands returned to his work with J. T. Way.

Newlands set up practice as an analytic chemist in 1864 and taught chemistry to supplement his income. In 1868 he was hired as the chief chemist in a sugar refinery owned by James Duncan. In the course of this work he developed a new system for cleaning sugar and applied his knowledge of chemistry to improve processing. After the refinery's business declined as a result of intense foreign competition, Newlands once again established himself as an independent analyst in 1886, this time in partnership with his brother, B.E.R. Newlands. The two wrote a treatise on sugar growing and refining.

While working as an industrial chemist, Newlands began to grapple with the idea of systematizing organic compounds. In 1863 he published his first paper in *Chemical News*; in it he summarized his concept that the elements shared a periodicity when they were arranged according to their atomic weights. His law showed the halogens in one group and the alkali metals in another. In a paper published in 1864 he detailed his law of octaves. Newlands compared this principle to the intervals of the musical scale. Unlike Dmitri Ivanovich Mendeleev, however, he did not leave gaps for undiscovered elements.

When Newlands presented his ideas to the Chemical Society, he was met with fierce criticism that his ordering was too arbitrary. The society rejected his paper for publication, and one member, G. C. Foster, a professor of physics, even derisively inquired whether he had thought of arranging the elements alphabetically. Newlands's law of octaves was not accepted until after Dmitri Ivanovich Mendeleev's periodic table revolutionized the study of chemistry. The significance of his work was finally recognized, however, and in 1884 he published his many papers in a collection, *On the Discovery of Periodic Law*. He was awarded the Davy Medal of the Royal Society in 1887. Newlands died of influenza in 1898 and was survived by his wife and their three children.

Newton, Sir Isaac
(1642–1727)
English
Mathematician, Physicist

Sir Isaac Newton's discoveries are considered some of the most influential and important in the history of science. He made profound contributions to the fields of physics, mathematics, and optics. His formulation of three laws of motion laid the foundation for classical mechanics, explained planetary motion, and provided the foundation for his general theory of gravitation. In the realm of mathematics he invented the calculus (independently of Leibniz) and conceived the binomial theorem. His research in optics led him to propose his theory of light—that white light was a mixture of various colors in the spectrum.

Newton was born prematurely on December 25, 1642. His father, who owned the manor of Woolsthorpe in Lincolnshire, England, had died three months before he was born. His mother, Hannah Ayscough Newton, remarried when Newton was three and left him in the care of his grandmother at Woolsthorpe. Upon the death in 1653 of her second husband, the Reverend Barnabas Smith, she returned to Woolsthorpe. Although the family owned quite a bit of land, they were neither wealthy nor noble.

Sir Isaac Newton, whose extraordinary contributions to the fields of physics, mathematics, and optics included his theory of light, which proposed that white light was a mixture of various colors in the spectrum. *AIP Emilio Segrè Visual Archives.*

After Newton had attended the King's School in Grantham for some time, his mother withdrew him with the intention of making him a farmer. He showed little skill or interest in this vocation, however. With the encouragement of his uncle Newton was prepared for the university instead. In 1661 he was admitted to Trinity College, Cambridge, where he earned his bachelor of arts degree in 1665.

Newton was forced to return to his home in Woolsthorpe in 1665 when Cambridge closed because of an outbreak of plague. His time away from the university proved to be fruitful. Newton himself later termed this period his *annus mirabilis* (miraculous year), because he began to ponder the forces behind natural phenomena he witnessed and to research the subjects that would later make him famous—gravity, optics, and light. When he returned to Cambridge in 1667, he was elected a fellow at Trinity, and in 1669 he was named Lucasian Professor of Mathematics at the age of 26.

Newton's work in mechanics and gravity earned him most fame. In 1687 he published his seminal work, *Mathematical Principles of Natural Philosophy*—known simply as the *Principia*—which discussed his three laws of motion. The first law posited that a body at rest or in uniform motion will continue in that state unless a force is applied; the second defined force as equaling the mass of a body multiplied by its acceleration; the third stated that if a body exerted a force (action) on another, there would be an equal but opposite force (reaction) on the first body. These laws provided the basis for Newton's law of gravitation, which proposed that any particle of matter in the universe attracts any other with a force determined by the product of their masses but decreases by the square of their distance apart. These theories, elaborated in the *Principia*, explained a diverse range of natural phenomena—from the motion of the planets and their moons to the tides—and became essential building blocks for classical mechanics and future scientific endeavors.

Newton's contributions to mathematics and optics were no less significant. In 1665 he first laid out the binomial theorem, which was a general formula for writing any power of a binomial without multiplying out. He invented the calculus in 1669, although he did not publish his findings until 1674. He published *Opticks*, in which he described his theory of light and included other mathematical research, in 1704.

But Newton's career was not devoted exclusively to science. From 1689 to 1690 and from 1701 to 1702 he served as a member of Parliament for the university. He was appointed warden of the Mint in 1696 and master of the Mint in 1699. After resigning his professorship at Cambridge in 1701, he was knighted for his services to the Treasury in 1705. After this time he dedicated little time to science, although he did remain president of the

Royal Society until his death. His legacy as a scientist remains unsurpassed. Newton died in London in 1727 at the age of 85.

Nice, Margaret Morse

(1883–1974)
American
Zoologist

The Austrian zoologist Konrad Lorenz, usually credited with cofounding ethology, the study of the behavior of animals in nature, once said that this honor really belonged to the bird researcher Margaret Morse Nice. She was born Margaret Morse on December 6, 1883, in Amherst, Massachusetts, the fourth of the seven children of Anson Daniel and Margaret Morse. Her father was a professor of history at Amherst College. By the time she was nine, Margaret was taking notes on the behavior of birds.

Morse attended college at Mount Holyoke in Massachusetts, from which she earned a bachelor's degree in 1906. She then became a "dutiful daughter-at-home" until lectures by a professor from Clark University in Worcester, Massachusetts, revived her interest in studying nature professionally. She went to Clark in 1907, studied there for two years, and finally was awarded a master's degree in ornithology (bird study) in 1915. While at Clark, Morse met and fell in love with the physiologist Leonard Blaine Nice. They married in 1909, and Margaret Nice gave up plans for further academic study. They had five daughters: Constance, Marjorie, Barbara, Eleanor, and Janet. Observing them, Margaret wrote several papers on language development and related subjects in child psychology.

In 1919, when the Nices were living in Norman, Oklahoma, Margaret decided to return to her childhood "vision of studying nature and trying to protect the wild things of the earth." Her first project, in which she enlisted her husband and daughters (she later praised Constance's "unquenchable zeal for climbing . . . trees"), was a study of mourning doves. This work prevented the dove hunting season from being extended by showing that the birds continued nesting through September and even into October.

Nice's most important research focused on the song sparrows near her home in Columbus, Ohio, where the family moved in 1928. Previous scientists had not tried to distinguish one bird from another, but Nice banded her sparrows so she could identify specific birds. Her careful and lengthy observations allowed her to outline the life history of the species at a level of detail never achieved before. The renowned ornithologist Ernst Mayr said that Nice "almost single-handedly initiated a new era in American ornithology. . . . She early recognized the importance of a study of bird *individuals* . . . [as] the only method to get reliable life history data."

Nice's work became a book, *Population Study of the Song Sparrow,* published in 1937. (A second volume appeared in 1943.) The book won the Brewster Medal in 1942. By then Nice was already greatly respected in birding circles; she had been made a member of the American Ornithological Union in 1931, only the fifth woman to be so honored, and a fellow of the union in 1937. Earlier days of being denigrated as "just a housewife" had long passed. Nice later received two honorary degrees as well, and a Mexican subspecies of song sparrow was named after her.

In 1936 the Nices left Ohio for Chicago, where opportunities for bird observation were limited. Margaret turned some earlier experiences into a book called *The Watcher at the Nest* (1939), but devoted herself mainly to "rais[ing] . . . friends for wildlife" and educating the public about the need for conservation. She died on June 26, 1974, at the age of 90.

Nicolle, Charles Jules Henri

(1866–1936)
French
Bacteriologist

Charles Nicolle received the 1928 Nobel Prize in physiology or medicine for his identification of the cause of the tropical disease typhus, which he discovered was transmitted by lice on the human body.

Nicolle was born on September 21, 1866, in Rouen, France. His father, Eugène Nicolle, was a medical doctor at the municipal hospital and a professor of natural history at the École des Sciences et des Arts.

Nicolle's brother, Maurice, a renowned bacteriologist himself, encouraged Nicolle to follow in his footsteps; Nicolle did so, conducting doctoral research at the Pasteur Institute in Paris on *Hemophilus ducreyi,* or Ducrey's bacillus, the bacterium responsible for the venereal disease of soft chancre. He received his medical degree in 1893 and returned to Rouen as a staff physician at the hospital there. He married Alice Avice, and together the couple had two sons, Marcelle and Pierre, both of whom went on to become physicians.

When it became apparent that the hospital at Rouen would never develop into a major biomedical research center, as he had hoped, he moved to Tunis, Tunisia, to take over the directorship of the Pasteur Institute there. Nicolle remained in Tunis the majority of his career, transforming a ramshackle clinic into a major research center for the investigation of tropical diseases. In 1932 the College de France appointed Nicolle to a professorship.

In 1909 Nicolle witnessed an outbreak of typhus; he observed that it spread readily in the general population but little in hospitals. Nicolle realized that the main difference between people on the streets and patients in the beds was that stripping of clothing and thorough cleaning followed hospital admission. Furthermore the hospital's laundry personnel typically contracted the disease, whereas the patients and medical staff inside the hospital did not contract it from these new patients. Nicolle guessed that the initial cleansing rid patients of lice, which he hypothesized to be the most likely carrier of the disease. He therefore immediately commenced a study by infecting a chimpanzee's blood with typhus, then transfusing this infected blood to a macaque monkey. When this monkey contracted typhus, Nicolle planted 29 human body lice from healthy people on the macaque monkey. Nicolle then transferred these lice to numerous other healthy monkeys, all of which contracted typhus, thus confirming lice as the causative agent of the spread of the disease. Halting the spread of typhus thus amounted to wiping out lice.

Later in his career Nicolle identified the phenomenon of inapparent infection, whereby the disease carrier exhibits no symptoms but spreads the disease nonetheless. This discovery helped to explain how diseases reappeared after the end of one epidemic to start another epidemic. Nicolle also researched African infantile leishmaniasis as well as rinderpest, brucellosis, measles, diphtheria, trachoma, and tuberculosis. Besides the Nobel Prize, Nicolle was honored as a French Commander in the Legion of Honor and as a member of the French Academy of Medicine. Nicolle died on February 28, 1936, in Tunis, Tunisia.

Nirenberg, Marshall Warren

(1927–)
American
Biochemist

Marshall Warren Nirenberg played an important role in deciphering the genetic code. He discovered which portion of deoxyribonucleic acid (DNA) is responsible for the synthesis of proteins. Specifically he determined which DNA sequence patterns code for which amino acids, the building blocks of proteins. Nirenberg's work is credited with fueling a better understanding of genetically determined diseases.

Nirenberg was born on April 10, 1927, in New York City. When he was 10, he moved with his parents, Harry Edward and Minerva Bykowsky Nirenberg, to Florida. In 1961 at the age of 34 he married a fellow biochemist, Perola Zaltzman.

At the University of Florida he received his undergraduate education; there he majored in biology and worked as a teaching assistant in that field. He received his M.Sc. from the University of Florida in 1952, then attended the University of Michigan for his doctoral studies. His dissertation explored the uptake of hexose (a sugar molecule) by ascites tumor cells; he was awarded his Ph.D. in 1957.

Upon completing his doctorate, Nirenberg undertook research as an American Cancer Society (ACS) fellow at the National Institutes of Health (NIH) in Bethesda, Maryland. In 1960 he joined the NIH permanently as a research scientist in biochemistry. He completed his most noteworthy work at the NIH.

Nirenberg's research built upon the legacy of JAMES DEWEY WATSON and FRANCIS HARRY COMPTON CRICK, who together had discovered the structure of DNA in 1953. At the time Nirenberg began his research, it had been established that different combinations of three nucleotides—those molecules that when linked form the building blocks of DNA or ribonucleic acid (RNA)—each coded for a specific amino acid. The four nucleotide bases in DNA, adenine, cytosine, guanine, and thymine (in RNA the thymine base is replaced by uracil), could combine as triplets in 64 permutations called codons. The sequence of nucleic acids in the codon dictates the structure of proteins. At that time no one had solved the ultimate puzzle of which of these codons coded for each of the 20 amino acids.

Nirenberg used a technique developed by SEVERO OCHOA to synthesize an RNA molecule consisting entirely of uracil nucleotides. After Nirenberg discerned that his synthesized molecule made a protein composed solely of the amino acid phenylalanine, he was able to conclude that his uracil triplet coded for phenylalanine. In short, Nirenberg's achievement was to crack the genetic code. He announced his findings in 1961 at a meeting of the International Congress of Biochemistry.

Nirenberg's research paved the way for the study of genetically determined diseases. He was awarded the Molecular Biology Award from the National Academy of Sciences in 1962 and the National Medal of Science in 1965. He won the Nobel Prize in physiology or medicine with HAR GOBIND KHORANA and Robert Holley in 1968. Upon completing his work on the genetic code, he continued to serve at the NIH in various capacities. From 1962 until 1966 he was head of the Section for Biochemical Genetics at the National Heart Institute of the NIH. Thereafter he worked as chief of the Laboratory of Biochemical Genetics at the NIH's National Heart, Lung, and Blood Institute. Motivated by his own experiences, Nirenberg became a strong and public advocate of government support for scientific research.

Nobel, Alfred Bernhard

(1833–1896)
Swedish
Chemist

Although he was recognized during his own lifetime as the inventor of dynamite, Alfred Bernhard Nobel is known today for the trust he established to award an eponymous prize each year to the finest minds in several fields. In addition to inventing dynamite and creating the Nobel Prize, Nobel patented his discoveries of blasting gelatin (a more powerful form of dynamite) and smokeless blasting powder. Nobel also perfected various forms of detonating cap technologies.

Nobel was born on October 21, 1833, in Stockholm, Sweden, to Immanuel and Andrietta Ahlsell Nobel. Through his mother he was related to Olof Rudbeck, an important Swedish scientist of the 17th century. His father was a builder, industrialist, and inventor, who taught his son engineering. When Alfred Nobel was nine, the family moved to St. Petersburg, Russia, where his father supervised the manufacture of a submarine mine he had conceived.

Nobel was educated primarily by tutors. By the time he was 16 he was fluent in English, French, German, Russian, and Swedish and had developed a background in chemistry. In 1850 he left Russia to travel. After studying chemistry in Paris, he spent four years in the United States, working under John Ericsson, a Swedish expatriate who built an ironclad warship. Nobel returned to St. Petersburg and remained there until his father's factory went bankrupt in 1859.

Upon his return to Sweden, Nobel immediately set out to manufacture the liquid explosive nitroglycerin. Shortly after it had begun production, Nobel's factory blew up in 1864, killing five people, including Nobel's youngest brother, Emile. Although the Swedish government forbade Nobel to rebuild his plant, he continued his experiments on nitroglycerin in an effort to make the substance safer. He eventually discovered that nitroglycerin could be absorbed completely in Kieselguhr, a dry organic packing material. This Kieselguhr/nitroglycerin mixture, which could be handled safely, he called *dynamite*. After devising a detonating cap, Nobel patented dynamite in various countries, including Britain (1867) and the United States (1868).

After this discovery Nobel continued his work with explosives. In 1867 he patented blasting gelatin, a more potent form of dynamite, which was also less sensitive to shock and moisture and was water-resistant as well. He produced Ballistite (which he also called Nobel's blasting powder) in 1887. This explosive, which was an amalgam of nitroglycerin, nitrocellulose, and camphor, was nearly smokeless. His inventions in the realm of explosives were complemented by the detonating caps he created to withstand simple firings. Nobel's later inventions were in the fields of electrochemistry, optics, biology, and physiology.

His numerous discoveries and patents made Nobel a tremendously wealthy man. He never married and was a committed pacifist throughout much of his life. When he died on December 10, 1896, he left almost his entire estate to a foundation dedicated to rewarding, as he put it, "those who, during the preceding year, shall have conferred the greatest benefit on mankind." Beginning in 1901, the Nobel Prize was awarded annually in five categories: peace, literature, physiology or medicine, chemistry, and physics. In 1968 the Bank of Sweden funded the addition of a sixth prize category, economics. The Royal Swedish Academy of Sciences chooses the winner each year in physics and chemistry; the Royal Caroline Medical Institute selects the medicine or physiology Nobel Prize recipient; the Swedish Academy determines the literature winner; and the Norwegian parliament chooses the peace prize recipient. Element 102 was named *nobelium* in his honor.

Noether, Emmy

(1882–1935)
German
Mathematician

Emmy Noether helped to explain ALBERT EINSTEIN's theory of relativity and the behavior of particles inside the atom and developed a new form of algebra that united many fields. The mathematician Norbert Wiener wrote shortly before Noether's untimely death that she was "one of the ten or twelve leading mathematicians of the present generation in the entire world."

Amalie Noether, always known as Emmy, was born on March 23, 1882, in Erlangen, Germany. Her father, Max, taught mathematics at Erlangen University, and her brother, Fritz, also became a mathematician. Around age 18 Emmy decided to do so as well, even though women were barred by law from enrolling at most German universities, including Erlangen. (Its academic senate had concluded in 1898 that the presence of women "would overthrow all academic order.") For two years she attended lectures given by Erlangen professors who were family friends, then audited other classes for a semester at Göttingen University. When Erlangen opened its doors to women in 1904, she returned home and enrolled there.

At Erlangen, Noether mainly studied with her father and a friend of his, Paul Gordan. Gordan specialized in invariants, or constants, which became the subject of Noether's thesis (which she later called "a jungle of formulas"). She was awarded her doctorate, with highest honors, in December 1907. For eight years afterward she helped her father (even sometimes giving his lectures after his health failed) without pay or title, since as a woman she could not join a university faculty. Meanwhile she published papers about invariants that began to attract international attention to her work.

Noether's work drew the attention of David Hilbert and Felix Klein, two renowned mathematicians at Göttingen, and she accepted an invitation to join them there in

1916. Her expertise in invariants helped them work out the mathematics behind Albert Einstein's general theory of relativity. The mathematician Hermann Weyl, a friend of Noether's, later said, "For two of the most significant sides of the theory of relativity, she gave . . . the genuine and universal mathematical formulation."

Hilbert and Klein wanted Noether to have a faculty position, but other professors objected. One asked, "What will our soldiers [then fighting in World War I] think when they return to the university and find that they are expected to learn at the feet of a woman?" Hilbert snapped back, "I do not see that the sex of the candidate is an argument against her admission. . . . The [academic] senate is not a public bathhouse."

The minister of education finally permitted Noether to lecture as Hilbert's assistant, but she was not given a title or paid a salary. Only in 1919, after women's legal and social position had been somewhat liberalized, was she made a *Privatdozent,* the lowest faculty rank. She could then lecture under her own name, but she still was not paid. In 1922 she was made an "unofficial extraordinary professor." Eventually she was given a tiny stipend, but she never became a regular professor at Göttingen.

Professor or not, Noether made important advances. In 1918, for instance, she formulated Noether's theorem, which showed that the conservation laws of physics are identical with the laws of symmetry and therefore are independent of time and place. This theorem proved extremely important in the new field of quantum physics, which describes the behavior of subatomic particles.

Mathematicians were equally impressed with Noether's work in another new field, abstract algebra—a field that, indeed, she helped to found. Instead of focusing on problems and formulas, Noether's algebra dealt with concepts or ideals. She showed that the same basic rules underlie many fields of mathematics. "She saw connections between things that people hadn't realized were connected before," says the algebraist Martha K. Smith. In 1932 Noether received the Teubner Memorial Prize.

When the National Socialists (Nazis) took control of Germany in 1933, Noether became a target because she was an independent Jewish woman with pacifist and leftist political sympathies. She lost her job in April 1933, one of the first six Göttingen professors the Nazis fired. She just calmly moved her small classes into her apartment. Hermann Weyl later said, "Her courage, her frankness, . . . her conciliatory spirit were, in the midst of all the hatred and meanness, despair, and sorrow . . ., a moral solace." Still, Noether's friends convinced her that she must leave Germany while she still could. After much searching they found her a post at Bryn Mawr, a respected women's college near Philadelphia, and she went there in fall 1933.

On April 10, 1935, Emmy Noether underwent what should have been a routine operation. Four days later, probably because of an infection, she developed a high fever and lost consciousness. She died on April 14, 1935, at age 53. In a letter sent to the *New York Times* soon after her death, Albert Einstein wrote that Noether was "the most significant creative mathematical genius thus far produced since the higher education of women began."

Noguchi, Constance Tom

(1948–)
American
Medical Researcher

Constance Tom Noguchi's research has improved the understanding and treatment of sickle-cell disease, a severe inherited blood disorder that usually strikes people of African descent. She was born in Kuangchou (Canton), China, on December 8, 1948. Her father, James Tom, a Chinese-American engineer, had married a Chinese woman, Irene Cheung, while working in China. They had three daughters before returning to the United States (when Connie was seven months old) and added a fourth later.

Constance Tom Noguchi, a pioneering research scientist whose work has improved the understanding and treatment of sickle-cell disease. *Courtesy Constance Tom Noguchi.*

Connie Tom's childhood on the edge of San Francisco's large Chinese community blended Chinese and American elements. Some of her father's many books stirred her interest in science, as did a high school class in which students designed, researched, and carried out their own experiments.

She attended the University of California at Berkeley, planning to become a physician, but she changed her major to physics. In her senior year she married a medical student, Phil Noguchi, and they later had two sons. She graduated in 1970.

Connie Noguchi continued her studies at George Washington University in Washington, D.C., and earned a Ph.D. in theoretical nuclear physics in 1975. She then joined the National Institutes of Health (NIH), the group of large government research institutes in Bethesda, Maryland, where she has been ever since. (She worked in the laboratory of chemical biology in the National Institute of Diabetes, Digestive, and Kidney Diseases in the 1990s.) Her work has won such awards as the Public Health Special Recognition Award (1993) and the NIH EEO Recognition Award (1995). Most of Noguchi's research has concerned sickle-cell disease, which is caused by a single defective gene. In people who inherit the gene from both parents an abnormal form of hemoglobin, the red pigment that carries oxygen in the blood, is produced. This substance causes red blood cells, normally disk-shaped, to curve in a sickle shape. The curved cells sometimes clog small blood vessels, depriving tissues of oxygen and causing pain and illness.

Noguchi and her coworkers have studied hydroxyurea, a drug that increases the quantity of an alternate form of hemoglobin that normally exists in the blood in only tiny amounts. This form is not damaged by the defective gene, so increasing it increases the number of healthy red cells. "It's like doing gene therapy without having to add new genes," Noguchi says. Noguchi has investigated the possibility of combining hydroxyurea with other drugs, such as a hormone that produces new red cells, so that it can be used in smaller amounts and thus cause fewer side effects.

Today Noguchi is studying the way genes interact to produce hemoglobin and other red blood cell chemicals in both normal people and those with sickle-cell disease. She hopes that this basic research will lead to new treatments for the illness, perhaps even repair of the gene that causes it.

Norrish, Ronald George Wreyford

(1897–1978)
English
Physical Chemist

Ronald Norrish collaborated with GEORGE PORTER from 1949 to 1965, pioneering techniques for studying reaction kinetics, or the rates of chemical reactions. Together they developed the methods of flash photolysis, or chemical reactions elicited by light bursts, and kinetic spectroscopy. They honed their skills and techniques to the point where they could isolate and analyze chemical reactions in intermediate stages lasting a mere thousandth of a millionth of a second. For these achievements the pair won the 1967 Nobel Prize in chemistry, which they shared with the German scientist MANFRED EIGEN, who did similar work, though Eigen utilized the relaxation technique of measuring the return of chemicals to a state of equilibrium after an induced disturbance.

Born on November 9, 1897, in Cambridge, England, to Amy and Herbert Norrish, Ronald Norrish in 1926 married Anne Smith, a lecturer at the University of Wales. Together the couple had two children, twin daughters.

Norrish attended Emmanuel College of Cambridge University on a natural sciences scholarship; World War I interrupted his studies, as he served as a lieutenant in the Royal Field Artillery in France. In 1918 German soldiers captured him, and he spent a year as a prisoner of war. Upon his return he took up where he left off at Emmanuel and graduated in 1921 with a B.S. in chemistry. Norrish remained at Emmanuel to work on his doctorate in physical chemistry under Eric Rideal studying chemical kinetics and photochemistry. He earned his Ph.D. in 1924 and was subsequently appointed a fellow of the college, where he remained the rest of his career. The college appointed him a demonstrator in chemistry in 1925 and the Humphrey Owen Jones Lecturer in Physical Chemistry in 1930. He became a professor of physical chemistry in 1937, serving as the director of his department as well. Norrish retired in 1965, though he maintained his relationship with the college.

Norrish wrote articles on his work with reaction kinetics throughout his career; he published an early article from his collaboration with Porter, "Chemical Reactions Produced by Very High Light Intensities," in a 1949 edition of the British journal *Nature*. He also collaborated with Porter on a 1954 article, "The Application of Flash Techniques to the Study of Fast Reactions," in the journal *Discussions of the Faraday Society*. He continued writing through the year of his retirement, when he published "The Kinetics and Analysis of Very Fast Reactions" in the journal *Chemistry in Britain*. Norrish also distinguished himself by modifying Draper's law. In the 19th century John Draper had posited that photochemical changes could be expressed as a ratio to the light intensity times the length of time the reaction takes. Norrish tweaked this equation by demonstrating that the ratio should be to the square root of the light intensity.

Besides the Nobel Prize, Norrish received the Liverside Medal from the Chemical Society and the Davy Medal from the Royal Society, both in 1958, and in 1964 the

Combustion Institute awarded him the Bernard Lewis Gold Medal. Norrish died on June 7, 1978, in Cambridge.

Northrop, John Howard

(1891–1987)
American
Biochemist

John Howard Northrop shared the 1946 Nobel Prize in chemistry with JAMES BATCHELLER SUMNER and WENDELL MEREDITH STANLEY for their roles in establishing that enzymes are proteins. Northrop contributed to this effort by purifying and crystallizing the enzymes pepsin, trypsin, and chymotrypsin, as well as ribonuclease and deoxyribonuclease (related to ribonucleic acid [RNA] and deoxyribonucleic acid [DNA]).

Northrop was born on July 5, 1891, in Yonkers, New York. His mother, Alice Belle Rich, taught biology at Hunter College, and his father, John Isaiah, taught zoology at Columbia University, where he died in a laboratory fire before his son's birth. Northrop attended Columbia University throughout, earning his B.S. in biochemistry in 1912, his M.A. in 1913, and his Ph.D. in 1915. In June 1918 Northrop married Louise Walker, and together the couple had two children, Alice Havemeyer and John.

Upon receiving his doctorate in 1915, Northrop worked in the laboratory of Jacques Loeb conducting research on fruit flies. This appointment commenced a relationship with the Rockefeller Institute for Medical Research (RIMR) for what would be his entire career, as he became a professor emeritus some 45 years later, in 1961. In 1949 RIMR closed the facility where Northrop worked in Princeton, prompting him to accept a visiting professorship at the University of California at Berkeley, during which he retained his affiliation with RIMR.

During World War I the U.S. Army Chemical Warfare Service commissioned Northrop as a captain. He aided the war effort by developing a fermentation process for acetone, a key to explosives. During World War II Northrop worked with the U.S. Office of Scientific Research and Development to invent the Northrop titrator (in a portable version as well), a device that measured concentrations of mustard gas in the air at some distance from the source. This invention proved an important prototype for similar anti–chemical warfare devices used later.

In 1929 Northrop collaborated with M. L. Anson to develop the diffusion cell, which simplified the isolation process. In 1931 he worked with Moses Kunitz to study the purity of substances with the phase rule solubility method, testing for the homogeneity of dissolved substances. Northrop utilized this method to prove conclusively Sumner's assertion that enzymes are in fact proteins. Northrop reported this discovery in the 1939 text *Crystalline Enzymes,* which he wrote in conjunction with Kunitz and Roger Herriott.

The year prior to the publication of this important text Northrop used similar methods to isolate a bacteriophage, demonstrating that it too was a nucleoprotein. In 1941 Northrop used the process again to prepare a diphtheria antibody, the first crystalline version. He later collaborated with W. F. Gobel to prepare an antibody for pneumococcus.

Besides the Nobel Prize, Northrop won the W. B. Cutting Traveling Fellowship as well as the Stevens Prize of the College of Physicians and Surgeons of Columbia University. For 62 years Northrop served on the editorial board of the *Journal of General Physiology.* Northrop died on May 27, 1987, in Wickenberg, Arizona.

Novello, Antonia Coello

(1944–)
American
Physician

Antonia Novello was the first woman and the first Hispanic to become U.S. surgeon general. She was born in Fajardo, Puerto Rico, on August 23, 1944. Her father, Antonio Coello, died when she was eight years old. She and her brother were raised by their mother, Ana Delia Flores, a junior high school principal, and stepfather, Ramon Rosario, an electrician.

Tonita, as she was called as a child, was sick much of the time because of a birth defect. (It was finally corrected by surgery when she was 18.) Admiration for the doctors who cared for her, as well as determination to prevent other children from suffering as she had, encouraged her to become a physician. She earned her bachelor's degree in 1965 and her M.D. in 1970 from the University of Puerto Rico.

In 1970 Coello married a flight surgeon, Joseph Novello, and both continued their studies at the University of Michigan Medical Center in Ann Arbor. Antonia was named Intern of the Year by the center's department of pediatrics in 1971, the first woman to be so honored. From 1976 to 1978 she had a private pediatric practice in Virginia, but she quit because she found dealing with seriously ill children too painful. "When the pediatrician cries as much as the parents [of the children] do, . . . it's time to get out," she told *People* magazine.

In 1978 Novello joined the U.S. Public Health Service and the National Institutes of Health (NIH), the large conglomerate of federal research institutes in Bethesda, Maryland. She also trained in health services administration at the Johns Hopkins School of Public Health in Baltimore and earned a master's degree in 1982. She worked with Congress, for instance, in helping to draft the National

Organ Transplant Act of 1984. In 1986 she became deputy director of the National Institute of Child Health and Human Development, one of the institutes at NIH, and a professor of pediatrics at the Georgetown University School of Medicine.

Nominated by President George Bush, Novello became the country's 14th surgeon general on March 9, 1990. She said her motto would be "Good science and good sense." During her term in office she emphasized the health concerns of children and young people, women, and Hispanics, "speak[ing] up for the people who are not able to speak for themselves." She criticized companies that marketed cigarettes and alcohol to youths.

Novello stepped down from the surgeon general's job in 1993. Later she served as a special representative for the United Nations Educational, Scientific, and Cultural Fund (UNICEF) and as a professor at Johns Hopkins University Medical School.

O

Ocampo-Friedmann, Roseli

(1937–)
American
Botanist

Roseli Ocampo-Friedmann and her husband, Imre Friedmann, have found living microorganisms inside Antarctic rocks and under the permanently frozen ground of the Far North. She was born Roseli Ocampo in Manila, the Philippines, on November 23, 1937, to Eliseo A. and Generosa Ocampo.

Ocampo earned a B.S. in botany from the University of the Philippines in 1958. In 1963 she went to Hebrew University in Jerusalem to do graduate work; there she met the Hungarian-born Imre Friedmann, who two years earlier had discovered that microscopic algae and cyanobacteria (sometimes called blue-green algae) could live under the surface of rocks in inhospitable deserts. Ocampo grew and studied some of these microorganisms in the laboratory as her master's degree project and discovered that she was outstandingly successful at getting them to thrive in test tubes. She had what Friedmann called a "blue-green thumb."

Ocampo earned her master's degree in 1966, returned to the Philippines, and worked for a while for the National Institute of Science and Technology in Manila. In 1968 she joined Friedmann at Florida State University in Tallahassee, where she completed her Ph.D. in 1973. She married Friedmann in 1974.

As a scientist couple working together, Roseli and Imre Friedmann traveled to deserts all over the world, looking for algae and other microorganisms. Roseli Ocampo-Friedmann added new organisms each year to her growing collection of living cultures. In the mid-1970s Friedmann found microorganisms inside rocks from Antarctica, and after many weeks of work Ocampo-Friedmann succeeded in culturing them in the laboratory.

As the Friedmanns continued their Antarctic research, Imre went to Antarctica and Roseli remained in Tallahassee to make cultures from the rock samples he sent. After two years, however, they realized that chances of contamination and damage would be reduced if she traveled there to start cultures on the spot. Roseli therefore made five trips with Imre to the southern continent. Since then the two have extended their research to Siberia, where they have studied bacteria in the permanently frozen ground, or permafrost, that underlies the Arctic tundra. The Friedmanns' research attracted media attention in the late 1970s and again in 1996 because scientists believe that the microorganisms the couple have studied inside frozen rocks might be similar to those that could once have existed on Mars.

Roseli Ocampo-Friedmann continued to use her "blue-green thumb" in research with her husband at Florida State University in Tallahassee. In 1987 she became a full professor at Florida A&M University, teaching general biology and microbiology. She received a resolution of commendation from the state government of Florida in 1978 and the U.S. Congressional Antarctic Service Medal from the National Science Foundation in 1981. Her collection of cultures of microorganisms from extreme environments all over the world numbered close to 1,000 different types in the late 1990s.

Ochoa, Severo

(1905–1993)
Spanish/American
Biochemist

Severo Ochoa was the first scientist to synthesize ribonucleic acid (RNA), the first molecular chain created outside a living organism. For this accomplishment Ochoa received the 1959 Nobel Prize in physiology or medicine

Severo Ochoa, the first scientist to synthesize RNA. *Bernard Becker Medical Library, Washington University School of Medicine.*

with ARTHUR KORNBERG, who used Ochoa's technique to synthesize deoxyribonucleic acid (DNA).

Ochoa was born on September 24, 1905, in Luarca, Spain, the youngest son in his family. He was named after his father, a lawyer. On July 8, 1931, he married Carmen Garcia Cobain, daughter of a Spanish lawyer and businessman. He became a citizen of the United States in 1956.

Ochoa attended Málaga College, where he earned his B.A. in 1921. He entered the University of Madrid Medical School in 1923 and earned his M.D. in 1929. He then conducted postdoctoral research at the Kaiser Wilhelm Institute in Berlin under OTTO FRITZ MEYERHOF in biochemistry and the physiology of the muscle. Between 1931 and 1933 he conducted research at the National Institute for Medical Research in London with Henry Hallett Dale under a University of Madrid fellowship.

In 1933 Ochoa returned to the University of Madrid as a lecturer in physiology and biochemistry. When the university established its new Institute for Medical Research, it appointed Ochoa as the director of the physiology department. However, the Spanish Civil War intervened in 1936, sending him back to Germany to research under Meyerhof; then to Plymouth, England, where he filled a six-month fellowship at the Marine Biological Laboratory; after which he served as a research assistant at the University of Oxford. World War II then intervened, and as a foreigner in England he could not participate in the scientific efforts that were devoted exclusively to the war, so he left for the United States in 1940.

There Ochoa teamed up with Carl Cori and GERTY THERESA RADNITZ CORI at Washington University in St. Louis. Then in 1941 New York University prevailed upon him to take up a position as research associate in medicine at Bellevue Psychiatric Hospital. The New York University Medical School appointed him as an assistant professor of biochemistry in 1945 and the next year made him chairman of the department of pharmacology, a position he held until 1954 except for a one-year sabbatical which he spent as a visiting professor at the University of California. In 1954 he became chairman of the department of biochemistry at NYU, a position he held for the next two decades.

In 1955 Ochoa discovered an enzyme in sewage that he called polynucleotide phosphorylase. He used this enzyme to synthesize RNA. In fact, he later discovered, the enzyme does not synthesize but rather degrades RNA, although laboratory conditions reverse this process, so that the enzyme does in this instance generate RNA. Ochoa later demonstrated the role of adenosine triphospate (ATP) in the storage and release of energy.

In 1974 Ochoa retired from NYU but took up a position at the Roche Institute of Molecular Biology in New Jersey. In 1985 he returned to his homeland as a professor of biology at the University Autonoma in Madrid. Besides the Nobel Prize, he received the 1951 Carl Neuberg Medal in biochemistry and the 1955 Charles Leopold Mayer prize. Ochoa died on November 1, 1993, in Madrid.

Ohm, Georg Simon

(1789–1854)
German
Physicist

Georg Simon Ohm, the importance of whose work was not recognized until late in his life, formulated the basic law of electric current flow. He postulated a theory of electricity, now known as Ohm's law, that holds that the amount of current flowing in a circuit made up of pure resistances is directly proportional to the total resistance of the circuit. He profoundly influenced the future study of electricity and was also an early champion of applying a mathematical approach to physics.

Georg Simon Ohm, who developed the basic law of electric current flow. *AIP Emilio Segrè Visual Archives, E. Scott Barr Collection.*

Born on March 16, 1789, in Erlangen, Bavaria, Ohm was the eldest son of Johann Wolfgang and Maria Elisabeth Beck Ohm. Of his parents' seven children only three (including Ohm) survived into adulthood. Ohm's mother was the daughter of a tailor, and his father, who exerted a considerable influence on Ohm, was a master locksmith. Johann Ohm was a self-taught man, who tutored his children in mathematics, physics, chemistry, and philosophy.

Ohm attended the Erlangen Gymnasium from 1800 to 1805, then enrolled at the University of Erlangen. After three semesters at the university, however, his father forced him to withdraw, after learning that his son participated in dancing and billiards. Ohm then settled in rural Switzerland, where from 1806 to 1809 he taught mathematics at a school in Bern canton. From 1809 until 1811 he served as a private tutor in Neuchâtel. In the spring of 1811 Ohm returned to the University of Erlangen and was awarded his Ph.D. after passing the required exams that fall.

Although Ohm dearly wanted a post at a university, he proceeded from Erlangen to Bamberg, where he worked as a mathematics instructor at a poorly funded *Realschule* until 1816. In 1817 Ohm was offered a teaching position at the Jesuit Gymnasium in Cologne, where he remained until 1827. While in Cologne he began to apply himself avidly to the study of physics and was inspired by the work of Fourier and HANS CHRISTIAN ØRSTED, who discovered electromagnetism in 1820. Ohm embarked on his own research into electricity and magnetism in 1825. After publishing a few experimental papers, he wrote his famous work, *The Galvanic Current Investigated Mathematically,* in 1827. In this text he elaborated his theory of electricity, in which he clearly established the relationships among current, electromotive force, and resistance. He discovered, in what is now known as Ohm's law, that the current in an electrical circuit is proportional to the voltage. Specifically he showed that the amount of current is directly proportional to the total resistance of the circuit. Ohm's law is frequently expressed by the formula $I = E/R$, where I equals the current in amperes, E is the electromotive force, and R is the resistance (the property of an object of opposing the flow of electrical current).

Despite his scientific breakthrough, Ohm was unable to find a faculty position at a university. He left Cologne in 1827 and became a professor of physics at the Polytechnic in Nuremberg. Eventually he received belated recognition for his work. He was awarded the Royal Society's Copley Medal in 1841 and was made a member of the Bavarian Academy in 1845. He became a full professor at the University of Munich in 1848 and was given the chair of the physics department there in 1852. Ohm never married. He died in Munich on July 6, 1854. In his honor the basic unit of electrical resistance was named the *ohm.*

Olden, Kenneth

(1938–)
American
Cellular Biologist, Biochemist

Kenneth Olden has devoted more than two decades to the study of possible links between the properties of cell-surface molecules and cancer. He is recognized as a leading authority on the structure and function of the extracellular matrix glycoprotein fibronectin—one of a family of proteins that play a role in the interactions between cells and the supporting structure around them. As a result of his numerous discoveries and his considerable experience overseeing research teams, Olden was appointed the director of the National Institute of Environmental Health Sciences and the National Toxicology Program at the National Institutes of Health in 1991. In this capacity he has investigated swainsonine, an anticancer drug.

On July 22, 1938, Olden was born in Parrottsville, Tennessee. His parents, Mack and Augusta Christmas Olden, supported his ambition to pursue a career in science. Olden married Sandra White, with whom he has four children—Rosalind, Kenneth, Stephen, and Heather.

After attending Knoxville College, Olden received his bachelor's degree in biology in 1960. He earned a master of science degree in genetics from the University of Michigan in 1964, and a doctorate in biology and biochemistry from Temple University in 1970. While still in graduate school Olden worked as a research assistant at Columbia University's Department of Biological Chemistry, and as a biology instructor at the Fashion Institute of Technology.

Olden's first appointment was as a research fellow at Harvard University Medical School in 1970. After the fellowship period concluded in 1973, he remained at Harvard for another year as a physiology instructor. In 1974 he was named a senior staff fellow at the Laboratory of Molecular Biology, Division of Cancer Biology and Diagnosis, of the National Cancer Institute at the National Institutes of Health (NIH). Olden advanced to the position of Expert in Biochemistry in 1977 and was promoted to research biologist the following year. A pair of articles he published during this period, "The Role of Carbohydrates in Protein Secretion and Turnover" and "Fibronectin Adhesion Glycoprotein of Cell Surface and Blood," were among the 100 most cited papers of 1978 and 1979. Olden solidified his reputation with his insightful investigations into extracellular matrix glycoprotein fibronectin, which is believed to be related to the spread of cancer. Olden was the first to demonstrate that the metastasis of some malignant cells can be prevented if the interaction between the fibronectin and the glycoprotein receptor around the cell is blocked.

Olden left the NIH in 1979, to become a professor of oncology at the Howard University Cancer Center. After becoming the center's deputy director in 1982, he was named its director of research in 1984. He also became chairman of Howard's oncology department in 1985. In 1991 he embarked on a new path once more, when he returned to the NIH as the director of the National Institute of Environmental Health Sciences and the National Toxicology Program. This appointment was particularly significant because Olden was the first African American to serve as a director of one of the NIH's 17 institutes. He has made the study of swainsonine a top priority.

In addition to being recognized by his peers as a key contributor to the field of oncology, Olden was appointed to the National Cancer Advisory Board by President Bush in 1991. The author or coauthor of over 108 publications, Olden has been a member of the American Society of Cell Biology, the American Society of Biological Chemistry, and the American Association of Cancer Research, where he has served on the board of directors.

Oort, Jan Hendrik

(1900–1992)
Dutch
Astronomer

Although Jan Hendrik Oort is best known for his theory on comets, he devoted most of his career to pioneering research into the structure and dynamics of the Milky Way galaxy. Using mathematical calculations, he showed that the outer regions of the galaxy actually rotated more slowly than the inner parts, a finding contrary to the prevailing theory that the galaxy rotated uniformly, as a wheel does. He also established that the Earth's solar system was not at the center of the Milky Way, as was then believed, but at the galaxy's outer edges. In the 1950s Oort helped develop radio astronomy, which revolutionized the study of the Milky Way's structure. In his most publicized work Oort postulated the existence of what came to be known as the Oort cloud, a mass of icy objects surrounding the solar system. He proposed that comets were pieces detached from this cloud.

Oort was born on April 28, 1900, in Franeker, a small town in the Netherlands. His parents, Ruth Faber Oort and Abraham Oort, a physician, had five children. When Oort was three, his family moved from Franeker to Wassenaar. He eventually married, and he and his wife, Mieke, had three children of their own.

After graduation from the gymnasium in Leiden in 1917 Oort matriculated at the University of Groningen, where he studied under the renowned astronomer Jacobus Kapteyn, the first scientist to measure and compute the position of stars in the Milky Way. In 1920 Oort's astronomical gifts were recognized when he was awarded the Bachiene Foundation Prize for a paper he wrote about stars of the spectral types F, G, K, and M. Upon his graduation from Groningen in 1922 Oort served as a research assistant at the Yale University observatory. At Yale he was influenced by the work of the Harvard astronomer HARLOW SHAPLEY, who demonstrated that the size of the galaxy was larger than previously supposed. Oort returned to the Netherlands in 1924 and received his Ph.D. in 1926 from the University of Groningen.

In 1926 Oort became an instructor at the University of Leiden, where he remained for the rest of his career. In 1927 he investigated the theories of the astronomer Bertil Lindblad and affirmed that the Milky Way did indeed rotate. Oort employed complex mathematical equations to prove that the galaxy exhibited differential rotation, whereby stars near the center of the Milky Way moved faster than those farther away. These calculations allowed him to deduce the mass and size of the galaxy, which he measured to be 100,000 light years across and 20,000 light years deep. Even more importantly, Oort demonstrated that our solar system was not at the center of the galaxy, but rather lingered on the outside, some 30,000 light years from the center. This insight forever changed scientists' conception of the Earth in space. After being appointed a professor of astronomy at Leiden, Oort postulated the existence of dark matter (or missing matter), mass that could be detected neither visually nor by contemporary calculations.

After World War II, during which Oort had been forced into hiding, he was appointed director of the Leiden Observatory. In the 1950s he collaborated with Hendrik van de Hulst in the development of radio astronomy, which offered a degree of precision in studying the galaxy, because radio waves, unlike visible light, are unaffected by dust or gas. In 1951 the pair discovered the 21-centimeter hydrogen line that cuts through the Milky Way. They were able to use this information to confirm that the galaxy completes a full rotation once every 225 million years. Oort also proposed the existence of a cloud of icy objects that surrounds the Sun, accounting for the appearance of comets. This phenomenon was later named the *Oort cloud* in his honor.

During his career Oort headed several astronomical groups, including the International Astronomical Union from 1958 to 1961. He cofounded the European Southern Observatory in 1962. Although he retired in 1970, Oort continued to research and publish papers on the Milky Way and other topics in astronomy. He died at the age of 92 on November 5, 1992.

Oppenheimer, J. Robert

(1904–1967)
American
Physicist

J. Robert Oppenheimer exerted a profound influence over both scientific and political developments. He made breakthrough findings in the field of quantum mechanics, the study of the energy of atomic particles. He discovered, for example, the positron. He also presided over the laboratory in Los Alamos, New Mexico, where the first atomic bomb was created. Despite his best efforts to influence U.S. postwar policy to limit the proliferation of nuclear weapons, Oppenheimer was unable to do so. His vocal opposition to the production of the hydrogen bomb led critics to accuse him of disloyalty.

Oppenheimer was born on April 22, 1904, in New York City. His father, Julius Oppenheimer, was an emigrant from Germany who became a successful businessman in the textile industry. His mother, Ella Friedman Oppenheimer, was a painter. He married Katherine Harrison in 1940. Oppenheimer's considerable intellectual abilities were apparent from an early age. At 11 he became the youngest person ever admitted to the New York Mineralogical Society.

J. Robert Oppenheimer, who presided over the laboratory in Los Alamos, New Mexico, where the first atomic bomb was created. He is shown here at home with his children. *Mrs. J. Robert Oppenheimer, courtesy AIP Emilio Segrè Visual Archives.*

After entering Harvard College in 1922, Oppenheimer graduated summa cum laude in only three years. In 1925 he went to the Cavendish Laboratory at Cambridge. In 1926 Oppenheimer left Cavendish for the University of Göttingen in Germany, where he worked with the physicist MAX BORN, studying variations in the vibration, rotation, and electronic properties of molecules. The duo conceived the Born-Oppenheimer method, which was essentially a quantum mechanics at the molecular, rather than the atomic, level. Oppenheimer received his doctorate from Göttingen in 1927.

From 1929 to 1942 Oppenheimer taught and conducted research at both the University of California at Berkeley and the California Institute of Technology; he also continued his work in particle physics. In 1930 he disproved conventional wisdom by demonstrating that the proton was the antimatter equivalent of the electron. Five years later he determined that it was possible to accelerate deuterons (consisting of a proton and a neutron) to much higher energies than neutrons alone. As a result, deuterons could be used to bombard atomic nuclei at high energies, allowing further research into atomic particles.

After ALBERT EINSTEIN warned of Nazi efforts to create an atomic weapon, the U.S. Army marshaled British and U.S. scientists to harness the awesome power of nuclear energy for military purposes. In 1942 Oppenheimer was chosen to direct this so-called Manhattan project. He assembled a group of theoretical physicists at the project's remote site in Los Alamos, New Mexico. The group succeeded in producing a fissionable form of plutonium, and on July 16, 1945, Oppenheimer and his colleagues observed the detonation of the first atomic bomb in Alamogordo, New Mexico.

After the war Oppenheimer expressed doubts about his role in the formation of so destructive a weapon. He cowrote the *Acheson-Lilienthal Report,* which argued (in vain) against the international proliferation of nuclear weapons. From 1946 to 1952 he served as the chairman of the General Advisory Committee of the Atomic Energy Commission. In this capacity he argued against the development of a hydrogen bomb and urged restraint in the nascent arms race. At the same time he presided over the Atomic Energy Commission, Oppenheimer was appointed director of the Institute for Advanced Studies at Princeton University. He occupied this position from 1947 until his retirement in 1966. At the institute he influenced the path physics would take in the future through his interactions with students and academics alike.

Oppenheimer's efforts came under attack in 1953, when the Eisenhower administration withdrew his security clearance, thereby effectively relieving him of his post at the Atomic Energy Commission. In an era dominated by the virulent politics and frenzied anticommunism of Senator Joe McCarthy, Oppenheimer's past associations with Left-leaning ideologues and the fact that he was not an enthusiastic supporter of the hydrogen bomb program were considered liabilities. Although a panel affirmed that he was a "loyal citizen," Oppenheimer did indeed lose his clearance and was forced to leave the Atomic Energy Commission. When the shameful years of the McCarthy era ended, Oppenheimer's many accomplishments were at last recognized. In 1963 President Lyndon Johnson gave Oppenheimer the Enrico Fermi Award. He died of throat cancer in 1967.

Ørsted, Hans Christian

(1777–1851)
Danish
Physicist, Chemist

Hans Christian Ørsted discovered the inherent relationship between electricity and magnetism when a wire carrying an electrical current deflected a magnetic compass needle. This discovery led to the development of electromagnetic theory.

Ørsted was born on August 14, 1777, in Rudkøbing, Denmark. He was the eldest son of Karen Hermansen and Søren Christian Ørsted, an apothecary who sent him and his brother, Anders Sandøe, to Germany when they were youngsters. They received little formal education there, but both passed the University of Copenhagen's entrance examination with honors. In 1797 Ørsted received his pharmaceutical degree with high honors, and in 1799 he received his Ph.D. His dissertation defended the philosphy of Immanuel Kant, which formed the foundation of his scientific beliefs. Although subsequent Kantian philosophers contended that Ørsted certainly misunderstood Kant's meaning, this influence on Ørsted's thinking was significant because many other contemporary scientists were misreading Kant in the exact same way at the time. Ørsted and others believed that Kant was saying that scientific law was more a perception of human reason than an immutable system. What lent scientific law its power was its discovery by human reason, which coincided with Divine Reason.

During the summer of 1801, after managing a pharmacy, Ørsted embarked on a journey through Europe that solidified his scientific knowledge. He learned from Johann Ritter in Göttingen and from Henrik Steffens and Franz von Baader in Berlin. His sojourn was not completely successful, as he championed the chemical theories of Ritter and J. J. Winterl in Paris on the strength of results that the two men had not substantiated experimentally. This experience hardened Ørsted, reinforcing the importance of empirical evidence (as opposed to philosophical belief) for scientific conclusions.

Upon Ørsted's return to Denmark in 1804 he did not receive the nomination for a professorship at the University of Copenhagen as he had hoped, but he earned this appointment by 1806 on the strength of his popular scientific lectures to the public. In 1829 he was appointed the director of the Polytechnic Institute in Copenhagen, a position he held the rest of his life.

In April 1820 Ørsted demonstrated during a lecture the disruption of a magnetic needle by an electrical current flowing through a wire that he passed beneath the compass. Though this effect had been observed in 1802 by Gian Domenico Romagnosi, it was not recognized until Ørsted published his finding in a paper, dated July 21, 1820, that redirected science to the study of electromagnetism. That year Ørsted also discovered piperine, the component that gave pepper its "heat." In 1822 he calculated the first accurate value for the compressibility of water, and in 1825 he prepared metallic aluminum.

Ørsted acted as a great popularizer of science, and he worked hard to raise the status of Danish science to the level of scientific thought throughout Europe at the time. In 1824 he founded the Society for the Promotion of Natural Science; since 1908 this organization has awarded the Ørsted Medal for furthering the physical sciences in Denmark. Ørsted died on March 9, 1851, in Copenhagen. In 1932 the physical unit of magnetic field strength was dubbed the *ørsted* in his honor.

P

Pappus of Alexandria

(c. A.D. 320)
Greek
Mathematician

Pappus of Alexandria's lasting accomplishment was his collection of eight books, the *Synagoge* (or Collection). This work, which provided a systematic account of the seminal works in ancient Greek mathematics, was apparently written as a handbook to be read with the original mathematical sources it discussed—such as writings by EUCLID and CLAUDIUS PTOLEMAEUS. The *Synagoge* incorporated texts that would otherwise have been totally lost to modern scholars, including Euclid's *Porisms*. A question posed in Pappus' *Synagoge* inspired RENÉ DU PERRON DESCARTES, Sir ISAAC NEWTON, and BLAISE PASCAL. In Book VII of the *Synagoge* Pappus presented a problem regarding the fact that for any hexagon inscribed in a conic the intersections of opposite pairs of sides are collinear. Moreover, one of Pappus' own theorems is recognized today as the foundation for modern projective geometry.

Almost no information about Pappus' birth, family, education, or life is known. The estimate that he lived around A.D. 320 was gleaned from a reference to a solar eclipse that he made in his commentary on Claudius Ptolemaeus. The eclipse was determined to have occurred in A.D. 320. Because Pappus dedicated the seventh and eighth books of the *Synagoge* to his son Hermodorous, scholars know that he had at least one child.

Pappus' vast *Synagoge* not only detailed almost the entire field of Greek geometry up to that point (complete with historical annotations), but also suggested ways to improve upon some of the existing theorems it cited. Of the original eight books of the *Synagoge* the first and part of the second are lost. All current translations of the text are derived from a single 12th-century manuscript, *Codex Vaticanus Graecus 218.*

Book I apparently dealt with arithmetic. The surviving fragment of Book II proposed a system of continued multiplication paired with the expression of large numbers in terms of powers above 10,000 (tetrads). Book III explored plane and solid geometry. In it Pappus gave an original answer to the problem of finding two mean proportionals between two given lines. He also provided a general theory of means, in which he differentiated among 10 different kinds. Book IV covered various theorems relating to a circle that circumscribes three given circles tangent to one another. Pappus also presented such topics as the spiral of ARCHIMEDES, the quadratrix of Hippias, and the conchoid of Nicomedes. In addition, Book IV dealt with the construction of a curve of double curvature (Pappus called this the *helix on a sphere*). Book V discussed areas of different plane figures and the volumes of different solids. Book VI detailed problems of geometry and astronomy that were originally presented by Euclid, ARISTARCHUS OF SAMOS, and others. Book VII not only listed all the works of Euclid, Apollonius of Perga, and others, but also distinguished between theorems and problems. Book VIII was concerned with mechanics.

Pappus also wrote commentaries on books by Claudius Ptolemaeus, including *Almagest, Planisphaerium,* and *Harmonia;* on *Elements* by Euclid; and on the *Analemma* by Diodorus. Moreover, his own ideas exerted a potent impact on future mathematicians.

Pascal, Blaise

(1623–1662)
French
Mathematician, Physicist, Philosopher

As one of the eminent scientific minds of his era, Blaise Pascal made profound contributions to mathematics and physics. In the field of hydrostatics (the laws of fluids at rest) Pascal formulated what came to be known as Pascal's law, which established that fluids transmit pressure equally in all directions. He also proved experimentally that increases or decreases in atmospheric pressure affected the level of a mercury column in a barometer. With Pierre de Fermat, Pascal devised the mathematical theory of probability. His later work on the "theory of indivisibles" proved to be a precursor to the methods of integral calculus. He used this theory to analyze problems involving infinitesimals. A profoundly religious man, Pascal retired from scientific life and wrote *Thoughts on Religion and Other Subjects,* which was published posthumously.

Pascal was born in France on June 19, 1623. His mother, Antoinette Bergon, died three years after his birth; his father, Étienne, devoted a great deal of attention to Pascal's upbringing. The elder Pascal, who was both a respected mathematician and a tax officer, took complete charge of his son's education.

Pascal never received formal schooling at any level. However, after his family moved to Paris in 1631, he came into contact with prominent intellectuals and scientists. The young Pascal showed early scientific gifts. He read EUCLID's *Elements* when he was 11 years old. At the age of 16, he made his first great discovery when he formulated one of the basic theorems of projective geometry. He described this theorem, now known as Pascal's theorem, in his 1639 *Essay on Conics.* He followed his father back to provincial France in 1640, when the family settled in Rouen, and remained there for the next seven years. Pascal did not remain idle in this relative solitude, though. Inspired by a desire to help his father with the time-consuming mathematical calculations that were part of his job, Pascal invented the first mechanical calculation machine in 1642. Upon his return to Paris in 1648 Pascal provided the first purely geometric solution to a famed problem posed by PAPPUS OF ALEXANDRIA in the fourth century.

Pascal next occupied himself with experiments in the field of hydrostatics (fluid statics). In 1646 he verified by experimentation TORRICELLI OF SALERNO's hypothesis that air pressure decreased with altitude. To do so, Pascal used barometers to measure air pressure at various altitudes—including in Paris and at the summit of the peak of Puy de Dôme. In 1647 he developed Pascal's law, which posited that pressure applied to a confined fluid is transmitted equally in all directions and to all parts of the enclosing vessel.

Pascal collaborated with Pierre de Fermat in 1654 to conceive the calculus of probabilities. Pascal was the first to apply the principle of the arithmetical triangle to combinatorial analysis and the problem of stakes. He published his finding in the 1654 *Traité du triangle arithmétique.* In 1659 he focused on his "theory of indivisibles," which was a forerunner of integral calculus. Pascal used his theory to solve problems involving infinitesimals, those variable nonzero numbers that are smaller in absolute value than any preassigned nonzero number. He also worked on the calculations of areas and volumes and the determination of centers of gravity.

Pascal had entered a Jansenist community at Port Royal in 1654, and he remained there until his death in 1662. The Jansenists practiced a rigorous form of Catholicism, and in this new environment Pascal devoted himself mostly to philosophical and spiritual writings. His famous treatise on religious matters, *Thoughts on Religion and Other Subjects,* was published posthumously in 1670. Pascal left a considerable legacy. He carried out elemental work in projective geometry, integral calculus, and the calculus of probability, and his rigorous adherence to empirical experimentation also did much to advance scientific methodology.

Pasteur, Louis

(1822–1895)
French
Chemist, Microbiologist

Although Louis Pasteur was educated as a chemist, his work had a tremendous impact on medicine. To combat the souring of alcohol or dairy products caused by microorganisms, he invented the technique (now called pasteurization) of heating the beverage, thereby killing any bacteria. Pasteur also disproved the erroneous theory of spontaneous generation, which held that substances could produce germs of their own accord. Pasteur showed that germs instead were introduced into a substance from the outside. His research on fermentation and spontaneous generation led him to propose the germ theory of diseases, which laid the foundation for the field of microbiology. Later in his life Pasteur developed vaccines for chicken cholera, anthrax, and rabies.

Born on December 27, 1822, Pasteur was the only son of Jean-Joseph and Jean-Étiennette Pasteur. When he was four his family moved from his birthplace of Dôle, France, to the town of Arbois, where he spent the rest of his childhood.

Pasteur attended the Collège d'Arbois and the Collège Royal de Besancon from 1831 until 1840. In preparing for his entrance exams to the École Normale in Paris he took classes at the Sorbonne with the famed chemist Jean-Baptiste Dumas. He entered the École Normale in 1843; there he received his doctorate with a focus on physics and chemistry in 1847.

Pasteur became a professor of chemistry at Strasbourg in 1849. His early work with *d*-tartrate crystals ultimately led to the discovery of molecular asymmetry. In 1854 he took a faculty position at Lille, where he began to explore the fermentation process that occurs in alcohol and milk. His endeavors in this field had the practical outcome of helping to preserve the French beer- and wine-making industry. Moreover his research demonstrated that the production of alcohol during fermentation was caused by yeast and that souring was due to the presence of additional organisms (namely, bacteria) in the beverage. He found that when alcohol—and likewise milk—was heated, this undesirable souring would not take place because the higher temperatures destroyed the bacteria. In 1857 Pasteur was named the director of the École Normale in Paris.

Pasteur applied the results of his fermentation work to the hotly contested topic of spontaneous generation. In 1862 he published his renowned text *Note on Organized Corpuscles That Exist in the Atmosphere,* which demonstrated that germs can be introduced to a substance only from an external source. He derived his conclusion from experiments involving a sterilized, fermentable fluid kept in two different kinds of vessels. One container, a swan-necked flask, did not allow dust or microorganisms to enter; the other container did. The fluid in the swan-necked flask remained free of germs, whereas its counterpart was quickly contaminated. These experiments led him to propose his germ theory of disease: Illness arises as a result of germs that attack the body from outside.

After working at the Sorbonne from 1867 to 1874, Pasteur returned to the École Normale, where he undertook pioneering research into human diseases. In 1880 he inadvertently found a vaccine for chicken cholera. He injected chickens with cholera bacilli that had been accidentally left out for some time. Not only did the chickens not become ill; they were also able to withstand exposure to fresh cholera bacilli. By 1882 he had developed a vaccine against anthrax, which he publicly and successfully tested at Pouilly-le-Fort. In 1885 Pasteur used his newly created rabies vaccine to save a nine-year-old boy, Joseph Meister, who was badly infected. Like his chicken cholera vaccine, Pasteur's rabies vaccine inoculated against infection and even treated the disease after it had been contracted. In eradicating the threat of rabies, Pasteur became a national hero. The French government funded the founding of the Pasteur Institute, and Pasteur was made its director.

Pasteur received countless honors and awards for his work. In 1881 he was given the Grand Cross of the Legion of Honor, and he was elected to the French Academy in 1882. Pasteur's contributions to the field of medicine were profound and lasting: His rabies vaccines motivated the next generation to seek other vaccines to prevent other diseases, and his germ theory paved the way for Lister's advances in antiseptic surgery and for other systematic research to combat infectious diseases.

Patrick, Ruth

(1907–)
American
Botanist, Ecologist

Ruth Patrick studied the effects of pollution on freshwater ecology and invented a sensitive tool for evaluating water pollution. She was born on November 26, 1907, in Topeka, Kansas, but grew up in Kansas City, Missouri. Her father, Frank Patrick, was a lawyer, whose hobby was studying diatoms, one-celled algae (plantlike living things) that are the base of the food chain in fresh water. When Ruth was good, he let her look through his microscope.

Patrick attended Coker College in South Carolina and graduated in 1929. During a summer at Cold Spring Harbor Laboratory in New York she met a fellow biology student, Charles Hodge IV. They married in 1931, but Patrick continued to use her maiden name professionally. They had one son. Patrick earned a master's degree in 1931 and a Ph.D. in botany in 1934, both from the University of Virginia.

Patrick has spent most of her professional life at the Academy of Natural Sciences in Philadelphia, where she worked part-time in 1937 and full-time in 1945. She was the chairperson of its board of trustees from 1973 to 1976 and is now its honorary chairperson. In 1947 she established a department of limnology (freshwater ecology) at the academy, now called the Environmental Research Division, and directed it until 1973; she is still its curator. She has also been an adjunct professor at the University of Pennsylvania since 1970.

Patrick's first favorite study subject, like her father's, was diatoms. With Charles Reimer she produced a monumental two-part work on the subject, *Diatoms of the United States,* published in 1966. She then expanded her research to include general ecology and biodiversity in rivers. She is considered the cofounder of the discipline of limnology.

Patrick was concerned with pollution's effect on ecology long before RACHEL LOUISE CARSON made the issue fashionable. Patrick's research showed that diatoms are sensitive indicators of pollution in fresh water, and she invented the diatometer, which determines the presence and kind of water pollution by measuring numbers of different species of diatoms. Patrick was also a pioneer in pointing out that scientists needed to study ecological communities, not just single species, to determine the effects of pollution.

Unlike Carson, Patrick has had a relatively cordial relationship with industry, for which she has often worked as a consultant. She has even been a director of Pennsylvania Power and Light Company and of du Pont—the first woman and the first environmentalist ever on the board of the latter. "We have to develop an atmosphere where the industrialist trusts the scientist and the scientist trusts the industrialist," she has said.

Patrick was elected to the National Academy of Sciences in 1970 and was president of the American Society of Naturalists from 1975 to 1977. She is also a member of the American Academy of Arts and Sciences. In 1975 Patrick won the John and Alice Tyler Ecology Award, the highest-paying award in science (it outpays even the Nobel Prize), and used the money to prepare the multivolume *Rivers of the United States,* first published in 1994. In 1995 she married Lewis H. Van Dusen, Jr. Her recent awards include the Benjamin Franklin Award from the American Philosophical Society (1993), the American Society of Limnology and Oceanography's Lifetime Achievement Award, and the National Medal of Science, the latter both received in 1996. She has continued to study the diverse life in rivers and to develop ways to monitor its health.

Patterson, Francine ("Penny")

(1947–)
American
Psychologist

By teaching American Sign Language to Koko and Michael, captive-born lowland gorillas, Francine Patterson says that she, like other scientists who have done similar work with chimpanzees, has "helped to dismantle an intellectual barrier long thought to separate humans from the great apes: the ability to use language."

Francine, or Penny as she is always called, was born in Chicago in 1947, one of four boys and three girls. Her father, C. H. Patterson, was a professor of educational psychology at the University of Illinois, so it was not surprising that she, too, attended that university and majored in psychology. She earned her B.A. in 1970.

While doing graduate work at Stanford University, Patterson attended a lecture by Allen and Beatrix Gardner, who had taught American Sign Language (Ameslan or ASL) to a chimpanzee named Washoe. Apes cannot make most sounds in human languages, but they can learn the gestures that make up ASL, which is used by the hearing-impaired. Washoe learned a number of Ameslan "words" and constructed meaningful sentences with them. Patterson decided to teach ASL to a gorilla as her doctoral project.

In 1972, with the permission of the San Francisco Zoo, Patterson began teaching ASL to Koko, a year-old female gorilla that had been born at the zoo. "As . . . Koko began to use words that revealed her personality, I . . . recognize[d] sensitiveness, strategies, humor, the stubbornness with which I could identify," Patterson wrote. "The realization that I was dealing with an intelligent and sensitive intellectual . . . sealed my commitment to Koko's future."

Ongoing conflicts with zoo officials led Patterson and her partner, Ronald Cohn, to establish a nonprofit organization called the Gorilla Foundation in 1976 and launch a public appeal for donations to buy Koko from the zoo so that their work could continue. With the help of individuals and the National Geographic Society they acquired not only Koko but a second gorilla, a male they named Michael. He, too, has learned ASL.

In 1979, after Patterson completed her Ph.D. in developmental psychology at Stanford, she and Cohn moved Koko and Michael to the nearby community of Woodside, where they still live. They added a third gorilla, a male named Ndume, to the group in 1991. Ndume, on loan from the Cincinnati Zoo, is a potential mate for Koko but is not part of the ASL project. The Gorilla Foundation, meanwhile, expanded to support not only Patterson's ongoing work but "improved care and welfare of captive gorillas . . . [and] efforts to . . . preserve gorillas in their natural habitat."

Patterson's awards include the Rolex Award for Enterprise (1978), the PAWS Award for Outstanding Professional Service (1986), and the Kilby International Award (1997). In the late 1990s she was an adjunct professor of psychology at the University of Santa Clara as well as the president and research director of the Gorilla Foundation. She said that her 26 years of research "shed light on human origins" and "have yielded . . . remarkable access to the gorilla mind."

Pauli, Wolfgang

(1900–1958)
Austrian/Swiss
Physicist

Wolfgang Pauli made significant contributions to the fledgling science of quantum mechanics, the theory of physics that uses the concept of the quantum (a fixed elemental unit of energy) to describe the dynamic properties of subatomic particles. As a student he wrote the article "Theory of Relativity," which summarized and analyzed developments in the emerging field of relativity theory. More importantly, he formulated what came to be known as the Pauli exclusion principle, which established that only two electrons can simultaneously occupy the same energy level in an atom. As part of this theory he introduced a new quantum number, *s*, in addition to the three quantum numbers (*n*, *l*, and *m*) that NIELS HENDRIK DAVID BOHR had established. This exclusion principle explained many aspects of atomic behavior that had previously baffled scientists. His second important achievement was to solve a vexing problem about beta decay. When electrons were emitted by atomic nuclei, they appeared as "missing" energy, contravening the conservation of energy principle. Pauli proposed that this energy did not vanish but was carried off by an undetected neutral subatomic particle. ENRICO FERMI named this particle the *neutrino,* and its existence was proved in 1956.

Pauli was born in Vienna on April 25, 1900. He inherited his scientific acumen from his father, Wolfgang

Joseph Pauli, who was a medical doctor and biochemist, and eventually became a professor at the University of Vienna. His mother, Bertha Schülz Pauli, was an author. The famed physicist ERNST MACH was his godfather. As a gifted student often bored by his subjects in school, the young Pauli took to reading treatises on modern physics during class.

After graduation from high school in 1918 Pauli enrolled in the University of Munich. After only two years there a professor encouraged him to write a paper on the subject of relativity. The 250-page document Pauli presented was praised by ALBERT EINSTEIN for its clarity and scope. Eventually the paper was published as a small book, *Theory of Relativity,* in 1921. Pauli received his Ph.D. from the university in 1922 (university policy had prevented him from completing it sooner) for a thesis on the hydrogen molecule ion. He was 22 years old. He married twice—to Kate Depner and then to Franciska Bertram—but never had children.

Pauli taught at the universities of Göttingen (1921–1922), Copenhagen (1922–1923), and Hamburg (1923–1928) before settling at the Federal Institute of Technology in Zurich for virtually the remainder of his career. He made the acquaintance of Niels Bohr while at Göttingen. Bohr had postulated the accepted model of the atom—a nucleus surrounded by rings of charged electrons in different energy levels—but the laws of classical mechanics could not explain why the electrons were distributed in these different energy levels. This issue piqued Pauli's interest. In 1925 he presented his exclusion principle, which proposed the existence of a fourth quantum number in addition to the three already known. Pauli asserted that only with this fourth quantum number could the electron's energy state be understood. Pauli's exclusion principle maintained that no two electrons in an atom could have the same four quantum numbers, or, in other words, occupy the same energy level. This principle explained the existence of the different electronic shells orbiting the nucleus.

In 1931 Pauli hypothesized the existence of the neutrino, the small neutral particle that accounted for the energy discrepancy when an electron was ejected from an atomic nucleus. When the nucleus lost an electron, he theorized, it also lost a neutrino, which effectively "carried off" some of the energy. Pauli's concept was proved correct in 1956, when the neutrino was first observed.

Except for three stints as a visiting professor at the Institute for Advanced Studies at Princeton University, Pauli remained at the Federal Institute of Technology. His work on the exclusion principle was finally awarded a Nobel Prize in 1945. His contributions to the emerging field of quantum mechanics were also recognized with the Max Planck Medal of the German Physical Society in 1958 and the Lorentz Medal of the Royal Dutch Academy of Sciences in 1930. He died suddenly on December 15, 1958, in Zurich.

Pauling, Linus Carl

(1901–1994)
American
Chemist

Linus Pauling's most significant scientific achievement, for which he was awarded the Nobel Prize in 1954, was his application of quantum mechanics to the study of molecular structures. After becoming acquainted with the new discipline of quantum mechanics—the branch of physics concerning the dynamic properties of subatomic particles—Pauling applied its underlying principles to the study of chemical bonds. Pauling determined that hybrid bonds formed by electrons in a carbon atom manifested an energy configuration inconsistent with then-accepted theories. He also established that most atoms formed chemical bonds that bore attributes of both ionic and covalent bonding. Ionic bonds are chemical bonds in which electrons are lost by one atom and gained by another to form a more stable molecule, whereas covalent bonds occur when atoms share pairs of electrons equally.

Linus Pauling, the only person ever to win two Nobel Prizes, shown picketing the White House as part of a mass demonstration opposing the resumption of U.S. atmospheric nuclear tests. *NARA, courtesy AIP Emilio Segrè Visual Archives.*

Pauling was born on February 28, 1901, in Portland, Oregon, to Herman and Lucy Darling Pauling. His father was a druggist who died of a perforated ulcer in 1910. His mother then strove to keep the family afloat.

Pauling enrolled at Oregon Agricultural College (now Oregon State University) in 1917. As a student he became interested in the nascent field of quantum mechanics. Upon graduating in 1922, Pauling pursued graduate work at the California Institute of Technology (Caltech) and was awarded his Ph.D. in chemistry summa cum laude in 1925. He and his wife, Ava Miller, whom he married in 1923, then left for Europe. While in Zurich, Pauling studied with seminal figures in quantum mechanics—NIELS HENRIK DAVID BOHR, ERWIN SCHRÖDINGER, and Arnold Sommerfeld. In 1927 he accepted a position as assistant professor of theoretical physics at Caltech.

Pauling employed tenets from quantum mechanics to solve problems in chemical theory. His work in bond hybridization, the process whereby electrons change character when forming chemical bonds with other atoms, was renowned. He turned his attention to the puzzle posed by the fact that, although carbon had two types of bonding electrons (referred to as 2*s* and 2*p*), the atom did not form two kinds of bonds (one with the 2*s* electrons and the other with 2*p* electrons). Instead four identical bonds were formed. Pauling recognized that the electrons of the carbon atom changed their nature during bonding, forming an energy configuration that hybridized the 2*s* and 2*p* energy levels. He named this hybrid *sp3*. Pauling also proposed an intermediate bonding type between ionic and covalent bonds. He published the results of his work on chemical bonds frequently; *The Nature of the Chemical Bond and the Structure of Molecules and Crystals,* published in 1939, provided a comprehensive summary of his efforts in this area.

In the mid-1930s Pauling focused on the structure of biological molecules. After considerable effort Pauling created a molecular model of hemoglobin, the substance that transports oxygen in the bloodstream. Pauling became more concerned with political issues after the outbreak of World War II and devoted much of his energy to the cause of disarmament. In 1945 he proved that sickle-cell anemia is caused by a single amino acid in the hemoglobin molecule. After witnessing the horrors of World War II, Pauling again changed course and tirelessly campaigned for disarmament. His 1958 book *No More War!* detailed the military threat facing the world. In 1964 he severed his relationship with Caltech and accepted a position at the Study of Democratic Institutions in Santa Barbara, California, where he remained for four years. He moved to the University of California at Santa Barbara in 1968, and two years later he became a professor at Stanford University. After his retirement in 1974 he founded the Institute of Orthomolecular Medicine (later named the Linus Pauling Institute of Science and Medicine).

In addition to the Nobel Prize in chemistry he won in 1954, Pauling received the Nobel Peace Prize in 1963, making him the only person ever to win two unshared Nobel Prizes. His work in model building influenced FRANCIS HARRY COMPTON CRICK and JAMES DEWEY WATSON in their ground-breaking study of the deoxyribonucleic acid (DNA) molecule. His pioneering application of quantum mechanics to chemical problems advanced the field of chemistry. He died in 1994 and was survived by his four children.

Pavlov, Ivan Petrovich

(1849–1936)
Russian
Physiologist

Although Ivan Petrovich Pavlov earned the Nobel Prize in 1904 for his pioneering studies of the digestive system, it was his research on conditioned reflexes, conducted after 1904, that won him international fame. This research contributed to the development of a physiologically oriented school of psychology that studied the significance of conditioned reflexes to learning and behavior. The expression *Pavlov's dog* refers to the famous experiments in which Pavlov taught a dog to salivate at the sound of a bell by associating the bell with the sight of food.

Pavlov's intense devotion to his work was probably inspired by his father, Pyotr Dmitrievich Pavlov, an influential priest in the town of Ryazan in central Russia, where Pavlov was born on September 26, 1849. A reader and scholar himself, Pyotr taught his son to read all good books at least twice so that he would understand them better. Varvara Ivanova, Pavlov's mother, also was from a family of clergy.

Pavlov studied, in accordance with his family's wishes, at the Ecclesiastical High School and the Ecclesiastical Seminary in Ryazan. While at the seminary he encountered the work of Dmitri Pisarev, who taught many of CHARLES ROBERT DARWIN's evolutionist theories. Not long afterward he left the seminary to enter the University of St. Petersburg, where he studied physiology and chemistry. Pavlov received his M.D. from the Imperial Medical Academy in St. Petersburg in 1879, and in 1881, he married Seraphima Vasilievna Karchevskaya, a friend of the writer Fyodor Dostoevsky. The couple eventually had four sons and a daughter. Pavlov completed his dissertation, for which he received a Gold Medal, in 1883.

Between 1884 and 1886 he studied in Leipzig, Germany, under the direction of Carl Ludwig, a cardiovascular physiologist, and Rudolf Heidenhain, a gastrointestinal physiologist. Another mentor, Sergei Botkin, asked Pavlov to run an experimental physiological laboratory. It was under Botkin's guidance that Pavlov first developed his interest in "nervism," the pathological influence of the central nervous system on reflexes.

Pavlov's first independent research when he returned to Russia, conducted between 1888 and 1890, was on the physiological characteristics of the circulatory system. Working on unanesthetized dogs, he would introduce a catheter into the femoral artery and record the influence of various pharmacological and emotional stimuli on the dogs' blood pressure. This work earned him his first professorship at the Military Medical Academy, then in 1895 Pavlov became the chair of the physiology department at the St. Petersburg Institute for Experimental Medicine, where he spent most of his career.

From 1890 to 1900 and to a lesser degree until about 1930 Pavlov studied the physiological processes of the digestive system. In one famous experiment he slit the gullet of a dog he had just fed, causing the food to drop out before it reached the dog's stomach. He was able to show that the sight, smell, and swallowing of food were enough to start secretion of digestive juices. For this work he received the Nobel Prize in 1904.

Pavlov next developed the laws of conditioned reflex by focusing on neural influences in digestion. He noted that his dogs would sometimes salivate at the sight of a lab assistant who often fed them. Through a series of repeated experiments he demonstrated that if a bell is rung each time a dog is given food, the dog eventually develops a "conditioned" reflex to salivate at the sound of the bell, even when food is not present. Pavlov's conditioned reflex theory became the subject of much research in the fields of psychiatry, psychology, and education.

In addition to his influential career in science, Pavlov was a political activist who frequently spoke out against communism and protested the actions of Soviet officials. In addition to the Nobel Prize, Pavlov was awarded the Order of the Legion of Honor of France and the Corey Medal of the Royal Society of London. He remains one of the fathers of modern science.

Payne, Katharine Boynton ("Katy")
(1937–)
American
Zoologist

Katharine Payne has helped to develop the field of bioacoustics, which studies the sounds that animals use to communicate, and has made key discoveries about sounds made by whales and elephants. Katy, as she is usually known, was born Katharine Boynton in Ithaca, New York, in 1937. Her father was a professor at Cornell University. While attending Cornell, where she majored in music, Katy met Roger Payne, a biology graduate student, and they married in 1960. During the next few years she and their growing family, which came to include four children, accompanied him on field studies of humpback whales and right whales.

After hearing recordings of vocalizations that humpbacks made under water, Roger Payne realized in 1966 that the sounds formed long, complex, repeating patterns; in short, they were songs, like bird songs. All the male whales in any given area of the ocean sing the same song at a particular time, but, as Katy Payne discovered, the song of each population gradually changes with time, and all the singing whales keep abreast of the changes. Katy spent more than a decade documenting and analyzing this fascinating example of cultural evolution—the passing on of learned traits—in whales. She found an intriguing set of similarities between the gradual changes in whale songs and those in human languages.

Payne became a research associate at Cornell University in 1984. In that same year, while observing elephants at the Metro Washington Park Zoo in Portland, Oregon, she noticed a throbbing in the air near the elephants' cages and suspected that it might be caused by powerful infrasonic vocalizations—sounds pitched too low for human ears to hear. Payne, working with William Langbauer and Elizabeth Marshall Thomas, used recording equipment that could pick up deep sounds to confirm her guess. No one had ever suspected that land animals could use infrasound to communicate, but low-frequency sound travels through air exceptionally well, and Payne thought that elephants' use of infrasound might give them a long-distance communication system.

Payne and various associates have been following up this notion ever since. They have not only demonstrated elephants' use of infrasound but also gained an understanding of how the animals use these unusual calls to help coordinate their widespread and highly complex societies. For instance, in Amboseli National Park in Kenya in 1985 and 1986 Payne and Joyce Poole, working with the same elephants CYNTHIA MOSS has studied, recorded an extensive vocabulary of elephant calls and found them rich in infrasound. The existence of these calls helps to explain how widely separated elephant groups can move in parallel over long periods as they feed and how male elephants find females during the brief periods when the latter are in breeding condition. In 1998, working with several meteorologists, Payne was trying to determine the effect of the atmosphere on infrasound communications used by elephants, lions, and certain other animals. In addition to her scientific work, Payne writes, she is an "active conservationist with a strong concern for the health and survival of wild animals and wild places."

Pert, Candace Beebe

(1946–)
American
Medical Researcher

Candace Pert discovered molecules in the brain that attach to natural chemicals that resemble opiates (narcotics). She has also done pioneering research on chemical links between the brain and the rest of the body and on a drug that may be a valuable treatment for acquired immunodeficiency syndrome (AIDS) and other diseases.

Pert was born Candace Dorinda Beebe on June 26, 1946, in New York City and grew up in Wantagh, Long Island. Her parents were Robert and Mildred Beebe. Robert Beebe held a variety of jobs, from selling ads to arranging band music, and Mildred was a court clerk.

Beebe majored in English at nearby Hofstra University, but while there she met an Estonian psychology student, Agu Pert, who interested her in science. She and Pert married in 1966, before she finished her bachelor's degree, and soon had the first of their three children. Candace finished her degree at Bryn Mawr, where Agu went to do further graduate work, and this time she took classes in chemistry and psychology.

Candace Pert began doing doctoral research at Johns Hopkins Medical School under Solomon Snyder, an expert on brain chemistry, in 1970. He suggested that she look for the opiate receptor, a substance whose existence was strongly suspected but had not been proved. Like many other chemicals that act on the body and brain, opiates become active only when they attach to receptor molecules on the surfaces of cells. Each chemical or group of chemicals has its own receptor, which appears only on the kinds of cells that chemical affects. At the time Pert started her research, receptors for only a few body chemicals had been identified.

In September 1972 after many attempts Pert succeeded in "tagging" opiate receptors in animal brain tissue with radioactive material. These were the first receptors located in the brain and the first that bound to a substance that did not exist naturally in the body. Publication of Pert's results in early 1973 not only earned her a Ph.D. (in 1974) but also made her famous.

In late 1975 two researchers in Scotland identified the first of the endorphins, the group of opiatelike natural brain chemicals to which Pert's receptors normally attach. Pert meanwhile adapted her technique to map receptors for opiates and other chemicals in different parts of the brain. She and Michael Kuhar of Johns Hopkins found opiate receptors in brain areas involved in perception of both pain and pleasure.

In 1975 Candace and Agu Pert joined the National Institute of Mental Health (NIMH), one of the government-sponsored National Institutes of Health in Bethesda, Maryland. Candace soon was heading her own laboratory. She won the Arthur S. Fleming Award for outstanding government service in 1978. She was angry, however, that she did not receive the Albert and Mary Lasker Award for Biomedical Research, which Snyder and the two Scottish researchers shared in that same year. Opinions differ about whether she was denied the award because she was still a graduate student when she did her opiate receptor research or because she was a woman.

In 1983 Pert became chief of a new section on brain biochemistry within the clinical neuroscience branch of NIMH, the only woman section chief in the institute at the time. She studied peptides, small molecules similar to proteins, many of which carry messages from one part of the body to another. From the early 1980s she and others found that many message-carrying peptides have receptors both in the brain and on cells of the immune system, which defends the body against disease.

Pert's research on peptides and receptors took a new turn in 1984 when she found that one kind of receptor on immune system cells called CD-4 cells also appeared in the brain. Other scientists showed that human immunodeficiency virus (HIV), which causes AIDS, enters CD-4 cells by way of this receptor. Pert believed that AIDS might be prevented or halted by a drug that blocked the receptors, and she began working with another NIH researcher, Michael Ruff, on this idea. Candace and Agu Pert had divorced in 1982, and in 1986 she married Ruff. That same year the two began publishing accounts of their research on a substance they called *peptide T,* which was similar to the part of the virus that binds to the CD-4 receptor.

Pert (who has continued to use the name under which she did her earlier research) and Ruff felt that the NIH was not supporting the peptide T research fully, and at the end of 1987 they left NIH and started their own company to pursue it. Although subsequent work with the drug ran into difficulties, recent research has given a boost to the couple's belief that it may combat some effects of AIDS, especially wasting and brain damage, and work against other kinds of diseases as well by blocking a second receptor, the chemokine receptor. Tests of peptide T continue.

Pert also focused during the 1990s on the links between brain and body, especially between the brain and the immune system, that her peptide and receptor research revealed. (She calls peptides "the molecules of emotion.") She and other researchers who believe that these links allow physical and mental health to influence each other, have formed a new field, called psychoneuroimmunology. Pert, who won the Kilby Award in 1993, and Ruff have continued their "search for truth" as research professors at Georgetown University Medical Center in Washington, D.C.

Piazzi, Giuseppe

(1746–1826)
Italian
Astronomer

Although Giuseppe Piazzi did not begin his study of astronomy until he was middle-aged, he made several significant observations in the field. In 1801 he discovered a planet, which he named Ceres in honor of the patron goddess of Sicily. He also devoted 10 years to the tedious task of making precise determinations of the coordinates of fixed stars. The result of this hard work was the 1814 publication of his catalog detailing the positions of 7,646 stars. In addition, he founded and served as director of the Observatory of Palermo. Later King Ferdinand I named him director general of the observatories in both Naples and Sicily.

Little is known about Piazzi's origins. He was born at Ponte in Valtellina, Italy, which is now in Switzerland. He entered a monastery of the Theatine order in Milan when he was a young man. After completing studies in Milan and in Rome, he attained a doctorate in philosophy and mathematics.

Piazzi taught mathematics from 1769 until 1779 in a number of Italian cities. In 1780 however, the prince of Caramanico offered him the prestigious chair of higher mathematics at the Academy of Palermo. At some point Piazzi apparently developed an interest in astronomy, for he communicated his interest in building an observatory in Palermo. After receiving permission to do so, Piazzi departed for England in the late 1780s in order to obtain the finest equipment for the observatory. While in England he met prominent figures in the field of astronomy, most notably Sir WILLIAM HERSCHEL, who had discovered the planet now known as Uranus. After publication of his first astronomical work in 1789 Piazzi's observatory finally opened in 1790. He was appointed director.

The first project Piazzi undertook in his new position was the exact determination of the astronomical coordinates for principal stars. This work required regular observations. On January 1, 1801, Piazzi sighted a celestial body that was not mentioned in any of his books. He continued to watch it and tentatively announced his discovery in the next months. He hypothesized that it was a planet, and further study of the body's orbital patterns confirmed this theory. In 1802 Piazzi named his discovery Ceres. He entered into a debate about Ceres with his acquaintance Herschel. Herschel insisted that Ceres be called an *asteroid,* not a planet, because he did not think it worthy to be included with the likes of Saturn and Jupiter. Piazzi obviously disagreed, and was subsequently vindicated when the term *asteroid* fell out of favor.

Throughout this period Piazzi continued his efforts to obtain precise coordinates. The catalog he produced in 1803 was far more accurate than previous ones and was admired for its scope and detail. Piazzi thereupon began to investigate the ascension of stars by situating them in direct relation to the Sun. He published another catalogue, which listed the mean position of 7,646 stars, in 1813. These accomplishments were not unnoticed. In 1817 King Ferdinand I summoned Piazzi to supervise the construction of an observatory in Naples and later appointed him director of both observatories. In these capacities Piazzi split his time between Naples and Sicily. In 1824 he once more settled permanently in Naples, where he was commissioned by Ferdinand to oversee the reform of the system of weights and measures.

Piazzi's achievements were recognized on a number of occasions. He was awarded the Lalande Prize of the Institut de France in 1803 and in 1813. Moreover he was elected president of the Neapolitan Academy of Sciences. Later astronomers built upon his careful records and his discoveries.

Pimental, David

(1925–)
American
Entomologist, Ecologist

David Pimental commenced his career as an insect ecologist, but the more he researched pest control, the more he realized its interrelationship to the broader issue of sustainable agricultural practices and public policy implementation. He thus broadened the scope of his research and writings to address environmentalism more generally, and his ideas have influenced important policy decisions on local, national, and international levels.

Pimental was born on May 24, 1925, in Fresno, California. His parents, Marion Silva and Frank Freitas Pimental, were grape and vegetable farmers who moved to a farm in Middleborough, Massachusetts, when Pimental was six years old. In 1949 Pimental married Maria Hutchins, a nutritionist who later served on the same faculty at Cornell University as her husband. Together the couple had three children—two daughters and one son.

Pimental attended St. John's University in Collegeville, Minnesota, where he received pilot and officer training until 1943, when he served in the United States Army Air Force for two years. He transferred to the University of Massachusetts at Amherst in 1945; he received his B.S. there in 1948 after studying at Clark University in Worcester, Massachusetts, during the summer of 1946. Pimental then proceeded to conduct doctoral research at Cornell University, where he received his Ph.D. in 1951.

When the Air Force called him up from reserve status to active duty, he transferred his assignment to the United States Public Health Service (USPHS), which appointed him chief of its Tropical Research Laboratory in San Juan, Puerto Rico, where he studied mongooses and snails as disease carriers from 1951 through 1955. During the springs and summers of 1954 and 1955 he served in Savannah, Georgia, as a project leader at the USPHS Technical Development Laboratory, and in the winters he conducted postdoctoral research on insect ecology at the University of Chicago. In 1955 he became an assistant professor of insect ecology at Cornell University, where he remained for the rest of his career. Cornell promoted him to full professor status in 1963, when it also appointed him chairperson of the department of entomology and limnology.

That year Pimental published a paper proposing new criteria for choosing pest control agents, and his ideas were largely adopted. This represented the beginning of his influential role in determining how human populations would interact with the environment. Between 1964 and 1966 he served on the President's Advisory Council, and in 1969 he served on a commission that promulgated the banning of the chemical pesticide DDT and also led to the formation of the Environmental Protection Agency.

Pimental also continued to exert his influence through his writings; for example, at the height of the energy crisis in 1973 he published a paper that traced the amount of energy consumed to produce food, the first study of its kind. Throughout his career he produced over 400 publications, including 17 books, such as *Food, Energy, and Society,* which he coauthored with his wife in 1979; *The Pesticide Question: Environment, Economics, and Ethics,* which he coedited with H. Lehman in 1993; and *World Soil Erosion and Conservation,* which he edited in 1993. Pimental's positive influence on the environment was recognized in 1992 with the Award for Distinguished Service to Rural Life.

Planck, Max

(1858–1947)
German
Physicist

As the originator of quantum theory Max Planck revolutionized physics. His work provided the foundation for a great many discoveries in modern physics. Planck's primary achievement, which ultimately won him the Nobel Prize, was his radical reworking of the conception of energy. Classical physics postulated, and indeed the entire scientific community believed, that energy was always transmitted in a continuous form, such as a wave. Planck's research led him to conclude that energy could exist in discrete units, which were later given the name *quanta.* In further elucidating this theory he discovered a universal constant in nature—of equal magnitude to other universal constants such as the speed of light—which came to be known as Planck's constant.

Max Planck, who is recognized as the creator of quantum theory. *AIP Emilio Segrè Visual Archives.*

Planck was born in Kiel, Germany, on April 23, 1858, to Johann von Planck and Emma Patzig Planck. Max was his parents' fourth child and also had two step-siblings from his father's previous marriage. When Planck was nine, the family moved from Kiel to Munich.

After graduating from the Königliche Maximillian Gymnasium in Munich in 1874, Planck entered the University of Munich. His education was interrupted by a severe illness in 1875 that forced him to withdraw. In 1877 he completed two semesters at the University of Berlin, as he continued to recuperate. He returned to Munich in 1878 and passed the state examination for higher-level teaching in mathematics and physics that year. He wrote his dissertation on the second law of thermodynamics and received his Ph.D. from the University of Munich in 1879.

Unfortunately his personal life was filled with tragedy. He married his childhood sweetheart, Marie Merck, in 1885, and the couple had three children, but after his wife died in 1909, both his daughters died while giving birth. His son was then killed during World War I. Merck eventually remarried, but his son from this union was executed in 1944 for plotting to overthrow Hitler. Planck's home was destroyed during an air raid in 1945, just two years before his death.

Prior to attaining his first university appointment in 1885 at the University of Kiel, Planck was a *Privatdozent* at the University of Munich. He was offered an assistant professorship at the University of Berlin in 1888; he remained there until 1926. He turned his attention to the topic of blackbody radiation in 1896. A blackbody is an ideal body or surface that absorbs all radiant energy when heated and gives off all frequencies of radiation when cooled. Physicists had tried to formulate a mathematical law that applied to the blackbody's radiation of heat, but no scientist could adequately explain blackbody radiation across the entire spectrum of frequencies. Planck solved this problem in 1900 with his ground-breaking theory of the *quantum* (Latin for "how much"). He proposed that energy moved in a stream of discrete units. Planck determined that the energy of each quantum is equal to the frequency of the radiation multiplied by the universal constant he discovered. The numerical value of Planck's constant is 6.62×10^{-27}.

Planck then decided to investigate the application of his quantum theory to the emerging field of relativity. He retired from the University of Berlin in 1926 and four years later became the president of the Kaiser Wilhelm Society (subsequently renamed the Max Planck Society) in Berlin. He increasingly dedicated his energies to philosophical and spiritual questions. In 1935 he published *Die Physik im Kampf um die Weltanschauung,* which explored the connections among science, art, and religion. His posthumously published *Philosophy of Physics* addressed the importance of finding general themes in physics.

Planck's contribution to the field of physics was immense. Not only did he launch a new subfield, quantum mechanics; his work also made future endeavors, such as the discovery of atomic energy, possible. It provided a foundation for the theories of WERNER KARL HEISENBERG, WOLFGANG PAULI, and later ALBERT EINSTEIN. Planck was awarded the Nobel Prize in physics in 1918 and achieved a status as one of the greatest scientific minds of his era. He died in 1947. On his headstone was engraved the mathematical expression of Planck's constant.

Playfair, John

(1748–1819)
Scottish
Mathematician, Geologist

John Playfair's scientific contributions ranged across several disciplines. After serving for a time as a minister, Playfair became a professor of mathematics. He published his *Elements of Geometry,* in which he presented an alternative version to EUCLID's fifth postulate, which is now known as Playfair's axiom. His more famous work, however, was his *Illustrations of the Huttonian Theory of the Earth,* which explained some of the theories of his friend the renowned geologist JAMES HUTTON. Playfair's *Illustrations* made Hutton's often abstruse concepts accessible to a wider scientific and public audience. For instance, Playfair cogently laid out, and then expanded upon, Hutton's uniformitarian ideas.

Born on March 10, 1748, near Dundee, Scotland, Playfair was the eldest son of the Reverend James Playfair. The details of his family life and childhood are not known. He was sent to the University of St. Andrews when he was 14; there he revealed an aptitude for mathematics. He left St. Andrews in 1769 and qualified for the ministry.

Upon his father's death in 1769 Playfair inherited the church benefice of Benvie, where he remained until 1782. After serving as a private tutor from 1782 until 1785, Playfair became a professor of mathematics at the University of Edinburgh. During his tenure there Playfair edited the periodical *Transactions of the Royal Society.* In 1795 he published his *Elements of Geometry,* which discussed Euclid's first six books in detail. Playfair also added his own insights, especially into Euclid's fifth postulate. Playfair asserted, in what is now termed *Playfair's axiom,* that "two straight lines, which intersect one another, cannot be parallel to the same straight line."

Playfair's reputation, however, was established not by his professional projects as a mathematician but by his work in geology. When his friend James Hutton died in 1797, Playfair began to analyze and build on Hutton's 1795 book *Theory of Earth.* In 1802 Playfair published the results of this project in *Illustrations on the Huttonian Theory of the Earth.* Although Hutton had been a pioneering figure in the nascent field of geology, his presentations of his ideas were often garbled and difficult to understand. Playfair's *Illustrations* did much to remedy this defect.

In a particularly significant passage Playfair described Hutton's 1785 theory of uniformitarianism, which held that the history of the Earth can be interpreted solely on the basis of geological processes evident to the modern observer. Hutton argued that geological processes had been operating for millions of years to create the current landscape. In *Illustrations* Playfair also proposed some of his own theories, the most important of which was that a river carves out its own valley. Playfair's career took another turn in 1805, when he was named a professor of natural philosophy at the University of Edinburgh. In this capacity he researched and lectured on physics and astronomy. In 1814 he published his *Outlines of Natural Philosophy.*

Of his diverse accomplishments Playfair's forays into geological studies were the most influential. His interpretation of Hutton appeared at a time when the field was at a crossroads. In effect Playfair's *Illustrations* summed up the first period of British geology and paved the way for the next generation of geologists, which included such individuals as William Smith, John Farey, and Thomas Webster. Playfair died in Edinburgh on July 20, 1819.

Poincaré, Jules-Henri

(1854–1912)
French
Mathematician

In the more than 500 papers that he published during his academic career Jules Henri Poincaré explored a range of topics that demonstrated the breadth of his abilities. He contributed to the fields of number theory, functions theory, and topology (the geometry of functions). His work in this latter subject helped lay the foundation for modern algebraic topology. Poincaré also sought to apply mathematics to the study of physical phenomena, and he did so in the fields of celestial mechanics, thermodynamics, and relativity theory. His research into the three-body problem—the question of how three bodies, such as planets, act on each other—resulted in significant advances toward its solution. His proposals that absolute motion did not exist and that no object could travel faster than the speed of light proved to be fundamental elements of relativity theory.

Jules-Henri Poincaré, whose work in topology (the geometry of functions) helped lay the foundation for modern algebraic topology. *AIP Emilio Segrè Visual Archives, Physics Today Collection.*

Poincaré was born on April 29, 1854, in Nancy, France, where his family remained throughout his childhood. His father, Léon Poincaré, a physician and professor of medicine at the University of Nancy, was well versed in the sciences. Despite the fact that Jules Poincaré suffered from poor health as a boy—diphtheria paralyzed his larynx for a time—he demonstrated mathematical ability at an early age.

During his time at the lycée in Nancy he won first prize in a national student competition. In 1873 he entered the École Polytechnique at the head of his class. It was reputed that he excelled in all his math classes without ever taking notes in class or reading the required texts. After graduation Poincaré matriculated at the École des Mines with the goal of becoming an engineer. His love of mathematics won out, however. He received his doctorate in mathematics upon completing his dissertation in 1879. With his wife, Jeanne Louise Marie Poulain D'Andecy, Poincaré had four children.

Poincaré taught at the University of Caen from 1879 to 1881. He then accepted a position as a professor of mathematical analysis at the University of Paris, where he stayed for the remainder of his career. His early work involved a set of functions, which he named Fuchsian functions but came to be known as automorphic functions. He used these functions, involving sets that correspond to themselves, to integrate linear differential equations with rational algebraic coefficients and to describe the coordinates of any point on an algebraic curve as uniform functions of a single algebraic variable. Poincaré also made breakthrough discoveries in the mathematical fields of topology and probability. His insights in the fledgling area of topology allowed him to formulate some of the basic tenets of modern algebraic topology. His work in probability was sufficiently dynamic that his papers on the topic garnered a readership among the general public.

In addition to his achievements in mathematics Poincaré used math to gain an understanding of physical phenomena. Chief among these efforts was his work on celestial mechanics. He helped move science closer to resolving the three-body problem and used new mathematical techniques to explain planetary orbits. In 1893 he published *The New Methods of Celestial Mechanics,* a summary of his application of new mathematical methods to astronomy. Some of his theories became relevant to the eventual development of relativity theory. Later in his life he published works on the philosophy of science for the general public. He explored topics such as the role convention (or the arbitrary choice of concepts) played in scientific method in *Science and Hypothesis* (1903) and *Science and Method* (1908).

Poincaré's multifaceted achievements were honored during his lifetime. In 1887 he was elected at the age of thirty-two to the French Academy of Sciences. He was named a foreign member of the Royal Society of England in 1894. He was praised as the last of the great mathematical universalists, and his work provided the foundation for many future mathematical endeavors—both pure and applied. He died at the age of 58 in 1912.

Porter, Sir George

(1920–)
British
Chemist

Sir George Porter received the 1967 Nobel Prize in chemistry with his collaborator RONALD GEORGE WREYFORD NORRISH for their development of flash photolysis to analyze the intermediate stages of very fast chemical reactions. The pair shared this prize with the German scientist MANFRED EIGEN, who did similar work, though Eigen utilized the relaxation technique of measuring the return of chemicals to a state of equilibrium after an induced disturbance.

Porter was born on December 6, 1920, at Stainforth in West Yorkshire, England. His parents were Alice Ann Roebuck and John Smith Porter. In 1938 he received the Ackroyd Scholarship to attend Leeds University. He earned his bachelor of science degree in chemistry in 1941. Having studied radio physics and electronics at Leeds, Porter spent World War II working as a radar specialist with the Royal Navy Volunteer Reserve. After the war he resumed his education, conducting doctoral work at Emmanuel College of Cambridge University. In 1949 he earned his Ph.D., and that year he also married Stella Brooke. Together the couple had two sons, John Brooke and Christopher.

After receiving his doctorate Porter remained at Cambridge as a demonstrator in chemistry. In 1952 the university promoted him to the position of assistant director of research in the department of physical chemistry. In 1954 he filled this same position at the British Rayon Research Association. The next year Sheffield University appointed him as a professor of physical chemistry. In 1963 the university named him the Firth Professor and promoted him to head the chemistry department. In 1966 the Royal Institution in London named him its Fullerian Professor of Chemistry; concurrently the Davy Faraday Research Laboratory appointed him as its director. Two decades later in 1987 Imperial College appointed him as a professor, and in 1990 the college named him chairman of its Center for Photomolecular Sciences.

Porter utilized the flash photolysis technique in multiple diverse applications throughout his career. He first focused on investigating free radicals that were liberated by the initial light burst and then illuminated by a subsequent flash of light for observation. He later managed to stabilize these free radicals with a method known as matrix isolation. He then applied the flash photolysis technique to gases and solids. He even extended his flash photolysis investigations to animals, analyzing the interactions between oxygen and hemoglobin, and to plants, studying chloroplasts, the properties of chlorophyll, and the process of photosynthesis. Porter eventually studied reactions occurring in a thousandth of a millionth of a second.

Porter gained increasing recognition for his efforts to reach the general public, especially children, with scientific knowledge. He participated in the production of several British Broadcasting Company television shows geared to children, including *Young Scientist of the Year, The Laws of Disorder,* and *Time Machines.* Besides the Nobel Prize, Porter received two honors in 1978—the Robertson Prize of the U.S. National Academy of Sciences and the Rumford Medal of the Royal Society. Porter was knighted in 1972, and in 1990 he became Baron Porter of Luddenham.

Pregl, Fritz

(1869–1930)
Austrian
Analytical Chemist

Fritz Pregl received the 1923 Nobel Prize in chemistry for his advances in the microanalysis of organic material. Although Pregl did not distinguish himself with original theories but rather built on the work of others, the advances he made were tremendous, opening up a whole new field of chemical study with his innovative improvements in the precision of his instrumentation.

Pregl was born on September 3, 1869, in Laibach, Austria (now Ljubljana in the Republic of Slovenia) to Raimund, the treasurer of a bank in Krain (now Kranj), who died when Pregl was young, and Friderike Schlacker, who moved him in 1887 to Graz, where he entered the University of Graz, the institution at which he spent the majority of the rest of his career. Pregl studied medicine and worked as a laboratory assistant to Alexander Rollett. Pregl obtained his M.D. in 1893.

Pregl then practiced medicine, specializing in ophthalmology, while still working in Rollett's laboratory. The University of Graz appointed him as an assistant lecturer in physiology and histology before naming him as a university lecturer in 1899. In 1904 Pregl traveled to Germany to study chemistry with Friedrich Wilhelm Ostwald in Leipzig and EMIL HERMANN FISCHER in Berlin. Pregl returned to the University of Graz in 1905 as an assistant at the medical and chemical laboratory. To these duties he added the post of forensic chemist for central Styria, where Graz was located. In 1910 he accepted the position of professor and head of the medical chemistry department at the University of Innsbruck. Pregl returned to Graz in 1913 as a full professor at the University of Graz's Medico-Chemical Institute. The university promoted him to the position of dean of the medical school in 1916 and in 1920 named him the vice-chancellor of the university.

Pregl's early studies on the properties of bile and urine convinced him of the necessity of improving the techniques and tools of microanalysis, as the amounts of bile required to conduct his studies were astronomical and the time required to analyze these samples was prohibitive. Pregl followed the same logic established by Justus von Liebig some 70 years earlier, when he founded the practice of microanalysis; Pregl simply improved upon von Liebig's methods. Pregl applied many of the microanalytical techniques of his contemporary, Friedrich Emich, but he continually tweaked them to allow for ever-smaller samples. Pregl improved upon the microbalance recently innovated by W. H. Kuhlman, which could measure to .01 milligram, refining it until it could measure .001 milligram. In 1911 he presented his improvements to the German Chemical Society in Berlin, and again in 1913 to a scientific congress in Vienna. In 1917 he published the book that summarized his work, *Die quantitative organische Mikroanalyse,* which went through multiple editions.

Besides the Nobel Prize, Pregl received the Lieben Prize. A bachelor throughout his life he endowed the Vienna Academy of Sciences in 1929 with funds to establish the Fritz Pregl Prize, awarding a year-long stipend to promising science students. Pregl died on December 13, 1930, in Graz, Austria.

Priestley, Joseph

(1733–1804)
English
Chemist

Although he received little formal training in the physical sciences, Joseph Priestley became one of the most important experimental chemists of his era. His most famous accomplishment—the discovery of the element oxygen—occurred during his broader examination of gases, including ammonia and nitrogen. His findings in this area offered considerable inspiration to ANTOINE-LAURENT LAVOISIER. In addition to his contributions to chemistry, Priestley published work on electricity, theology, politics, and linguistics. Employed for most of his life as a minister and teacher, Priestley championed the rights of the individual in government and ultimately rejected many of the tenets of Christianity.

Priestley was born the eldest of six children on March 13, 1733, near Leeds in Yorkshire, England. His father, Jonas Priestley, was a cloth dresser. After his mother, Mary Swift, died when Joseph was six, he was sent to live with his aunt, Sarah Priestley Keighley, with whom he stayed until he was 19. The loss of his mother and separation from his immediate family instilled in him a sense of independence that deeply informed his life and work.

Priestley's family encouraged him to pursue the ministry and advocated his membership in the Dissenting church (the appellation for congregations that did not conform to the Church of England). After his early education at parish schools Priestley entered the Dissenting Academy at Daventry in 1752.

After remaining at Daventry for three years, Priestley was appointed an assistant minister to the independent Presbyterian community in Needham Market, Suffolk, in 1755. His spiritual move toward the more rational doctrines of Unitarianism (he rejected the Trinity), however, made him unpopular, and in 1758 he moved to a more sympathetic congregation in Nantwich, Cheshire. Upon being ordained a Dissenting minister in 1762, he married Mary Wilkinson, with whom he had four children. It was during this period that Priestley became interested in science.

Priestley's public recognition as a scientist occurred in 1766 with the publication of his treatise on electricity *The History and Present State of Electricity,* which not only encapsulated contemporary thinking on the subject but also proposed original experiments. In 1767 he was named minister of Mills Hill Chapel in Leeds, which afforded him more time for his study and research. He became drawn to matters of chemistry. At one of the many breweries in Leeds he observed carbon dioxide (which he referred to as "fixed air") and hydrogen. His curiosity about gases (or "airs," as he called them) was piqued, and he embarked on a series of laboratory experiments that resulted in his discovery of 10 gases. He discovered four gases while he was at Leeds: nitric acid, nitrogen dioxide, nitrous oxide, and hydrogen chloride. His true genius lay in his design of laboratory equipment that allowed him to isolate so many gases. Upon leaving Leeds, he accepted the earl of Shelburne's offer in 1773 to serve as his personal librarian and intellectual companion in Wiltshire. Priestley's greatest discovery occurred there in August of 1774, when he heated red mercuric acid and produced a colorless, odorless gas—oxygen. He published his findings about gases in a series of volumes, *Experiments and Observations on Different Kinds of Air,* between 1774 and 1786.

His development as a chemist was paralleled by the evolution of his political ideals. His 1769 treatise *Essay on the First Principle of Government, and on the Nature of Political, Civic, and Religious Liberty,* emphasized individualism, especially at the political level. In 1779 he left his position with Shelburne to become a minister in Birmingham. By publicly refuting Edmund Burke's condemnation of the French Revolution in 1790 and attacking what he considered to be irrational aspects of Christianity, he became unpopular. In 1791 a mob destroyed his home, library, and laboratory. He immigrated to the United States in 1794 and remained there until his death on February 6, 1804, in Northumberland, Pennsylvania.

Priestley's legacy was tremendous. He influenced Lavoisier's conclusion that oxygen was an essential part of the atmosphere and a factor in both respiration and combustion. (It was Lavoisier who actually termed Priestley's discovery *oxygen*). Priestley was awarded the Copley Medal of the Royal Society in 1774.

Prigogine, Ilya

(1917–)
Russian/Belgian
Theoretical Chemist

Ilya Prigogine received the 1977 Nobel Prize in chemistry for his pioneering work in the field of nonequilibrium thermodynamics, a branch of science devoted to explaining systems as they actually exist, with processes that are irreversible in time. Prigogine coined the term *dissipative structures* to define systems far from equilibrium that spring up into states of organization spontaneously, such as traffic patterns in response to changing conditions and the organization of termite mounds, which seem to create order out of chaos.

Prigogine was born on January 25, 1917, in Moscow, Russia, on the eve of the Bolshevik Revolution. His mother was Julia Wichman; his father, Roman Prigogine, was a chemical engineer and factory owner. His brother, four years his elder, set the precedent of becoming a chemist. The family left Soviet Russia in 1921 for Lithuania and then Berlin, where they remained until 1929, when the rise of Nazism drove the Jewish émigrés to Brussels, Belgium.

In 1935 Prigogine entered the Free University of Brussels, the institution where he spent the remainder of his career. He earned his master's degree in 1939, then conducted doctoral research under Théophile De Donder. His dissertation "The Thermodynamic Study of Irreversible Phenomena" earned him his Ph.D. in 1941.

By 1947 Prigogine had been appointed full professor at Free University. In 1959 he added the duty of director of the Instituts Internationaux de Physiques et de Chimie, otherwise known as the Solvay Institute. In 1961 he married Marina Prokopowicz, an engineer. Between then and 1966 he spent time at the University of Chicago; from 1967 on he spent three months of the year at the Ilya Prigogine Center for Statistical Mechanics and Thermodynamics at the University of Texas.

Prigogine's work involved the structure and organization of systems through time. In classical thermodynamics the second law states that in closed systems order tends to lead toward disorder as time passes. In other words, the trend toward entropy is irreversible. However, many systems are open, or exchange energies with outside systems. These systems are unstable in that they can transform into new systems at points called moments of choice or bifurcation points. Prigogine's development of the Brusselator, with the assistance of G. Nicolis and Réné Lefever, helped to quantify and explain these theoretical structures, thus verifying their validity.

Prigogine published extensively throughout his career. In 1957 he published *The Molecular Theory of Solutions,* pointing out that at a low temperature liquid helium will spontaneously separate into two phases, helium-3 and helium-4. This hypothesis was later experimentally confirmed. In 1967 the third edition of *Thermodynamics of Irreversible Processes* appeared, explaining Prigogine's theories of nonequilibrium and irreversible time. In the 1980s he addressed these complex topics to a larger audience; the results were the 1980 text *From Being to Becoming: Time and Complexity in the Physical Sciences,* and the 1984 text *Order Out of Chaos: Man's New Dialog with Nature,* coauthored by Isabelle Stengers. This book won Prigogine the title of Commandeur de l'Ordre des Arts et des Lettres, an honor usually bestowed upon authors of literary works.

Proust, Joseph-Louis

(1754–1826)
French
Chemist

Joseph Louis Proust's most important achievement was his formulation of the law of constant proportions, which held that the elements making up any compound—regardless of how that compound is prepared—are present in a fixed proportion of weight. The significance of this concept became apparent in 1808, when JOHN DALTON published his atomic theory that served as a foundation for modern physical science. Proust's discovery was aided by his analytical skills and persistent adherence to quantitative measures.

Born on September 26, 1754, in Angers, France, Proust was the second son of Joseph and Rosalie Sartre Proust. His father was an apothecary, who encouraged young Joseph to pursue the same career. Proust received his early education from his godparents. He married Anne Rose Châtelain Daugbigné in 1798; the couple had no children.

Proust attended the local Oratorian college in Angers, then served as an apprentice to his father. In 1774, against his parents' wishes, Proust opted to continue his education in Paris, where he studied chemistry with Hilaire-Martin Rouelle. While there he also absorbed the teaching of Clérembourg, an apothecary.

In 1776 Proust was named chief apothecary at the Salpêtrière Hospital in Paris. He moved to Spain in 1778 to become a professor of chemistry at the recently founded Real Seminario Patriótico Vascongado in Vergara. But he remained there for only a short time before returning to France in 1780. He taught chemistry in France then returned to Spain, where he ultimately was appointed a professor of chemistry at the Royal Artillery College in Segovia in 1788. Eleven years later he was selected to head a new chemical laboratory in Madrid. That same year Proust discovered glucose, which he called "grape sugar" because he distilled it from that fruit. This finding was significant, as glucose proved to be a cheaper and more reliable source of sugar than cane, which had to be imported from the West Indies.

As head of the laboratory in Madrid, Proust had access to a first-rate scientific facility, which provided him the opportunity to refine his law of constant proportions. Whereas his counterpart Claude-Louis Berthollet maintained that chemical compounds could be formed from elements in a wide range of proportions, Proust firmly asserted that they contained elements only in certain proportions, for example, though he did not conceive of it in these terms, Proust's law predicts that a molecule of water must always comprise two hydrogen and one oxygen atom, rather than some indeterminate or variable quantity of those two elements. Berthollet and Proust had a long and good-natured dispute about the issue of proportions.

After Proust returned to France in 1806, Napoleon's army invaded Spain. In the ensuing political dislocation Proust's laboratory was sacked. In 1810 the French government commissioned him to establish a "grape sugar" factory, and in 1820 Proust took over his brother's pharmacy in Angers. He remained in France until his death.

The significance of Proust's law of proportion was confirmed by JOHN DALTON's atomic theory, which postulated that a definite number of atoms joined to form molecules. Late in his life Proust received official recognition. In 1816 he was elected to the Institut de France and three years later became a member of the Legion of Honor. He was granted a pension by Louis XVIII in 1820.

Ptolemaeus, Claudius (Ptolemy)

(c. 100–c. 178)
Greek
Astronomer, Geographer

Claudius Ptolemaeus, commonly known as Ptolemy, was best known for developing the geocentric theory of the motion of the planets. The theory, also referred to as the Ptolemaic theory, proposed that the Earth was the center of the universe, and the Sun, Moon, and planets revolved around it. Ptolemy also made great advances in the field of geography. He introduced principles of cartography and attempted to determine accurate positions and locations through latitudinal and longitudinal positions, straying from the traditionally employed spatial approach to geography.

Little is known of the life of Ptolemy, including the place of his birth. It is generally assumed that Ptolemy was born to Greek parents sometime around 100 and grew up in Alexandria, Egypt. Around the second century a town called Ptolemais Hermii was situated on the Nile River; because it was frequently the custom to name children after their place of birth, some scholars believe Ptolemy's name provides proof of his birthplace. Other scholars, however, contend that Ptolemy was born in Greece and later moved to Alexandria.

Alexandria was the center of intellectual learning during the second century, and Ptolemy had access to the city's vast array of libraries. He was particularly interested in science, chiefly astronomy, as well as philosophy and mathematics. It is alleged that Ptolemy erected an observatory on the upper floor of a temple to gain the best possible perspective on the sky.

Almagest is most likely Ptolemy's best-known work, and in it Ptolemy presented his astronomical observations and his explanation for the motion of the universe. In Ptolemy's view the Earth was stationary, perfectly spherical, and the center of the universe. Surrounding the Earth were the Moon, Mercury, Venus, the Sun, Mars, Jupiter, and Saturn, in that order; the stars surrounded the planets and enclosed the universe. Ptolemy was greatly inspired by ARISTOTLE, who believed in an Earth-centered universe; Plato; and the Greek astronomer HIPPARCHUS. Influenced by Aristotle's proposal that all heavenly objects and their motion were perfect, he believed that planets moved in a uniform pattern around the Earth at uniform speeds. Because Ptolemy's observations were not consistent with this belief, he devised a complex geometrical system to explain the irregularities. His calculations of planetary motions used spherical trigonometry and were accurate enough to convince the majority of scientists of his day to adopt the geocentric theory.

Another of Ptolemy's important works was *Geography,* which included a map of the world as well as numerous regional maps. Ptolemy explained principles of cartography, including the idea of assigning coordinates to geographic locations and the concepts of latitude and longitude. Ptolemy based his system on his theory that the Earth was a perfect sphere, using degrees of a circle to calculate distances on land. Over water, however, Ptolemy could only guess at the distances. He also had to guess longitudinal positions, as they could not be measured by the method used for latitudinal positions, which were calculated through observations of the Sun and stars. As a result Ptolemy's maps were not particularly accurate and his Earth was considerably smaller than its actual size.

Though Ptolemy's geocentric theory was incorrect, it persisted until the 15th century, when NICOLAUS COPERNICUS proposed his heliocentric theory. And though Ptolemy's maps may have been somewhat inaccurate, his geographical theory was the dominant view until the Renaissance. In addition, it is said Ptolemy's *Geography* inspired Christopher Columbus to undertake his journey over the Earth and that Columbus based his maps on Ptolemy's calculations. Ptolemy was also known for his work *Tetrabiblos,* which discussed his views on astrology (a field that was closely tied to astronomy during his day), and his work *Optics,* which detailed the basic principles of optics.

Purcell, Edward Mills

(1912–1997)
American
Physicist

Both Edward Purcell and FELIX BLOCH independently but simultaneously developed methods for measuring nuclear magnetic resonance, or NMR, the frequency at which atomic particles spin. For this joint achievement the two shared the 1952 Nobel Prize in physics.

Purcell was born on August 30, 1912, in Taylorville, Illinois. His parents were Mary Mills, a high school teacher, and Edward A. Purcell, a former country schoolteacher and the general manager of an independent telephone company. In 1937 Purcell married Beth C. Busser. Together the couple had two sons, Frank and Dennis.

Purcell attended Purdue University, studying electrical engineering to earn his bachelor's degree in 1933. He then spent a year abroad as an international exchange student, studying at the Technische Hochschule in Karlsruhe, Germany. Upon his return he entered Harvard University, where he earned his master's degree in physics in 1935 and his Ph.D. in 1938.

Purcell continued on at Harvard as a physics instructor until 1941, when World War II loomed large and almost all scientific activity focused on the war. Purcell contributed to the war effort as the leader of the Fundamental Developments Group at the Massachusetts Institute of Technology (MIT) Radiation Laboratory developing sensitive radar systems for nighttime navigation. After the war in 1946 Purcell returned to Harvard as an associate professor. Harvard promoted him to the status of full professor in 1949. In 1958 the university named him the Donner Professor of Science and in 1960 promoted him to the title of Gerhard Gade University Professor, a chair that he retained for the next two decades, until his 1980 retirement.

Purcell commenced studying nuclear magnetic moments, or the force of rotation of nuclei in a magnetic field. He utilized research methods similar to those employed by his colleague ISIDOR ISAAC RABI, during his MIT tenure. Purcell used a combination of electromagnets activated by radio waves, which allowed him to determine the signature frequencies of different atoms, or their nuclear magnetic resonance. In 1946 Purcell developed his own method of NMR detection, improving upon Rabi's so-called atomic-beam method. In his Nobel acceptance speech, Purcell reported his sense of wonder and delight at witnessing these atoms' identifying their own special tracks of motion.

Purcell later identified the phenomenon of chemical shift, which is the change in NMR signatures due to the transformation of substances such as crystals and liquids in response to their surroundings. Purcell also made a significant contribution to radio astronomy in 1951 as the first scientist to report his observation of a 21-centimeter wavelength of microwave radiation, energy that originated in interstellar neutral hydrogen. This observation confirmed the 1944 theoretical prediction by the Dutch astronomer H. C. van de Hulst. Subsequently these microwaves were used to help map the galaxy and to measure the temperature and motion of interstellar gas.

Purcell was inducted into the National Academy of Sciences in 1951, and in 1978 he received the National Medal of Science from the National Science Foundation. Purcell died on March 7, 1997, in Cambridge, Massachusetts.

Pythagoras

(C. B.C. 580–C. B.C. 500)
Greek
Mathematician, Philosopher

Pythagoras is renowned for his formulation of mathematical principles that influenced future mathematicians and proved seminal in the development of Western rational philosophy. Curiously the impetus for Pythagoras' work arose from his spiritual beliefs. He founded a sect centered around the creed that all things could be reduced to numerical relationships. In the course of exploring these relationships Pythagoras devised a theory concerning right triangles that now bears his name. He also explained musical tones as a function of the length of the strings of the instrument making the sound, and he introduced the concept of irrational numbers. There is, however, some question as to whether he personally made many of the discoveries attributed to him or whether they were made by various of his followers and imputed to him.

As with the pedigree of his inventions, the historical record is quite spotty on the details of Pythagoras' life. It is believed with some certainty, however, that he was born on the Greek island of Samos in approximately B.C. 580. At some point thereafter he moved to Croton (now Crotona) in southern Italy, apparently to escape a repressive government on Samos. He founded his ethicopolitical academy at Croton and attracted a sizable following to his religious brotherhood. This mystical sect strove to achieve spiritual purification, which they believed would culminate in the soul's union with the divine. The group gained adherents and flourished for a time until it became associated with democratic political movements, which earned it the enmity of the state. Historians surmise that Pythagoras was exiled around B.C. 500 to Metapontum, where he remained until his death. His followers were persecuted and forced into hiding. In the middle of the fifth century the Pythagoreans' meeting houses were burned, and by the middle of the fourth century his community had been obliterated.

Theories of astronomy, geometry, and especially numbers were the pillars of Pythagoras' doctrines. The fundamental tenet of the group was that reality was mathematical in its nature: Everything could be reduced to, and explained by, mathematical relationships. This worldview inspired the tremendous mathematical accomplishments achieved by the Pythagoreans. Pythagoras and/or his followers formulated what came to be known as the Pythagorean theorem. This geometrical concept posited that for any right triangle the sum of the squares of the two shorter sides equals the square of its longest size, known as the hypotenuse. This theorem is expressed by the equation $A^2 + B^2 = C^2$. The Pythagoreans also proved the incommensurability of the side and diagonal of a square (that is, that they had no common divisor). Another conceptual breakthrough that emerged from the group was the concept of irrational numbers, or those that are not expressible as a perfect fraction. Pythagoras himself is credited with launching the science of acoustics. His work on musical tones led him to postulate an early expression of mathematical physics.

Pythagoras' impact on subsequent scientific and philosophical movements was profound. Classical Greek philosophy absorbed many of his group's teachings, especially the notion that existing objects are composed of forms, not just material substances. Not only were Plato and Aristotle influenced by the Pythagoreans, but so too were many medieval Christian philosophers. Each of the Pythagoreans' mathematical discoveries advanced the development of science. Indeed, fundamental Pythagorean astronomical concepts held that Earth was spherical and that it (and other planets and stars) moved in a spherical motion (or orbit). NICOLAUS COPERNICUS claimed that these postulates played a role in the formulation of his hypothesis that the Earth and other planets rotate in orbits around the Sun.

Pytheas of Massilia

(fl. 330 B.C.)
Greek
Navigator, Astronomer

Pytheas of Massilia solidified his place in history by providing the first description of the seas of northern Europe. Although there are no extant copies of his work *On the Ocean,* it is described in detail in the writings of the Greek historian Polybius, which have survived. According to Polybius, Pytheas sailed up the east coast of Britain to a northern land he called Thule, which historians now speculate was likely either Iceland or Norway. Although many of Pytheas' contemporaries, as well as later commentators, claimed that this feat was impossible, the precision of Pytheas' descriptions seems to confirm his achievement.

Few specific facts about Pytheas' life are known. He was most likely born in the Greek colony of Massilia, which is now known as Marseilles, France. Around the time he embarked on his voyage new scientific concepts related to geography and navigation were being discovered. It was probably through the writings of Eudoxus of Cnidus, who lived sometime around 400 B.C. and explored both the Arabian Sea and the east coast of Africa, that Pytheas became acquainted with the theory that the Earth was a sphere.

Information about the particulars of Pytheas' northern voyage is as sparse as his biography, but we do know that from Massilia he most likely sailed from the Mediterranean Sea through the Strait of Gibraltar and into the Atlantic Ocean. After stopping at the Phoenician city of Gades (now Cádiz, Spain), it is believed, he followed the European coastline to Brittany. He eventually reached southern England, where he visited the famed tin mines of Cornwall. Pytheas claimed to have explored Britain on foot and to have circumnavigated the island. At some point he then proceeded on to countries in northern Europe. Some historians speculate that he reached the mouth of the Vistula River on the Baltic Sea. In any event, he braved the frigid northern seas to arrive at a land he called Thule, which he reported was a six-day sail from England. The name *Thule* had not existed in the classical world prior to his time and Pytheas' evocative descriptions brought this arctic realm alive for his contemporaries. The sea near Thule, he explained, was "neither land nor sea nor air but a mixture of all, like a sea-lung, in which sea and land and everything swing and which may be the bounds of all, impassable by foot or boat."

On his journey Pytheas made a number of discoveries. For instance, he noted that the days lengthened the farther north he progressed, culminating in midsummer when the Sun never set. He also astutely recorded that the polestar was not at the true pole and that the Moon influenced the tides. He correctly described the shape of Britain and also made impressively accurate calculations of its size. His estimates of the island's circumference and its distance from Marseilles were close to the actual figures. *On the Ocean* also gave its readers the first account of Britain's inhabitants and their habits. He devoted a passage to the natives' drink of cereal and honey, and he compared some of their farming methods to the Greeks' methods.

Although Pytheas' report was viewed with skepticism by some geographers of his time, such as Strabo, others, including ERATOSTHENES and HIPPARCHUS, accepted the validity of his claims. In any event, even allowing for a bit of hyperbole, Pytheas ranks as one of the boldest and most successful explorers of antiquity. His voyage marked the first recorded history of Britain, and long after his odyssey the name *Thule* remained cemented in the vocabulary and imagination of the ancient world.

Quinland, William Samuel

(1885–1953)
West Indian/American
Pathologist

Although William Samuel Quinland was offered a teaching position at Harvard University's Medical College, he turned down the prestigious opportunity so that he could better serve the African-American community as a professor at the black medical school Meharry. This decision typified Quinland's career. As both a pathologist and an educator he sought to focus on African-American patients. He conducted 28 studies on disease among African Americans, a group often overlooked by cancer researchers. Quinland was also the first African American to be elected to the American Association of Pathologists and Bacteriologists.

Quinland was born on October 12, 1885, in All Saints, Antigua (then a part of the British West Indies). The son of William Thomas and Floretta Williams Quinland, he received his secondary education in the West Indies before launching his medical career. In 1923 he married Sadie Lee Watson, and the couple had two children.

After teaching in Antiguan public schools Quinland served as a laboratory assistant for three years in the Ancon Hospital in the Canal Zone, Panama. He then spent four years in a similar post at the Candelaria Hospital in Brazil before immigrating to the United States. He augmented one year at Howard University (1914) with classes at Oskaloosa College in Iowa, where he completed his B.S. degree in 1918. The following year he received his medical degree from Meharry Medical College in Nashville, Tennessee. After graduation he was named a Rosenwald Fellow in pathology and bacteriology at Harvard Medical School, where he remained until 1922. During this period he obtained his certificate in pathology and bacteriology (in 1921) and published his first professional article, on carcinoma. During his final year at Harvard, Quinland served as an assistant in pathology at the Peter Bent Brigham Hospital in Boston.

Harvard offered him a post as a professor, but Quinland chose instead to return to Meharry when his fellowship ended. In 1922 he was appointed professor and head of the pathology department at Meharry, where he would work until 1947. Quinland did more than fulfill his teaching obligations at Meharry: From 1931 until 1937 he was the associate medical director of the college's George W. Hubbard Hospital and worked as a staff pathologist at the Millie E. Hale Hospital. To keep apprised of developments within his field, he completed postgraduate studies as a fellow at the University of Chicago in 1941 and 1942. During his tenure at Meharry, Quinland published studies on tuberculosis, syphilis, and heart disease, as well as his seminal investigation of carcinoma among African Americans. In *Primary Carcinoma in the Negro* he recorded and analyzed the occurrence of 300 different types of carcinoma among male and female African Americans of different ages. In 1947 Quinland left Meharry to head the laboratory service of the Veterans Administration Hospital in Tuskegee, Alabama, where he continued to investigate tumors.

Quinland remained active at the Veterans Hospital until his death on April 6, 1953. Despite the considerable discrimination facing African-American scientists during this period, he was not only elected to the American Association of Pathologists and Bacteriologists in 1920, but also appointed to the American Board of Pathology in 1937 and the College of American Pathologists in 1947. Moreover, he served as the editor of the *Journal of the National Medical Association* and was honored with a post as a reserve surgeon for the United States Public Health Service. His carcinoma studies that took into account the race of the patient were a powerful impetus for future research into diseases afflicting African Americans.

R

Rabi, Isidor Isaac

(1898–1988)
Austrian/American
Physicist

I. I. Rabi's most important contribution to physics was his invention of the atomic- and molecular-beam magnetic resonance methods of observing spectra. These methods allowed for the precise measurement and description of atomic molecules, as well as the confirmation of the theories of quantum electrodynamics. For this achievement he received the 1944 Nobel Prize in physics.

Rabi was born on July 29, 1898, in Rymanow, Galicia (now Poland). His family moved in 1899 to New York City, where he grew up speaking Yiddish at home and English on the streets. His mother was Janet Teig and his father was David Rabi, a jack-of-all-trades who made women's blouses in a sweatshop, according to Rabi. In 1926 Rabi married Helen Newmark, and together the couple had two daughters, Nancy Elizabeth and Margaret Joella.

After attending Manual Training High School in Brooklyn, Rabi entered Cornell University in 1916; he graduated in 1919 with a B.S. in chemistry. Three years later he entered graduate school at Cornell, but in 1923 he transferred to Columbia University as a doctoral student in physics. Rabi supported himself while pursuing his doctorate by teaching physics at City College in New York. He wrote his dissertation on the effects of magnetic fields on matter to earn his Ph.D. in 1927.

After receiving the doctorate Rabi conducted a two-year tour of Europe, visiting the major centers of scientific thought—Munich, Copenhagen, Hamburg, Leipzig, and Zurich—and consulted such luminary scientists as OTTO STERN, NIELS HENRIK DAVID BOHR, WOLFGANG PAULI, and WERNER KARL HEISENBERG. Rabi was most impressed by Stern's work, especially the experiment Stern devised in 1922 with Walther Gerlach demonstrating that a single molecular beam divides into dual beams when passed through a magnetic field. Upon his return to Columbia as a lecturer in physics in 1929 Rabi devoted himself to investigating the effects of magnetism on atomic molecules in more depth. As Columbia promoted him to the posts of assistant professor in 1930 and associate professor in 1932, Rabi was busy designing methods of determining the magnetic moments of nuclei. By the time he was appointed a full professor in 1937, Rabi had invented the means to identify the nuclear magnetic resonance of atomic molecules. That year he also discovered the possibility of reversing the spin of a nucleus by means of a radio-frequency signal.

During World War II Rabi worked at the Massachusetts Institute of Technology on microwave radar technology; he later returned to New York to accept the position of chairman of the physics department at Columbia, which named him its first University Professor in 1964. Also after the war Rabi was named to the General Advisory Committee of the Atomic Energy Commission, and he succeeded J. ROBERT OPPENHEIMER as the chairman of the committee from 1952 to 1956. As a member of the United Nations Educational, Scientific, and Cultural Organization (UNESCO) Rabi championed the creation of the European Center for Nuclear Research (CERN) in Geneva, Switzerland.

In addition to the Nobel Prize, Rabi received numerous prestigious awards: the 1942 Elliott Cresson Medal from the Franklin Institute; the 1948 U.S. Medal of Merit; the Niels Bohr International Gold Medal and the Atoms for Peace Award, both in 1967; and the 1985 Franklin Delano Roosevelt Freedom Medal. Rabi died on January 11, 1988, in New York City.

Isidor Isaac Rabi, whose invention of the atomic-beam and molecular-beam magnetic resonance methods of observing spectra won him the 1944 Nobel Prize in physics. *Brookhaven National Laboratory.*

Rajalakshmi, R.

(1926–)
Indian
Biochemist, Psychologist, Nutritionist

Though lacking a formal education in biochemistry, R. Rajalakshmi has made significant contributions to the field, particularly in her work on nutrition. Since the 1960s Rajalakshmi has examined the link between nutrition and brain function, studied human pregnancy and lactation, and researched diet and nutrition, especially in regard to low-income people in India.

Born Lakshmi in 1926 at Quilon, in Kerala, India, to a middle-class family, Rajalakshmi added the prefix *Raja* in 1931 when entering school. The third of seven children, she grew up in Madras, India, where her maternal grandparents resided. Rajalakshmi's father worked as an accountant in the postal audit office, and her mother was a homemaker. At the age of 12 Rajalakshmi began to help take care of her younger siblings. Because the education of her siblings had been irregular, she tutored them to bring them up to appropriate class levels.

A good student, Rajalakshmi was determined to attend college, and in 1941 she entered Wadia College, in Poona, on a merit scholarship. Wadia College was at that time affiliated with the University of Bombay. Four years later Rajalakshmi received her undergraduate degree in mathematics, with a minor in physics. Though she intended to pursue graduate studies in mathematics at Ferguson College, also in Poona, her plans were delayed when her parents fell ill and she became their caretaker. Because the school term was half over when Rajalakshmi's parents recovered, Rajalakshmi opted to work as a teacher in Kancheepuram, where she stayed for three years. Rajalakshmi later earned her teaching degree from Lady Willingdon College, in Madras. In 1953 she earned a master's degree in psychology from Banaras Hindu University, and in 1958 she was awarded a doctorate in psychology from McGill University, in Montreal, Canada.

After working as a high school teacher for a number of years, Rajalakshmi turned to university teaching and became increasingly interested in biochemistry and nutrition, largely because her husband, C. V. Ramakrishnan, was involved in biochemistry; he became the director of the newly established biochemistry department at Maharaja Sayajirao University of Baroda in 1955. Though Rajalakshmi did not have a job upon her return from Canada in 1958, she began to help doctoral students prepare their dissertations and thus became increasingly familiar with subjects in biochemistry. In the early 1960s Rajalakshmi studied the effects of protein deficiency on learning performance with her husband. In addition, while supervising two Ph.D. theses on nutrition and human lactation, she observed that nursing mothers on low-calorie diets maintained satisfactory lactation and that the ascorbic acid content in the milk was greater than that in the diet. These findings inspired Rajalakshmi to apply for the position of an honorary research worker so that she could work on her own projects. Rajalakshmi proposed two projects—one on the effects of nutritional deficiency on psychological performance and brain biochemistry, the other on ascorbic acid secretion during lactation. She applied for grants from the Human Nutrition Division of the U.S. Department of Agriculture and became the first person in the university to obtain funding from the department.

In 1964 Rajalakshmi became reader in Psychometrics and Biometrics at Baroda, and three years later she transferred to the biochemistry department to teach advanced nutrition. In 1976 Rajalakshmi became a full professor. In addition to working on lactation, Rajalakshmi was instrumental in proposing practical dietary and nutrient allowances for low-income Indian families. Rajalakshmi was also among the first to propose that malnutrition was closely tied to emotional and environmental factors at different stages of development, and that proper nutrition of babies born with low birth weights leads to normal development.

Raman, Sir Chandrasekhara Venkata

(1888–1970)
Indian
Physicist

The Raman effect was named after the man who discovered it, C. V. Raman. The effect occurs when light passes through a transparent substance—solid, liquid, or gas—and diffracts. Although most light waves retain the same frequencies upon exiting the substance, a few emerge with wavelengths that differ. These different wavelengths became known as Raman frequencies or lines. For this discovery Raman received the 1930 Nobel Prize in physics, the first Indian and indeed the first Asian so honored in this field.

Chandrasekhara Venkata Raman, who won the 1930 Nobel Prize in physics, the first Indian to be awarded a Nobel Prize in this field. *AIP Emilio Segrè Visual Archives.*

Raman was born on November 7, 1888, in Trichinopoly, India, the second of eight children. His mother, S. Parvati Ammal, was the daughter of Saptarshi Sastri, a renowned Sanskrit scholar. His father, Ramanathan Chandrasekaran Iyer, was a lecturer in physics, mathematics, and physical geography at A.V.N. College in Vizagapatam.

Raman won a scholarship to Presidency College of the University of Madras and graduated first in his class in 1904, at the age of 16, with gold medals in physics and English. He then commenced graduate study, during which he published his first paper, "Unsymmetrical Diffraction Bands Due to a Rectangular Aperture," in the British journal *Philosophical Magazine* in 1906, at the age of 18. In 1907 he earned his master's degree in physics with highest honors. That year he married the artist Lokasundari.

Despite his proven acumen for physics, Raman was unable to pursue the discipline as a career, as his frail health prevented him from traveling to England, where he could have pursued advanced studies. Instead he entered the Indian Civil Service, securing a coveted post as an accountant by scoring highest on the examination. On his own he continued to experiment and research in physics. In 1917 Calcutta University offered him the newly established Palit Chair for Physics, which he accepted even at half his current pay. In 1928 he observed and reported on the Raman effect; that work proved useful scientifically in analyzing the degree to which the Raman lines diverged from the spectrum, revealing information about the structure of the molecule.

In 1934 Raman founded the Indian Academy of Sciences and its journal, the *Proceedings of the Indian Academy of Sciences,* which he contributed to regularly. The previous year Raman was appointed president of the Indian Institute of Science in Bangalore, a position he held until 1937. He remained the head of the physics department from his arrival until his departure in 1948 to found the Raman Research Institute.

Besides the Raman effect, several scientific phenomena were named after him, including the Raman curve, which described the frequency response of a violin, and the Raman-Nath theory, which explained what happened to a light beam passing through a liquid agitated by sound waves. Raman was knighted in 1929. He died on November 20, 1970, in Bangalore, and his ashes were strewn about the woods on the campus of the research institute named after him. India revered its native son, Raman, for modernizing the study of science there and giving generously of his time and expertise to his students, who dispersed to spread Raman's influence throughout the country.

Ramanujan, Srinivasa Iyengar

(1887–1920)
Indian
Mathematician

S. I. Ramanujan was a mathematical genius who compensated for his lack of formal education with intuitive brilliance concerning numbers and their properties. Since much of his work lacked disciplined explanations for the complex theoretical proofs, it provided fodder for generations of mathematicians to contemplate his originality and attempt to recreate his thought processes. Long after his death computer scientists and polymer chemists found applications in their fields for Ramanujan's mathematical theories.

Ramanujan was born on December 22, 1887, in Erode, India, into the Brahman class. His mother, Komalatammal, sang devotional songs in the local temple, and his father, K. Srinivasa Iyengar, worked as a bookkeeper for a cloth merchant in Kumbakonam. Ramanujan's parents arranged his marriage with the parents of Janaki, the teenage girl who became his wife in 1909.

In 1897 Ramanujan placed first in the Tanjore district's primary examination. By that time he was stumping his mathematics teachers with questions they scarcely understood. In 1903 he borrowed a text from a college student, George Shoobridge Carr's *Synopsis of Elementary Results in Pure and Applied Mathematics,* a list of solutions without explanations or proofs. Ramanujan set out to provide the proofs and solutions to these equations, a daunting task for even an advanced mathematician. The next year Ramanujan graduated from high school with a special prize in mathematics and a scholarship to Government College of the University of Madras. However, Ramanujan's singular obsession with mathematics caused him to fail college, even after attempting to retake the fine arts examination four times between 1904 and 1907.

Without a degree Ramanujan experienced difficulty securing a job, especially since his mind remained singularly focused on mathematics. In 1910 he published his first paper, "Some Properties of Bernoulli's Numbers," in the *Journal of the Indian Mathematical Society.* Ramachandra Rao, a professor of mathematics at Presidency College of University of Madras, recognized his genius and supported him financially before securing a modest clerk's job at the Madras Port Trust for Ramanujan in 1912.

On January 16, 1913, Ramanujan wrote a letter to the number theorist Godfrey Harold Hardy of Cambridge University containing original solutions to problems that Hardy himself could not conceive—the sheer brazenness of the proofs convinced Hardy of their authenticity. Hardy arranged for a two-year fellowship from the University of Madras to support Ramanujan to travel to Cambridge, only after the goddess Namagiri appeared in a dream to his mother, absolving him of his religious duties at home. Ramanujan spent the next five years at Trinity College, learning the rudiments of modern mathematics while stimulating the most learned mathematicians with almost-incomprehensible mathematical theories. During this time he published such articles as "Modular Equations and Approximations to pi" and "Highly Composite Numbers."

In 1918 Trinity College appointed Ramanujan as a fellow, and that year the Royal Society also elected him as its youngest fellow, at the age of 30. The year before, however, he contracted what was probably a case of tuberculosis, and in 1919 he returned home to India. His religious beliefs prevented him from accepting medical aid, though those beliefs also sustained his interest until his final days in mathematics, which he believed to be an expression of God's thoughts. Ramanujan died on April 26, 1920, in Chatput, India.

Ramón y Cajal, Santiago

(1852–1934)
Spanish
Histologist

A pioneer in the field of histology, Santiago Ramón y Cajal was awarded the 1906 Nobel Prize in physiology or medicine for his work on the structure of the nervous system. Ramón y Cajal successfully established that the neuron, or individual nerve cell, was the basic unit of structure of the nervous system. His research greatly advanced scientific understanding of the nerve impulse and of the role of the neuron in nervous function. Ramón y Cajal also studied the brain's cellular structures.

Born in the small village of Petilla de Aragon in Spain on May 1, 1852, to Antonia Cajal and Justo Ramón y Casasús, a barber-surgeon, Ramón y Cajal grew up in a poor environment. The family later moved to Zaragoza, where Ramón y Cajal's father earned his M.D. from the university and became a professor of anatomy. Ramón y Cajal developed an early interest in art, which was contrary to his strict father's desire that he study medicine. As a result, Ramón y Cajal became rebellious and did poorly in school. He was then forced to endure apprenticeships with a barber and then a shoemaker, during which he also studied anatomy with his father.

After completing the disciplinary apprenticeships, Ramón y Cajal began medical studies at the University of Zaragoza. He received his degree in 1873 and entered the army medical service. Malaria led to his discharge, and while recuperating Ramón y Cajal earned his M.D. in 1879. He married Silveria Fananas Garcie in 1880 and the couple had six children.

Interested in research in anatomy, Ramón y Cajal chose an academic research career. He headed the University of Zaragoza's anatomical museum from 1879 to 1883, then taught anatomy at the University of Valencia, beginning in 1883. Ramón y Cajal was professor of histology at the University of Barcelona from 1887 to 1892 and professor of histology and pathologic anatomy at the University of Madrid from 1892 to 1922.

After studying anatomical tissues and becoming a knowledgeable histologist, Ramón y Cajal began to investigate the structure of the nervous system. Improving upon the process of staining nervous tissue developed by the Italian histologist CAMILLO GOLGI, Ramón y Cajal studied embryonic tissue samples and determined that the neuron was the basic structural unit of the nervous system. He also supported the neuron theory, which proposed that nerve fibers did not come into contact with nerve cells. This hypothesis contradicted the common theory that the nervous system was an interconnected system.

Ramón y Cajal investigated the retina of the eye and the tissues of the inner ear. He also studied the connections of nerve cells in the gray matter of the brain. He created a gold stain in 1913 for studying the structure of nervous tissue in the brain and spinal cords of embryos. The stain became useful in the diagnosis of brain tumors. Ramón y Cajal also researched the deterioration of tissue in the nervous system and the regeneration of severed nerve fibers.

In addition to the 1906 Nobel Prize, which he shared with Golgi, Ramón y Cajal received the Rubio Prize in 1897, the Moscow Prize in 1900, and the Royal Academy of Berlin's Helmholtz Gold Medal in 1905. Ramón y Cajal was a prolific writer and published numerous works. King Alfonso XIII of Spain commissioned the building of the Cajal Institute in Madrid in 1920, and Ramón y Cajal worked there until his death in 1934.

Richards, Ellen Henrietta Swallow

(1842–1911)
American
Chemist

Ellen Swallow Richards, the first woman graduate of the Massachusetts Institute of Technology (MIT), was one of the first scientists to use chemistry to check the purity of air, water, and food. She taught homemakers how to use science to improve their families' health and was one of the founders of ecology, which studies interactions in the environment of living things.

Ellen Henrietta Swallow was born on December 3, 1842, in Dunstable, Massachusetts, the only child of Peter and Fannie Swallow. Her father was a schoolteacher, farmer, and storekeeper, and her mother also taught school. She wrote later that she was "born with a desire to go to college," but she had to endure years of what she called "purgatory" before she saved enough money to enroll in 1868 at Vassar College in Poughkeepsie, New York. Her only complaint about Vassar was that, because of the common belief that women were too frail for arduous academic work, "they won't let us study enough." She was inspired by the astronomy classes of MARIA MITCHELL and the chemistry classes of Charles Farrar, and Farrar's insistence that chemistry be used to solve practical problems made her decide to major in that subject. She earned her B.A. in 1870.

Swallow decided to continue her education at the new Institute of Technology in Boston, which later became MIT. The college had never had a woman student, but officials told Swallow in December 1870 that she would be admitted as a special student in chemistry at no charge. She thought the college was recognizing her financial need, but she learned later that she had received free tuition so that the president "could say I was not a student, should any of the trustees or students make a fuss about my presence. Had I realized upon what basis I was taken," she added, "I would not have gone." Nonetheless, she was pleased to become "the first woman to be accepted at any 'scientific' school." She earned a B.S. in chemistry from MIT in 1873, as well as a master's degree from Vassar. She went on to complete the requirements for a doctorate, but MIT would not grant the degree because, she wrote, "the heads of the department did not wish a woman to receive the [university's] first D.S. in chemistry."

Swallow's friends on MIT's chemistry faculty included William R. Nichols, whom she assisted in an examination of the state's water supply for the Massachusetts State Board of Health that began in 1872, and Robert Hallowell Richards, the young head of the metallurgical and mining engineering laboratory, whom she married in June 1875. Ellen Richards used her chemistry skills to help her husband, and he in turn encouraged her to continue her scientific career. For her analysis of metals in ores she was made the first woman member of the American Institute of Mining and Metallurgical Engineers in 1879.

In 1876 a chemistry laboratory for women opened at MIT, with John Ordway as its head and Richards his unpaid assistant. She was made a paid instructor in chemistry two years later. When the lab closed in 1883 because women students were then allowed to use the same facilities as men, Richards feared she would no longer "have anything to do or anywhere to work," but the university opened the country's first institute in sanitary chemistry soon afterward, and she again became an assistant to William Nichols, who headed it. Her title became instructor in sanitary chemistry.

As scientists were recognizing the link between illness and pollution of air, water, and food, governments were starting to demand chemical analysis of these vital elements as a safeguard to public health. With A. G. Woodman she summarized her teachings on this subject in *Air, Water, and Food for Colleges*, published in 1900.

Richards supervised the laboratory during a new survey of Massachusetts water directed by Nichols's successor, Thomas M. Drown, between 1887 and 1889. She effectively established the world's first water purity tables and the first state water quality standards. She taught that "water rightly read is the interpreter of its own history," containing traces of every substance it has encountered.

From about 1890 on Richards concentrated on the practical education of homemakers. She believed that many medical and social problems could be solved or prevented if housewives learned how to make food, water, and air clean and healthful. She redesigned her own home as a demonstration laboratory. In 1890 she and others also opened the New England Kitchen, which sold nutritious, inexpensive meals and let people watch the meals being prepared.

Richards and others who shared her interest developed the idea of scientifically based household management into a new discipline called home economics. In 1908 they formed the American Home Economics Association, dedicated to "the improvement of living conditions in the home, the institutional household and the community." They chose Richards as the association's first president, an office she held until 1910. She wrote many books, articles, and lectures on the subject.

Richards's understanding of the way interactions between the living and nonliving environment shape the life of living things led her to call in 1892 for "the christening of a new science," which she termed *oekology,* from the Greek word for "household." Ernst Haeckel and some other European biologists proposed a similar idea at about the same time. The modern discipline of ecology echoes the latter scientists' focus on plants and animals, but Richards was more interested in the environment of human beings. The claim that Richards "founded ecology" is somewhat exaggerated, but both the environmental and consumer education movements descend in part from her work.

Ellen Richards died on March 30, 1911, of heart disease at the age of 68. Many prizes, funds, and buildings were named for her, including the Ellen H. Richards Fund for research in sanitary chemistry at MIT and Pennsylvania College's Ellen H. Richards Institute for research to improve standards of living. Her best memorial, however, was the improved health in American homes and cities that resulted from her reforms.

Richter, Charles Francis

(1900–1985)
American
Seismologist

Charles Francis Richter, along with Beno Gutenberg, developed the standard scale for measuring the power of earthquakes. Although it was Gutenberg who suggested that they make the scale logarithmic by degrees of 10, his name was dropped from the designation. It was Richter who suggested that the maximum amplitude of earthquake waves measured on a seismograph, after adjusting for the distance from the epicenter of the quake, be called its *magnitude,* borrowing this term from astronomers. The Richter scale became so ubiquitous that even earthquake measurements according to subsequently devised scales were reported as "Richter scale" readings.

Richter was born on April 26, 1900, on a farm near Hamilton, Ohio. His parents divorced when he was young, and his maternal grandfather moved him to Los Angeles in 1909. Richter attended the University of Southern California his freshman year before transferring to Stanford University, where he received his A.B. in physics in 1920. He conducted his doctoral work in theoretical physics at the California Institute of Technology (Caltech) and earned his Ph.D. in 1928. That year he married Lillian Brand, a creative writing teacher.

By then the Caltech president and Nobel laureate Robert A. Millikan had already tapped Richter to work as a research assistant at the newly established Seismological Laboratory in Pasadena, then managed by the Carnegie Institution of Washington. There Richter teamed up with Gutenberg to devise an objective, quantitative scale for measuring and comparing earthquakes to replace the subjective, qualitative scale devised by the Italian priest and geologist Giuseppe Mercalli in 1902. The pair published their new scale in 1935; it was rapidly accepted and used worldwide. Meanwhile Richter and Gutenberg constantly tracked and recorded earthquakes throughout California. They reported their findings in the 1941 text *Seismicity of the Earth,* which was reprinted in 1954.

The year after the publication of the scale Richter returned to Caltech, which promoted him to the status of professor of seismology in 1952. Richter remained there the rest of his career, save the one year he served as a Fulbright Scholar at the University of Tokyo, 1959–1960. In 1958 he had collected all of his accumulated knowledge on earthquakes into one book, *Elementary Seismology,* the definitive text on the subject. Richter continued to study and track earthquakes even after his retirement by establishing a seismic consulting firm.

Richter devoted his life singularly to seismology. He maintained an antialarmist attitude, pointing out that highway travel was riskier than earthquakes. He filtered his energy into preparing for earthquakes by reviewing building codes; he opposed buildings higher than 30 stories. He remained skeptical of earthquake predictions because he understood the limitations of tracking seismic activity, though he did organize the Southern California Seismic Array, a network of interconnected instrumentation that recorded and interpolated seismic information. Richter's obsession with seismology extended into his personal life, as he installed a seismograph in his living room to track earthquake activity at all hours. Richter died of congestive heart failure on September 30, 1985, in Pasadena.

Robbins, Frederick Chapman

(1916–)
American
Virologist, Pediatrician

While conducting experiments on the mumps virus, Frederick Chapman Robbins and his colleague THOMAS HUCKLE WELLER took advantage of some extra test tubes of human embryonic tissue to try to incubate poliomyelitis. At that time polio was known to grow only in neural and brain tissue, which acted as toxins when researchers tried to prepare a polio vaccine. Surprisingly, the polio cultured in the embryonic tissue, an important discovery that led to the development of a polio vaccine by JONAS EDWARD SALK. Robbins, Weller, and JOHN FRANKLIN ENDERS received the 1954 Nobel Prize in physiology or medicine for their roles in the breakthrough that led to the polio vaccine.

Robbins was born on June 21, 1916, in Auburn, Alabama, the eldest of three sons, to Christine F. Chapman, a Wellesley graduate and researcher in botany, and Dr. William Jacob Robbins, a plant physiologist who had formerly directed the New York Botanical Garden and was a professor of botany at the University of Missouri. In 1948 Robbins married Alice Havemeyer Northrop, and together the couple had two daughters, Alice and Louise.

Robbins earned his B.A. in 1936 and his B.S. in premedicine in 1938 from the University of Missouri. He then proceeded to Harvard Medical School, where he earned his M.D. in 1940. He served his internship and residency at Children's Hospital in Boston until World War II interrupted. Between 1942 and 1946 Robbins served as the chief of the section of the Fifteenth Medical General Laboratory in North Africa and Italy studying viruses and bacterial diseases. He was awarded a Bronze Star after the war for this work.

In 1946 Robbins returned to Children's Hospital as an assistant resident, and in 1948 he was named chief resident. He then received a two-year senior fellowship from the National Research Council, working in Ender's laboratory. Concurrently he held an associate professorship of pediatrics at Harvard, and in 1950 he was named the associate director of Isolation Services at Children's Hospital. He moved in 1952 to take on the directorship of the pediatrics and contagious diseases department at the Cleveland City Hospital and to serve as a professor of pediatrics at Case Western Reserve Medical School. In 1965 the medical school named him its dean, a position he filled until his 1980 retirement.

Robbins's attention became increasingly focused on health policy, as he was appointed president of the Society for Pediatric Research in 1961 and 1962. From 1973 through 1974 he served as the president of the American Pediatric Society. In 1979 he was named the chairman of the advisory council for the congressional Office of Technology Assessment. After retirement from Case Western Reserve in 1980 he was the president of the Institute of Medicine until 1985, and in that same period he was appointed a distinguished professor of pediatrics at Georgetown University.

Although Robbins will be remembered for the contribution to virology that earned him the Nobel Prize, he perhaps had an even larger impact through his work on health policy.

Robinson, Julia Bowman

(1919–1985)
American
Mathematician

Julia Bowman Robinson made advances in logic and number theory and helped to find the answer to what an eminent mathematician called one of the greatest unsolved problems in mathematics. She was born in St. Louis, Missouri, on December 8, 1919, the second daughter of Ralph Bowers and Helen Hall Bowman, who owned a machine tool and equipment company. She grew up in Arizona and San Diego, California, with her sisters, father, and stepmother (her mother died when Julia was two).

At the University of California at Berkeley, Bowman for the first time found others who shared her interest in mathematics. One was a young professor, Raphael M. Robinson, with whom she took long walks. By the time of their marriage in late 1941 Julia had obtained a bachelor's (1940) and a master's degree (1941). She would have liked to raise a family, but a childhood illness that seriously damaged her heart made pregnancy dangerous for her. Her husband reminded her that "there was still mathematics," and she resumed her studies and obtained a Ph.D. in 1948.

As a consultant for the Rand Corporation in 1949–50 Robinson proved that in a "zero sum game"—a game in which there are two players, one of whom must win and one lose—a situation in which each player follows a strategy that is the average of the values of the previous two moves would converge toward a solution of the game (that is, one of the players would win). The mathematician David Gale told Robinson that her theorem was the most important one in elementary game theory.

Robinson's chief achievement, however, concerned the 10th in a list of unsolved mathematical problems made by the German mathematician David Hilbert around 1900. The problem was to find out whether one could determine whether a diophantine equation had a solution that could be expressed in integers (whole numbers). Working with Martin Davis and Hilary Putnam, Robinson proposed in 1961 that this problem could be approached by studying exponentiation (numbers increasing by a power) and polynomials (equations that use both multipliers and powers).

Early in 1970 Robinson learned that a young Russian, Yuri Matijasevich, had used her work to solve the problem. As she had suspected, the solution was negative: One cannot determine whether a diophantine equation has an integer solution. Far from being angry that Matijasevich had arrived at the solution first, she wrote to him: "It is beautiful, it is wonderful. If you really are 22 [he was], I am especially pleased to think that when I first made the conjecture [that the answer to the problem was no], you were a baby and I just had to wait for you to grow up!"

Robinson was awarded many honors for her part in solving Hilbert's problem. In 1975 she became the first woman mathematician elected to the National Academy of Sciences, and Berkeley made her a full professor even though her health prevented her from taking a full teaching load. She was also elected a member of the American Academy of Arts and Sciences, received a MacArthur Genius Fellowship in 1982, and in 1983 became the American Mathematical Society's first woman president. Her heart was repaired in 1961; she died of leukemia on July 30, 1985.

Robinson once wrote: "Rather than being remembered as the first woman this or that, I would prefer to be remembered, as a mathematician should, simply for the theorems I have proved and the problems I have solved."

Rohrer, Heinrich

(1933–)
Swiss
Physicist

In 1986 the Swedish Academy of Arts and Sciences split its Nobel Prize in physics, awarding half to Ernst Ruska for the invention of the electron microscope in 1931 and the other half to Heinrich Rohrer and GERD BINNIG for their invention of the scanning tunneling microscope (STM), which they had tested successfully a mere five years earlier and were still refining. Whereas the electron microscope magnified images as small as 5 angstroms (atoms typically measure 1 to 2 angstroms), the new microscope magnified an image as small as 0.1 angstrom, thus allowing for the study of the atomic structure of the surface of solids.

Rohrer was born on June 6, 1933, in Buchs, St. Gallen, Switzerland. His mother was Katharina Ganpenbein; his father, Hans Heinrich Rohrer, was a manufactured goods distributor. In 1961 Rohrer married Rose-Marie Eggar, and together the couple had two daughters.

In 1951 Rohrer entered the Federal Institute of Technology in Zurich to study physics. He earned his degree in 1955 and conducted doctoral research on superconductivity to earn his Ph.D. in 1960. Rohrer continued on at the institute as a research assistant until 1961, when he commenced a two-year stint conducting postdoctoral research on superconductivity at Rutgers University in New Jersey. When he returned to Zurich he joined the International Business Machines (IBM) Research Laboratory in Roschliken, Switzerland, where he remained for the rest of his career save one year when he was a visiting scholar at the University of California at Santa Barbara studying nuclear magnetic resonance in 1974 and 1975.

He commenced studying magnetic fields and critical phenomena at IBM, but his attention was eventually drawn to the desire to view the atomic structure of solid surfaces, a feat not possible within the constraints of the most powerful electron microscope. Binnig's arrival at IBM in 1978 acted as a catalyst for Rohrer, and together they embarked on research into developing a spectroscopic probe. They hit upon the idea of tunneling, a quantum phenomenon whereby electrons leaked from adjacent surfaces would leap back and forth between overlapping "clouds" of electrons. Binnig and Rohrer proposed building a probe that would act as the adjacent surface to the surface being studied to create a tunneling effect between the two. The surface of the spectroscopic probe would be outfitted with a tungsten electron field emitter tip to measure the appearance of the electrons on the surface, guided by precision piezoelectric transducers that maintained a constant distance from the surface, thereby tracing it to create three-dimensional representations of atomic structure.

Rohrer and Binnig continued to improve the design of the STM, most notably by stabilizing it against disruption from sound and physical vibrations by means of a stone table placed on top of inflated rubber tires and surrounded by magnets drawn to copper surfaces, reducing vibration to 0.1 angstrom vertically and 6 angstroms laterally. The pair published reports on their advancements in the article "The Scanning Tunneling Microscope" in *Scientific American* and in "Scanning Tunneling Microscopy: From Birth to Adolescence" in the *Review of Modern Physics*.

Röntgen, Wilhelm Conrad

(1845–1923)
German
Physicist

Wilhelm Conrad Röntgen wrote 48 papers reporting his research on the specific heats of gases, the heat conductivity of crystals, the Faraday and Kerr effects, and the compressibility of solids and liquids, among other topics. He wrote only two papers on the research that earned him lasting fame: his 1895 discovery of X rays, for which he received the first Nobel Prize in physics in 1901.

Röntgen was born on March 27, 1845, in Lennep, Germany. His mother was Charlotte Constanze Frowein, who was originally from Lennep before her family moved to Amsterdam, and his father, Friedrich Conrad, was a cloth manufacturer and merchant. When Röntgen was three years old, his family moved to Apeldorn, Holland. On January 19, 1872, Röntgen married Anna Bertha Ludwig, the daughter of a German revolutionary who had immigrated to Switzerland. Although the couple never had children of their own, they adopted Ludwig's niece, Josephine Bertha, in 1887, when she was six years old.

In December 1862 Röntgen matriculated at Utrecht Technical School, which expelled him two years later when he took responsibility for a teacher's caricature drawn by a classmate. The University of Utrecht admitted him as an "irregular" student in January 1865, and in November of that year he transferred to the Swiss Federal Institute of Technology in Zurich, where he received his diploma in mechanical engineering in August 1868. On the advice of his mentor, August Kundt, Röntgen conducted his doctoral research in physics; he wrote his thesis "Studies about Gases" to earn his Ph.D. in June 1869.

Röntgen served as Kundt's assistant, following him to the University of Würzburg in Germany in 1870 and then to the University of Strasbourg in France in 1872. The Hohenheim Agricultural Academy appointed him as a professor of physics in 1875, but he returned to Strasbourg as an associate professor of physics the next year. He served as a professor of physics at the University of Giessen in Germany between 1879 and 1888, then returned to Würzburg as a physics professor and director of its Physical Institute. Between 1894 and 1900 he served as rector for the university.

While experimenting on November 8, 1895, with Sir WILLIAM CROOKES's cathode rays, which caused the glass of a vacuum tube to luminesce from the discharge of an electrical current, Röntgen discovered that a screen covered in barium platinocyanide crystals from the tube glowed. Since the cathode rays traveled only a matter of centimeters, Röntgen surmised that another energy force must have caused the glow. He experimented another seven weeks, noticing that these mysterious rays could expose photographic paper and travel through glass, wood, and even his hand, but not through certain metals, such as lead. On December 22, 1895, he produced the first X-ray photograph, using his wife's hand as subject. He reported his findings to the editors of the journal of the Physical and Medical Society of Würzburg on December 28, 1895, and on January 13, 1896, he demonstrated the fascinating discovery before the Prussian court.

The kaiser awarded Röntgen the Prussian Order of the Crown, Second Class, but Röntgen declined the offer of the title of *von*, as well as the patent for his discovery. He even donated his monetary award from the Nobel Prize to the University of Würzburg, earmarking it for scientific research. He won countless awards, notably the 1896 Rumford Medal of the Royal Society of London, but he remained exceedingly humble concerning his discovery of X rays. After a short illness due to intestinal cancer Röntgen died on February 10, 1923, at his country home in Weilheim, near Munich.

Rubin, Vera Cooper

(1928–)
American
Astronomer

Vera Rubin has made several discoveries that produced major changes in astronomers' view of the structure of the universe. She was born Vera Cooper in Philadelphia on July 23, 1928, one of Philip and Rose Cooper's two daughters. Her father was an electrical engineer. From childhood on, she has said, "I just couldn't look at the sky without wondering how anyone could do anything but study the stars."

Cooper studied astronomy at Vassar College in New York and graduated in 1948. There she met Robert Rubin, a young physicist studying at Cornell. They married just after her graduation, and she followed him to Cornell for her master's degree work even though its astronomy department was small.

For her thesis project Rubin studied the motion of galaxies. She found that galaxies at about equal apparent distances from Earth were moving faster in some parts of the sky than in others, a result that no astronomical theories could explain. Rubin's work earned her degree in 1951, but when she presented it at a meeting of the American Astronomical Society, she says she was all but politely hooted off the stage. She and others have since confirmed it.

Soon afterward the Rubins moved to Washington, D.C. Vera stayed home at first to take care of her young son (she later had a daughter and two more sons), but she missed astronomy so much that she burst into tears each time she read an issue of the *Astrophysical Journal*. Her husband urged her to begin studying for her doctorate at nearby Georgetown University.

Rubin's project this time used mathematics to determine the distribution of galaxies in the local part of the universe. Its result was just as unsettling as her earlier conclusion. Galaxies were supposed to be distributed evenly in space, but Rubin found that this was not so. More recently astronomers such as MARGARET JOAN GELLER have confirmed by observation that the universe has a "lumpy" structure, in which clusters of galaxies are separated by nearly empty voids. The reason for this distribution still is not known.

Rubin received her Ph.D. from Georgetown in 1954 and remained on its faculty for 11 years, meanwhile raising her family. In 1965 she joined the Department of Terrestrial Magnetism (DTM), part of the Carnegie Institution of Washington, where she still works. Despite its name, the DTM sponsors projects of many types, including astronomical ones.

Around 1970 Rubin and another DTM astronomer, Kent Ford, measured how fast stars in different parts of Andromeda, the nearest full-size galaxy to ours, were rotating around the galaxy's center. They assumed that most of the galaxy's mass was near its center, as most of its light was. If this was so, gravity should cause stars near the center of the galaxy to move faster than those near the edge. Rubin and Ford found, however, that stars near the edges of the galaxy were moving as fast as, sometimes faster than, those near the center.

Rubin and Ford showed in the mid-1970s that other galaxies follow this same pattern. Indeed, some stars in the outer parts of galaxies move so fast that they ought to escape the galaxies' gravity and fly off into space, yet that does not happen. This suggests that the galaxies contain a large amount of mass that no one can see. Rubin and others have since confirmed that 90 percent of the universe is made up of what has come to be called *dark matter.* No one knows what this material is.

Rubin's startling and important discoveries have won her election to the National Academy of Sciences and, in 1993, the National Medal of Science, the U.S. government's highest science award. In 1996 she also was awarded the Gold Medal of London's Royal Astronomical Society. Rubin was the first woman to receive this medal since CAROLINE LUCRETIA HERSCHEL in 1828.

Russell, Bertrand Arthur William

(1872–1970)
British
Mathematician, Philosopher

Although best known for his innovative work in mathematical logic, Bertrand Russell had legendary intellectual accomplishments. He published more than 40 books across the disciplines of philosophy, mathematics, science, ethics, sociology, education, history, religion, and politics. Much of his work was aimed at reducing the pretensions of human knowledge, thereby discarding many of the assumed (but unproven) axioms upon which much scientific thought rested. He devoted other works to unifying mathematics and logic, a task that required establishing a small number of logical principles from which all of mathematics originated. A committed pacifist, he was twice imprisoned for his stance against war and nuclear proliferation.

Russell was born on May 18, 1872, in Trelleck, Monmouthshire, England, to politically active parents. His father, Lord Amberly, had been elected to Parliament but lost his seat because he championed birth control and voting rights for women. His mother, Kate Stanley, was the daughter of a liberal politician. Both parents died when Russell was only four. He was subsequently raised by his grandmother, a devout Presbyterian who prized virtue and duty above all else. Russell later credited her with providing him with a strong moral compass and an indifference to the judgments of others.

The roots of Russell's adult character and concerns were evident in his youth. While studying the work of EUCLID as a child, Russell became disillusioned when he realized that, in order to remain intact, the seemingly flawless discipline of mathematics required that certain assumptions be taken on faith. He enrolled at Trinity College, Cambridge, in 1890, after being schooled at home by tutors. He initially embraced mathematics but turned his attention to philosophy in 1893. He received his degree in moral sciences in 1894 and submitted a dissertation on the foundations of geometry the following year.

Russell's long and varied career consisted mainly of lectureships at many different universities, including Cambridge and Harvard Universities and Bryn Mawr College; peripatetic travel; and prolific writing. More than anything, Russell's publications defined his work and made him famous. His books on mathematics and logic were particularly influential.

In 1903 Russell published *The Principles of Mathematics,* which strove to distill mathematical knowledge to core logical principles. He collaborated with his former tutor, Alfred North Whitehead, in his three-volume masterpiece *Principia Mathematica* (1910, 1912, and 1913), to pursue this objective further. In a similar vein but an entirely different field, Russell sought to reduce language to what he called its "atomic facts," in order to foster greater precision in the description of objects. He presented this theory of description in *The Analysis of Matter* (1927) and *The Analysis of Mind* (1921). Another important work was the *Introduction to Mathematical Philosophy* (1919), which he wrote when he was imprisoned for protesting Britain's involvement in World War I. Russell was passionately opposed to war and devoted a large portion of his considerable energies to this topic. He spoke eloquently against the proliferation of nuclear weapons and was president of both the 1957 Pugwash Conference of Scientists and the Campaign for Nuclear Disarmament.

Russell's personal life was as varied as his professional one. He was married four times—to Alys Pearsall Smith (1894), Dora Black (1921), Patricia Spence (1936), and Edith Finch (1952). He had three children. His rejection of the accepted morality of the time on occasion led to his ostracism and the loss of academic appointments.

The influence Russell exerted on the development of logic, philosophy, and the philosophy of science was enormous. Not only were his works widely read, but also in his teaching capacity he shaped the minds of his students, who included the philosopher Ludwig Wittgenstein. Russell was elected to the Royal Society in 1913 and awarded the Nobel Prize in literature in 1958. He died in Penyrhyndeudraeth, Merioneth, England, on February 2, 1970.

Rutherford, Ernest

(1871–1937)
New Zealander
Physicist

Although it was Ernest Rutherford's explanation of radioactivity that earned him a Nobel Prize, his most significant contribution to science was made later, when he established that the atom consisted of a dense nucleus orbited by electrons. While researching radioactivity, he discovered alpha and beta radiation and came to understand that radioactivity was the result of spontaneous disintegration of the atoms of one element into the atoms of a totally different element. He also concluded that alpha particles consisted of the nuclei of helium atoms.

Ernest Rutherford, who confirmed that the atom consists of a dense nucleus orbited by electrons. *Smith Collection, Rare Book & Manuscript Library, University of Pennsylvania.*

Rutherford was born near Spring Grove, New Zealand, on August 30, 1871. His father, James Rutherford, pursued a number of jobs, including logging, construction, and farming. His mother, Martha Thompson Rutherford, was a schoolteacher. Before he left New Zealand, Ernest Rutherford met his wife, Mary Newton. After their marriage in 1900 the couple had one daughter.

Rutherford received both his B.A. (in 1892) and his master of arts degree in mathematics and mathematical physics (in 1893) from Canterbury College at Christchurch, New Zealand. In 1895 he won a scholarship to Cambridge University in England and became the physicist J. J. Thomson's first research assistant at the Cavendish Laboratory.

In 1897 Rutherford began his pioneering studies of radioactivity, the phenomenon that ANTOINE-HENRI BECQUEREL had discovered the previous year. Rutherford's initial research with uranium led him to conclude that uranium emitted two types of radiation—alpha rays (which were less penetrating) and beta rays (the more penetrating of the two). He continued to investigate radioactivity after he left the Cavendish for a position at McGill University in Montreal in 1898. In 1901 Rutherford collaborated with FREDERICK SODDY to study the seemingly erratic behavior of thorium. They had noticed that radioactive emanations from thorium led to the production of a new radioactive substance. After careful analysis they concluded in 1903 that radioactivity is the physical manifestation of an element's decay. Decay occurs because some elements are not permanent. When an element such as thorium releases radiation, its atoms are actually being transformed into a new element—one with a lower atomic number. In other words, the former element has shed some of its protons. Rutherford also formulated the concept of what is now called the half-life of a radioactive element, which is the time half a substance takes to decay. Rutherford discerned that this decay occurred geometrically over time. In 1903 he concluded that alpha particles, which had a positive charge, were the nuclei of helium atoms.

Rutherford accepted the chair of the physics department at Manchester University in 1907. In 1911 he discovered that when a stream of alpha particles are fired at gold foil, some of the particles are deflected back. On the basis of this observation he concluded that the accepted model of an atom—a mass of positively charged particles in which negatively charged electrons are embedded—was inaccurate. Rutherford made the revolutionary proposal that electrons orbited a small, dense nucleus in rings. Rutherford's colleague at Manchester, NIELS HENRIK DAVID BOHR, employed this insight in conjunction with the quantum theory elaborated by MAX PLANCK to prove that electrons have stable orbits around the nucleus. In 1919 Rutherford was named director of the Cavendish. In the 1920s he predicted the existence of the neutron, which was later confirmed by Sir JAMES CHADWICK. In 1934 Rutherford was part of a collaboration that achieved the first fusion reaction, which was created by bombarding deuterium with deuterons (the nuclei of deuterium atoms).

Rutherford's accomplishments were amply recognized during his lifetime. In addition to receiving the Nobel Prize in chemistry in 1908, he was knighted in 1914 and made a baron in 1931. His work had a profound and lasting impact. He meticulously laid what became the foundation for nuclear physics. His conception of the atom fueled Bohr's use of quantum mechanics to explain atomic behavior. Moreover, his leadership helped launch the new field of particle physics.

S

Sabin, Alfred Bruce

(1906–1993)
Polish/American
Virologist

Alfred Bruce Sabin developed an oral vaccine for polio that was eventually accepted as the best way to combat the dreaded disease. Although JONAS EDWARD SALK's injected polio vaccine was the first to be widely used and thus won for him the name recognition that Sabin never achieved, Sabin's attenuated live-virus vaccine ultimately proved to have significant advantages over Salk's. Sabin also created vaccines against dengue fever and Japanese B encephalitis. Moreover, he gained insight into the human immune system's mechanisms for resisting viruses and later turned his attention to the role of viruses in cancer.

Sabin was born on August 26, 1906, in Bialystok, Poland (at the time, Russia). His parents, Jacob and Tillie Sabin, suffered from crushing poverty under the czarist regime. In 1921 the family immigrated to the United States. After settling in Paterson, New Jersey, Jacob Sabin worked in the textile business.

In 1923 Sabin enrolled at New York University (NYU), thanks to an uncle who agreed to pay for Sabin's education if he would attend dental school. After Sabin became enchanted with the science of virology and switched majors in 1925, the uncle withdrew his support, and Sabin had to struggle to pay his tuition. Sabin completed his B.S. in 1928 and entered NYU's College of Medicine. After receiving his M.D. in 1931, he fulfilled his residency at New York's Bellevue Hospital. He completed his internship in 1934 and accepted a position at the Rockefeller Institute for Medical Research. He married Sylvia Tregillus in 1935, and the couple had two children. He was later married to Jane Warner, and finally to Heloisa Dunshee De Abranches, whom he wed in 1972.

Sabin focused on the fight against polio early in his career. The paralytic disease, which is caused by a virus that embeds itself in the brain stem and attacks the central nervous system, had reached epidemic proportions in the United States, cruelly wreaking most devastation among children. In its most severe form polio resulted in complete loss of control over all muscles. In 1936 Sabin succeeded in growing the polio virus in central nervous tissue obtained from human embryos, but he needed huge amounts of the virus to develop a vaccine, and this production method was far too costly to pursue. Sabin served in the United States Army Medical Corps during World War II. The research he conducted into encephalitis and dengue while stationed in the Pacific theater led to his creation of vaccines for both.

Upon his return he accepted a position at the University of Cincinnati College of Medicine, where he made a substantial breakthrough by proving that polio existed in both the digestive tract and the nervous system. This insight showed that the virus could be cultivated in non–nervous system tissue, thereby saving both time and money. In 1949 JOHN FRANKLIN ENDERS, FREDERICK CHAPMAN ROBBINS, and Thomas Sweller grew the first polio virus in human nonnervous tissue.

Jonas Salk beat Sabin in the race for a polio vaccine. In 1953 Salk announced that he had developed an effective vaccine made from dead (killed) viruses. Sabin believed that the killed virus vaccine would not be as safe as one formulated from a weakened, live virus, so he continued his efforts. When Salk's initial vaccine infected some of its recipients with polio, Sabin appeared vindicated. But Salk then took extra precautions, and his vaccine became the accepted standard in the United States. In 1957 Sabin isolated three strains of the virus that, when administered orally in a vaccine, were strong enough to stimulate the immune system but weak enough to present no danger. Sabin lobbied tirelessly for the implementation of his vaccine on a wide scale. He argued that it was safer, easier to administer, and more effective over the long term than Salk's. After Sabin's version was used with good results in massive trials in the Soviet Union and was embraced by the World Health Organization, the U. S. Department of Health approved its manufacture in the United States in 1960.

Sabin continued his investigations into other aspects of virology during and after his formulation of a polio vaccine. He addressed the potential role of viruses in cancer, as well as the probability of developing a human immunodeficiency virus (HIV) vaccine. He died in 1993.

Sabin, Florence Rena

(1871–1953)
American
Medical Researcher

Florence Sabin succeeded in three different careers: teaching, research, and—after her retirement, in her 70s—public health reform. She was born in Central City, Colorado, on November 9, 1871. Her mother, the former Serena Miner, had been a teacher. Florence spent her first years in small mining towns where her father, George K. Sabin, worked as an engineer, and the family moved to Denver in 1875. Three years later Serena Sabin died, and Florence and her older sister, Mary, went to live with relatives in Chicago and, later, Vermont. Florence attended Smith College, from which she graduated with a B.S. in mathematics and zoology in 1893.

After teaching and saving her money for three years, Sabin enrolled in Johns Hopkins Medical School, one of the few good medical schools that admitted women, and earned her M.D. in 1900. She wanted to do research, and Franklin Paine Mall, the head of Hopkins's anatomy department, became her mentor. While still a medical student she prepared a three-dimensional model of a baby's mid- and lower brain more accurate than any made before. Her 1901 laboratory manual based on this model, *An Atlas of the Medulla and Midbrain,* became a standard textbook.

In 1902 Sabin became Mall's assistant, the first woman member of the Hopkins medical faculty. She was made an associate professor of anatomy in 1907. She became a full professor in 1917, the first woman to achieve that rank at Johns Hopkins, but this title was a consolation prize for not making her head of the anatomy department on Mall's death, a position that Mall's wife believed that he himself would have wanted her to have. Mall's wife said that "only the lingering prejudice against women . . . prevented Sabin's well-merited advancement."

Sabin's most important research at Johns Hopkins concerned the lymphatic system, a network of vessels that carries a milky fluid called lymph. Lymph contains chemicals and cells that belong to the immune system. Anatomists had thought that the lymphatic system developed before birth from spaces in the tissues, separately from the circulatory system, but Sabin showed that lymph vessels budded from veins. She summarized her findings in a 1913 book, *The Method and Growth of the Lymphatic System.*

During her years at Hopkins Sabin also used a technique for staining living cells that she had learned in Germany to investigate the origin of blood cells and blood vessels. She once stayed up all night watching the blood system develop in a chick embryo, ending with the embryo heart's making its first beat. She called it "the most exciting experience of my life."

Sabin was as well known for her teaching as for her research. She taught that "it is more important for the student to be able to find out something for himself than to memorize what someone else has said." She saw research as a way to improve teaching: "No one can be a really great educator unless he himself is an investigator," she said. Simon Flexner, scientific director of the prestigious Rockefeller Institute (later Rockefeller University) in New York City, invited Sabin to join the institute in 1925 and set up a new department of cellular studies. He called her "the greatest living woman scientist and one of the foremost scientists of all time." She accepted, the first woman to receive full membership in the institute. There she did research on cells of the immune system and on the immune response to the microbe that caused tuberculosis. She also learned which chemicals in the microbe damaged the body.

Sabin retired from the Rockefeller Institute in 1938 at age 68, and went to live with her sister in Denver. Her third career began in 1944, when John Vivian, then governor of Colorado, invited her to head the subcommittee on health of the state's Post-War Planning Committee. Vivian may have chosen Sabin because he believed a report that described her as a "nice little old lady with her hair in a bun . . . who has spent her entire life in a laboratory, doesn't know anything about medicine on the outside, and won't give any trouble," but if so, he received a surprise. Later, Governor W. Lee Knous more accurately called her "Florence the atom bomb."

Sabin in fact had always been interested in public health. Far from rubber-stamping Colorado's existing health laws, she launched into a thorough investigation that took her all over the state. She found that both the laws and their enforcement were woefully inadequate. The "little doctor," as she came to be known, then drafted a legislative program for public health reform, assembled piles of evidence to show the need for it, and hounded legislators and officials until they passed most of it in 1947. She made similar reforms in Denver when the city's mayor made her head of the Interim Board of Health and Hospitals in 1947 and later of the Department of Health and Welfare and of the permanent Board of Health and Hospitals.

Florence Sabin was one of the best known and most honored women scientists of her time. In 1924 she became the first woman president of the American Association of Anatomists, and a year later she received the even greater honor of becoming the first woman elected to the prestigious National Academy of Sciences. In the 1930s she received a host of honorary degrees and other awards from various colleges, including the National Achievement Award (1932) and the M. Carey Thomas Prize (1935). Finally in 1951 she won the Albert and Mary Lasker Award for medical research, one of the highest awards in American science. Sabin stopped her public activities at the end of 1951 and spent the last years of her life caring for her ailing sister. Sabin died of a heart attack on October 3, 1953, just short of her 82nd birthday.

Sagan, Carl Edward

(1934–1996)
American
Astronomer, Exobiologist

Carl Sagan proposed a number of theories on scientific topics ranging from the effects of nuclear war on the planet's climate to the potential for the discovery of life outside the Earth. He was unique in his extensive knowledge of biology and genetics, which he applied to his primary field of astronomy. He maintained a lifelong, passionate interest in exobiology—the discipline devoted to investigating the possibility of extraterrestrial life. His greatest contribution, however, was his capacity to make scientific concepts available and intriguing to a popular audience. His television show *Nova,* for example, used a synthesis of special effects and accurate science to captivate a mass audience.

Sagan was born in New York City on November 9, 1934, to Samuel and Ruth Gruber Sagan. His father was a Russian emigrant who worked as a cutter in a clothing factory. As a child Sagan was drawn to the study of astronomy and devoured science fiction novels. Sagan was married three times—to LYNN ALEXANDER MARGULIS in 1957, Linda Salzman in 1968, and Ann Druyan. He had five children.

In 1955 Sagan received his B. S. from the University of Chicago, where he remained to pursue his doctorate. As a graduate student he became enchanted with exobiology and was fortunate to receive support in this typically unrespected pursuit from a number of respected scientists, including HERMANN JOSEPH MULLER, GERARD PETER KUIPER (one of the very few astronomers at that time who was

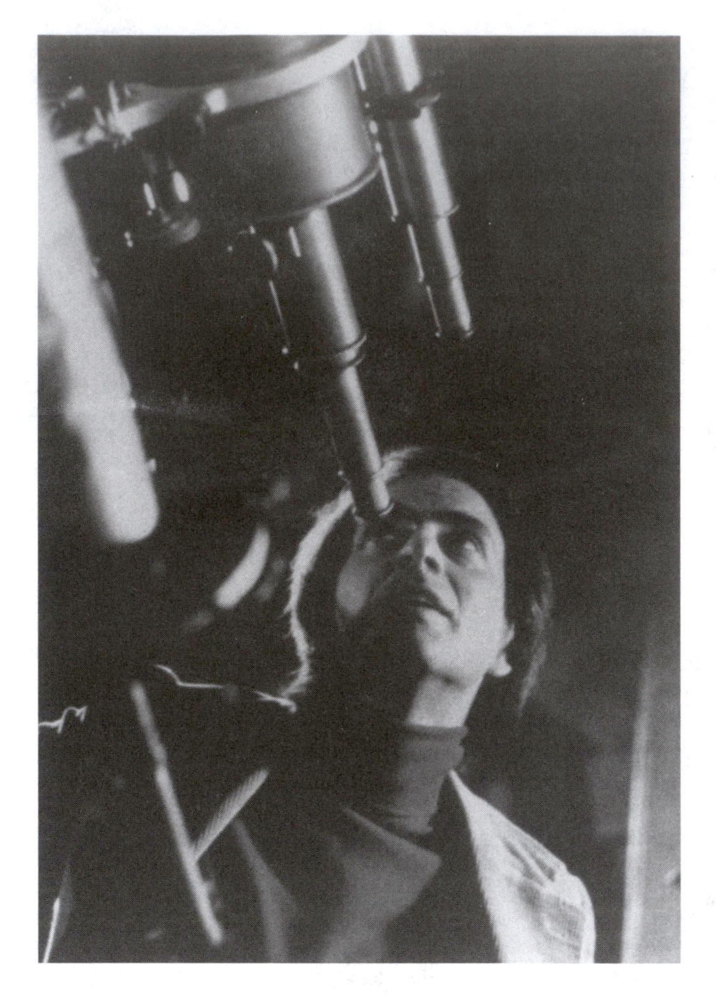

Carl Sagan, who is best known for his popular television show *Nova,* which made scientific concepts accessible and intriguing to a broad, general audience. *Cornell University, courtesy AIP Emilio Segrè Visual Archives.*

also a planetologist), and HAROLD CLAYTON UREY, who had advised STANLEY LLOYD MILLER in his experiments into the origins of life. Sagan's dissertation, "Physical Studies of the Planets," earned him a Ph.D. in astronomy and astrophysics in 1960.

Sagan was the Miller Residential Fellow at the University of California at Berkeley from 1960 to 1962. He then accepted a faculty position at Harvard University, where he remained until 1968. During this time at Harvard he postulated two theories that were subsequently confirmed. He hypothesized that the color variations on Mars were caused by dust shifted in windstorms, and that Venus's surface was extremely hot because the atmosphere trapped the Sun's heat. In 1968 he became an assistant professor at the Center for Radiophysics and Space Research at Cornell University, where he stayed until his retirement. He was named the David Duncan Professor of Astronomy and Space Science at Cornell in 1977, after he served as associate director of the center.

His later theoretical work focused on the possibility that Jupiter's moon Titan could sustain life, and also on the potential effects of nuclear war on the Earth's atmosphere. In 1983 he coauthored a paper, "Nuclear Winter: Global Consequences of Multiple Nuclear Explosions," which galvanized both political leaders and the public. The paper asserted that even a "small" nuclear war would wreak havoc on the planet and ultimately cause a substantial temperature drop, a phenomenon termed *nuclear winter.*

A significant facet of Sagan's work occurred outside academia. He advised the National Aeronautics and Space Administration (NASA) on many missions, including the *Mariner* Venus mission, the *Mariner 9* and *Viking* missions to Mars, and those involving the *Pioneer 10, Pioneer 11,* and *Voyager* spacecraft. He helped design a plaque affixed to some of these crafts that bore messages describing human life in case extraterrestrials encountered the space ships. Sagan also was dedicated to spreading his love of science to a broad audience. He published several popular science books, such as *The Cosmic Connection* (1973), *The Dragons of Eden* (1977), and *Broca's Brain: Reflections on the Romance of Science* (1979). He frequently appeared on Johnny Carson's *Tonight Show,* serving as a sort of science adviser. In 1980 he launched the television show *Cosmos,* which introduced a huge audience to such formerly esoteric concepts as black holes.

Sagan's enthusiasm and receptiveness to new ideas were his true legacy. His books were widely read, and he moved exobiology closer to mainstream scientific thought. He was awarded the Pulitzer Prize for literature in 1978 and received numerous other awards, including the Rittenhouse Medal in 1980 and the UCLA Medal in 1991.

Salk, Jonas Edward

(1914–1995)
American
Microbiologist

Jonas Salk's major accomplishment was his development of the first safe and effective vaccine against polio. With his collaborator, Thomas Francis, Jr., Salk pioneered the use of a "killed virus" as a vaccine antigen. Salk's discovery made him a national hero. He also helped create a killed-virus flu vaccine and conducted epidemiological research. In 1960 he founded the Salk Institute for Biological Studies in La Jolla, California, a facility dedicated to cutting-edge research.

Born on October 28, 1914, in New York City, Salk was the eldest son of Orthodox Jewish emigrants from Poland. His father, Daniel Salk, was a garment worker with a talent for and a love of drawing. Dora Press, Salk's mother, took care to cultivate the considerable intellectual gifts the young Salk displayed.

Enrolled at the City College of New York, Salk received his bachelor's degree in science in 1933, when he was only 19. Although he had originally planned to pursue law, Salk had been captivated by biology during his undergraduate studies. In 1933 he entered New York University's School of Medicine, where he met his future collaborator, Dr. Thomas Francis, Jr. He obtained his M.D. in 1939 and continued working for Francis while fulfilling a two-year internship at Mount Sinai Hospital in New York. He also married Donna Lindsay in 1939, and the couple had three sons. After his first marriage ended, Salk wed Françoise Gilot (Pablo Picasso's first wife) in 1967.

Upon finishing his internship, Salk accepted a National Research Council fellowship and followed Francis to the University of Michigan, where the two developed a killed-virus vaccine for the flu in 1943. This achievement was significant. The killed virus was able to destroy live flu viruses at the same time that it stimulated the body to produce antibodies to ward off future infection. In 1946 Salk accepted a position as assistant professor of epidemiology at the University of Michigan, but he left for the University of Pittsburgh's Virus Research Laboratory the following year. He published papers on polio, some of which were read by Daniel Basil O'Connor, the director of the National Foundation for Infantile Paralysis, a well-funded organization devoted to eradicating polio. O'Connor contributed considerable money to Salk's fledgling attempts to devise a polio vaccine.

At the University of Pittsburgh Salk joined researchers from other institutions to classify the more than 100 strains of polio virus. In 1951 he felt certain that all polio viruses could be arranged into three categories. He built on the work of ALBERT BRUCE SABIN, JOHN FRANKLIN ENDERS, THOMAS HUCKLE WELLER, and FREDERICK CHAPMAN ROBBINS, each of whom played a part in making cultivation of the polio virus easier in a lab. Salk grew samples of the three types in cultures of monkey kidneys, then exposed the virus in formaldehyde for 13 days, thereby killing it. He felt certain that the killed-virus technique would be safer than the live-virus method advocated by Sabin and others. In 1952 Salk began to administer his killed-virus vaccine to children—first to those who already had the disease, later to those who had never contracted it. In both cases he found that the vaccine definitively elevated antibody levels. These initial positive results were published in the *Journal of the American Medical Association* in 1953. The following year Salk spearheaded the implementation of a massive inoculation program, with 1 million children participating. On April 12, 1955, Salk's vaccine was pronounced both potent and safe. A tragedy occurred that June, however, when 200 children contracted polio from the vaccine because one of the participating labs had used the wrong strain of the virus. This problem was quickly corrected and did not recur. In 1960 Salk founded his eponymous institute.

Salk's vaccine accounted for a 96 percent reduction in the number of cases of polio in the United States by the summer of 1961, and he was hailed as a hero. His chief scientific competitor, Sabin, created a live-virus vaccine, which gradually supplanted Salk's innovation because it could be administered orally. In addition to many other awards, Salk received the Presidential Medal of Freedom in 1977. He died of heart failure in 1995.

Sanchez, Pedro Antonio

(1940–)
Cuban/American
Soil Scientist

Pedro Antonio Sanchez spent the majority of his career practicing soil science in the field, focusing on tropical soil management as a means of promoting sustainable agriculture. He actively opposed rain forest destruction and soil erosion, promoting instead agroforestry and improvement of agriculture through nutrient cycling.

Sanchez was born on October 7, 1940, in Havana, Cuba, the oldest of four children. His mother, Georgina San Martin, was a pharmacist and high school teacher, and his father, after whom he was named, was a farmer and fertilizer salesperson. Sanchez married for the first time in 1965 and had three children—Jennifer, Evan, and Juliana. Then in 1990 he married Cheryl Palm, a fellow soil scientist.

Sanchez attended college in the United States at Cornell University, where he received his bachelor of science in agronomy in 1962 and his master of science in soil science in 1964. While working toward his doctorate, he served as a graduate assistant in soil science at the University of the Philippines–Cornell Graduate Education Program in Los Baños, Philippines. He received his Ph.D. in soil science in 1968.

North Carolina State University (NCSU) hired Sanchez as an assistant professor of soil science in 1968, and he remained affiliated with NCSU throughout his career, which he conducted mostly in South America. He spent his first three years as coleader of the National Rice Program of Peru, an NCSU-sponsored program promoting Peruvian self-sufficiency through sustainable agricultural practices. Between 1971 and 1976 he led the Tropical Soils Program, conducting soil research in the Cerrado of Brazil, the Amazon of Peru, and in Central America. NCSU promoted him to associate professor in 1973. He acted as coordinator of the Beef–Tropical Pastures Program of the Centro Internacional de Agricultural Tropical in Cali, Colombia, between 1977 and 1979. NCSU promoted him to full professor in 1979.

In 1982 Sanchez returned to South America as the chief of the NCSU Mission in Lima, Peru. He expanded across the Atlantic and Pacific in 1984 as the coordinator of NCSU's Tropical Soils Program in Bolivia, Indonesia, and Madagascar. Between 1990 and 1991 he directed the Center for World Environment and Sustainable Development, a joint initiative involving Duke University, the University of North Carolina at Chapel Hill, and NCSU. In 1991 Sanchez scaled back to professor emeritus status while he served as the third director general of the International Center for Research in Agroforestry in Nairobi, Kenya.

Sanchez has authored and edited 10 books, notably *Properties and Management of Soils in the Tropics* in 1976, as well as more than 125 articles, including "Soil Fertility Dynamics after Clearing a Tropical Rainforest in Peru," which he cowrote in 1983 for the *Soil Science Society of America Journal*. He received international recognition for his scientific efforts: The Instituto Colombiano Agropecuario awarded him its 1979 Diplome de Honor; the Peruvian government bestowed him with its 1984 Order de Merito Agricola; and in 1993 the Soil Science Society of America granted Sanchez its International Soil Science Award and the American Society of Agronomy bequeathed him its International Service in Agronomy Award. These awards acknowledged that he practiced his science to benefit his fellows.

Saruhashi, Katsuko

(1920–)
Japanese
Geologist, Chemist

Katsuko Saruhashi studied carbon dioxide levels in seawater long before people began to suspect that this gas might increase temperatures on Earth. She also tracked the spread of radioactive debris from atomic bomb tests.

Saruhashi was born in Tokyo, Japan, on March 22, 1920. While she was a student at Toho University, from which she graduated in 1943, she met the government meteorologist Yasuo Miyake, who became her friend and mentor. After World War II ended, he hired her as a research assistant in his new Geochemical Research Laboratory, part of the Japanese Transport Ministry's Meteorological Research Institute.

Around 1950 Miyake suggested that Saruhashi measure the concentration of carbon dioxide (CO_2) in seawater. "Now everyone is concerned about carbon dioxide, but at the time nobody was," Saruhashi says. Indeed, she had to design most of her own techniques for measuring the gas. Her project earned her a doctor of science degree from the University of Tokyo in 1957, the first doctorate in chemistry that the university awarded to a woman.

In the early 1950s the United States, the Soviet Union, and several other nations tested nuclear bombs at remote sites, filling the air with radioactive debris. Concern about such fallout led the Japanese government to ask Miyake's laboratory in 1954 to measure the amount of radioactive material reaching Japan in rain and the amount found in seawater off the country's coast. Miyake put Saruhashi in charge of this project, which she says was the first of its kind. She found that fallout from an American bomb test on the Pacific island of Bikini reached Japan in seawater a year and a half after the test. She later measured fallout in other parts of the world as well. Evidence gathered by Saruhashi and others helped protesters persuade the United States and the Soviet Union to stop above-ground testing of nuclear weapons in 1963.

Saruhashi also continued her measurements of carbon dioxide in seawater, finding that water in the Pacific releases about twice as much CO_2 into the atmosphere as it absorbs from the air. (There is about 60 times more CO_2 dissolved in seawater than in air.) This result suggests that the ocean is unlikely to weaken possible global warming by absorbing excess CO_2.

Saruhashi was made director of the Geochemical Research Laboratory in 1979. She retired from this post a year later. In 1990, after Miyake's death, she became executive director of the Geochemistry Research Association in Tokyo, which Miyake had founded in 1972. She still held this post in 1998. Among the many honors she has received are election to the Science Council of Japan in 1980 (she was its first woman member), the Miyake Prize for geochemistry in 1985, and the Tanaka Prize from the Society of Sea Water Sciences in 1993.

When Saruhashi retired from the directorship of the Geochemical Laboratory, her coworkers gave a gift of 5 million yen (about $50,000). She used the money as an initial fund to establish the Saruhashi Prize, given each year since 1981 to a Japanese woman making important contributions to the natural sciences. Its first recipient was the population geneticist Tomoko Ohta. Saruhashi says the prize "highlight[s] the capabilities of women scientists. . . . Each winner has been not only a successful researcher but . . . a wonderful human being as well."

Schiaparelli, Giovanni Virginio

(1835–1910)
Italian
Astronomer

Giovanni Virginio Schiaparelli made a number of discoveries significant to the study of astronomy. He spent most of his career as director of the Brera Observatory in Milan, where he discovered the asteroid Hesperia and proposed a theory on the origin of meteors. His belief that meteors were the products of the dissolution of comets was later shown to be correct. Curiously, one of Schiaparelli's observations gained a great deal of notoriety as a result of a mistranslation. While studying the surface of Mars, Schiaparelli noted the existence of straight lines, which he called *canali*. This word, which means "channels" in Italian, was translated into English as "canals," sparking the notion that canals had been constructed on the Red Planet. He devoted the latter part of his life to the history of astronomy.

Schiaparelli was born on March 14, 1835, in the town of Savigliano in the Cuneo province of Italy. The details of his family life as a child and an adult are sparse. After he graduated with a degree in civil engineering from the University of Turin in 1854, he studied astronomy, first under Johann Encke in Berlin, then with Friedrich Struve at the Pulkovo Observatory in St. Petersburg in 1856.

In 1860 Schiaparelli was appointed astronomer at the Brera Observatory. In 1862 he was named its director, a position he maintained until he retired in 1900. His primary interest was in the solar system. His first major finding occurred in 1861 when he discovered the asteroid Hesperia. Schiaparelli devoted a great deal of his early career to studying comets. After undertaking some initial work in 1860 on different types of comet tails, he postulated that comets give rise to meteors. He carefully observed meteor swarms in August 1866 and became convinced that the showers resulted from the disintegration of comets. In 1871 he published a detailed summation of this theory, *Entwurf einer astronomischen Theorie der Sternschnuppen*, which was substantially confirmed by future research.

Schiaparelli shifted his focus in 1877, when he began to study Mars carefully. His work was aided by the fact that Mars made one of its closest recorded approaches to Earth that very year. In the course of these observations Schiaparelli discerned a great number of features that had previously been undetected, such as a southern polar ice cap and a series of dark, straight lines traversing the planet. Although the American astronomer PERCIVAL LOWELL, working from a poor translation of Schiaparelli's term *canali*, erroneously posited that the topographical features were built by an advanced Martian civilization, Schiaparelli did not advocate this view. The *Mariner* space probes eventually established that the "canali" were natural phenomena, not the product of construction. Schiaparelli also posited theories about the rotation of Mercury and Venus, which he believed rotated on their axes at the same rate at which they revolved around the Sun, thereby always keeping one side facing the Sun. This view came to be the conventional wisdom until the radar technology of the 1960s disproved it.

After he retired from the observatory, Schiaparelli collaborated on translations of some of the ancient treatises on astronomy, such as those by CLAUDIUS PTOLEMAEUS and al-Battani. His 1903 book *Astronomy in the Old Testament* was considered an excellent addition to the field of astronomical history. He became a senator of the Kingdom of

Italy in 1889 and was member of the Lincei. He twice won the Lalande Prize of the Institut de France.

Schrieffer, John Robert

(1931–)
American
Physicist

J. Robert Schrieffer shared the 1972 Nobel Prize in physics with John Bardeen and LEON NEIL COOPER. Together the three men developed a theory of superconductivity that they named by combining their initials, the BCS theory.

Schrieffer was born on May 31, 1931, in Oak Park, Illinois, to Louise Anderson and John Schrieffer. Schrieffer studied electrical engineering and physics at the Massachusetts Institute of Technology and received his B.S. in 1953. He received his M.S. the next year from the University of Illinois and commenced his doctoral research there under Bardeen on electrical conduction on semiconductor surfaces before switching his topic to superconductivity. He earned his Ph.D. in 1957.

After receiving his doctorate, Schrieffer moved to the University of Chicago as an assistant professor. He returned to the University of Illinois in 1960 as an associate professor. That same year he married Anne Grete Thomsen, and together the couple had two daughters and one son. In 1962 he moved to Philadelphia as a professor at the University of Pennsylvania, which named him the Amanda Wood Professor of Physics in 1964. In 1980 Schrieffer transferred to the University of California at Santa Barbara, which named him the Essan Khashoggi Professor of Physics in 1985. In addition he served as the director of the Institute of Theoretical Physics from 1984 through 1989. In 1992 Florida State University in Tallahassee appointed Schrieffer as a professor and as the chief scientist at the National High Magnetic Field Laboratory.

Schrieffer conducted the work that earned his Nobel while still a doctoral candidate. He and his colleagues discovered the process whereby certain metals and alloys, when supercooled to the vicinity of absolute zero (zero degrees kelvin), lose all electrical resistance, thus creating a perfectly efficient system in which no energy is lost. The trio announced their theory in a letter published in the November 15, 1956, issue of *Physical Review.* The trio followed up in the December 1, 1957, issue of *Physical Review* with a substantial article, "The Theory of Superconductivity," explaining their theory. Schrieffer continued to write about superconductivity, publishing the article "Recent Advances in the Theory of Superconductivity" in 1960, as well as a book-length treatment of the topic, *Theory of Superconductivity,* published in 1964.

Other scientists followed up on the BCS theory by trying to raise the temperature at which superconductivity could occur past the threshold of 77.4 degrees Kelvin, or the temperature at which nitrogen liquifies. Until then superconductivity remained a theory in terms of practical use, since the materials and conditions required were prohibitively expensive. Liquid nitrogen, on the other hand, is plentiful and thus cheap. In 1987 PAUL CHING-WU CHU discovered a combination of materials that could conduct electricity at 93 degrees Kelvin, far past the threshold temperature for liquid nitrogen.

In addition to studying superconductivity, Schrieffer conducted research on particle physics, metal impurities, spin fluctuations, and chemisorption. In addition to the Nobel Prize, he won the 1968 Comstock Award from the National Academy of Sciences and the 1984 National Medal of Science.

Schrödinger, Erwin

(1887–1961)
Austrian
Physicist

Erwin Schrödinger's most significant scientific contribution was his formulation of a mathematical equation that accurately described the behavior of electrons in atoms. After encountering the work of the French physicist Prince LOUIS-VICTOR-PIERRE-RAYMOND DE BROGLIE, who suggested that particles of matter might possess properties that could be expressed by wave functions, Schrödinger realized that electrons' orbits could be described that way. Subsequently Schrödinger turned his attention to more philosophical topics, including the theoretical underpinnings of physics and the relationship between biological and physical sciences.

Schrödinger, an only child, was born on August 12, 1887, in Vienna, Austria. His mother was the daughter of Alexander Bauer, a Viennese professor of chemistry. Schrödinger's father, Rudolf Schrödinger, ran an oilcloth factory, but his true passions were painting and botany.

After graduating from the Akademische Gymnasium in Vienna in 1906, Schrödinger entered the University of Vienna, where he first developed an interest in theoretical physics under the tutelage of Friedrich Hansenöhrl. Upon receiving his Ph.D. in 1910, Schrödinger took a position at the university's Second Physics Institute as a laboratory assistant to the experimental physicist Franz Exner.

The outbreak of World War I took Schrödinger away from Vienna, while he served as an artillery officer on the Italian front. After the war Schrödinger returned to the university; he then took a succession of teaching positions before arriving with his wife (he had married Annemarie Bertel in 1920) at the University of Zurich in 1921.

In Zurich in 1925 Schrödinger first encountered Prince de Broglie's work, when it was cited in a paper by

Erwin Schrödinger, who formulated a mathematical equation that accurately described the behavior of electrons in atoms. *AIP Meggers Gallery of Nobel Laureates.*

ALBERT EINSTEIN, and applied its insights to his own contemplation of the movement of electrons in an atom. Thirteen years earlier NIELS HENRIK DAVID BOHR's empirical observations had suggested that electrons inhabited specific and defined orbits within an atom. Schrödinger decided that the electron behavior posited by Bohr could be described by a wave function. After much effort Schrödinger hit on the formula that came to be known as the Schrödinger wave equation, which he revealed in a series of papers published in 1926. The import of this discovery was immediately recognized by his peers. Schrödinger's equation not only explained the phenomena observed by Bohr but also provided a theoretical underpinning for the model of the atom that scientists had previously been able to justify only by empirical means.

In 1927 Schrödinger assumed the chair of theoretical physics at the University of Berlin from the retiring MAX PLANCK. He remained in Berlin until 1933, when he left Germany for England after the accession to power of Hitler's National Socialist Party. In the same week that he began teaching at Oxford, Schrödinger learned that he had been awarded the Nobel Prize for physics, which he shared with the English physicist PAUL ADRIEN MAURICE DIRAC, who was recognized for predicting the existence of the positron. Schrödinger remained at Oxford until 1936, when he returned to Austria to take a position at the University of Graz. He was summarily dismissed shortly after Hitler invaded in 1938. Aware of Schrödinger's difficulties, the prime minister of Ireland, who was also a mathematician, arranged to have an Institute for Advanced Studies established at the University of Dublin in order to provide Schrödinger a professional and personal home. Schrödinger traveled to Ireland in 1939 and remained there for the next 17 years.

Schrödinger's work took a somewhat contemplative turn in Ireland. In 1944 he published *What Is Life?* which explored the connections between theoretical physics and human biology. He hypothesized that biologists inquiring into the nature of the gene might find an analogy with the method physicists used to unlock the secrets of the atom. The book proved a powerful influence on both physicists and biologists and is credited with influencing the seminal figures of modern molecular biology, including FRANCIS HARRY COMPTON CRICK.

Schrödinger's accomplishments were recognized during his lifetime. In addition to receiving the Nobel Prize, he was elected to the Royal Society of England, the Prussian Academy of Sciences, the Austrian Academy of Sciences, and the Pontifical Academy of Sciences. In 1956, after the occupation of Austria ended, he acceded to what had been a decade of entreaties and returned to the country of his birth.

Schwinger, Julian Seymour

(1918–1994)
American
Physicist

Julian Schwinger shared the 1965 Nobel Prize in physics with RICHARD PHILIP FEYNMAN and SHINICHIRO TOMONAGA, each of whom had independently formulated theories of quantum electrodynamics, which reconciled quantum mechanics with ALBERT EINSTEIN's special theory of relativity.

Schwinger was born on February 12, 1918, in the Jewish Harlem section of New York City, the younger of two brothers. His mother was Belle Rosenfled,;his father, Benjamin Schwinger, was a garment manufacturer. Schwinger was a child prodigy intellectually, graduating from high school at the age of 14 after skipping three grades. While in high school he read articles by PAUL ADRIEN MAURICE DIRAC in the *Proceedings of the Royal Society of London,* material geared to scientists with advanced degrees. While attending the College of the City of New York as an undergraduate Schwinger published a paper on quantum mechanics in the journal *Physical Review,* a discussion that greatly impressed ISIDOR ISAAC RABI, who arranged for a scholarship to be granted to Schwinger from his institution, Columbia University. In 1936 at the age of 17 Schwinger graduated from Columbia, and in 1939 at the age of 20 he received his Ph.D. from Columbia as well.

Schwinger proceeded that year to the University of California at Berkeley as a research associate working in J. ROBERT OPPENHEIMER's laboratory. In 1941 he moved to Purdue University as an instructor, and by the time he left in 1943, he had been promoted to assistant professor. World War II then intervened, and Schwinger divided his wartime scientific efforts between the Manhattan Project designing the atomic bomb in the Metallurgical Laboratory at the University of Chicago and developing improved microwave radar systems at the Massachusetts Institute of Technology. After the war in 1945 he moved to Harvard University as an associate professor, and in 1947 at the age of 29 he was made the youngest full professor in Harvard's long history. That same year Schwinger married Clarice Carrol. In 1972 he was appointed professor of physics by the University of California at Los Angeles, a post he retained until his death.

At the 1948 meeting of the American Physical Society, Schwinger announced his reconciliation of quantum physics with the theories of electromagnetism proposed by Einstein, WERNER KARL HEISENBERG, WOLFGANG PAULI, and most importantly for Schwinger, Dirac. Dirac's work was coming into question at that time, because his mathematical predictions were not borne out by experimentation, specifically because they would require the atomic particle to have infinite mass. Schwinger's theory corrected Dirac's theory in a way that left it standing.

Schwinger published several books on his theories, including *Discontinuities in Waveguides* in 1968, *Particles and Sources* in 1969, and *Particle Kinematics and Dynamics* in 1970. In most of his scientific writing Schwinger addressed a professional audience using the vocabulary of the laboratory. Later in his career he wrote *Einstein's Legacy: The Unity of Space and Time,* addressed to a more general audience in lay terms to help them understand quantum physics.

In addition to the Nobel Prize, Schwinger received the 1964 National Medal of Science. Schwinger died of cancer on July 16, 1994, in Los Angeles.

Scott, Charlotte Angas

(1858–1931)
English/American
Mathematician

Charlotte Scott taught mathematics to generations of Bryn Mawr students and made important contributions to algebraic geometry. She was born in Lincoln, England, on June 8, 1858, to the Reverend Caleb Scott and his wife, Eliza Ann. Her father was the principal of a small college as well as a Congregationalist minister.

Scott enrolled in Girton College, a women's college of Cambridge University, in 1876. She completed her studies with honors in 1880, scoring eighth highest in the difficult tripos examination, though her score was not officially announced because she was a woman. She taught mathematics at Girton for several years afterward, meanwhile continuing her own studies at the University of London. She earned a bachelor of science degree in 1882 and a doctorate in 1885.

In the year Scott obtained her doctorate, M. Carey Thomas, the first head of Bryn Mawr College in Pennsylvania, invited her to head the college's mathematics department. Scott accepted, becoming the only woman on Bryn Mawr's six-person faculty. She taught there for 40 years, earning the department a worldwide reputation and attracting such mathematical stars as ANNA JOHNSON PELL WHEELER and EMMY NOETHER.

Scott also made important contributions as a theoretical mathematician. In an early text, *An Introductory Account of Certain Modern Ideas in Plane Analytical Geometry* (1894), she treated point and line coordinates together to show the principles they had in common. She later helped to develop the new field of algebraic geometry by showing the "geometrical reality" beneath singularities and intersections of plane algebraic curves. Scott proved that complicated singularities could be broken down into clusters of simpler ones.

Scott helped to found the American Mathematical Society in 1894, serving on its council (the only woman member) from 1894 to 1897 and as its first woman vice president in 1906. Her peers ranked her 14th among the world's top 93 mathematicians. Scott retired from Bryn Mawr in 1925 and returned to England. She died at Cambridge on November 8, 1931.

Seaborg, Glenn Theodore

(1912–1999)
American
Nuclear Chemist

Glenn Seaborg's work on isotopes and radioactive elements led to his discovery of several transuranium elements, most importantly plutonium, which was used to advance nuclear technology. He also identified radioisotopes, which led to advances in radiological imaging technology and in radiotherapy. For these achievements, he received the 1951 Nobel Prize in chemistry with EDWIN MATTISON MCMILLAN.

Seaborg was born on April 19, 1912, in Ishpeming, Michigan. His parents, Selma O. Erickson and Herman

Theodore Seaborg, a machinist, emigrated from Sweden in 1904. In 1942 he married Helen L. Griggs, and together the couple had four sons and two daughters.

Seaborg's parents moved to Los Angeles to give him a better opportunity for an education, and he took advantage of this by becoming the valedictorian of his high school class. In 1929 he entered the University of California at Los Angeles to study chemistry; he earned his degree in 1934. He transferred to the University of California at Berkeley for graduate work, studying under the chemist Gilbert Newton Lewis and the physicist ERNEST ORLANDO LAWRENCE. In 1937 Seaborg earned his Ph.D. with a dissertation on the interaction of "fast" neutrons with lead.

Seaborg remained at Berkeley as an assistant in Lewis's laboratory. During this time he discovered the radioactive isotopes of iodine-131, iron-59, and cobalt-60. In 1941 he discovered the 94th element, plutonium-238, which proved to be important for the nuclear reactions that fueled the atomic bomb. On April 19, 1942, Seaborg departed for Chicago to participate in the Manhattan Project, where he utilized ultramicrochemical analysis to separate plutonium from uranium. By 1944 he had prepared enough plutonium for the detonation devices of two bombs.

Seaborg then shifted his focus to discovering more transuranic elements. He came up with the actinide concept, suggesting a shift in the periodic table that had been devised by DMITRI IVANOVICH MENDELEEV in 1869, by isolating actinium and 14 other elements in a separate section of the table. Seaborg correctly assumed that these elements were ripe for investigation in search of radioactive, transuranic elements, which he found in the form of element 95, americium, and element 96, curium.

After Seaborg returned from his wartime duties, he was appointed the associate director of the Lawrence Radiation Laboratory at Berkeley in 1948, and he assembled a group of outstanding scientists who made a number of important discoveries there. This group discovered elements 97 through 102—berkelium, californium, einsteinium, fermium, mendelevium, and nobelium. In 1958 Seaborg turned his attention more toward administrative duties, serving as the vice-chancellor of the University of California at Berkeley. Between 1961 and 1971 he was the chairman of the Atomic Energy Commission. The University of California at Berkeley then named him a University Professor of Chemistry in 1971.

Seaborg also made his mark through publications, which tended toward the academic in his earlier career, but later turned to a wider audience. In 1981 he published his account of the crisis over nuclear weapons, *Kennedy, Krushchev and the Test Ban*. Then in 1987 he published *Stemming the Tide: Arms Control in the Johnson Years*. In tribute to his contributions to the discovery of elements 93 through 102, a group working at the Lawrence Laboratory named element 106, discovered in 1974, seaborgium. Glenn Seaborg died in 1999.

Seibert, Florence Barbara

(1897–1991)
American
Microbiologist, Biochemist

Florence Seibert showed how even "pure" water could cause illness. She also purified the substance used in the skin test by which doctors screened people for tuberculosis, greatly increasing the test's accuracy. She was born on October 6, 1897, in Eaton, Pennsylvania, to George and Barbara Seibert, the middle of their three children. Her father owned a rug business.

A childhood attack of infantile paralysis, or polio, left Florence with a severe limp. When she enrolled at Goucher College in Baltimore, her father stayed there for a week, thinking he might need to take her home again, but Florence thrived. "I learned . . . that I was not an invalid but was able to stand on my own two feet with a chance to make a contribution to the world," she wrote in her autobiography, *Pebbles on the Hill of a Scientist*.

Seibert studied mathematics, biology, and chemistry at Goucher and graduated in 1918. After working briefly for a paper mill, she went on to study biochemistry at Yale. As her doctoral project she investigated the puzzling fact that injection of certain supposedly harmless substances sometimes caused fevers in humans or animals. At first she deepened the mystery when she found that even injections of distilled water, which should have been perfectly pure, could cause fever. This could pose a serious health problem, since medicines were often dissolved in distilled water before being injected.

Seibert learned that even though the heat of distillation killed bacteria in the water, poisons made by the microbes remained in the steam and sometimes dripped back into the water. These substances caused the fevers. A trap that prevented steam droplets from reentering the water solved the problem. Seibert won the Ricketts Prize in 1924 and the John Elliott Memorial Award from the American Association of Blood Banks in 1962 for this early work.

After earning her Ph.D. in 1923, Seibert transferred to the University of Chicago, where she worked in the Ricketts Laboratory and eventually became an assistant professor in biochemistry. She joined the laboratory of Esmond Long, working part-time with him and part-time at the Sprague Memorial Institute, and moved with him in 1932 to the Henry Phipps Institute in Philadelphia, part of the University of Pennsylvania.

Most of Seibert's research during her years with Long was devoted to purifying tuberculin, a substance made by tuberculosis bacteria that was used in a skin test to screen people for that then-common disease. "Old tuberculin," as the crude bacterial preparation was called, contained many impurities, and the amount of actual tuberculin in it varied from batch to batch. The skin test could be dependable only if tuberculin were purified, and Florence Seibert devoted 35 years to this painstaking task. Two new techniques for separating compounds in a mixture that she learned about in Sweden, high-speed centrifugation and electrophoresis, eventually allowed her to perfect a purified protein derivative (PPD) of tuberculin that became the standard for the United States in 1941 and for the world in 1952.

In 1938 Seibert became the first woman to receive the Trudeau Medal of the National Tuberculosis Association. Other awards included the First Achievement Award of the American Association of University Women (1943) and the Garvan Medal of the American Chemical Society. Seibert became a full professor at the University of Pennsylvania in 1955 and retired three years later to St. Petersburg, Florida, with her younger sister, Mabel. She continued research in her home on microorganisms associated with cancers in mice and rats.

Seibert was inducted into the National Women's Hall of Fame in 1990, when she was 92 years old. She died the following year.

Harlow Shapley, who was called a latter-day Copernicus for his discoveries about the Earth's solar system in relation to the Milky Way galaxy. *AIP Emilio Segrè Visual Archives, Shapley Collection.*

Shapley, Harlow

(1885–1972)
American
Astronomer

For discovering that the Earth's solar system was not at the center of the Milky Way galaxy the astronomer Harlow Shapley was called a latter-day NICOLAUS COPERNICUS. His contributions to astronomy and humanitarian causes were numerous. Shapley concluded that the Milky Way was much larger than had previously been assumed and devoted considerable effort to the study of nebulae (a term that at that time referred to any celestial body not identified as a star). He worked tirelessly to save Jewish scientists from Nazi Germany and after the war became committed to the cause of peace. During his tenure as the director of the Harvard University Observatory he played a significant role in making that observatory the preeminent center for astronomical research in the United States.

Shapley and his twin brother, Horace, were born on November 2, 1885, in Nashville, Missouri. Their father, Willis Shapley, was a farmer and a teacher who died when the boys were young. Their mother, Sarah Stowell Shapley, raised the family singlehandedly thereafter. In 1914 Harlow Shapley wed Martha Betz, a classmate who not only was a mathematician and astronomer herself, but also coauthored several scientific papers with her husband. The couple had five children.

Shapley enrolled at the University of Missouri at the age of 20 and obtained his B.A. in mathematics and physics in 1910 and his M.A. in 1911. He then moved to the Princeton University Observatory, where he received his Ph.D. in astronomy in 1913. His dissertation on eclipsing binary stars, which he later expanded and published, was considered an influential text.

Shapley's first position was at the Mount Wilson Observatory in Pasadena, California, where he worked under GEORGE ELLERY HALE from 1913 to 1920. He made some of his more profound discoveries during this period. He built upon the work of the astronomer HENRIETTA SWAN LEAVITT, who had demonstrated that the rate at which a Cepheid (a type of star) pulsed was directly related to its brightness. Shapley used this insight to determine a star's absolute brightness and then compare that to its observed brightness. Employing this method, he was able to calculate the distance from Earth of any given Cepheid. With this measuring system he also determined the placement of Earth's solar system within the Milky Way. In 1915 he estimated the galaxy's center to be some 50,000 light-years from the Sun, in the constellation Sagittarius. It was later established that this "Shapley Center" was 33,000 light-years from the Sun. Shapley also proposed that the diameter of the Milky Way was 10 times greater than previously thought.

In 1921 Shapley left California for the Harvard Observatory, where he served as director until 1952. He continued to pursue his research, but his more significant contribution at Harvard was improving the reputation of the observatory itself. Not only did he create a graduate program in astronomy, but he also actively courted the best scientists from around the world to the observatory. In fact, he spearheaded a movement to rescue Jewish scientists from the horrors of the Nazi regime. He saw to it that dozens of these persecuted intellectuals took positions at Harvard instead of perishing in concentration camps.

After the war Shapley continued to focus on international politics. He was a key figure in the creation of the United Nations Educational, Scientific, and Cultural Organization (UNESCO), and he represented the United States at the writing of its charter in 1945. He strongly advocated scientific cooperation with the Soviet Union and strove to undermine the House Committee on Un-American Activities, which accused him of sympathizing with communists. He chaired the Committee of One Thousand, an organization devoted to First Amendment rights, which counted ALBERT EINSTEIN among its members. Shapley taught at Harvard until 1956 and then lectured on astronomy, politics, and philosophy across the country.

Shapley's findings about the size and structure of the Milky Way shifted the current conception of the Earth's solar system and its place in the galaxy. His tenure as director of Harvard's Observatory helped shape the development of astronomy in the United States. He saved a great many lives and provided a passionate example of how science and politics could work together for positive causes. Shapley died on October 20, 1972, in Boulder, Colorado.

Sherrington, Charles Scott

(1857–1952)
English
Neurophysiologist

Charles Scott Sherrington helped establish the field of neurophysiology, devoting his career to the study of the transmission of nerve impulses between the central nervous system and individual muscles. Sherrington created a new paradigm of neurophysiology when he suggested the integrative theory of nerve interrelationship, whereby individual nerves are separated by gaps that are bridged by what Sherrington called *synapses*. For this work he received the 1932 Nobel Prize in physiology or medicine with Edgar Douglas Adrian.

Sherrington was born on November 27, 1857, in London, England. His father was James Norton Sherrington, who died when his son was young; his mother was Anne Brookes, who later married Caleb Rose, Jr., an Ipswich physician. Sherrington married Ethel Mary Wright, and together the couple had one son, Charles E.R. Sherrington, born in 1897.

Sherrington commenced his medical training at St. Thomas's Hospital in London in 1875. He then attended Caius College at Cambridge University studying physiology under Michael Foster for his bachelor's degree in medicine in 1884. He followed up on this with a three-year tour of Europe, where he conducted research on physiology, histology, and pathology in a series of laboratories.

Sherrington returned to teach systematic physiology at his alma mater, St. Thomas's. In 1891 he accepted the appointment as a professor and the superintendent of the Brown Institute for Advanced Physiology and Pathological Research. In 1895 he accepted the physiology chair at the University of Liverpool, a post he held the next 13 years, until he was named the Waynflete Professor of Physiology at Oxford University in 1913. He held this chair until his 1936 retirement, two dozen years later.

Sherrington's research suggested a kind of compromise between the two major theories of neurophysiological function: On the one hand, the reticular theory posited that the nervous system was completely interconnected, whereas the neuron theory (popularly known as the neuron doctrine) held that nerves functioned independently. Sherrington's experimentation confirmed both theories mutually, so Sherrington integrated the theories, positing that the nerves were separated by gaps that were somehow bridged by what Sherrington coined *synapses*. Alternate messages of excitation and inhibition travel across these bridges in paths interlinking individual nerves into a network. Sherrington identified what he called *reciprocal innervation*, or the phenomenon whereby excitation and inhibition messages can travel simultaneously through the system. Sherrington also split the reactions of the nerves into two categories, higher and lower, the former referring to cerebrally controlled reactions and the latter to reactions controlled by the muscles. Sherrington gathered all his accumulated knowledge concerning the central nervous system into one text, *The Integrative Action of the Nervous System*, which was published in 1906.

Sherrington served as the president of the Royal Society of London from 1920 through 1925. In 1922 he received the Knight Grand Cross of the British Empire, and two years later he received the Order of Merit. Sherrington died of heart failure on March 4, 1952, in Eastbourne, England.

Shoemaker, Eugene Merle

(1928–1997)
American
Geologist, Astrogeologist

Eugene Shoemaker conducted important research into the geology of Earth, the Moon, and other celestial bodies, an unconventional scope of work for a geologist. He demonstrated that the Barringer Crater (also known as the Meteor Crater) in Arizona was formed not by volcanic action, as had been speculated, but by a meteorite's striking the Earth. He worked for an astrogeological unit of the United States Geological Survey (USGS) and in the Manned Space Sciences Division of the National Aeronautics and Space Administration (NASA), where he helped to map the Moon's surface and formulate mission objectives for space voyages. He founded the Palomar Planet-Crossing Asteroid Survey (later called the Palomar Asteroid and Comet Survey), which conducted the first systematic search for asteroids. He also propounded theories about meteorological events that may have led to the extinction of the dinosaurs by virtue of the climatic upheavals they caused.

Shoemaker was the eldest child of George Estel and Muriel May Scott Shoemaker. After his birth in Los Angeles on April 28, 1928, the family moved frequently before returning to Los Angeles when Shoemaker was 14. George Shoemaker worked at an assortment of jobs; Muriel, a teacher, encouraged her son's early interest in geology.

After receiving a bachelor's degree in geology from the California Institute of Technology (Caltech) in 1947, Shoemaker worked with the USGS for two years. From 1950 to 1951 and again from 1953 to 1954 he attended Princeton University, where he received his M. A. in 1954. In 1960 he was awarded his Ph.D. from Princeton for his dissertation on the presence of coesite in the Barringer Crater. Given that a volcano exerts insufficient force to create coesite (a form of silica produced when quartz is squeezed at massive pressure), Shoemaker and his collaborator Edward Chao were able to show that the crater must have been made by a meteor.

After completing his doctorate Shoemaker turned his attention to the skies. He worked for a USGS astrogeological unit in 1960 and at NASA in 1962 and 1963. In the mid-1960s he was in charge of television experiments on Project Ranger, and later Project Surveyor, both of which were uncrewed space voyages. He devised the technique of photoclinometry to analyze the photos taken on these flights and computed the steepness of lunar craters by examining the shadows they projected on the ground. He also assisted NASA in establishing scientific objectives for its crewed missions. The insight he provided into the Moon's surface convinced NASA that a probe could indeed be landed there.

After resigning from the USGS in 1966, Shoemaker returned to the study of impact craters on Earth. In 1973 he founded the Palomar Planet-Crossing Asteroid Survey to search for asteroids. In 1982 Shoemaker's wife, Carolyn Spellman Shoemaker, joined his project. Together the couple, who had married in 1951, discovered more than 300 asteroids. They also had three children together.

Shoemaker was a professor at Caltech from 1969 to 1985 and chairman of the school's Division of Geological and Planetary Sciences from 1969 until 1972. In 1980 Shoemaker focused his attention on the work of LUIS WALTER ALVAREZ, who asserted that the extinction of the dinosaurs 65 million years ago was caused by the cataclysmic impact of an asteroid or meteor that stirred up a cloud of dust that blocked the Sun's rays and lowered the Earth's temperatures. Alvarez's team proposed the 120-mile-wide Chicxulub crater in the Yucatan as the likely spot of impact. Shoemaker argued that a smaller crater in Iowa dating from the same era indicated that multiple impacts rocked the Earth and caused the change in climate. He proposed that the barrage of meteors was caused by the breakup of a giant comet pelting the planet with debris.

Shoemaker received many significant honors, including the National Medal of Science and the NASA Medal for Scientific Achievement. He is a fellow of the American Academy of Arts and Sciences and a member of the National Academy of Sciences. His work is recognized for the insight it shed on the impact history of the planet. Shoemaker died in 1997.

Siegbahn, Karl Manne Georg

(1886–1978)
Swedish
Physicist

Karl Siegbahn devoted his career to the study of X-ray spectroscopy, expanding knowledge in the field by improving on instrumentation and techniques. He managed to measure the length of X-ray wavelengths and received the 1924 Nobel Prize in physics for this pioneering work in X-ray spectroscopy.

Siegbahn was born on December 3, 1886, in Orebro, Sweden, to Emma Sofia Mathilda Zetterberg and Nils Reinhold Georg Siegbahn, a stationmaster for the Swedish national railway system. Siegbahn married Karin Högbom in 1914, and together the couple had two sons, Bo and Kai Manne Börge. Kai won the 1981 Nobel Prize in physics for work that built on his father's in spectroscopy.

Siegbahn commenced study of astronomy, chemistry, mathematics, and physics at the University of Lund in 1906. The summer after he received his bachelor's degree in 1908 he studied in Göttingen, and the next summer he studied in Munich. In 1910 he earned his licentiate, the equivalent of a master's degree. Siegbahn then worked as a research assistant in Johannes Rydberg's laboratory studying electromagnetics while pursuing his doctorate. He received his Ph.D. in 1911 with a dissertation that took measurements of magnetic fields. As with his transition between undergraduate and graduate work Siegbahn spent his summer between his doctorate and his professional career traveling to Paris and Berlin for further study.

Siegbahn continued in Rydberg's lab and assumed his mentor's teaching responsibilities as the elder fell ill. When Rydberg died in 1920 Siegbahn inherited his chair as the head of the physics department at the University of Lund. In 1923 Uppsala University offered him a professorship in physics, which he accepted and held for the next 14 years, when the University of Stockholm appointed him professor of physics. Siegbahn simultaneously accepted the directorship of the newly founded Nobel Institute of Physics, which he ran until 1975, making it a world-class facility. From 1939 to 1964 Siegbahn was a member of the International Committee on Weights and Measures.

In 1924 Siegbahn discovered by means of spectroscopy the phenomenon of X-ray refraction by a prism, as with visible light splitting into the colors of the spectrum. He confirmed the existence of the hypothesized K and C series of spectral lines corresponding to two distinct "shells" of electrons within atoms, the former "hard" and the latter "soft," depending on the atomic weight of the molecule under study. In 1916 he discovered yet another shell that was creating yet another set of spectral lines, the M series lines; he later discovered the N series of spectral lines. The next year he published his findings in his classic text *The Spectroscopy of X-Rays.*

Besides the Nobel Prize, Siegbahn received two medals from the Royal Society of London—the Hughes Medal in 1934 and the Rumford Medal in 1940. He was also awarded the Duddell Medal of the London Physical Society. Siegbahn died on September 26, 1978, in Stockholm.

Simpson, Joanne Malkus

(1923–)
American
Meteorologist

Joanne Simpson, the first woman to earn a Ph.D. in meteorology, has chased hurricanes in planes, studied ways to increase rain and calm storms, and worked out valuable models of clouds, storms, and heat balance in the tropics. She was born Joanne Malkus on March 23, 1923, in Boston, Massachusetts, to a newspaper editor father and a reporter mother. In a 1973 speech she described her childhood as "intellectual but unhappy," making her determined to "get somewhere and be somebody."

Joanne learned meteorology at the University of Chicago as part of a special training program during World War II. She taught the subject to military recruits while continuing her own education and earned a B.A. in 1943 and an M.S. in 1945. Unlike other women who learned meteorology during the war, she refused to leave this "men's field" when the conflict ended. In 1949 she became the first woman to gain a doctorate in meteorology (from the University of Chicago). While working on her Ph.D., she also worked as an instructor at the Illinois Institute of Technology, married, and became a mother. She divorced soon after obtaining her Ph.D.

Joanne remained at the Illinois Institute of Technology until 1951; eventually she became an assistant professor. She then worked at the Woods Hole Oceanographic Institution until 1960. There her main research subjects were winds, especially those involved in fierce tropical storms such as hurricanes, and the way that the movement of heat in the tropical atmosphere contributes to such storms. She married again and had three more children, though in time this marriage also failed. In 1960 she became a full professor at the University of California at Los Angeles (UCLA).

Joanne became the director of the National Oceanic and Atmospheric Administration's experimental meteorology laboratory in Coral Gables, Florida, in 1964 and remained there until 1974. She has said she forged this laboratory "out of nothing . . . on a shoestring." She married a fellow meteorologist, Robert Simpson, in 1965 and hoped to do research with him on the project, but government nepotism rules blocked that plan.

In the late 1960s and early 1970s Simpson did research on cumulus clouds, the large, thick clouds that produce rain. She created the first mathematical one-dimensional computer model of such clouds and tested it by observation. She also experimented with cloud seeding, a technique for increasing rainfall, and continued her study of hurricanes, often personally flying research planes into their centers.

Simpson became the Corcoran Professor of Environmental Sciences at the University of Virginia in 1974 and held that post until 1979. During that time she also worked with her husband in a private company, Simpson Weather Associates. She used extensive new data to refine her hypothesis about heat flow in the tropical ocean and atmosphere.

Since 1979 Simpson has worked for the Goddard Space Flight Center in Greenbelt, Maryland, first as head of the center's Severe Storms Branch and, since 1988, as the chief scientist of its Meteorology and Earth Science Directorate. She was also project scientist for the Tropical Rainfall Measuring Mission from 1986 until the mission was launched in 1997. She has continued her studies of tropical storms and extended her cloud modeling to show how clouds merge into systems that produce over 90 percent of tropical rainfall.

Simpson has won a number of awards during her career, including the Meisinger Award of the American Meteorological Society (1962), a gold medal from the Department of Commerce (1972), NASA's Exceptional Scientific Achievement Medal (1982), the first Nordberg

Award for Earth Sciences (1994), and a NASA Outstanding Leadership Medal. She was president of the American Meteorological Society in 1989.

Sithole-Niang, Idah

(1957–)
Zimbabwean
Biochemist, Geneticist

Idah Sithole-Niang has made a reputation for herself in her native Zimbabwe by developing ways to use genetic engineering to improve food crops there. She was born in that southeast African country on October 2, 1957. She won scholarships to study at the University of London, where she learned biochemistry and earned a B.S. in 1982, and at Michigan State University in Lansing, where she earned a Ph.D. in 1988. Sithole then married Cheikh I. Niang, and they had one son.

In Michigan, Sithole did genetic studies of both viruses and plants. After she became a lecturer in biochemistry at the University of Zimbabwe in 1992, she combined these two subjects, focusing on viruses that infect plants, in particular the potyvirus, which affects cowpeas, a legume that is one of Zimbabwe's chief food crops. She has identified and worked out the structure of genes that code for the chemicals that make up the potyvirus, and she hopes to insert some of these genes into cowpeas and thereby make them resistant to the virus.

In another set of experiments Sithole has studied a gene that could be introduced into cowpeas to make them resistant to a common family of herbicides called atrazines. Maize (corn) is naturally resistant to these herbicides, which are sprayed on cornfields to control weeds. If cowpeas were also resistant to atrazine, they could be grown in fields with maize, both enriching the soil and providing a cheap source of food.

She has received a number of fellowships and awards, including the U.S. Agency for International Development Fellowship (1983–88) and a Rockefeller Foundation Biotechnology Career Fellowship (1992–95). In the late 1990s Sithole-Niang, as she began calling herself in 1996, broadened her interests to include policy issues such as the safety of genetically engineered organisms. As a member of the Interim Biosafety Board she has been updating biosafety guidelines and regulations for Zimbabwe.

Snyder, Solomon Halbert

(1938–)
American
Neuroscientist

Solomon Halbert Snyder distinguished himself as a researcher by identifying the location in the brain of opiate receptors and by isolating opiatelike substances manufactured in the body. He also pioneered the techniques necessary to investigate these receptors. In other areas he identified a protein responsible for the detection of odor. In his laboratory he demonstrated the reproduction of adult brain cells; he also discovered what may be a new class of neurotransmitters.

Snyder was born on December 26, 1938, in Washington, D.C., the second of five children born to Patricia Yakerson, a real estate broker, and Samuel Simon Snyder, a cryptanalyst who broke codes during World War II for the National Security Agency. When Snyder was a mere nine years old, his father taught him to program computers. In 1955 Snyder entered Georgetown University as a premed student, and in 1958, before he had even received his bachelor's degree, Georgetown University Medical School admitted him. In 1962 at the age of 23, Snyder received his M.D. On June 10 of that year, he married Elaine Borko, a psychotherapist. Together the couple had two children, Judith and Deborah.

Snyder served his internship the next year at the Kaiser Foundation Hospital in San Francisco. He then worked for two years as a research assistant in the laboratory of the Nobel laureate Julius Axelrod at the National Institute of Mental Health. In 1965 he commenced his residency in psychiatry at Johns Hopkins Hospital in Baltimore, where he excelled, and Johns Hopkins University School of Medicine hired him as an assistant professor of pharmacology while he was still a resident. He remained at Johns Hopkins the rest of his career; by 1980 he had attained the status of Distinguished Service Professor in neuroscience, pharmacology, and psychiatry, while also serving as the director of the department of neuroscience.

In 1972 Snyder collaborated with a graduate student, CANDACE BEEBE PERT, on the most significant research of his career, searching for the site of opiate receptors in the brain in order to understand better how opiates kill pain as well as to understand the physiological processes of addiction. Snyder and Pert published their results in a 1973 *Science* article, "Opiate Receptor: Demonstration in Nervous Tissue." Snyder published numerous other articles and books in his career, including *Madness and the Brain* in 1974, *Opiate Receptor Mechanisms: Neurochemical and Neurophysiological Processes in Opiate Drug Action and Addiction* with Steven Matthysse in 1975, *The Troubled Mind: A Guide to Release from Distress* in 1976, and *Biological Aspects of Mental Disorder* in 1980.

In recognition of the significance of Snyder's research he received the 1978 Albert Lasker Medical Research Award, one of the most prestigious awards in his field. A 1991 *Scientific American* article called Snyder "one of the country's most prolific and creative neuroscientists." Snyder attributed much of his creativity to his wide reading in search of seemingly unrelated developments in the medical community that he might be able to apply to neuroscience.

Soddy, Frederick

(1877–1956)
English
Chemist

Frederick Soddy's greatest scientific achievements involved the study of radioactivity. From 1901 to 1903 he collaborated with ERNEST RUTHERFORD, and the two produced ground-breaking work on the decay of radioactive elements. In 1913 Soddy discovered the existence of isotopes (various forms of a chemical element possessing different atomic weights but identical chemical properties), for which he was subsequently awarded the Nobel Prize. Becuase of the impact of World War I his interests during the postwar years shifted from chemistry to political and economic matters.

Soddy, the youngest of his parents' seven children, was born to a prosperous family in Eastbourne, England, on September 2, 1877. His father, Benjamin Soddy, was a successful corn merchant. His mother, Hannah Green Soddy, died when he was only 18 months old.

Frederick Soddy, who won the Nobel Prize in chemistry for his discovery of isotopes. *Smith Collection, Rare Book & Manuscript Library, University of Pennsylvania.*

Soddy's scientific talents were evident from his youth and were later nurtured by R. E. Hughes, who taught Soddy at Eastbourne College. Hughes also encouraged Soddy to study chemistry at Oxford, and he did so in 1896, after an intermediate year at University College in Aberystwyth, Wales. He graduated with a first-class-honors degree in 1898 and remained at Oxford conducting independent research until 1900.

Soddy then became a junior demonstrator at McGill University in Montreal, where he met Rutherford. In 1901 the two began to investigate the curious behavior of radioactive elements. They had noticed that the element thorium transmuted into an entirely different substance in the course of producing radioactive emanations. Over two years of study they came to realize that radioactivity is the physical manifestation of an element's decay. In other words, by emitting radioactive particles, a radioactive element such as thorium sheds protons, thereby altering its atomic weight and becoming an entirely different element. This discovery engendered a complete reworking of the scientific understanding of radioactivity.

In 1903 Soddy returned to England, where he engaged in a brief partnership with Sir William Ramsey, during which they demonstrated that the disintegration of radium invariably produces helium. In 1904 Soddy became a lecturer at the University of Glasgow, where he remained for the next 10 years. It was there that he began the work that led to his Nobel Prize–winning discovery. During the first decade of the 20th century many scientists were striving to identify the elements produced by the decay of radium and uranium. They were troubled by the fact that more new elements appeared than could be accommodated by the gaps DMITIRI IVANOVICH MENDELEEV had left in the periodic table. Moreover, many of these substances shared identical chemical properties. Soddy's experiments led him to hypothesize that these substances might simply be variants of a smaller number of elements, and he published a paper suggesting this theory in 1910. Over the next three years he developed the theory that these puzzling substances could represent various forms of the same chemical element that possessed different atomic weights. In 1913 Soddy attached the term *isotope* to this concept in a paper he published in *Chemical News*. The scientific community recognized the validity of this explanation.

In 1914 Soddy left Glasgow for the University of Aberdeen, where he conducted military research to aid the war effort. In 1918 he took a post at Oxford, where he remained for the next 18 years. But the devastation of the war was profoundly disillusioning to him. Feeling that advances in science had led to no concomitant advances in human civilization, Soddy abjured original research in favor of political, social, and economic endeavors. He became particularly involved with issues relating to the status of Ireland and the female suffrage movement.

Upon the sudden death of his wife, Winifred Moller Beilby Soddy, whom he had married in 1908, Soddy retired from Oxford and spent the remainder of his life speaking and writing about contemporary political issues. Having no children, he bequeathed much of his estate to a trust designed to ameliorate social inequities in various regions of England, but Soddy was never able to achieve the success in the public arena that he had in chemistry. Along with the Nobel Prize, he received the Cannizzaro Prize and was made a member of the Royal Society of England and the Chemical Society, and a foreign member of the Italian, Russian, and Swedish Academies of Science.

Somerville, Mary Fairfax

(1780–1872)
British
Mathematician

In spite of determined attempts to block her education, Mary Somerville learned enough mathematics and science to explain complicated theories to a popular audience. She was born to Sir William and Margaret Fairfax on December 26, 1780, in Jedburgh, Scotland, but grew up in the seaside village of Burntisland. Her father, a vice-admiral in the British navy, was often away from home. The large Fairfax family had little money.

As a child Mary was free to roam the shores and fields around her home except during one "utterly wretched" year in a girls' boarding school, where she felt equally confined by torturous clothing and boring memorization. She led an active social life in her teen years, even becoming known as the "Rose of Jedburgh" because of her beauty. A yearning for something more was stirred in her one day when, as she recounted in her memoirs, she spotted a mathematical puzzle in a fashion magazine that contained "strange-looking lines mixed with letters, chiefly *X*s and *Y*s." A friend explained, "It's a kind of arithmetic; they call it Algebra." From there she found her way to geometry but had to ask her brother's tutor to buy her books on these subjects because it was not considered proper for a young woman to enter a bookstore.

After Mary obtained her books, her parents tried to stop her from studying them because, like many people of their time, they feared that too much intellectual effort could harm a woman's physical and mental health. ("We shall have Mary in a straightjacket one of these days," her father worried.) Like MARIE SOPHIE GERMAIN's parents, they tried to prevent their daughter from studying at night by taking away her supply of candles. Mary, however, had memorized the formulas she wanted to work on and simply lay in the dark, solving problems in her head.

In 1804 Mary married Samuel Greig, a distant cousin, who was a captain in the Russian navy. They had two sons. Greig died three years after the marriage, and Mary and the boys returned to Burntisland. Her family still considered her "eccentric and foolish," but as a widow she now had more money and independence and could pursue learning more openly.

Mary married another cousin, William Somerville, in 1812. Unlike Greig, Somerville, an army surgeon, supported his wife's educational ambitions. After he became head of the Army Medical Department in 1816 and they moved to London, he introduced her to leading scientists of the day. They had three daughters, one of whom died young.

Mary Somerville did a few experiments in physics and wrote scientific papers during this period. One paper, which her husband presented to the prestigious Royal Society (since women could not attend its meetings) in 1826, associated magnetism with ultraviolet light from the Sun. Though later shown to be incorrect, Somerville's papers were well received. Still, she concluded that she lacked originality as a scientist, a condition she unfortunately believed was true of all women. "I have perseverance and intelligence but no genius," she wrote. "That spark from heaven is not granted to the [female] sex."

In 1827 Lord Henry Brougham, head of an educational group called the Society for the Diffusion of Useful Knowledge, asked Mary Somerville to write a translation and explanation of PIERRE SIMON DE LAPLACE's *Mécanique céleste* (Celestial mechanics), an influential but difficult work that summarized what was known about the mathematics of gravity. Published in 1831 as *The Mechanisms of the Heavens,* the book that resulted was a great success. Laplace himself said that Somerville was the only woman, and one of the few people of either gender, who understood his work. The book's lengthy preface, written by Somerville and republished separately in 1832, explained the mechanical principles that governed the solar system and the mathematics behind Laplace's ideas. This included differential and integral calculus, for which she provided original proofs and diagrams. The book remained a college text for nearly 100 years.

In 1834 Somerville published a second book, *On the Connexion of the Physical Sciences,* a summary of current scientific ideas that showed the relationships among seemingly different fields. "In all [physical sciences] there exists such a bond of union, that proficiency cannot be attained in any one without knowledge of others," she wrote in the preface. The book went through 10 revised editions. A remark in one, suggesting that unusual features in the orbit of the planet Uranus might be caused by an as-yet-unknown planet that lay beyond it, led John

Couch Adams to calculate the planet's orbit, a process that led to the discovery of Neptune.

Somerville's third book, *Physical Geography,* was published in 1848. It described new ideas in geology, including the belief that the Earth was extremely old. It became the most popular of her books, though some critics denounced Somerville as "godless" for agreeing with scientists who said that the Earth was much older than the age calculated from the Bible.

Somerville wrote one more book, *Molecular and Microscopic Science,* which discussed advances in chemistry, physics, and biology. It was published in 1869, when she was 89 years old. The science in this book was outdated, however, and reviews were "kindly rather than laudative."

Somerville's books made her the best-known woman scientist of the 19th century. A bust of her was placed in the great hall of the British Royal Society in 1831, and she and CAROLINE LUCRETIA HERSCHEL were both made honorary members of the Royal Astronomical Society—the first women to achieve this status—in 1835. In that same year Somerville received a lifetime pension from the king. The British Royal Geographical Society and the Italian Geographic Society both gave her gold medals. The Somervilles had moved to Italy in the 1840s in the hope of improving William Somerville's health, and Mary stayed there after his death in 1865. She was still working on mathematics problems on the day she died, at Naples in 1872, at the age of 92. In 1879 a women's college at Oxford University, Somerville College, was named for her.

Sperry, Roger Wolcott

(1913–1994)
American
Neurobiologist

Roger Wolcott Sperry's pioneering experiments into the functional specialization of the brain not only garnered him the 1981 Nobel Prize, but also influenced later developments in the fields of neurobiology and experimental psychobiology. Prior to his work little was known about the two nearly identical hemispheres of the brain, which were connected by bands of nervous tissues, the corpus callosum (or the great cerebral callosum). Decades of research led Sperry to conclude that these left and right hemispheres perform distinct functions and that the corpus callosum integrates the two by transmitting information.

Sperry was born in Hartford, Connecticut, on August 20, 1913. His father, Francis Bushnell Sperry, a banker, died when Roger was 11. To support the family in the wake of her husband's death, Sperry's mother, Florence Kramer Sperry, returned to school and later worked in high school administration. In 1949 Sperry married Gay Deupree, with whom he had two children.

Sperry matriculated at Oberlin College, where he earned a B.A. in English in 1935 and an M.A. in psychology in 1937. He then enrolled at the University of Chicago. Under the guidance of the biologist Paul Weiss, Sperry explored nerve connections. Eventually he disproved Weiss's accepted theory on the subject. In his dissertation Sperry concluded that the location and function of nerves were intricately related. He was awarded his Ph.D. in zoology in 1941.

Sperry worked as a postdoctoral fellow at Harvard University until 1942, when he accepted a position at the Yerkes Laboratories of Primate Biology in Orange Park, Florida. Sperry's early research concentrated on nerve regeneration. At this time it was commonly held that nerve function preceded form—that the brain and nervous system "learn" to function properly. To test this theory, Sperry cut the optic nerve in salamanders (which, like other amphibians, can regenerate nerve tissue) and rotated their eyeballs 180 degrees. He found that the nerve did not form new pathways to accommodate the new eye position. Instead, the optic nerve followed the original pathway. He surmised that nerves have discrete, specialized functions based on genetically predetermined differences in nerve cell chemicals.

After a three-year stint at the Office of Scientific Research and Development to fulfill his military duties during World War II, Sperry returned to the University of Chicago in 1946 as an assistant professor of anatomy. In 1954 he was named Hixon Professor of Psychobiology at the California Institute of Technology. Sperry then began to explore split-brain functions. Although scientists had long known that the cerebrum of the brain comprised two hemispheres that controlled muscle movements—the left hemisphere responsible for muscles on the right side of the body, and the right hemisphere for those on the left—scientists believed that the right hemisphere was less developed than the left hemisphere. (This theory derived from observations of patients who had endured trauma to the left hemisphere of the brain and thereafter suffered from speech impediments and other cognitive dysfunctions.) Moreover, it was assumed that the corpus callosum had no function other than preventing the brain hemispheres from sagging.

Sperry's research overturned these hypotheses. After severing the corpus callosum in animals, he found that behavioral responses learned by the left side of the brain were not transferred to the right side. The animal with the split brain behaved as if it had two separate brains. This observation led Sperry to suggest that the corpus callosum was responsible for transmitting information between the two hemispheres. Sperry also worked with human patients whose corpus callosum had been surgically cut to prevent seizures. He determined that the right hemisphere of the brain was by no means inferior to the left. Indeed the right half of the brain carried out concrete thinking and was responsible for spatial relations, emotions, and musical ability.

In addition to receiving the 1981 Nobel Prize in physiology or medicine (which he shared with TORSTEN NILS WIESEL and DAVID HUNTER HUBEL), Sperry was named a member of numerous scientific societies, including the National Academy of Sciences. Sperry's insights into the differences between the two hemispheres of the brain and the essential functions performed by the right brain would influence later experiments into different areas of the brain.

Stanley, Wendell Meredith

(1904–1971)
American
Biochemist

Wendell Meredith Stanley's most significant achievement was his work in isolating, purifying, and crystallizing viruses, thereby demonstrating their molecular structures. His experiments on the tobacco mosaic virus (TMV) led him to conclude that the virus was rod-shaped and contained a protein component. During World War II Stanley developed a vaccine for viral influenza, a discovery that spared countless lives. His research provided the foundation for the fledgling field of molecular biology.

Stanley was born on August 16, 1904, in Ridgeville, Indiana. His parents, James and Claire Plessinger Stanley, published a local newspaper, and as a child Stanley was often pressed into service to deliver papers, search out stories, and set type.

In January 1926 Stanley graduated with a B.A. in chemistry from Earlham College in Indiana. His goal was to become a football coach, but on a trip to Illinois State University with an Earlham chemistry professor he met Roger Adams, an organic chemist. Stanley was so inspired by Adams's field that he decided to abandon his coaching ambitions and entered graduate school at Illinois State in 1926. After studying various chemicals to treat leprosy, Stanley was awarded his M.A. in 1927 and his Ph.D. in 1929. In 1929 he married a fellow chemist, Marian Staples Jay, with whom he collaborated on a paper. The couple eventually had four children together. After receiving his Ph.D., Stanley embarked on a postdoctoral fellowship at the University of Munich in Germany, where he deepened his knowledge of experimental biochemistry.

Stanley returned to the United States in 1931 to become a research assistant at the Rockefeller Institute. Work on a marine plant, *Valonia,* introduced him to problems in biological chemistry and augmented his understanding of biophysical systems. In 1932 Stanley transferred to the Rockefeller Institute's Division of Plant Pathology, where he began his investigation into TMV, a disease that caused discoloration in tobacco plants. Although it was already known that the infectious agent of TMV was a virus, little was understood about its workings. Stanley sought to understand its structure. After infecting several Turkish tobacco plants with TMV, Stanley ground the plants into a liquid. He tested the liquid with more than 100 different chemicals and thereby determined that various enzymes were able to inactivate the virus. Even more significant was the fact that TMV was inactivated by the same enzymes that typically inactivated proteins. Stanley concluded that TMV was proteinlike. His next challenge was to purify the virus. After two and one-half years of painstaking distillation, he became the first person to isolate a virus in its crystalline form and found that TMV crystals were still capable of causing infection. In 1938 two researchers in England confirmed Stanley's insight when they discovered that TMV was a nucleoprotein, 94 percent protein and 6 percent nucleic acid.

During World War II Stanley was recruited to join the effort at the Office of Scientific Research and Development to create a vaccine against viral influenza. With his team he found that formaldehyde, a biological preservative, not only stimulated the human body to produce antibodies, but also inactivated the virus. In 1948 Stanley accepted a position at the University of California, Berkeley, where he directed the virology laboratory and chaired the department of biochemistry. He recruited a team of scientists who made stunning contributions to the study of virology. In 1960 Stanley oversaw an effort to determine the entire amino acid sequence of TMV protein.

Stanley's achievement in crystallizing TMV and discerning its structure was recognized in 1946 when he was awarded the Nobel Prize in chemistry (which he shared with JOHN HOWARD NORTHROP and JAMES BATCHELLER SUMNER). In 1948 he received the Presidential Certificate of Merit for his role in producing the vaccine. His work was essential to the field of molecular biology. Furthermore, by crystallizing TMV, Stanley revealed that a crystalline substance with chemical properties, such as a virus, could also be a living entity. In addition to publishing more than 150 articles, Stanley was known for defending scientific colleagues against the politics of McCarthyism. Stanley died of a heart attack on June 15, 1971.

Stark, Johannes

(1874–1957)
German
Physicist

Although Johannes Stark's early career was marked by stunning accomplishments, he later embraced Adolph Hitler's National Socialist movement and called for German science to purge itself of Jewish contributions. Stark was recognized, however, as a brilliant physicist. In addition to founding the acclaimed *Yearbook of Radioactivity and Electronics,* Stark discovered both the Doppler effect in canal rays and what came to be known as the Stark effect, achievements for which he was awarded the Nobel Prize in 1919. Though his work affirmed ALBERT EINSTEIN's relativity theory, he later renounced it as a product of "Jewish" science.

Stark was born on April 15, 1874, at his family's farm in Schickenhof, Bavaria. Little is known about his parents or early childhood. Later in life he married Louise Uepter, with whom he had five children. After attending public schools in Bayreuth and Regensberg, Stark enrolled as a science major at the University of Munich in 1894. He was awarded his Ph.D. in 1897. For the next three years he served as an assistant to Eugene Lommel at Munich.

In 1900 Stark accepted a position at the University of Göttingen, where he made his first major discovery. Like other physicists, Stark was familiar with the principle of the Doppler effect, a theory proposed in 1842 by JOHANN CHRISTIAN DOPPLER. The Doppler effect is the apparent variation in frequency of an emitted wave as that wave changes position relative to an observer. Wavelengths shorten as they approach (producing a higher pitch) and lengthen as they move away (resulting in a lower one). An example of this phenomenon occurs when a train approaches a station. Its whistle sounds higher-pitched the closer the train moves toward the passengers waiting on the platform. EDWIN POWELL HUBBLE observed this same effect in the lines of the spectrum of luminous bodies, such as stars. Stark's insight was to use canal rays (beams of positively charged particles produced in a vacuum tube) to observe the Doppler effect. In 1905 he made an important breakthrough when he employed this process to investigate the Doppler effect in the spectral lines of hydrogen atoms.

Stark was offered a post as lecturer at the Technical College in Hannover in 1906, but he disagreed with his superior, and left after three years. In 1909 he was appointed a professor at the Technical College in Aachen, where he remained for eight years. There he addressed a problem posed in 1896 by the Dutch physicist Pieter Zeeman, who had noted that an element's spectral lines split in the presence of a magnetic field. Although no scientist had previously been able to replicate this effect, Stark succeeded in 1913, when he split the spectral lines in an electric field by generating canal rays in a vacuum tube containing electrodes, a cathode, and 20,000 volts. The phenomenon was named the *Stark effect* in his honor.

After accepting a position at the conservative University of Greifswald in 1917, Stark moved to the more liberal University of Würzburg in 1920. But he abandoned his physics research, bickered with his colleagues, and resigned in 1922. Thereafter he immersed himself in Nazi politics and sought to establish conservative scientific organizations, such as the Fachgemeinschaft der deutschen Höchschuleher der Physik. Although he later attempted to return to academia, he had burned too many bridges with his hateful proposals and flagrant efforts to undermine his peers. In the 1930s he worked as an administrator of science in posts obtained as a result of his loyalty to the Nazis but was never able to realize his goal of overseeing the creation of German scientific policy. Stark's earlier scientific accomplishments had been recognized. In 1910 he won the Nobel Prize in physics, as well as the Baumgartner Prize of the Vienna Academy of Sciences. By the onset of World War II, however, Stark was isolated, having failed in his bids to claim academic and policy-making power. He retired to his estate in Eppenstatt. In 1947 a German denazification court sentenced him to four years in a labor camp. He returned to Eppenstatt after serving his time and died on June 21, 1957.

Stein, William Howard

(1911–1980)
American
Biochemist

For his ground-breaking work with his colleague STANFORD MOORE, William Howard Stein was awarded the Nobel Prize in 1972. In their decades-long collaboration, Stein and Moore researched the structure and function of proteins, especially the digestive enzyme ribonuclease. The pair also invented technology that made such analysis possible. The fraction collector and automated system devised by Stein and Moore allowed them to collect and separate amino acids (the building blocks of proteins) and subsequently became standard laboratory equipment. Their techniques and results facilitated future investigations of proteins.

Born on June 25, 1911, in New York City, Stein was the second child of Fred and Beatrice Borg Stein. Though a successful businessman, his father retired early to devote himself to public service. While Fred Stein volunteered at community health care services, Beatrice Stein lobbied for better educational facilities for underprivileged children. The couple encouraged William to pursue science.

Upon earning his B.A. from Harvard University in 1933, Stein began graduate studies in chemistry there. After a year, however, he realized that his true interest lay in biochemistry. He thereupon transferred to the College of Physicians and Surgeons of Columbia University, where he received his Ph.D. in 1938 for a dissertation exploring the amino acid composition of elastin, a protein in the walls and veins of arteries. Stein married Phoebe Hockstader on June 22, 1936, and the couple had three sons.

In 1938 Stein accepted a position at the Rockefeller Institute (now Rockefeller University) under Max Bergmann. At the institute he developed his collaborative relationship with Moore. Stein's initial project was to conceive a methodology for analyzing the amino acids glycine and leucine. World War II interrupted his efforts, as Bergmann's lab was commissioned to find a counteragent for mustard gas.

After the war Bergmann's team at the Rockefeller Institute was asked to establish the institute's first program of protein chemistry. Although scientists were well aware that proteins performed essential roles as enzymes, hormones, oxygen carriers, and antibodies, virtually nothing was known about their chemical makeup. In an effort to improve the process of separating amino acids and obtain more fundamental information about proteins' structures Stein and Moore built upon the filtering technique of partition chromatography developed by A.J.P. Martin and Richard Synge. Stein and Moore eventually invented the automatic fraction collector and an automated system to analyze amino acids. These discoveries reduced the time required to differentiate and examine amino acids from two weeks to a few hours.

Stein and Moore applied their new technology to the study of ribonuclease, an enzyme secreted by the pancreas that catalyzes the breakdown of nucleic acids, thereby assisting in the digestion of food. The duo found that ribonuclease's amino acid sequence was a three-dimensional chain that caused a catalytic reaction by folding and bending. This crucial discovery became the jumping-off point for investigation by later scientists into enzyme catalysis. Stein was appointed a professor of biochemistry at Rockefeller in 1954. With Moore he then detailed the structure of deoxyribonuclease, a molecule twice as complex as ribonuclease.

Stein's active participation in research was cut short in 1969, however, when he contracted Guillain-Barré syndrome, a disease that left him a quadriplegic, confined to a wheelchair. But the techniques invented by Stein and Moore revolutionized the study of proteins. Stein's accomplishments were recognized not only with his 1972 Nobel Prize in chemistry (shared with Moore and CHRISTIAN BOEHMER ANFINSEN) but also with the 1972 Theodore Richard Williams Medal of the American Chemical Society. In addition to serving as chairperson of the U.S. National Committee for Biochemistry from 1968 to 1969, Stein edited the *Journal of Biological Chemistry.* He led the *Journal* to preeminence in the field of biochemistry publications. Stein died on February 2, 1980.

Stern, Otto

(1888–1969)
German/American
Physicist

Otto Stern is best remembered for devising methods of using a molecular beam to determine the magnetic moment of atoms and nuclei, as well as other physical constants. Most significantly, Stern's work bore out the predictions of quantum theory concerning the spatial quantization of atoms. Quantum theory holds that atoms are particles with electric charges that rotate in space. *Spatial quantization* is the concept that the magnetic moment of an atom could have only specific and finite properties. (A *magnetic moment* is the magnetic field that attends an atom.) Stern's work aided the further development of quantum theory.

Stern, the oldest of his parents' five children, was born on February 17, 1888, in Sohrau, Upper Silesia (which is now Zory, Poland). His parents, Oskar and Eugenie Rosenthal Stern, were Jewish traders, who encouraged their children's intellectual curiosity. Otto himself never married.

Upon graduation from the Johannes Gymnasium in Breslau (now Wroclaw, Poland) in 1906 Stern attended the universities of Freiberg im Breisgau, Munich, and Breslau; he received his Ph.D. in physical chemistry from Breslau in 1912. Stern then served for two years as a research assistant to ALBERT EINSTEIN, an experience that stimulated Stern's interest in the fields of quantum theory and thermodynamics. Stern served in the Meteorology Corps of the German army during World War I, though he still found the time to author two significant papers applying quantum theory to statistical thermodynamics. At war's end Stern took a position at the University of Frankfurt-am-Maim. There he became a research assistant to MAX BORN and began to use the recently developed technique of employing molecular beams to investigate the characteristics of molecules and atoms.

In 1911 the French physicist Louis Dunoyer had demonstrated that molecules or atoms placed in a high-vacuum chamber travel in fixed patterns. When vapor or gas is introduced into the chamber, beams of particles—not dissimilar to rays of light—become visible. These are known as molecular beams. Stern initially employed this method in 1920, to provide the first empirical confirmation of JAMES CLERK MAXWELL's theoretical description of the distribution of molecular velocities in gas.

Two years later, after taking a position as an associate professor of theoretical physics at the University of Rostock in 1921, he conducted his most famous work. Along with Walther Gerlach, Stern discovered that when a beam of silver atoms passes through a nonuniform magnetic field, it splits into two discrete parts. Now known as the Stern-Gerlach experiment, this process demonstrated the accuracy of previous predictions about the spatial quantization of atoms. It also permitted Stern to ascertain the precise magnetic moment of silver.

Continuing his peripatetic ways, Stern moved to the University of Hamburg in 1923 to become a professor of physical chemistry and director of the Institute for Physical Chemistry. In Hamburg he employed molecular beams to prove that particles have wavelike properties, as LOUIS-VICTOR-PIERRE-RAYMOND DE BROGLIE had earlier hypothesized. Outraged by the anti-Semitic programs of Adolph Hitler's new government, Stern resigned his position in 1933 and moved to the United States. He became a research professor in physics at the Carnegie Institute in Pittsburgh and was naturalized in 1939. He was awarded the Nobel Prize in physics in 1943.

After World War II Stern retired to Berkeley, California, where he died in a movie theater in 1969. In addition to receiving the Nobel Prize, he was named a member of the National Academy of Sciences, the American Philosophical Society, and the Royal Danish Academy. The University of California at Berkeley and the Eidgenössische Technische Hochschule in Zurich conferred honorary doctorates on him.

Stevens, Nettie Maria

(1861–1912)
American
Geneticist

Nettie Stevens was one of two scientists who discovered what determines whether a living thing will be male or female. She was born in Cavendish, Vermont, to Ephraim and Julia Stevens on July 7, 1861. Her father was a carpenter. After earning a credential from Westfield Normal School in 1883, she spent the first part of her adult life as a teacher and librarian. Eventually, however, she decided on a career in research. She enrolled at Stanford University in California in 1896 and earned a bachelor's degree in 1899 and a master's degree in physiology a year later.

While doing doctoral research at Bryn Mawr, Stevens met the genetics pioneer THOMAS HUNT MORGAN and became interested in the study of genetics. Morgan helped her win a fellowship to study overseas. She earned her doctorate in 1903 and then joined the faculty of Bryn Mawr, rising to the rank of associate professor and becoming beloved as a teacher. She told one student, "How could you think your questions would bother me? They never will, so long as I keep my enthusiasm for biology; and that, I hope, will be as long as I live."

Geneticists were just beginning to associate the "factors" that controlled inheritance of traits, first described by the Austrian monk JOHANN GREGOR MENDEL in 1866, with threadlike bodies called chromosomes in the nucleus, or central body, of cells. In her most important research Stevens observed that although all unfertilized eggs of the common mealworm contained the same 10 chromosomes, that was not true of this insect's sperm. One chromosome in some sperm cells, which she called *X*, resembled one seen in the egg. In other sperm cells this chromosome was replaced by another, smaller one, which she termed *Y*. She speculated that if an egg were fertilized by a sperm carrying an X chromosome, the resulting offspring would be female. If the egg were fertilized by a sperm carrying a Y, the offspring would be male.

Stevens's discovery confirmed the link between chromosomes and inheritance as well as showing how gender is determined. She also proved that inheritance of gender followed the rules that Mendel had worked out. She published her results in 1905. So did Edmund B. Wilson, a better-known male scientist, who had made the same finding independently. The two are properly given equal credit for the discovery, although some books have ignored or downplayed Stevens's contribution.

Stevens later found similar chromosome differences in other insects, and other scientists confirmed her conclusions as well. She also made the important discovery that chromosomes exist in pairs. Edmund Wilson wrote that she was "not only the best of the women investigators, but one whose work will hold its own with that of any of the men of the same degree of advancement." Unfortunately Stevens died of breast cancer on May 4, 1912, at the age of 51, limiting her late-blooming career to only 12 years.

Stewart, Alice

(1906–)
English
Medical Researcher

Finding that x-raying pregnant women often led to leukemia in their children, Alice Stewart was the first scientist to show that low-level radiation could harm human health. Stewart was born into the large family of two physician parents in England, on October 4, 1906. (Being one of many children, she told interviewer Gail Vines, meant that she "learn[ed] not to mind about battles.") She earned a master's degree in 1930 and an M.D. in 1934 from Cambridge University.

Aided by the demand for women workers created by World War II, Stewart, then married and the mother of two children, joined the Nuffield Department of Clinical Medicine at Cambridge's rival university, Oxford. What she calls her "semi-ingenuity in thinking up things" helped her devise ways to protect the health of workers in several war-related industries. As a result she became the youngest woman ever to be made a fellow of the Royal College of Physicians.

Soon after the war ended Stewart found herself the head of Oxford's new department of social medicine (public health)—just as the university cut off most of the department's funding. "If I'd been a man, I would never have stood it," she told Vines, "but being a woman I didn't have all that number of choices." Looking for a high-profile project that might restore support, she decided to find out why unusually large numbers of young children were dying of leukemia. She asked medical officers all over England to interview mothers of children with cancer and of healthy children of the same age. The interviews showed that the mothers of sick children were twice as likely to have been x-rayed during pregnancy as the mothers of healthy ones. "That set the jackpot going and . . . has kept me in the business of low-level radiation ever since."

When Stewart issued her first report in 1956, she says, "I was unpopular . . . because . . . X-rays were a favourite toy of the medical profession." She extended her work into what became the Oxford Survey of Childhood Cancers, which eventually listed 22,400 childhood cancer deaths all over Britain between 1953 and 1979. When the larger sample confirmed her first findings, she told a *Ms.* interviewer, Amy Raphael, "my research practically brought prenatal X-rays to a halt."

Stewart retired from Oxford around 1976 and moved to the University of Birmingham, where she is still a senior research fellow. In the mid-1980s she and her collaborator George Kneale, supported by a $2 million grant from the Three Mile Island Public Health Fund, investigated the effects of radiation on American nuclear plant workers. They found, among other things, that the chance of development of cancer after low-level exposure increases greatly once workers pass the age of 50. Some authorities have questioned Stewart's findings, but in 1986 she won the Right Livelihood Award, sometimes called the "alternative Nobel Prize," and the *New York Times* once called her "the Energy Department's most influential and feared scientific critic." Stewart told Gail Vines that her secret weapon is that "I know that I am going to be right." To Amy Raphael she added, "My epitaph would have to be . . . 'She stuck with the job.'"

Stewart, Sarah

(1906–1976)
American
Microbiologist, Cancer Researcher

Sarah Stewart was one of the first to show that viruses could cause cancer in mammals. She was born to George and Maria Andrade Stewart on August 16, 1906, in Tecalitlan, Mexico, where her father, an engineer, owned several mines. An uprising when Sarah was five years old forced her family to flee to the United States, and they eventually settled in New Mexico.

Stewart graduated from New Mexico State University in 1927 with bachelor's degrees in home economics and general science. She obtained a master's degree in bacteriology from the University of Massachusetts at Amherst in 1930, then worked as an assistant bacteriologist at the Colorado Experimental Station for three years. She studied for a Ph.D. at the University of Colorado School of Medicine and later at the University of Chicago; she earned her degree from the latter in 1939.

In 1936 Stewart joined the Microbiology Laboratory at the National Institutes of Health (NIH), the group of large government-sponsored research institutes in Bethesda, Maryland. From then until 1944 she did research on bacteria that can survive without air and on ways to protect people against diseases they cause, such as botulism. Meanwhile she grew interested in viruses and the possibility that they might cause cancer.

The Microbiology Laboratory did not study cancer, so Stewart transferred to the National Cancer Institute at NIH in 1947. Researchers there, however, regarded her belief that viruses could cause cancer in mammals (they were known to do so in birds) as "eccentric" at best. She thought they were ignoring her because she was not a medical doctor, so she went back to school at Georgetown University and at age 39 earned an M.D. (the first Georgetown had granted to a woman). This led to the loss of her job at the Cancer Institute, although the institute's director found her a laboratory in the Public Health Service hospital in Baltimore.

In 1953 a New York researcher, Ludwik Gross, found a virus that appeared to produce cancer in mice. While attempting to verify Gross's discovery, Stewart found a second virus that seemed to cause a mouse leukemia. In 1956 she asked BERNICE EDDY, a friend and fellow NIH researcher, to help her grow this virus in mouse cells in laboratory dishes. Stewart and Eddy found that their virus caused cancer in every kind of animal into which they injected it. They named it the SE polyoma virus: *SE* for "Stewart and Eddy," and *polyoma* meaning "many tumors."

Most other researchers were skeptical when Stewart and Eddy first announced their findings in 1957, but their work persuaded others to start looking into cancer-causing viruses. "The whole place just exploded after Sarah found polyoma," Alan Rabson of the National Cancer Institute told the writer Edward Shorter. "It was a major, major discovery." These scientists confirmed the women's results, and Stewart's ideas became accepted. President Lyndon Johnson gave her the Federal Women's Award in 1965 for her work on cancer-causing viruses.

Stewart continued this work for the rest of her career. She and others showed that viruses could produce cancer in birds, mice, cats, and hamsters. She also isolated viruses from several human cancers, though she could not prove that the viruses caused the disease. She died—of cancer, ironically—on November 27, 1976. BERNICE EDDY said of her, "She was a forceful individual who did not let anything stand in [her] way if she could help it."

Stokes, William

(1804–1878)
Irish
Physician

Although it was John Cheyne's paper that identified periodic respiration, it was William Stokes who drew it to the attention of the greater medical community, in his 1854 text *The Diseases of the Heart and Aorta,* and thus periodic respiration became known as Cheyne-Stokes breathing. Stokes's significance extends much further than this one contribution, though, as his influence was felt in many diverse areas. He is attributed with modernizing the practice of clinical medicine in Britain by raising its standards. For example, he noted that the classical methods of treatment by bleeding and purging were far from scientific remedies and were in fact counterproductive. Along these lines he also introduced the notion of teaching medicine in the same setting where it is practiced—in a clinical environment. His abdominal studies resulted in the first identification of the signs and symptoms of abdominal aneurysm. In a related development he reported the beneficial effects of opium as a treatment for acute abdominal diseases in the text *Clinical Observations on the Use of Opium.*

Stokes was born on October 1, 1804, in Dublin, Ireland. His mother was Mary Anne Picknell, and his father was Whitley Stokes. Stokes studied chemistry and medicine at Glasgow before transferring to the University of Edinburgh, where he received his M.D. in 1825. In 1828 he married Mary Black, and together they had several children. Stokes earned a second M.D. from the University of Dublin in 1839.

Stokes commenced his medical career in 1825 at the Dublin General Dispensary and at the Meath Hospital; he retained the latter appointment for the next two decades. Also in 1825 he published the first in a succession of significant medical texts, *An Introduction to the Use of the Stethoscope.* Besides publishing books and papers, he served as the editor of the *Dublin Quarterly Journal of Medical Science* from 1834 on. In 1837 he published *A Treatise on the Diagnosis and Treatment of Diseases of the Chest.*

In 1843 the University of Dublin appointed Stokes as the Regius Professor of Medicine. Three years later in 1846 he published an important paper, "Observations on Some Cases of Permanently Slow Pulse." In it he described the phenomenon that came to be known as Stokes-Adams attacks. In 1861, the year the Royal Society of London elected him a member, he served as the physician to the queen of England. He capped off his career with the publication in 1874 of *Lectures on Fever.*

Stokes was honored for his medical innovations and discoveries with the Prussian Order of Merit. Numerous universities recognized Stokes by bestowing honorary degrees on him: The University of Edinburgh honored him in 1861, and both Oxford University and Cambridge University granted him honorary degrees in 1863. He helped found the Pathological Society of Dublin, and he was a member of the Royal Medical Association. He served as the president of the Royal Irish Academy. Stokes died on January 10, 1878, in Dublin.

Sumner, James Batcheller

(1887–1955)
American
Biochemist

James Batcheller Sumner became the first person to crystallize an enzyme. His efforts to crystallize urease, which catalyzes the metabolic reaction in which urea breaks down into ammonia and carbon dioxide, convinced him that enzymes were, in fact, proteins. Although it was then commonly believed that enzymes were not proteins but low-molecular-weight substances, Sumner persevered in his theories. He was vindicated in 1930, when JOHN HOWARD NORTHROP isolated the enzyme pepsin in the form of a crystalline protein. In 1946 Sumner was awarded the Nobel Prize in chemistry for his discoveries.

Born on November 19, 1887, in Canton, Massachusetts, Sumner was the son of Charles and Elizabeth Rand Sumner. His wealthy family farmed but was also involved in manufacturing. Sumner was an avid hunter and outdoorsman as a child. Unfortunately a hunting accident resulted in severe damage to his left arm. Although the arm was paralyzed, the left-handed Sumner taught himself to use his right arm in activities ranging from tennis to writing.

Sumner enrolled at Harvard University in 1906, intending to study electrical engineering. He soon shifted his focus to chemistry, however, and graduated in 1910. After a one-year stint at his uncle's textile plant, and as a chemistry teacher at both Mount Allison University and the Worcester Polytechnic Institute, Sumner returned to Harvard in 1911. He received his A.M degree in 1913, and was awarded his Ph.D. in biological chemistry in 1914 for his dissertation "The Formulation of Urea in the Animal Body." In 1914 Sumner accepted a position as an assistant professor of biochemistry at the Ithaca division of the Cornell University Medical College. He was promoted to full professor in 1929. Sumner's private life was turbulent, however. With Bertha Louise Ricketts, whom he married in 1915, he had five children but divorced in 1930. In 1931 he wed Agnes Pauline Lundquist, but this union also ended in divorce. He married Mary Morrison Beyer in 1943, and the couple had two children.

Sumner began his important experiments on urease in 1917, motivated in part by his graduate research on the topic. After nine years of painstaking work he achieved his goal—a crystalline globulin with urease activity. In 1926 he published the first of many papers on this subject. He strenuously asserted that urease was indeed a protein, mainly because he could not separate the protein from the crystals. Nevertheless, his colleagues cleaved to the teachings of the Nobel laureate Richard Willstätter, who held that enzymes were absorbed into carrier proteins (but were not themselves proteins). Between 1926 and 1930 Sumner published a series of papers shoring up his claims. The importance of his work was not recognized, however, until Northrop succeeded in crystallizing additional proteins.

After Sumner's department at Cornell University Medical College was terminated in 1938, he moved to Cornell's School of Agriculture, were he taught in both the zoology and biochemistry departments. In 1937 he crystallized catalase and proved it to be a protein; he later studied peroxidases, lipoxidase, and other enzymes. In 1947 the school formed a laboratory of enzyme chemistry and appointed Sumner its director. He continued to conduct research with enzymes there.

In 1946 Sumner's discovery regarding the fundamental nature of enzymes was acknowledged when he was awarded the 1946 Nobel Prize (shared with Northrop and WENDELL MEREDITH STANLEY). His findings laid the foundation for future work in enzyme chemistry. Sumner died on August 12, 1955, in Buffalo, New York.

Svedberg, Theodore

(1884–1971)
Swedish
Chemist

Theodore Svedberg, known affectionately as "The Svedberg" to his colleagues, conducted influential experiments in colloid chemistry, which earned him the 1926 Nobel Prize in chemistry. The field of colloid chemistry had languished before Svedberg's involvement in it. Colloids are minuscule substances that are dispersed, but do not dissolve, in a medium. Both the particles and the medium may be solid, liquid, or gaseous. Because these particles are so small, they are unaffected by gravity and thus remain suspended in the solution indefinitely. Moreover, colloids cannot be observed under a microscope. Svedberg's invention of the ultracentrifuge, a device that greatly increased the force of gravity and thus broke the suspension, made studying colloids an easier task. Along with facilitating breakthroughs in colloid chemistry, Svedberg's ultracentrifuge proved invaluable to other disciplines as well.

Born on August 30, 1884, in Flerång, Sweden, Svedberg was the only child of Elias and Augusta Alstermark Svedberg. His father, a civil engineer at an ironworks, often took his son on long trips through the Swedish countryside. Svedberg himself had 12 children from his four marriages to Andrea Andreen (in 1909), Jan Frodi Dahlquist (in 1916), Ingrid Blomquist Tauson (in 1938), and Margit Hallen Norback (in 1948).

After receiving his B.A. in chemistry from the University of Uppsala in January 1904, Svedberg began graduate studies. He was awarded his doctorate from Uppsala in 1907 for his dissertation on colloids. In his thesis Svedberg proposed a new method for creating colloidal solutions of metals. Previously an electric arc was passed between metal electrodes submerged in liquid; Svedberg's innovation was to use an alternating current and an induction coil immersed in liquid. His apparatus produced very pure colloidal mixtures and allowed results to be reproduced, thereby making quantitative analyses possible.

In 1907 Svedberg accepted a position as a lecturer in physical chemistry at the University of Uppsala; he was named chair of the institution's newly created physical chemistry department in 1912. He remained at Uppsala for the rest of his career. Svedberg initially turned his attention to the phenomenon of Brownian motion, the continuous and random movement of particles suspended in a liquid medium first noted by the British botanist Robert Brown. Brownian motion had previously captured the attention of such notable scientists as ALBERT EINSTEIN and Jean Perrin. Like all other colloids, the particles involved in Brownian motion were too small to be affected by forces of gravity. Since the particles could not be seen under the direct light of a microscope, Svedberg tried an ultramicroscope, which used refracted light, to examine specimens. Although he found that the colloidal solution obeyed the laws of Newtonian physics, he was unable to determine the sizes of the smallest particles because the constant collision of the particles with water molecules kept the particles suspended. Svedberg tried a number of techniques to force the particles to settle out. Eventually he hit on centrifugation, a technique that mimics the action of gravity. Although centrifuges were then being used to separate red blood cells from plasma, colloid particles were too tiny to be handled by available centrifuges.

After accepting a guest professorship at the University of Wisconsin in 1923, Svedberg collaborated on the creation of the first ultracentrifuge, a device that spun so rapidly that it exerted a gravitational force thousands of times greater than the Earth's. This ultracentrifuge enabled Svedberg to study proteins, which retained their colloidal properties in solutions. He then applied his centrifuge to the examination of carbohydrate molecules, specifically the sugars of the Lileaceae family (lilies and irises).

Svedberg's 1926 Nobel Prize bolstered Swedish science and led to the founding of the Institute of Physical Chemistry at Uppsala in 1930. With Svedberg at its helm the institute conducted innovative research in the field of colloid chemistry. Despite his reaching the mandatory retirement age, the Swedish government appointed Svedberg lifetime director of the Gustav Werner Institute for Nuclear Chemistry. His contributions to colloid chemistry proved useful to a number of scientific fields; his ultracentrifuge became an essential tool in protein chemistry and he reinvigorated the study of colloid chemistry. The centrifugation coefficient unit was named the *svedberg unit* in his honor. He died in Stockholm on February 25, 1971.

Sydenham, Thomas

(1624–1689)
English
Physician

Dubbed the "English HIPPOCRATES," Thomas Sydenham made notable advances in the theory and practice of medicine. He was an ardent advocate of using concrete clinical observation—rather than speculative theories—as the basis for treatment, a position he set forth eloquently in his masterful 1676 publication *Medical Observation*. He proposed that diseases, much like plant species, could be separated into groups based on certain characteristics. He was one of the first to describe scarlet fever, as well as what came to be known as Sydenham's chorea, and he invented liquid laudanum (derived from opium) as a painkiller. He was a pioneer in the use of quinine.

Sydenham was born on September 10, 1624, at Wynford Eagle, Dorset, England. His parents, William and Mary Sydenham, had several children, though there is no clear record extant as to exactly how many. William, a squire, was from a family of prominence. Sydenham's eldest brother, also named William, was a close confidant of Oliver Cromwell, the English revolutionary leader who established the Commonwealth after prevailing in the English Civil War. Sydenham wed Mary Gee in 1655 and had three sons—William, who also became a physician; Henry; and James.

Sydenham enrolled at Oxford in 1642, but his education was interrupted by the civil war that erupted in 1644. He left school for a time and fought on the side of Cromwell's Parliamentarians. After returning to Oxford, he obtained his bachelor of medicine degree in 1648. The war continued to rage, and Sydenham returned to battle in 1651, when he was severely wounded in the battle of Worcester. There is some evidence that his valor in battle made him popular with Cromwell's party, thereby making his election as fellow to All Souls College a mere formality. He obtained his license to practice medicine in 1663 and was admitted as a member of Pembroke College, Cambridge, in 1676. He received an M.D. that year as well.

Sydenham established himself in London and devoted the rest of his life to private practice. In this capacity he eschewed unfounded theoretical approaches to medicine and maintained that science (and indeed all human understanding) should be bounded by the limits of observation and reasoning. His close association with ROBERT BOYLE and John Locke undoubtedly influenced this epistemological outlook. In 1669 Sydenham carefully began to record the many diseases he encountered in his practice. He continued to make these clinical observations for the next five years. From this record he produced his most important work, *Observayiones medica* (Medical observations), in 1676. In it he postulated his view of sound medical practice: He believed it essential to link medical treatments to clinical observation, to delineate the different kinds of diseases, and to make many different trials of a treatment before declaring it efficacious.

Sydenham also made significant contributions to the field of pharmacology during his career. He devised a liquid form of laudanum and pioneered the prescription of iron compounds to treat anemia. He identified scarlet fever and Sydenham's chorea. Above all, he was renowned for alleviating the suffering of the sick. His legacy was a scientific approach to medicine and the treatment of infections as specific entities.

T

Tamm, Igor Evgenievich
(1895–1971)
Russian
Physicist

Although he worked on a diverse array of projects during his career, Igor Tamm is best remembered for his theoretical explanation, devised with ILYA MIKHAILOVICH FRANK, of the cause of the Cherenkov effect. The effect, named after its discoverer, PAVEL ALEKSEYEVICH CHERENKOV, is the blue radiation glow that occurs when gamma rays pass through water. Cherenkov had identified this phenomenon in 1934, but no satisfactory explanation had accounted for it until Tamm and Frank posited their solution in 1937. The pair deduced that the effect was akin to a sonic boom caused by an object moving faster than the speed of sound. In this case, however, the particles were moving faster than the speed of light. (Although a fundamental tenet of physics holds that nothing may exceed the speed of light in a vacuum, an object may break that barrier when traveling through another medium.) This theory laid the groundwork for the development of measuring devices that remain significant to the operation of particle accelerators.

Tamm was born to Evgeny Tamm and Olga Davydova in Vladivostok, Russia, on July 8, 1895. His father was a civil engineer. When Igor was five, the family moved to Elizavetgrad (later called Kirovograd) in the Ukraine. Igor married Natalie Shiuskaya in 1917, and the couple had two children.

Igor graduated from the Elizavetgrad Gymnasium in 1913. After spending a year at the University of Edinburgh, Tamm returned to Russia and matriculated at Moscow State University, where he studied physics and math. He received his bachelor's degree in physics in 1918.

Tamm accepted a post at the Crimean University in 1919 and remained there until 1921. He then took a position in Simferopol at the Odessa Polytechnic Institute. In 1922 Tamm moved to Moscow and began teaching at the J. M. Sverdlov Communist University. While at that institution Tamm also received appointments at the Second Moscow University in 1923 and Moscow State University in 1924. In 1925 he left the Sverdlov and became a professor of theoretical physics at Moscow State, where he was named physics department head in 1930 and received his doctorate in 1933. In 1934 Tamm was appointed director of the theoretical section of the P. N. Lebedev Physical Institute in Moscow, where he remained for the rest of his life.

Influenced by a colleague in Simferopol, Tamm began conducting research in the field of crystal optics during the 1920s. Tamm also identified particular properties possessed by electrons on the surface of crystalline solids. These Tamm surface levels, as they came to be known, proved important to the future development of solid-state devices, particularly those involving semiconductors (solids that conduct electricity better than insulators, but not as well as pure conductors).

It was also at Moscow State that Tamm engaged in the work that later gained him the Nobel Prize. By the early 1930s he had developed an interest in atomic nuclei. After Frank joined the faculty in 1935, he and Tamm embarked on the work that eventually led to their explanation of the Cherenkov effect. The importance of this discovery was quickly recognized and resulted in the invention of the Cherenkov detector, a measuring device that made photoelectric analysis of Cherenkov radiation possible. Constructed of glass or some other transparent substance through which the high-velocity particles could pass, Cherenkov detectors are now widely used to study particles produced in cyclotrons and other accelerators.

Tamm shifted his focus back to elementary particles and nuclear physics after his collaboration with Frank. In 1945 he developed an innovative means of calculating the interaction of molecules, which came to be known as the Tamm-Dankov method. (This method remains important for the theoretical examination of certain subatomic particles including mesons and nucleons.) In the 1950s he turned his attention to plasma physics. He also became very active in the movement for the peaceful application of nuclear technology and was prominent in the Pugwash movement of the 1950s and 1960s. In addition to sharing the 1958 Nobel Prize in physics with Frank and Cherenkov, Tamm twice received the Order of Lenin, as well as the Order of the Red Badge of Labor and the Order of State Prize, First Degree. He was also elected to the Soviet Academy of Sciences.

Tarski, Alfred

(1901–1983)
Polish/American
Mathematician, Logician

Alfred Tarski is best known for his accomplishments in the disciplines of mathematics and logic. His research in mathematics focused on algebra and set theory; his work in logistics concentrated on semantics. Tarski produced a mathematical definition of truth in language and a corresponding proof demonstrating that such a definition results in contradictions. Additionally he developed an algorithm that determined whether basic statements in Euclidean geometry were true or false. This contribution later proved significant in the development of computer science, as it helped make machine calculations possible.

Born Alfred Tajtelbaum on January 14, 1901, Tarski was the first son of Ignacy Tajtelbaum and Rose Iuussak Tajtelbaum. The family, which also included a younger brother, lived in Warsaw, Poland, where his father was a shopkeeper. Tarski was an outstanding student in secondary school, where his favorite subject was biology. Upon matriculation at the University of Warsaw he intended to major in biology. However, at the urging of his professor, he changed his emphasis to mathematics after solving a complex problem in set theory. In 1924 Tarski received his Ph.D. from the University of Warsaw. Around this time, possibly concerned that anti-Semitism would affect his career, Alfred changed his last name to Tarski, believing that name sounded more Polish. He also met his future wife, Maria Witkowski, whom he would marry in 1929. The couple had two children, Jan and Ina.

After obtaining his degree, Tarski began teaching mathematics and logic at the University of Warsaw. To supplement his university income, which was inadequate to support his family, he also became a full-time instructor at Zeromski's Lycee. He remained in both of these positions until 1939. Tarski later stated that because of his teaching load during those years, he was unable to devote as much time as he would have liked to his research. Nevertheless, the papers he did publish between 1925 and 1939 were sufficient to establish his reputation as one of the most respected logicians of the 20th century.

It was during these years, when he was working in conjunction with Stefan Banach, that the Banach-Tarski paradox was formulated. The paradox demonstrates the limitations of mathematical theories that break down space into pieces. In the early 1930s Tarski began investigating semantics, the study of the meaning of symbols. This research produced his definition of truth in language and its accompanying proof of contradiction.

These pioneering developments in studying models of linguistic communications became a new field of study known as model theory. Model theory is concerned with the mathematical properties of grammatical sentences in comparison to various linguistic communication models. During his career Tarski conducted additional research concerning undecidable theories, Euclidean geometry, Boolean algebra, and metamathematics.

In 1939 Tarski went to the United States for a lecture tour. Upon the German invasion of Poland, which precipitated the outbreak of World War II, Tarski found himself stranded in the United States, while his wife and children were unable to leave Poland. He was not to be reunited with his family until the war ended. Between 1939 and 1942 Tarski held positions at several institutions, including Harvard University, Princeton University, and the City College of New York. In 1942 Tarski assumed a lecturer's position at the University of California at Berkeley. He spent the rest of his career there and became a full professor in 1946 and professor emeritus in 1968. In addition to teaching some of the most important 20th-century mathematicians and logicians, Tarski established Berkeley's prestigious Group in Logic and the Methodology of Science. He also acted as a visiting professor at numerous institutions in the United States and abroad before retiring in 1973. After retirement Tarski continued his research and acted as a doctoral adviser to students until his death in 1983 of a smoking-related lung condition.

Tarski was a member of the American Mathematical Society, the Polish Logic Society, and the International Union for the History and Philosophy of Science. He held numerous honorary degrees and was awarded a fellowship in the British Academy and election to the National Academy of Sciences. In 1981 Tarski received Berkeley's highest faculty honor, the Berkeley Citation.

Tartaglia, Niccoló

(1500–1557)
Italian
Mathematician

Remembered primarily for his work on the solution of third-degree equations, Tartaglia also made significant contributions to fundamental principles of arithmetic, numerical calculations, rationalization of denominators, extraction of roots, and other mathematical problems. In addition to being the first to publish the pattern now known as Pascal's triangle, which appeared in his *General trattato* in 1556, he was first to publish translations of EUCLID into Italian and of ARCHIMEDES into Latin.

A speech impediment resulting from a sword wound Tartaglia suffered in 1512 during the French attack on Brescia earned him the nickname *Tartaglia,* which means "stammerer," but records indicating that he had a brother with the surname *Fontana* have led some historians to attribute that name to Niccoló as well. His father, a postal courier named Michele, died about 1506, leaving his wife and children in poverty.

Tartaglia did not begin to learn to write until he was 14 and because of lack of money was unable to continue his studies long enough to master the alphabet. Much of what he learned, he accomplished on his own. Little is known about his mathematical studies, but sometime between 1516 and 1518 he moved to Verona and was employed as a teacher of the abacus. Certain records dating from 1529–1533 indicate that he had a family, that he lived in poverty, and that he was in charge of a school. He was a professor of mathematics in Venice, where he moved in 1534, and gave public lessons in the Church of San Zanipolo. He spent the remainder of his life in Venice except during a brief period when he returned to Brescia between 1548 and 1549.

Tartaglia's discovery in 1535 of a method for solving third-degree equations was actually a rediscovery of an unpublished rule formulated by Scipione Ferro in the early part of the 16th century. Tartaglia left the method unpublished but explained it to his friend Girolamo Cardano in 1539. Although Cardano had repeatedly asked Tartaglia to reveal the rule and had sworn to keep it secret, he published it in his *Ars magna* in 1545, crediting both Ferro and Tartaglia. This angered Tartaglia, who proceeded to publish his own research and to deride Cardano for his breach of promise. Their conflict grew in proportion as other mathematicians rose to Cardano's defense, and Tartaglia eventually lost the argument. In geometry Tartaglia was a pioneer in determining how to calculate the volume of a tetrahedron from the lengths of its sides.

In addition to Tartaglia's translations of the great classical scientists, which played a major role in disseminating scientific knowledge throughout Europe, his three-volume *Trattato di numeri et misure,* an encyclopedia on elementary mathematics, made a significant contribution to the field. He also made notable inventions in the fields of military science and topography. His *Quesiti et inventione diverse,* published in 1546, dealt with such topics as the firing of artillery, cannonballs, gunpowder, topographical surveying, and statistics. He played an instrumental role in the study of projectiles, in the development of firing tables in ballistics, and in the application of the compass to surveying. A man of wide-ranging interests, he also conducted experiments and wrote about diving suits, weather forecasting, and raising of sunken ships.

Taussig, Helen

(1898–1986)
American
Physician

Helen Taussig designed an operation that saved thousands of "blue babies," children born with a heart defect that prevents most of their blood from reaching the lungs, slowly depriving the child of an adequate supply of oxygen. She also helped to prevent the epidemic of birth defects in Europe caused by a drug called thalidomide from occurring in the United States. She was born in Cambridge, Massachusetts, on May 24, 1898, to Frank Taussig, a Harvard economics professor, and his wife, Edith, also a teacher. Helen was the youngest of their four children. Her mother died when she was 11 years old.

Helen began college at Radcliffe, then transferred to the University of California at Berkeley after two years to gain a "broader experience." When she graduated in 1921, she had decided to become a doctor and wanted to take her medical training at Harvard, but its medical school did not admit women. She received permission to take a few courses there, and she took others at Boston University. Alexander Begg, her anatomy professor at Boston, was one of the few who encouraged her. One day he thrust a beef heart into her hand, saying, "Here, it wouldn't hurt you to become interested in a major organ of the body." Thus began Taussig's study of the heart.

Taussig followed Begg's advice to go to Johns Hopkins Medical School, from which she earned her M.D. in 1927. She specialized in pediatric cardiology, or heart diseases of children. In 1930 Edwards Park, the school's chairman of pediatrics, put her in charge of his new pediatric cardiology clinic. She was devoted to her patients, whom she called her "little crossword puzzles" because of their often mysterious ailments. Most doctors identified heart defects by sounds heard through a stethoscope, but Taussig, left somewhat deaf by a childhood illness, used her eyes instead. She examined children with a fluoroscope, which produced real-time X-ray images, and observed the movements of their chests as they breathed.

Taussig became especially interested in heart problems caused by birth defects. One set of four defects that occurred together was called tetralogy of Fallot, after the French physician who had first described it. The two most important defects were a narrowing of the pulmonary artery, which takes blood from the heart's right ventricle to the lungs to receive oxygen, and a hole in the wall of muscle that separates the right half of the heart from the left. Taussig was the first to realize that these defects prevented most blood from reaching the lungs, thus slowly starving the children of oxygen. The oxygen shortage made their skin look bluish, earning them the nickname "blue babies." They rarely lived past childhood.

Another common defect, the ductus arteriosus, was a short vessel that connected the aorta, the main vessel that carries blood into the body, to the pulmonary artery. The ductus normally exists in a fetus but seals off at birth. If this does not happen, pressure from the aorta pushes too much blood into the lungs, damaging their delicate tissue. Some hapless children had both a tetralogy of Fallot and an open ductus arteriosus. Neither defect could be repaired directly because heart surgery was still very primitive.

In 1939 the Boston surgeon Robert Gross devised an operation for closing an open ductus arteriosus. This usually restored the health of children in whom the ductus was the only defect, but Taussig noticed that if children also had tetralogy of Fallot, the operation made their condition worse. It occurred to her that the ductus arteriosus actually helped children with tetralogy of Fallot by allowing more blood to reach their lungs. Their lungs were not harmed because so little blood traveled to them from the pulmonary artery. Why not, then, make an artificial ductus in these children?

Gross was not interested in Taussig's idea, but Alfred Blalock, chief of surgery at Johns Hopkins, was more willing to take a chance on it when she described it to him in 1943. Experimenting on dogs, he perfected an operation in which he joined the subclavian artery, which carries blood to the arms in humans, to the pulmonary artery. Blalock first tried the surgery on a human baby, Eileen Saxon, on November 29, 1944. "It was like a miracle," the child's mother told an interviewer. Soon afterward, Taussig herself witnessed the "miracle" during an operation on what she described as a "small, utterly miserable, six-year-old boy. . . . When the clamps were released [after Blalock had joined the blood vessels] the anesthesiologist said suddenly, 'He's a lovely color now!' I walked around to the head of the table and saw his normal, pink lips. From that moment the child was healthy, happy, and active."

Blalock and other surgeons went on to perform this miracle on some 12,000 other children. Furthermore, one noted surgeon says, the Blalock-Taussig "blue baby" operation showed that extremely sick children could survive surgery and thus "prompted surgeons to venture where they had not dared to venture previously. The result is much of present-day cardiac surgery." The operation continued to be used until surgery advanced enough for the heart defects themselves to be repaired.

Taussig meanwhile went on treating her young "crossword puzzles" and training doctors in pediatric cardiology, the specialty she had helped to develop. She became a professor of pediatrics, Johns Hopkins's first woman full professor, in 1959. In January 1962 a young West German doctor who was studying under Taussig told her that in some parts of Europe there had been a sudden increase in the number of children born with the severe birth defect *phocomelia*, or "seal limbs." These children's hands and feet were attached directly to their trunks, giving a flipperlike appearance. They often had internal defects as well. A German physician had published his belief that phocomelia might result when pregnant women took a popular drug called thalidomide (Contergan), often prescribed to control their morning nausea.

Taussig immediately flew to Europe to study the problem. When she returned, she presented her findings to FRANCES OLDHAM KELSEY of the U.S. Food and Drug Administration (FDA), who had already blocked a drug company's request to sell thalidomide in the United States. Taussig's report confirmed Kelsey's doubts. By this time the American company had withdrawn its FDA application, but thousands of samples of thalidomide had been distributed to physicians for "test" purposes. Taussig's urgent warnings prevented most of these doctors from giving the drug to their patients and thus helped to prevent most cases of phocomelia in America. Kelsey and Taussig also successfully campaigned for new FDA rules requiring testing of drugs on pregnant animals.

Helen Taussig retired from Johns Hopkins in 1963, but she continued to work at the cardiology clinic and do research. In 1965 she became the first woman president of the American Heart Association. President Lyndon Johnson gave her the Medal of Freedom, the highest award the United States can give a civilian. A 1977 article called Taussig "probably the best-known woman physician in the world." She moved to Pennsylvania in the late 1970s and died as a result of a car accident on May 21, 1986.

Taylor, Stuart Robert

(1937–)
American
Biologist, Cell Physiologist

Stuart Taylor's foremost achievement was his invention of a high-speed supercomputer imaging system, computer-assisted measurements of excitation-response activities (CAMERA). Taylor applied this system to the study of the minute activities of muscle fibrils (slender fibers). His research overturned prior theories and led to a better understanding of the mechanism of a muscle cell's contractions. He also investigated the role of calcium in instigating muscle contraction.

Taylor was born in Brooklyn, New York, on July 15, 1937, to Rupert Robert, a physician, and Enid Hansen

Taylor. Taylor's aptitude for science became evident in his youth, when he was accepted at the prestigious science-focused Stuyvesant High School in Manhattan. Taylor married in 1963 and had three children. He was later divorced.

After receiving his B.A. in zoology from Cornell University in 1958, Taylor earned his M.A. in zoology in 1961 from Columbia University, where he served as a lecturer. In 1966 he was awarded his Ph.D. in biology from New York University for his dissertation "Electro-Mechanical Coupling in Skeletal Muscle." Taylor's interest in muscle contraction was piqued by a 1964 lecture given by Sir ANDREW FIELDING HUXLEY, a British physiologist and Nobel laureate. Fascinated by Huxley's broad discussion of topics ranging from muscle contraction to electrical activity during contraction, Taylor accepted a position as a postdoctoral fellow at Huxley's laboratory at University College, London, in 1967.

Taylor returned to the United States in 1970 to take a position as an instructor (later an assistant professor) of pharmacology at the Downstate Medical Center in Brooklyn. In 1971 he was appointed a staff member in pharmacology, physiology, and biophysics at the Mayo Foundation, where he jointly held a professorship at the University of Minnesota. He became a professor at Mayo Medical School and Graduate School of Medicine in 1980 and remained there for over a decade.

During his tenure at the Mayo Foundation, Taylor conceived and developed his cell-imaging computer, CAMERA. Whereas prior attempts to film muscle contractions had been limited to 30 frames per second, CAMERA was able to produce 5,000 images per second. This innovation was significant because it afforded Taylor the opportunity to observe individual muscle cells at work (as CAMERA's enhanced imaging capacity allowed Taylor to record the act of a muscle contracting). Scientists had previously believed that sarcomeres—the individual contractile units of each muscle cell fibril—contracted and relaxed in unison to prevent the muscle cell from stretching and tearing apart. However, Taylor found that the links between regions were weak, indicating that this theory was erroneous. He concluded that a muscle cell's twitch was not the result of the movement of individual sarcomeres, but rather the average response of many sarcomeres. Taylor also utilized CAMERA to explore the relationship of calcium to muscle contractions. He found that calcium prompts contractions and that a feedback mechanism between calcium and contraction exists.

Since the early 1990s Taylor has been a Distinguished Professor at Hunter College of the City University of New York. In addition to serving on the editorial boards of several prominent scientific journals, Taylor has been active in recruiting minority students to the sciences. A fellow of the American Association for the Advancement of Science, he is recognized as an authority on muscle cell contraction.

Teller, Edward

(1908–)
Hungarian/American
Physicist

Often referred to as "the father of the hydrogen bomb," Edward Teller conducted ground-breaking work in the field of nuclear physics, focusing particularly on fusion reactions. Unlike fission reactions (such as those that powered the first generation of atomic bombs), in which energy is derived from the splitting of atoms, fusion is the process of melding atoms together. The military applications of his work were tremendous, and Teller reveled in them. In later years he influenced Ronald's Reagan's decision to advance the Strategic Defense Initiative (SDI, or "Star Wars" to its critics), a missile defense system intended to shoot down incoming projectiles before they neared their targets.

Born in Budapest on January 15, 1908, Teller was the son of a prosperous lawyer, Max Teller, and Ilona Deutch Teller. At age 18 Teller won a prestigious math contest for Hungarian high school students. He hoped to study mathematics at the university but was dissuaded by his father, who believed it would not provide a secure career. Instead, he earned his undergraduate degree in chemical engineering from the University of Budapest and his doctorate in physics from the University of Leipzig in 1930.

In 1934 Teller received a Rockefeller Foundation fellowship and traveled to Copenhagen's Institute for Theoretical Physics to work with NIELS HENDRIK DAVID BOHR. At this time he married Augusta Maria Harkanyi. The pair lived for a year in London, where Teller was a lecturer at the University of London, before moving to Washington, D.C., where he assumed a full professorship in physics at George Washington University. While at George Washington, Teller worked with GEORGE GAMOW to formulate the Gamow-Teller selection rules for beta decay of radioactivity. At this early stage of his career Teller, a prolific writer, had already written about molecular vibrations, stellar energy, and magnetic cooling processes, but his passion for fusion led to his involvement in the Manhattan Project, beginning in 1940.

Working at Los Alamos National Laboratory in New Mexico, Teller contributed to the development of the atomic bomb, but he primarily explored the possibility of developing thermonuclear weapons. Success in this pursuit proved elusive: Initial efforts were plagued by erroneous calculations and design flaws. Moreover, scientific and governmental support for the project waned after the detonation of an atomic bomb over Hiroshima. After World War II ended, Teller left Los Alamos to teach theoretical physics at the University of Chicago, where he remained until 1949. That year the Soviet Union successfully exploded an atomic device, and U.S. government interest in the fusion bomb was rekindled. Teller returned to Los Alamos and resumed work on the project alongside Stanislaw Ulam. The two developed the Teller-Ulam configuration, which relied on the radiation from an initial atomic fission explosion to ignite the thermonuclear core and precipitate the fusion reaction. A test device was successfully detonated on November 1, 1952.

When prominent Los Alamos scientists, including J. ROBERT OPPENHEIMER, voiced opposition to further thermonuclear weapons research, Teller used his considerable political influence to lobby for the creation of a second nuclear research facility. Lawrence Livermore Laboratory was established in 1954, with Teller as its associate director from 1958 to 1960. During Teller's tenure Livermore was especially noted as a center for fusion research. Teller had begun teaching at the University of California at Berkeley in 1953 and remained a professor there until retiring in 1975.

Teller's political influence reached its peak with Ronald Reagan's election in 1980. Teller obtained government funding to research an active nuclear defense system, SDI. The project was supposed to work by means of nuclear-powered lasers that would destroy enemy missiles in space. However, it failed to yield results and proved extremely costly, attracting sharp criticism. SDI funding was cut off in 1993, though Congress has periodically considered restoring it.

Throughout the 1990s Teller remained active, conducting personal research on superconductivity and astrophysics. Teller is a member of the National Academy of Science, the American Nuclear Society, and the American Academy of Arts and Sciences, among other organizations. He has also received numerous awards, including the Fermi Award (1962), the National Science Medal (1983), the Presidential Citizen Medal (1989), and the Order of Banner with Rubies of the Republic of Hungary.

Theophrastus

(c. 372 B.C.–c. 287 B.C.)
Greek
Botanist

Theophrastus is considered the father of scientific botany because of his extensive writing on the topic and in-depth consideration of plant physiological and etiological characteristics. One of his main strengths was his ability to draw images with his words. Theophrastus is also significant as a conduit of ARISTOTLE's teachings, as his own writings reflected a clear connection to Aristotle's philosophies.

Theophrastus was born in approximately 372 B.C., in Eresus, on the island of Lesbos. He reportedly studied under Plato at the Academy in Athens before traveling to Asia Minor in 347 B.C., to Lesbos from 344 B.C. to 342 B.C., and on to Macedonia from 342 B.C. to 335 B.C. He presumably met Aristotle in one of these locations, as he returned to Athens in 335 B.C. to teach at Aristotle's school, the Lyceum. He acted as Aristotle's chief assistant until Aristotle's retirement in 323 B.C., at which point Theophrastus inherited the helm of the Lyceum and led the school into its most successful period. Two thousand pupils reportedly passed through the Lyceum during the 35 years of his tenure.

Theophrastus was a prolific writer, covering topics as diverse as philosophy, history, law, literature, music, poetics, and politics. Most of this work, however, did not survive. Of his extant writings his work on botany far outweighs his treatment of other topics. Thus his reputation as a botanist is in part due to the availability of that material. The two most extensive works on botany by Theophrastus that survived are the nine volumes of his *Inquiry into Plants* and the six volumes of his *Etiology of Plants*. The former work systematically described and classified plants, drawing on personal observations as well as the observations of his students. He believed in the power of perception over the power of abstract reason when it came to botany. The latter work concerned itself mostly with plant physiology.

In these works Theophrastus cataloged more than 500 species of plants. He appreciated the interrelationships among fruit, flower, and seed and understood the distinction between monocotyledons and dicotyledons. He also distinguished between angiosperms, or flower-bearing plants, and gymnosperms, or cone-bearing plants. Theophrastus described with a surprising degree of accuracy the process of plant germination. His students aided him in some of his studies; hailing from distant regions of the Hellenistic world, they collected and took to Athens plant samples or sketches for him to examine. From the diversity of plant life he observed Theophrastus concluded that plant distribution depended to a large degree on soil and climate variation.

Though Theophrastus is best known as a botanist because of the unavailability of his other works, he also contributed to the history and criticism of the discipline of science. His thinking followed the philosophies of Aristotle closely, and in fact his own theories served to reduce the influence of Platonic thinking on Aristotelian concepts. Theophrastus died in approximately 287 B.C. in Athens, Greece.

Ting, Samuel Chao Chung

(1936–)
American
Physicist

Samuel Ting is best known for his 1974 discovery of a new subatomic particle, J/psi. This important finding was

Samuel Ting, whose discovery of the subatomic particle J/psi provided the first confirmation of the existence of "charm" quarks. *Brookhaven National Laboratory.*

awarded the 1976 Nobel Prize, which Ting shared with Burton Richter, who had independently made the same discovery nearly simultaneously (though utilizing a different experimental process). J/psi, a heavy particle with a long lifespan, is thought to be composed of a "charm" quark. Quarks are hypothetical particles that are believed to constitute the elementary building blocks of the universe. At the time of Ting's research physicists had assumed the existence of three quark types—"up," "down," and "strange." Although the concept of "charm" quarks had been proposed in 1970, J/psi provided the first confirmation that they did indeed exist.

Born in Ann Arbor, Michigan, on January 27, 1936, Ting was the son of Kuan Hai Ting and Tsun-Ying Wang. After his father completed his engineering studies at the University of Michigan, the family returned to China when Samuel was only two months old. Since both his parents were professors (his father of engineering, his mother of psychology), Ting was raised by his maternal grandmother. The onset of World War II prevented Ting from beginning school until he was 12 years old. When the war ended, the family moved to Taiwan.

Ting, like his father, enrolled at the University of Michigan in 1956. Upon receiving his bachelor's degree in mathematics and physics in 1959, he began graduate work at Michigan. He married an architect, Kay Louise Kune, in 1960, and the couple had two daughters. Ting was awarded his Ph.D. in physics in 1962, then spent a year as a Ford Foundation Fellow at the European Center for Nuclear Research in Geneva (CERN). His research at CERN utilized a proton synchrotron, which accelerates protons to high speeds, allowing for analysis and measurement.

After a brief stint in 1965 as a professor at Columbia University, Ting worked at the German synchrotron project in Hamburg, Germany, where he was part of a team that devised a double-arm spectrometer, an apparatus that analyzes and measures particle emissions. With this device Ting was able to calculate the masses of particles and their combined energy, thereby simplifying the identification of particles. Eventually Ting affirmed the accuracy of quantum electrodynamic theory (the theory concerned with the interaction of matter with electromagnetic radiation) as it applied to the production of electron and positron pairs by photon radiation.

In 1969 Ting returned to the United States to become a professor at the Massachusetts Institute of Technology (MIT). Intrigued by heavy photons—particles of radiation that he had investigated at Hamburg—Ting launched a project at the Brookhaven National Laboratory in Long Island, New York, in 1971 to determine their properties. (He was still a professor at MIT during this period.) Ting fired streams of protons at a beryllium target and in 1974 discovered an unexpectedly sharp spike of high-energy electron-positron pairs. He had found evidence of a new particle, one that was heavier than known elementary particles and also lasted a comparatively long time. After he reported the discovery to the Frascati Laboratory in Italy, his results were almost immediately confirmed. Ting published the stunning breakthrough in *Physical Review Letters;* interestingly, Burton Richter had noted the same phenomenon at almost the exact same time. This new particle was named J/psi.

Ting and Richter shared the 1976 Nobel Prize, and Ting won the E. O. Lawrence Award that year. Ting is a fellow of the American, European, and Italian Physical Science Societies. His discovery fueled the drive to discern additional subatomic particles, and he has been regarded as a bold thinker and careful experimental practitioner.

Tiselius, Arne Wilhelm Kaurin

(1902–1971)
Swedish
Biochemist

For his pioneering research using electrophoresis in the study of serum proteins, as well as for his work on adsorption chromatography, Arne Wilhelm Kaurin Tiselius was awarded the Nobel Prize in chemistry in

1948. Although electrophoresis, the migration of molecules through a solution under the influence of an electric field, had been recognized for some time, it became a practical and useful technique for the analysis of chemical substances only with Tiselius's investigations. Tiselius also made advances in chromatography and adsorption procedures of analyzing polypeptides.

Born on August 10, 1902, in Stockholm, Sweden, Tiselius developed an interest in science during his grammar school years. Tiselius's father, Hans Abraham J. Tiselius, had earned a college degree in mathematics from the University of Uppsala and worked for an insurance company. His mother, Rosa Kaurin, was the daughter of a Norwegian cleric. After Tiselius's father died in 1906, the family moved to Göteborg to be closer to relatives.

After completing studies at the gymnasium in Göteborg, Tiselius entered the University of Uppsala in 1921. There he studied chemistry, physics, and mathematics, working under the noted physical chemist THEODORE SVEDBERG. Tiselius received his master's degree in 1924 and continued to work as Svedberg's research assistant. He earned his doctorate in 1930 for his dissertation on electrophoresis and was appointed docent in the chemistry department at the University of Uppsala, where he remained for the duration of his professional career. Also in 1930 Tiselius married Ingrid Margareta Dalén, and the couple eventually had two children.

Tiselius expanded his studies to include biochemistry and the possibility of studying proteins by using electrophoresis after earning his doctorate. He became concerned with the possibilities of contaminated materials—even substances sufficiently centrifuged sometimes revealed impurities during electrophoresis experiments. Tiselius found this to occur most frequently with serum proteins. The problem led Tiselius to investigate chromatographic analysis, the process of separating mixtures by using light of a certain frequency. These studies eventually prompted Tiselius to develop new electrophoretic instruments. Tiselius used one version of his device to discover that blood plasma consisted of a mixture of various elements and separated blood serum proteins into four different groups—the globulins, which he labeled alpha, beta, and gamma, and albumins.

After his breakthrough discovery of the four groups of proteins Tiselius continued to refine his electrophoretic methods and used them in an attempt to study polypeptides. When polypeptides were broken down, however, the resulting individual peptides were too similar for the methods to provide sufficient information. Tiselius thus began to investigate methods of adsorption, the accumulation of molecules on the surface of a liquid or solid. In 1943 he made an important advance in adsorption methods of analysis that were used to measure the concentration of the product of elution, the extraction of one material from another. One problem with the method was that despite the success of elution in separating a mixture, it also caused part of the solution to be corrupted by molecules from the other part. This effect was known as *tailing*. Tiselius proposed an adaptation that prevented tailing. This later came to be called displacement analysis.

In the 1940s Tiselius continued his research in electrophoresis. He also was instrumental in the development of science in Sweden, helping to establish the Science Advisory Council to the Swedish government. Tiselius served as vice president of the Nobel Foundation a year before he was awarded the prize. In addition to the Nobel Prize in chemistry, Tiselius received a number of honors, including the 1926 Bergstedt Prize of the Royal Swedish Scientific Society, the 1956 Franklin Medal of the Franklin Institute, and the 1961 Paul Karrer Medal in Chemistry of the University of Zurich.

Todd, Alexander Robertus, Baron

(1907–1997)
British
Chemist

A versatile chemist, Baron Alexander Robertus Todd focused his research on the chemistry of biologically important natural products. Todd successfully synthesized vitamins, including B_1, E, and B_{12}, which had commercial and industrial significance, and he studied the active ingredients in cannabis. During World War II, Todd worked on the development of chemical weapons. Todd is perhaps best known for his investigations of nucleotides, nucleosides, and nucleotide coenzymes. It was for this pioneering work that he was awarded the 1957 Nobel Prize in chemistry.

Born on October 2, 1907, in Glasgow, Scotland, Todd was one of three children born to Alexander Todd and Jane Lowrie. Though Todd's father was a businessman, financial resources were scarce. Todd's interest in chemistry was sparked before he reached the age of 10, and he often experimented with a chemistry set given to him.

After completing studies in 1924 at the Allan Glen's School, a high school focused on the sciences, Todd entered the University of Glasgow to study chemistry. In 1928 he earned his bachelor's degree and was awarded a Carnegie research scholarship that allowed him to extend his studies under T. E. Patterson for an additional year. Todd then pursued graduate work at the University of Frankfurt am Main in Germany. There he began his investigations of natural products, first studying apocholic acid. After receiving his doctorate in 1931, he returned to England and entered Oxford University for further studies. Working under Robert Robinson, who would win a Nobel Prize in 1947, Todd earned another doctorate in 1934.

Todd left Oxford after completing his doctoral studies and went to the University of Edinburgh on a Medical Research Council grant. There he studied the chemistry of vitamins, beginning with thiamine, or vitamin B_1, which he successfully synthesized, though not before other scientific teams had. Todd also investigated the chemistry of vitamin E, a study he continued after moving to the Lister Institute of Preventive Medicine in London in 1936. At Lister, Todd undertook research on the cannabis plant (marijuana). Todd was able to extract the compound cannabinol from the cannabis plant resin. Todd married Alison Dale, whose father was the Nobel Prize winner Henry Hallett Dale, in 1937. The couple eventually had three children.

Todd moved again in 1938, this time to the University of Manchester. There he continued his vitamin E and cannabis studies but turned to the development of chemical agents with the beginning of World War II. Directing a team of chemists, Todd created a method for producing a sneeze gas and designed a plant for the production of blistering agents.

It was at Manchester that Todd began the work that would lead to the Nobel Prize. His vitamin studies prompted Todd to investigate nucleosides, compounds that make up the structural units of nucleic acids (deoxyribonucleic acid [DNA] and ribonucleic acid [RNA]). Nucleosides consisted of a sugar and a purine or pyrimidine base. A nucleotide was made up of a nucleoside combined with a phosphate group. Todd was able to develop a method for synthesizing nucleosides. He then attached a phosphate group to them and formed nucleotides. Todd learned more about the structure of nucleotides after moving to Cambridge University in 1944 and successfully synthesized them. This work led Todd to synthesize adenosine, adenosine triphosphate (ATP), and adenosine diphosphate (ADP) in 1949. These compounds worked in energy storage and production in the muscles of living organisms and in plants.

Further work involved the study of vitamin B_{12} as well as the chemistry of plant and insect pigments. Todd was knighted in 1954 and named Baron Todd of Trumpington in 1962. In 1977 he was made a member of the Royal Order of Merit. Todd served as president of the Royal Society from 1975 to 1980 and was the master of Cambridge University's Christ's College from 1963 to 1978.

Tombaugh, Clyde William

(1906–1997)
American
Astronomer

As an astronomical observer continuing the project that PERCIVAL LOWELL commenced in 1905 of searching for a trans-Neptunian planet, Clyde Tombaugh discovered Pluto, the ninth planet in our solar system, in 1930. Though this discovery earned Tombaugh lasting fame, he also distinguished himself by the sheer number of astronomical observations he made. He was also a master telescope builder, grinding some 36 telescopic mirrors and lenses in his lifetime.

Tombaugh was born on February 4, 1906, in the farming community of Streator, Illinois. He was the eldest of six children born to Muron Tombaugh, a poor farmer, and Adella Chritton. In 1934 Tombaugh married Patricia Irene Edson, and together the couple had two children, Alden and Annette.

At the time of his discovery of Pluto, Tombaugh had no formal education. He had educated himself, however, specifically in astronomy. He hand-constructed his first telescope, which was based on a 1925 article in *Popular Astronomy*, in a seven-foot rectangular box, grinding the mirror on a fence post. He later constructed a telescope with a nine-inch reflector made from parts of a 1910 Buick. In 1928 he sent drawings he made of Jupiter and Mars to Lowell, who immediately offered him a job at the Lowell Observatory in Flagstaff, Arizona, as a professional observer on the planet survey.

Tombaugh used a blink comparator, a device that alternates up to 10 times a second two photographs of the same area taken at different times to reveal moving objects, to track down the planet according to Lowell's calculations. On February 18, 1930, Tombaugh located the new planet, which coincided closely with Lowell's predictions, and on March 13, 1930, Tombaugh announced his discovery. Two years later he entered the University of Kansas as an undergraduate. In 1937 he discovered what he called the "Great Perseus-Andromeda Stratum of Extra-Galactic Nebula," a cluster of 1,800 galaxies. This lesser-known discovery challenged the firmly held belief of the regular distribution of galaxies. He returned to the University of Kansas to earn his master's degree between 1938 and 1939; he wrote his thesis on his restoration of the university's 27-inch reflecting telescope.

In 1943 Tombaugh augmented his work on the planet survey by teaching physics at the Arizona State Teacher's College in Flagstaff, where he also taught navigation for the U.S. Navy. In 1944 he taught astronomy and the history of astronomy at the University of California at Los Angeles. In 1946 the Lowell Observatory inexplicably dismissed him from the planet survey. During his tenure there Tombaugh observed 90 million star images, cataloging 29,548 galaxies, 3,969 asteroids (775 of which were previously unreported), two previously undiscovered comets, one nova, and of course Pluto.

In 1946 Tombaugh acted as an optical physicist and astronomer for the U.S. Army, improving its missile-tracking capabilities at the White Sands Proving Grounds near Las Cruces, New Mexico. He joined the faculty in the department of earth sciences at New Mexico State University in 1955 and was promoted to the status of associate professor in 1961 and full professor in 1965. He retired to professor emeritus status in 1973. In 1980 he published *Out of Darkness: The Planet Pluto,* coauthored by Patrick Moore. Tombaugh refused to relinquish his handmade telescopes to the Smithsonian Institution's historical collections until he was finished using them. That day came on January 17, 1997, when he died in Las Cruces, New Mexico.

Tomonaga, Shinichiro

(1906–1979)
Japanese
Physicist

Shinichiro Tomonaga, one of the founders of quantum electrodynamics. *AIP Emilio Segrè Visual Archives.*

Recognized as one of the founders of quantum electrodynamics (QED), the theoretical physicist Shinichiro Tomonaga developed a quantum theory that was consistent with the theory of special relativity. The theory of QED applied quantum mechanics and relativity theory to explain the actions of particles and their behavior in interacting with energy. When Tomonaga began his research, the theory of QED was not fully consistent with special relativity theory, and no quantum theory that could be applied to subatomic particles with high energies existed. Tomonaga's advances were recognized with the 1965 Nobel Prize in physics, which he shared with the American physicists RICHARD PHILIP FEYNMAN and JULIAN SEYMOUR SCHWINGER.

Born on March 31, 1906, in Tokyo, Japan, Tomonaga was raised in an academic environment. Tomonaga's mother was Hide Tomonaga; his father was Sanjuro Tomonaga, an educator. When Tomonaga was a young child the family moved to Kyoto, where Tomonaga's father had taken a job as professor of philosophy at Kyoto Imperial University. In 1940 Tomonaga married Ryoko Sekiguchi, whose father directed the Tokyo Metropolitan Observatory. They had three children.

After completing studies at Kyoto's esteemed Third High School, Tomonaga entered Kyoto Imperial University to study physics. In high school and college he was a classmate of HIDEKI YUKAWA, who would later become the first Japanese Nobel laureate in physics. After completion of his undergraduate work in 1929 Tomonaga stayed at the university to work as a research assistant. Three years later he took a research assistant position at the Institute of Physical and Chemical Research in Tokyo, where he stayed for five years. Tomonaga then journeyed to the University of Leipzig to further his studies. There he completed a paper on the atomic nucleus that resulted in his receiving a doctorate from Kyoto Imperial University in 1939. In 1941 he accepted a position as professor of physics at Bunrika University in Tokyo (now Tokyo University of Education).

A significant portion of Tomonaga's work in QED took place in the early 1940s. QED had come about in the 1920s as scientists realized the study of elementary particles required more than classical laws of physics. The British physicist PAUL ADRIEN MAURICE DIRAC advanced QED when he developed new formulations of the atomic theory in the late 1920s. Dirac's theory was not faultless, and his proposal that particles would have both infinite mass and infinite charge under particular circumstances was troublesome. Though many physicists rejected this prediction, the physical unfeasibility of which was referred to as *divergence difficulty,* Tomonaga believed there was a way to preserve Dirac's basic method while also settling the divergence difficulties. Using a mathematical approach known as renormalization, Tomonaga showed that infinite mass and infinite charge were actually elemental aspects of Dirac's theory and presented the idea that two particles interacted through the exchange of a third virtual particle.

Though Tomonaga presented his findings in a paper in 1943, World War II prevented his ideas from reaching an international audience until 1947. Schwinger and Feynman independently reached similar conclusions, and the three were awarded the 1965 Nobel Prize for solving the issues of divergence difficulties and for advancing the theory of QED. Tomonaga received a number of additional honors, including the 1948 Japan Academy Prize, the 1952 Order of Culture of Japan, and the 1964 Lomonosov Medal of the Soviet Academy of Sciences. Tomonaga left the Tokyo University of Education in 1949 for a job as visiting scholar at the Institute for Advanced Studies in Princeton, New Jersey. In 1951 Tomonaga returned to Japan to direct the Institute for Scientific Research. He helped establish the University of Tokyo's Institute for Nuclear Studies and became the university president in 1956. After retiring in 1962, Tomonaga served as president of the Science Council of Japan and headed the Institute for Optical Research.

Tonegawa, Susumu

(1939–)
Japanese
Molecular Biologist

For his pioneering research of the immune system Susumu Tonegawa was awarded the 1987 Nobel Prize in physiology or medicine, the first Japanese scientist to receive the Nobel Prize in that field. Tonegawa discovered how particular cells of the immune system were able to rearrange themselves genetically in order to produce a wide array of antibodies. Before Tonegawa's discoveries biologists had been unaware of how antibody-producing cells, which had a limited number of genes, could generate millions of different antibodies. Tonegawa's work greatly advanced the understanding of the immune system.

Born on September 5, 1939, in Nagoya, Japan, Susumu Tonegawa was one of four children. His mother was Miyoko Masuko; his father was Tsutomu Tonegawa, an engineer. Because Tonegawa's father's work made it necessary to move on a regular basis, Tonegawa's parents decided to send him and his brother to live with an uncle in Tokyo. During his high school years Tonegawa developed an interest in science, particularly chemistry.

After completing studies at the top-ranked Hibiya High School, Tonegawa entered the University of Kyoto in 1959. He received his undergraduate degree in chemistry in 1963, then traveled to the United States to pursue graduate work in molecular biology at the University of California at San Diego. There he studied under Masaki Hayashi in the biology department, researching the genetic transcription in bacteriophages, viruses that attack certain bacteria. After earning his doctorate in biology in 1968, Tonegawa began his postdoctoral work, first at the University of California at San Diego, then at the Salk Institute, located near San Diego.

In 1971 Tonegawa applied for a position as a molecular biologist at the Institute of Immunology in Basel, Switzerland. Though he had received little training in immunology, Tonegawa was offered the job. It was in Basel that Tonegawa embarked upon his breakthrough research on the immune system. At the time it was known that organisms were capable of producing millions of different antibodies and that each of these antibodies responded to different antigens, or foreign molecules. What was not known, however, was how this was accomplished, and this is what Tonegawa, along with his collaborator Nobumichi Hozumi, set out to research. Tonegawa's many experiments showed that part of the B-lymphocyte, the antibody-producing cell, could be rearranged and shuffled in numerous ways into different sequences, making possible the generation of about 1 billion different types of antibodies, each of which was designed to react to a specific antigen. This revolutionary work won Tonegawa the Nobel Prize in physiology or medicine in 1987.

After a decade in Basel, Tonegawa in 1981 returned to the United States to take a position as professor of biology at the Massachusetts Institute of Technology (MIT) Center for Cancer Research and Development. In 1992 Tonegawa and his research team discovered a particular gene that influences the ability to learn. Tonegawa has received a number of honors in addition to the Nobel Prize, including the 1981 Genetics Grand Prize of the Japanese Genetics Promotions Foundation, the 1983 V. D. Mattia Award of the Roche Institute of Molecular Biology, the 1986 Robert Koch Prize, and Brazil's Order of the Southern Cross in 1991. Tonegawa married Mayumi Yoshinari in 1985, and they had three children. In the late 1990s Tonegawa continued to work as a biology professor at MIT.

Trotula of Salerno

(fl. 11th century)
Italian
Physician

Trotula wrote a text on the diseases of women that was used for centuries. She probably belonged to the aristocratic di Ruggiero family and lived during the 11th century in Salerno, a town near Naples in southern Italy. Salerno's medical school was considered the best in Europe at the time and was the only one that admitted Muslims, Jews, and women. Indeed, its "women of Salerno," a group of women physicians, were famous.

Trotula is thought to have taught at the medical school as well as treating patients. She was married to another physician, Joannes Platearius, and had two sons,

Johannes and Matthias, who were also physicians and writers. The four wrote a medical encyclopedia called *Practica Brevis*.

Trotula's chief surviving book is the *Passionibus Mulierum Curandorum* (Diseases of women), sometimes called *Trotula Major.* It includes material on normal birth and the care of newborns. (A second book, *Trotula Minor,* deals with cosmetics and skin diseases.) In the preface to her major work Trotula explained how she had come to practice medicine. "Women on account of modesty . . . dare not reveal the difficulties of their sicknesses to a male doctor. Wherefore I, pitying their misfortunes . . ., began to study carefully the sicknesses which most frequently trouble the female sex." She probably wrote her book at least partly to tell male doctors facts about women's bodies that their patients might not want or be able to communicate.

Trotula's book quoted the best ancient medical authorities, such as GALEN and HIPPOCRATES OF COS, and presented some ideas that were ahead of its time. For instance, Trotula wrote that "conception is hindered as often by a defect of the man as of the woman," a radical notion in a day when infertility was almost always blamed on the woman. Trotula showed understanding of her patients' psychological as well as physical needs. Victorian historians were surprised at her outspokenness about sexual matters, but feminist scholars say that such an attitude would have been expected in a woman doctor.

Trotula's medical advice was more sensible and less based on magic than that of most of her contemporaries. She told doctors to diagnose illness by observing, for example, urine, pulse, and color of the face. She urged cleanliness, a healthy diet, and freedom from stress as treatments for many conditions. If more was needed, she recommended mild treatments, usually herbs and oils, rather than harsh medications, bleeding, or surgery. She was said to approach patients with gentleness and optimism. It was little wonder that midwives and physicians relied on her text for so long.

Tsiolkovsky, Konstantin Eduardovich

(1857–1935)
Russian
Astrophysicist

Konstantin Tsiolkovsky was a pioneer in the field of astronautics. Most of his work remained in the theoretical realm, as his ideas were too revolutionary to receive funding, even if the technologies existed. Nevertheless, he took into consideration every aspect of cosmic flight he could conceive, including the shapes of fuselages and wings, the cantilever of wings, and the use of internal combustion engines fueled by a mixture of liquid hydrogen and oxygen. He went so far as to design instruments for measuring the impact of gravitational acceleration on the human body and devised mathematical formulas to describe the movement of spacecraft. His basic rocket equation, or the fundamental relationship between the velocity and mass of a rocket, combined with its exhaust velocity, saw continued use. He even foresaw the need for what he called "cosmic rocket trains," or propulsion in stages.

Tsiolkovsky was born on September 17, 1857, in the village of Izhevskoye in the province of Ryazan. His father, Eduard Tsiolkovsky, was a forester, teacher, and minor government official; his mother was Maria Yumasheva. In 1867 at the age of 10 he fought a bout of scarlet fever that left him deaf. He turned to reading for solace and thus educated himself.

Between 1873 and 1876 he lived in Moscow, where he studied science and attended lectures with the aid of an ear trumpet, even though he could not afford to attend the university officially. In 1879 he passed the teacher's examination, and the next year he commenced teaching arithmetic and geometry at the Borovsk Uyzed School in Kaluga. In 1892 he transferred to a high school in Kaluga, where he taught until his retirement in 1920.

In 1881 Tsiolkovsky submitted his second paper, "The Theory of Gases," to the Russian Physico-Chemical Society, which realized that he had replicated the work of two decades earlier of JAMES CLARK MAXWELL, though Tsiolkovsky had no knowledge of Maxwell or the background scientific work upon which Maxwell founded his theory. Tsiolkovsky's work impressed the society sufficiently that it elected him as a member in 1883, upon his next submission, "On the Theoretical Mechanics of a Living Organism."

Tsiolkovsky commenced his theorizing on space travel during his Moscow sojourn in the mid-1870s. He compiled many of his theories and ideas concerning space travel in an important paper he started writing in 1896 and finished in 1903, though it wasn't until 1911 that "Investigations of Outer Space by Reaction Devices" was published in the journal *Herald of Aeronautics*. He further summarized these ideas in the 1920 book *Beyond Earth*.

In 1897 Tsiolkovsky constructed the first wind tunnel in Russia, which he explained in the article "Air Pressure on Surfaces Introduced into an Artificial Air Flow." The Russian Academy of Sciences awarded him 470 rubles (the equivalent of $235 at the time) to construct a larger wind tunnel, which he accomplished in May 1900.

Unfortunately this was the only governmental support he received until the October Revolution of 1917, after which the Communist government recognized his significance by electing him a member of the Socialist Academy. The Council of the Peoples' Commissariats of the Russian Federation granted him a pension, which eased his poverty. Tsiolkovsky continued to work into old age, producing three-quarters of his 500-plus papers after the revolution, as opposed to the 60 years between 1857 and 1917. Tsiolkovsky died on September 19, 1935, in Kaluga, Russia. The Soviet government attempted to launch *Sputnik I* in Tsiolkovsky's honor on the 100th anniversary of his birth and the 22nd anniversary of his death, though the actual launch date was postponed. A huge crater on the dark side of the moon was named for him.

Tsui, Lap-Chee

(1950–)
Chinese/Canadian
Geneticist, Biologist

A molecular geneticist, Lap-Chee Tsui made his major contribution to the scientific community with his discovery of the gene that causes cystic fibrosis, a hereditary and fatal disease, the most common genetic disease among whites. Tsui's work greatly furthered the understanding of cystic fibrosis, and Tsui continued to study the gene through the late 1990s with the hope of improving treatments for cystic fibrosis patients, as well as gaining additional insight into the disease and its development.

Born on December 21, 1950, in Shanghai, China, Tsui was the son of Jing-Lue Tsui and Hui-Ching Hsue. Tsui grew up in the small village of Dai Goon Yu, which was located on the Kowloon side of Hong Kong. As children Tsui and his friends often conducted simple experiments with tadpoles, fish, and silkworms, but Tsui's ambition did not involve science—he hoped to become an architect. Tsui married Lan Fong Ng, and the couple had two sons.

Awarded the Yale-in-China Scholarship in 1968, Tsui attended the Chinese University of Hong Kong and studied biology. After earning his bachelor's degree in 1972, he continued his studies and received his master's degree in biology in 1974. Tsui left Hong Kong for the United States in 1974 to pursue his doctorate in biological sciences at the University of Pittsburgh. After Tsui was awarded his Ph.D. in 1979, he held several postdoctoral positions. From 1979 to 1980 he was a postdoctoral investigator at the Oak Ridge National Laboratory in Tennessee. Though he had not been much involved in genetics, Tsui journeyed to Toronto, Ontario, Canada, in 1981 to assume the position of postdoctoral fellow in the department of genetics at the Hospital for Sick Children. Tsui remained with the hospital through the 1990s.

Tsui began his breakthrough research on cystic fibrosis while at the Hospital for Sick Children. Though Tsui published his first paper dealing with cystic fibrosis in 1985, it was not until May 1989 that Tsui and his research team uncovered the gene that causes cystic fibrosis. The genetic disease, which primarily affects the sweat glands, respiratory system, and the pancreas, causes an excess buildup of mucus, resulting in respiratory problems and other ailments. Those born with cystic fibrosis often died at an early age, and few lived past the age of 30. Through a series of experiments Tsui and his team discovered a type of genetic mutation in a deoxyribonucleic acid (DNA) sequence—the genetic information needed to identify that one amino acid in a protein chain was missing. To confirm that they had discovered the cystic fibrosis gene, Tsui isolated a region of the human chromosome 7 on which the cystic fibrosis gene was located and compared DNA sequences of cystic fibrosis patients with those of individuals who did not have the disease. By September Tsui and his staff were confident that they had isolated the gene.

In following years Tsui continued to investigate the gene to learn more about its actions. He learned that the function of the DNA sequence that accompanied the genetic mutation was to create a protein called cystic fibrosis transmembrane conductance regulator (CFTR), which was involved in generating mucus. It was discovered that cystic fibrosis patients lacked one amino acid in the CFTR protein, a condition that caused mucus to accumulate. Tsui's research provided greater understanding of how the disease developed and simplified the diagnosis process.

Tsui continued to study the molecular genetics of cystic fibrosis, and in the late 1990s he conducted molecular cloning experiments to isolate the gene and worked on defining mutations in the gene. Tsui received numerous awards and honors for his achievements, including the Pharmaceutical Manufacturers Association of Canada's Gold Medal of Honor, the Cystic Fibrosis Foundation's Paul di Sant'Agnese Distinguished Scientific Achievement Award, the Genetic Society of Canada's Award of Excellence, the Cystic Fibrosis Foundation's Doris Tulcin Cystic Fibrosis Research Achievement Award, and the Franklin Institute's Cresson Medal. Tsui was awarded the 1996 Canadian Medical Association Medal of Honor and was elected a fellow of the Royal Society of Canada (1990) and the Royal Society of London (1991).

Tull, Jethro

(1674–1741)
British
Agriculturist

Considered one of the founding fathers of modern agriculture, Jethro Tull was responsible for developing a variety of agricultural machinery. Tull's inventions helped spark the agricultural revolution, considerably changing farming methods by allowing farmers to work more efficiently and more productively in the field. Tull's methods were not immediately accepted by his fellow farmers but were later adopted into general use.

Baptized on March 30, 1674, Tull was born at Basildon, in Berkshire, England. Tull's father owned a farm in Oxfordshire, which Tull later inherited. Tull attended St. John's College, Oxford, then entered Gray's Inn in London to pursue a career in law. Though Tull was called to the bar in 1699, he chose to return to the country to run his father's farm. Tull's career change was due in part to poor health, from which he had suffered for much of his life.

Tull's first contribution to modern agriculture was the seed drill, which he developed in 1701. The drill mechanized and simplified the process of sowing seeds, which traditionally were broadcast by hand. The drill was capable of performing three functions, which had previously been handled separately: drilling, sowing, and covering the seeds with soil. The seed drill, which was designed to be pulled by a horse, was made up of two hoppers, funnel-shaped vessels into which seed was placed before being dispensed. The openings of the hoppers were controlled by a spring-loaded mechanism. As the seed drill was pulled forward, a wooden gear turned, thereby allowing the seed to drop out at regular intervals. A plow and harrow, a sharp instrument used to break up the ground, were also part of the machine, employed to cut the groove, or drill, in the soil for the seed and to turn the soil to cover the seed. Tull's mechanical seed drill sowed in rows, a revolutionary practice at the time, and was capable of sowing three rows of seed at one time.

After creating the seed drill, Tull moved on to develop machinery to facilitate plowing. Recognizing the existence of different types of soil and the need for equipment specialized for particular types of soil, Tull created first a two-wheeled plow designed for lighter soils such as those that could be found in southern England and in Midland areas. The plow was capable of cutting to a depth of about seven inches. Tull also developed a swing plow more appropriate for heavy clay soils. These soils were typically found in the eastern regions of England. Another breakthrough instrument created by Tull was a horse-drawn hoe. At the time farmers were often forced to resort to using a breast plow to cut through top layers of grass and weeds before the primary plowing could take place. Tull was able to develop a hoe using four coulters, blades that created vertical cuts in the soil, that were arranged in such a manner as to pull up the grass and weeds and permit them to dry on the surface.

In 1709 Tull moved to Berkshire, where he purchased his own farm. In 1733 after more than three decades working in agriculture Tull published a book on farming, *The New Horse Hoeing Husbandry: Or an Essay on the Principles of Tillage and Vegetation*. The French philosopher Voltaire practiced Tull's farming methods on his estate, but in general, many farmers were skeptical of Tull's principles and were slow to adopt them. Despite his initial struggle for acceptance of his agricultural techniques, Tull greatly influenced farming for centuries to follow, and his inventions helped to increase farm productivity. Tull died at Prosperous Farm in Berkshire in 1741.

Turing, Alan Mathison

(1912–1954)
British
Mathematician

Alan Turing set the theoretical foundation for modern computers while working on the practical application of his pioneering computer theory. In a very influential paper written while he was still pursuing his doctorate, he hypothesized an automatic machine that foreshadowed the modern computer. He applied these ideas in World War II to building machines that could crack German codes. After the war he devised a simple test, called the *imitation game* at first and later renamed the *Turing test*, to determine artificial intelligence.

Turing was born on June 23, 1912, in Paddington, England. His father, Julius Mathias Turing, spent most of his time in India as a member of the British civil service. His mother, Ethel Sara Stoney, accompanied her husband, leaving Turing and his older brother, John, in boarding school.

Turing did not distinguish himself at Sherbourne School in Dorset, except in science and mathematics. Trinity College rejected him twice, but King's College, Cambridge University, accepted him in 1931. In 1935 he submitted his master's thesis, "On the Gaussian Error Function," which won the Smith's Prize in 1936. That year he attended Princeton University to pursue his doctorate under Alonzo Church. He received the Proctor Fellowship in his second year, and in 1938 he earned his Ph.D. The next year he returned to King's College on a fellowship.

World War II interrupted his academic work, as the communications department of the Foreign Office recruited him to the Government Code and Cypher School in Bletchley, Buckinghamshire. There he developed the "Bombe," a machine for deciphering German "Enigma" codes. He probably also oversaw construction of the "Collosus," a decoding machine that used vacuum tubes, though information about this project remained classified by the British government. In June 1945 he declined a position at Cambridge to work for the Mathematics Division of the National Physical Laboratory (NPL) planning the automatic computing engine (ACE). Then in 1948 the Royal Society Computing Laboratory at Manchester College appointed him deputy director of the Manchester Automatic Digital Machine (MADAM) project.

Turing earned his enduring reputation mostly on the strength of several important papers. In 1936 he wrote "On Computable Numbers, with an Application to the Entscheidungs Problem" and presented it to the London Mathematical Society in 1937. It addressed the logical problem posed by David Hilbert whether all mathematical problems could be solved and concluded in agreement with Kurt Gödel that some problems could not be solved. However, Turing hypothesized in a footnote that an automatic machine, later known as the Turing machine, could identify these "undecidable" problems if provided the correct algorithms. Turing published his second most influential paper, "Computing Machinery and Intelligence," in 1950. In it he proposed the imitation game, later renamed the Turing test, whereby a person interrogated a computer and a human and ascertained whether the answers revealed human intelligence. Turing predicted that by the year 2000 computers would exist with sufficient artificial intelligence to fool the human interrogator within five minutes on 70 percent of the questions.

For his wartime work Turing was awarded the Order of the British Empire although he did not serve in the military and was an avowed homosexual and atheist. In 1952, however, he was convicted of "gross indecency" when his homosexuality, which was illegal in England at the time, came to the attention of the police. His reduced sentence included estrogen treatment. Although his professional career apparently did not suffer from this prosecution, he nevertheless committed suicide by eating a cyanide-laced apple on June 7, 1954.

Turner, Charles Henry

(1867–1923)
American
Entomologist

Charles Henry Turner contributed significantly to the understanding of insect behavior despite the fact that he chose to teach at the high school level and thus lacked institutional backing for his research. As an African American, Turner confronted racism in trying to publish his results. His findings were not taken as seriously as those of white men. The validity and importance of his scientific discoveries could not be ignored, though, as his observations spoke for themselves, revealing interesting aspects of insect life, such as the turning movement of ants, which was named *Turner's circling* in his honor.

Turner was born on February 3, 1867, in Cincinnati, Ohio. His father, Thomas Turner, was born a free man in Canada and worked as a church custodian. His mother, Eat Campbell, was a freed woman from Kentucky who worked as a nurse. Turner married Leontine Troy while he was in college; she died in the mid-1890s, leaving him with their five children. Turner eventually remarried.

Turner attended the University of Cincinnati, where he earned his B.S. in 1891 and his M.S. a year later. Uncertain of whether he wanted to teach at the high school or the college level, he tried both. He remained at the University of Cincinnati as an assistant professor in biology between 1892 and 1893. Upon the recommendation of Booker T. Washington he was offered a position as a professor of biology from 1893 through 1895 at Clark College, which was founded by the Freedmen's Aid Society of the African Methodist Episcopal (AME) church.

Turner then tried his hand at high school teaching. Between 1896 and 1908 he taught high school in Illinois, Ohio, and Tennessee. He also taught at the Haines Normal and Industrial Institute in Augusta, Georgia, and did doctoral work at the University of Chicago. In 1907 he received his Ph.D., graduating magna cum laude on the strength of his dissertation, "The Homing of Ants: An Experimental Study of Ant Behavior." In 1908 he finally found his calling, teaching biology and psychology at Sumner High School in St. Louis, Missouri, where he remained for the rest of his career.

Turner financed most of his experimentation personally on a limited salary. Though this constraint proved a hardship throughout his life, it also prompted him to devise simple experiments that displayed an appreciable degree of elegance. For example, he used a piece of paper and a watermelon rind to try to camouflage a burrowing bee's hole. Turner showed that insects can distinguish pitch; that honeybees recognize color, as they are led to flowers not only by odor but also by sight; that ants follow light and not odor back to their colony; and that wasps use their sight, not a hypothetical sixth sense, to relocate their nests. He researched the hearing of moths and the ability of cockroaches to be trained.

Turner died on February 14, 1923, in Chicago, Illinois. After his death the St. Louis Board of Education named the Charles H. Turner School for the physically disabled in his honor.

U

Urey, Harold Clayton

(1893–1981)
American
Physical Chemist

Harold Urey won the 1934 Nobel Prize in chemistry for his discovery of deuterium, an isotope of hydrogen with twice its atomic weight. He also found isotopes of carbon, nitrogen, oxygen, and sulfur.

Urey was born on April 29, 1893, in Wallerton, Indiana, to Samuel Clayton Urey, a schoolteacher and lay minister in the Church of the Brethren, who died when Urey was six years old, and Cora Reinoehl, who continued the child's religious upbringing. On June 12, 1926, in Lawrence, Kansas, Urey married Freida Daum, a bacteriologist. Together the couple had four children—Gertrude Elizabeth, Freida Rebecca, Mary Alice, and John Clayton.

Since Urey could not afford college, he taught in Indiana between 1911 and 1912 and in Montana between 1912 and 1914; there he earned enough money for in-state tuition at the Montana State University, where he matriculated in 1914. He graduated in 1917 with a B.S. in zoology and was immediately called upon to support the war effort by helping to manufacture high explosives at the Barrett Chemical Company in Philadelphia, a task that ran contrary to his pacifist upbringing. After the war he returned to teach briefly at Montana State, then moved to the University of California at Berkeley to pursue a doctorate in physical chemistry under Gilbert Norton Lewis. For his dissertation Urey performed spectroscopic research on the heat capacity and entropy, or degree of randomness, in gases. Urey earned his Ph.D. in 1923, after which he pursued postdoctoral studies with NIELS HENDRIK DAVID BOHR at the Institute for Theoretical Physics at the University of Copenhagen.

Urey served as an associate professor of chemistry at Johns Hopkins University in Baltimore between 1925 and 1929 and at Columbia University in Manhattan from 1929 to 1933, when he was named the Ernest Kempton Adams Fellow. The next year Columbia promoted him to full pro-

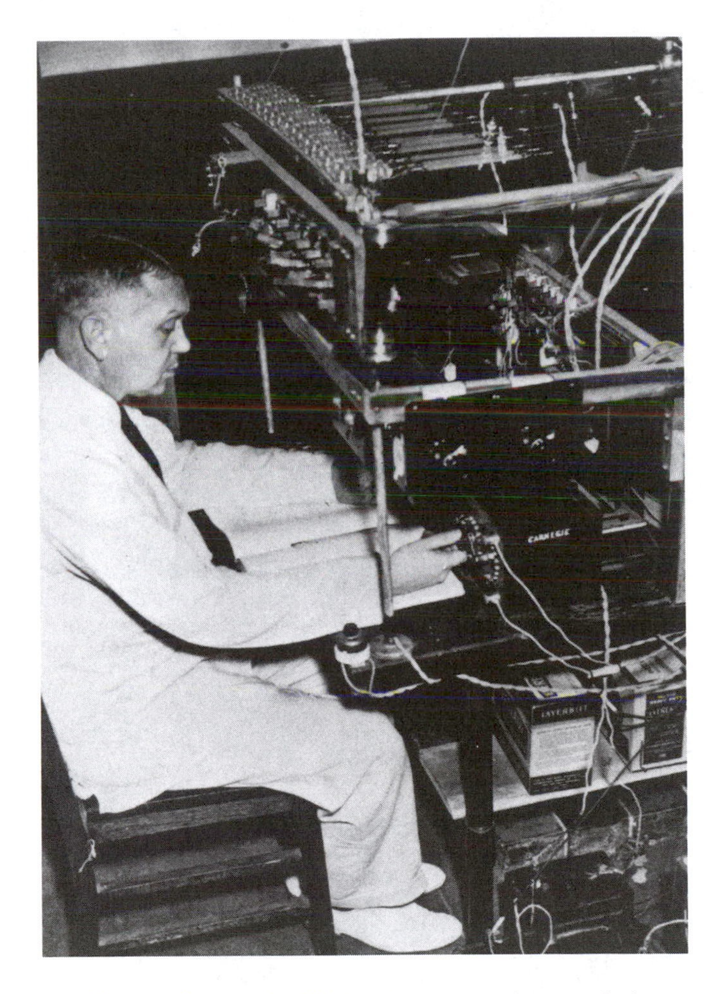

Harold Urey, who won the 1934 Nobel Prize in chemistry for his discovery of the isotope deuterium. *Argonne National Laboratory, courtesy AIP Emilio Segrè Visual Archives.*

fessor. These promotions recognized his important discovery of deuterium, which he achieved in 1931 with the assistance of his colleagues Ferdinand Brickwedda and George M. Murphy by slowly evaporating four liters of hydrogen, leaving one milliliter of "heavy" hydrogen that had a different spectroscopic line reading from normal hydrogen's. Urey had also published his first book in 1930 with A. E. Ruark, the seminal text in physical chemistry, *Atoms, Molecules, and Quanta,* and he served as the first editor of the *Journal of Physical Chemistry.*

Urey again assisted in the war effort in 1940, this time serving on the Uranium Committee of the Manhattan Project and as the director of Columbia's Substitute Alloys Materials Laboratory from 1942, isolating the isotope uranium-235 for use in atomic reactions. After the war Urey moved to the Enrico Fermi Institute of Nuclear Studies at the University of Chicago, which named him the Martin A. Ryerson Distinguished Service Professor in 1952. Then in 1958 the University of California at San Diego in La Jolla named him a Professor at Large.

Urey spent the latter part of his career researching global temperature shifts as recorded by isotope concentration in fossils. He also calculated to an amazing degree of accuracy the abundance of elements in the universe, considered the origin of the solar system and of the Earth's moon, and conducted experiments on replicating the origin of life on Earth. Urey died of a heart attack on January 5, 1981, in La Jolla, California.

V

Van der Meer, Simon

(1925–)
Dutch
Physicist

The developer of a technique known as stochastic cooling, Simon Van der Meer is a specialist in particle accelerator engineering. His successful use of methods for accumulating and focusing particles in an accelerator resulted in the discovery of the charged W and neutral Z bosons, which are elementary particles. For this work Van der Meer shared the 1984 Nobel Prize in physics with his colleague Carlo Rubbia.

Born on November 24, 1925, in The Hague, the Netherlands, Van der Meer was an only son. His mother was the former Jetske Groeneveld; his father was Pieter van der Meer, a teacher. Van der Meer married Catharina M. Koopman in 1966, and they had two children.

Unable to enroll in college after graduating from the local gymnasium in 1943 because of occupation by the German army, Van der Meer studied at the gymnasium for two additional years. In 1945 he attended the technical university in Delft to study physics. After earning an engineering degree in 1952, Van der Meer accepted a position with the Phillips Research Laboratory in Eindhoven. There he worked on electron microscopy and high-voltage equipment. After four years and in the same year Van der Meer earned his doctorate from Delft, he left the laboratory for a position at the recently founded European Center for Nuclear Research (CERN), located in Geneva, Switzerland. CERN, a joint effort of 13 nations, was a research center focused on the study of elementary particles.

After joining CERN, Van der Meer worked on the development and design of the proton synchrotron (PS), an accelerator in which charged particles are accelerated and fixed to a path by an electric field and a magnetic field. Part of the strategy was to generate masses of charged particles and accelerate them to high speeds in a smaller accelerator. The particles would then be moved to the PS and accelerated to a higher degree. The electric and magnetic fields needed to be designed in such a way as to prevent the particles from scattering and colliding with one another. The particles had to be restricted to a fixed path. Van der Meer also worked on stochastic cooling, a technique for accumulating bunches of particles into a small area. He first developed the method in 1972 to reduce the arbitrary motion of particles.

In the mid-1970s Van der Meer began investigating the force-carrying W and Z bosons, the existence of which had been proposed in several theories. He worked with Carlo Rubbia, who hoped to find the bosons through a modified version of the PS accelerator. He hypothesized that beams of protons and beams of antiprotons, accelerated in reverse directions and allowed to collide, would generate sufficient energy to produce the bosons. Protons and antiprotons generally did not bunch together but repelled each other instead. Van der Meer thus used stochastic cooling to develop an electrical system that focused the beams and kept them in place. His design led to the creation of the super proton synchrotron (SPS) in 1979. In 1983 the SPS began operation, and soon both the W and Z bosons were discovered.

In addition to the 1984 Nobel Prize in physics, Van der Meer has received a number of honors and awards, including honorary degrees from universities in Amsterdam, Geneva, and Genoa. Van der Meer was appointed to the Royal Netherlands Academy of Arts and Sciences as well as the American Academy of Arts and Sciences. He retired from CERN in 1990.

Van Dover, Cindy Lee

(1954–)
American
Marine Biologist

Cindy Lee Van Dover is the first scientist and first woman certified as a submersible pilot. Her discovery that certain deep-sea shrimp are sensitive to light has led to new speculations about hot-water vents on the ocean floor and the strange communities that live around them. She was born on May 16, 1954, in Red Bank, New Jersey, to James K. and Virginia Van Dover.

In 1977, the same year Cindy Van Dover earned a B.S. in environmental science from Rutgers University, scientists' picture of the deep sea changed drastically. They had believed that very few organisms lived there because, as far as was known, all life ultimately depended on creatures' ability to make food from the Sun's energy through the process of photosynthesis, and sunlight did not penetrate the water below about 900 feet. Then, however, oceanographers in tiny craft called submersibles, exploring spots on the seafloor where the Earth's continental plates were spreading apart, found vents through which fountains of water spewed, black with mineral deposits and heated to 650°F by the magma below. These vents were surrounded by bizarre life-forms that included white mats of bacteria, eyeless shrimp, and eight-foot-tall red-tipped tube worms.

Van Dover saw her first hydrothermal vent community—so beautiful that it had been nicknamed the Rose Garden—in a dive in the submersible *Alvin,* owned by the Woods Hole Oceanographic Institution, in 1985, the same year she earned a master's degree in ecology from the University of California at Los Angeles. The following year, studying for a doctorate in a combined program of the Massachusetts Institute of Technology and Woods Hole, she began looking at a species of shrimp found only around the vents. In videotapes of living vent communities she noticed two bright lines on the shrimps' backs not visible on her preserved specimens. She dissected this area and found that the lines were flaps of tissue connected to large nerves. This finding suggested that the tissue was a sense organ. Van Dover then had what seemed at first an outlandish idea: Could it be a type of eye?

Other scientists to whom Van Dover sent samples of the tissue said that features of its structure and chemical characteristics supported her idea that it might sense light. No light around the vents had been reported, but, Van Dover thought, the vent water was so hot that some of its heat energy might shade into dim red light, just as a heater coil glows red. She asked a geologist taking a sensitive camera down on *Alvin* to turn off the sub's lights—that had not been done before—and see whether the camera spotted any light near the vents. He did so in June 1988 and sent back the startling message: "VENTS GLOW."

After receiving her doctorate in 1989, Van Dover spent nine months learning how to pilot *Alvin* herself. She was the first scientist and the first woman whom the navy certified as a submersible pilot commander. She piloted *Alvin* on 48 dives between 1990 and 1992.

Van Dover returned to scientific work in 1993 and is now pursuing another seemingly outlandish idea. As far as is known, all vent creatures depend on the bacteria there, which can make food from sulfur compounds in the water by a process completely different from photosynthesis. Some microorganisms, however, can carry on photosynthesis in light as dim as the vent light Van Dover helped to discover, and she is trying to find vent microbes that can photosynthesize. If she succeeds, she will have uncovered the first example of natural photosynthesis that does not require sunlight. Some scientists have speculated that life on Earth originated around undersea vents, and Van Dover's idea suggests that photosynthesis may have begun there, too.

Van Dover has remained affiliated with Woods Hole throughout her career. She has also been a Visiting Scholar at Duke University in North Carolina (1994–95), an associate professor at the University of Alaska at Fairbanks (1995–98), and science director of the West Coast National Undersea Research Center. In 1998 she became an assistant professor at the College of William and Mary in Virginia. *Ms.* magazine chose her as its Woman of the Year in 1988, and she has also won several scientific awards, including the Vetlesen Award from Woods Hole (1990) and the NOAA/MAB Research Award (1996).

In addition to studying vent light and its possible contribution to photosynthesis, Van Dover has researched other ecological aspects of vent communities and the effects of deep-sea waste disposal on underwater life. She has also worked to increase support for exploring the deep sea. "We actually know more about the surface of Mars and Venus . . . than we know about the topography of our own seafloor," she has said; yet the health of the communities of creatures that live there "may be critical to the balance of the world's oceans" and to life on land as well.

Virtanen, Artturi Ilmari

(1895–1973)
Finnish
Biochemist

Recognized as one of Finland's most important scientists, Artturi Ilmari Virtanen made important contributions in agricultural chemistry and nutrition. Virtanen's best-known work was in his development and improvement of preservation methods for livestock feed, or fodder, and silage, fodder prepared by storing protein-rich green plants in a silo. Silage often suffered from spoilage as a result of various fermentation processes, and this was a major problem in areas that experienced long, harsh winters, such as Virtanen's homeland. Virtanen's preservation methods and his discoveries in agricultural chemistry earned him the Nobel Prize in chemistry in 1945.

Born in Helsinki, Finland, on January 15, 1895, Virtanen was the child of Serafina Isotalo and Kaarlo Virtanen. He married Lilja Moisio in 1920, and the couple had two children, both boys.

After graduating from the classical lyceum in Viipuri (now Vyborg, Russia), Virtanen entered the University of Helsinki to study chemistry, physics, and biology. He earned his master's degree in 1916 and spent one year working as an assistant chemist in the Central Industrial Laboratory of Helsinki. Virtanen then pursued graduate studies at the University of Helsinki and received his doctorate in 1919. For his postdoctoral studies Virtanen journeyed across Europe, studying physical chemistry in Zurich, Switzerland, and bacteriology and enzymology in Sweden. It was also during his journeys that Virtanen was drawn to the field of biochemistry.

After his travels Virtanen in 1921 took a position directing the laboratory at the Finnish Cooperative Dairies Association in Valio. There his duties included overseeing the production of cheese, butter, and other dairy items. Virtanen also began teaching chemistry at the University of Helsinki in 1924.

Virtanen's research involved studying the enzyme activity of plant cells and their protein content to determine whether the proteins in plant cells were enzymes. This led to his investigation of nitrogen, which was plentiful in proteins and important in the nutrition of humans and animals. Virtanen researched legumes, which were capable of changing atmospheric nitrogen into nitrogenous substances appropriate for plant production, and soon realized that a large amount of nitrogenous material was lost during storage as silage. Animals that were fed this less nutritious fodder produced inferior dairy products; Virtanen therefore tackled the problem of preserving the quality of this fodder. He knew that lactic acid was a product of fermentation and that it made silage sufficiently acidic to stop harmful fermentation processes. Virtanen thus mixed a solution of dilute hydrochloric and sulfuric acids to increase the acidity of newly stored silage to a point at which damaging fermentation would be prevented. This method, known as the *AIV method* after Virtanen's initials, did not negatively affect the fodder, the cows, or the products of the cows that were fed the fodder. The AIV method was first used in Finland in 1929.

Virtanen headed Finland's Biochemical Research Institute in Helsinki from 1931 and taught biochemistry at both the Helsinki University of Technology and the University of Helsinki. He continued his biochemical research after receiving the 1945 Nobel Prize in chemistry, studying the nitrogen-fixing bacteria in the root nodules of legumes and the composition of plants. He also developed improved methods for preserving butter and discovered a red pigment in plant cells that was comparable in composition and purpose to hemoglobin. Virtanen served on the editorial boards of a number of scientific journals, was the Finnish representative on a United Nations Commission on Nutrition, and was president of Finland's State Academy of Science and Arts from 1948 to 1963. He died on November 11, 1973, in Helsinki.

Volta, Count Alessandro Giuseppe Antonio Anastasio

(1745–1827)
Italian
Physicist

Alessandro Giuseppe Antonio Anastasio Volta was best known for inventing the electric battery, which provided not only the first reliable source of electricity but also the first source of readily available, continuous current. For his pioneering work he was made a count by Napoleon and granted numerous honors and awards. Volta's discov-

Alessandro Giuseppe Anastasio Volta, who is best known as the inventor of the electric battery. *Smith Collection, Rare Book & Manuscript Library, University of Pennsylvania.*

eries greatly advanced the understanding of electricity and paved the way for further contributions in the field, such as William Nicholson's studies of electrolysis and Sir HUMPHRY DAVY's work in electrochemistry.

Volta was born on February 18, 1745, in Como, Italy, to Filippo Volta and Maddalena de' Conti Inzaghi, who raised him in an aristocratic and religious environment. Many members of his family were involved in the church, and most of the male members became priests, including his three paternal uncles and three brothers. Two of his sisters became nuns. After Volta's father died when Volta was about seven, one of his uncles directed his education.

Though Volta's family persuaded him to study law, a field in which many family members on his mother's side were involved, Volta at the age of 18 was already determined to investigate electricity. In 1757 he began study at the local Jesuit college. He later attended the Seminario Benzi.

In 1774 Volta took a job teaching physics at the Royal School of Como. A year later he invented the electrophorus, an instrument used to produce static electricity, which provided the most effective way to store electric charge at that time. Volta also in 1778 encountered and isolated methane gas. In 1779 he accepted the chair of physics at the University of Pavia. It was in Pavia that Volta conducted his breakthrough research on electricity. In 1780 his friend Luigi Galvani found that contact of a frog's leg muscle with two metals, copper and iron, generated an electric current. Galvani believed that the animal's tissues produced the electricity. Volta was interested in discovering the source of the electricity and began conducting experiments in 1794 using only metal. He discovered that he could produce electric current without animal tissue; instead he placed different metals in contact with one another, and this arrangement generated electricity.

Volta introduced his battery, the voltaic pile, in 1800. It was made up of alternating disks of silver and zinc separated by brine-soaked cardboard. The voltaic pile successfully generated electric current and resolved the controversy regarding the source of electricity—some believed it was produced by animal tissue, and others thought electric current was generated by metals. Volta exhibited his voltaic pile in 1801 in Paris and provided Napoleon with a demonstration of the battery. Napoleon was sufficiently impressed to make Volta a count and senator of Lombardy. Napoleon also awarded Volta the medal of the Legion of Honor.

After presenting his voltaic pile, Volta did little to improve upon the battery. In 1815 he began working at the University of Padua. Volta received many honors for his work, including the Royal Society of London's Copley Medal, which he received three years after becoming a member in 1791. He also was a member of the Paris Academy and was a correspondent of the Berlin Academy of Sciences. He died in Como, Italy, on March 5, 1827. The volt, the unit of electric potential, was named in Volta's honor.

Von Humboldt, Alexander
(1769–1859)
German
Naturalist

Having wide-ranging interests and a fervor for scientific knowledge, the naturalist and explorer Baron Alexander von Humboldt did much to popularize science. Humboldt's expedition to South America, Cuba, and Mexico advanced the science of ecology considerably. Not only did Humboldt gather and analyze large volumes of specimens, but he also studied volcanoes, theorizing that volcanic activity had played a significant role in the development of the Earth's crust. Humboldt also was the first to propose a canal through Panama and introduced isobars and isotherms on weather maps, thereby furthering comparative climatology.

A sickly child and unenthusiastic student, Humboldt apparently did not develop his tireless drive for knowledge until his college years. Humboldt's father, Alexander Georg von Humboldt, was a Prussian officer in the army of Frederick the Great. His mother, Marie Elisabeth Colomb Holwede Humboldt, was a member of a family of French Protestants and had inherited wealth after her first husband's death. Humboldt was born September 14, 1769, in Berlin, Germany; his father died when he was 10 years old. Thereafter he and his brother, Wilhelm, were raised by their strict and undemonstrative mother, who pressed them in their studies so that they would be prepared for political lives as adults.

Humboldt entered the University of Frankfurt an der Oder to study economics in 1787 but decided to study engineering instead. He thus went to Berlin to receive training in engineering but then became enamored of botany. Collecting and classifying plant specimens in the Berlin area, Humboldt soon began to dream of exploring faraway lands. He attended the University of Göttingen from 1789 to 1790 to study science. There he became interested in mineralogy and geology, prompting him to enroll in the School of Mines in Freiberg, Saxony. After intensive studies Humboldt left Freiberg in 1792 without a degree. Soon he began working as a mining engineer for the government and was sent to Ansbach-Bayreuth, where he reorganized and supervised mines.

In 1796 Humboldt received an inheritance, which enabled him to finance his own scientific explorations. He resigned from his mining position in 1797 and began preparing for his travels by studying geological measuring systems. The Napoleonic Wars interfered with his plans, but in 1799 Humboldt was finally able to depart on his first expedition. With permission from the Spanish government to visit the colonies in Central and South America, Humboldt and a French botanist, Aimé Bonpland, set sail. They traveled for five years, navigating the Orinoco River and visiting Peru, Venezuela, Ecuador, Cuba, and Mexico. The two covered more than 6,000 difficult miles on horseback, on foot, and in canoes and made abundant discoveries. Humboldt climbed numerous mountains, including the Chimorazo volcano to a height of more than 19,000 feet without oxygen. As a result of these climbs Humboldt was the first to attribute mountain sickness to lack of oxygen at high altitudes. He also studied the coastal currents of the Pacific; the Humboldt current (now the Peru current) was named for him.

Humboldt spent 1804 to 1827 in Paris, analyzing and publishing the information he and Bonpland gathered on the expedition. Significant among the findings were his observations of Andean volcanoes, which led him to conclude that volcanic action affected the geological history of the Earth, and the collection of meteorological data. Humboldt had recorded mean daily and nightly temperatures and marked his weather maps with isotherms, lines that connected points of equal temperature, and isobars, lines that connected points of equal barometric pressure. Also of importance was his analysis of the flora and fauna of various geographical regions.

Depleted of money, Humboldt returned to Berlin in 1827 and served as tutor to the crown prince. In 1829 he traveled to Siberia and took geological and meteorological measurements. In his final 25 years Humboldt worked on writing *The Cosmos*, an ambitious attempt to detail the structure of the universe. Humboldt died while working on the fifth volume of the work. The state of ecology was advanced considerably by Humboldt's enthusiasm and extensive studies.

Von Neumann, John Louis

(1903–1957)
Hungarian/American
Mathematician

John Von Neumann was considered a genius because of his ability to understand complex mathematical and scientific theories, abstracting them down to simpler expressions comprehensible to lesser minds. His career bridged the gap between pure mathematics and applied mathematics, which subsequently developed into separate disciplines. His formulation of Von Neumann architecture set the foundation for most applications of computer design. What became known as Von Neumann algebras were derived from quantum mechanics. Von Neumann was also instrumental in the development of the first atomic bomb, and he devoted the latter part of his career to implementing sane policies regarding nuclear power.

John Von Neumann, whose Von Neumann architecture created the foundation for most applications of computer design. *AIP Emilio Segrè Visual Archives.*

Von Neumann was born on December 28, 1903, in Budapest, Hungary. He was the eldest of three sons born to Margaret Von Neumann and Max Von Neumann, a successful banker. In 1929 Von Neumann married Mariette Kovesi of Budapest, and together the couple had one daughter, Marina, in 1935. They divorced in 1937, and in 1938 Von Neumann married Klara Dan, also of Budapest.

In 1921 Von Neumann commenced study of mathematics at the University of Budapest and at the University of Berlin, and in 1925 he received his degree in chemical engineering from the Swiss Federal Institute of Technology in Zurich. In 1926 he earned his Ph.D. in mathematics from the University of Budapest, where he wrote his dissertation on set theory. That year he received a Rockefeller grant for postdoctoral work at the University of Göttingen under David Hilbert, with whom he developed axiomatizations that satisfied both ERWIN SCHRÖDINGER's wave theory and WERNER KARL HEISENBERG's particle theory in quantum mechanics.

Von Neumann inaugurated his professional career as a *Privatdozent* at the University of Berlin from 1927 to 1929 and at the University of Hamburg from 1929 to 1930. Von Neumann then spent the rest of his career at Princeton University, first as a visiting professor, then as a professor of mathematics at Princeton's Institute for Advanced Study from 1933 until his early death in 1957. In 1932 he published his influential text *The Mathematical Foundations of Quantum Mechanics*, which represented the first distillation of quantum theories applied to mathematics. While at Princeton, Von Neumann became a naturalized citizen of the United States.

In 1937 Von Neumann began a lasting relationship with the United States government, which was anticipating a war, as a consultant to the Ballistics Research Laboratory of the Army Ordnance Department. In 1941 he became a consultant on the theory of detonation of explosives for the National Defense Research Council, and on mine warfare for the Navy Bureau of Ordnance. He joined the Los Alamos Scientific Laboratory in 1943 to work on the Manhattan Project; there he contributed the idea of implosion as the detonation technique for the atomic bomb. Von Neumann continued work on nuclear policy as a member of the Atomic Energy Commission from 1954.

In 1944 Von Neumann took a brief respite from his wartime activity to publish *The Theory of Games and Economic Behavior* on the minimax theorem that he had formulated in 1928, which stated that the rational choices of game players make it pointless to play, as the outcome is determined not by their efforts but by the rules of the game alone. His coauthor Oskar Morgenstern applied this idea to economics, though it also applied to politics and other human endeavors. That same year he continued his wartime work in conjunction with the Moore School of Engineering at the University of Pennsylvania transforming the electronic numerical integrator and computer (ENIAC) into the electronic discrete variable automatic computer (EDVAC) to improve calculations involving wartime technologies. Von Neumann's 1945 report on EDVAC contained the first statement of what became known as Von Neumann architecture, or computer memory storage. He also pioneered the technology for the central processing unit (CPU) and random access memory (RAM).

Von Neumann received the 1956 Enrico Fermi Science Award and the 1956 Medal of Freedom from President Eisenhower. The year before, however, he was diagnosed with bone cancer, and he died on February 8, 1957, in Washington, D.C.

Von Sachs, Julius

(1832–1897)
German
Botanist

Julius von Sachs made discoveries that were to form the foundation for significant portions of the science of botany. His extensive experimental work addressed nearly every aspect of plant physiological processes. He described water movement in plants and established that photosynthesis occurred within chloroplasts and that starch present in these chloroplasts was the first visible indication of photosynthesis. He studied the physiological characteristics of stimuli and invented laboratory equipment such as the clinostat to do so. The respect and admiration his work garnered provided the neglected field of plant physiology a much needed boost. Early in his career he became the first person to teach an entire course on the subject at a German university.

Born on October 2, 1832, in Breslau, Silesia (now Wroclaw, Poland), Sachs was one of nine children. His parents, Maria-Theresia and Christian Gottlieb Sachs, were poor. His father, an engraver, died in 1848, and his mother died the next year.

Without his parents' support Sachs had no way to finance his education and was forced to withdraw from the Gymnasium. Fortunately a physiologist named Johannes Purkinje invited Sachs to become his assistant at the University of Prague. Sachs eventually obtained his Ph.D. at Prague in 1856 and qualified as a lecturer in plant physiology the following year.

In 1859 Sachs became an assistant professor in plant physiology at the Agricultural and Forestry College in Tharandt. Two years later he left for Poppelsdorf to become a teacher at the Agricultural College there. He made an important discovery in 1861, when he proved that what became known as chloroplasts are the site of photosynthesis (the production of organic compounds from carbon dioxide and water) in plants. He also established that chlorophyll, an ingredient essential to the photosynthetic process, is not distributed throughout the plant but rather is contained within the chloroplasts. In 1862 he demonstrated that the starch found within these chloroplasts was the first visible product of photosynthesis. The iodine test he designed to detect the presence of starch is still used in laboratory classes today.

Sachs took a faculty position at the University of Würzburg in 1868, and he remained a professor of botany there for the rest of his career. He began to explore the physiological characteristics of stimuli in 1873, with his research into geotropism (the effect of gravity on a plant's growth), phototropism (a plant's movement toward or away from light), and hydrotropism (the effect of moisture on a plant's movement or growth). He also invented devices to facilitate his experiments. The most famous of these was the clinostat, which compensated for gravity. In 1874 he presented his imbibition theory, which asserted that absorbed water is carried through tubes in the walls of a plant with no cooperation from living cells and without entry into cell cavities. For the most part Sachs's experimental efforts were far more successful than his attempts to formulate theoretical explanations for the phenomena he witnessed. His *Textbook of Botany*, published in 1868, was a thorough and sound summation of botanical knowledge.

Sachs's most lasting contribution to botany was the interest in plant physiology his exacting research fueled. Moreover the conclusions he derived, especially those pertaining to photosynthesis, became some of the core principles of botany. He received many honors. In 1877 he was declared privy councillor and awarded the Order of Maximilian and the Order of the Bavarian Crown. He was allowed to add *von* to his name, an indication that he had been granted personal nobility. Many universities, including those at Bonn, Bologna, and London, conferred honorary doctorates on him. Sachs was also elected to the Royal Society of England.

Vrba, Elisabeth

(1942–)

South African/American

Paleontologist

CHARLES ROBERT DARWIN's theory of evolution claimed that species appeared or died out gradually as a result of random changes that gave some members of a species a competitive advantage over others. Elisabeth Vrba, however, believes that evolution sometimes makes sudden jumps and that climate change drives most of them, including the ones that led to modern humans.

Elisabeth was born on May 27, 1942, in Hamburg, Germany. Her father, a law professor, died when she was two years old, and her mother moved to Namibia in southern Africa. Elisabeth majored in zoology and statistics at the University of Cape Town, South Africa, and graduated with honors. She moved to Pretoria in 1967; married a civil engineer, George Vrba; and began teaching high school.

Vrba began work for the Transvaal Museum in 1968, first as an unpaid assistant. About 10 years later she became the museum's deputy director and was put in charge of its collection of fossil hominids, the ancestors of humans. On the basis of her studies of both hominid and antelope fossils she has devised a theory she calls the *turnover pulse hypothesis*.

Vrba noticed that many changes occurred in both hominid and antelope species about 5 million years ago, and she suspects that these changes resulted from a shift in climate that has been demonstrated by fossil and geologic evidence. Much of the world became drier and colder at that time, and large parts of Africa's formerly solid blanket of forests were replaced by grassland. Species of forest-living antelopes and apes either clung to the remaining forest lands or died out and were replaced by new species that could adapt to grassland. The forest apes eventually became chimpanzees, gorillas, and (in Asia) orangutans, whereas the grassland ape was *Australopithecus afarensis*, the first hominid. Vrba believes that two similar climate changes, one between 2.8 and 2.5 million years ago and the other about 900,000 years ago, led, respectively, to the appearance of the human genus (*Homo*) and the spread of the human ancestor *Homo erectus* out of Africa.

In the turnover pulse hypothesis Vrba has generalized these ideas to propose that changes in climate are the chief cause of rapid changes in species and perhaps of evolution itself. She is not the first to suggest that the pace of evolution sometimes speeds up and that changes in the environment, especially of climate, produce widespread changes in species. She is, however, one of the strongest believers in a link between these two ideas: "Making a new species requires a physical event to force nature off the pedestal of equilibrium," she told the *Discover* writer Ellen Shell.

Vrba, with her husband and daughter, moved to the United States in 1986. Shortly thereafter she joined the faculty of Yale University, where she became a tenured professor. The controversy her ideas stir does not bother her, she told Ellen Shell: "I'm interested in pushing out the frontiers of science, not sailing my boat through tranquil seas."

W

Wagner-Jauregg, Julius

(1857–1940)
Austrian
Physician, Psychiatrist

For his breakthrough discoveries regarding the use of high fever to treat mental illness Julius Wagner-Jauregg became in 1927 the first psychiatrist to receive the Nobel Prize in physiology or medicine. Wagner-Jauregg artificially induced malaria to treat general paresis, or paralysis, which was caused by the then-fatal disease of syphilis. His investigations of such stress treatments led to the development of shock therapy in treating mental disorders.

Born Julius Wagner in Wels, Austria, on May 7, 1857, Wagner-Jauregg was the child of Ludovika Ranzoni and Adolf Johann Wagner. Wagner-Jauregg's mother died when he was still quite young; his father worked as a government official and later joined the ranks of the nobility, after which the family name was changed to *Wagner von Jauregg*. The *von* was abandoned after the fall of the Austro-Hungarian Empire in 1918.

After completion of his studies at Vienna's Schotten-Gymnasium, Wagner-Jauregg entered the University of Vienna to study medicine. He was educated in experimental biology and became friends with SIGMUND FREUD, the founder of psychoanalysis. Wagner-Jauregg earned his medical degree in 1880 and tried unsuccessfully to find a position in general medicine. Though he had little training in psychiatry or mental illness, he settled for a post as assistant in the university's psychiatric clinic. He married Anna Koch, and the couple had two children.

As early as 1888 Wagner-Jauregg began to study what was to become his lifelong work—the use of fever to treat mental illness. Wagner-Jauregg had noticed cases in which such illnesses as scarlet fever, malaria, and smallpox caused a reduction of mania, melancholy, or paresis among many patients. In 1889 he became a professor of psychiatry at the University of Graz and began conducting experiments in which he injected tuberculin into mental patients to induce fever. However, tuberculin, which was used to treat tuberculosis, was found to be unsafe, and Wagner-Jauregg ceased his experiments. He left Graz in 1893 for the University of Vienna, where he became a full professor and the head of the Hospital for Nervous and Mental Diseases. There he furthered his investigations, injecting patients suffering from paresis with typhus vaccine and staphylococci. The results were unsatisfactory, however, and it was not until World War I that Wagner-Jauregg was able to begin his malaria experiments.

By World War I it was known that untreated syphilis, usually a venereal disease, caused paresis, which was considered fatal and brought about paralysis and insanity. In 1917 Wagner-Jauregg injected nine paretic patients with malaria. The patients were treated with quinine, the common treatment for malaria, after they had suffered several bouts of fever. Six of the nine patients were restored to good health.

Wagner-Jauregg's malaria treatment was not uncontroversial but gradually became widely accepted after further clinical trials demonstrated a low rate of mortality when mild strains of malaria were used. Malaria continued to be used to treat syphilis until penicillin was discovered during World War II. In addition to the Nobel Prize, Wagner-Jauregg received such honors as the University of Edinburgh's Cameron Prize in 1935 and the American Committee for Research on Syphilis Gold Medal in 1937.

Waldeyer-Hartz, Heinrich Wilhelm Gottfried von

(1836–1921)
German
Anatomist

Wilhelm Waldeyer, as he was known, was the first to describe cancer in modern terms, transforming the medical perception of this disease from an incurable malady to a treatable condition, if discovered early in its development. In the middle of his career he identified several anatomical features that now bear his name. Late in his career he coined the terms *chromosome* and *neuron*, and he proposed the neuron theory of the nervous system.

Waldeyer was born on October 6, 1836, in Hehlen, Germany. His mother, Wilhelmine von Hartz, was the daughter of a schoolteacher, and his father, Johann Gottfried Waldeyer, was an estate manager. In 1856 he entered the University of Göttingen to study the natural sciences. However, Friedrich Gustav Jacob Henle's lectures in anatomy influenced Waldeyer toward that specialty. He studied briefly at the University of Greifswald before conducting his doctoral work at the University of Berlin, where he wrote a dissertation on the structure and function of the clavicle that was published in 1862. He married in 1866, and four of his children survived him; none entered the medical profession.

After receiving his doctorate, he worked at the University of Königsberg as an assistant in the department of physiology and taught histology. He moved on to the University of Breslau, where he lectured in pathology and histology. In 1863 he reported his findings that cancer develops from a single cell, countering the contemporary belief in cancer as a systemic disease. He also accounted for the spread of cancer through the vascular and lymphatic systems. Then in 1864 he published a paper on the histological transformations that occur in muscles in the wake of typhoid fever; this publication secured his appointment as a professor of pathology and the director of the department of postmortem investigations in 1865.

In 1868 the University of Breslau named him the chairperson of its pathology department. That year he described the two duodenal fossae, thenceforth known as *Waldeyer's fossae*, that related to hernias. In 1870 he located in the ovaries what became known as Waldeyer's line, the boundary of the mesovarium at the helus. He also identified what was later referred to as *Waldeyer's ring*, or lymphoid tissue of the throat, the faucial and pharyngeal tonsils. In 1872 the University of Strasbourg hired Waldeyer as a professor of normal human anatomy in the course of replacing French professors with German ones, as the region had just exchanged hands. After 11 years there Waldeyer moved to the University of Berlin in 1883 as a professor of anatomy and the director of the university's anatomical institute.

One telling measure of Waldeyer's esteem as a physician was that in 1887 Emperor Frederick III called upon him to diagnose his tumor of the vocal cords. Waldeyer devoted his late career to teaching and administration, transforming the University of Berlin's anatomical institute from a poorly equipped facility to a world-class research laboratory. Waldeyer died after suffering a stroke on January 23, 1921.

Wallace, Alfred Russel
(1823–1913)
Welsh
Naturalist

Alfred Wallace developed a theory of natural selection independently of CHARLES ROBERT DARWIN, and the two delivered their ideas jointly, though Darwin's name attached itself singularly to the theory of evolution through propriety and the circumstance that his ideas were more fully developed and backed by evidence. Besides being one of the original proponents of the theory of evolution, Wallace made contributions to the scientific fields of ethnology, zoogeography, geology, vaccination, and astronomy.

Wallace was born on January 8, 1823, in Usk, Monmouthshire, Wales, the eighth of nine children born to Thomas Vere Wallace and Mary Anne Greenell. Wallace married Annie Mitten in April 1866. Together the couple had three children—Herbert, who died at the age of four; Violet; and William.

Wallace's formal education ended when he was 14; he traveled to London to apprentice as a surveyor under his brother William from 1838 to 1843, when business declined. Wallace then became a master, teaching English, arithmetic, surveying, and elementary drawing between 1844 and 1845 at Collegiate School in Leicester, where he met the entomologist Henry Walter Bates. Wallace enticed Bates into exploring the Amazon River in South America, which they did in tandem from 1848 to 1852. Unfortunately Wallace lost the specimens he had collected in a fire that sank his ship in the Atlantic on August 6, 1852. However, he recorded his impressions of the expedition in the 1853 book *A Narrative of Travels on the Amazon and Rio Negro.*

Wallace conducted an eight-year tour of the Malay Archipelago between 1854 and 1862, when he collected about 127,000 specimens. During this time he wrote the paper "On the Law Which Has Regulated the Introduction of New Species," which caught Darwin's attention upon its publication in 1855. Wallace argued for the evolution of species from closely allied species. It was in February 1858, however, while experiencing an attack of malaria at Ternate in the Moluccas and reading Thomas R. Malthus's *Essay on Population,* that Wallace first conceived the idea of survival of the fittest; he summarized the idea over three successive evenings and sent it to Darwin in the next mail, which left Ternate on March 9, 1858. Upon receipt of the manuscript Darwin recognized the coincidence of his own and Wallace's ideas about evolution.

On July 1, 1858, Wallace read to the Linnaean Society his paper "On the Tendency of Varieties to Depart Indefinitely from the Original Type," demonstrating species' struggle for existence, reproduction rates, and food supply dependence, all of which contributed to his theory. Darwin read an abstract of his larger theory of evolution, which he was preparing for publication in *On the Origin of Species.* Both scientists gained instant notoriety, though Darwin eclipsed Wallace as the proprietary theorist.

In the 1870 text *Contributions to the Theory of Natural Selection* Wallace posited his differences with Darwin, including Wallace's assertion that human intelligence did not evolve by natural selection alone but was derived in part from external influence such as a spiritual connection. In 1889, Wallace published *Darwinism,* and though it posited further differences with Darwin, it summarized Darwinian evolutionary theory very well.

In 1869 Wallace published *The Malay Archipelago,* in which he divided the archipelago into the western islands of Borneo and Bali, with affinities to Asia, and the eastern islands of Celebes and Lombok, with affinities to Australia. What's known as Wallace's line divides the islands, as mammals on the Asian side advanced further than their more primitive counterparts on the Australian side of the line. Wallace followed up on these ideas in 1876 with the two volumes of *Geographical Distribution of Animals* and in 1880 with *Island Life.*

Wallace received several awards from the Royal Society, which elected him a member in 1893, including the 1868 Royal Medal, the 1890 Darwin Medal, and the 1908 Copley Medal. From the Linnean Society of London he received the 1892 Gold Medal and the first Darwin-Wallace Medal in 1908. In 1910 he received the Order of Merit from the British government. He died on November 7, 1913, in Broadstone, Dorset.

Washington, Warren M.

(1936–)
American
Meteorologist

The first African American to become president of the American Meteorological Society, Warren Washington has made significant contributions to the atmospheric sciences. Warren has focused efforts on studying the Earth's climate through the development of computer models that analyze and predict weather and climate changes. Washington is the head of the Climate and Global Dynamics Division of the National Center for Atmospheric Research (NCAR).

Born on August 28, 1936, in Portland, Oregon, Washington was one of five boys born to Dorothy Grace Morton and Edwin Washington, Jr. A college graduate, Washington's father had moved to Portland in 1928 in search of a teaching position. Because of racial discrimination, however, he was unable to secure such a job and worked instead as a porter in Pullman cars for the Union Pacific Railroad. Washington's mother later became a nurse.

Though discouraged from entering college by his high school counselor, Washington had for years demonstrated an aptitude for science. His natural bent was supported by his teachers, and Washington thus entered Oregon State University in the small town of Corvallis, Oregon, to study physics. In a college with few African-American students, Washington found himself the sole African American majoring in physics. Washington's interest in meteorology developed when he took a job at a weather station near the university. Operating a weather radar that was designed to track storms approaching from the Pacific Ocean, he was intrigued by the as-yet little explored technology of weather radar. After earning his bachelor's degree in 1958, Washington continued with graduate studies at Oregon State University and received his master's degree in 1960. Because Oregon State University did not offer a doctoral program in meteorology, Washington traveled to Pennsylvania State University, which was known for its pioneering work in the use of computers to study weather. Washington was awarded his doctorate in meteorology in 1964, one of four African Americans who had such a doctorate.

After earning his doctorate, Washington took a job with NCAR in Boulder, Colorado. One of his early projects was to develop a mathematically based climate prediction program, on which he collaborated with his colleague Akira Kasahara. At that time there were few weather observatories, and weather data were not plentiful. Through the years however, the available weather data increased, as did the complexity and power of computers, and Washington was able to develop more complex climate models that consider such weather elements as the moisture of soil and the atmosphere, the volume of sea ice, and the effects of surface and air temperature. By the early 1970s, for example, he and his team produced images that showed the connection between the jet stream's crossing the Pacific Ocean and India's yearly monsoon.

Washington's research has contributed to the ability of meteorologists to make fairly accurate weather forecasts and has helped make possible the prediction of global climate changes. His work has also advanced understanding of the greenhouse effect, which results in global warming.

Washington has written more than 100 published articles and continues his research to increase the accuracy of climate models. He served on the President's National Advisory Committee on Oceans and Atmosphere from 1978 to 1984 and has served as the head of NCAR's Climate and Global Dynamics Division since 1987. A cofounder of the Black Environmental Science Trust (BEST), Washington has worked to advance opportunities for African Americans in environmental sciences. Among the societies in which Washington is a member are the American Association for the Advancement of Science and the African Scientific Institute. A widower, Washington has six children.

Watson, James Dewey

(1928–)
American
Geneticist, Biophysicist

James Watson collaborated with FRANCIS HARRY COMPTON CRICK to discover the key to human heredity by building a model of the molecular structure of deoxyribonucleic acid, or DNA. He won the 1962 Nobel Prize in medicine or physiology with Crick and MAURICE HUGH FREDERICK WILKINS for this revolutionary discovery. Watson spent the remainder of his career in genetic research, endeavoring to map all known genes, which number in the hundred thousands, for a period in the late 1980s.

Watson was born on April 6, 1928, in Chicago, Illinois, to Jean Mitchell and James Dewey Watson. He married Elizabeth Lewis in 1968 and together the couple had two sons, Rufus Robert and Duncan James. A child prodigy, Watson entered the University of Chicago at the age of 15 and received his bachelor's degree in zoology in 1947. He indulged an interest in ornithology at the University of Michigan's summer research station at Douglas Lake in 1946, but reading ERWIN SCHRÖDINGER's *What Is Life?* convinced him to pursue the genetic key to the mystery of the reproduction of life. In 1947 he commenced doctoral study under SALVADOR EDWARD LURIA at Indiana University, joining the Phage Group organized by Luria, MAX DELBRÜCK, and other prominent biologists. Watson wrote his disssertation on the effect of X rays on the rate of bacteriophage lysis, or bursting, to earn his Ph.D. in 1950.

In the next year he worked on a National Research Council fellowship grant to research protein molecular structure in Copenhagen, Denmark. He then moved to Cambridge, England, under the National Foundation for Infantile Paralysis, working at the Cavendish Laboratory from 1951 to 1953. There he partnered with Crick to search for the molecular structure of DNA. On the basis of X-ray diffraction photographs from Wilkins and ROSALIND ELSIE FRANKLIN and data on DNA's four organic nucleotides, or bases, from Erwin Chargaff, Watson and Crick followed the model-building strategy of LINUS CARL PAULING to construct a double-helix structure resembling two interlacing spiral staircases that conformed to existing information on DNA. They published their findings in the April–May 1953 issues of the British journal *Nature*. Watson won the coin toss to determine whose name would appear first.

In the summer of 1953 Watson commenced what would become a long-term relationship with the Cold Spring Harbor Laboratory of Quantitative Biology on Long Island in New York. He took over directorship of the lab in 1968 and shifted his full-time attention to this position in 1976 when he retired from Harvard University, where he had taught since 1955 and served as a professor of biology since 1961. In 1989 he became director of the Human Genome Project of the National Institutes of Health, mapping all known chromosomes, but he resigned within two years over policy disputes. Watson retired in 1993.

In 1965 Watson published *The Molecular Biology of the Gene,* the first widely adopted textbook on the topic. In 1968 he published his landmark personal and professional autobiography of the years 1951 through 1953, *The Double Helix: A Personal Account of the Discovery of the Structure of DNA.* He later collaborated with John Tooze and David Kurtz on the 1983 publication *The Molecular Biology of the Cell.* Aside from the Nobel Prize, recognition of Watson's career peaked in 1977 when Jimmy Carter presented him the Presidential Medal of Freedom.

Wegener, Alfred Lothar

(1880–1930)
German
Meteorologist, Geophysicist

Alfred Wegener articulated the first full-fledged theory of continental drift to explain the creation of continents and oceans, though Francis Bacon had noted the correspondence between the coasts of Africa and South America as early as 1620. Though Wegener presented compelling evidence of corresponding continental shelves, fossils, and flora and fauna between the continents separated by the Atlantic Ocean, he could not explain the mechanism controlling the drift. His hypothesis was the target of scientific derision until the 1960s, when the theory of plate tectonics confirmed the phenomenon of continental drift.

Wegener was born on November 1, 1880, in Berlin, Germany, to Anna and Richard Wegener. His father was a doctor of theology and the director of an orphanage. Wegener married Else Köppen, daughter of the eminent German meteorologist Wladimir Köppen, in 1912. Wegener attended the Universities of Heidelberg and Innsbruck before receiving his Ph.D. in astronomy in 1905 from the University of Berlin. His dissertation converted astronomical tables from the 13th century to decimal notation. He abandoned astronomy in favor of meteorology after graduation.

In 1906 Wegener fulfilled his life's dream of exploring Greenland, landing the position of official meteorologist on a Danish expedition studying polar air circulation that lasted until 1908. Upon his return he lectured in meteorology at the Physical Institute at the University of Marburg, where he remained until 1912. His clarity and candor—he readily admitted his distaste for mathematics, a sentiment many students likely shared—made his lectures popular. After World War I, during which he was wounded twice while serving in the German military as a junior officer, he took over his father-in-law's position at the German Marine Observatory at Gross Borstel near Hamburg as director of the meteorological research department. In 1924 the University of Graz created a special chair in meteorology and geophysics specifically for Wegener; he occupied it until his death in 1930.

In 1910 Wegener toyed with the idea of continental drift when, like Bacon, he noted the corresponding coastlines of Africa and South America. Evidence of paleontological correspondences that he happened upon in 1911 spurred his renewed interest, though these findings supported a land bridge theory that he rejected. He presented the seed of his own theory in the paper "The Geophysical Basis of the Evolution of the Large-Scale Features of the Earth's Crust (Continents and Oceans)" in Frankfurt on January 6, 1912, and again four days later in Marburg.

In 1915 Wegener published the book upon which his reputation rests, *The Origin of Continents and Oceans,* which fleshed out his theory of continental displacement, later termed *continental drift,* that originated in the late Paleozoic era (250 million years ago) when the one continent, which he called *Pangaea,* surrounded by one ocean, which he called Panthalassa, began to separate. Subsequent editions provided more evidence and reached more readers. However, the rejection of his theory demonstrated the degree to which the subjectivity of renowned scientists could eclipse the importance of scientific objectivity.

Wegener participated in three other expeditions to Greenland, in 1912 to 1913 to study glaciology and climatology, in 1929, and in 1930. He celebrated his 50th birthday on this final expedition, then set out on a trek from which he did not return. He died, apparently of heart failure, in November 1930 in Greenland.

Weinberg, Steven

(1933–)
American
Physicist

The physicist Steven Weinberg participated in developing a theory, which became known as the electroweak theory, that unified electromagnetic and weak interactions. Weinberg also predicted the existence in "neutral currents" of one of the three particles, or bosons, found in the weak force. When these bosons were detected in experiments conducted in the early 1980s, the electroweak theory was validated. For his participation in formulating the theory Weinberg shared the 1979 Nobel Prize in physics with his former classmate SHELDON LEE GLASHOW and the Pakistani physicist Abdus Salam, both of whom independently contributed to the electroweak theory.

Born on May 3, 1933, in New York City, Weinberg developed an interest in science at an early age. His parents, Frederick Weinberg and Eva Israel, encouraged Weinberg's scientific bent, as did his schoolteachers. Weinberg's father worked as a court stenographer. Weinberg married a law professor, Louise Goldwasser, in 1954; they had one child.

After finishing his high school education at the prestigious Bronx High School of Science in 1950, Weinberg entered New York's Cornell University to study physics. Glashow graduated from the same high school and attended Cornell as well. Weinberg earned his bachelor's degree in 1954 then spent one year studying at the Niels Bohr Institute in Copenhagen, Denmark, under the theoretical physicist Gunner Källén. Weinberg returned to the United States in 1955 and entered Princeton University to pursue graduate studies. After completing his dissertation on the subject of weak interactions, he earned his doctorate in 1957.

Weinberg developed his theory of electromagnetic and weak forces largely during the 1960s, when he worked at the University of California's Lawrence Berkeley Laboratories. It had been known for some time that there were four fundamental forces in nature—gravity, electromagnetism, the strong nuclear force, and the weak nuclear force. Efforts to combine these forces into a unifying theory persisted, and in the early 1960s these attempts were narrowed in focus to electromagnetic and weak forces. In 1967 Weinberg published *A Model of Leptons,* a paper that presented his theory. Four years later he suggested that when matter and neutrinos, electrically neutral subatomic particles, collided, a neutral current was generated. In this neutral current, Weinberg predicted, one of the three particles that produced the weak force could be found. In 1973 neutral weak currents were detected during experiments conducted at the European Center for Nuclear Research (CERN). Ten years later the physicist Carlo Rubbia observed the bosons predicted in the electroweak theory while working at CERN.

For his contributions to the electroweak theory Weinberg was awarded part of the Nobel Prize in physics in 1979. Weinberg continues to conduct research on a variety of topics, including muon physics, broken symmetries, and scattering theory. He has also devoted his time to the study of cosmology. Among his written works on cosmology is *Gravitation and Cosmology,* which he published in 1972. Weinberg left the Lawrence Berkeley Laboratories in 1969 to work at the Massachusetts Institute of Technology (MIT). In 1973 he joined the physics staff at Harvard University while also working at the Smithsonian Astrophysical Observatory. Since 1982 Weinberg has worked in the physics and astronomy departments at the University of Texas in Austin. In addition to the Nobel Prize, Weinberg has received numerous honors, including the American

Physical Society's 1977 Dannie Heineman Prize, the Franklin Institute's 1979 Elliott Cresson Medal, and the 1992 National Medal of Science. He was elected a member of the Royal Society of London in 1981.

Weller, Thomas Huckle

(1915–)
American
Physician, Virologist

A highly regarded virologist, Thomas Weller shared the 1954 Nobel Prize in physiology or medicine for cultivating the poliomyelitis, or polio, virus in tissue cultures. This discovery eventually led to the development of polio vaccines by JONAS EDWARD SALK. Weller's methods for growing cells in tissue cultures were widely adopted by scientists and greatly advanced laboratory studies of viruses. Weller made significant discoveries regarding many other viruses, including those that cause chicken pox and rubella.

Born on June 15, 1915, in Ann Arbor, Michigan, Weller was the son of Elisie A. Huckle and Carl V. Weller. Weller's father was the chair of the pathology department at the University of Michigan, in Ann Arbor. Weller married Kathleen Fahey in 1945 and had four children.

After attending the University of Michigan, where he earned his bachelor's degree in 1936 and his master's degree in 1937, Weller entered Harvard Medical School. There he met and developed a relationship with FREDERICK CHAPMAN ROBBINS, with whom he would later share the Nobel Prize. As a fellow of the Rockefeller Foundation in 1938 Weller traveled to Tennessee and Florida, where he studied public health and malaria, respectively. In 1940 Weller earned his medical degree with honors in parasitology. After fellowships in tropical medicine and bacteriology Weller went to the Children's Hospital in Boston for his internship. In 1941 he completed his internship and started his residency, hoping to become a pediatrics specialist. Weller's residency was delayed by the onset of World War II. He served in the U.S. Army's Medical Corps from 1942 to 1945, studying bacteriology and virology in Puerto Rico. In particular he researched pneumonia and schistosomiasis, a tropical disease affecting the intestines and many organs. After his discharge in 1945 Weller returned to the Children's Hospital, completed his residency, and started working with JOHN FRANKLIN ENDERS.

While working with the mumps virus in 1948, Weller had some leftover human tissue, so he and Enders attempted to cultivate the poliomyelitis virus in the tubes of tissue. This led to additional efforts by the team, which now included Robbins, to grow the virus. They used human foreskin and intestinal cells of a mouse, among other media, in their experiments and discovered substantial growth of the virus in the intestinal cells of a human. The use of antibiotics allowed Weller and his colleagues to isolate and analyze the virus. Not only did their work make the discovery of a polio vaccine possible, but it also made feasible, through Weller's techniques for cultivating cells, the increased detection and analysis of other types of viruses.

In addition to his breakthrough work growing the poliomyelitis virus, Weller was the first to cultivate the rubella, or German measles, virus in tissue cultures. He and his colleague Franklin Neva also grew the chicken pox virus, known as the varicella virus, from human tissue, and Weller also researched intestinal parasites. In 1949 he became the assistant director of the Children's Hospital's infectious diseases division. He also taught tropical medicine and tropical public health at Harvard University and in 1954 was appointed chair of Harvard's public health department. Weller remained at Harvard until 1985, when he became professor emeritus. In addition to the Nobel Prize, Weller received such honors as the Mead Johnson Prize in 1953, a prize he shared with Robbins. He joined the National Academy of Sciences in 1964 and served on numerous committees concerned with medical advancements, such as the World Health Organization and the National Institute of Allergy and Infectious Disease.

Werner, Abraham Gottlob

(1749–1817)
German
Mineralogist, Geologist

Abraham Werner became the foremost mineralogist and geologist in his day as the founder of the Neptunist school of thought, which supported the aqueous origin of all rocks, as opposed to the Plutonists' or Vulcanists' arguments for the igneous origin of rocks. Werner also rejected uniformitarianism, or the belief that geological evolution has been continuous and uniform, a theory that supplanted his theories after his death.

Werner was born on September 25, 1749, in Wehrau, Upper Lusatia (now Osiecznica, Poland). He had an older sister named Sophia, but he was the only son of Regina Holstein and Abraham David Werner, the inspector of the Duke of Solm's ironworks in Wehrau and Lorenzdorf. Werner developed an early interest in mineralogy, as his father educated him until he was nine years old. He then attended the Waisenschule in Bunzlau (now Boleslawiec, Poland) until his mother died in 1764, when his father took him on as his *hüttenschreiber*, or assistant, in the ironworks until 1769. That year Werner commenced two years of study at the Freiberg Mining Academy. He then pursued a traditional education at the University of Leipzig, starting in 1771, though his interest in mineralogy eclipsed his interest in law and language. He left the university without a degree in 1774.

That same year Werner published a book on fossils that prompted the Freiberg Mining Academy to offer him a position as a teacher of mining and as the curator of the mineral collection. He accepted and remained with the academy for the rest of his life, transforming it from a provincial school to an internationally acclaimed institution, mostly on the strength of his persuasive argument for the Neptunian theory of geological history, which commenced with a deluge of biblical proportions covering the whole of the Earth. This flood deposited five geological layers of different types of rock to form the surface of the Earth, according to Werner. However, Werner never traveled outside his home region, and although his system applied perfectly well to Saxony, by no means did it apply throughout the world, much less throughout Europe. Nevertheless the Neptunian theory held sway in his lifetime, in part because most of the prominent geologists of the day passed through the doors of his academy and fell under his influence.

As a mineralogist Werner developed a classification system that mitigated the dispute between categorization by external form versus chemical composition. Werner's proposal combined aspects of both sides by dividing minerals into four classes—earthy, saline, combustible, and metallic. His ideas spread not by publication, as he developed an aversion to writing later in his life, but rather by word of mouth in academic circles throughout Europe. Werner published only 26 scientific tracts, and late in life he took to leaving his mail unopened to avoid answering it. The Académie of Sciences elected him as a foreign member in 1812, though he did not learn of the honor until he read about it in a journal. Werner died on June 30, 1817, in Dresden, Germany.

Werner, Alfred

(1866–1919)
Swiss
Chemist

For his development of the coordination theory of chemistry, which greatly advanced the field of inorganic chemistry, Alfred Werner was awarded the 1913 Nobel Pize in chemistry. Werner's coordination theory allowed for the classification of inorganic compounds and presented ideas about how atoms and molecules were bonded together. The theory was formulated early in Werner's career, and he spent many subsequent years conducting experiments to verify and validate his work.

Born on December 12, 1866, in the small town of Mulhouse, France, Werner was the youngest of four children. Werner's father, Jean-Adam Werner, was an ironworker at the time of Werner's birth. His mother, Jean-Adam's second wife, was Salome Jeanette Tesche, the daughter of a prosperous German family. When Werner was four years old, Alsace, the province in which the Werners lived, was annexed to Germany as a result of the Franco-Prussian War. The Werners remained loyal to France, and the family spoke French in the home. When Werner was still a child, the family moved from the city to the country, where Werner's father took up dairy farming. Werner demonstrated an interest in chemistry from an early age, building a chemistry laboratory in the family's barn and studying chemistry at the Ecole Professionelle.

During his one year of compulsory military service Werner took some organic chemistry courses at the Technical University in Karlsruhe. After completing his military duty, Werner attended the Federal Institute of Technology in Zurich, Switzerland. He was an excellent chemistry student but did poorly in mathematics. After receiving a degree in technical chemistry in 1889, Werner continued his studies, and he earned his doctorate one year later. His dissertation, a significant work in organic chemistry, discussed the similarity in shape of nitrogen compounds and carbon compounds. Werner completed his postdoctoral work in Paris, where he studied thermochemical topics.

In 1892 Werner began teaching atomic theory at the Federal Institute of Technology. Though his career was just starting, it was in 1893 that Werner published his seminal work, *Contribution to the Construction of Inorganic Compounds,* which presented his coordination theory. The theory, which dealt with the large class of chemical compounds that involved the inorganic complexes of metals, suggested that single atoms or molecules could be joined and grouped around a central atom. The theory allowed for the prediction of the existence of unknown compounds and the explanation of the properties of known compounds. Werner's coordination theory catapulted him into the limelight, and in 1893 he became a professor of organic chemistry at the University of Zurich.

Werner subsequently spent nearly 25 years carrying out experiments to validate his theory. He prepared more than 8,000 compounds, and in 1907 he successfully prepared a compound predicted by his coordination theory—an ammonia-violeo salt. In 1911 Werner resolved a coordination compound into optical isomers, thereby providing further evidence to support his theory. Two years later he won the Nobel Prize in chemistry, the first Swiss chemist to receive the honor.

In addition to the Nobel Prize, Werner received numerous honors for his pioneering studies, including the French Chemical Society's Leblanc Medal. A prolific writer, he published more than 100 papers in organic and inorganic chemistry and completed several textbooks before his death in 1919. Werner married Emma Wilhelmine Giesker in 1894, the year he became a Swiss citizen. The couple had two children.

Wexler, Nancy Sabin

(1945–)
American
Psychologist, Medical Researcher

Nancy Wexler has turned a family tragedy into a new understanding of a deadly inherited disease. She was born on July 19, 1945, in Washington, D.C., to Milton Wexler, a psychoanalyst, and his wife, Leonore. In August 1968, when Nancy was 22 years old and her parents had been divorced for four years, her father told her and her older sister, Alice, that he had just learned that their mother was suffering from an inherited brain disorder, Huntington's disease (formerly Huntington's chorea). He explained that the disease, which usually does not reveal itself until middle age, produces a slow slide into insanity accompanied by uncontrollable twisting or writhing movements. There is no cure or treatment. Because the disease is caused by a single dominant gene, Nancy and Alice each had a 50–50 chance of developing it as well.

The news was devastating, but Milton Wexler said later that Nancy "went from being dismal to . . . wanting to be a knight in shining armor going out to fight the devils." Of a similar mind, Milton contacted the Committee to Combat Huntington's Chorea, an organization led by Marjorie Guthrie, who had been married to the folk singer Woody Guthrie, the disease's best-known victim. Milton opened a chapter of the group in Los Angeles, and Nancy set up another in Michigan, where she was studying psychology at the University of Michigan at Ann Arbor. She did her thesis research on the psychological effects of belonging to a family suffering from a hereditary disease and earned her Ph.D. in 1974.

Guthrie's organization was devoted mainly to improving care, but the Wexlers were more interested in research. In 1974 Milton Wexler, with Nancy's help, founded a new organization for this purpose, the Hereditary Disease Foundation. Nancy became the foundation's president in 1983.

The group agreed that identifying the gene that causes Huntington's was probably the best approach to a treatment or cure. At minimum doing so would create a test that showed who would get the disease before signs of it developed, to help people at risk plan their future. In October 1979 a researcher at the Massachusetts Institute of Technology, David Housman, told the foundation about a new technique that narrowed down the location of unknown genes by using known marker genes called restriction fragment length polymorphisms (RFLPs). The more often a certain form of an RFLP was inherited with a certain form of an unknown gene in a given family, the more likely the unknown gene was to be near the RFLP on the same chromosome.

The only problem was that at that time only one human RFLP was known. Finding a RFLP that happened to be near the Huntington's gene might take decades. Still the Hereditary Disease Foundation gave Housman a grant to try his idea, and Nancy Wexler arranged additional funding through the Congressional Committee for the Control of Huntington's Disease and Its Consequences, of which she had been made executive director in 1976.

To determine inheritance patterns, the researchers needed a large family in which some members had Huntington's disease. Luckily Nancy Wexler, investigating another aspect of the disease, had learned of such a family in Venezuela and visited them earlier in 1979. In 1981 she and an international research team made the first of what became yearly trips to collect skin and blood samples from the family for testing. The Venezuelans cooperated when they learned that Wexler's own family had the disease and she, too, had given samples. In return for the family's help Wexler's team gave them medical and social aid (the family is very poor), and Wexler personally provided what a Venezuelan team member called "immeasurable love."

At first the RFLP project had unusually good luck. Dr. James Gusella of Massachusetts General Hospital, who was in charge of the testing, found a RFLP inherited with the Huntington's gene in 1983. It was only the 12th RFLP he had tried. In addition to giving a great boost to the gene hunt, this identification made possible a test that would indicate with about 96 percent accuracy whether someone would develop Huntington's disease.

To improve the chances of locating the Huntington's gene, the Hereditary Disease Foundation, beginning in 1984, persuaded six laboratories in the United States and Britain to share their results and credit. John Minna, a scientist in the group, told Wexler's sister, Alice, "The person that made all of the HD [Huntington's disease] thing work . . . was Nancy. . . . It was her acting as . . . glue and go-between, doing whatever was necessary, that was the real key." The group finally found the Huntington's gene in 1993 and is now learning how it does its deadly work.

Nancy Wexler's other personal contribution has been tracing the ancestry of the Venezuelan family. Their family tree now spans 10 generations, including some 14,000 members, and covers both walls of a corridor outside Wexler's office at the Columbia University Medical Center, whose faculty she joined in 1985. (She became a full professor in 1992.) Wexler won the Albert Lasker Public Service Award in 1993 for her contributions to the search for the Huntington's gene and for "increasing awareness of all genetic disease."

In the 1990s Wexler's work focused on the implications of testing for inherited diseases. In 1989 she was made head of a committee that oversees research on the ethical, social, and legal questions raised by the Human Genome Project, which aims to work out the "code" of the complete collection of human genes. Information from this project eventually will make testing for any inherited disease or genetic weakness possible. Such testing could yield medical benefits, but test results could also be used to deny people medical insurance or employment. Wexler has tried to find ways to prevent such tragedies, just as she has worked to create hope out of the tragedy that struck her own life.

Wheeler, Anna Johnson Pell

(1883–1966)
American
Mathematician

Anna Johnson Pell Wheeler was an expert in linear algebra as well as an outstanding teacher and administrator. She was born Anna Johnson, the third child of Swedish immigrants, Andrew and Amelia Johnson, on May 5, 1883, in Hawarden, Iowa. She grew up in Akron, where her father was a furniture dealer and undertaker.

Anna Johnson earned her B.A. from the University of South Dakota in 1903. There a professor, Alexander Pell, encouraged her to do graduate study in mathematics. She earned master's degrees from the University of Iowa in 1904 and Radcliffe College in 1905. She then won a Wellesley fellowship to study at the renowned University of Göttingen, Germany, in 1906–1907. Pell joined her in Göttingen in 1907, and they were married there.

Anna Pell earned a Ph.D. from the University of Chicago in 1910. From 1911 to 1918 she taught at Mount Holyoke in South Hadley, Massachusetts, after which she moved to Bryn Mawr. In 1925, four years after Alexander Pell's death, Anna became a full professor and head of Bryn Mawr's mathematics department, succeeding CHARLOTTE ANGAS SCOTT. She married a classics scholar, Arthur Leslie Wheeler, in that same year and moved to Princeton to be with him, teaching part-time at Bryn Mawr until his death in 1932 and then returning to full-time work.

In addition to her fame as a teacher and administrator, Wheeler was well known as a research mathematician. Her specialty was linear algebra of infinitely many variables, a branch of what is now called functional analysis, and its applications to differential and integral equations. In 1927 the American Mathematical Society invited her to give its Colloquium Lectures; she was the only woman so honored until 1980. She served on the society's council and board of trustees.

Anna Wheeler retired in 1948 and died on March 26, 1966, just before her 83rd birthday. According to the science historians Judy Green and Jeanne Laduke, Wheeler "received more recognition from the mathematical community before World War II than perhaps any other American woman."

Whipple, Fred Lawrence

(1906–)
American
Astronomer

Fred Whipple asserted a theory of comet dynamics in 1950 that competed with several other theories until the appearance of Halley's comet in 1986 confirmed the validity of Whipple's model. Whipple's devotion to their study yielded the discovery of six previously unknown comets.

Whipple was born on November 5, 1906, in Red Oak, Iowa, to Celestia MacFarland and Henry Lawrence Whipple, both farmers. Whipple married Dorothy Woods in 1927, and they had one son, Earle Raymond, before their 1935 divorce. On August 20, 1946, Whipple married Babette Frances Samuelson, and they had two daughters, Dorothy Sandra and Laura.

Whipple attended Occidental College from 1923 to 1924 before transferring to the University of California at Los Angeles (UCLA), where in 1927 he received a degree in mathematics. He moved north to the University of California at Berkeley to pursue his doctorate in astronomy, which he received in 1931. He then accepted a position at the Harvard College Observatory. Harvard appointed him as a professor of astronomy in 1945, then named him the Philips Professor of Astronomy in 1950, a post he retained for a quarter of a century. He accepted a simultaneous appointment as the director of the Smithsonian Institution's Astrophysical Observatory in Cambridge in 1955 until his 1973 retirement, when he accepted the title of Senior Scientist.

In 1942 Whipple published *Earth, Moon, and Planets,* which went through three editions and was regarded as the quintessential text on the solar system. Whipple's influence was felt even more significantly through his theories about comets. To conduct research on comets, he devised a "two-station" method of photography using cameras with rotating shutters, a setup that allowed him to determine a comet's or a meteor's trajectory, atmospheric drag, velocity, and solar orbit. Using this method, he first theorized that meteors were visible debris related to comets. He then suggested a new theory of comets.

In 1950 Whipple advanced what became known as the "dirty snowball" theory of comet makeup. Essentially a comet is a ball of ice, Whipple hypothesized, surrounded by ammonia, methane, and cosmic dust—hence "dirty." When the comet is in the far reaches of space, it is frozen hard, but as it nears the Sun, it heats to the depth of the gases in the nucleus. This effect causes the gases to erupt, releasing a jet of energy from the nucleus of the comet. The energy affects the motion of the comet, depending on the direction of the eruption, increasing or decreasing the velocity of the comet or shifting its trajectory to one side or the other. Whipple's snowball theory elegantly accounted for the reasons why the flight of comets did not obey Newtonian laws.

The return of Halley's comet in 1986 confirmed Whipple's dirty snowball theory. The West African republic of Mauritania commemorated the event with a postage stamp bearing the likeness of Whipple. Whipple's contributions were previously recognized by a 1948 Presidential Certificate of Merit for his wartime work on "confusion reflectors," or tiny pieces of foil that jammed German radars. He received the American Astronautical Society's 1960 Space Flight Award for developing a method for tracking satellites in the wake of the Soviet's surprise launching of *Sputnik*. He was awarded the Donahue Medal not once but six times, once for each previously unknown comet he discovered.

Whipple, George Hoyt

(1878–1976)
American
Pathologist

An enthusiastic teacher and talented experimenter, George Hoyt Whipple shared the 1934 Nobel Prize in physiology or medicine for his hemoglobin research. Whipple's experiments demonstrated that the consumption of liver relieved the effects of anemia, and his work soon led to the development of a liver treatment of pernicious anemia.

Born on August 28, 1878, in Ashland, New Hampshire, Whipple was a member of a family with a medical legacy. His father, Ashley Cooper Whipple, was a doctor, as was his grandfather. Whipple's mother was the former Frances Anna Hoyt. Whipple's father died when Whipple was only two years old, and he was raised by the female members of the family.

After attending the prestigious Phillips Academy in Andover, Massachusetts, Whipple entered Yale College (now Yale University) to study medicine. An outstanding student, Whipple was also an exceptional athlete, and he participated in such sports as baseball, rowing, and gymnastics. Whipple earned his bachelor's degree in 1900 and in 1901 began medical school at Johns Hopkins University. Though Whipple had considered specializing in pediatrics, he chose to focus on pathology after earning his M.D. in 1905 and took a job as an assistant in pathology at the university. One of Whipple's early achievements took place while he was an assistant—during an autopsy Whipple identified an intestinal tissue condition that later became known as Whipple's disease.

Studies of liver damage and the function of bile eventually led Whipple to investigate the role of hemoglobin in 1911. He understood there were links among bile, hemoglobin, and the liver, and in 1913 he demonstrated that the breakdown of hemoglobin could produce bile pigments. Whipple extended his studies of hemoglobin in California, where he moved in 1914 to head the University of California in San Francisco's Hooper Foundation for Medical Research. He was accompanied by a medical student, Charles W. Hooper, from Johns Hopkins and joined by an assistant, Frieda Robscheit-Robbins. In a series of experiments Whipple and his team bled laboratory dogs to induce anemia. After feeding the dogs different foods to observe the effects on hemoglobin, they discovered that a diet of liver seemed to cause an increase in the production of hemoglobin.

In 1921 Whipple accepted a position at the University of Rochester in New York. There he and Robscheit-Robbins continued their hemoglobin research and honed their techniques. They were able to induce long-term anemia in dogs and became convinced that a diet of liver effectively boosted hemoglobin regeneration. In 1925 Whipple announced the findings, and soon a liver extract was available on a commercial basis. A year later two physicians, GEORGE RICHARDS MINOT and WILLIAM PARRY MURPHY, developed a liver treatment of pernicious anemia, then a fatal disease. Minot, Murphy, and Whipple shared the Nobel Prize in 1934.

Whipple's later studies included the investigation of the role of iron in the human body and research involving the use of tissue proteins and plasma to restore hemoglobin to treat anemia. A tireless researcher and educator, Whipple continued teaching at the University of Rochester until 1955, when he was 77 years old. During his career he published more than 200 works on such subjects as anemia, livery injury, and pigment metabolism. Among the honors Whipple received were the 1939 Kober Medal and the 1962 National Academy of Sciences's Kovalenko Medal. Whipple married Katharine Ball Waring in 1914 and had two children. He died on February 1, 1976, in Rochester, New York, at the age of 98.

White, Gilbert

(1720–1793)
English
Naturalist

Gilbert White's scientific knowledge stemmed almost exclusively from personal experience and observation of the gardens and forests surrounding his family home, The Wakes, in Selborne, England. White decided to forgo more prestigious positions to accept a modest post as curate in his home parish, freeing him to concentrate on his naturalist journal and letter writing to friends about the local flora and fauna. These friends prevailed upon him to edit his observations for publication, and he reluctantly agreed in 1789; the result, *The Natural History and Antiquities of Selborne*, was hailed as a poetic masterpiece. The text went through about 200 reprints in English and was translated into several other languages.

White was born on July 18, 1720, in Selborne, in Hampshire, England. His parents were Anne Holt, the daughter of a wealthy clergyman, and John White. White attended school in Farnham and Basingstoke before commencing study at Oriel College of Oxford in 1740, where he earned his bachelor's degree in 1743. The next year he was elected a fellow of Oxford, a relationship he maintained throughout his life. In October 1746 he earned his master's degree from Oxford, and in 1747 he received his deacon's orders as curate at Swarranton, the parish of his uncle, Reverend Charles White. In 1750 the bishop of Hereford ordained White into priesthood. However, White refused all appointments save that at Moreton Pinkney, Northhamptonshire, which was salaried by Oxford and was maintained by the neighboring curate, allowing him to remain at The Wakes.

White remained a bachelor, devoting himself singularly to his naturalist observations. His observations of birds exceeded those of most ornithologists of the time, who limited their interest to plumage and anatomy. White listened to bird songs, thus drawing the distinctions among three British leaf warblers that had identical plumage, which made them seem to be the same bird, though in fact they had three different calls. White noted the strange habit of cuckoos' laying their eggs in other birds' nests. He also followed the migration of swallows. He identified the harvest mouse as Britain's smallest mammal and recognized that the noctule bat, previously seen only in continental Europe, was an inhabitant of Britain. Other subjects he touched on included phrenology, agriculture, weather, archaeology, and astronomy—in short, all natural phenomena within his bounded geographic scope.

White corresponded with, among others, Thomas Pennant and Daines Barrington, who urged him to publish his observations. He resisted until four years before his death, when he finally relented by gathering together 110 letters and editing them into the text of *The Natural History and Antiquities of Selborne*. Other of his writings were published posthumously, namely, his "Garden Kalender," which appeared in 1900, and his "Calender of Flora, 1766," which appeared in 1911. Further excerpts from his "Naturalist's Journal" were published in 1931 as *Gilbert White's Journals*. Consistent with his humility in life, White's gravestone read simply, "G. W. June 26, 1793." He died where he was born and lived, in Selborne. Edmund Gosse eulogized White as "a man who has done more than any other to reconcile science with literature."

Wiesel, Torsten Nils

(1924–)
Swedish/American
Neurophysiologist

For important contributions that furthered the understanding of the organization and function of the brain, specifically the role of the visual cortex in mammals, Torsten Nils Wiesel shared the 1981 Nobel Prize in physiology or medicine with his colleague DAVID HUNTER HUBEL and the neurobiologist ROGER WOLCOTT SPERRY. Wiesel and Hubel studied the visual, or striate, cortex and found that the visual system was much more complex than had been traditionally believed. Their investigations also uncovered the developmental stages of the visual cortex and demonstrated that there was a critical phase in the development of the visual system.

Born on June 3, 1924, in Uppsala, Sweden, Wiesel was the child of Anna-Lisa Bentzer and Fritz S. Wiesel. Wiesel's father worked as the head psychiatrist at Stockholm's Beckomberga Mental Hospital, and Wiesel lived at the hospital during his childhood. He attended a private school, where he enjoyed sports more than academics. Wiesel married Teiri Stenhammer in 1956, but the couple divorced in 1970. Wiesel was remarried three years later to Grace Yee, and they had one child. In 1981 this second marriage dissolved in divorce.

In 1941 Wiesel entered the Karolinska Institute in Stockholm to study medicine. Studying both neurophysiology and psychiatry, he earned his medical degree in 1954. He then began teaching at the institute. Wiesel also took a position as an assistant at Karolinska Hospital's Department of Child Psychiatry. A year after graduation Wiesel traveled to the United States for postdoctoral work at Johns Hopkins University's School of Medicine in Baltimore, Maryland.

Wiesel's studies of the visual system began at Johns Hopkins, where he worked under Stephen Kuffler. Kuffler was interested in visual physiology and had researched the nerve activity of the retina in various animals, particularly cats. Kuffler's studies had demonstrated that there was a significant difference between the visual system of mammals and that of nonmammalian species, and this finding influenced Wiesel to undertake research of mammalian brains, where crucial visual processes occurred. Wiesel was joined in his research by Hubel in 1958, and a year later they followed Kuffler to Harvard Medical School. As part of their experiments Wiesel and Hubel used microelectrodes to assess the electrical impulses of cells in the visual cortex of cats. They were able to determine that different cortical cells responded to different patterns and levels of light. In further investigations Wiesel and Hubel used radioactive amino acids to trace the pathway of vision from the retina of the eye to the visual cortex, uncovering the important function of neurons in the visual system. Prior to the team's exhaustive studies the scientific community accepted that all cortical cells were basically the same. Wiesel and Hubel's work greatly advanced the understanding of how the deeply complex visual system processes information.

Wiesel and Hubel also investigated the development of the visual system and the effect of visual impairments on the stages of development. Experimenting on kittens, the team discovered that though mammals were born with a full visual cortex, impairments during the critical developmental stage negatively affected the visual system. Early impairments, if not corrected immediately, would lead to permanent cortical processing deficits. These findings greatly influenced the treatment of eye disorders, particularly in pediatric patients.

After more than 20 years of collaboration Hubel and Wiesel's partnership ended when Wiesel accepted a job directing Rockefeller University's neurobiology laboratory in 1984. Wiesel became a U.S. citizen in 1990, and two years later he became president of Rockefeller University. Since 1994 he has served as chairman of the National Academy of Science's Committee on Human Rights, and in 1995 he became chairman of the board of the Aaron Diamond AIDS Research Center. That same year he married Jean Stein.

In addition to the Nobel Prize, Wiesel received such honors as the 1972 Lewis S. Rosentiel Award, the 1975 Jonas S. Friedenwald Memorial Award, the 1977 Karl Spencer Lashley Prize, and the 1978 Louisa Gross Horowitz Prize.

Wiles, Andrew John

(1953–)
English
Mathematician

After spending years in seclusion working on Fermat's last theorem, the most famous of all mathematical problems, Andrew Wiles gained instant fame when he announced his discovery of a proof in 1993. The problem dates back to 1637, when the French number theorist Pierre de Fermat scribbled an equation in the margin of his copy of Diophantus's *Arithmetica,* along with the comment that a proof written in the margin of the book, as was his habit, would not fit. The equation was related to the Pythagorean theorem, replacing the exponent 2 with n. Fermat boldly stated that there existed no solution for n for whole numbers greater than 2. He demonstrated this fact for the number 4, and in 1780 LEONHARD EULER demonstrated it for the number 3. Mathematicians followed this method for centuries, essentially hoping to find a number that would solve the equation and thus disprove the theorem. However, when computer programs tested numbers up to 4 million, it became clear that disproving the theory by this method would not work, as the theorem could be correct, and other strategies would be required. Wiles devoted his career to one of these strategies.

Wiles was born in April 1953. His father was a professor of theology. At the age of 10 Wiles became fascinated with Fermat's last theorem, and he spent his adolescence toying with the equation until he later realized that the solution had stumped eminent mathematicians for over three centuries for good reason. Wiles studied mathematics, focusing on elliptical curves, at Clare College of Cambridge University; he earned a master's degree in 1977 and a Ph.D. in 1980. Elliptical curves turned out to be the key to the mystery of the theorem.

Upon earning his doctorate in 1980, Wiles moved to the United States to teach mathematics at Princeton University. In 1988 the London Mathematics Society awarded him the Whitehead Prize, allowing him to fill the position of Research Professor in Maths and Professorial Fellow at Merton College of Oxford. In 1990 he returned with his wife and two daughters to Princeton, where he continued his seven-year odyssey in search of the mathematical grail.

In 1954 the Japanese mathematician Yutaka Taniyama proposed the Taniyama conjecture, which suggested that for every elliptical curve there existed a related function with specific properties. The German mathematician Gerhard Frey proposed in 1985 that a solution to Fermat's equation would relate to a specific set of elliptical curves. Identifying this set, called Frey's curves, would disprove Fermat's last theorem. In 1986 the American mathematician Kenneth Ribert proved the viability of Frey's approach, focusing Wiles's ideas on the potential solution to Fermat's last theorem.

Wiles worked in seclusion in his attic, unaided by a computer. In 1991 he overcame his first major hurdle, then in May 1993 a paper by Harvard's Barry Mazur included a mathematical construction that proved to be the last key to Wiles's solution. Finally, on June 23, 1993, under a shroud of secrecy, Wiles announced at his third lecture to a small mathematics conference in Cambridge, England, his proof of Taniyama's conjecture, and, by extension, of Fermat's last theorem. He submitted his 200-page proof to a board of mathematical experts, who identified minor gaps in the theory that Wiles subsequently filled. After a year-long review Wiles's proof was accepted for publication, validating its legitimacy.

Wilkins, J. Ernest, Jr.

(1923–)
American
Physicist, Mathematician

A skilled mathematician and engineer, J. Ernest Wilkins, Jr., worked in numerous fields, including nuclear engineering. Early in his career Wilkins worked on the Manhattan Project, which was formed during the World War II years to develop a nuclear bomb. Wilkins worked in private industry as well as in academia, teaching at such schools as Howard University and Tuskegee Institute (now Tuskegee University). A prolific writer, Wilkins has published about 100 works on a diverse number of topics, including nuclear reactor design, applied mathematics, and optics. His work on gamma-ray penetration advanced nuclear reactor design and neutron absorption studies.

Born on November 27, 1923, in Chicago, Illinois, Wilkins grew up in a prominent Methodist family. Wilkins's mother, Lucile Beatrice Robinson, had a master's degree in education and worked as a teacher; his father, J. Ernest Wilkins, Sr., was a successful Chicago lawyer. In the early 1940s Wilkins's father was the president of the Cook County Bar Association, and in the early 1950s he was appointed assistant secretary of labor by President Eisenhower, making him the first African American to serve a subcabinet post in the federal government. Wilkins was a bright child, able to solve basic arithmetic problems by the age of three.

Though Wilkins's two brothers chose to study law, Wilkins was more interested in mathematics, and at the phenomenally young age of 13 he entered the University of Chicago. He earned his bachelor's degree in 1940 and a year later received his master's degree. In 1942 at the age of 19 Wilkins was awarded his doctorate. He then traveled to Princeton University to study at the Institute for Advanced Study. More than a decade later Wilkins returned to school and earned two additional degrees—a bachelor's and a master's degree in mechanical engineering from New York University.

After a brief period of teaching at Tuskegee, Wilkins returned to the University of Chicago and began working on the Manhattan Project through the school's Metallurgical Laboratory. In 1946 he began his career in industry, accepting a job as mathematician at the American Optical Company in New York. There he developed and refined optical instruments, working to improve the lenses of telescopes to enhance the view of distant stars. In 1947 he married Gloria Stewart, with whom he had two children. Wilkins married a second time in 1984.

In 1950 Wilkins once again took up nuclear physics when he began working for the Nuclear Development Corporation of America (later United Nuclear Corporation). The company designed nuclear power plants as well as propulsion systems for submarines and ships. Few in the nuclear power industry possessed knowledge of both nuclear physics and plant and operational design, and Wilkins considered this a detriment. Though he worked full-time, Wilkins chose to return to school, gaining degrees in mechanical engineering to further his understanding of how to design nuclear power plants. Shortly after earning his master's degree in 1960, Wilkins moved to San Diego, California, to work at the General Dynamics Corporation's department of theoretical physics.

In 1970 Wilkins returned to academia and began teaching mathematics and physics at Howard University in Washington, D.C. After six years he took a year's leave, which he spent at the University of Chicago's Argonne National Laboratory. In 1977 Wilkins revisited industry, accepting a job at EG&G Idaho, Inc., a firm that designed nuclear power plants and explored the uses of nuclear power. In 1984 he became an Argonne Fellow and spent one year at Argonne National Laboratory before retiring. After about five years of retirement Wilkins became Distinguished Professor of Applied Mathematics and Mathematical Physics at Clark Atlanta University.

Wilkins has been involved in numerous societies and organizations, including the American Nuclear Society, of which he was president from 1974 to 1975, and the National Academy of Engineering, which he joined in 1976. He also served on committees dedicated to finding peaceful uses of atomic power.

Wilkins, Maurice Hugh Frederick

(1916–)
British
Biophysicist

Maurice Wilkins won the 1962 Nobel Prize in medicine or physiology with JAMES DEWEY WATSON and FRANCIS HARRY COMPTON CRICK for their discovery of the molecular structure of deoxyribonucleic acid (DNA). Specifically Wilkins contributed the X-ray diffraction photography used to create the model of DNA.

Wilkins was born on December 15, 1916, in Pongoroa, New Zealand, to Eveline Constance Jane Whittaker and the physician Edgar Henry Wilkins, both Irish immigrants. Wilkins married Patricia Ann Chidgey in 1959, and together the couple had four children—two daughters and two sons.

From the age of six Wilkins attended King Edward's School in Birmingham, England, and at St. John's College at Cambridge University he earned his B.A. in physics in 1938. Wilkins advanced to the University of Birmingham, where he earned his Ph.D. in 1940 with a dissertation on the electron-trap theory of phosphorescence and thermoluminescence.

While he was pursuing his doctorate the Ministry of Home Security and Aircraft Production had already corraled Wilkins to contribute to the preparations for war with research on how to improve radar screens. Wilkins transferred his wartime efforts to the University of California at Berkeley, where he participated in the Manhattan Project separating uranium isotopes for the atomic bomb.

John T. Randall invited Wilkins to lecture in physics at St. Andrews University in Scotland after the war in 1945. Wilkins's reading of ERWIN SCHRÖDINGER's *What Is Life? The Physical Aspects of the Living Cell* prompted him to consider the intersection between physics and biology, or biophysics.

The next year Randall established the Medical Research Council Biophysics Unit at King's College in London with Wilkins's assistance. This assistant's role became official in 1955, when Wilkins was appointed deputy director of the unit. He then became director between 1970 and 1972. Two years later he served as the director of the Medical Research Council's Neurobiology Unit, a position he retained until 1980. Wilkins served simultaneously at King's College as a professor of molecular biology from 1963 to 1970, when he transferred to the biophysics department as its head; in 1981 he became a professor emeritus.

In 1946 researchers at the Rockefeller Institute in New York identified DNA as the genetic building blocks of life. Wilkins devoted himself wholeheartedly to the study of DNA, specifically working to determine its structure. Wilkins discovered that DNA lent itself perfectly to X-ray diffraction photography. In 1951 the X-ray diffraction expert ROSALIND ELSIE FRANKLIN joined Wilkins, though the pair disagreed over the apparent helical appearance of DNA. Some reports suggest that Wilkins's sharing of X-ray photographs with Watson and Crick occurred without Franklin's knowledge. Nevertheless Wilkins and Franklin's X-ray photographs proved vital to the discovery of the structure of DNA. After this discovery Wilkins devoted his efforts to understanding the structure of DNA's relative, ribonucleic acid, or RNA, which he showed to have a helical structure like DNA's.

Besides the Nobel Prize, Wilkins's achievements have been recognized with the 1960 Albert Lasker Award from the American Public Health Association as well as his 1959 election as a fellow of the Royal Society of King's College. More importantly, Wilkins served his political conscience as the president of the British Society for Social Responsibility in Science.

Wilkinson, Sir Geoffrey

(1921–1996)
English
Chemist

Best known for his work on inorganic complexes, Sir Geoffrey Wilkinson greatly advanced chemists' understanding of chemical structure through his studies of metallocenes, or sandwich molecules. Wilkinson investigated the structure of a compound that became known as ferrocene and determined that it consisted of a single iron atom sandwiched between two five-sided carbon-based rings, thereby uncovering a new class of chemical structure. This discovery led to the creation of numerous sandwich compounds and essentially founded a new field of chemistry. Wilkinson also made advances in nuclear chemistry, and his work on metal-to-hydrogen bonding had commercial and industrial applications, in addition to contributing to organic and inorganic chemistry. For his pioneering work on metallocenes Wilkinson shared the 1973 Nobel Prize in chemistry.

Born on July 14, 1921, in Todmorden, England, to Ruth Crowther and Henry Wilkinson, young Wilkinson was interested in chemistry; that interest was encouraged and nurtured by an uncle who owned a chemical firm in his hometown.

After graduating from Todmorden Secondary School, Wilkinson attended the University of London's Imperial College of Science and Technology. He earned his bachelor's degree in 1941 and immediately embarked on graduate studies. In 1942 Wilkinson traveled to Canada to work on the Atomic Energy Project, funded by the National Research Council. While studying the fission process and the products of atomic fission reactions, Wilkinson developed a new method, known as ion-exchange chromatography. With the technique he discovered many new isotopes. After earning his doctorate in 1946, Wilkinson took a research position at the University of California at Berkeley and continued his investigation of nuclear chemistry, focusing particularly on studying neutron-deficient isotopes. In 1950 he became a research associate at the Massachusetts Institute of Technology, where he shifted his research from nuclear chemistry to transition elements, the elements in the center of the periodic table. In 1951 he married Lise Solver Schau, and the couple had two children.

It was during his four years at Harvard University, which began in 1951, that Wilkinson uncovered the previously unknown chemical structure of sandwich molecules. He read about a newly synthesized compound that eventually was called ferrocene. With his colleague ROBERT BURNS WOODWARD he correctly proposed that the structure of the compound was made up of a metal atom sandwiched between two carbon rings of five sides each. Wilkinson went on to synthesize many other sandwich compounds.

In the 1960s, after returning to Imperial College in 1955 to head the inorganic chemistry department, Wilkinson continued his research of transition elements and began to study the complexes containing metal-hydrogen bonds. One of the compounds he investigated became known as Wilkinson's catalyst and was the first complex of its kind to be used as a homogeneous catalyst for hydrogenation. This class of compound was also used as a catalyst for hydroformylation and had industrial applications.

In 1988 Wilkinson became professor emeritus of the Imperial College. In addition to his 1973 Nobel Prize, which he shared with ERNST OTTO FISCHER, who had worked independently on similar research, Wilkinson received a number of honors and awards, including the 1968 French Chemical Society's Lavoisier Medal, the Royal Society's 1972 Transition Metal Chemistry Award, and the 1973 Galileo Medal presented by the University of Pisa in Italy. In 1976 Wilkinson was knighted. A prolific writer, Wilkinson published more than 400 works and wrote *Advanced Inorganic Chemistry,* a standard textbook. He died on September 26, 1996.

Williams, Anna Wessels

(1863–1954)
American
Microbiologist

Anna Williams made advances in identification and treatment of infectious diseases that saved thousands of lives, though men were often given credit for her work. The second child of six, she was born on March 17, 1863, in Hackensack, New Jersey, to William and Jane Williams. Her father was a teacher and school official. Anna wrote that she first became interested in science when she looked through a teacher's "wonderful microscope."

Williams earned a certificate from the New Jersey State Normal School in 1883 and began teaching school. In 1887, however, her sister almost died while giving birth to a dead baby. Wanting to prevent similar tragedies, Williams decided to become a physician. She earned an M.D. from the Women's Medical College of the New York Infirmary in 1891 and remained at the school as an instructor in pathology for several years.

In 1894 Williams joined the new New York City Department of Health Laboratory, the country's first city-run laboratory for diagnosis of disease. Its chief job was identifying disease-causing microbes in tissue, blood, food, or drink. A volunteer at first, Williams was hired in 1895 as an assistant bacteriologist. A biography of the laboratory's director, W. H. Park, said that Williams "became more indispensable . . . with each year of service."

The laboratory also carried out research on infectious diseases, and Williams made her greatest contributions in this area. Scientists had found a way to make an antitoxin that counteracted the toxin, or poison, produced by the bacteria that caused a serious disease called diphtheria, which killed thousands of children yearly. Making the antitoxin required toxin, however, and finding a strain of bacteria that produced large amounts of toxin dependably was hard. In 1894 Williams identified such a strain, which became known as the Park-Williams strain—or just the Park 8 strain, although Park had not even been in the laboratory when Williams found it. It became the standard strain used for diphtheria toxin production.

Williams's second important discovery involved rabies. At the time the only test for this fatal disease took two weeks. In 1904, while examining infected brain tissue under a microscope, Williams noticed certain cells that did not appear in normal brains. An Italian physician, Adelchi Negri, made the same discovery a short time later. The cells became known as Negri bodies rather than Williams bodies because Negri published his description before Williams did. They became the basis of a much faster test for rabies. Williams improved on Negri's version of the test, producing one that gave results in half an hour rather than several hours. Her form became standard.

Williams studied many other infectious diseases, including scarlet fever, typhoid fever, pneumonia, and influenza, and coauthored several books about disease-causing microorganisms with Park. She became assistant director of the New York laboratory in 1906. In 1932 she became chairperson of the Laboratory Section of the American Public Health Association, the first woman to hold such a post. She retired in 1934 and moved to New Jersey, where she lived until her death on November 20, 1954.

Williams, Oswald S.

(1921–)
American
Aeronautical Engineer

Oswald S. Williams, or "Ozzie," as he was known, helped pave the way for African-American scientists to excel within a segregated, racist society. In the 1950s he led the group that developed the first airborne radio beacons for locating crashed planes. In the 1960s he headed the effort at Grumman International to develop the reaction control subsystem that guided the Apollo lunar module. In the 1970s he filled a more administrative position at Grumman, negotiating trade and industrial relations with emerging African nations, specifically providing them with the technological tools necessary to harness solar and wind energy. In the 1980s he went to work as a professor of marketing at St. John's University in Queens, New York.

Williams was born on September 2, 1921, in Washington, D.C. His mother was Marie Madden; his father, after whom he was named, was a postal worker. The family moved to New York City when he was young, and in 1938 he graduated from Boys High School of Brooklyn. He matriculated at New York University, where the dean discouraged him from studying something as specialized as aeronautical engineering because of his race. Williams persisted despite this warning, earning his bachelor's degree in 1943. That year he married Doris Reid, and together the couple had three children: Gregory, who died in 1982; Bruce; and Meredith. Williams proceeded to earn his master's degree in aeronautical engineering in 1947.

During World War II Williams served as the senior aerodynamicist for the Republic Aviation Corporation. On the basis of wind tunnel experiments he calculated important formulas, such as those for calculating the lift of airplane wings, the propelling force necessary to fly, and the drag on the plane once it was airborne. He helped design the P47 Thunderbolt, which was instrumental in the war in escorting America's high-altitude bombers. After the war and the awarding of his master's degree he worked as a design draftsman for Babcock and Wilcox Co. in 1947. Between 1948 and 1950 he was a technical writer for the United States Navy Material Catalog Office.

In 1950 Greer Hydraulics, Inc., hired Williams as an engineer, and he served as the group project leader in developing the first experimental airborne radio beacon for pinpointing crashed planes. His challenge was to design an instrument that would send a signal strong enough to be detected wherever the plane landed: underwater, at high altitude, obscured by mountain ranges, and so on. In 1956 he joined the Reaction Motors Division of Thiokol Chemical Corp. to design small rocket engines.

Grumman International hired Williams in 1961 as a propulsion engineer. He headed the eight-year, $42 million effort to develop the reaction control subsystem for the Apollo lunar module. He then moved on to work in Grumman's marketing department. In 1974 Grumman appointed Williams as a vice president in charge of relations with emerging African nations. Williams not only helped Grumman establish business links in these new markets, but also helped these nations with emerging economies equip themselves to take care of their basic energy needs with renewable resources, such as solar and wind power.

In 1981 Williams earned an M.B.A. from St. John's University, where he went on to teach marketing. Thus at the end of his career Williams demonstrated his excellent business sense, although the focus of his early career had been more purely scientific, developing new technologies for aeronautical use. Besides doing ground-breaking aeronautical engineering work, Williams was a bold pioneer for African-American professionals.

Wilson, Charles Thomson Rees

(1869–1959)
Scottish
Physicist

C.T.R. Wilson won the 1927 Nobel Prize in physics with Arthur H. Compton for the invention of the Wilson cloud chamber, used in the study of radioactivity, X rays, cosmic rays, and the scattering of high-energy photons.

Wilson was born on February 14, 1869, near Glencorse, in Midlothian, Scotland. His father, John Wilson, who conducted important experiments in sheep farming, died when Wilson was four years old. His mother was the elder Wilson's second wife, Annie Clark Harper, from a Glasgow family of thread manufacturers. In 1908 at the age of 39 C.T.R. Wilson married Jessie Frank Dick, and together the couple had one son and two daughters.

At the age of 15 Wilson attended Owens College (later renamed Victoria College of Manchester); he graduated at 18 with a first-class degree in zoology. He won a scholarship to Sidney Sussex College at Cambridge University to study physics and chemistry. After teaching briefly at Bradford Grammar School in Yorkshire, Wilson returned to Cambridge in 1894, and in 1896 he won the Clerk Maxwell Studentship for three years.

In 1899 the Meteorological Council appointed Wilson as a researcher on atmospheric electricity. The next year he was elected a fellow of Sidney Sussex College and appointed as a university lecturer and demonstrator. In 1911 the Solar Physics Observatory named Wilson an observer in meteorological physics; he was promoted to the position of reader in electrical meteorology in 1918. Between 1925 and 1934 he was the Jacksonian Professor of Natural Philosophy at Cambridge. Wilson then spent an active retirement, continuing to conduct research well into his 80s. In fact, in 1956 he presented his paper "A Theory of Thundercloud Electricity" to the Royal Society and became the oldest fellow of the society ever to make a presentation.

In 1895 Wilson contructed his first cloud chamber, which was based on an idea that originated during a hike to the top of Ben Nevis, Britain's highest peak, in September 1894, when he witnessed spectecular optical phenomena through the clouds that he wished to recreate in the laboratory. When X rays were discovered in 1895, Wilson promptly applied them to his study of cloud formation. In 1900 he demonstrated that ions are present in air, and he was able to distinguish between positively charged and negatively charged ions in his cloud chamber. His crowning achievement with the chamber was his photography of the vapor trails left by positively charged alpha rays in 1911. It was mainly on the strength of this experiment that Wilson won the Nobel Prize.

Wilson's career was highly decorated with awards besides the Nobel Prize. In 1900 the Royal Society elected him as a fellow, and he received three subsequent prizes from the society: the 1911 Hughes Medal, the 1922 Royal Medal, and the 1935 Copley Medal. After his long and active career Wilson died on November 15, 1959, in Carlops, Pebbleshire, Scotland.

Wong-Staal, Flossie (Yee-ching Wong)

(1946–)
Chinese/American
Geneticist, Medical Researcher

Flossie Wong-Staal has been a leader in research on acquired immunodeficiency syndrome (AIDS) since the disease and the virus that causes it were first identified.

She was born Yee-ching Wong in Kuangchou, or Canton, China, on August 27, 1946. Her father, Sueh-fung Wong, was a cloth exporter-importer. In 1952, not long after Communists took over the Chinese government, the Wong family moved to Hong Kong. Yee-ching became Flossie after the American nuns at her school asked students' parents to select English names for them, and her father chose the one given to a typhoon (tropical storm) that had just swept through the city.

Flossie's teachers assigned her to study science, and she found she liked it. When she attended college at the University of California at Los Angeles (UCLA), she specialized in molecular biology, the study of the structure and function of the chemicals in living things. This relatively new field was filled with excitement as scientists deciphered the genetic code. She graduated in 1969.

In 1971, while working for her Ph.D. at UCLA, Wong married a medical student, Steven Staal. They later had two daughters, Stephanie and Caroline. Wong-Staal, as she called herself after her marriage, earned her degree in 1972, when she won the Woman Graduate of the Year award.

Both Staal and Wong-Staal found jobs at the National Institutes of Health (NIH), the group of government-sponsored research institutes in Bethesda, Maryland, in 1973. Wong-Staal worked in the laboratory of Robert Gallo at the National Cancer Institute. At the time Gallo was one of the few researchers investigating the possibility that viruses could cause cancer in humans. Wong-Staal found genes in human cells similar to ones in viruses that caused cancer in monkeys and apes. Gallo has written that she "evolve[d] into one of the major players in my group. Because of her insight and leadership qualities, she gradually assumed a supervisory role."

In 1981 Gallo's group identified the first virus proven to cause a human cancer, a rare form of leukemia. They named it *human T-cell leukemia virus* (HTLV); T cells are one type of immune system cell in the blood. They later found a second, similar virus, which they called *HTLV-2*.

Meanwhile other scientists were studying a strange illness that had appeared mostly among homosexual men in large cities. They eventually called it acquired immunodeficiency syndrome, or AIDS, because it devastated its victims' immune systems. Gallo and Wong-Staal were struck by the fact that the as-yet-unknown agent that caused AIDS chiefly attacked T cells and was transmitted through sexual contact or transfer of blood or from mother to fetus—just as HTLVs are.

While checking out the possibility that one of their viruses caused the disease, Gallo's group found a previously unknown virus similar to theirs. They named it HTLV-3. Scientists in France's Pasteur Institute isolated what proved to be the same virus at the same time, late 1983. The two labs have shared credit for discovering the virus, which came to be called the human immunodeficiency virus, or HIV.

Gallo's lab then turned to a full-time study of HIV. In 1984 Flossie Wong-Staal became the first to clone the virus's genes. She also worked out the chemical sequence of each gene and determined its function. Most of the genes controlled the speed of the virus's growth. Wong-Staal's work helped to explain why most people infected with HIV remain seemingly healthy for years before development of AIDS.

In 1990 Wong-Staal, by then divorced from Steven Staal, left Gallo's lab and moved with her younger daughter to the University of California at San Diego. There she heads a center devoted to AIDS research. One approach to treatment she has investigated inserts a gene coding for a chemical called a ribozyme into immune system cells. The ribozyme snips HIV's genetic material into several pieces, preventing it from reproducing.

One sign of Wong-Staal's fame is the number of times that researchers cite her work in their own papers. A 1990 survey conducted by the Institute of Scientific Information in Philadelphia showed that Wong-Staal's papers were cited more than those of any other woman scientist during the 1980s.

Woodward, Robert Burns

(1917–1979)
American
Chemist

An outstanding chemist who is largely recognized as the founder of modern organic synthesis, Robert Burns Woodward successfully synthesized a great number of complicated organic substances, including quinine, cortisone, cholesterol, and vitamin B_{12}. For his significant achievements Woodward received the Nobel Prize in chemistry in 1965. He also, with his colleague ROALD HOFFMANN, developed the theory of the conservation of orbital symmetry, which advanced the understanding of various chemical reactions.

Born on April 10, 1917, in Boston, Massachusetts, Woodward was the child of Arthur Woodward and Margaret Burns. Woodward's father died when Woodward was a young child. Woodward displayed an affinity for chemistry at a young age, and he often experimented with his home chemistry set. Woodward married twice—the first marriage was to Irji Pullman in 1938. Together the couple had two children. In 1946 Woodward married Eudoxia Muller, with whom he had two children.

After graduation from high school at the early age of 16 Woodward entered the Massachusetts Institute of Technology (MIT) in 1933 to study chemistry. Though his college career had an inauspicious start—he nearly flunked out because of lack of interest in some of the compulsory classes and rules—Woodward earned his doctorate after only four years, one year after receiving his bachelor's degree. After completion of his postdoctoral work at Harvard University, Woodward began teaching there.

In the early 1940s Woodward began his efforts at synthesizing organic compounds, and in 1944 he succeeded, with his colleague William E. Doering, in synthesizing quinine, a feat chemists had been attempting to accomplish for more than 100 years. Quinine was used to treat malaria, and its successful synthesis had positive implications during the World War II era. To synthesize quinine, Woodward and Doering developed a procedure with 17 steps. Because each step in the synthetic process rearranged the atomic order or changed the chemical groups, Woodward's experiments produced just a half-gram of quinine, though he started with five pounds of materials. The successes followed quickly after the quinine achievement.

In 1947 Woodward linked amino acids into a chain molecule to produce a synthetic protein. In 1951 he completed the first total synthesis of the steroids cortisone and cholesterol. Among the many substances synthesized by Woodward and his research team were reserpine, which was used in the treatment of mental illness; strychnine, a toxin often employed in the killing of rats; chlorophyll; and a tetracycline antibiotic. In 1971 Woodward completed the synthesis of vitamin B_{12}, a coenzyme vitamin. The synthesis required more than 100 steps and led Woodward and his colleague Hoffmann to propose the concept of the conservation of orbital symmetry. Their theory detailed the circumstances under which some organic reactions are able to occur, and the resultant Woodward-Hoffmann rules simplified the process of predicting chemical reactions.

Though Woodward spent his entire career at Harvard University, he was involved in numerous organizations and societies. In 1963 the CIBA-Geigy Corporation established the Woodward Institute in Zurich, Switzerland, in his honor, and Woodward traveled there frequently to work on projects. In addition to the 1965 Nobel Prize, Woodward received the Royal Society of Britain's 1959 Davy Medal and 1978 Copley Medal and the 1964 U.S. National Medal of Science. Woodward helped establish two journals in organic chemistry. There is speculation that had Woodward lived long enough, he would have shared the 1981 Nobel Prize in chemistry that was awarded to Hoffmann for the orbital theory.

Wrinch, Dorothy Maud

(1894–1976)
British/American
Mathematician, Chemist

Bridging barriers between scientific disciplines, Dorothy Wrinch helped to found molecular biology with her mathematics-based theory about the structure of protein molecules. She was born on September 12, 1894, in Rosario, Argentina, to an engineer, Edward Wrinch, and his wife, Ada. One account says that when Dorothy's father first took her to school in Rosario, he told her teachers, "This child is to be a mathematician." Her parents returned to England while she was still a child, and she grew up in London. Wrinch attended Girton, a women's college in Cambridge University, from 1913 to 1918 and earned a B.A. and an M.A. She studied both mathematics and philosophy. After graduation she taught mathematics at the University of London. She earned an M.S. in 1920 and a D.S. in 1922.

In 1922 Wrinch married John Nicholson, a mathematical physicist at Oxford, and they had a daughter, Pamela, in 1928. Unusual for the time, Wrinch continued to work under her maiden name. She obtained an M.A. from Oxford in 1924 and began tutoring full-time at Lady Margaret Hall, one of the university's colleges for women. In 1927 she was appointed a lecturer for three years, the first woman granted a university lectureship in mathematics at Oxford. She used a group of her papers in mathematical physics and applied mathematics as the basis for a second D.S. degree, the first that Oxford had awarded to a woman, in 1929. Wrinch separated from Nicholson, an alcoholic, in 1930, and they divorced in 1938.

Wrinch contributed to many fields during her life, including several branches of pure and applied mathematics, sociology, biology, and chemistry. She did her most important work, development of the first precise theory of the structure of proteins, in the late 1930s. At that time scientists thought that understanding the structure of proteins would reveal "the secret of life" because proteins were believed to carry inherited information. Attempts to understand protein structure led to establishment of molecular biology, the study of the structure and function of chemicals in living things, which now dominates much of biological laboratory research.

Wrinch proposed that proteins were made up of six-sided units called cyclols, which interlocked to form two-dimensional sheets. This theory—and Wrinch's sometimes strident insistence on it—aroused great controversy, with Nobel Prize–winning scientists taking both sides. It could explain much of what was known about the behavior of proteins, but, despite Wrinch's claims to the contrary, little direct evidence supported it.

Wrinch was a visiting scholar at Johns Hopkins University from 1939 to 1941 and hoped to remain in the United States, but the contention about her theory made finding a permanent position hard. Otto Glaser, a biologist at Amherst College in Massachusetts, helped her in her aim, and they married in 1941. Wrinch became a U.S. citizen in 1943 and began teaching full-time at Smith College in 1944.

Wrinch's later achievements included the book *Fourier Transforms and Structure Factors* (1946), which described the mathematics of X-ray crystallography. She spent most of her time, however, trying to prove her cyclol theory. Research in the early 1950s confirmed that cyclols existed in nature, but they did not prove to be the key to protein structure that Wrinch had hoped. Furthermore, by that time attention had shifted to nucleic acids, when they, rather than proteins, were shown to be the carriers of inherited information. Wrinch retired from Smith in 1971 and died in February 1976 at the age of 82.

Wu, Chien-Shiung

(1912–1997)
Chinese/American
Physicist

"This small, modest woman was powerful enough to do what armies can never accomplish," a reporter from the *New York Post* wrote of Chien-Shiung Wu in 1957. "She helped destroy a law of nature." In a difficult, painstaking experiment Wu proved that what had been considered a basic law of physics did not always hold true.

Chien-Shiung Wu was born in Liu-ho, a town near Shanghai, China, on May 29, 1912. Her father, Wu Zong-yee, was a school principal. Unlike most Chinese of his day, he believed that girls should receive the same education as boys.

When Chien-Shiung studied physics in high school, she says, "I soon knew it was what I wanted to go on with." She continued her education at the National Central University in Nanjing, from which she graduated in 1936. She decided to do her graduate work in the United States and enrolled in the University of California at Berkeley, then "the top of the world," as she put it, for anyone interested in the interior of atoms. Wu earned her doctorate in 1940. In 1942 she married Luke Yuan, a fellow physicist whom she had met at Berkeley, and they had a son, Vincent, in 1945.

Wu's research specialty became beta decay, in which a neutron in the nucleus of a radioactive atom spontaneously breaks apart, releasing a fast-moving electron (a beta particle) and a second particle called a neutrino, which has no mass or charge. A proton is left behind, automatically converting the atom into an atom of a different element.

After teaching physics at Smith College for a year, Wu (who continued to use her maiden name professionally) moved to Columbia University in March 1944 and joined the Manhattan Project, the research project that led to the creation of the atomic bomb. Her work consisted chiefly of looking for ways to produce more of the radioactive form of uranium. She also improved the design of Geiger counters, which detect radiation.

Wu remained at Columbia after the war, rising to the rank of associate professor in 1952 and continuing her studies of beta decay. She became a U.S. citizen in 1954. She was known as a fair but exacting teacher (she once said, "In physics . . . you must have total commitment. It is not just a job. It is a way of life") and a meticulous experimentalist. Indeed a fellow scientist said of her, "She has virtually never made a mistake in her experiments." This latter reputation drew two other Chinese-born physicists, CHEN NING YANG and TSUNG-DAO LEE, to her late in 1956, to ask for help in finding out whether a groundbreaking idea they had was correct.

Some materials in nature are either "right-handed" or "left-handed"; in other words, they are mirror images of each other. For instance, particles in the atomic nucleus can spin either clockwise or counterclockwise. Since 1925 physicists had believed that physical reactions would be the same whether the particles involved in them were right-handed or left-handed. This was called the law of conservation of parity. When one form of radioactive decay appeared to violate the parity law, most physicists thought that an error must have occurred, but Yang and Lee made the shocking suggestion in mid-1956 that the parity law might not hold true in weak nuclear interactions, which include radioactive decay.

Wu decided to test Yang and Lee's idea with an experiment that used radioactive cobalt, or cobalt-60. A strong electromagnetic field could make the cobalt atoms line up, just as iron filings do near a magnet, and spin along the same axis. Wu could then count the number of beta particles thrown from their nuclei in different directions as the atoms decayed. If the parity law held true for weak interactions, the number of particles thrown off in the direction of the nuclei's spin would be the same as the number thrown the opposite way. If the law did not hold, the numbers would differ.

There was only one problem: At normal temperatures the cobalt atoms would move around too much to line up under the magnet. The experiment therefore had to be done at almost absolute zero (-459.67°F or -273.15°C), the temperature at which all atomic motion due to heat stops. Wu took her work to the National Bureau of Standards in Washington, D.C., the only laboratory in the country where material could be cooled to such a low temperature. During the testing as she repeated the experiment many times, Wu often got only four hours of sleep a night. "It was . . . a nightmare," she said later. "I wouldn't want to go through [it] again."

At the end of the nightmare, however, the results were clear: Far more beta particles flew off in the direction opposite the nuclei's spin than in the direction that matched the spin. The law of parity therefore did not apply to weak interactions. This finding, which Wu announced on January 16, 1957, caused a great sensation. Lee and Yang won a Nobel Prize in 1957 for their theory. Wu was not so honored, but in 1958 she became a full professor at Columbia and was elected to the National Academy of Sciences.

Wu remained for the rest of her career at Columbia, where she was, as Emilio Segrè, one of her former professors at Berkeley, put it, the "reigning queen of nuclear physics." She continued to make important discoveries and tests of others' theories and to win awards for her work, including the Comstock Award of the National Academy of Sciences and the National Medal of Science, the country's highest science award. She retired in 1981.

Soon after her parity experiment Wu told a group of students, "It is the courage to doubt what has long been established, the incessant search for its verification and proof, that pushed the wheel of science forward." Wu has made major contributions to pushing that wheel. She died on February 16, 1997 at St. Luke's–Roosevelt Center in New York City after a stroke.

Y

Yalow, Rosalyn Sussman
(1921–)
American
Physicist, Medical Researcher

Rosalyn Yalow and her research partner, Solomon Berson, invented a technique called radioimmunoassay that measures substances in body fluids so accurately that reporters have said it could detect a lump of sugar dropped into Lake Erie. For this advance Yalow won a Nobel Prize in 1977.

Yalow was born Rosalyn Sussman on July 19, 1921, in the South Bronx area of New York City, to Simon and Clara Sussman, who had grown up in the city's immigrant community. Simon Sussman owned a small paper and twine business, which earned just enough money for his family to live on. "If you wanted something, you worked for it," Yalow recalled.

By the time Rosalyn was eight, she had decided that she was going to be a "big deal" scientist and marry and have a family to boot. She went to Hunter College, which charged no tuition to New York City residents. There her attention turned to physics because, as she says, "In the late thirties . . . nuclear physics was the most exciting field in the world." She graduated with high honors in January 1941.

As a Jewish woman with little money Sussman had three strikes against her in trying to gain admission into a graduate school or medical school. At first she thought her only hope was to work as a secretary at Columbia University Medical School, which would allow her to take classes there free. Impending war drained universities of men and created new openings for women, however, and Sussman obtained a teaching assistantship in physics at the University of Illinois in Champaign-Urbana. On her first day of classes in the fall of 1941 she met another Jewish New Yorker, a rabbi's son from Syracuse named Aaron Yalow. They married on June 6, 1943, and later had two children. From the start, Yalow supported his wife's career.

Once the Yalows finished their Ph.D.'s in 1945—Rosalyn was only the second woman to earn a physics doctorate from Illinois—they returned to New York. Rosalyn became an engineer at the Federal Telecommunications Laboratory, the first woman to hold this position; when the laboratory moved away a year later, she began teaching at Hunter. Hunter had no research facilities, however, and she wanted to do research. Aaron suggested that she look into the new field of medical physics, which focused mostly on radioisotopes, radioactive forms of elements that were usually made in the atom-smasher machines invented during the war.

Rosalyn Yalow consulted Edith Quimby, a pioneer researcher in medical physics at Columbia, and Quimby in turn introduced Yalow to her boss, Gioacchino Failla. On Failla's recommendation the Bronx Veterans Administration (VA) Hospital hired Yalow in December 1947. Her laboratory, one of the first radioisotope laboratories in the United States, was a former janitor's closet, and she had to design and build most of her own equipment. Until 1950 she continued to teach at Hunter.

In the fall of 1950 Yalow found her ideal professional partner in a young physician named Solomon A. Berson. Their collaboration lasted 22 years. One coworker told Sharon McGrayne that Yalow and Berson had "a kind of eerie extrasensory perception. Each knew what the other was thinking. . . . Each had complete trust and confidence in the other."

In one of their first studies, published in 1956, Yalow and Berson used radioisotope tagging to show that the immune systems of diabetics, who must have daily injections of the hormone insulin to make up for their body's lack of it, formed substances called antibodies in response to the insulin they took, which was from cows or pigs and thus was slightly different from human insulin.

This finding was startling enough—researchers had believed that insulin molecules were too small to produce such a response—but more importantly, Berson and Yalow realized that they could turn their discovery on its head to create a very sensitive way of measuring insulin or almost any other biological substance in body fluids. They injected the substance they wanted to test for into laboratory animals, making the animals produce antibodies to it. They then mixed a known amount of these antibodies with a known amount of the radioactively tagged substance and a sample of the fluid to be tested.

Antibodies attach to molecules of the substance that caused their formation. The nonradioactive substance in the sample attached to some of the antibodies, preventing the radioactive substance from doing so. After a certain amount of time Yalow and Berson measured the radioactive material not attached to the antibodies. The more substance had been in the sample, the more radioactive material would be left over. This test, the radioimmunoassay, can detect as little as a billionth of a gram of material. In 1978 *Current Biography Yearbook* called it "one of the most important postwar applications of basic research to clinical medicine."

Yalow accumulated many honors for her work, including the American Medical Association's Scientific Achievement Award and election to the National Academy of Sciences in 1975. In 1976 she became the first woman to win the Albert Lasker Basic Medical Research Award, often considered a prelude to a Nobel Prize in physiology or medicine. A year later she finally won the "Big One"—the Nobel Prize itself. It was the first time that the surviving member of a partnership was honored for work done by both. (She shared the prize with two other researchers who had made discoveries about hormones in the brain.) Yalow was the second woman (after GERTY THERESA RADNITZ CORI) to win a Nobel Prize in physiology or medicine and the first American-born woman to win any science Nobel Prize. In 1988 she also won the National Medal of Science, the highest science award in the United States.

Yalow retired in 1991, but in 1998 she was still working at her office at the Veterans Administration Hospital and doing some research, although her health had begun to fail. During the 1990s she gave many lectures on such subjects as nuclear power, which she feels is unjustly feared; the need for better science education in the United States; and the need for more women scientists. "The world cannot afford the loss of the talents of half its people," she said in her Nobel Prize acceptance speech.

Yang, Chen Ning

(1922–)
Chinese/American
Physicist

A theoretical physicist who worked on the theory of weak interactions, Chen Ning Yang is best known for disproving the principle of conservation of parity, a law that had long been accepted by physicists as universal and fundamental. Yang's hypothesis, which he developed with his collaborator TSUNG-DAO LEE in the mid-1950s, included suggested experiments for proving his ideas. These experiments were carried out by CHIEN-SHIUNG WU and validated Yang and Lee's work. For their important theoretical breakthrough Yang and Lee shared the 1957 Nobel Prize in physics.

Born on September 22, 1922, in Hofei, in the Anhwei province in China, Yang was the child of Meng Hwa Loh and Ke Chuan Yang, a mathematics professor. When Yang was a child, the family moved to Peking, where Yang's father had been offered a teaching position at Tsinghua University. Influenced by a biography of Benjamin Franklin that he had read, Yang took on "Franklin" as his first name. In 1937 with the invasion of the Japanese army Tsinghua University was merged with National Southwest Associated University and moved to Kunming. Yang's family thus moved once again, to Kunming.

For his undergraduate studies Yang attended the National Southwest Associated University, where he earned his physics degree in 1942. Yang then enrolled at Tsinghua University, where his father still taught, and gained his master's degree in 1944. Before embarking on his doctoral studies, Yang taught high school for one year. A fellowship carried Yang to the United States in 1946, as there were no doctoral programs in physics in China at that time. Hoping to study under the physicist ENRICO FERMI, Yang attended the University of Chicago, where he studied nuclear physics under EDWARD TELLER. It was in Chicago that Yang began his long collaboration with Lee, a fellow student from National Southwest Associated University with whom he had not been well acquainted. After earning his doctorate in 1948, Yang was an assistant to Fermi for a year before going to Princeton, New Jersey, to work at Princeton University's Institute for Advanced Study. In 1950 he married Chih Li Tu, and together they had three children.

Yang and Lee began developing their theories about parity conservation in the 1950s. For some time Yang had been interested in weak interactions, the forces believed to cause some decay of elementary particles. Yang and Lee decided to explore the subject further, influenced by information about the decay of the K-meson, a subatomic particle. The principle of parity conservation, which arose in 1925, held that the physical laws of nature remained unaltered in mirror-image systems. The K-meson, however, seemed to decay into two different configurations. Some believed there were two types of K-mesons, but Yang and Lee proposed that there was only one class of K-meson, and that it decayed in two ways—one that conserved parity and one that did not. They published their ideas in 1956 in *Question of Parity Conservation in Weak Interactions*, and soon Columbia University's Wu began conducting the experiments outlined in Yang and Lee's paper. By 1957 their hypothesis was confirmed.

Yang also made advances in statistical mechanics and developed a theory, known as the Yang-Mills theory after Yang and his collaborator R. L. Mills, that described interactions for elementary particles and quantum fields. In 1965 Yang took a position at the State University of New York at Stony Brook. In addition to the 1957 Nobel Prize, he has received the 1957 Einstein Award, the 1980 Rumford Medal of the American Academy of Arts and Sciences, the 1986 Liberty Award, and the National Medal of Science.

Yener, Kutlu Aslihan

(1946–)
Turkish/American
Archaeologist

Aslihan Yener's discovery of mines and metal-refining sites has changed researchers' understanding of metalworking and trade in the ancient Middle East. She has also developed a new method of determining the chemical composition of artifacts. She was born on July 21, 1946, in Istanbul, Turkey, and her parents, the businessman Reha Turkkan and his wife, Eire Guntekin, took her to the United States when she was just six months old. She and her younger sister grew up in New Rochelle, New York, where, she says, she "almost lived at the Natural History Museum."

Yener studied chemistry at Adelphi University in Garden City, New York, but after a few years she "got the travel bug" and went to Turkey. She transferred to Robert College (later Bosphorus University) in Istanbul in 1966 and changed her major to archaeology. After her graduation in 1969 she studied at Columbia University from 1972 to 1980 and earned a Ph.D.

Yener first did chemical analyses of lead and silver in objects from the ancient Middle East to determine which mines they came from. This knowledge helped her form a picture of trade in the area. She then turned to tin, an important metal because it was both relatively rare and essential for making bronze. Bronze, an alloy of copper and tin, was used to make most metal objects between about B.C. 3000 and 1100, a period now called the Bronze Age.

In 1987, while Yener was an associate professor in history at Bosphorus University (a post she held from 1980 to 1988) as well as working for the Turkish Geological Research and Survey Directorate, she found the remains of a Bronze Age tin mine in the Taurus Mountains, which in ancient times were part of an area called Anatolia. Lead and silver mines had been found there before, but ancient writings had claimed that tin was imported from much farther east. Yener's 1989 discovery of a city where tin ore had been refined, at a site called Goltepe near her mine, provided confirming evidence that Anatolia had been an important source of tin. Her findings have changed archaeologists' understanding of trade in this period, when complex civilizations and international trade were starting to appear.

Yener has worked for the Oriental Institute of the University of Chicago since 1994 and is now an associate professor there. In early 1998 she announced a new technique for analyzing the chemical composition of ancient objects, which uses Argonne Laboratories' Advanced Photon Source, a device that passes high-energy X rays through objects and acts as a "chemical microscope." The technique can determine the composition of different parts of the same object, revealing how objects were made and mended. "This is the most important scientific development in archaeology since the discovery of radiocarbon dating," the most widely used method of estimating the ages of ancient artifacts, Yener claimed in a press release.

York, James Wesley, Jr.

(1939–)
American
Theoretical Physicist

James Wesley York, Jr., focused his research on gravity and particle physics, specifically considering the complex gravitational fields of black holes. He applied statistical mechanics to these self-gravitating systems and utilized quantum theory to explain their properties. Along with colleagues he discovered a new form of the Einstein equations and applied them to his work in gravity physics, particle physics, and relativistic astrophysics.

York was born on July 3, 1939, in Raleigh, North Carolina. He married Betty Mattern in 1961, and together the couple had two children, a daughter named Virginia and a son named Guilford. In 1962 he earned his bachelor of science degree from North Carolina State University (NCSU), where he remained to conduct his doctoral work. While he was writing his dissertation in 1965, NCSU hired York as an assistant professor of physics. The next year he earned his Ph.D. in physics, and he remained on the faculty for two more years.

Princeton University hired York in 1968 as a research associate in physics; the university promoted him the next year to a lectureship, a position he held for one year. In 1970 Princeton again promoted York, to the position of assistant professor, which he held until 1973. York served as a visiting assistant professor at the University of Maryland concurrently in 1972. In 1973 he returned to his home state as an associate professor at the University of North Carolina at Chapel Hill (UNC), where, with the exception of numerous visiting professorships, he remained throughout the rest of his career.

From 1974 on, York served as a principal investigator for the National Science Foundation. In 1976 he traveled to the Sorbonne as a visiting professor at the University of Paris. The next year he worked as a visiting scientist at the Harvard-Smithsonian Center for Astrophysics. York twice relocated to the University of Texas as a visiting professor, in 1979 and in 1987. From 1984 to 1990 he directed the Institute of Field Physics. Meanwhile at UNC he rose from the status of associate professor to full professor between 1973 and 1988. In 1989 UNC named York the Agnew H. Bahnson Jr. Professor of Physics, a chair he retained thereafter.

York published extensively in his career. In 1999 he collaborated with Yvonne Choquet-Bruhat on *The Cauchy Problem of General Relativity.* A glimpse at the titles of some of his journal articles reveals his degree of specialization, as he wrote not for a general audience but for fellow physicists: "Gravitational Wave Extraction and Outer Boundary Conditions by Perturbative Matching," published in a 1998 edition of *Physical Review Letters,* for example, or "Conformal 'Thin-Sandwich' Data for the Initial-Value Problem of General Relativity," also published in *Physical Review Letters* in 1999.

York advanced the understanding of physics by reconciling the vast with the minuscule. By understanding the phenomena that apply to black holes across the universe, York sought to understand further the phenomena that apply to particles throughout the universe, especially those that compose the earthly environment.

Thomas Young, whose principle of the interference of light boldly questioned, and subsequently replaced, Newton's corpuscular theory of light. *AIP Emilio Segrè Visual Archives.*

Young, Thomas

(1773–1829)
English
Physicist

Thomas Young was so bold as to challenge Newtonian physics with his principle of the interference of light, which reestablished the wave theory of light from the previous century and called into question Newton's corpuscular theory. Though scientists scoffed at the notion of refuting Newton, Young had devised simple yet elegant experiments to back up his views. Nevertheless Young's theory was not accepted until confirmed by the French physicists Augustin J. Fresnel and François Arago. Young also proposed the notion of tricolor vision, later accepted as the Young-Helmholtz three-color theory, after L. F. Von Helmholtz.

Young was born on June 13, 1773, in Milverton, Somerset, in southwest England. His father was a banker, but Young was raised by his grandfather. He attended medical school at the Universities of Edinburgh, Göttingen, and Cambridge. On June 13, 1773, Young married Eliza Maxwell, of Scottish aristocracy. In 1799 he established a medical practice in London that met with moderate success, though his mind focused more on his research than on his patients. Young had begun experimentation early, and he made important conjectures in some of his first publications.

Young published his paper "Observations on Vision" as early as 1793, when he was merely 20 years old. In it he argues that the eye lens is a muscle, adjusting its shape to focus vision. In his 1796 doctoral dissertation he retracted this hypothesis in response to new research by Everard Home, John Hunter, and Jesse Ramsden, which suggested that focus was controlled not by the lens, as subjects with removed lenses could still focus, but by changes in the curvature of the cornea and the elongation and shortening of the eyeball itself. However, Young reverted to his original hypothesis in 1800, as reported in his article "On the Mechanism of the Eye," which stated that accommodation, or the change of focus, was achieved by changes in the curvature, not of the cornea, but of the lens. In order to demonstrate one of his points, Young described and measured his own astigmatism. Still working with optics, Young proposed in 1801 that the eye creates vision of all colors through three "principle colors," the primary colors red, yellow, and blue. Young restated this theory in 1807; later Helmholtz confirmed it.

In December 1809 Young was elected into the Royal College of Physicians, and in January 1811 he was appointed a physician at St. George's Hospital, a position that would allow him to continue his experimentation on scientific issues broader than just medicine or physiology. Outside optics Young made many contributions to scientific understanding. In hydrodynamics he measured the surface tension of liquids, and in physics he measured the size of molecules and described the function of the constant in the equation describing elasticity in Hooke's law; this refinement earned the constant the title *Young's modulus*.

Young demonstrated the interference of light by passing light onto a screen through two pinholes set close to one another. He observed alternating bands of light and darkness, thus proving the wave theory of light. Young followed up on this assertion of the wave theory by proposing in 1817 the transversal nature of light waves, vibrating perpendicularly to the direction of travel, as opposed to the accepted longitudinal view of light waves as vibrating in the direction of travel. Young died on May 10, 1829, in London.

Yukawa, Hideki

(1907–1981)
Japanese
Physicist

For mathematically predicting the existence of mesons, subatomic particles that possess masses between those of the electron and the proton, Hideki Yukawa became the first Japanese to receive a Nobel Prize. Yukawa's theory, which he developed in the mid-1930s by using quantum theory, was substantiated more than a decade later, when the physicist Cecil Frank Powell discovered the predicted particle in cosmic rays. For his meson theory and his pioneering studies of elementary particles Yukawa was awarded the 1949 Nobel Prize in physics.

Born Hideki Ogawa on January 23, 1907, in Tokyo, Japan, Yukawa changed his name in 1932, when he married and adopted the family name of his wife, Sumi Yukawa. The couple had two children.

Hideki Yukawa was one of seven children born to Koyuki and Takuji Ogawa. His father worked at Tokyo's Geological Survey Bureau when Yukawa was born but soon became a geology professor at Kyoto Imperial University (later Kyoto University). Yukawa was heavily influenced by his grandmother, a former teacher at a samurai school, who taught Yukawa to read and write at an early age. After graduating from the prestigious Third High School in Kyoto, where he was a classmate of SHINICHIRO TOMONAGA, who would later become a Nobel laureate as well, Yukawa entered Kyoto University to study physics. Yukawa developed his interest in physics during his high school years and went so far as to teach himself the German language so he could understand the works of the German physicist MAX PLANCK. Yukawa stayed at Kyoto University after receiving his master's degree in 1929, first working as a laboratory research assistant then becoming a lecturer. In 1933 he moved to Osaka Imperial University (later Osaka University) to teach and pursue his doctorate, which he earned in 1938.

At Osaka University, Yukawa developed his groundbreaking meson theory after investigating the force that held neutrons and protons together in the atomic nucleus. WERNER KARL HEISENBERG, one of the founders of quantum theory, had determined that only two particles, protons and neutrons, made up the nucleus. Since only protons possessed a positive charge, it seemed a nucleus with more than one proton would be unstable, as protons would repel one another and the nucleus would be prone to come apart. The existence of some other force that held the protons and neutrons together was thought to be the answer, and Yukawa's investigations of the force led to the meson theory. He suggested that the force that held the nucleons (protons and neutrons) together was a particle that exchanged some unit of energy between the nucleons. Using quantum theory, Yukawa predicted that the mass of this particle would be about 200 times that of the electron.

Yukawa publicized his meson theory in 1934, and in 1937 his theory was confirmed when CARL DAVID ANDERSON uncovered a particle similar to the one Yukawa had predicted. The particle, later named the *muon,* had the expected mass, but when it was found not to interact with nucleons often enough to be the bonding force of nucleons, Yukawa's theory was somewhat disregarded. In 1947, however, the theory was finally validated when Powell found the predicted particle in a cosmic ray shower. The discovered meson was later called the pi-meson, and Yukawa was promptly awarded the Nobel Prize in physics in 1949.

Yukawa spent most of his career at Kyoto University, where he returned in 1939 after several years at Osaka University. In 1949 he was a visiting professor at Princeton University's Institute for Advanced Study, and he taught at Columbia University in New York for several years before returning to Kyoto University in 1953. He retired in 1970. In addition to the Nobel Prize, Yukawa received such honors as the 1940 Imperial Prize of the Japan Academy, the 1964 Lomonosov Gold Medal of the Soviet Academy of Sciences, the 1964 Federal Republic of Germany's Order of Merit, and the 1977 Order of the Rising Sun.

Z

Ziegler, Karl

(1898–1973)
German
Chemist

Best known for his research on polymers, Karl Ziegler was a versatile chemist who greatly advanced the plastics industry. He discovered the Ziegler catalysts, made up of organometallic compounds, which not only revolutionized the plastics industry but also found applications in the development of rubbers, fibers, and films. For his achievements Ziegler shared the Nobel Prize in chemistry in 1963 with the Italian chemist GIULIO NATTA, who expanded upon Ziegler's findings.

Born on November 26, 1898, in Helsa, near Kassel, Germany, Ziegler was the son of Luise Rall and Karl Ziegler, a Lutheran minister. Ziegler demonstrated an avid interest in chemistry at an early age and carried out experiments in his home laboratory.

In 1916 Ziegler entered the University of Marburg, his father's alma mater, to study chemistry. His knowledge and skills were such that he was able to obtain his doctorate in 1920. Ziegler remained at the University of Marburg for five years, working as a lecturer, before traveling to the University of Frankfurt. In 1926 he assumed a post at the University of Heidelberg, and a decade later he joined the staff at the University of Halle.

Ziegler's early work, which covered the span from 1923 to 1943, focused on structural chemistry; he investigated free radicals, highly reactive atoms or groups of atoms with at least one unpaired electron. He also researched the reactions of organometallic compounds with double bonds as well as cyclic compounds. Free radicals generally had very short lives, but Ziegler discovered that if reactive species were not present, some free radicals could stay alive and be handled as regular compounds. His studies of free radicals led him to research lithium, an organic derivative of reactive metals, and he succeeded in producing highly reactive compounds. Ziegler's work on cyclic carbon compounds, the synthesis of large rings of long-chain molecules, found practical applications in the synthesis of musk, a perfume base.

In 1943 Ziegler was offered a position as head of the Kaiser Wilhelm Institute (later the Max Planck Institute) for Coal Research, located in Mülheim. There he made his breakthrough discovery while studying the reactions of organoaluminum compounds. One experiment produced unexpected results, and upon further investigation Ziegler found that an organometallic compound was responsible.

Ziegler then experimented with various metal salts and discovered that some of these metals resulted in the formation of polyethylene. Polyethylene was an early plastic, made by the polymerization of the ethylene molecule into long chains. The problem with this process, however, was that branches would develop and weaken the plastic. In addition, production of the plastic required high temperatures and high atmospheric pressures. Ziegler's metal catalysts, on the other hand, did not cause branching and produced a durable plastic. Ziegler's process was a low-temperature, low-pressure one as well. The commercial and industrial implications of Ziegler's catalysts were recognized immediately, and his work greatly advanced chemical research.

The production and quality of plastics were improved significantly by Ziegler's work, and he became wealthy as a result of his research. Ziegler remained with the Institute for Coal Research until his retirement in 1969. An active member of the scientific community, he reestablished the German Chemical Society after World War II in 1949 and acted as its president. At the age of 70 Ziegler founded the Ziegler Fund for Research. He married Maria Kurz in 1922, and the couple had two children.

Zinder, Norton David

(1928–)
American
Molecular Geneticist

Focused primarily on the study of the molecular genetics of bacteriophages, viruses that attack certain bacteria, Norton David Zinder is best known for his discovery of bacterial transduction, the transfer of genetic material from one bacterial cell to another by a bacteriophage. This finding advanced the understanding of the behavior and location of bacterial genes.

Born on November 7, 1928, in New York City, Zinder was one of two sons born to Jean Gottesman and Harry Zinder, a manufacturer. After graduation from the esteemed Bronx High School of Science, Zinder entered Columbia University. There he studied biology and earned his bachelor's degree in 1947. Upon the recommendation of a former professor Zinder then attended the University of Wisconsin to pursue graduate studies. He earned a master's degree in genetics in 1949, the same year that he married Marilyn Zinder. The couple eventually had two children. Three years later he earned his doctorate in medical microbiology, then accepted a position as associate professor at the Rockefeller Institute for Medical Research (later Rockefeller University) in his hometown of New York City.

Zinder began his studies of microbial genetics at the University of Wisconsin, where he studied under the geneticist Joshua Lederberg. Lederberg, who in 1958 received a Nobel Prize, had discovered mating, or genetic conjugation, in *Escherichia coli,* a bacterium found in the intestinal tract that caused dysentery and other ailments, in 1946. Zinder planned to extend Lederberg's work and study the genus *Salmonella,* which was closely related to *Escherichia coli. Salmonella* caused such diseases as food poisoning and typhoid fever. To study the bacteria, Zinder set out to obtain a large number of mutant bacteria for his experiments. To accomplish this, Zinder developed a new technique to obtain mutant bacteria; he then began his conjugation, or exchange of genetic material, experiments. What he discovered, however, was not conjugation but genetic transduction, a different method of genetic exchange. Zinder found that hereditary material was carried from one bacterial cell to another via a phage, which attacked the bacterial cell. The phage then took control of the cell's genetic material, replicated, and destroyed the bacterial cell. Zinder's discovery had many implications in genetic studies and eventually led a group of scientists to demonstrate that bacterial genes that affect particular physiological processes were grouped in what became known as operons.

In later research Zinder uncovered the F2 phage, the only phage known to contain ribonucleic acid (RNA) as its genetic material. Since 1993 Zinder has served as the dean of graduate and postgraduate studies at Rockefeller University. He has received a number of honors and awards for his pioneering studies, including the 1962 Eli Lilly Award in Microbiology and Immunology from the American Society of Microbiology, the 1966 United States Steel Award in Molecular Biology presented by the National Academy of Sciences, and the 1969 Medal of Excellence from Columbia University. Zinder not only has published many scientific articles, but also has served as editor for such journals as *Virology* and *Intervirology.* Zinder was elected to the American Academy of Arts and Sciences, the National Academy of Sciences, the American Society of Microbiology, the American Society of Virology, and the Genetics Society of America.

BIBLIOGRAPHY

Abir-Am, P. G., and Dorinda Outram, eds. *Uneasy Careers and Intimate Lives: Women in Science 1789–1979.* New Brunswick, N.J.: Rutgers University Press, 1987.

American Men and Women of Science, 1995–96: A Biographical Directory of Today's Leaders in Physical, Biological, and Related Sciences. New York: R. R. Bowker, 1994.

Asimov, Isaac. *Asimov's Biographical Encyclopedia of Science and Technology: The Lives and Achievements of 1510 Great Scientists from Ancient Times to the Present Chronologically Arranged.* New York: Doubleday, 1982.

Association of Women in Mathematics. "Women in Mathematics." Available online. URL: http://www.awm-math.org.

Bailey, B. *The Remarkable Lives of 100 Women Healers and Scientists.* Holbrook, Mass.: Bob Adams, 1994.

Bailey, M. J. *American Women in Science.* Santa Barbara, Calif.: ABC-CLIO, Inc., 1994.

Barnhart, John H. *Biographical Notes upon Botanists.* Boston: G. K. Hall, 1965.

Barrett, Eric C., and David Fisher, eds. *Scientists Who Believe: Twentiy-one Tell Their Own Stories.* Chicago: Moody Press, 1984.

The Biographical Dictionary of Scientists. New York: P. Bedrick Books, 1983–85.

Bridges, Thomas C., and Hubert H. Tiltman. *Master Minds of Modern Science.* New York: L. MacVeagh, 1931.

Cassutt, Michael. *Who's Who in Space: The First 25 Years.* Boston: G. K. Hall, 1987.

Concise Dictionary of Scientific Biography. New York: Scribner, 1981.

Current Biography. New York: H. W. Wilson, 1940– .

Daintith, John, Sarah Mitchell, Elizabeth Tootill, and Derek Gjertsen, eds. *Biographical Encyclopedia of Scientists.* Philadelphia: Institute of Physics Publishing, 1994.

Darrow, Floyd L. *Masters of Science and Invention.* New York: Harcourt, 1923.

Dash, Joan. *The Triumph of Discovery: Women Scientists Who Won the Nobel Prize.* Englewood Cliffs, N.J.: Julian Messner, 1991.

Debus, A. G., ed. *World Who's Who in Science.* Chicago: Marquis Who's Who, 1968.

Defries, Amelia D. *Pioneers of Science.* London: Routledge and Sons, 1928.

Dewsbury, Donald A., ed. *Studying Animal Behavior: Autobiographies of the Founders.* Chicago: University of Chicago Press, 1989.

Dictionary of American Biography. New York: Scribner's, 1928– .

Dictionary of Scientific Biography. New York: Scribner's, 1970–1980.

Elliott, Clark A. *Biographical Dictionary of American Science: The Seventeenth through the Nineteenth Centuries.* Westport, Conn.: Greenwood Press, 1979.

Engstrand, Iris W. *Spanish Scientists in the New World: The Eighteenth-Century Expeditions.* Seattle: University of Washington Press, 1981.

Eric's Treasure Trove of Scientific Biographies. Available online. URL: http://www.treasure-troves.com

Feldman, Anthony. *Scientists and Inventors.* New York: Facts On File, 1979.

Gaillard, Jacques. *Scientists in the Third World.* Lexington: University Press of Kentucky, 1991.

Herzenberg, Caroline L. *Women Scientists from Antiquity to the Present: An Index.* West Cornwall, Conn.: Locust Hill, 1986.

Howard, Arthur Vyvyan. *Chamber's Dictionary of Scientists.* New York: Dutton, 1961.

Hsiao, T.C., ed. *Who's Who in Computer Education and Research: U.S. Edition.* Latham, N.Y.: Science and Technology Press, 1975.

Hutchings, D., and E. Candlin. *Late Seventeenth-Century Scientists.* Oxford, N.Y.: Pergamon Press, 1966.

Ireland, Norma O. *Index to Scientists of the World from Ancient to Modern Times: Biographies and Portraits.* Boston: Faxon, 1962.

Jones, Bessie Zaban, ed. *The Golden Age of Science: Thirty Portraits of the Giants of 19th-Century Science by Their Scientific Contemporaries.* New York: Simon and Schuster, 1966.

Kass-Simon, G., and Patricia Farnes. *Women of Science: Righting the Record.* Bloomington: Indiana University Press, 1990.

Kessler, James H., Jerry S. Kidd, Renee A. Kidd, Katherine Morin, and Tracy More, eds. *Distinguished African American Scientists of the 20th Century.* Phoenix, Ariz.: Oryx Press, 1996.

Kohler, Robert E. *Partners in Science: Foundation Managers and Natural Scientists, 1900–1945.* Chicago: University of Chicago Press, 1991.

Krapp, Kristine M. *Notable Black American Scientists.* Detroit, Mich.: Gale Research. 1998.

Makers of Modern Science: A Twentieth Century Library Trilogy. New York: Scribner, 1953.

McGraw-Hill Modern Men of Science: 426 Leading Contemporary Scientists. New York: McGraw-Hill, 1966–68.

McGraw-Hill Modern Scientists and Engineers. New York: McGraw-Hill, 1980.

McGrayne, Sharon Bertsch. *Nobel Prize Women in Science: Their Lives, Struggles, and Momentous Discoveries.* Seacaucus, N.J.: Carol Publishing Group, 1993.

McMurray, Emily J., and Donna Olendorf, eds. *Notable Twentieth Century Scientists.* Detroit, Mich.: Gale Research, 1995.

Millar, David, Ian Millar, John Millar, and Margaret Millar, eds. *The Cambridge Dictionary of Scientists.* New York: Cambridge University Press. 1996.

Murray, Robert H. *Science and Scientists in the Nineteenth Century.* New York: Macmillan, 1925.

The Nobel Foundation, "Nobel Laureates." Available online. URL: http://www.nobel.se

North, J. *Mid-Nineteenth-Century Scientists.* Oxford, N.Y.: Pergamon Press, 1969.

Olby, Robert C., ed. *Late Eighteenth-Century European Scientists.* Oxford, N.Y.: Pergamon Press, 1966.

A Passion to Know: 20 Profiles in Science. New York: Scribner, 1984.

Pelletier, Paul A. *Prominent Scientists: An Index to Collective Biographies.* New York: Neal-Schuman, 1994.

Porter, Roy, ed. *The Biographical Dictionary of Scientists.* New York: Oxford University Press, 1994.

St. Andrews, History of Science. Available online: URL: http://www-history.mcs.st-andrews.ac.uk/history/.

Schlessinger, Bernard S., and June H. Schlessinger. *Who's Who of Nobel Prize Winners.* Phoenix, Ariz.: Oryx Press, 1991.

Scott, Michael Maxwell. *Stories of Famous Scientists.* London: Barker, 1967.

Shearer, B. F., and B. S. Shearer, eds. *Notable Women in the Life Sciences.* Westport, Conn.: Greenwood Press, 1996.

———. *Notable Women in the Physical Sciences.* Westport, Conn.: Greenwood Press, 1997.

Siedel, Frank, and James M. Siedel. *Pioneers in Science.* Boston: Houghton, 1968.

Uglow, Jennifer S. *International Dictionary of Women's Biography.* New York: Continuum, 1989.

University of California-Los Angeles. "Women in Science and Engineering." Available online. URL:http://www.physics.ucla.edu/~cwp/Phase2/.

Unterburger, Amy L., ed. *Who's Who in Technology.* Detroit, Mich.: Gale Research Inc., 1989.

Van Sertima, Ivan, ed. *Blacks in Science: Ancient and Modern.* New Brunswick, N.J.: Transaction Books, 1983.

Van Wagenen, Theodore F. *Beacon Lights of Science: A Survey of Human Achievement from the Earliest Recorded Times.* New York: Thomas Y. Crowell, 1924.

Weisberger, Robert A. *The Challenged Scientists: Disabilities and the Triumph of Excellence.* New York: Praeger, 1991.

Who's Who in Science and Engineering, 1994–1995. New Providence, N.J.: Marquis Who's Who, 1994.

Who's Who in Science in Europe. Essex, England: Longman, 1994.

Who's Who of British Scientists, 1980–81. New York: St. Martin's, 1981.

Youmans, William J., ed. *Pioneers of Science in America: Sketches of Their Lives and Scientific Work.* New York: Arno Press, 1978.

Yount, Lisa. *A to Z of Women in Science and Math.* New York: Facts On File, 1999.

Zuckerman, Harriet. *Scientific Elite: Nobel Laureates in the United States.* New York: Free Press, 1979.

ENTRIES BY FIELD

PHYSIOLOGY

PLANT BREEDING

PSYCHOLOGY

SEISMOLOGY

SOIL SCIENCE

VIROLOGY

ZOOLOGY

Entries by Country of Birth

ENTRIES BY COUNTRY OF MAJOR SCIENTIFIC ACTIVITY

NETHERLANDS

NIGERIA

PERSIA

POLAND

PORTUGAL

RUSSIA

SCOTLAND

SOUTH AMERICA

SPAIN

SWEDEN

SWITZERLAND

UNITED STATES

ZIMBABWE

ENTRIES BY YEAR OF BIRTH

1850–1899

CHRONOLOGY

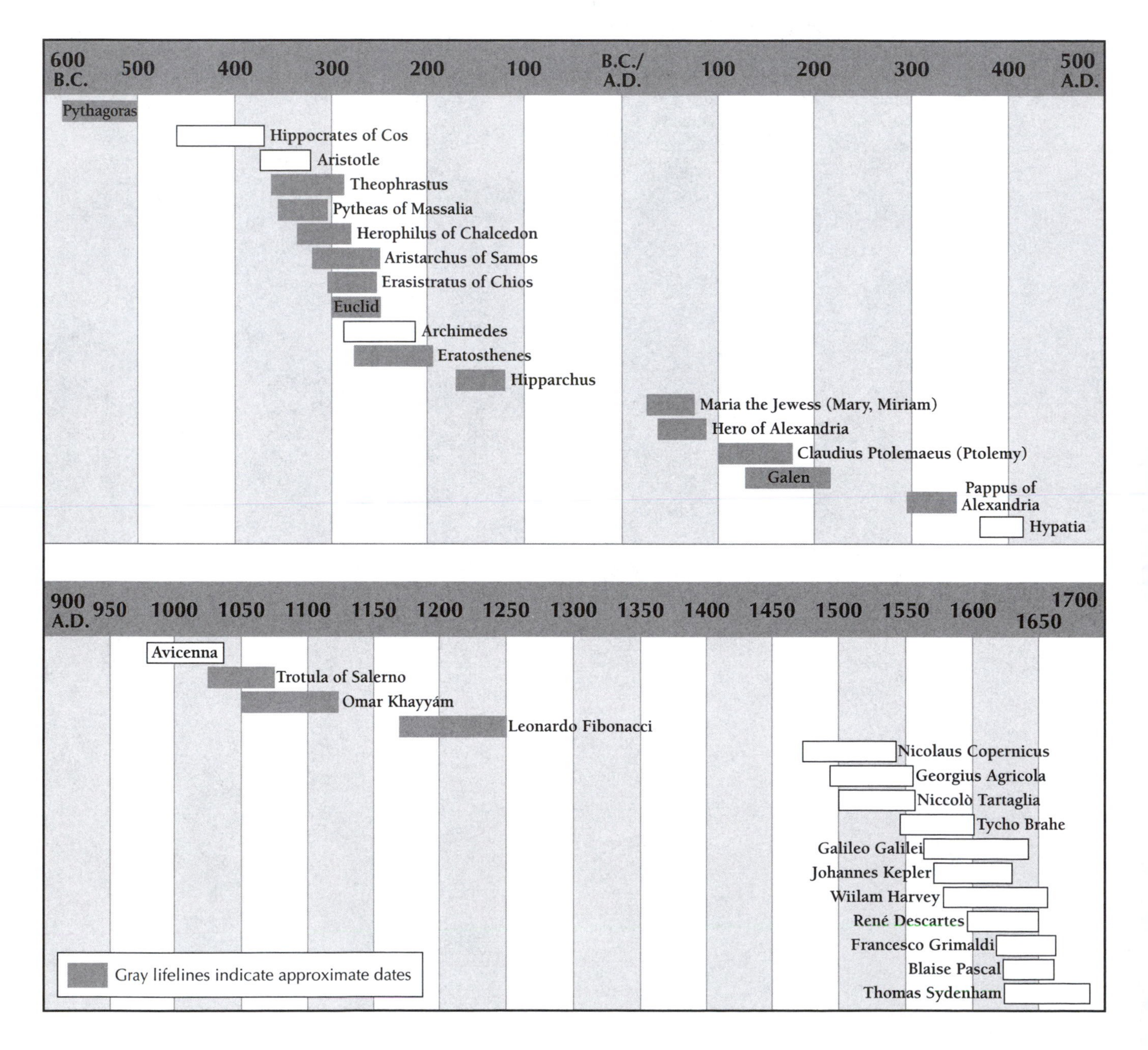

600 B.C.
500
400
300
200
100
B.C./ A.D.
100
200
300
400
500 A.D.
Pythagoras
Hippocrates of Cos
Aristotle
Theophrastus
Pytheas of Massalia
Herophilus of Chalcedon
Aristarchus of Samos
Erasistratus of Chios
Euclid
Archimedes
Eratosthenes
Hipparchus
Maria the Jewess (Mary, Miriam)
Hero of Alexandria
Claudius Ptolemaeus (Ptolemy)
Galen
Pappus of Alexandria
Hypatia
900 A.D.
950
1000
1050
1100
1150
1200
1250
1300
1350
1400
1450
1500
1550
1600
1650
1700
Avicenna
Trotula of Salerno
Omar Khayyám
Leonardo Fibonacci
Nicolaus Copernicus
Georgius Agricola
Niccolò Tartaglia
Tycho Brahe
Galileo Galilei
Johannes Kepler
Wiilam Harvey
René Descartes
Francesco Grimaldi
Blaise Pascal
Thomas Sydenham
Gray lifelines indicate approximate dates

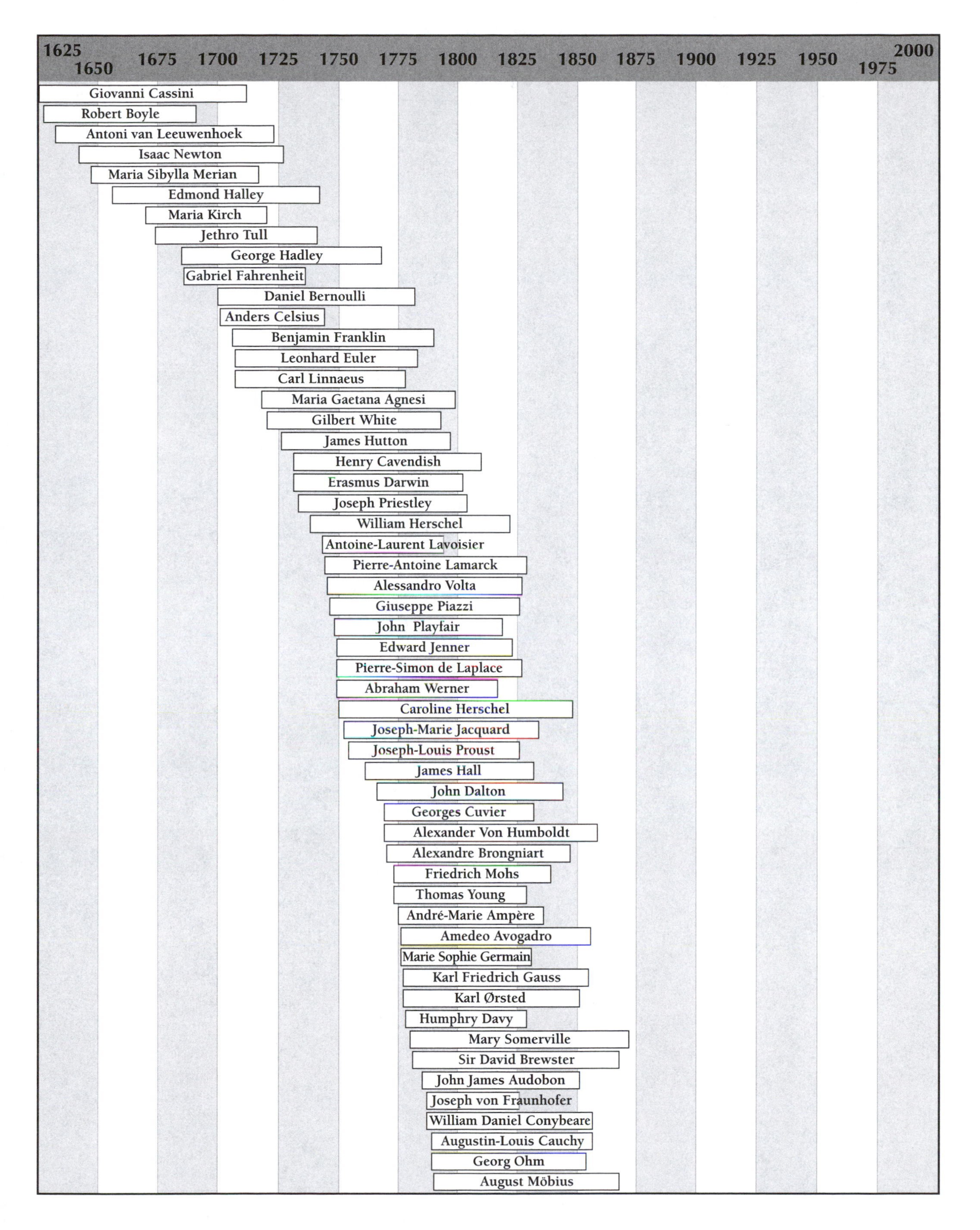
1625
1650
1675
1700
1725
1750
1775
1800
1825
1850
1875
1900
1925
1950
1975
2000
Giovanni Cassini
Robert Boyle
Antoni van Leeuwenhoek
Isaac Newton
Maria Sibylla Merian
Edmond Halley
Maria Kirch
Jethro Tull
George Hadley
Gabriel Fahrenheit
Daniel Bernoulli
Anders Celsius
Benjamin Franklin
Leonhard Euler
Carl Linnaeus
Maria Gaetana Agnesi
Gilbert White
James Hutton
Henry Cavendish
Erasmus Darwin
Joseph Priestley
William Herschel
Antoine-Laurent Lavoisier
Pierre-Antoine Lamarck
Alessandro Volta
Giuseppe Piazzi
John Playfair
Edward Jenner
Pierre-Simon de Laplace
Abraham Werner
Caroline Herschel
Joseph-Marie Jacquard
Joseph-Louis Proust
James Hall
John Dalton
Georges Cuvier
Alexander Von Humboldt
Alexandre Brongniart
Friedrich Mohs
Thomas Young
André-Marie Ampère
Amedeo Avogadro
Marie Sophie Germain
Karl Friedrich Gauss
Karl Ørsted
Humphry Davy
Mary Somerville
Sir David Brewster
John James Audobon
Joseph von Fraunhofer
William Daniel Conybeare
Augustin-Louis Cauchy
Georg Ohm
August Möbius

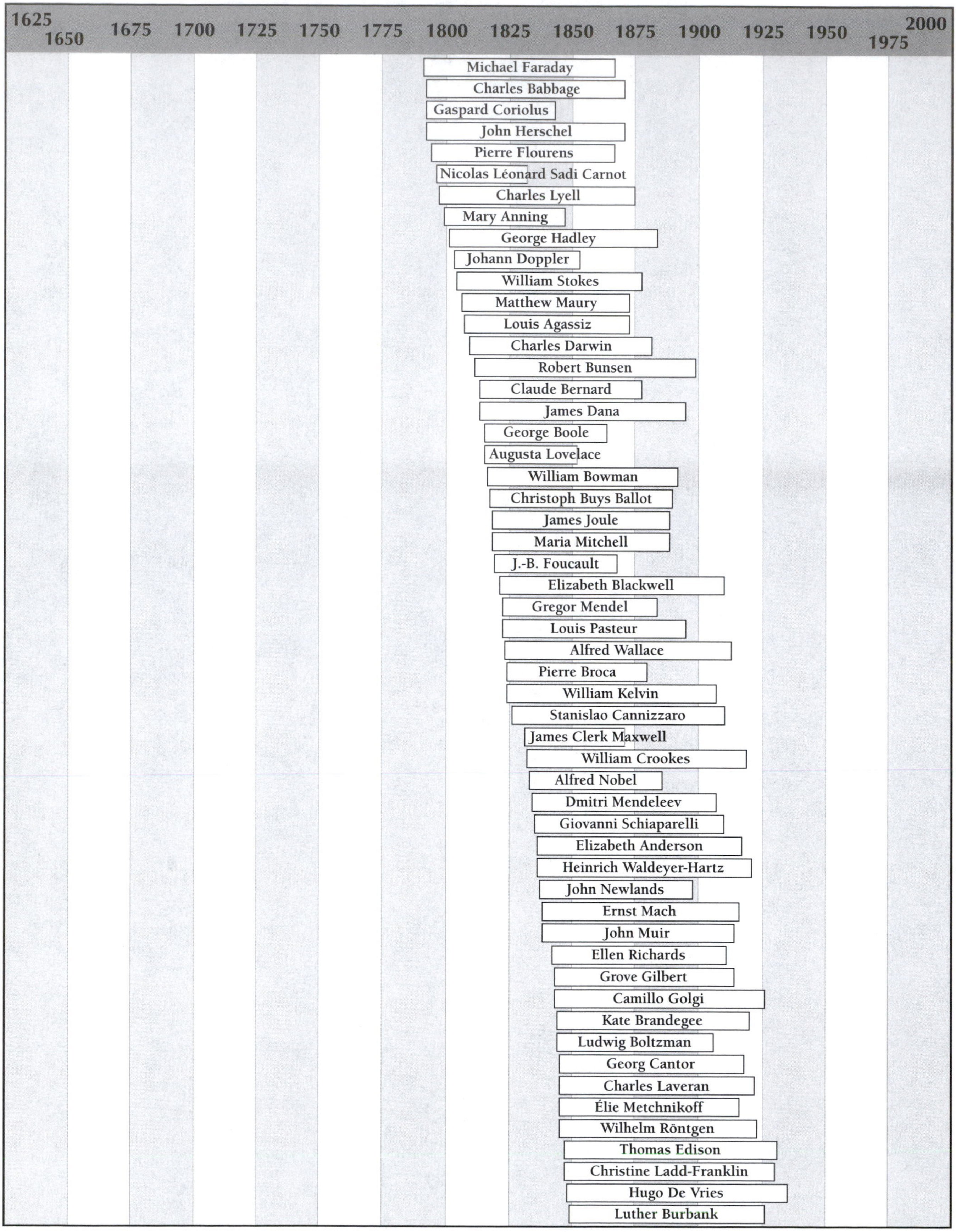

1625 1650 1675 1700 1725 1750 1775 1800 1825 1850 1875 1900 1925 1950 1975 2000

Michael Faraday
Charles Babbage
Gaspard Coriolus
John Herschel
Pierre Flourens
Nicolas Léonard Sadi Carnot
Charles Lyell
Mary Anning
George Hadley
Johann Doppler
William Stokes
Matthew Maury
Louis Agassiz
Charles Darwin
Robert Bunsen
Claude Bernard
James Dana
George Boole
Augusta Lovelace
William Bowman
Christoph Buys Ballot
James Joule
Maria Mitchell
J.-B. Foucault
Elizabeth Blackwell
Gregor Mendel
Louis Pasteur
Alfred Wallace
Pierre Broca
William Kelvin
Stanislao Cannizzaro
James Clerk Maxwell
William Crookes
Alfred Nobel
Dmitri Mendeleev
Giovanni Schiaparelli
Elizabeth Anderson
Heinrich Waldeyer-Hartz
John Newlands
Ernst Mach
John Muir
Ellen Richards
Grove Gilbert
Camillo Golgi
Kate Brandegee
Ludwig Boltzman
Georg Cantor
Charles Laveran
Élie Metchnikoff
Wilhelm Röntgen
Thomas Edison
Christine Ladd-Franklin
Hugo De Vries
Luther Burbank

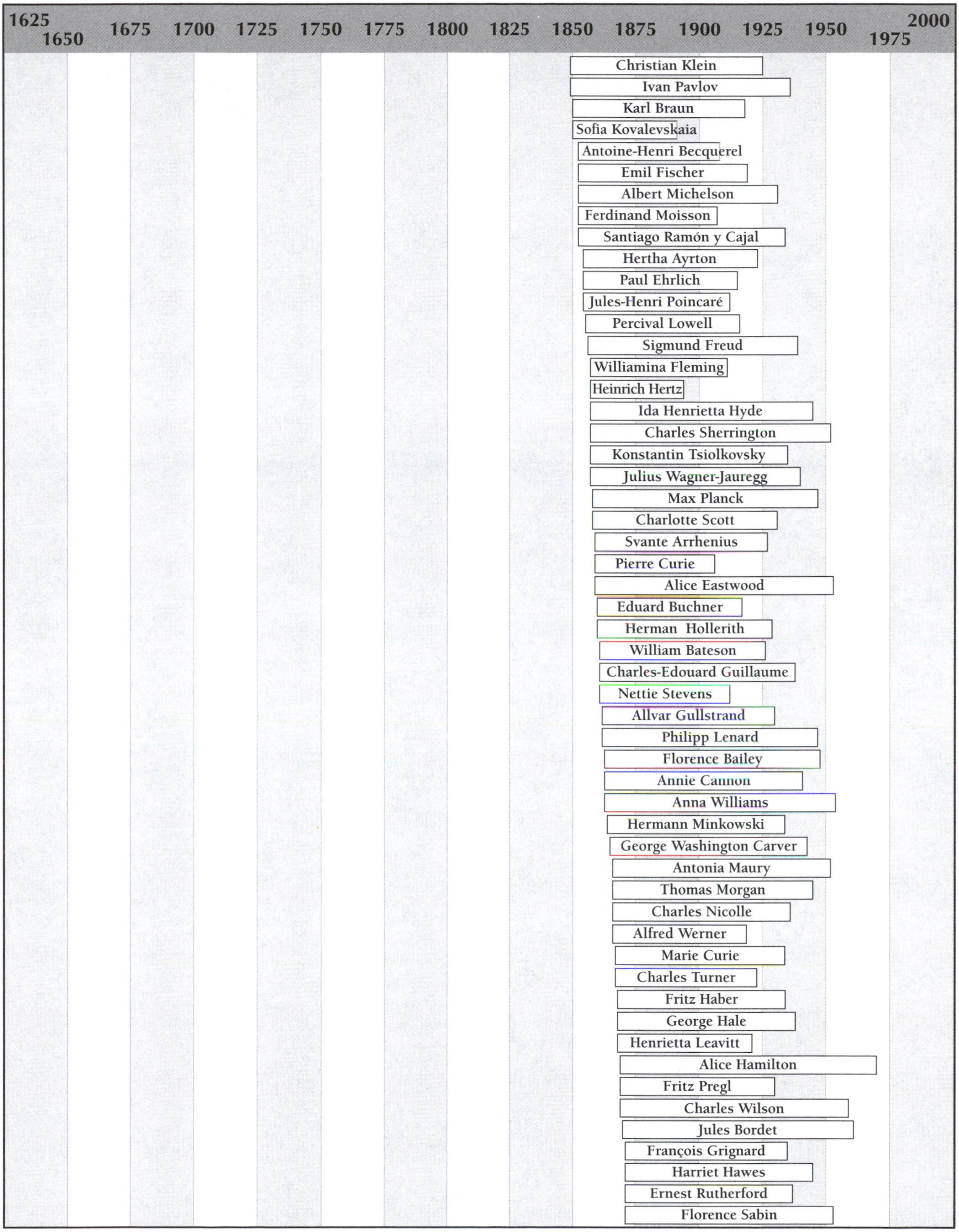

1625 1650 1675 1700 1725 1750 1775 1800 1825 1850 1875 1900 1925 1950 1975 2000

Christian Klein
Ivan Pavlov
Karl Braun
Sofia Kovalevskaia
Antoine-Henri Becquerel
Emil Fischer
Albert Michelson
Ferdinand Moisson
Santiago Ramón y Cajal
Hertha Ayrton
Paul Ehrlich
Jules-Henri Poincaré
Percival Lowell
Sigmund Freud
Williamina Fleming
Heinrich Hertz
Ida Henrietta Hyde
Charles Sherrington
Konstantin Tsiolkovsky
Julius Wagner-Jauregg
Max Planck
Charlotte Scott
Svante Arrhenius
Pierre Curie
Alice Eastwood
Eduard Buchner
Herman Hollerith
William Bateson
Charles-Edouard Guillaume
Nettie Stevens
Allvar Gullstrand
Philipp Lenard
Florence Bailey
Annie Cannon
Anna Williams
Hermann Minkowski
George Washington Carver
Antonia Maury
Thomas Morgan
Charles Nicolle
Alfred Werner
Marie Curie
Charles Turner
Fritz Haber
George Hale
Henrietta Leavitt
Alice Hamilton
Fritz Pregl
Charles Wilson
Jules Bordet
François Grignard
Harriet Hawes
Ernest Rutherford
Florence Sabin

1625 1650 1675 1700 1725 1750 1775 1800 1825 1850 1875 1900 1925 1950 1975 2000

Bertrand Russell
Alexis Carrel
Carl Bosch
Joseph Erlanger
Guglielmo Marconi
António Egas Moniz
Johannes Stark
Ferdinand Moisson
Carl Jung
Robert Bárány
Harriet Brooks
John Macleod
Francis Aston
James Jeans
Frederick Soddy
Lillian Gilbreth
Lise Meiner
George Whipple
Albert Einstein
Alfred Wegener
Alice Evans
Hans Fischer
Alexander Fleming
Walter Hess
Irving Langmuir
Max Born
James Franck
Julia Gardner
Robert Goddard
Melanie Klein
Emmy Noether
Victor Hess
Elmer Imes
Margaret Nice
Anna Wheeler
Friedrich Bergius
St. Elmo Brady
Peter Debye
Otto Meyerhof
Theodore Svedberg
Neils Bohr
Elizabeth Hazen
Karen Horney
George Minot
William Quinland
Harlow Shapley
Archibald Hill
Aldo Leopold
Karl Siegbahn
Ruth Benedict
Gustav Hertz
Srinivasa Ramanujan
Erwin Schrödinger
James Sumner

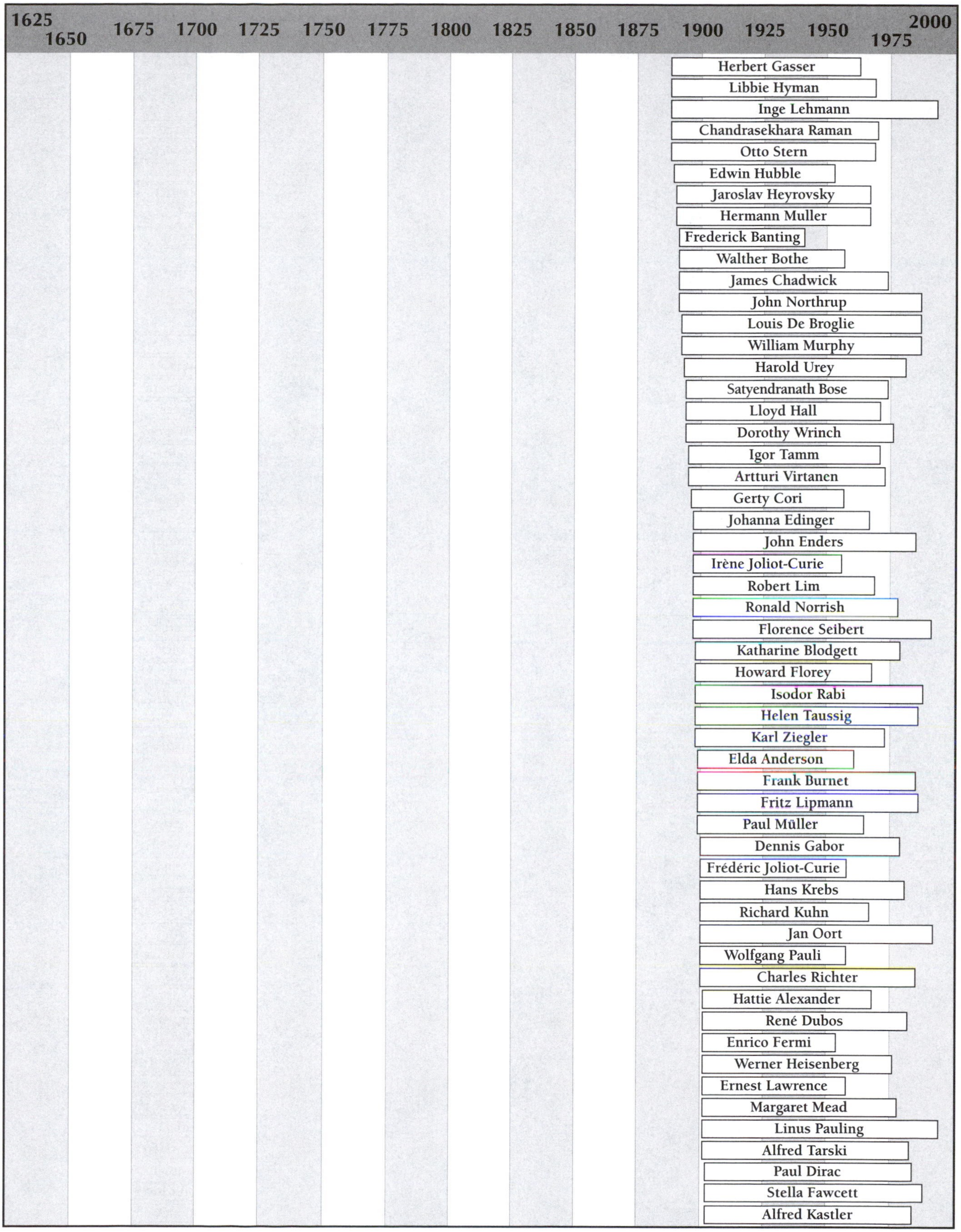

1625 1650 1675 1700 1725 1750 1775 1800 1825 1850 1875 1900 1925 1950 1975 2000

Herbert Gasser
Libbie Hyman
Inge Lehmann
Chandrasekhara Raman
Otto Stern
Edwin Hubble
Jaroslav Heyrovsky
Hermann Muller
Frederick Banting
Walther Bothe
James Chadwick
John Northrup
Louis De Broglie
William Murphy
Harold Urey
Satyendranath Bose
Lloyd Hall
Dorothy Wrinch
Igor Tamm
Artturi Virtanen
Gerty Cori
Johanna Edinger
John Enders
Irène Joliot-Curie
Robert Lim
Ronald Norrish
Florence Seibert
Katharine Blodgett
Howard Florey
Isodor Rabi
Helen Taussig
Karl Ziegler
Elda Anderson
Frank Burnet
Fritz Lipmann
Paul Müller
Dennis Gabor
Frédéric Joliot-Curie
Hans Krebs
Richard Kuhn
Jan Oort
Wolfgang Pauli
Charles Richter
Hattie Alexander
René Dubos
Enrico Fermi
Werner Heisenberg
Ernest Lawrence
Margaret Mead
Linus Pauling
Alfred Tarski
Paul Dirac
Stella Fawcett
Alfred Kastler

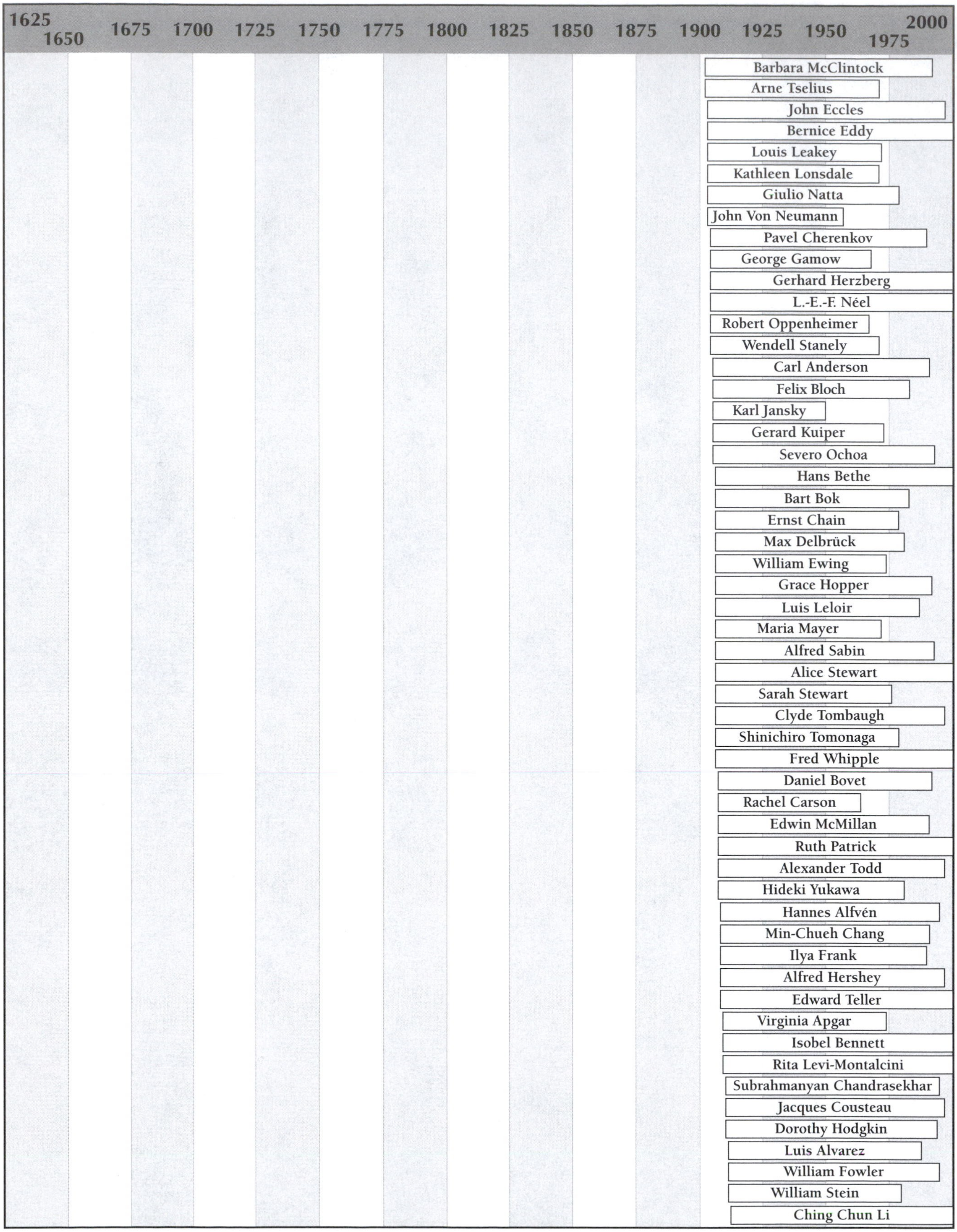

1625 1650 1675 1700 1725 1750 1775 1800 1825 1850 1875 1900 1925 1950 1975 2000

Barbara McClintock
Arne Tselius
John Eccles
Bernice Eddy
Louis Leakey
Kathleen Lonsdale
Giulio Natta
John Von Neumann
Pavel Cherenkov
George Gamow
Gerhard Herzberg
L.-E.-F. Néel
Robert Oppenheimer
Wendell Stanely
Carl Anderson
Felix Bloch
Karl Jansky
Gerard Kuiper
Severo Ochoa
Hans Bethe
Bart Bok
Ernst Chain
Max Delbrück
William Ewing
Grace Hopper
Luis Leloir
Maria Mayer
Alfred Sabin
Alice Stewart
Sarah Stewart
Clyde Tombaugh
Shinichiro Tomonaga
Fred Whipple
Daniel Bovet
Rachel Carson
Edwin McMillan
Ruth Patrick
Alexander Todd
Hideki Yukawa
Hannes Alfvén
Min-Chueh Chang
Ilya Frank
Alfred Hershey
Edward Teller
Virginia Apgar
Isobel Bennett
Rita Levi-Montalcini
Subrahmanyan Chandrasekhar
Jacques Cousteau
Dorothy Hodgkin
Luis Alvarez
William Fowler
William Stein
Ching Chun Li

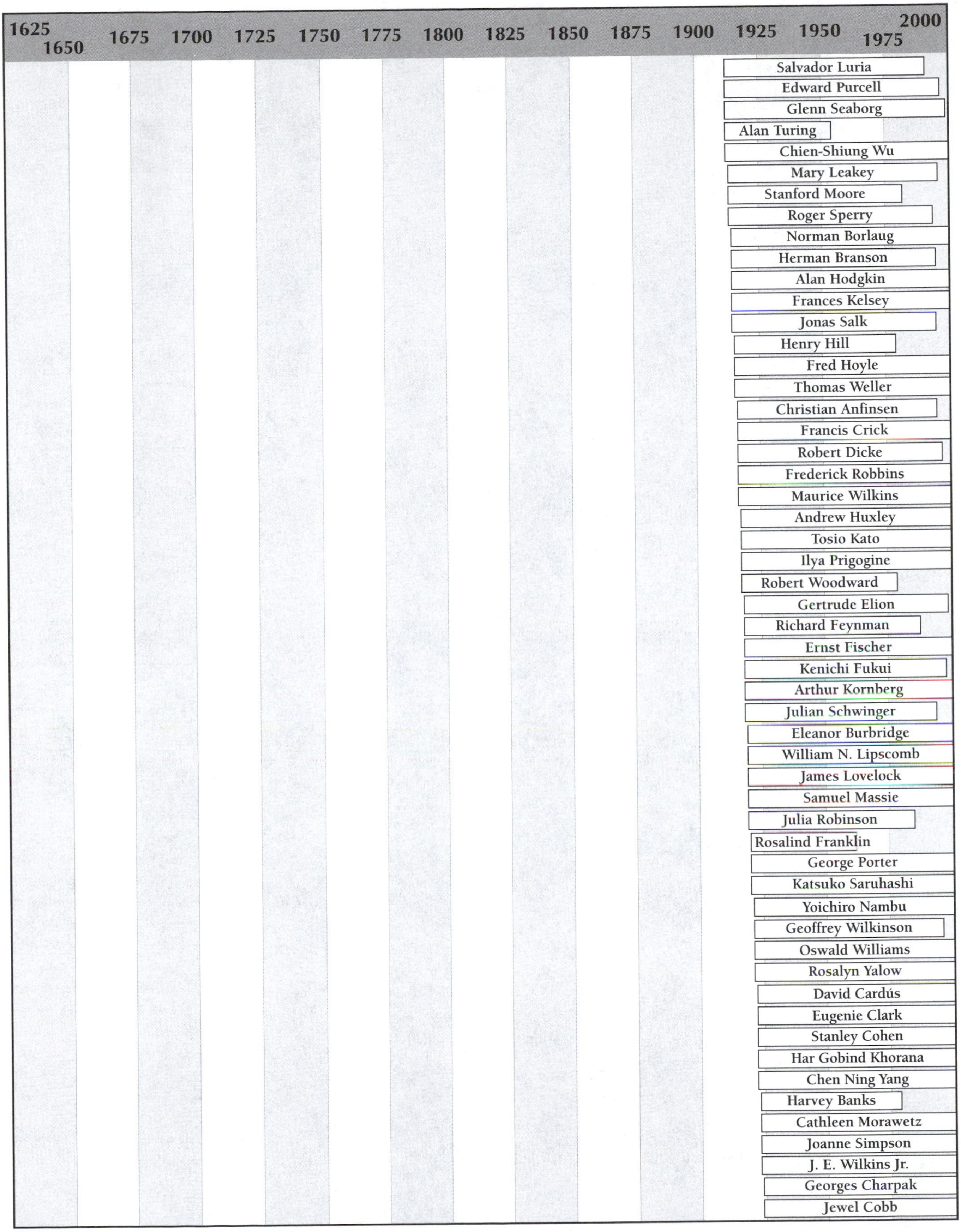

1625 1650 1675 1700 1725 1750 1775 1800 1825 1850 1875 1900 1925 1950 1975 2000

Salvador Luria
Edward Purcell
Glenn Seaborg
Alan Turing
Chien-Shiung Wu
Mary Leakey
Stanford Moore
Roger Sperry
Norman Borlaug
Herman Branson
Alan Hodgkin
Frances Kelsey
Jonas Salk
Henry Hill
Fred Hoyle
Thomas Weller
Christian Anfinsen
Francis Crick
Robert Dicke
Frederick Robbins
Maurice Wilkins
Andrew Huxley
Tosio Kato
Ilya Prigogine
Robert Woodward
Gertrude Elion
Richard Feynman
Ernst Fischer
Kenichi Fukui
Arthur Kornberg
Julian Schwinger
Eleanor Burbridge
William N. Lipscomb
James Lovelock
Samuel Massie
Julia Robinson
Rosalind Franklin
George Porter
Katsuko Saruhashi
Yoichiro Nambu
Geoffrey Wilkinson
Oswald Williams
Rosalyn Yalow
David Cardús
Eugenie Clark
Stanley Cohen
Har Gobind Khorana
Chen Ning Yang
Harvey Banks
Cathleen Morawetz
Joanne Simpson
J. E. Wilkins Jr.
Georges Charpak
Jewel Cobb

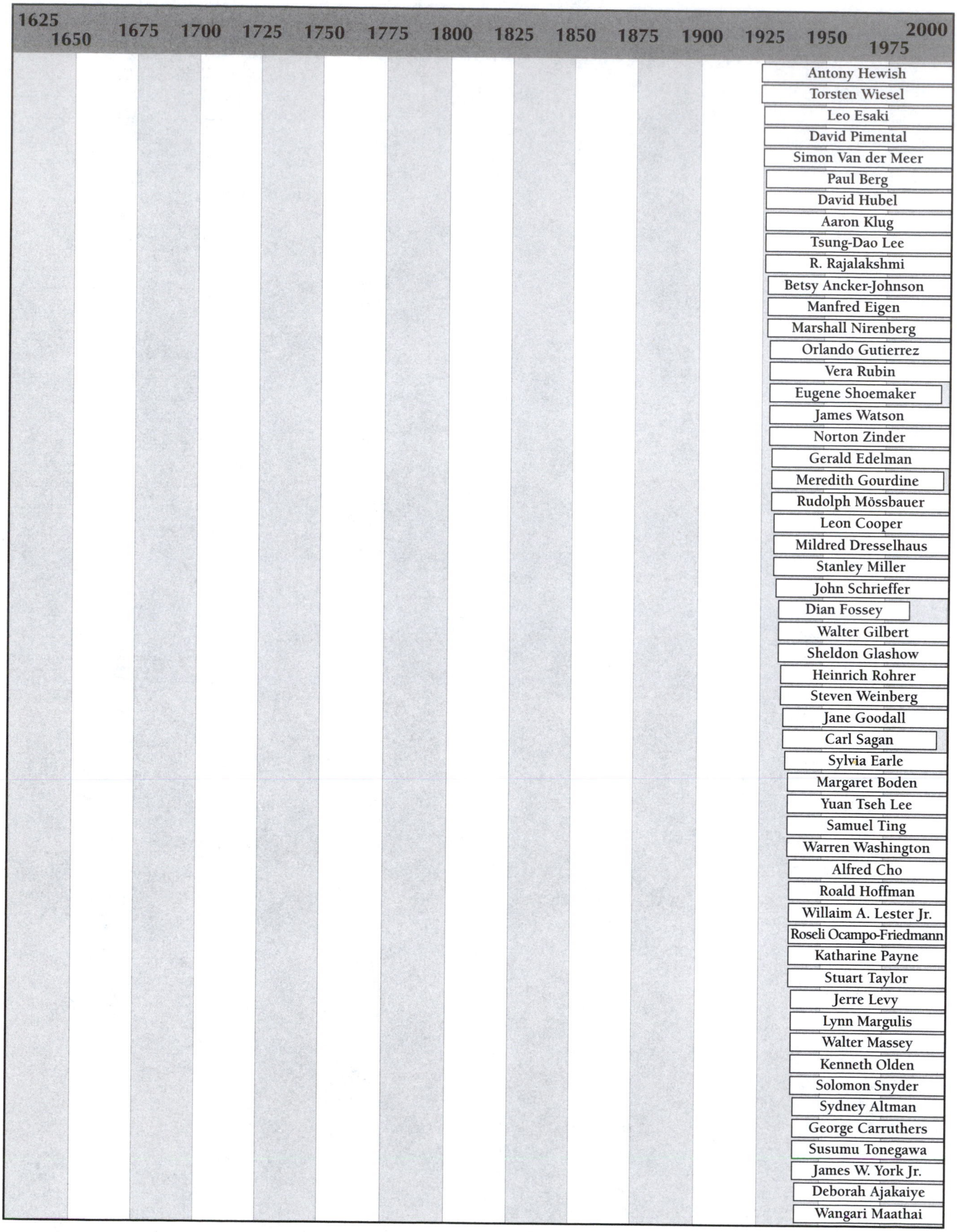

1625 1650 1675 1700 1725 1750 1775 1800 1825 1850 1875 1900 1925 1950 1975 2000

Antony Hewish
Torsten Wiesel
Leo Esaki
David Pimental
Simon Van der Meer
Paul Berg
David Hubel
Aaron Klug
Tsung-Dao Lee
R. Rajalakshmi
Betsy Ancker-Johnson
Manfred Eigen
Marshall Nirenberg
Orlando Gutierrez
Vera Rubin
Eugene Shoemaker
James Watson
Norton Zinder
Gerald Edelman
Meredith Gourdine
Rudolph Mössbauer
Leon Cooper
Mildred Dresselhaus
Stanley Miller
John Schrieffer
Dian Fossey
Walter Gilbert
Sheldon Glashow
Heinrich Rohrer
Steven Weinberg
Jane Goodall
Carl Sagan
Sylvia Earle
Margaret Boden
Yuan Tseh Lee
Samuel Ting
Warren Washington
Alfred Cho
Roald Hoffman
Willaim A. Lester Jr.
Roseli Ocampo-Friedmann
Katharine Payne
Stuart Taylor
Jerre Levy
Lynn Margulis
Walter Massey
Kenneth Olden
Solomon Snyder
Sydney Altman
George Carruthers
Susumu Tonegawa
James W. York Jr.
Deborah Ajakaiye
Wangari Maathai

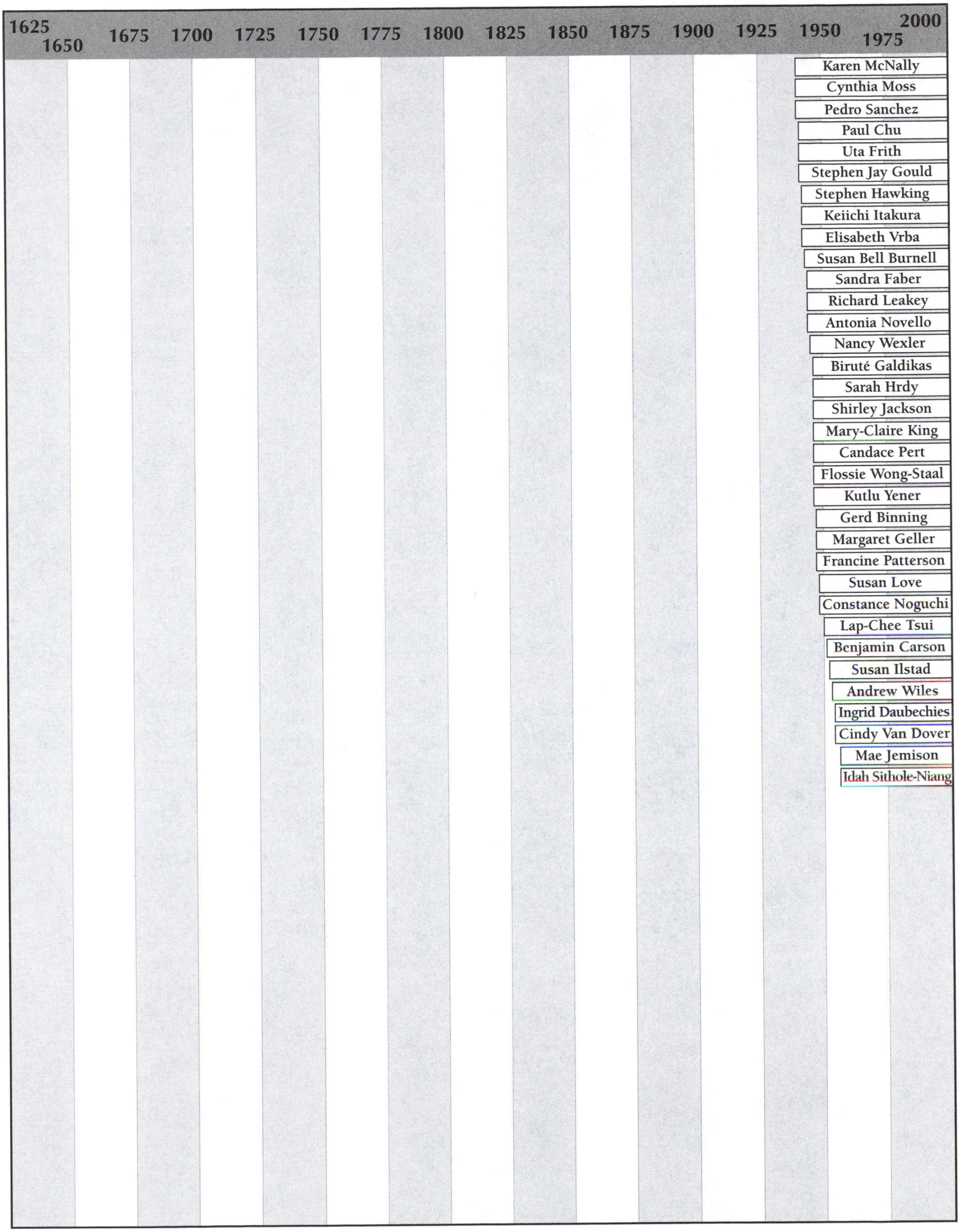

1625 1650 1675 1700 1725 1750 1775 1800 1825 1850 1875 1900 1925 1950 1975 2000

Karen McNally
Cynthia Moss
Pedro Sanchez
Paul Chu
Uta Frith
Stephen Jay Gould
Stephen Hawking
Keiichi Itakura
Elisabeth Vrba
Susan Bell Burnell
Sandra Faber
Richard Leakey
Antonia Novello
Nancy Wexler
Biruté Galdikas
Sarah Hrdy
Shirley Jackson
Mary-Claire King
Candace Pert
Flossie Wong-Staal
Kutlu Yener
Gerd Binning
Margaret Geller
Francine Patterson
Susan Love
Constance Noguchi
Lap-Chee Tsui
Benjamin Carson
Susan Ilstad
Andrew Wiles
Ingrid Daubechies
Cindy Van Dover
Mae Jemison
Idah Sithole-Niang

INDEX

Boldface page numbers denote main entries; *italic* page numbers indicate illustrations.

D

G

H